LALLEMENT—Semigroups and Combinatorial Applications
LAMB—Elements of Soliton Theory
LAY—Convex Sets and Their Applications
LINZ—Theoretical Numerical Analysis: An Introduction to Advanced Techniques
LOVELOCK and RUND—Tensors, Differential Forms, and Variational Principles
MARTIN—Nonlinear Operators and Differential Equations in Banach Spaces
MELZAK—Companion to Concrete Mathematics
MELZAK—Invitation to Geometry
NAYFEH—Perturbation Methods
NAYFEH and MOOK—Nonlinear Oscillations
ODEN and REDDY—An Introduction to the Mathematical Theory of Finite Elements
PASSMAN—The Algebraic Structure of Group Rings
PETRICH—Inverse Semigroups
PRENTER—Splines and Variational Methods
RIBENBOIM—Algebraic Numbers
RICHTMYER and MORTON—Difference Methods for Initial-Value Problems, 2nd Edition
RIVLIN—The Chebyshev Polynomials
RUDIN—Fourier Analysis on Groups
SAMELSON—An Introduction to Linear Algebra
SCHUMAKER—Spline Functions: Basic Theory
SHAPIRO—Introduction to the Theory of Numbers
SIEGEL—Topics in Complex Function Theory
Volume 1—Elliptic Functions and Uniformization Theory
Volume 2—Automorphic Functions and Abelian Integrals
Volume 3—Abelian Functions and Modular Functions of Several Variables
STAKGOLD—Green's Functions and Boundary Value Problems
STOKER—Differential Geometry
STOKER—Nonlinear Vibrations in Mechanical and Electrical Systems
STOKER—Water Waves
TURÁN—On A New Method of Analysis and Its Applications
WHITHAM—Linear and Nonlinear Waves
WOUK—A Course of Applied Functional Analysis
ZAUDERER—Partial Differential Equations of Applied Mathematics

INVERSE SEMIGROUPS

INVERSE SEMIGROUPS

MARIO PETRICH

A Wiley-Interscience Publication
JOHN WILEY & SONS
New York • Chichester • Brisbane • Toronto • Singapore

Library of Congress Cataloging in Publication Data

Petrich, Mario.
Inverse semigroups.

(Pure and applied mathematics)
Includes bibliographical references and index.
1. Inverse semigroups. I. Title. II. Series:
Pure and applied mathematics (John Wiley & Sons)
QA171.P434 1984 512'.2 83-16910
ISBN 0-471-87545-7

Printed in the United States of America

10 9 8 7 6 5 4 3 2 1

PREFACE

In its half-century history, the theory of semigroups has grown from modest attempts to generalize both group theory and ring theory in the direction of a single binary operation obeying only the associative law to the present stage of an independent branch of algebra with a large body of important results.

The first books on semigroup theory represented attempts to systematize most of the knowledge on the subject, whereas later works have tended to cover only some particular areas. One of the areas of semigroup theory which showed great popularity almost from its beginning is that of inverse semigroups. The present book represents the first attempt to systematize the principal parts of the theory of inverse semigroups. Not all results in this area are included, but the main subjects are treated in complete detail from elementary definitions to the frontier of the existing state of the art.

Most of the material presented in this text stems from research papers, a small amount has appeared in book form, and a few things are new. The main effort is focused on the systematization and presentation in a form suitable for class work and easy reference. Exercises are designed to test the understanding of the material, and a few open problems ought to challenge the interested reader.

The choice of the material and the form of its presentation is primarily geared to exhibit the method of investigation rather than to achieve the greatest possible generality in the shortest possible way. This entails treating some subjects in more detail than they seem to deserve from the point of view of their intrinsic value. An instance of this situation is well illustrated by the structure theorems and properties of monogenic inverse semigroups or of ω-regular semigroups. The very fact that one is able to describe their structure in such detail and to answer almost any question about them should serve as a model for investigation of other classes of semigroups.

The case may be made for a survey of all significant results in the field of inverse semigroups rather than a thorough study of selected topics. The aim here is to develop ideas that have proved themselves useful in the study of the structural or other aspects of the theory of inverse semigroups and to present results that hopefully form a basis for further investigation of the subject. The selection was also guided by universality and simplicity of the underlying ideas to the detriment of technically difficult, and sometimes deeper, results. The effort to systematize the often diverging tendencies of approaches to problems, and the very diversity of the questions posed and answered, is channeled here in the direction of completeness of smaller parts of the theory rather than a global attack on the amorphous whole. It is hoped that a unified theory does emerge from this attempt.

The text is independent of the exercises. However, exercises often extend the scope of the theory in the main text and afford an opportunity for a better understanding of the material as well as a possibility for constructing alternative proofs for many of its assertions. Results, examples or problems that are credited to an author or authors but are not given a numbered reference were communicated to me personally and appear here for the first time. The source of exercises is not recorded.

The writing of the manuscript was performed during the author's odyssey: Université de Montpellier (France), Matematički Institut SANU (Yugoslavia), Universität Oldenburg (F. R. Germany), Simon Fraser University (Canada), and Universität Wien (Austria).

Peter R. Jones, Francis Pastijn, and Norman R. Reilly have corrected a great many errors, small and large, and supplied new proofs, examples, and results. Their generous help made the final product a much better piece of work. T. E. Hall, E. I. Kleiman, G. Kowol, D. Krgović, J. Meakin, and H. Mitsch also contributed towards improving the manuscript in various ways. Many colleagues at the above-mentioned institutions and numerous other persons helped in bringing this enterprise to fruition. However, without Norman Reilly and the hospitality of Simon Fraser University the final preparation of the manuscript may have taken another decade.

To all these persons, I offer here my sincere gratitude.

MARIO PETRICH

November 1983

CONTENTS

I

PRELIMINARIES

1. A historical survey of the theory of inverse semigroups shows that there have been two foci of the origin and development of inverse semigroups: the Soviet and the Western schools, initiated by Wagner and Preston, respectively.

2. A list of needed concepts from the theory of partially ordered sets, and in particular lattices, is collected first and is used freely throughout the text. Simple properties of some of these concepts are also proved if they will be needed later.

3. Definitions related to semigroups in general such as identities and zeros, subsemigroups, idempotents, generation, orthogonal sum, and so on, are listed next. The semigroups of partial and full transformations are also introduced.

4. Homomorphisms and their close relatives congruences are introduced together with several ramifications. An explicit expression for the congruence generated by an arbitrary relation on a semigroup is derived.

5. Ideals and their variants are introduced, as well as the kernel of a semigroup. Certain simple properties of these concepts are also established including a characterization of 0-simple semigroups.

6. Green's relations are introduced and their most important properties are established. These include the structure of a $\mathcal{D}$-class and the behavior of idempotents in relation to $\mathcal{D}$-classes.

7. Regular elements and semigroups admit several characterizations; so do completely regular elements. Several auxiliary results are proved, including Lallement's lemma.

8. Concepts related to the translational hull are introduced, and a few of their properties are established.

9. A modest portion of the theory of ideal extensions is discussed, including general extensions, strict and pure extensions, and dense extensions. The relationship with the translational hull plays here an essential role.

10. Free semigroups and free semigroups with involution are defined and constructed, the latter by a construction which is part of the construction of a free group.

11. Identities and varieties are discussed in some detail. The expression for the join of two varieties, and for the variety generated by a semigroup, are explicitly found. The relationship of fully invariant congruences on a free semigroup with varieties is established.

12. The concepts related to amalgamation are introduced including the weak, special, and strong amalgamation properties. Two useful lemmas concerning these notions are proved.

13. The free product and the amalgamated free product of semigroups are discussed in some detail. Several universal properties of these concepts are established.

14. A short list of needed definitions from category theory is given.

I.1 INTRODUCTION

Inverse semigroups were introduced by Wagner in 1952 as regular semigroups with commuting idempotents. In 1953 Liber proved that Wagner's definition is equivalent to the requirement that every element has a unique inverse. Wagner called inverse semigroups "generalized groups" and he and some of his followers have used this term since that time. The term "inverse semi-groups" was introduced by Preston who independently discovered this class of semigroups in 1954.

From their inception to the present day, inverse semigroups have attracted a wide attention among workers in semigroups. Their popularity has several objective and subjective reasons.

In the first place, the closeness of inverse semigroups to groups made it possible to search for structure theorems vaguely modeled on those in group theory. Even though this approach had only a limited success, groups still play a decisive role in important structure theorems for various classes of inverse semigroups. Although the similarity of inverse semigroups and groups is not as substantial as it may appear on the first examination, there is an important analogy between them. Inverse semigroups represent an abstraction of the properties of sets of one-to-one partial transformations closed under composition and inversion just as groups play that type of role for permutation groups. This fact has actually been the leitmotiv and the focus of attention of the Saratov school of inverse semigroups (or should we say "generalized groups"?) headed and inspired by Wagner and by Schein.

In the second place, the simple and esthetically pleasing axioms for inverse semigroups have exercised a certain charm upon many researches in the field of semigroups. Many structure theorems and concepts for various classes of regular semigroups have much simpler formulations for the corresponding classes of inverse semigroups. Such inverse semigroups as the bicyclic semigroup (which has been rediscovered many times) and Brandt semigroups enjoy properties of great value in the study of other classes besides inverse semigroups. It is this abstract approach that was adopted by Preston and pursued by Munn, McAlister, Reilly, and others. The valuable contributions by Clifford antedate these authors and concern the structure of inverse semigroups belonging to certain special classes.

These are some of the principal movers of the theory and some of the objective reasons for the attention paid to inverse semigroups. The subjective reasons for the considerable development of the theory of inverse semigroups can be found in the magnetic personalities of the prime movers: Wagner and Schein created a following in the Soviet school of inverse semigroups, Preston and Munn in the school of the West.

Research activity in inverse semigroups has been both intensive and extensive. On the intensive side, deep structure theorems abound for special classes of inverse semigroups. Let us recall only a few jewels: Clifford's theorems for semilattices of groups and for Brandt semigroups, Reilly's theorem for bisimple ω-semigroups and McAlister's theorem for E-unitary inverse semigroups. In this special category belong the Wagner and the Munn representations. On the extensive side, the number of papers dealing entirely or partly with inverse semigroups is a large one. The bibliography at the end of this text represents an attempt to collect all the items dealing primarily with or bearing upon inverse semigroups.

The following is a concise discussion of the topics covered in various chapters.

I. An extensive collection of concepts, and some of their properties, concerning semigroups in general make up a chapter on preliminaries. In particular, the following topics are discussed: semigroups, congruences and homomorphisms, ideals, Green's relations, regularity, the translational hull, ideal extensions, free semigroups, varieties, amalgamation, and free products. The chapter starts with the needed definitions from the theory of partially ordered sets and ends with a list of needed concepts from category theory.

II. Some of the important, and widely researched, classes of inverse semigroups include Clifford semigroups, Brandt semigroups, strict inverse semigroups, Bruck semigroups over monoids, and Reilly semigroups. These classes harbor some of the most important constructions of the theory and provide suitable examples exhibiting various phenomena discussed in the succeeding chapters.

III. Congruences on inverse semigroups were first described by Preston; a different approach was later devised by Scheiblich. In the study of the congruence lattice, the initial steps of Reilly and Scheiblich play a significant role. Of all the classes of semigroups for which congruences have been investigated, the study of congruences on inverse semigroups has been most profitable. In fact, all the important structure theorems for inverse semigroups are based on various special congruences.

IV. Inverse semigroups admit an analogue of the Cayley theorem in group theory, namely the Wagner representation by one-to-one partial transformations. The Munn representation is a homomorphism of an inverse semigroup into the inverse semigroup of isomorphisms between principal ideals of its semilattice of idempotents. Congruence-free inverse semigroups admit several interesting characterizations. The general theory of representations of inverse semigroups by one-to-one partial transformations on a set, due to Schein, exhibits many features akin to those of group representations by permutations.

V. The translational hull of an inverse semigroup is again an inverse semigroup, a result first proved by Ponizovskiĭ. Related to the subject of the translational hull are the two hulls $C(S)$ and $\hat{S}$ designed by Schein and McAlister, respectively. The translational hull of a Clifford semigroup admits a suitable Clifford representation. For Brandt semigroups the translational hull is sufficiently transparent so that one may construct ideal extensions of these semigroups in great detail.

VI. Just as in the case of group extensions, it is natural to consider conjugate extensions of inverse semigroups. Treating these extensions, one is led to the conjugate hull of an inverse semigroup, analogous to the automorphism group of a group. Normal extensions of inverse semigroups represent a close analogue of the Schreier group extensions. They were initially studied by Petrich, but a general solution was furnished by Allouch. The theory of normal extensions runs somewhat parallel to the Schreier theory with the added complication of a partition of idempotents.

VII. The McAlister structure theorem for E-unitary inverse semigroups in terms of P-semigroups certainly dominates most treatments of general or special E-unitary inverse semigroups. An alternative construction for these semigroups was offered by Petrich, Reilly, and Žitomirskiĭ. The structure of F-inverse semigroups was described by McFadden and O'Carroll. This relatively new field is already rich in significant achievements.

VIII. Even though Wagner proved the existence of free inverse semigroups rather early, it was Scheiblich who provided for it a concrete construction. His work caused a burst of activity which produced improvements in his construction as well as other descriptions of free inverse semigroups. The underlying ideas of the McAlister P-theorem have much in common with Scheiblich's work. Jones established some remarkable properties of free inverse semigroups.

IX. Free monogenic inverse semigroups were first described by Gluskin. Much later, alternative descriptions followed, the simplest one being by Scheiblich as a subdirect product of two copies of the bicyclic semigroup. Congruences and various properties of these semigroups were investigated in some detail.

X. Bisimple inverse monoids exhibit many features reminiscent of groups. The first construction of these semigroups was offered by Clifford in an early paper. Since then they attracted the attention of many researches. Munn and McAlister provided alternative constructions, and Reilly modified Clifford's construction to describe bisimple inverse semigroups.

XI. Inverse semigroups whose idempotents form an ω-chain are said to be ω-regular. The main contributors to deciphering the structure of these semigroups were Reilly, Kočin, and Munn. Their structure is so well elucidated that one is able to answer many questions concerning these semigroups. These results stimulated much interest in inverse semigroups with some restrictions on idempotents.

XII. Varieties of inverse semigroups is a subject of recent origin but it already includes some deep results. The joins and meets of an inverse semigroup variety with the variety of groups provide the first insight into the structure of the lattice of inverse semigroup varieties. These results and the structure of the lowest three levels of the lattice are due to Kleiman. Djadčenko investigated the so-called small varieties and Reilly the completely semisimple varieties.

XIII. Almost all important results on amalgamation of inverse semigroups are due to Hall. He proved, in several different ways, that the class of inverse semigroups has the strong amalgamation property. In fact, he characterized precisely, up to group varieties, which inverse semigroup varieties have the (weak) strong amalgamation property. An alternative approach to the treatment of amalgamation of inverse semigroups was contributed by Howie.

XIV. One of the first attempts to "construct" all inverse semigroups is that of Schein by means of a Croisot groupoid, a partial order on it, and partial products. Meakin devised a similar approach based on a Croisot groupoid, a semilattice structure on its idempotents and "structure mappings" among some of the $\mathcal{R}$-classes of the groupoid. These constructions have theoretical, rather than practical, value showing to what extent some of the ingredients of an inverse semigroup determine the semigroup itself.

Various chapters may be grouped as follows. Chapter I consists of preliminaries. Chapter II provides basic special classes and constructions and concerns the structure of the semigroups in these classes. Chapters III and IV treat special aspects concerning all inverse semigroups; similarly Chapters V and VI concern several hulls and extensions of general inverse semigroups. Chapters VII to XI contain studies of the structures of inverse semigroups belonging to some special classes (hence are of the same general character as Chapter II). Chapters XII to XIV can be characterized as global analysis from three different points of view; they again concern all inverse semigroups.

Chapters are denoted by Roman numerals, sections by the chapter number and an Arabic numeral, and statements by yet another Arabic numeral. References within chapters indicate only the section and statement numbers, say 2.3; references to other chapters bear full information, say VII.2.3. The bibliography includes some papers dealing only marginally or not at all with inverse semigroups, but whose content may be of interest in our development. It excludes all announcements and conference reports, with a few exceptions when these items are of particular interest.

I.2 PARTIALLY ORDERED SETS

This is a brief compendium of concepts and simple properties related to partially ordered sets and more particularly lattices.

I.2.1 Definition

If X is any set, then any subset of the Cartesian product $X \times X$ is a *relation* on X. A *partially ordered set* $(X, \leq)$, to be simply denoted by X, is a pair where X is a nonempty set and $\leq$ is a reflexive, antisymmetric, and transitive relation on X.

Now let X be a partially ordered set. If the greatest lower bound (respectively least upper bound) of two elements α and β of X exists, we denote it by $\alpha \wedge \beta$ (respectively $\alpha \vee \beta$) and call this element the *meet* (respectively the *join*) of α and β. If any two elements of X have a lower bound in X, then X is *lower directed*. If any two elements of X have a meet, then X is a *lower semilattice*.

If Y is a nonempty subset of X which is a semilattice under the order induced on it by the order of X, then Y is a *subsemilattice* of X. If any two elements of X have a meet and a join, then X is a *lattice*. If Y is a subset of X which has a greatest lower bound, the latter is denoted by $\wedge Y$ or $\wedge_{\alpha \in Y} \alpha$ and is called the *meet* of Y; analogously for the *join* $\vee Y$ or $\vee_{\alpha \in Y} \alpha$.

Further, X is *linearly* (or *totally*) *ordered* if for any $x, y \in X$, either $x \leq y$ or $y \leq x$; in such a case X is a *chain*.

I.2.2 Lemma

Let X be a partially ordered set. If for some $\alpha, \beta, \gamma \in X$, $(\alpha \wedge \beta) \wedge \gamma$ and $\alpha \wedge (\beta \wedge \gamma)$ exist, then they are equal.

Proof. The proof of this lemma is left as an exercise.

For functions on partially ordered sets, we have the following concepts.

I.2.3 Definition

Let X and X' be partially ordered sets. A function $\varphi : X \to X'$ *preserves order* (or is *order preserving*) if for any $\alpha, \beta \in X$, $\alpha \leq \beta$ implies $\alpha\varphi \leq \beta\varphi$; φ *inverts order* (or is *order inverting*) if for any $\alpha, \beta \in X$, $\alpha \leq \beta$ implies $\beta\varphi \leq \alpha\varphi$. A bijection φ of X onto X' is an *order isomorphism* if both φ and φ^{-1} preserve order; in the case that $X = X'$, φ is an *order automorphism* of X. A bijection φ of X onto X' is an *order antiisomorphism* if both φ and φ^{-1} invert the order.

The following subsets of a partially ordered set are of particular interest.

I.2.4 Definition

Let X be a partially ordered set. A nonempty subset Y of X is an (*order*) *ideal* of X if for any $\alpha \in Y$, $\beta \in X$, $\beta \leq \alpha$ implies $\beta \in Y$. For any $\alpha \in X$, the set

$$[\alpha] = \{\beta \in X | \beta \leq \alpha\}$$

is the *principal* (*order*) *ideal of* X *generated by* α. If Y is an ideal of X such that for every $\alpha \in X$, $Y \cap [\alpha]$ is a principal ideal, then Y is a *p-ideal*. An ideal Y of X is *essential* if for any $\alpha \in X$, there exists $\beta \in Y$ such that $\beta \leq \alpha$. For any $\alpha, \beta \in X$, $\alpha \leq \beta$, the set

$$[\alpha, \beta] = \{\gamma \in X | \alpha \leq \gamma \leq \beta\}$$

is an *interval* of X. If $\alpha, \beta \in X$ are such that $\alpha < \beta$ and $\alpha < \gamma < \beta$ for no $\gamma \in Y$, then β *covers* α, or α *is covered by* β, which we denote by $\alpha \prec \beta$. We denote by X^1 the partially ordered set obtained from X by adjoining to it an element which plays the role of the greatest element of X^1.

We now turn to lattices.

I.2.5 Definition

Let L be a lattice. Then L is *distributive* if

$$\alpha \wedge (\beta \vee \gamma) = (\alpha \wedge \beta) \vee (\alpha \wedge \gamma) \qquad (\alpha, \beta, \gamma \in L);$$

we may equivalently interchange the signs $\wedge$ and $\vee$. A weaker condition is: L is *modular* if

$$\alpha \leq \gamma \Rightarrow \alpha \vee (\beta \wedge \gamma) = (\alpha \vee \beta) \wedge \gamma \qquad (\alpha, \beta, \gamma \in L).$$

The lattice L is *complete* if every nonempty subset of L has a meet and a join. A sublattice V of L which is complete under the order induced on it by the order of X and whose meets and joins coincide with those of X is a *complete sublattice* of L. A subset V of L is a *complete* $\wedge$*-sublattice* of L if V is a complete lattice whose meets coincide with those in L. A *complete* $\vee$*-sublattice* has an analogous meaning.

There is a simple criterion for completeness which is often useful.

I.2.6 Lemma

Let X be a partially ordered set. If X has a greatest element and each nonempty subset of X has a meet, then X is a complete lattice.

Proof. Let Y be a nonempty subset of X. The set Z of all upper bounds of Y is nonempty since the greatest element of X is an element of Z. By hypothesis, $\alpha = \wedge Z$ exists, and thus $\alpha = \vee Y$, that is, Y has a join.

For mappings on a lattice, we have the following concepts.

I.2.7 Definition

Let L and L' be lattices. A mapping $\varphi : L \to L'$ is a *homomorphism* if

$$(\alpha \wedge \beta)\varphi = \alpha\varphi \wedge \beta\varphi, \qquad (\alpha \vee \beta)\varphi = \alpha\varphi \vee \beta\varphi \qquad (\alpha, \beta \in L);$$

in such a case, we say that φ *preserves meets and joins*, and this definition can be extended to arbitrary meets and joins in an obvious way. If φ is also a bijection of L onto L', it is an *isomorphism* of L onto L'. If L and L' are complete lattices and $\varphi : L \to L'$ preserves arbitrary meets (respectively joins), then φ is a *complete* $\wedge$*-homomorphism* (respectively *complete* $\vee$*-homomorphism*); the conjunction of the two conditions makes a *complete homomorphism*.

An equivalence ρ on L is a *congruence* on L if

$$\alpha\rho\beta \Rightarrow (\alpha \wedge \gamma)\rho(\beta \wedge \gamma), \quad (\alpha \vee \gamma)\rho(\beta \vee \gamma) \qquad (\alpha, \beta, \gamma \in L).$$

Note that the correspondence of congruences and homomorphisms in lattices is the same as in any universal algebra. The following lemma will be useful.

I.2.8 Lemma

If φ is an order isomorphism of a lattice L onto a lattice L', then φ is a (lattice) isomorphism.

Proof. Let φ be as in the statement of the lemma, and let $a, b \in L$. Then $a \wedge b \le a$ implies $(a \wedge b)\varphi \le a\varphi$ and analogously $(a \wedge b)\varphi \le b\varphi$ so that $(a \wedge b)\varphi \le a\varphi \wedge b\varphi$. Further, $a\varphi \wedge b\varphi \le a\varphi$ which yields $(a\varphi \wedge b\varphi)\varphi^{-1} \le a$ and symmetrically $(a\varphi \wedge b\varphi)\varphi^{-1} \le b$. Hence $(a\varphi \wedge b\varphi)\varphi^{-1} \le a \wedge b$ and thus $a\varphi \wedge b\varphi \le (a \wedge b)\varphi$. Consequently, $(a \wedge b)\varphi = a\varphi \wedge b\varphi$. A dual argument shows that $(a \vee b)\varphi = a\varphi \vee b\varphi$.

We now discuss binary relations.

I.2.9 Definition

Let X be any set. We denote by $\mathfrak{B}(X)$ the *set of all relations on* X. Then $\mathfrak{B}(X)$ is a complete lattice under inclusion with least element the *empty relation* $\varnothing$ and greatest element the *universal relation* ω on X. The *equality* (or *identical*) *relation* ε is the least reflexive relation on X. Further, $\mathcal{B}(X)$ is provided with a multiplication defined by

$$x\alpha\beta y \Leftrightarrow x\alpha z,\ z\beta y \qquad \text{for some } z \in X.$$

The notation ε and ω will be used consistently; only in the case of possible confusion, we will write ε_X and ω_X instead of ε and ω, respectively. Simple verification shows that the multiplication of binary relations is associative. We thus may write products without parentheses and define α^n as the nth iterate $\alpha\alpha \cdots \alpha$.

I.2.10 Definition

Let $\rho \in \mathcal{B}(X)$. The relation ρ^{-1} defined by

$$x\rho^{-1}y \Leftrightarrow y\rho x \qquad (x, y \in X)$$

is the *inverse relation* of ρ. The relation $\rho^t = \cup_{n=1}^{\infty}\rho^n$ is the *transitive closure* of ρ; explicitly

$$x\rho^t y \Leftrightarrow \text{there exist } z_1, z_2, \ldots, z_n \in X \text{ such that}$$

$$x = z_1, \qquad z_i\rho z_{i+1}, \qquad i = 1,2,\ldots,n-1, \qquad z_n = y.$$

We then have the following simple result.

I.2.11 Lemma

For any $\rho \in \mathcal{B}(X)$, the following statements hold.

(i) ρ^t is the least transitive relation on X containing ρ.
(ii) $(\rho \cup \rho^{-1} \cup \varepsilon)^t$ is the least equivalence relation on X containing ρ.

Proof. The proof of this lemma is left as an exercise.

The intersection of equivalence relations on X is evidently an equivalence relation. Since also ω is an equivalence relation, 2.6 implies that the partially ordered set of all equivalence relations on X is a complete $\cap$-sublattice of $\mathcal{B}(X)$. The join of a nonempty family $\mathcal{F}$ of equivalence relations on X is of the form $(\cup\mathcal{F})^t$ according to 2.11. The following special case is of particular interest.

I.2.12 Lemma

Let α and β be equivalence relations on a set X. Then $\alpha\beta$ is an equivalence relation if and only if $\alpha\beta = \beta\alpha$, in which case $\alpha\beta = \alpha \vee \beta$ in the lattice of equivalence relations on X.

Proof. Since $\alpha \cup \beta \subseteq \alpha\beta \subseteq \alpha \vee \beta$, if $\alpha\beta$ is an equivalence relation, then $\alpha\beta = \alpha \vee \beta$ and

$$\alpha\beta = (\alpha\beta)^{-1} = \beta^{-1}\alpha^{-1} = \beta\alpha.$$

Conversely, if $\alpha\beta = \beta\alpha$, then

$$(\alpha\beta)^{-1} = \beta^{-1}\alpha^{-1} = \beta\alpha = \alpha\beta,$$
$$\alpha\beta\alpha\beta = \alpha\alpha\beta\beta = \alpha\beta,$$

so $\alpha\beta$ is symmetric and transitive, and it is obviously reflexive, so it is an equivalence relation.

We will need the following simple result.

I.2.13 Lemma

If L is a lattice of commuting equivalence relations on a set X, then L is modular.

Proof. In view of 2.12, for any $\alpha, \beta \in L$, we have $\alpha \vee \beta = \alpha\beta$. Now let $\alpha, \beta, \gamma \in L$ be such that $\alpha \subseteq \gamma$. Let $a[(\beta\alpha) \cap \gamma]b$. Then $a\beta\alpha b$ and $a\gamma b$ and thus $a\beta c$, $c\alpha b$ for some $c \in S$. Since $\alpha \subseteq \gamma$, we get $c\gamma b$. But then $a\gamma b$ and $b\gamma c$ which implies $a\gamma c$. Now $a\beta c$ and $a\gamma c$ yield $a(\beta \cap \gamma)c$, which together with $c\alpha b$ gives $a(\beta \cap \gamma)\alpha b$. Consequently, $(\alpha\beta) \cap \gamma \subseteq \alpha(\beta \cap \gamma)$. In the lattice notation this reads $(\alpha \vee \beta) \wedge \gamma \leq \alpha \vee (\beta \wedge \gamma)$; since $\alpha \leq \gamma$, the opposite inclusion is true in any lattice. Therefore L is modular.

The following construction will be needed.

I.2.14 Definition

Let P and Q be partially ordered sets. On $P \times Q$ introduce a relation $\leq$ by

$$(p,q) \leq (p',q') \text{ if } p = p', q \leq q' \quad \text{or} \quad p < p'.$$

One verifies easily that $\leq$ is a partial order in $P \times Q$; $\leq$ is the *lexicographic order* on $P \times Q$. The partially ordered set $(P \times Q, \leq)$ is the *ordinal product* of P and Q, to be denoted by $P \circ Q$.

I.2.15 Exercises

(i) Let Y be a lower semilattice and I be an order p-ideal of Y. For every $\alpha \in Y$, define $\bar{\alpha}$ by the condition $[\alpha] \cap I = [\bar{\alpha}]$. What can be said about the mapping $\alpha \to \bar{\alpha}$ $(\alpha \in Y)$?

(ii) Show that the lattice of all normal subgroups of any group is modular. Give an example of a group in which the lattice of all subgroups is not modular.

(iii) Show that any isomorphism of complete lattices is a complete isomorphism.

(iv) Let X be a nonempty set. Show that the complete lattices of equivalence relations on X and of partitions of X are isomorphic.

(v) Let X be a set, τ a relation on X, and $\mathfrak{F}$ a nonempty family of relations on X. Prove

(α) $\tau(\bigcup_{\rho\in\mathfrak{F}}\rho) = \bigcup_{\rho\in\mathfrak{F}}\tau\rho$,

(β) $(\bigcup_{\rho\in\mathfrak{F}}\rho)\tau = \bigcup_{\rho\in\mathfrak{F}}\rho\tau$,

(γ) $\tau(\bigcap_{\rho\in\mathfrak{F}}\rho) \subseteq \bigcap_{\rho\in\mathfrak{F}}\tau\rho$,

(δ) $(\bigcap_{\rho\in\mathfrak{F}}\rho)\tau \subseteq \bigcap_{\rho\in\mathfrak{F}}\rho\tau$.

(vi) Prove the following statements concerning a relation ρ on a set.

(α) ρ is reflexive if and only if $\rho \supseteq \varepsilon$.

(β) ρ is symmetric if and only if $\rho = \rho^{-1}$.

(γ) ρ is transitive if and only if $\rho^2 \subseteq \rho$.

(δ) ρ is a partial order if and only if $\rho \cap \rho^{-1} = \varepsilon$ and $\rho^2 \subseteq \rho$.

(ε) ρ is a linear order if and only if $\rho \cap \rho^{-1} = \varepsilon$, $\rho^2 \subseteq \rho$, and $\rho \cup \rho^{-1} = \omega$.

(vii) Let ρ and τ be relations on a set. Show that $(\rho\tau)^{-1} = \tau^{-1}\rho^{-1}$, $(\rho^{-1})^{-1} = \rho$, and deduce that $\rho\rho^{-1}$ is symmetric.

For a comprehensive treatment of lattices, consult Grätzer [2].

I.3 SEMIGROUPS

We compile here a list of concepts and notation concerning semigroups most of which will be used constantly.

I.3.1 Notation

For any sets A and B, we write $A \setminus B = \{a \in A | a \notin B\}$. The singleton set $\{x\}$ is often denoted by x. The empty set is denoted by $\varnothing$. The cardinality of A is denoted by $|A|$. The additive group of integers is denoted by $\mathbb{Z}$, the set of natural numbers $\{1, 2, \dots\}$ by $\mathbb{N}$, the set of nonnegative integers by N.

I.3.2 Definition

A set S together with a (binary) operation, usually called *multiplication*, is a *groupoid*. If the operation is not defined for all pairs of elements of S, it is a *partial groupoid*. A groupoid S satisfying the *associative law*

$$a \cdot (b \cdot c) = (a \cdot b) \cdot c \qquad (a, b, c \in S)$$

is a *semigroup*. A semigroup having only one element is *trivial*.

We will generally omit the symbol for multiplication and denote the product by juxtaposition ab, except when there are several operations present, in which case we may write $a * b$ or $a \circ b$. The associative law implies the *general associative law* which says that the product of any number of elements does not depend on the way of performing the multiplication of the elements in given order. For nonempty subsets $A_1, A_2, \ldots, A_n$ of a semigroup, we have the *complex multiplication*

$$A_1 A_2 \cdots A_n = \{a_1 a_2 \cdots a_n | a_i \in A_i, i = 1, 2, \ldots, n\},$$

$$A^n = AA \cdots A \qquad (n \text{ times}).$$

For the rest of this section, let S denote an arbitrary semigroup.

I.3.3 Definition

An element e of S is a *left* (respectively *right*) *identity* of S if $es = s$ (respectively $se = s$) for all $s \in S$. Further, e is a *two-sided identity* (or simply an *identity*) of S if it is both a left and a right identity of S. A semigroup with an identity is a *monoid*. If S is a monoid, the maximal subgroup of S whose identity is the identity of S is the *group of units* of S; its elements are the *invertible elements* (or *units*) of S.

One may always *adjoin an identity* to a semigroup S by letting $e \notin S$ and declaring on $S \cup \{e\}$ the multiplication in S and

$$es = se = s \qquad (s \in S \cup \{e\}).$$

The following notation comes in quite handy.

I.3.4 Notation

Let $S = S^1$ if S has an identity, otherwise let S^1 be the semigroup S with an identity adjoined. The identity of any monoid is usually denoted by 1. We denote by the symbol 1 any trivial (semi)group.

I.3.5 Definition

An element z of S is a *left* (respectively *right*) *zero* of S if $zs = z$ (respectively $sz = z$) for all $s \in S$. Further, z is a *two-sided zero* (or simply a *zero*) of S if it is both a left and a right zero of S. If S has a zero and all products are equal to zero, S is a *null semigroup*.

A zero can be adjoined to any semigroup in the same way as an identity. The following notation will be used repeatedly.

I.3.6 Notation

Let S be a semigroup with zero 0. Then $S^* = S \setminus \{0\}$ denotes the partial groupoid in which only the products ab are defined where $ab \neq 0$ in S.

We will generally denote by 0 the zero in any semigroup (that has a zero), but occasionally other symbols may be used.

I.3.7 Definition

A nonempty subset T of S is a *subsemigroup* of S if T is closed under the multiplication of S; if also T is a group under the induced operation, it is a *subgroup* of S.

In terms of complex multiplication, T is a subsemigroup of S if and only if $T^2 \subseteq T$. In such a case, T is a semigroup in its own right.

I.3.8 Definition

If A is a nonempty subset of S, then the intersection of all subsemigroups of S containing A is the *subsemigroup T of S generated by A*; if $T = S$, S is *generated by A*, and A is a *set of generators for S*. The subsemigroup of S generated by a singleton $\{s\}$ is the *cyclic* (or *monogenic*) *semigroup generated by s*, to be denoted by $[s]$. An element s of S is of *finite order* if $[s]$ is a finite semigroup, otherwise s is of *infinite order*.

Now let S be a monoid. A subsemigroup of S containing the identity of S is a *submonoid* of S. A subset A of S *generates S as a monoid* if no proper submonoid of S contains A.

The symbol $[\alpha]$ is given a different meaning in 2.4. Since the notation $[s]$ will be used rarely, there should arise no confusion as to the meaning of these symbols. Clearly, if $s \in S$, then $[s]$ consists of all powers of s. For example, the set $\mathbb{N}$ of natural numbers under addition is a cyclic semigroup generated by the element 1.

I.3.9 Definition

An element e of S is *idempotent* if $e^2 = e$. A semigroup is *idempotent*, or is a *band*, if all its elements are idempotent. Two elements a and b of S *commute* if

$ab = ba$; S is *commutative* if any two elements of S commute. A commutative idempotent semigroup is a *semilattice*. The two-element semilattice $\{0, 1\}$ with $01 = 0$ is denoted by Y_2.

The following concepts will be used extensively.

I.3.10 Definition

The set of all idempotents of a subset A of S is denoted by E_A. The *natural* (or *canonical*) *ordering of idempotents of* S is given by

$$e \leq f \Leftrightarrow e = ef = fe \qquad (e, f \in E_S).$$

(It is readily verified that $\leq$ is a partial ordering.) If S has no zero, $e \in E_S$ is a *primitive idempotent* if it is minimal relative to the natural ordering. If S has a zero, then $e \in E_{S^*}$ is a *primitive idempotent* if it is minimal in E_{S^*}.

We have defined a lower semilattice in 2.1 as a special partially ordered set. The notions of a lower semilattice and a semilattice, which is a semigroup, are usually identified in view of the following strong relationship of the two notions.

I.3.11 Lemma

If Y is a semilattice, then Y is a lower semilattice under the partial order: $\alpha \leq \beta$ if $\alpha = \alpha\beta$. Conversely, if Y is a lower semilattice, then Y is a semilattice under the operation $\alpha\beta = \alpha \wedge \beta$.

Proof. The proof of this lemma is left as an exercise.

I.3.12 Definition

If A is a nonempty subset of S, then

$$c_S(A) = \{s \in S | sa = as \text{ for all } a \in A\}$$

is the *centralizer* of A in S; $Z(S) = c_S(S)$ is the *center* of S.

I.3.13 Definition

Let S be a semigroup with zero and let there be given a system of subsemigroups $\{S_\alpha\}_{\alpha \in A}$ such that $S_\alpha \cap S_\beta = S_\alpha S_\beta = 0$ if $\alpha \neq \beta$ and $S = \cup_{\alpha \in A} S_\alpha$. In such a case, S is an *orthogonal sum* (or 0-*direct union*) of semigroups S_α, to be denoted by $S = \Sigma_{\alpha \in A} S_\alpha$.

This definition may be considered internal. Externally we may proceed as follows. Let $\{S_\alpha\}_{\alpha \in A}$ be a set of pairwise disjoint semigroups with zero. On the set $S = (\cup_{\alpha \in A} S_\alpha^*) \cup 0$, where 0 is an extra symbol, define a multiplication by

$$a * b = ab \quad \text{if} \quad a, b \in S_\alpha, \quad ab \neq 0_\alpha \text{ in } S_\alpha$$

for some $\alpha \in A$, and all other products are equal to 0. It is obvious that this makes S a semigroup, and if we consider S_α as a subsemigroup of S, we have $S = \Sigma_{\alpha \in A} S_\alpha$.

We now introduce concrete semigroups of partial and of full transformations on a set.

I.3.14 Definition

Let X be a set. A function α mapping a subset Y of X into X is a *partial transformation* of X; Y is the *domain* of α, denoted by $\mathbf{d}\alpha$, and the set

$$\{x \in X | y\alpha = x \quad \text{for some} \quad y \in Y\}$$

is the *range* of α, denoted by $\mathbf{r}\alpha$; the cardinal number of $\mathbf{r}\alpha$ is the *rank* of α denoted by rank α. For convenience the *empty transformation*, denoted by $\varnothing$, is the mapping with $\mathbf{d}\varnothing = \mathbf{r}\varnothing = \varnothing$.

Let $\mathcal{F}(X)$ denote the set of all partial transformations of X together with the following multiplication: for $\alpha, \beta \in \mathcal{F}(X)$,

$$x(\alpha\beta) = (x\alpha)\beta \quad \text{if} \quad x \in \mathbf{d}(\alpha\beta) = \{y \in \mathbf{d}\alpha | y\alpha \in \mathbf{d}\beta\}.$$

Let $\mathcal{F}_0(X) = \{\alpha \in \mathcal{F}(X) | \text{rank } \alpha \leq 1\}$.

It is easy to verify that the multiplication in $\mathcal{F}(X)$ is associative. We have written the functions on the right of the argument and composed them as such.

I.3.15 Definition

The semigroup $\mathcal{F}(X)$ is the *semigroup of all partial transformations of X written on the right*. Dually, the *semigroup $\mathcal{F}'(X)$ of all partial transformations of X written on the left* has the multiplication defined by: for $\alpha, \beta \in \mathcal{F}'(X)$,

$$(\alpha\beta)x = \alpha(\beta x) \quad \text{if} \quad x \in \mathbf{d}(\alpha\beta) = \{y \in \mathbf{d}\beta | \beta y \in \mathbf{d}\alpha\}.$$

The symbols $\mathcal{F}_0'(X)$, $\mathbf{d}\alpha$, $\mathbf{r}\alpha$, and rank α carry over from $\mathcal{F}(X)$.

If the elements of $\mathcal{F}(X)$ are considered as binary relations on X, then the multiplication in $\mathcal{F}(X)$ coincides with that of binary relations on X defined in 2.9. It follows that $\mathcal{F}(X)$ is a semigroup; by reversing the order of multiplication, we may also deduce that also $\mathcal{F}'(X)$ is a semigroup.

I.3.16 Definition

A partial transformation of a set X is *full*, or is a *transformation* of X, if its domain coincides with X. The *semigroup of all full transformations* of X is denoted by $\mathfrak{T}(X)$ if they are written on the right, and by $\mathfrak{T}'(X)$ if they are written on the left. The *identity mapping* of X, written on either side, is denoted by ι_X. The *group of all permutations* of X is denoted by $\mathfrak{S}(X)$ and $\mathfrak{S}'(X)$ depending on whether the functions are written on the right or on the left.

It is clear that ι_X and $\varnothing$ are the identity and the zero of $\mathfrak{F}(X)$, $\mathfrak{F}'(X)$, and $\mathfrak{B}(X)$, respectively.

I.3.17 Exercises

(i) Prove the general associative law: in a semigroup S, the product of the elements $a_1, a_2, \ldots, a_n$ in this order does not depend on the configuration of parentheses indicating how the product was computed.

(ii) Let S be a semigroup with a left identity e and a right identity f. Show that $e = f$ and deduce that e is the unique identity of S. Prove an analogous statement for zeros.

(iii) Let α and β be partial transformations of a set X. Show that
$$\operatorname{rank}(\alpha\beta) \leq \min\{\operatorname{rank}\alpha, \operatorname{rank}\beta\}.$$

(iv) For any nonempty set X, characterize primitive idempotents of $\mathfrak{F}(X)$, $\mathfrak{T}(X)$, and $\mathfrak{B}(X)$.

(v) Let X be any nonempty set.

- (α) Find the center of $\mathfrak{F}(X)$.
- (β) Characterize the idempotents of $\mathfrak{F}(X)$.
- (γ) Determine the natural order of idempotents in $\mathfrak{F}(X)$.
- (δ) Find all left zeros and all right zeros of $\mathfrak{T}(X)$.

(vi) Prove that the semigroup S of natural numbers under multiplication has a generating set contained in every generating set of S. Is this true for the semigroup of natural numbers under addition?

(vii) Show that in any infinite cyclic semigroup $[x]$, $x^m = x^n$ implies $m = n$.

(viii) Prove that every finite semigroup contains an idempotent.

(ix) Let A and B be nonempty sets, $C \subseteq B \times A$, and let 0 be a symbol not contained in $A \times B$. On the set $S = (A \times B) \cup \{0\}$ define a multiplication by
$$(a, b)(a', b') = (a, b')$$
if $(b, a') \in C$, and all other products are equal to 0. Show that S is a semigroup; find its idempotents and their natural order.

For an extensive treatment of general semigroups, consult the texts by Ljapin [7], Clifford-Preston [1], Petrich [8], and Howie [4]. The subject has arisen as an effort of generalizing both ring theory and group theory as well as an abstraction of properties of full and partial transformations of a set.

I.4 HOMOMORPHISMS AND CONGRUENCES

Besides the definitions of the concepts related to homomorphisms and congruences, we discuss also the least congruence containing a given relation on a semigroup and the join of congruences.

I.4.1 Definition

Let S and S' be semigroups. A mapping $\varphi: S \to S'$ is a *homomorphism* if it respects the multiplication, that is to say,

$$(ab)\varphi = (a\varphi)(b\varphi) \qquad (a, b \in S).$$

If φ is also one-to-one, it is a *monomorphism* or *embedding*; if φ maps S onto S' it is an *epimorphism* and S' is a *homomorphic image* of S. A homomorphism φ which is also a bijection of S onto S' is an *isomorphism* of S onto S' and we write $S \cong S'$; if also $S = S'$, then φ is an *automorphism*. A homomorphism of S into itself is an *endomorphism*. Under the composition of mappings written on the right, endomorphisms of S form a semigroup, to be denoted by $\mathcal{E}(S)$. Its group of units is the *group* $\mathcal{A}(S)$ *of all automorphisms of* S. If an endomorphism φ of S fixes all elements of $S\varphi$, then φ is a *retraction* and $S\varphi$ is a *retract* of S.

If S and S' are monoids, then a homomorphism $\varphi: S \to S'$ is a *monoid homomorphism* if φ maps the identity of S onto the identity of S'. If G is a group, for every $g \in G$, let

$$\varepsilon_g : x \to g^{-1}xg \qquad (x \in G)$$

be the *inner automorphism of* G *induced by* g. Denote by $\mathcal{IA}(G)$ *the group of all inner automorphisms of* G.

A homomorphism φ of S is *trivial* if $S\varphi$ is a trivial semigroup.

Most of these concepts are applicable to any universal algebra. There are further variants of some of these notions, and we now list some of them.

I.4.2 Definition

A bijection of a semigroup S onto a semigroup S' is an *antiisomorphism* if φ reverses the multiplication in the sense that

$$(ab)\varphi = (b\varphi)(a\varphi) \qquad (a, b \in S).$$

If also $S = S'$, then φ is an *antiautomorphism* of S; if in addition $\varphi^2 = \iota_S$, then φ is an *involution*.

I.4.3 Definition

For any semigroup S and nonempty set X, a homomorphism $\varphi : S \to \mathfrak{T}(X)$ [or $\mathfrak{T}'(X)$] is a *representation* of S by transformations of X; if φ is one-to-one, it is *faithful*.

There is a representation of any semigroup of particular importance as follows.

I.4.4 Definition

For every $s \in S$, define ρ_s by

$$x\rho_s = xs \qquad (x \in S).$$

Then the mapping

$$\rho : s \to \rho_s \qquad (s \in S)$$

is the *right regular representation* of S.

There is often a need for functions mapping sets with operations which are only partially defined. Among the whole spectrum of possibilities, we will mainly encounter the following one.

I.4.5 Definition

Let A and B be sets each with a binary operation defined for some pairs of elements. A mapping $\varphi : A \to B$ is a *partial homomorphism* if for any $a, b \in A$ such that ab is defined in A, $(a\varphi)(b\varphi)$ is defined in B, and $(ab)\varphi = (a\varphi)(b\varphi)$.

We now turn to congruences.

I.4.6 Definition

Let S be a semigroup. An equivalence relation ρ on S is a *left* (respectively *right*) *congruence* on S if for any $a, b, c \in S$, $a\rho b$ implies $ca\rho cb$ (respectively $ac\rho bc$); ρ is a *congruence* on S if it is both a left and a right congruence on S. A congruence ρ on S different from ω and ε is *proper*.

I.4.7 Lemma

Let ρ be an equivalence relation on a semigroup S. Then ρ is a congruence on S if and only if for any $a, b, c, d \in S$, $a\rho b$ and $c\rho d$ imply $ac\rho bd$.

Proof. The proof of this lemma is left as an exercise.

This simple result makes it possible to set up the following concepts.

I.4.8 Definition

Let ρ be a congruence on a semigroup S. Denote the ρ-class containing element s of S by $s\rho$. The set S/ρ of all ρ-classes with the multiplication

$$(a\rho)(b\rho) = (ab)\rho$$

is the *quotient semigroup* induced by ρ. The mapping

$$\rho^{\#} : s \to s\rho \qquad (s \in S)$$

is the *natural homomorphism* of S onto S/ρ.

If $\mathcal{C}$ is a class of semigroups and $S/\rho \in \mathcal{C}$, then ρ is a *$\mathcal{C}$-congruence* on S.

The converse situation occurs quite often.

I.4.9 Definition

If $\varphi : S \to S'$ is a homomorphism of semigroups S and S', then φ *induces the congruence* $\overline{\varphi}$ defined by

$$a\overline{\varphi}b \Leftrightarrow a\varphi = b\varphi \qquad (a, b \in S).$$

There is a special situation which occurs with semigroups with zero.

I.4.10 Definition

Let S be a semigroup with zero. A congruence ρ on S is *pure* if $\{0\}$ constitutes a ρ-class. A homomorphism $\varphi : S \to S'$ is *pure* if φ induces a pure congruence on S.

Since the intersection of an arbitrary nonempty family of congruences on a semigroup S is again a congruence, and the universal relation ω is a congruence, the set of all congruences on S forms a complete lattice under inclusion (see 2.6).

I.4.11 Notation

For any semigroup S, we denote by $\mathcal{C}(S)$ the *lattice of all congruences on S ordered by inclusion*.

We will often work with the congruence generated by a relation on a semigroup, and it is handy to have an explicit expression for it.

I.4.12 Definition

Let ρ be a relation on a semigroup S. Define a relation ρ^e on S by

$$a\rho^e b \Leftrightarrow a = xcy, \quad b = xdy \quad \text{for some} \quad x, y \in S^1, c\rho d.$$

The passage from a to b or vice versa is an *elementary ρ-transition.*

I.4.13 Lemma

For any relation ρ on a semigroup S, the relation

$$\rho^* = \left((\rho \cup \rho^{-1} \cup \varepsilon)^e\right)^t$$

is the least congruence on S containing ρ.

Proof. Since ρ^* contains ε, it is reflexive. Clearly, $\rho \cup \rho^{-1} \cup \varepsilon$ is symmetric, whence it follows easily that $(\rho \cup \rho^{-1} \cup \varepsilon)^e$ and thus also ρ^* is symmetric. Transitivity of ρ^* is obvious. Further, since $(\rho \cup \rho^{-1} \cup \varepsilon)^e$ is compatible with the multiplication so is ρ^*. Hence ρ^* is a congruence. If τ is any congruence on S containing ρ, then $\{\rho, \rho^{-1}, \varepsilon\} \subseteq \tau$ and thus $(\rho \cup \rho^{-1} \cup \varepsilon)^e \subseteq \tau$ and finally also $\rho^* \subseteq \tau$.

Explicitly, we have

$$a\rho^* b \Leftrightarrow a = b$$

or

$$\begin{aligned} a &= x_1 c_1 y_1 \\ x_1 d_1 y_1 &= x_2 c_2 y_2 \\ x_2 d_2 y_2 &= x_3 c_3 y_3 \\ &\vdots \\ x_n d_n y_n &= b \end{aligned}$$

for some $x_i, y_i \in S^1$, $c_i, d_i \in S$ such that either $c_i \rho d_i$ or $d_i \rho c_i$, $i = 1, 2, \ldots, n$.

As the first application of this result, we have the following useful statement.

I.4.14 Lemma

Let $\mathcal{F}$ be a nonempty family of congruences on a semigroup S. In the lattice $\mathcal{C}(S)$, $\vee \mathcal{F} = (\cup \mathcal{F})^t$. Explicitly, $a \vee \mathcal{F} b$ if and only if there exist $c_1, c_2, \ldots, c_{n-1} \in S$ and $\rho_1, \rho_2, \ldots, \rho_n \in \mathcal{F}$ such that

$$a\rho_1 c_1, c_1 \rho_2 c_2, \ldots, c_{n-1} \rho_n b.$$

Therefore $\mathcal{C}(S)$ is a complete sublattice of the lattice of equivalence relations on S.

Proof. The proof of this lemma is left as an exercise.

The following construction will be useful.

I.4.15 Lemma

For any congruences ρ and τ on a semigroup S such that $\rho \subseteq \tau$, define a relation τ/ρ on S/ρ by

$$(a\rho)(\tau/\rho)(b\rho) \Leftrightarrow a\tau b.$$

Then τ/ρ is a congruence on S/ρ and $(S/\rho)/(\tau/\rho) \cong S/\tau$.

Proof. It is verified easily that τ/ρ is a congruence on S/ρ. Define a mapping φ by

$$\varphi : s\rho \to s\tau \qquad (s \in S).$$

It follows easily that φ is a homomorphism of S/ρ onto S/τ. Furthermore, for any $a, b \in S$,

$$(a\rho)\varphi = (b\rho)\varphi \Leftrightarrow a\tau = b\tau \Leftrightarrow (a\rho)(\tau/\rho)(b\rho)$$

whence the desired isomorphism.

The next two concepts are of a universal algebraic nature.

I.4.16 Definition

If $\{S_\alpha\}_{\alpha \in A}$ is a family of semigroups, their *direct product* is the semigroup defined on the Cartesian product $\prod_{\alpha \in A} S_\alpha$ with coordinatewise multiplication. The notation for the direct product is $\prod_{\alpha \in A} S_\alpha$ except when A is finite, say $A = \{1, 2, \ldots, m\}$, in which case we write $S_1 \times S_2 \times \cdots \times S_m$. Any semigroup isomorphic to a direct product of semigroups S_α is itself a *direct product* of S_α, $\alpha \in A$.

A more general construction follows.

I.4.17 Definition

Let $\{S_\alpha\}_{\alpha \in A}$ be a family of semigroups, let $S = \prod_{\alpha \in A} S_\alpha$ and π_α denote the projection homomorphism $\pi_\alpha : S \to S_\alpha$. Any semigroup S' isomorphic to a subsemigroup V of S such that $V\pi_\alpha = S_\alpha$ for all $\alpha \in A$ is a *subdirect product* of semigroups S_α, $\alpha \in A$. A semigroup S is *subdirectly irreducible* if it has the property: whenever $S \subseteq \prod_{\alpha \in A} S_\alpha$ is a subdirect product, then one of the projection homomorphisms π_α is one-to-one.

The following result will prove useful.

I.4.18 Lemma

Let $\{\rho_\alpha\}_{\alpha \in A}$ be a family of congruences on a semigroup S such that $\bigcap_{\alpha \in A} \rho_\alpha = \varepsilon$. Then S is a subdirect product of semigroups S/ρ_α, $\alpha \in A$.

Proof. Define a mapping χ by

$$\chi : a \to (a\rho_\alpha)_{\alpha \in A} \qquad (a \in S).$$

The verification that χ is an isomorphism of S onto a subdirect product of semigroups S/ρ_α, $\alpha \in A$, is left as an exercise.

I.4.19 Exercises

(i) Show that every semigroup S can be embedded into $\mathfrak{T}(X)$ for some set X.

(ii) Characterize the semigroup of all endomorphisms of an infinite cyclic semigroup.

(iii) Let ρ be an equivalence relation on a semigroup S. Construct the greatest congruence on S contained in ρ.

(iv) Find all endomorphisms of the group of rational numbers under addition.

(v) Show that every nontrivial semilattice is a subdirect product of two-element chains. Deduce that every nontrivial subdirectly irreducible semilattice is a two-element chain.

(vi) Show that a nontrivial semigroup S is subdirectly irreducible if and only if S has a congruence different from ε contained in all congruences different from ε.

(vii) Let s be an element of a semigroup S. If s is of infinite order, prove that $[s]$ is isomorphic to the semigroup of natural numbers under addition.

(viii) If $S = \Sigma_{\alpha \in A} S_\alpha$ is an orthogonal sum, show that S can be embedded into the direct product $\Pi_{\alpha \in A} S_\alpha$.

(ix) Let S, A, and B be semigroups, $\alpha : S \to A$ and $\beta : S \to B$ be epimorphisms. Let

$$C = \{(a, b) \in A \times B \mid a = s\alpha,\ b = s\beta \text{ for some } s \in S\}.$$

Show that C is a subdirect product of A and B.

(x) Let ρ be a right congruence on a semigroup S. For every $s \in S$, define ρ^s on the set S/ρ of equivalence classes of ρ by $\rho^s : x\rho \to (xs)\rho$ $(x \in S)$. Show that $\bar{\rho} : s \to \rho^s$ $(s \in S)$ is a homomorphism of S into $\mathfrak{T}(S/\rho)$. If S has an identity, show that the congruence induced by $\bar{\rho}$ is the greatest congruence on S contained in ρ.

Most of the concepts discussed here were adapted to semigroups from general algebra and belong to the folklore.

I.5 IDEALS

We introduce here several concepts related to ideals and establish some of their simple properties.

I.5.1 Definition

Let S be a semigroup. A nonempty subset I of S is a *left* (respectively *right*) *ideal* of S if for any $s \in S$, $a \in I$, we have $sa \in I$ (respectively $as \in I$). Further, I is a *two-sided ideal* (or simply an *ideal*) of S if it is both a left and a right ideal of S. A left (right, two-sided) ideal I of S is *proper* if $I \neq S$.

A nonempty intersection of ideals of a semigroup S is again an ideal of S. Hence we may introduce the following notions.

I.5.2 Definition

Let s be an element of a semigroup S. The intersection of all left ideals of S containing s is the *principal left ideal of S generated by s*, to be denoted by $L(s)$. *Right* and *two-sided principal ideals of S generated by s* are defined analogously and are denoted by $R(s)$ and $J(s)$, respectively.

It is easy to verify that with the notation in 3.4, we have

$$L(s) = S^1 s, \qquad R(s) = sS^1, \qquad J(s) = S^1 s S^1.$$

With every ideal of a semigroup one can associate a congruence in the following way.

I.5.3 Definition

Let I be an ideal of a semigroup S. The relation ρ_I on S defined by

$$a \rho_I b \quad \text{if} \quad a, b \in I \quad \text{or} \quad a = b$$

is the *Rees congruence* on S relative to the ideal I. The quotient semigroup S/ρ_I is the *Rees quotient semigroup* and is denoted by S/I.

One verifies readily that a Rees congruence is indeed a congruence. There are various conditions on a semigroup which can be expressed in terms of ideals. One of the simplest is the following.

I.5.4 Definition

A semigroup is *simple* if it has no proper ideals. A semigroup S with zero is 0-*simple* if $S^2 \neq 0$ and S has no proper nonzero ideals.

The following lemma provides a convenient criterion for 0-simplicity.

I.5.5 Lemma

Let S be a semigroup with zero and assume that $S \neq 0$. Then S is 0-simple if and only if $SaS = S$ for all $a \in S^*$.

Proof

Necessity. Let $I = \{a \in S | SaS = 0\}$. Then I is an ideal of S and hence either $I = 0$ or $I = S$. In the latter case, we have $S^3 = 0$. Since S^2 is an ideal of S and by hypothesis $S^2 \neq 0$, we must have $S^2 = S$. But then $S = S^2 = S^3 = 0$, a contradiction. Hence $I = 0$, and thus for any $a \in S^*$, SaS is a nonzero ideal of S and $SaS = S$.

Sufficiency. Since S contains a nonzero element a, we have $S = SaS \subseteq S^2$ and thus $S^2 \neq 0$. Let I be an ideal of S and let $0 \neq a \in I$. Then $S = SaS \subseteq I$ and hence $I = S$. Therefore, S is 0-simple.

I.5.6 Corollary

A semigroup S is simple if and only if $SaS = S$ for all $a \in S$.

Proof. The proof of this corollary is left as an exercise.

I.5.7 Lemma

The following conditions on an ideal I of a semigroup S are equivalent.

(i) I is contained in every ideal of S.
(ii) I is the intersection of all ideals of S.
(iii) I is a simple semigroup.
(iv) $I = J(a)$ for all $a \in I$.

Proof. Items (i) and (ii) are equivalent since a nonempty intersection of ideals is again an ideal. If I satisfies (i), then for any $a \in I$, we have $I \subseteq J(a)$; since the opposite inclusion always holds, (iv) is satisfied. It is clear that (iv) implies (iii). Suppose that (iii) holds and let K be an ideal of S. Then

$KI \subseteq K \cap I$ so that $K \cap I \neq \varnothing$ and is thus an ideal of I. Since I is simple, we must have $K \cap I = I$, that is $K \supseteq I$. Hence (iii) implies (i).

In the light of the above lemma, it is natural to introduce the following important concept.

I.5.8 Definition

The intersection, if nonempty, of all ideals of a semigroup S is the *kernel* of S.

Hence the kernel, if it exists, is characterized by any of the conditions in 5.7. Item (i) in 5.7 means that I is the least element in the partially ordered set of all ideals of S ordered by inclusion. Since the intersection of a finite number of ideals is an ideal, every finite semigroup has a kernel. The semigroup of all natural numbers under addition does not have a kernel.

The next concept is also of basic importance.

I.5.9 Definition

A simple (respectively 0-simple) semigroup having a primitive idempotent is *completely simple* (respectively *completely* 0-*simple*).

There are several special kinds of ideals which play a role in certain situations. We introduce only the following ones.

I.5.10 Definition

Let I be an ideal of a semigroup S. Then I is *prime* (respectively *completely prime*) if for any $a, b \in S$, $aSb \subseteq I$ (respectively $ab \in I$) implies that either $a \in I$ or $b \in I$. Further, I is *categorical* if for any $a, b, c \in S$, $abc \in I$ implies that either $ab \in I$ or $bc \in I$.

If S has a zero, then S is *categorical at zero* if 0 is a categorical ideal.

It is the last concept that will be of greater importance in our considerations. The definition will be used in its contrapositive formulation: S is categorical at zero if and only if for any $a, b, c \in S$, $ab \neq 0$ and $bc \neq 0$ imply $abc \neq 0$.

The next concept is sometimes useful.

I.5.11 Definition

If T is a subsemigroup of a semigroup S, then the *idealizer* of T in S is the largest subsemigroup of S containing T as an ideal, to be denoted by $i_S(T)$.

This definition is justified by the following fact: with the notation of 5.11, one proves easily that

$$i_S(T) = \{s \in S | st, ts \in T \text{ for all } t \in T\}.$$

Note that if T contains the identity of S (if it exists), then $i_S(T) = T$.

I.5.12 Exercises

(i) Let I be an ideal of a semigroup S. Show that I is a prime ideal if and only if for any ideals A and B of I, $AB \subseteq I$ implies that either $A \subseteq I$ or $B \subseteq I$.

(ii) Let S be a nontrivial semigroup with zero and assume that S is categorical at zero and $aSa \neq 0$ for every $a \in S^*$. Show that S is uniquely an orthogonal sum of semigroups whose zero is a prime ideal.

(iii) For any nonempty set X, characterize all ideals of $\mathcal{F}(X)$ and $\mathcal{T}(X)$.

(iv) For ideals I and J of a semigroup S such that $I \subseteq J$, show that $S/J \cong (S/I)/(J/I)$.

(v) Let I and J be ideals of a semigroup S. Show that $I \cup J$ and $I \cap J$ are ideals of S and that $(I \cup J)/J \cong I/(I \cap J)$.

(vi) Let I be an ideal of a semigroup S. Show that I is categorical if and only if S/I is categorical at zero.

(vii) Show that an orthogonal sum of at least two semigroups S_α does not have 0 as a prime ideal. Also show that if each S_α is categorical at zero, so is their orthogonal sum. Deduce that a categorical ideal need not be prime.

(viii) Let X be a countably infinite set and I be the ideal of $\mathcal{T}(X)$ consisting of all transformations of finite rank. Show that I is prime but not categorical (cf. the preceding exercise).

(ix) Give an example of a 0-simple semigroup which is not completely 0-simple.

(x) Let S be a semigroup and $Y_2 = \{0, 1\}$ be a two-element chain. Show that a mapping φ on S onto Y_2 is a homomorphism if and only if $0\varphi^{-1}$ is a proper completely prime ideal of S.

(xi) Show that a semigroup S has no proper semilattice congruences if and only if it has no proper completely prime ideals.

For a fuller discussion of the subject of this section consult the standard texts on semigroups.

I.6 GREEN'S RELATIONS

After introducing these relations, we establish several results related to them. We then discuss briefly principal factors.

I.6.1 Definition

For a semigroup S, the relations $\mathcal{L}$, $\mathcal{R}$, and $\mathcal{J}$ defined on S by

$$a\mathcal{L}b \Leftrightarrow L(a) = L(b),$$

$$a\mathcal{R}b \Leftrightarrow R(a) = R(b),$$

$$a\mathcal{J}b \Leftrightarrow J(a) = J(b),$$

are *Green's relations* (or *equivalences*) on S.

It is easy to see that $\mathcal{L}$ (respectively $\mathcal{R}$) is a right (respectively left) congruence on S.

I.6.2 Lemma

In any semigroup S, the relations $\mathcal{L}$ and $\mathcal{R}$ commute as binary relations.

Proof. Let $a\mathcal{L}c$ and $c\mathcal{R}b$ for $a, b, c \in S$. By definition, we have $a = uc$, $b = cv$, $c = wa = bz$ for some $u, v, w, z \in S^1$. Let $d = av$; then $d = ucv = ub$, $a = uc = ubz = dz$, $b = cv = wav = wd$. Consequently, $a\mathcal{R}d$ and $d\mathcal{L}b$ so that $\mathcal{L}\mathcal{R} \subseteq \mathcal{R}\mathcal{L}$. The opposite inclusion is established similarly.

In view of 2.12 and 6.2, we have that $\mathcal{L}\mathcal{R}$ is an equivalence relation on S, and we may enlarge the set of Green's relations as follows.

I.6.3 Definition

The relations $\mathcal{D} = \mathcal{L}\mathcal{R}$ and $\mathcal{H} = \mathcal{L} \cap \mathcal{R}$ on any semigroup are also called *Green's relations* (or *equivalences*).

I.6.4 Notation

For any element a of a semigroup S and $\mathcal{T} = \mathcal{L}, \mathcal{R}, \mathcal{J}, \mathcal{D}, \mathcal{H}$, the $\mathcal{T}$-class containing a is denoted by T_a.

The above definition leads naturally to the following concepts.

I.6.5 Definition

A semigroup is *bisimple* (or *$\mathcal{D}$-simple*) if it consists of a single $\mathcal{D}$-class. A semigroup S with zero is 0-*bisimple* if $S^2 \neq 0$ and S^* constitutes a $\mathcal{D}$-class of S.

In 5.4 and 6.5, $S^2 \neq 0$ serves the purpose of excluding two-element null semigroups. We now establish a few basic results concerning Green's relations.

I.6.6 Lemma

For any elements a and b of a semigroup S, we have $ab \in R_a \cap L_b$ if and only if $R_b \cap L_a$ contains an idempotent. In such a case,

$$aH_b = H_a b = H_a H_b = H_{ab} = R_a \cap L_b.$$

Proof

Necessity. The hypothesis implies that $a = abx$ and $b = yab$ for some $x, y \in S^1$. Hence

$$b = yab = y(abx)b = (yab)xb = bxb.$$

Let $e = bxya$; then

$$e^2 = (bxya)(bxya) = (bxb)xya = bxya = e,$$

$$b = bxb = bx(yab) = (bxya)b = eb,$$

so that e is an idempotent and $e\mathcal{R}b$. One shows similarly that also $e\mathcal{L}a$ and thus $e \in R_b \cap L_a$.

Sufficiency. Let e be an idempotent in $R_b \cap L_a$. Then $a = ue$ and $e = bv$ for some $u, v \in S^1$. Hence $a = ue = (ue)e = (ab)v$ which implies that $a\mathcal{R}ab$. One shows similarly that $b\mathcal{L}ab$ and thus $ab \in R_a \cap L_b$.

Suppose that $R_b \cap L_a$ contains an idempotent. If $a' \in H_a$, $b' \in H_b$, then $a'b' \in R_a \cap L_b = H_{ab}$ since $R_b \cap L_a$ contains an idempotent, and thus $H_a H_b \subseteq H_{ab}$. In order to prove all the remaining inclusions, by symmetry, it suffices to show that $H_{ab} \subseteq aH_b$. Hence let $c \in H_{ab}$. Then $c = abx = yab$, $ab = cw = zc$, $a = abu$, $b = vab$ for some $x, y, w, z, u, v \in S^1$ since $ab \in R_a \cap L_b$. Let $d = bx$. Then

$$b = vab = vcw = vabxw = bxw = dw,$$

$$d = bx = vabx = vc = (vya)b,$$

$$b = vab = vzc = vzabx = (vza)d,$$

so that $d\mathcal{H}b$. Hence $c = ad \in aH_b$, as required.

I.6.7 Lemma

Let e and f be idempotents of a semigroup S. For every $a \in R_e \cap L_f$, there exists a unique $b \in R_f \cap L_e$ such that $ab = e$ and $ba = f$.

Proof. Let $a \in R_e \cap L_f$. Then $a = ea = af$, $e = ax$, $f = ya$ for some $x, y \in S^1$. Let $b = fx$. Then

$$ab = afx = ax = e,$$

$$ba = fxa = yaxa = yea = ya = f,$$

$$b = fx = f(fx) = fb,$$

$$b = fx = yax = ye,$$

which shows that $b \in R_f \cap L_e$. If c satisfies all the conditions imposed on b, then $c = fc = bac = be = b$, which establishes uniqueness.

I.6.8 Corollary

If e is an idempotent of a semigroup S, then H_e is a group.

Proof. By 6.6, we have that $H_e^2 = H_e$ so H_e is closed under multiplication. Further, e is the identity of H_e, and every element of H_e has a (unique) inverse by 6.7.

The next result is known as *Green's lemma.*

I.6.9 Lemma

Let a and b be $\mathcal{R}$-related elements of a semigroup S. By hypothesis there exist $s, s' \in S^1$ such that $as = b$ and $bs' = a$. Then the mappings

$$\sigma : x \to xs \qquad (x \in L_a),$$

$$\sigma' : y \to ys' \qquad (y \in L_b),$$

are mutually inverse, $\mathcal{R}$-class preserving bijections of L_a and L_b.

Proof. If $x \in L_a$, then $xs \mathcal{L} as$ and thus $xs \in L_b$ in view of $b = as$. Hence σ maps L_a into L_b; analogously σ' maps L_b into L_a. For any $x \in L_a$, we further have $x = ta$ for some $t \in S^1$ and thus

$$x\sigma\sigma' = xss' = (ta)ss' = t(as)s' = tbs' = ta = x.$$

Hence $\sigma\sigma'$ is the identity mapping on L_a; analogously $\sigma'\sigma$ is the identity mapping on L_b. If $x \in L_a$, then $x\sigma = xs$ and $x = (x\sigma)s'$ so that $x \mathcal{R} x\sigma$. Hence σ is $\mathcal{R}$-class preserving; analogously σ' is also.

I.6.10 Corollary

Let a, b, c be elements of a semigroup S such that $a \mathcal{R} b$ and $b \mathcal{L} c$. By hypothesis there exist $s, s', t, t' \in S^1$ such that $as = b$, $bs' = a$, $tb = c$, and

$t'c = b$. Then the mappings

$$\xi : x \to txs \qquad (x \in H_a),$$

$$\xi' : z \to t'zs' \qquad (z \in H_c)$$

are mutually inverse bijections of H_a and H_c.

Proof. In view of the properties of σ in 6.9, we see that $\sigma|_{H_a}$ is a bijection of H_a onto H_b; the dual of 6.9 provides us with a function $\tau : y \to ty$ of R_b onto R_c such that $\tau|_{H_b}$ is a bijection of H_b onto H_c. Obviously, ξ is a composition of these two mappings. A dual argument is valid for ξ'; that ξ and ξ' are mutually inverse follows from the corresponding property of σ and σ', τ and τ'.

I.6.11 Lemma

Let e and f be $\mathcal{D}$-related idempotents of a semigroup S. For $a, b \in S$ such that $ab = e$ and $ba = f$ (these exist by 6.7), the mappings σ and τ defined by

$$\sigma : x \to bxa \qquad (x \in eSe),$$

$$\tau : y \to ayb \qquad (y \in fSf)$$

are mutually inverse, $\mathcal{D}$-class preserving, isomorphisms of the semigroups eSe and fSf.

Proof. If $x \in eSe$, then

$$(bxa)f = (bxa)(ba) = bx(ab)a = bxea = bxa$$

and analogously $f(bxa) = bxa$ so that $x\sigma \in fSf$. If $x, x' \in eSe$, then

$$(x\sigma)(x'\sigma) = (bxa)(bx'a) = bx(ab)x'a = bxex'a = bxx'a = (xx')\sigma.$$

Hence σ is a homomorphism of eSe into fSf. For any $x \in eSe$, we further have

$$x\sigma\tau = a(bxa)b = (ab)x(ab) = exe = x$$

and analogously $y\tau\sigma = y$ for all $y \in fSf$. It follows that the mappings σ and τ are mutually inverse; in particular they are both isomorphisms.

For any $x \in eSe$, we have $x = xe = (xa)b$ so that $x\mathcal{R}xa$; also $xa = exa = a(bxa)$ which gives $xa\mathcal{L}bxa$. Consequently, $x\mathcal{D}x\sigma$ for any $x \in eSe$.

The following notation will come in handy.

I.6.12 Notation

If e is an idempotent of a semigroup S, we write $M_e = \{f \in E_S | f \leq e\}$.

I.6.13 Corollary

Let e and f be $\mathfrak{D}$-related idempotents of a semigroup S. Then $H_e \cong H_f$. If D is a $\mathfrak{D}$-class of S, then either $M_e \cap D = M_f \cap D = \varnothing$ or $M_e \cap D$ and $M_f \cap D$ are isomorphic as partially ordered sets. In particular, if S is 0-bisimple, then M_e and M_f are isomorphic for any $e, f \in E_{S^*}$.

Proof. The first assertion follows from 6.11 since H_e and H_f are groups of units of eSe and fSf, respectively. It also follows from 6.11 that $M_e \cap D$ and $M_f \cap D$ are simultaneously nonempty, and if so, then $\sigma|_{M_e \cap D}$ is the required isomorphism.

The next lemma will prove very useful. It represents the first inkling into the behavior of idempotents in a (0-)simple semigroup which is not completely (0-)simple.

I.6.14 Lemma

If a (0-)simple semigroup S is not completely (0-)simple, then for every nonzero idempotent e of S there exists an idempotent g such that $e > g$ and $e \mathfrak{D} g$.

Proof. We consider the case with zero; the case without zero has virtually the same proof. Since S is not completely 0-simple, e is not primitive and hence $0 \neq f < e$ for some $f \in E_S$. In view of 5.5, we have $SfS = S$ and thus $e = xfy$ for some $x, y \in S$. Let $a = exf$, $b = fye$, $g = ba$. Then $ab = (exf)(fye) = e$, and thus

$$g^2 = b(ab)a = bea = ba = g,$$

$$gf = baf = ba = g, \qquad fg = fba = ba = g,$$

and thus $g \leq f < e$. Further, $a = e(xf)$ and $e = ab$ imply $a \mathfrak{R} e$. Also

$$a = exf = ab(exf) = aba = ag,$$

which together with $g = ba$ implies $a \mathfrak{L} g$. Consequently, $e \mathfrak{D} g$ as required.

The following notation will be useful.

I.6.15 Notation

If a is an element of a semigroup S, let

$$I(a) = J(a) \setminus J_a = \{s \in J(a) | J(s) \neq J(a)\}.$$

I.6.16 Lemma

If a is an element of a semigroup S and $I(a) \neq \varnothing$, then $I(a)$ is an ideal of S.

Proof. Let $x \in I(a)$ and $y \in S$. Then $xy \in J(a)$ since $J(a)$ is an ideal of S. If $xy \in J_a$, then $J(x) \subseteq J(a) = J(xy) \subseteq J(x)$ and the equality prevails, which yields $x \in J_a$, a contradiction. Thus $xy \in I(a)$ and analogously also $yx \in I(a)$.

This result makes it possible to introduce the following basic notion.

I.6.17 Definition

Let a be an element of a semigroup S. The Rees quotient semigroup $J(a)/I(a)$ [where $J(a)/\varnothing = J(a)$] is a *principal factor* of S.

I.6.18 Lemma

Each principal factor of a semigroup is 0-simple, simple, or null.

Proof. Let a be an element of a semigroup S. First let B be an ideal of S such that $I(a) \subset B \subseteq J(a)$. Then for any $b \in B \setminus I(a)$, we have $J(b) = J(a)$. Hence if $x \in J(a)$, then $x \in J(b) \subseteq B$ and thus $B = J(a)$. Assume that $I(a) \neq \varnothing$; the case $I(a) = \varnothing$ is treated similarly. It follows that $Q = S/I(a)$ has no nonzero ideals properly contained in $I = J(a)/I(a)$. Note that I is an ideal of Q. Let $0 \neq c \in I$. Then Q^1cQ^1 is a nonzero ideal of Q contained in I and thus $Q^1cQ^1 = I$.

Now suppose that $J(a)/I(a)$ is not null. Then $I^2 \neq 0$ and hence $I^2 = I$; consequently,

$$I = I^3 = I(Q^1cQ^1)I = (IQ^1)c(Q^1I) \subseteq IcI.$$

It follows that $I = IcI$ which by 5.5 implies that I is 0-simple.

Using the concept just introduced, we can now single out a large class of semigroups as follows.

I.6.19 Definition

A semigroup all of whose principal factors are completely 0-simple or completely simple is *completely semisimple.*

According to 6.18, the principal factors of any semigroup S are 0-simple, null, and at most one can be simple. The possible simple principal factor is the kernel of S by 5.7. Hence in a completely semisimple semigroup, the only completely simple principal factor is its kernel if it exists.

I.6.20 Exercises

(i) Characterize Green's relations on $\mathcal{F}(X)$ and $\mathcal{T}(X)$ for any nonempty set X.

(ii) Find necessary and sufficient conditions on $\alpha, \beta \in \mathcal{F}(X)$ in order that $\alpha\beta \in R_\alpha \cap L_\beta$ (in terms of α and β as partial transformations of X).

(iii) Let X be a nonempty set, $\alpha \in \mathcal{F}(X)$, $\varphi, \psi \in E_{\mathcal{F}(X)}$. If $\alpha \in R_\varphi \cap L_\psi$, construct the unique $\beta \in R_\psi \cap L_\varphi$ (as a function) such that $\alpha\beta = \varphi$, $\beta\alpha = \psi$.

(iv) Let S be a semigroup, λ be a right congruence on S contained in $\mathcal{L}$, and ρ be a left congruence on S contained in $\mathcal{R}$. Show that λ and ρ commute and that their product $\lambda\rho$ is an equivalence relation contained in $\mathcal{D}$. Deduce that the lattice of all congruences on S contained in $\mathcal{H}$ is modular.

(v) Let e be an idempotent of a semigroup S. Show that $L_e R_e = D_e$.

(vi) Prove that the following conditions on a semigroup S are equivalent.

(α) No principal factor of S is null.

(β) For every ideal I of S, $I^2 = I$.

(γ) For every $a \in S$, $a \in SaSaS$.

(Such semigroups are called *semisimple*.)

(vii) For any idempotents e and f of a semigroup S, show that

(α) $e\mathcal{L}f \Leftrightarrow ef = e, fe = f$,

(β) $e\mathcal{R}f \Leftrightarrow ef = f, fe = e$.

(viii) Let ρ be a congruence on a semigroup S such that $\rho \subseteq \mathcal{L}$. Show that for any $a, b \in S$, $a\mathcal{L}b$ if and only if $(a\rho)\mathcal{L}(b\rho)$.

A comprehensive discussion of Green's relations can be found in Clifford–Preston [1].

I.7 REGULARITY

Besides the introduction of (completely) regular elements and semigroups, we prove several useful lemmas concerning these concepts.

I.7.1 Definition

An element a of a semigroup S is *regular* if $a = axa$ for some $x \in S$; S is *regular* if all its elements are regular.

The next result shows that $\mathcal{D}$-classes containing regular elements admit some interesting characterizations.

I.7.2 Lemma

The following conditions on a $\mathcal{D}$-class D of a semigroup S are equivalent.

(i) D contains an idempotent.
(ii) Every $\mathcal{L}$- and $\mathcal{R}$-class contained in D contains an idempotent.
(iii) D contains a regular element.
(iv) Every element of D is regular.

Proof. (i) *implies* (ii). Let e be an idempotent in D and L be an $\mathcal{L}$-class contained in D. Then for any $c \in L$, there exists an element a such that $e\mathcal{R}a$ and $a\mathcal{L}c$. Hence $a = eu$, $e = av$ for some $u, v \in S^1$. Letting $f = va$, we obtain $a = e(eu) = ea = a(va) = af$ so that $a\mathcal{L}f$ and $f = va = v(ea) = v(av)a = (va)^2 = f^2$, as required. A similar argument shows that every $\mathcal{R}$-class contained in D contains an idempotent.

(ii) *implies* (iii). This is trivial.

(iii) *implies* (iv). Let $a, b \in D$ with $a = axa$. Then $a\mathcal{L}c$ and $c\mathcal{R}b$ for some $c \in S$ and thus $a = uc$, $c = va = bz$, $b = cw$ for some $u, v, w, z \in S^1$. Consequently,

$$b = cw = vaw = (va)xaw = bzx(uc)w = b(zxu)b$$

and b is regular.

(iv) *implies* (i). If $a = axa \in D$, then $a\mathcal{L}xa$ so that $xa \in D$ and xa is evidently an idempotent.

In view of this lemma, it is natural to introduce the following concept.

I.7.3 Definition

A $\mathcal{D}$-class of a semigroup is *regular* if it contains a regular element.

The next definition provides a concept stronger than that of a regular element.

I.7.4 Definition

An element a of a semigroup S is *completely regular* if $a = axa$ and $xa = ax$ for some $x \in S$; S is *completely regular* if all its elements are completely regular.

For this concept, the following result is a close analogue of 7.2.

I.7.5 Lemma

The following conditions on an $\mathcal{H}$-class H of a semigroup S are equivalent.

(i) H contains an idempotent.
(ii) H is a maximal subgroup of S.
(iii) Every element of H is completely regular.
(iv) H contains a competely regular element.

Proof. (i) *implies* (ii). That H is a subgroup of S is stated in I.6.8. If G is a subgroup of S having e as an identity, then it follows at once that for any $g \in G$, we have $g\mathcal{H}e$ so that $G \subseteq H$.

The implications (ii) *implies* (iii) and (iii) *implies* (iv) are trivial.

(iv) *implies* (i). If $a \in H$ and $a = axa$, $xa = ax$, then it is easy to verify that $ax \in H$ and that ax is an idempotent.

I.7.6 Corollary

If G is a maximal subgroup of a semigroup S, then G is an $\mathcal{H}$-class of S. A semigroup S is a union of its (maximal pairwise disjoint) subgroups if and only if S is completely regular.

Proof. The proof of this corollary is left as an exercise.

The next concept is of basic importance.

I.7.7 Definition

If a and b are elements of a semigroup S, then b is an *inverse* of a if $a = aba$ and $b = bab$.

The relationship between regularity and inverses is elucidated by the following simple result.

I.7.8 Lemma

Every regular element of a semigroup has an inverse.

Proof. For if $a = axa$, then xax is an inverse of a.

The following lemma has many useful applications.

I.7.9 Lemma

Let T be a regular subsemigroup of a semigroup S. The Green's relations $\mathcal{L}$, $\mathcal{R}$, and $\mathcal{H}$ on T are the restrictions of those on S.

Proof. It suffices to consider $\mathcal{L}$. Let $\mathcal{L}_S$ and $\mathcal{L}_T$ be the corresponding $\mathcal{L}$-relations, $a, b \in T$, and $a\mathcal{L}_S b$. Then $a = ub$ and $b = va$ for some $u, v \in S^1$. Letting a' and b' be inverses of a and b, respectively, in T, we obtain $a = (ub)b'b = ab'b$ and $b = (va)a'a = ba'a$. But then $a\mathcal{L}_T b$ which proves that $\mathcal{L}_S|_T \subseteq \mathcal{L}_T$; the opposite inclusion is trivial.

The next two lemmas are of general interest for homomorphisms of regular semigroups. The first is known as *Lallement's lemma* and is of fundamental importance.

I.7.10 Lemma

Let φ be a homomorphism of a regular semigroup S onto a semigroup T. If $e \in E_T$, then $e\varphi^{-1}$ contains an idempotent.

Proof. Let $a \in S$ be such that $a\varphi = e$. Let x be an inverse of a^2 and $f = axa$. Then $f \in E_S$ and

$$f\varphi = (axa)\varphi = (a\varphi)(x\varphi)(a\varphi) = (a\varphi)^2(x\varphi)(a\varphi)^2$$

$$= (a^2xa^2)\varphi = a^2\varphi = (a\varphi)^2 = e^2 = e,$$

as required.

I.7.11 Lemma

Let φ, S, and T be as in 7.10. If $e, f \in E_T$ are such that $e \geq f$, then for every $g \in E_S$ such that $g\varphi = e$, there exists $h \in E_S$ such that $g \geq h$ and $h\varphi = f$.

Proof. By 7.10, there exist $g, h' \in E_S$ such that $g\varphi = e$ and $h'\varphi = f$. Let x be an inverse of $h'gh'$ and let $h = gh'xh'$. Then

$$h^2 = gh'x(h'gh')xh' = gh'xh' = h,$$

and since $x\varphi$ is an inverse of $(h'gh')\varphi = fef = f$, we get

$$h\varphi = ef(x\varphi)f = f.$$

I.7.12 Exercises

(i) If T is a regular subsemigroup of a semigroup S, show that neither $\mathcal{D}$- nor $\mathcal{J}$-relation on T need be the restrictions of those on S.

(ii) Prove that $\mathcal{F}(X)$ is a regular semigroup and characterize its completely regular elements. Find suitable isomorphic copies of maximal subgroups of $\mathcal{F}(X)$. Do the same for $\mathcal{T}(X)$.

(iii) Show that each of the following conditions on a semigroup S is equivalent to regularity.

(α) Every principal left ideal is generated by an idempotent.

(β) Every principal left ideal has a right identity.

(γ) Every $\mathcal{L}$-class contains an idempotent.

(δ) For every right ideal R and every left ideal L, $R \cap L = RL$.

(iv) Show that if a regular $\mathcal{D}$-class of a semigroup S is a subsemigroup of S, then it must be bisimple.

(v) Prove that the following conditions on an element a of a semigroup S are equivalent.

(α) a is completely regular.

(β) $a \mathcal{H} a^2$.

(γ) $a \in a^2S \cap Sa^2$.

(δ) $a \in a^2Sa^2$.

(vi) Show that a semigroup S is completely regular if and only if for every $a \in S$, $a \in a^2Sa$.

(vii) Give an example of a regular semigroup S and a subsemigroup T of S such that $\mathcal{L}_S|_T \neq \mathcal{L}_T$ and $\mathcal{H}_S|_T \neq \mathcal{H}_T$.

(viii) Let V be a vector space over a division ring and L be the semigroup of all linear transformations on V written as right operators.

(α) Show that L is regular.

(β) Characterize idempotents of L. Which ones are primitive?

(γ) Characterize completely regular elements of L.

(δ) Determine Green's relations on L.

(ε) Determine all ideals of L.

(ix) Show that every $\mathcal{D}$-class of a regular semigroup S with linearly ordered idempotents is a subsemigroup of S.

For fuller discussions of regularity, we refer to the standard texts on semigroups.

I.8 THE TRANSLATIONAL HULL

We present here a list of concepts concerning translations, and in particular the translational hull, of arbitrary semigroups and establish a few simple properties of these concepts.

I.8.1 Definition

Let S be a semigroup and let x and y stand for arbitrary elements of S. A transformation λ (respectively ρ) on S is a *left* (respectively *right*) *translation* of S if $\lambda(xy) = (\lambda x)y$ [respectively $(xy)\rho = x(y\rho)$]; if also $x(\lambda y) = (x\rho)y$, then λ and ρ are *linked* and the pair (λ, ρ) is a *bitranslation* of S. A bitranslation $\omega = (\lambda, \rho)$ can be considered as a "bioperator" by defining $\omega x = \lambda x$, $x\omega = x\rho$. The set $\Lambda(S)$ [respectively $\mathrm{P}(S)$] of *all left* (respectively *right*) *translations* of S is a semigroup under the composition of functions $(\lambda\lambda')x = \lambda(\lambda' x)$ [respectively $x(\rho\rho') = (x\rho)\rho'$]. Bitranslations multiply componentwise:

$$(\lambda, \rho)(\lambda', \rho') = (\lambda\lambda', \rho\rho').$$

It is easy to verify that the product of two bitranslations of a semigroup, considered as elements of $\Lambda(S) \times \mathrm{P}(S)$, is again a bitranslation. We thus arrive at the following basic concept.

I.8.2 Definition

The subsemigroup of the direct product $\Lambda(S) \times \mathrm{P}(S)$ consisting of all bitranslations of a semigroup S is the *translational hull* of S, to be denoted by $\Omega(S)$.

The following special kinds of translations and bitranslations are of particular importance.

I.8.3 Definition

Let s be an element of a semigroup S. Then the function λ_s (respectively ρ_s), defined by $\lambda_s x = sx$ (respectively $x\rho_s = xs$) for all $x \in S$, is the *inner left* (respectively *right*) *translation of S induced by s*; the pair $\pi_s = (\lambda_s, \rho_s)$ is the *inner bitranslation induced by s*. The set $\Pi(S) = \{\pi_s | s \in S\}$ is the *inner part of* $\Omega(S)$; analogously $\Gamma(S) = \{\lambda_s | s \in S\}$ and $\Delta(S) = \{\rho_s | s \in S\}$ are the *inner parts of* $\Lambda(S)$ *and* $\mathrm{P}(S)$, respectively.

The next lemma comes in very handy in manipulation with translations; it will be used often without express reference.

I.8.4 Lemma

In any semigroup S, we have

(i) $\lambda\lambda_s = \lambda_{\lambda s}, \quad \rho_s\rho = \rho_{s\rho} \quad [s \in S, \lambda \in \Lambda(S), \rho \in \mathrm{P}(S)]$,

(ii) $\omega\pi_s = \pi_{\omega s}, \pi_s\omega = \pi_{s\omega} \quad [s \in S, \omega \in \Omega(S)]$.

Proof. With the notation as in the statement of the lemma, we obtain for any $x \in S$,

$$(\lambda\lambda_s)x = \lambda(\lambda_s x) = \lambda(sx) = (\lambda s)x = \lambda_{\lambda s}x,$$

proving that $\lambda\lambda_s = \lambda_{\lambda s}$. The proof of the second part of (i) is symmetric. If $(\lambda, \rho) \in \Omega(S)$, then

$$x(\rho\rho_s) = (x\rho)\rho_s = (x\rho)s = x(\lambda s) = x\rho_{\lambda s}$$

proving that $\rho\rho_s = \rho_{\lambda s}$, which together with the first part of (i) shows that $\omega\pi_s = \pi_{\omega s}$. The second part of (ii) follows similarly.

I.8.5 Corollary

For any semigroup S, $\Gamma(S)$ is a left ideal of $\Lambda(S)$; $\Delta(S)$ is a right ideal of $P(S)$; $\Pi(S)$ is an ideal of $\Omega(S)$.

We need some more concepts.

I.8.6 Definition

For any semigroup S, the mapping

$$\pi : s \to \pi_s \qquad (s \in S)$$

is the *canonical homomorphism of* S *into* $\Omega(S)$ [or *onto* $\Pi(S)$] (it is easy to verify that π is indeed a homomorphism). The semigroup S is *weakly reductive* if π is one-to-one (equivalently, $ax = bx$ and $xa = xb$ for all $x \in S$ implies $a = b$).

I.8.7 Lemma

If S is a weakly reductive semigroup and $\omega, \omega' \in \Omega(S)$ are such that $\omega\pi_s = \omega'\pi_s$ and $\pi_s\omega = \pi_s\omega'$ for all $s \in S$, then $\omega = \omega'$.

Proof. For any $s \in S$, we have $\pi_{\omega s} = \pi_{\omega' s}$ and $\pi_{s\omega} = \pi_{s\omega'}$ by 8.6, and thus $\omega s = \omega' s$ and $s\omega = s\omega'$ since S is weakly reductive. Hence $\omega = \omega'$ as asserted.

I.8.8 Definition

A set T of bitranslations of a semigroup S is *permutable* if for any $(\lambda, \rho), (\lambda', \rho') \in T$, we have $(\lambda s)\rho' = \lambda(s\rho')$ for all $s \in S$.

I.8.9 Lemma

For a weakly reductive semigroup S, $\Omega(S)$ is permutable.

Proof. For any $(\lambda, \rho), (\lambda', \rho') \in \Omega(S)$ and $x, y \in S$, we obtain

$$x[(\lambda y)\rho'] = [x(\lambda y)]\rho' = [(x\rho)y]\rho' = (x\rho)(y\rho') = x[\lambda(y\rho')],$$

$$[(\lambda y)\rho']x = (\lambda y)(\lambda' x) = \lambda[y(\lambda' x)] = \lambda[(y\rho')x] = [\lambda(y\rho')]x,$$

which by weak reductivity implies $(\lambda y)\rho' = \lambda(y\rho')$.

The next result provides an isomorphism of translational hulls of two isomorphic semigroups.

I.8.10 Lemma

Let θ be an isomorphism of a semigroup S onto a semigroup T. For every $(\lambda, \rho) \in \Omega(S)$, define $\bar{\lambda}$ and $\bar{\rho}$ by

$$\bar{\lambda}t = [\lambda(t\theta^{-1})]\theta, \qquad t\bar{\rho} = [(t\theta^{-1})\rho]\theta \qquad (t \in T).$$

Then the mapping

$$\bar{\theta}: (\lambda, \rho) \to (\bar{\lambda}, \bar{\rho}) \qquad [(\lambda, \rho) \in \Omega(S)]$$

is an isomorphism of $\Omega(S)$ onto $\Omega(T)$ satisfying

$$\pi_s\bar{\theta} = \pi_{s\theta}, \qquad (\omega s)\theta = (\omega\bar{\theta})(s\theta), \qquad (s\omega)\theta = (s\theta)(\omega\bar{\theta})$$

$$[s \in S, \omega \in \Omega(S)].$$

Proof. The straightforward verification is left as an exercise.

The translational hull of an orthogonal sum of semigroups satisfying a mild condition is given by the following result.

I.8.11 Lemma

Let $S = \Sigma_{\alpha \in A} S_\alpha$ and assume that $S_\alpha^2 = S_\alpha$ for every $\alpha \in A$. Then $\Omega(S) \cong \Pi_{\alpha \in A} \Omega(S_\alpha)$.

Proof. Let η be a function defined by

$$\eta: (\lambda^\alpha, \rho^\alpha)_{\alpha \in A} \to (\lambda, \rho) \qquad [(\lambda^\alpha, \rho^\alpha) \in \Omega(S_\alpha)]$$

where $\lambda 0 = 0\rho = 0$ and $\lambda a = \lambda^\alpha a$, $a\rho = a\rho^\alpha$ if $a \in S_\alpha^*$. The proof that η is an isomorphism of $\prod_{\alpha\in A}\Omega(S_\alpha)$ into $\Omega(S)$ is left as an exercise. To prove that η is onto, let $(\lambda, \rho) \in \Omega(S)$. Let $a \in S_\alpha$ and assume that $\lambda a \neq 0$. By hypothesis $a = xy$ for some $x, y \in S_\alpha$. Hence, $\lambda a = \lambda(xy) = (\lambda x)y \neq 0$, and thus $\lambda x \in S_\alpha$ since $y \in S_\alpha$ so that $\lambda a \in S_\alpha$. It follows that λ maps S_α into itself; analogously ρ has the same property. Hence letting $\lambda^\alpha = \lambda|_{S_\alpha}$ and $\rho^\alpha = \rho|_{S_\alpha}$ for all $\alpha \in A$, we obtain $(\lambda^\alpha, \rho^\alpha)_{\alpha\in A}\eta = (\lambda, \rho)$ and thus η is onto.

We will also need the following notions.

I.8.12 Definition

Let S be a semigroup. Then S is *left* (respectively *right*) *cancellative* if $xa = xb$ (respectively $ax = bx$) in S implies that $a = b$; S is *cancellative* if it is both left and right cancellative.

Further, S is *left* (respectively *right*) *reductive* if $xa = xb$ (respectively $ax = bx$) for all $x \in S$ implies that $a = b$; S is *reductive* if it is both left and right reductive.

I.8.13 Lemma

Let S be a reductive semigroup. Then the projections

$$\sigma : (\lambda, \rho) \to \lambda \qquad [(\lambda, \rho) \in \Omega(S)],$$

$$\tau : (\lambda, \rho) \to \rho \qquad [(\lambda, \rho) \in \Omega(S)]$$

are both one-to-one.

Proof. Let $(\lambda, \rho), (\lambda, \rho') \in \Omega(S)$. Then for any $x, y \in S$,

$$(x\rho)y = x(\lambda y) = (x\rho')y.$$

Since y is arbitrary, we have $x\rho = x\rho'$. This holds for all $x \in S$, so that $\rho = \rho'$. Thus σ is one-to-one, and the proof for τ is analogous.

I.8.14 Exercises

(i) Show that every bitranslation of a semigroup S is inner if and only if S has an identity.

(ii) Show that for a weakly reductive semigroup S,

$$i_{\Lambda(S)\times P(S)}(\Pi(S)) = \Omega(S).$$

(iii) Show that if S is a cancellative semigroup, so is $\Omega(S)$.

(iv) Prove that a weakly reductive semigroup in which every left translation is linked to at most one right translation must be right reductive.

(v) Prove that in a right reductive semigroup, every left translation is linked to at most one right translation.

(vi) Show that a regular semigroup is weakly reductive but need not be either left or right reductive.

(vii) Characterize semigroups in which every transformation is a left translation. Also characterize semigroups in which the identity transformation is the only left translation.

(viii) Show that a cancellative regular semigroup is a group.

(ix) Let θ be an antiisomorphism of a semigroup S onto a semigroup T. For every $(\lambda, \rho) \in \Omega(S)$, define $\bar{\lambda}$ and $\bar{\rho}$ by

$$\bar{\lambda}t = \left[(t\theta^{-1})\rho\right]\theta, \qquad t\bar{\rho} = \left[\lambda(t\theta^{-1})\right]\theta \qquad (t \in T).$$

What kind of mapping is $(\lambda, \rho) \to (\bar{\lambda}, \bar{\rho})$?

(x) Let S be a weakly reductive semigroup. Show that every automorphism of $\Pi(S)$ can be uniquely extended to an automorphism of $\Omega(S)$.

(xi) Show that a cancellative simple semigroup containing an idempotent is a group.

A comprehensive discussion of the translational hull can be found in Petrich [8]; see also Clifford–Preston [1].

I.9 IDEAL EXTENSIONS

One of the simplest ways of building a larger semigroup out of two given semigroups is the formation of ideal extensions. We present here a minimum of the theory of ideal extensions.

I.9.1 Definition

If S is an ideal of a semigroup V, then V is an *ideal extension* of S. More precisely, V is an ideal extension of S by the Rees quotient semigroup V/S. This extension is *proper* if $V \neq S$.

If no other type of extension is being considered, one simply refers to an *extension* rather than an ideal extension. The *extension problem* consists of starting with a semigroup S and a semigroup Q with zero and constructing all semigroups V such that S is an ideal of V and $V/S \cong Q$. In view of the definition of the Rees quotient V/S, we may take $V = S \cup Q^*$, where $Q^* = Q \setminus \{0\}$ and Q is assumed disjoint from S, and find all multiplications on V which agree with the given multiplication on S and the products ab in Q for which $ab \neq 0$.

If S is an ideal of V, then each element $v \in V$ induces a left translation on S by the multiplication on the left $s \to vs$, and a right translation by $s \to sv$. This is formalized in the following basic result.

I.9.2 Theorem

Let S be an ideal of a semigroup V. For each $v \in V$, let

$$\lambda^v s = vs, \qquad s\rho^v = sv \qquad (s \in S).$$

Then the mapping

$$\tau = \tau(V: S) : v \to \tau^v = (\lambda^v, \rho^v) \qquad (v \in V)$$

is a homomorphism of V onto a semigroup of permutable bitranslations of S which extends the canonical homomorphism $\pi : S \to \Omega(S)$. If S is weakly reductive, then τ is the unique extension of π to a homomorphism of V into $\Omega(S)$.

Proof. That τ is a homomorphism of S onto a set of permutable bitranslations of S is an obvious consequence of the associative law in V and its verification is left as an exercise. The definition of the canonical homomorphism π (see 8.6) shows that τ extends π. Assume that $\omega : v \to \omega^v$ is another homomorphism of V into $\Omega(S)$ which extends π. Then for any $v \in V$ and $s \in S$, we have

$$\omega^v \pi_s = \omega^v \omega^s = \omega^{vs} = \pi_{vs} = \tau^{vs} = \tau^v \tau^s = \tau^v \pi_s$$

and dually $\pi_s \omega^v = \pi_s \tau^v$. Since this holds for all $s \in S$, if S is weakly reductive, 8.7 implies that $\omega^v = \tau^v$, and thus $\omega = \tau$.

In view of the above result, it is useful to introduce the following concepts.

I.9.3 Definition

The mapping $\tau(V: S)$ is the *canonical homomorphism of V into* $\Omega(S)$. The image of V under $\tau(V: S)$ is the *type of the extension V* of S and will be denoted by $T(V: S)$.

For a solution of the extension problem for weakly reductive semigroups, it is convenient to have the following notion.

I.9.4 Definition

Let S and Q be semigroups, and assume that Q has a zero. A partial homomorphism $\varphi: Q^* \to \Omega(S)$ such that $(a\varphi)(b\varphi) \in \Pi(S)$ if $a, b \in Q^*$, $ab = 0$ is an *extension function.*

The next result provides a general solution for the extension problem for weakly reductive semigroups.

I.9.5 Theorem

Let S be a weakly reductive semigroup and Q be a semigroup with zero disjoint from S. Let $\varphi: Q^* \to \Omega(S)$, in notation $\varphi: a \to \varphi^a = (\lambda^a, \rho^a)$ be an extension function. On $V = S \cup Q^*$ define a multiplication $*$ by

$$a * b = \begin{cases} a\rho^b & \text{if } a \in S,\ b \in Q^* \\ \lambda^a b & \text{if } a \in Q^*,\ b \in S \\ c & \text{if } a, b \in Q^*,\ ab = 0 \text{ in } Q,\ \varphi^a\varphi^b = \pi_c, \end{cases}$$

and the remaining products are those in S and Q. Then V is an ideal extension of S by Q, to be denoted by $V = \langle S, Q; \varphi\rangle$. Conversely, every ideal extension of S by Q can be so constructed.

Proof. The proof of the direct part consists in verifying the associative law in V, which is left as an exercise. Note that weak reductivity of S provides the unique element c in the third part of the definition of $a * b$ above. The converse is proved easily by applying 9.2.

The next concept will be useful here and later.

I.9.6 Definition

For S a subsemigroup of semigroups V and V', a homomorphism φ of V into V' which leaves the elements of S fixed is an *S-homomorphism*. Further, if φ is also an isomorphism of V onto V', it is an *S-isomorphism*. Finally, if $V' = S$, then φ is an *S-endomorphism of V.*

If V and V' are S-isomorphic ideal extensions of S, then V and V' are *equivalent ideal extensions* of S.

A simple criterion for equivalence of ideal extensions of weakly reductive semigroups is provided by the next result.

I.9.7 Proposition

Two ideal extensions $V = \langle S, Q; \varphi \rangle$ and $V' = \langle S, Q'; \varphi' \rangle$ of a weakly reductive semigroup S are equivalent if and only if there exists an isomorphism ψ of Q onto Q' such that $\varphi = \psi\varphi'$.

Proof. First let θ be an S-isomorphism of V onto V'. Define a mapping ψ on Q by $\psi|_{Q^*} = \theta|_{V \setminus S}$ and $0\psi = 0'$. If $a, b \in Q^*$ and $ab = 0$ in Q, then $ab \in S$ in V and hence

$$(a\psi)(b\psi) = (a\theta)(b\theta) = (ab)\theta = ab \in S.$$

Thus $ab = 0$ in Q implies $(a\psi)(b\psi) = 0'$ in Q', which evidently implies that ψ is an isomorphism of Q onto Q'. For any $a \in Q^*$ and $s \in S$, we obtain

$$\lambda^a s = as = (as)\theta = (a\theta)(s\theta) = (a\psi)s = \lambda'^{a\psi} s$$

and analogously $s\rho^a = s\rho'^{a\psi}$, and thus $\varphi = \psi\varphi'$.

For the converse, we define θ on V by $\theta|_S = \iota_S$, $\theta|_{V \setminus S} = \psi|_{Q^*}$. It is left as an exercise to verify that θ is a homomorphism of V onto V'; the remaining properties of θ are obvious.

The type $T(V\colon S)$ of an ideal extension V of a semigroup S evidently contains $\Pi(S)$ since the canonical homomorphism $\tau(V\colon S)$ maps S onto $\Pi(S)$. The following concepts describe ideal extensions with extremal properties of the type.

I.9.8 Definition

An ideal extension V of a semigroup S is *strict* if $T(V\colon S) = \Pi(S)$; V is a *pure* extension if $\tau^a(V\colon S) \in \Pi(S)$ implies $a \in S$.

In other words, V is a strict extension of S if and only if every element of V induces an inner bitranslation of S, and V is a pure extension of S if and only if only the elements of S induce inner bitranslations of S. The relevance of these concepts is justified by the following result.

I.9.9 Theorem

Let V be an ideal extension of a semigroup S. The complete inverse image K of $\Pi(S)$ under $\tau(V\colon S)$ is the greatest strict extension of S contained in V, and V is a pure extension of K.

Proof. Note that $\tau(V\colon S)$ is a homomorphism of V onto $T(V\colon S)$ which contains $\Pi(S)$ as an ideal, which implies that K is an ideal of V. In addition,

$\tau(V: S)$ maps S onto $\Pi(S)$ so that $K \supseteq S$ and S is an ideal of K. It is obvious that K is a strict extension of S. If V' is an extension of S and a subsemigroup of V, then $\tau(V': S)$ must be the restriction of $\tau(V: S)$ to V'. Hence if V' is a strict extension of S, then we must have $V' \subseteq K$.

Let $v \in V$ be such that $\tau^v(V: K) \in \Pi(K)$. Then for some $a \in K$, we have $vk = ak$ and $kv = ka$ for all $k \in K$. In particular, $vs = as$ and $sv = sa$ for all $s \in S$, and thus $\tau^v(V: S) \in \Pi(S)$. But then $v \in K$, which proves that V is a pure extension of K.

I.9.10 Corollary

Every ideal extension of S by Q is a pure extension of a strict extension. In particular, if Q has no proper nonzero ideals, the extension is either strict or pure.

We now consider some special types of extensions. For this we first need some standard concepts.

I.9.11 Definition

Let S be a subsemigroup of a semigroup V. Then S is a *retract* of V if there exists an S-endomorphism of V onto S, which is then called a *retraction*. If, in addition, S is an ideal of V, then S is a *retract ideal* of V and V is a *retract ideal extension* of S.

I.9.12 Proposition

Every retract ideal extension V of a semigroup S is strict. The converse holds if S is weakly reductive, in which case V has only one S-endomorphism.

Proof. Let φ be an S-endomorphism of V. For any $v \in V$ and $s \in S$, we obtain

$$vs = (vs)\varphi = (v\varphi)(s\varphi) = (v\varphi)s$$

and analogously $sv = s(v\varphi)$, which implies that $\tau^v(V: S) = \tau^{v\varphi}(V: S) \in \Pi(S)$ since $v\varphi \in S$. This means that V is a strict extension of S.

Now assume that S is weakly reductive and let V be a strict extension of S. Then the canonical homomorphism $\pi : S \to \Omega(S)$ is an isomorphism of S onto $\Pi(S)$, and thus $\tau(V: S)\pi^{-1}$ is an S-endomorphism of V. The proof of the last statement of the proposition is left as an exercise.

A particularly simple way of forming ideal extensions uses the following concept.

I.9.13 Definition

Let S and Q be semigroups, assume that Q has a zero, and let $\varphi : Q^* \to S$ be a partial homomorphism. Assume that S and Q are disjoint, and on $V = S \cup Q^*$ define a multiplication by

$$a * b = \begin{cases} a(b\varphi) & \text{if } a \in S, b \in Q^* \\ (a\varphi)b & \text{if } a \in Q^*, b \in S \\ (a\varphi)(b\varphi) & \text{if } a, b \in Q^*, ab = 0 \text{ in } Q, \end{cases}$$

and let the other products be as originally given in S and Q. Then the multiplication in V is *determined by the partial homomorphism* φ.

I.9.14 Proposition

With the notation as in 9.13, V is a retract extension of S by Q. Conversely, every retract extension of S by Q is determined by a partial homomorphism of Q^* into S.

Proof. Straightforward verification shows that the multiplication defined in 9.13 is associative. It is obvious that S is an ideal for it. The mapping χ defined by

$$\chi : a \to \begin{cases} a\varphi & \text{if } a \in Q^* \\ a & \text{if } a \in S \end{cases}$$

is easily seen to be a retraction of V onto S.

For the converse, if ψ is a retraction of V onto S, then an easy verification shows that $\varphi = \psi|_{Q^*}$ is a partial homomorphism of Q^* into S which determines the multiplication in V.

The details of this proof are left as an exercise.

The next two lemmas will be useful.

I.9.15 Lemma

Every retract ideal extension of a semigroup S by a semigroup Q is a subdirect product of S and Q.

Proof. Let V be a retract extension of S by Q. Let φ be a retraction of V onto S and ψ be the natural homomorphism of V onto Q associated with the Rees congruence relative to the ideal S of V. Define a mapping χ by

$$\chi : v \to (v\varphi, v\psi) \qquad (v \in V).$$

Then χ is a homomorphism of S into $S \times Q$ with both projections of $S\chi$ being onto. Assume that $a\chi = b\chi$ so that $a\varphi = b\varphi$ and $a\psi = b\psi$. By the latter condition, either $a, b \notin S$ and $a = b$ or $a, b \in S$. In the case that $a, b \in S$, the first condition yields $a = b$. Therefore χ is one-to-one and V is a subdirect product of S and Q.

I.9.16 Lemma

Every ideal extension of a monoid is a retract extension.

Proof. Let V be an ideal extension of a monoid S. Define a mapping φ on V by

$$\varphi : v \to v1 \qquad (v \in V)$$

where 1 is the identity of S. Simple verification shows that φ is a retraction of V onto S.

For a different kind of extension, we introduce the following concepts.

I.9.17 Definition

Let S be a subsemigroup of a semigroup V. A congruence ρ on V is an *S-congruence* if its restriction to S is the equality relation on S.

An ideal extension V of a semigroup S is *dense* if the equality relation is the only S-congruence on V. If, in addition, V is under inclusion a maximal dense extension, then S is a *densely embedded ideal* of V.

The next result essentially describes all dense extensions of weakly reductive semigroups.

I.9.18 Theorem

Let V be an ideal extension of a weakly reductive semigroup S. Then the congruence $\mathfrak{T}(V: S)$ induced by $\tau(V: S)$ is the greatest S-congruence on V. In particular, V is a dense extension of S if and only if $\tau(V: S)$ is one-to-one. Finally, S is a densely embedded ideal of V if and only if $\tau(V: S)$ is an isomorphism of V onto $\Omega(S)$.

Proof. Let ρ be an S-congruence on V, and let $a\rho b$. Then for any $s \in S$, we have $as = bs$ and $sa = sb$ since the restriction of ρ to S is the equality relation. But then $\tau^a(V: S) = \tau^b(V: S)$, which shows that $\rho \subseteq \mathfrak{T}(V: S)$.

Note that by weak reductivity $\pi = \tau(V: S)|_S$ is one-to-one, and hence $\mathfrak{T}(V: S)$ restricted to S is the equality relation. It now follows from the first part of the theorem that V is a dense extension of S if and only if $\mathfrak{T}(V: S)$ is the equality relation.

Now assume that S is a densely embedded ideal of V. Then $\tau(V: S)$ is an isomorphism of V onto the type $T(V: S)$. If $T(V: S) \neq \Omega(S)$, we could define a multiplication on $V \cup (\Omega(S) \setminus T(V: S))$ in a natural way making it a dense extension of S strictly containing V, contradicting the maximality of V.

Conversely, let $\tau(V: S)$ be an isomorphism of V onto $\Omega(S)$. Let V' be any extension of S strictly containing V as a subsemigroup and let $a \in V' \setminus V$. Then $\tau^a(V': S) \in \Omega(S)$ and since $\tau(V: S)$ maps V onto $\Omega(S)$, there exists $b \in V$ such that $\tau^a(V': S) = \tau^b(V: S)$. But $\tau(V: S)$ is the restriction of $\tau(V': S)$ to V, so that $\tau^a(V': S) = \tau^b(V': S)$. Since $a \neq b$, this shows that $\tau(V': S)$ is not one-to-one, and hence V' is not a dense extension of S. Consequently, S is a densely embedded ideal of V.

To conclude, we introduce the following notion.

I.9.19 Definition

An isomorphism φ of a semigroup S into a semigroup T is a *dense embedding* if $S\varphi$ is a densely embedded ideal of its idealizer in T.

I.9.20 Exercises

(i) Let S be an ideal of a semigroup V, and let φ be an S-endomorphism of V. Show that φ written on the left (respectively right) is a left (respectively right) translation of V and that the two translations are linked and permutable.

(ii) Show that two equivalent ideal extensions of any semigroup have the same type. Also show that two dense ideal extensions of a weakly reductive semigroup having the same type are equivalent. Deduce that the classes of equivalent dense extensions of a weakly reductive semigroup S can be put into one-to-one correspondence with subsemigroups of $\Omega(S)/\Pi(S)$ containing its zero (in particular, if S is a densely embedded ideal of V and V', then they are S-isomorphic).

(iii) Show that a dense extension of a left reductive right cancellative semigroup is left reductive and right cancellative. Deduce that for any ideal S of a semigroup V, V is cancellative if and only if S is cancellative and V is a dense extension of S.

(iv) Show that a dense extension of a commutative reductive semigroup is commutative and reductive. Deduce that a dense extension of a semilattice is a semilattice. Also deduce that for an ideal S of a semigroup V, V is commutative and cancellative if and only if S is commutative and cancellative and V is a dense extension of S.

(v) Let V be an extension of a commutative reductive semigroup S. Show that S is contained in the center of V.

(vi) Give an example of a noncommutative semigroup which is an extension of a commutative semigroup by a commutative semigroup.

(vii) Let S be any semigroup and V be the semigroup obtained by the adjunction of an identity to S. Show that V is either a retract extension or a dense extension of S.

(viii) Show that the following conditions on a semigroup S are equivalent.

(α) S has an identity.
(β) Every extension of S is strict.
(γ) Every extension of S is a retract extension.

Ideal extensions are systematicaly treated in Petrich [8].

I.10 FREE SEMIGROUPS

We construct a free semigroup on a nonempty set X as the set of all nonempty words with letters in X under concatenation. Free monoids, free semigroups with involution, and free groups admit similar constructions. We also introduce defining relations.

Even though we will mainly speak about semigroups, the concepts involved are of a universal algebraic nature. For example, in the first definition below, an algebra means a set with possibly several operations with various arities. The class considered is allowed to contain only algebras of the same kind, and homomorphisms among these algebras respect all operations.

We start with a list of standard concepts.

I.10.1 Definition

Let $\mathcal{C}$ be a class of algebras, F an algebra in $\mathcal{C}$, X a nonempty set, and $\varphi: X \to F$ a mapping. The pair (F, φ) is a *free $\mathcal{C}$-algebra on X* if for any $C \in \mathcal{C}$ and any mapping $\psi: X \to C$ there exists a unique homomorphism $\bar{\psi}: F \to C$ making the diagram

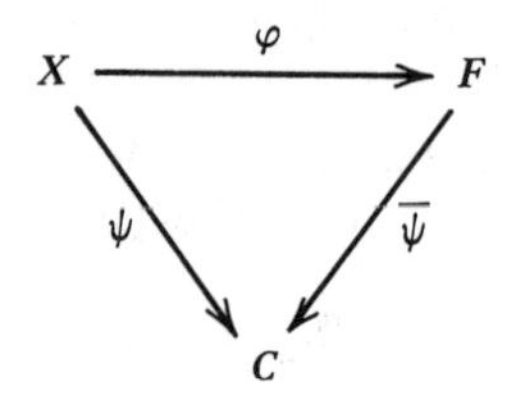

commutative.

Note that uniqueness in the above definition is assured if $X\varphi$ generates F. The mapping φ is often omitted, and one speaks of F as $\mathcal{C}$-*free*, or as a *free object in* $\mathcal{C}$ (*on* X), or as a *relatively free object* (*in* $\mathcal{C}$).

We now construct the first free object.

I.10.2 Notation

Let X be a nonempty set. The finite sequences of elements of X, written $x_1x_2 \cdots x_n$, are *words* over the *alphabet* X; the *empty word* is the sequence $\varnothing$ consisting of the empty set alone; the elements of X are *letters* (or *variables*). The set X^* of all words over X is provided with the multiplication of *juxtaposition* (or *concatenation*):

$$(x_1x_2 \cdots x_m)(y_1y_2 \cdots y_n) = x_1 \cdots x_my_1 \cdots y_n,$$

$$(x_1x_2 \cdots x_m)\varnothing = \varnothing(x_1 \cdots x_m) = x_1 \cdots x_m, \qquad \varnothing\varnothing = \varnothing,$$

and it evidently forms a monoid. Its subsemigroup consisting of all nonempty words is denoted by X^+. One-letter words are identified with letters.

I.10.3 Proposition

For any nonempty set X, (X^*, ι_X) [respectively (X^+, ι_X)] is a free monoid (respectively semigroup) on X.

Proof. Let ψ be a mapping of X into a monoid M. Define

$$\bar{\psi}: x_1x_2 \cdots x_n \to (x_1\psi) \cdots (x_n\psi), \qquad \bar{\psi}: \varnothing \to 1 \qquad (x_i \in X).$$

It follows at once that $\bar{\psi}$ is a homomorphism of X^* into M making the requisite diagram commute. Its uniqueness follows from the obvious fact that $X\iota_X$ generates X^* (as a monoid). The case of X^+ requires practically the same argument.

We will consistently write X^* and X^+ instead of (X^*, ι_X) and (X^+, ι_X), respectively. Note that in this case $X^* \setminus \{\varnothing\} = X^+$, that is to say, the free semigroup is obtained from the free monoid by removing its identity.

I.10.4 Definition

Let S be a semigroup and $s \to s^{-1}$ be an involution of S. The algebra $(S, \cdot, ^{-1})$, where $\cdot$ denotes the multiplication in S, is a *semigroup with involution*. If S has an identity, $(S, \cdot, ^{-1})$ is a *monoid with involution*.

Note that both homomorphisms and congruences of a semigroup with involution must be compatible both with the multiplication and the involution.

We construct next a free monoid with involution on a nonempty set.

I.10.5 Construction

Let X be a nonempty set, φ be a bijection of X onto some set X' disjoint from X, and let $Y = X \cup X'$. Define a unary operation $^{-1}$ on Y^* by

$$x^{-1} = \begin{cases} x\varphi & \text{if } x \in X \\ x\varphi^{-1} & \text{if } x \in X', \end{cases} \qquad \varnothing^{-1} = \varnothing,$$

and

$$(x_1 x_2 \cdots x_n)^{-1} = x_n^{-1} x_{n-1}^{-1} \cdots x_1^{-1}.$$

Clearly, $^{-1}$ is an involution on Y^*. Let $Z = (Y^*, \cdot, ^{-1})$ be the monoid Y^* with involution $^{-1}$.

I.10.6 Proposition

The pair (Z, ι_X) is a free monoid with involution on X.

Proof. Let $\psi: X \to M$ where M is a monoid with involution. Define a mapping α by

$$x\alpha = \begin{cases} x\psi & \text{if } x \in X \\ (x^{-1}\psi)^{-1} & \text{if } x \in X', \end{cases}$$

$$\varnothing\alpha = 1, \qquad \text{the identity of } M,$$

$$(x_1 x_2 \cdots x_n)\alpha = (x_1\alpha)(x_2\alpha) \cdots (x_n\alpha) \qquad \text{if } x_i \in Y.$$

Since the elements of Y^* can be uniquely written as products of elements of Y, α is well defined with domain Y^*. It is clear that α is a homomorphism, hence is the unique homomorphism for which the diagram

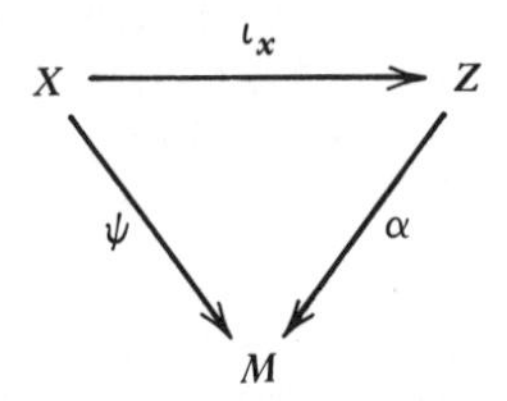

is commutative since X generates Z as a monoid with involution.

I.10.7 Definition

For any $w = x_1 x_2 \cdots x_n \in Y^*$, the transition $w \to x_1 x_2 \cdots x_{i-1} x_{i+2} \cdots x_n$, where $x_{i+1} = x_i^{-1}$, is a *reduction*. A word which cannot be further reduced is said to be *reduced*.

I.10.8 Lemma

For every word $w \in Y^*$, there is a finite number of reductions giving a reduced word. This reduced word does not depend on the order of reductions performed.

Proof. The proof of this lemma is left as an exercise.

I.10.9 Notation

For any $w \in Y^*$, we denote by $r(w)$ the reduced word obtained from w by a sequence of reductions. Let

$$G_X = \{ w \in Y^* | r(w) = w \}$$

with multiplication

$$w \cdot u = r(wu) \qquad (w, u \in G_X).$$

It is well known that the pair (G_X, ι_X) is a *free group on X*; we denote it simply by G_X. This form of multiplication implies that $w \to r(w)$ is a homomorphism of Z onto G_X. It is clear that the induced congruence is the least group congruence on Z.

There are a number of concepts closely related to those introduced above.

I.10.10 Definition

Let X be a nonempty set, $u_\alpha, v_\alpha \in X^+$ for $\alpha \in A$, and let

$$\rho = \{ (u_\alpha, v_\alpha) | \alpha \in A \}.$$

Then X^+/ρ^* is the *semigroup generated by X subject to the defining relations* $\{u_\alpha = v_\alpha\}_{\alpha \in A}$.

The same type of definition applies to monoids, semigroups with involution, and so on. For example, for monoids we take X^*, so some of the relations may be of the form $u = 1$. All defining relations in a group can be reduced to one of this form.

I.10.11 Exercises

(i) Describe free cyclic (α) semigroups, (β) monoids, (γ) semigroups with involution, and (δ) groups.

(ii) Show that any cyclic semigroup can be given in terms of a free cyclic semigroup with a single defining relation.

(iii) Let X be a generating set for a semigroup S. Show that there is a set ρ of defining relations such that $S \cong X^+/\rho^*$.

(iv) Let S be the semigroup generated by two elements a and b subject to the following defining relations:

$$a^2 = a, \qquad b^2 = b, \qquad a = aba, \qquad b = bab.$$

(α) Construct the multiplication table for S.

(β) Show that all elements of S are primitive idempotents.

The free semigroup with involution was constructed by Wagner [8]. The rest of this section amounts to an adaptation to semigroups of the universal algebraic folklore.

I.11 VARIETIES

Besides introducing the concepts related to varieties we establish here the correspondence of fully invariant congruences on a free semigroup on a countably infinite set and the lattice of all varieties of semigroups. In addition, we prove several useful lemmas.

I.11.1 Definition

A pair of elements of a free semigroup X^+, which are written as products of generators (here called *variables*) is a (*semigroup*) *identity*, to be denoted by $u = v$. A semigroup S *satisfies the identity* $u = v$ if for any homomorphism φ: $X^+ \to S$, we have $u\varphi = v\varphi$. Equivalently, for any substitution of variables in u and v by elements of S, the resulting elements of S are equal. In such a case, $u = v$ is a *law in* S. The identity $u = v$ is *homotypical* if u and v contain the same variables and *heterotypical* otherwise.

For the present, we will be discussing only semigroup identities, which then will be referred to simply as identities. We will encounter later other types of identities.

I.11.2 Definition

Let $\mathcal{F}$ be a nonempty family of identities. The class $\mathcal{V}$ of all semigroups satisfying each identity in $\mathcal{F}$ is *the variety determined by the identities* $\mathcal{F}$, or simply a *variety*. In such a case, $\mathcal{F}$ is a set of *defining identities for* $\mathcal{V}$ and we write $\mathcal{V} = [\mathcal{F}]$; if $\mathcal{F} = \{u = v\}$, we write simply $\mathcal{V} = [u = v]$. If $\mathcal{U}$ is a variety contained in $\mathcal{V}$, it is a *subvariety* of $\mathcal{V}$; it is a *proper subvariety* of $\mathcal{V}$ if $\mathcal{U} \neq \mathcal{V}$.

It is often useful to have a characterization of a variety of algebras of the same kind independent of the definition involving a family of identities. For this purpose, we have the following *theorem of Birkhoff*.

I.11.3 Theorem

Let $\mathcal{V}$ be a variety of algebras and $\mathcal{C}$ be a subclass of $\mathcal{V}$. Then $\mathcal{C}$ is a subvariety of $\mathcal{V}$ if and only if $\mathcal{C}$ is closed under direct products, subalgebras, and homomorphic images.

Note that the nontrivial part of this theorem is its converse, for the proof of which we refer to Grätzer [1].

I.11.4 Definition

Two sets of identities are *equivalent* if they determine the same variety. An identity equivalent to $u = u$ is *trivial*. If an identity $u = v$ is a law for every semigroup in a variety $\mathcal{V}$, then $u = v$ is *valid in* $\mathcal{V}$ or is a *law in* $\mathcal{V}$. A family $\mathcal{F}$ of laws in a variety $\mathcal{V}$ is a *basis for the identities in* $\mathcal{V}$ if $\mathcal{F}$ determines the variety $\mathcal{V}$.

We start with the following simple result.

I.11.5 Lemma

The set of all (semigroup) varieties has the cardinality at most that of the continuum. It is a complete lattice under inclusion.

Proof. Since every identity $u = v$ contains a finite number of variables, any family $\mathcal{F}$ of identities is equivalent to a countable family $\mathcal{C}$ of identities since at most a countably infinite set of variables is needed for the variables occurring in the identities in $\mathcal{F}$. Hence the cardinality of the set of all varieties is at most that of the continuum. It is clear that the intersection of any nonempty family of varieties is a variety. Now 2.6 implies that the set of all varieties forms a complete lattice under inclusion.

I.11.6 Definition

Let $\mathcal{C}$ be a nonempty class of semigroups. The intersection of all varieties containing all semigroups in $\mathcal{C}$ is the *variety generated by* $\mathcal{C}$. If $\mathcal{C}$ consists of a single semigroup S, then we speak of the *variety generated by* S.

The next result provides a useful characterization of the semigroups making up the variety generated by a class $\mathcal{C}$ of semigroups. It represents the first consequence of Birkhoff's theorem.

I.11.7 Proposition

Let $\mathcal{C}$ be a nonempty class of semigroups. Then the variety $\mathcal{V}$ generated by $\mathcal{C}$ consists of homomorphic images of subsemigroups of direct products of semigroups in $\mathcal{C}$.

Proof. Let $\mathcal{D}$ be the class of all semigroups which are homomorphic images of subsemigroups of direct products of semigroups in $\mathcal{C}$. Then $\mathcal{C} \subseteq \mathcal{D} \subseteq \mathcal{V}$ where the first inclusion is trivial and the second follows from (the easy part of) 11.3. It is obvious that $\mathcal{D}$ is closed under homomorphic images and it follows easily that it is closed for taking of subsemigroups. The verification that $\mathcal{D}$ is closed for formation of direct products is straightforward and is left as an exercise. Now (the hard part of) 11.3 yields that $\mathcal{D}$ is a variety, so by minimality of $\mathcal{V}$ it must coincide with $\mathcal{V}$.

There are two special cases of particular interest. Note that S^K is the direct product of K copies of a semigroup S.

I.11.8 Corollary

The variety $\mathcal{V}$ generated by a semigroup S consists of homomorphic images of subsemigroups of S^K for nonempty sets K.

Proof. The proof of this corollary is left as an exercise.

I.11.9 Corollary

Let $\mathcal{U}$ and $\mathcal{V}$ be varieties of semigroups. Then the join $\mathcal{U} \vee \mathcal{V}$ consists of semigroups S for which there exist $U \in \mathcal{U}$, $V \in \mathcal{V}$, a subdirect product $T \subseteq U \times V$, and a homomorphism φ of T onto S.

Proof. This follows from 11.7 in view of the fact that the varieties $\mathcal{U}$ and $\mathcal{V}$ are closed for the formation of subdirect products. The details are left as an exercise.

For the correspondence between fully invariant congruences and varieties, we need some preparation.

I.11.10 Definition

A congruence ρ on a semigroup S is *fully invariant* if it is invariant under all endomorphisms of S, that is, if $a\rho b$ and $\varphi \in \mathcal{E}(S)$, then $(a\varphi)\rho(b\varphi)$. The set of all fully invariant congruences on S, to be denoted by $\mathcal{FI}(S)$, is easily seen to be a complete sublattice of the lattice of all congruences on S.

Recall the definition of an order antiisomorphism in 2.3. It is easy to see that if $\varphi\colon L \to L'$ is an order antiisomorphism of complete lattices, then

$$\Big(\bigvee_{\alpha\in A} a_\alpha\Big)\varphi = \bigwedge_{\alpha\in A}(a_\alpha\varphi), \qquad \Big(\bigwedge_{\alpha\in A} a_\alpha\Big)\varphi = \bigvee_{\alpha\in A}(a_\alpha\varphi)$$

over any index set A.

The desired result can now be proved.

I.11.11 Theorem

Let X be a countably infinite set and $\mathcal{L}$ be the lattice of all (semigroup) varieties. Define a mapping ρ by

$$\rho\colon \mathcal{V} \to \rho(\mathcal{V}) \qquad (\mathcal{V} \in \mathcal{L})$$

where $\rho(\mathcal{V})$ is the relation defined on X^+ by

$$u\rho(\mathcal{V})v \Leftrightarrow u = v \text{ is a law in } \mathcal{V}.$$

Also define a mapping $\mathcal{V}$ by

$$\mathcal{V}\colon \rho \to \mathcal{V}(\rho) \qquad [\rho \in \mathcal{FI}(X^+)]$$

where $\mathcal{V}(\rho)$ is the variety determined by the set of identities $u = v$ where $u\rho v$.

Then ρ and $\mathcal{V}$ are mutually inverse order antiisomorphisms of $\mathcal{L}$ and $\mathcal{FI}(X^+)$.

Proof. 1. Let $\mathcal{V} \in \mathcal{L}$. It is easy to see that $\rho(\mathcal{V})$ is a congruence on X^+. Let $\alpha \in \mathcal{E}(X^+)$ and $u\rho(\mathcal{V})v$. Let $S \in \mathcal{V}$; then S satisfies $u = v$. Hence for all homomorphisms β of X^+ into S, we have $u\beta = v\beta$. For any such homomorphism β, we have that $\alpha\beta$ is a homomorphism of X^+ into S, so that $u\alpha\beta = v\alpha\beta$. Since β is arbitrary, we deduce that $u\alpha = v\alpha$ is valid in S, and thus $(u\alpha)\rho(\mathcal{V})(v\alpha)$. Consequently, $\rho(\mathcal{V})$ is fully invariant.

Now let $\mathcal{V}, \mathcal{V}' \in \mathcal{L}$ be such that $\mathcal{V} \subseteq \mathcal{V}'$ and let $u\rho(\mathcal{V}')v$. Then $u = v$ is a law in $\mathcal{V}'$ and thus $\mathcal{V}' \subseteq [u = v]$. The hypothesis then implies that $\mathcal{V} \subseteq [u = v]$

and so $u = v$ is a law in $\mathcal{V}$ and hence $u\rho(\mathcal{V})v$. Consequently, $\rho(\mathcal{V}') \subseteq \rho(\mathcal{V})$ and thus ρ inverts the inclusion relation.

2. Next let $\rho \in \mathcal{FI}(X^+)$. Then $\mathcal{V}(\rho)$ is trivially a variety of semigroups, and the function $\mathcal{V}$ obviously inverts inclusion.

In order to complete the proof, it suffices to show that the two composition functions $\rho\mathcal{V}$ and $\mathcal{V}\rho$ are identity mappings on their respective domains.

3. We show first that

$$X^+/\rho \in \mathcal{V}(\rho) \qquad [\rho \in \mathcal{FI}(X^+)]. \tag{1}$$

For this, it suffices to show that X^+/ρ satisfies all the laws $u = v$ corresponding to pairs (u, v) for which $u\rho v$. So let $u\rho v$, and let α: $X^+ \to X^+/\rho$ be any homomorphism.

For each $x \in X$, we have $x\alpha \in X^+/\rho$. Choose any $y \in X^+$ for which $x\alpha = y\rho^\#$ (this is possible since $\rho^\#$ maps X^+ onto X^+/ρ). Let α' be the mapping α': $x \to y$ and extend it to a homomorphism of X^+ into X^+ (this is possible since X^+ is free). It is now clear that the diagram

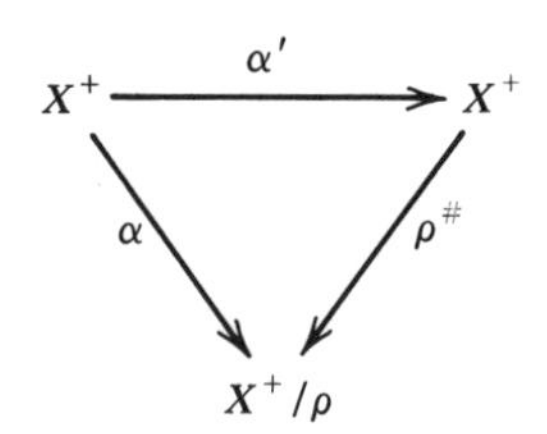

is commutative. Since ρ is fully invariant and α' is an endomorphism of X^+, we obtain $(u\alpha')\rho(v\alpha')$ whence

$$u\alpha = u\alpha'\rho^\# = v\alpha'\rho^\# = v\alpha$$

which proves (1).

4. Let $\mathcal{V} \in \mathcal{L}$. First let $u = v$ be a law in $\mathcal{V}$. Then $u\rho(\mathcal{V})v$ and thus $u = v$ is a law in $\mathcal{V}(\rho(\mathcal{V}))$. Conversely, let $u = v$ be a law in $\mathcal{V}(\rho(\mathcal{V}))$. By (1), $u = v$ is a law in $X^+/\rho(\mathcal{V})$. Therefore $u\rho(\mathcal{V})v$ and hence $u = v$ is a law in $\mathcal{V}$. Consequently, $\mathcal{V} = \mathcal{V}(\rho(\mathcal{V}))$.

5. Now let $\rho \in \mathcal{FI}(X^+)$. If $u\rho v$, then $u = v$ is a law in $\mathcal{V}(\rho)$ and thus $u\rho(\mathcal{V}(\rho))v$. Hence $\rho \subseteq \rho(\mathcal{V}(\rho))$. Assume $u\not\rho v$. Clearly, $u = v$ is not a law in X^+/ρ. Since $X^+/\rho \in \mathcal{V}(\rho)$ by (1), $u = v$ is not a law in $\mathcal{V}(\rho)$. Hence $(u, v) \notin \rho(\mathcal{V}(\rho))$. By contrapositive we get $\rho(\mathcal{V}(\rho)) \subseteq \rho$. Therefore, $\rho = \rho(\mathcal{V}(\rho))$.

The next result provides useful information concerning $\mathcal{V}$-free semigroups (see 10.1).

I.11.12 Proposition

Let $\mathcal{V} \in \mathfrak{L}$ and X be a nonempty set. Then $\rho(\mathcal{V})$ is the least congruence ρ on X^+ for which $X^+/\rho \in \mathcal{V}$; in addition, $X^+/\rho(\mathcal{V})$ is $\mathcal{V}$-free. Moreover, every semigroup in $\mathcal{V}$ is a homomorphic image of some $\mathcal{V}$-free semigroup.

Proof. By 11.11, $\rho(\mathcal{V}) \in \mathcal{F}\mathcal{I}(X^+)$. It follows from part 4 of the proof of 11.11 that $X^+/\rho(\mathcal{V}) \in \mathcal{V}(\rho(\mathcal{V})) = \mathcal{V}$.

Let τ be a congruence on X^+ such that $X^+/\tau \in \mathcal{V}$. If $u\rho(\mathcal{V})v$, then X^+/τ satisfies the identity $u = v$, so that in X^+, $u\tau v$. Thus $\rho(\mathcal{V}) \subseteq \tau$, which proves the minimality of $\rho(\mathcal{V})$. This proves the first assertion of the proposition.

Let $T \in \mathcal{V}$ and φ: $X \to T$ be any function. By the freeness of X^+, there exists a unique homomorphism φ': $X^+ \to T$ rendering the diagram

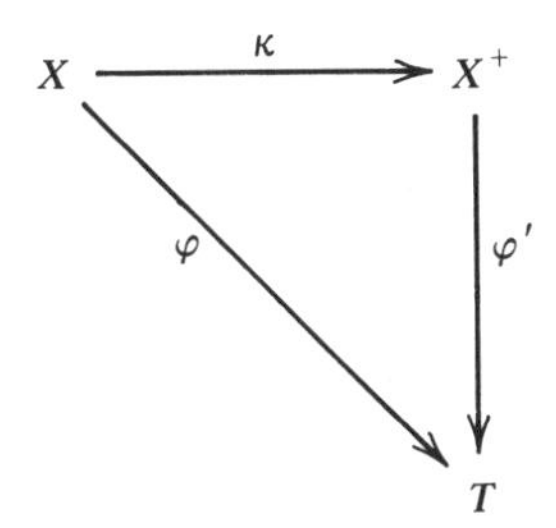

commutative, where κ is the canonical injection. We can extend this diagram as follows:

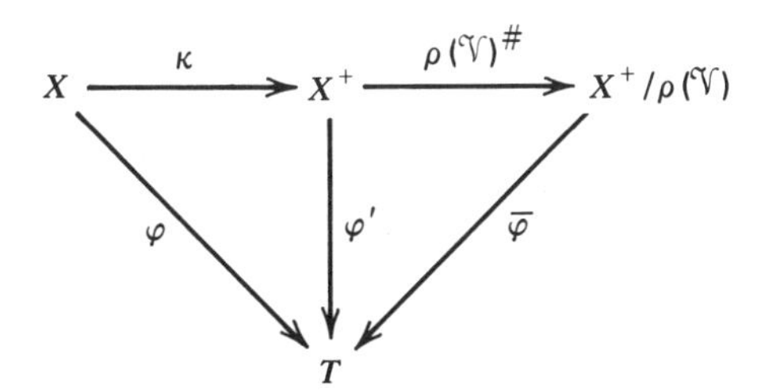

Now, since $T \in \mathcal{V}$, and by the above, $\rho(\mathcal{V})$ is the least congruence on X^+ for which $X^+/\rho(\mathcal{V}) \in \mathcal{V}$, we must have $\rho(\mathcal{V}) \subseteq \tau$, where τ is the congruence induced by φ'. But then φ' factors uniquely through $\rho(\mathcal{V})^\#$ giving the unique $\bar{\varphi}$ needed in the definition of $\mathcal{V}$-freeness. This proves the second statement of the proposition.

To see that the third statement holds, we let $S \in \mathcal{V}$. Then S is a homomorphic image of some free semigroup X^+. By minimality of $\rho(\mathcal{V})$ this homomorphism factors through $\rho(\mathcal{V})^\#$ and hence S is a homomorphic image of $X^+/\rho(\mathcal{V})$, which is $\mathcal{V}$-free by the second part of the proposition.

Concepts more general than those of an identity and a variety are provided by the following.

I.11.13 Definition

An *implication* is an ordered pair $(\mathfrak{F}, u = v)$ where $\mathfrak{F}$ is a finite family of identities, to be denoted by $\mathfrak{F} \Rightarrow u = v$. A semigroup S *satisfies the implication* $\mathfrak{F} \Rightarrow u = v$ if the following takes place: whenever in each $w = z$ in $\mathfrak{F}$ all variables are substituted by some elements of S and the elements resulting from such w and z are equal, then the elements of S obtained from u and v by this substitution are also equal. The class of all semigroups satisfying a fixed set of implications is a *quasivariety*.

I.11.14 Lemma

Let Q be a quasivariety of semigroups. Then the partially ordered set of all Q-congruences on a semigroup S under inclusion is a complete lattice with greatest element ω and meet equal to the intersection.

Proof. Let the quasivariety Q be given by a system $\mathfrak{I}$ of implications, and let $\mathfrak{F}$ be a family of Q-congruences on a semigroup S. Let $\{u_\alpha = v_\alpha\}_{\alpha \in A} \Rightarrow w = z$ be an implication in $\mathfrak{I}$. Now let u_α and v_α stand for an evaluation in S of the words u_α and v_α, and assume that $u_\alpha(\cap \mathfrak{F})v_\alpha$ for all $\alpha \in A$. Then $u_\alpha \rho v_\alpha$ for all $\rho \in \mathfrak{F}$, $\alpha \in A$. Hence $w\rho z$ for all $\rho \in \mathfrak{F}$ and thus $w(\cap_{\rho \in \mathfrak{F}} \rho)z$. Hence $S/\cap \mathfrak{F} \in Q$ and $\cap \mathfrak{F}$ is evidently the meet of $\mathfrak{F}$. Trivially ω is a Q-congruence. Therefore, the partially ordered set of all Q-congruences on S is a complete lattice by 2.6.

Except for Birkhoff's theorem, we have phrased all concepts and statements for semigroups. All the concepts are, however, of universal algebraic nature, and the proofs of the statements in this more general context follow almost verbatim those we have reproduced. For monoids, one speaks of *monoid identities* which may involve identities of the form $u = 1$ (as in the case of groups). Identities for semigroups with involution may involve u^{-1} (again as in the case of groups).

I.11.15 Exercises

(i) For $n > 1$, show the equivalence of $\{x = x^n, xy = yx\}$ and $\{x = x^n, x^n y = yx^n\}$, and of $\{x = x^{n+1}, x^n y = yx^n\}$ and $\{x = x^{n+1}, x^{2n} y = yx^{2n}\}$.

(ii) Find all (semigroup) identities in one variable.

(iii) Is every set of (semigroup) identities in one variable equivalent to a single identity in one variable?

(iv) Show that the set $\{x^2 = x, xy = yx\}$ is not equivalent to any set of (semigroup) identities consisting of a single identity.
(v) Find all quasivarieties of semilattices.
(vi) Find all subvarieties of the variety determined by the identity $x = xyx$.

The material of this section represents an adaptation of the universal algebraic concepts to semigroups and is part of the algebraic folklore. Consult B. H. Neumann [1] for a universal algebraic setting and H. Neumann [1] for varieties of groups. Varieties of semigroups are surveyed in Aĭzenštat–Boguta [1] and Evans [1]. A wealth of information concerning varieties of algebraic systems can be found in Malcev [2].

I.12 AMALGAMATION

We state here formal definitions of the concepts concerning the subject of amalgamation as it pertains to classes of semigroups. Two useful lemmas are proved here: one asserts that the discussion of amalgamation can be reduced to a consideration of only two semigroups to be amalgamated, and the other one that the weak and the special amalgamation properties together imply the strong amalgamation property.

I.12.1 Definition

Let $\{S_i\}_{i \in I}$ be a nonempty family of semigroups, U be a semigroup, and suppose that all these semigroups are pairwise disjoint. For each $i \in I$, let φ_i be an embedding of U into S_i. This system, denoted by $\mathfrak{A} = [S_i; U; \varphi_i]_{i \in I}$, is an *amalgam of semigroups* S_i *with core* U.

We will abbreviate the notation for an amalgam by omitting some symbols when there is no doubt as to its meaning, and for $|I| = 2$, write $[S_1, S_2; U; \varphi_1, \varphi_2]$ or simply $[S_1, S_2; U]$.

When considering an amalgam of the form $[S_1, S_2; U]$, we will often assume that $U \subseteq S_1 \cap S_2$ or even $U = S_1 \cap S_2$; this is permissible since isomorphic copies of S_1 and S_2 satisfying these conditions can always be found. This will simplify the arguments in several proofs.

I.12.2 Definition

Let $\mathfrak{A} = [S_i; U; \varphi_i]_{i \in I}$ and assume that there exists a semigroup S and for each $i \in I$, an embedding $\lambda_i\colon S_i \to S$ and an embedding $\lambda\colon U \to S$. If

$$\text{(i)} \qquad \varphi_i \lambda_i = \lambda \qquad (i \in I),$$

then the amalgam $\mathfrak{A}$ is *weakly embedded* into the semigroup S. If, in addition,

$$\text{(ii)} \qquad S_i\lambda_i \cap S_j\lambda_j = U\lambda \qquad (i, j \in I, i \neq j),$$

then $\mathfrak{A}$ is *strongly embedded* into S.

We also speak of (*weak*, *strong*) *embedding* of amalgams.

Note that condition (i) above means commutativity of the diagram

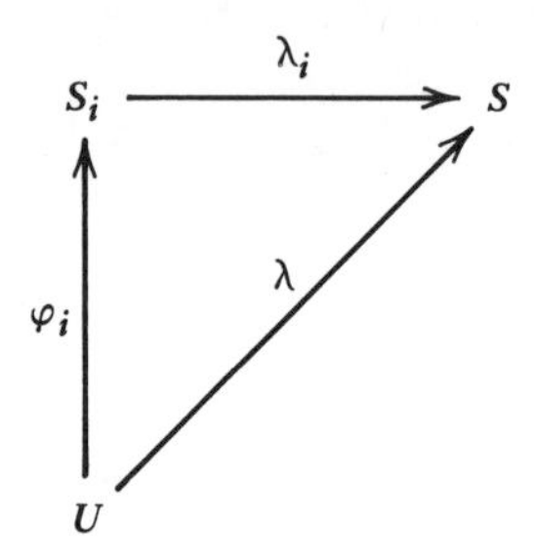

for all $i \in I$. The inclusion $S_i\lambda_i \cap S_j\lambda_j \supseteqq U\lambda$ in condition (ii) is automatically satisfied because of condition (i).

I.12.3 Definition

Let $\mathcal{C}$ be a class of semigroups. An *amalgam* $[S_i; U; \varphi_i]_{i\in I}$ *is in* $\mathcal{C}$ if all semigroups S_i and U are in the class $\mathcal{C}$; it is (*weakly*, *strongly*) *embeddable in* $\mathcal{C}$ if it is so embeddable in a semigroup in class $\mathcal{C}$. If every amalgam in $\mathcal{C}$ is (strongly, weakly) embeddable in $\mathcal{C}$, $\mathcal{C}$ has the (*weak*, *strong*) *amalgamation property*.

Our first result here reduces the consideration of embedding of arbitrary amalgams to the case of embedding of amalgams consisting of two semigroups. It is of universal algebraic nature.

I.12.4 Lemma

Let $\mathcal{C}$ be a class of semigroups which is closed for taking the union of a directed subfamily of $\mathcal{C}$. If any amalgam in $\mathcal{C}$ of two semigroups is (weakly, strongly) embeddable in $\mathcal{C}$, then $\mathcal{C}$ has the (weak, strong) amalgamation property.

Proof. First assume the validity of weak embedding of any amalgams in $\mathcal{C}$ consisting of two semigroups. The weak amalgamation property will be proved by transfinite induction.

Let $[S_i; U; \varphi_i]_{i \in I}$ be an amalgam in $\mathcal{C}$. We shall suppose that I is well ordered of type τ, and we identify I with the well-ordered chain of ordinals that precede τ. We thus rewrite the amalgam as $[S_\alpha; U; \varphi_\alpha]_{\alpha<\tau}$.

Let us formulate our induction hypothesis for the ordinal $\alpha(<\tau)$: for all $\beta \leq \alpha$, the amalgam $\mathfrak{A}_\beta = [S_\xi; U; \varphi_\xi]_{\xi<\beta}$ is weakly embeddable in a semigroup T_β in $\mathcal{C}$ and there exist embeddings

$$\lambda_\beta \colon U \to T_\beta,$$

$$\lambda_{\gamma,\beta} \colon S_\gamma \to T_\beta, \qquad \gamma < \beta,$$

such that

$$\lambda_\beta = \varphi_\gamma \lambda_{\gamma,\beta} \qquad \text{for all } \gamma < \beta. \tag{2}$$

Moreover, if $\gamma < \beta \leq \alpha$, then there exists an embedding $\sigma_{\gamma,\beta} \colon T_\gamma \to T_\beta$ such that

$$\sigma_{\gamma,\zeta}\sigma_{\zeta,\beta} = \sigma_{\gamma,\beta} \qquad \text{if } \gamma < \zeta < \beta \leq \alpha, \tag{3}$$

and

$$\lambda_{\gamma,\zeta}\sigma_{\zeta,\beta} = \lambda_{\gamma,\beta} \qquad \text{if } \gamma < \zeta < \beta \leq \alpha. \tag{4}$$

Obviously, our hypothesis is satisfied for the ordinals 0, 1, and 2. Let $\alpha < \tau$ and let us suppose that the hypothesis is satisfied for all ordinals $\beta < \alpha$. We must show that the hypothesis is satisfied for α, that is, we must find a semigroup T_α and embeddings $\lambda_\alpha \colon U \to T_\alpha$, $\lambda_{\gamma,\alpha} \colon S_\gamma \to T_\alpha$ for all $\gamma < \alpha$, such that

$$\lambda_\alpha = \varphi_\gamma \lambda_{\gamma,\alpha} \qquad \text{for all } \gamma < \alpha. \tag{5}$$

We must moreover establish the existence of embeddings $\sigma_{\gamma,\alpha} \colon T_\gamma \to T_\alpha$, $\gamma < \alpha$, such that

$$\sigma_{\gamma,\beta}\sigma_{\beta,\alpha} = \sigma_{\gamma,\alpha} \qquad \text{if } \gamma < \beta < \alpha, \tag{6}$$

and

$$\lambda_{\gamma,\beta}\sigma_{\beta,\alpha} = \lambda_{\gamma,\alpha} \qquad \text{if } \gamma < \beta < \alpha. \tag{7}$$

Let us first consider the case that α is a limit ordinal. Then $[T_\gamma, \gamma < \alpha; \sigma_{\gamma,\beta}, \gamma < \beta < \alpha]$ is a directed subfamily of $\mathcal{C}$. We consider the direct limit T_α in $\mathcal{C}$ of this directed family, and for all $\beta < \alpha$ we let $\sigma_{\beta,\alpha} \colon T_\beta \to T_\alpha$ be the natural embedding of T_β into T_α. If $\gamma < \alpha$, then there exists β such that $\gamma < \beta < \alpha$, and we define $\lambda_{\gamma,\alpha} = \lambda_{\gamma,\beta}\sigma_{\beta,\alpha}$. One verifies easily that $\lambda_{\gamma,\alpha}$ is independent of the choice of β. The embedding λ_α may be introduced by (5) and we may verify that $\lambda_\alpha = \varphi_\gamma\lambda_{\gamma,\alpha}$ is independent of the choice of γ. In other words, (5), (6), and (7) are satisfied and the hypothesis holds for α.

We now suppose that α is not a limit ordinal, that is $\alpha = \beta + 1$. Consider the amalgam $[S_\beta, T_\beta; U; \varphi_\beta, \lambda_\beta]$. By our assumption this amalgam is weakly embeddable in $\mathcal{C}$. Hence there exists T_α in $\mathcal{C}$ and embeddings

$$\lambda_{\beta,\alpha}: S_\beta \to T_\alpha, \qquad \sigma_{\beta,\alpha}: T_\beta \to T_\alpha$$

such that

$$\varphi_\beta\lambda_{\beta,\alpha} = \lambda_\beta\sigma_{\alpha,\beta} = \lambda_\alpha.$$

For any $\gamma < \beta$, we put

$$\lambda_{\gamma,\alpha} = \lambda_{\gamma,\beta}\sigma_{\beta,\alpha}, \qquad \sigma_{\gamma,\alpha} = \sigma_{\gamma,\beta}\sigma_{\beta,\alpha}.$$

It is routine to check that the conditions (5), (6), and (7) are satisfied for α.

We have proved that the hypothesis holds for all $\alpha < \tau$. We can easily deduce from this that the amalgam $[S_\alpha; U; \varphi_\alpha]_{\alpha<\tau}$ is weakly embeddable in a semigroup in $\mathcal{C}$.

We now assume the validity of strong embedding of any amalgams in $\mathcal{C}$ of two semigroups. In the induction hypothesis for α in the above proof, we must in addition suppose that for all $\beta \le \alpha$, the amalgam $\mathfrak{A}_\beta$ is strongly embeddable in T_β. Let $\alpha < \tau$ be an ordinal and suppose that the hypothesis holds for all ordinals less than α. We define the embeddings $\sigma_{\beta,\alpha}$, $\lambda_{\beta,\alpha}$, λ_α, and the semigroup T_α as above, thereby taking care of the fact that $[S_\beta, T_\beta; U; \varphi_\beta, \lambda_\beta]$ is strongly embedded in T_α in case $\alpha = \beta + 1$. We must show that

$$U\lambda_\alpha = S_\zeta\lambda_{\zeta,\alpha} \cap S_\gamma\lambda_{\gamma,\alpha} \qquad \text{for all } \zeta, \gamma < \alpha. \tag{8}$$

The inclusion

$$U\lambda_\alpha \subseteq S_\zeta\lambda_{\zeta,\alpha} \cap S_\gamma\lambda_{\gamma,\alpha}$$

always holds. Let us first assume that there exists ξ such that $\zeta, \gamma < \xi < \alpha$. If in this case $x \in S_\zeta\lambda_{\zeta,\alpha} \cap S_\gamma\lambda_{\gamma,\alpha}$, then $x\sigma_{\xi,\alpha}^{-1} \in S_\zeta\lambda_{\zeta,\xi} \cap S_\gamma\lambda_{\gamma,\xi}$, and from the induction hypothesis we then have $x\sigma_{\xi,\alpha}^{-1}\lambda_\xi^{-1} = x\lambda_\alpha^{-1} \in U$. In this case (8) holds. If there is no ordinal ξ such that $\zeta, \gamma < \xi < \alpha$, then $\alpha = \beta + 1$ and $\zeta = \beta$ or $\gamma = \beta$. Statement (8) then follows from the fact that $[S_\beta, T_\beta; U; \varphi_\beta, \lambda_\beta]$ is strongly embeddable in T_α.

This lemma considerably simplifies the arguments in any proof establishing the (weak, strong) amalgamation property of some class $\mathcal{C}$ of semigroups.

We can relax the requirement that the semigroups S_i and U, which are ingredients of an amalgam, be pairwise disjoint. We can let U be a subsemigroup of some (plural) S_i, with φ_i the inclusion mapping, or simply $\varphi_i = \iota_U$. We may even allow $\cap_{i \in I} S_i = U$. A special type of strong embedding is provided by the following concept.

I.12.5 Definition

Let U be a subsemigroup of a semigroup S, and ψ be an isomorphism of S onto a semigroup S'. Consider the amalgam $\mathfrak{A} = [S, S'; U; \iota_U, \psi|_U]$. If in a class $\mathcal{C}$ of semigroups every such amalgam is strongly embeddable in $\mathcal{C}$, then $\mathcal{C}$ has the *special amalgamation property.*

It is trivial that the strong amalgamation property implies both the weak and the special amalgamation properties; the converse of this is provided by the following useful result.

I.12.6 Lemma

Let a class $\mathcal{C}$ of semigroups have the weak and the special amalgamation properties. Then it also has the strong amalgamation property.

Proof. Let $[S, T; U]$ be an amalgam in $\mathcal{C}$. We may suppose that $U \subseteq S \cap T$. By the weak amalgamation property, there exists a semigroup P in $\mathcal{C}$ and embeddings φ: $S \to P$ and ψ: $T \to P$ such that $\varphi|_U = \psi|_U$. Thus, we may suppose that S and T are subsemigroups of P with the property that $U \subseteq S \cap T$.

Let P' be a semigroup disjoint from P and θ be an isomorphism of P onto P'. The amalgam $[P, P'; U; \iota_U, \theta|_U]$ is strongly embeddable into a semigroup V by the special amalgamation property. Hence we may suppose that P and P' are subsemigroups of V such that $U = P \cap P'$. But then S and T are subsemigroups of V such that $U = S \cap T$, that is to say, they are strongly embedded in V. The assertion now follows by 12.4.

This lemma makes it possible to split the arguments in the proof that some class $\mathcal{C}$ has the strong amalgamation property into those showing the weak and the special amalgamation properties each of which may be easier to establish. In addition, the introduction of the special amalgamation property provides another ramification of the properties some given class of semigroups may satisfy.

This section represents an adaptation to the case of semigroups of concepts and results pertaining to a general universal-algebraic setting; for a survey, see Jónsson [1]. Result 12.4 can be found in Howie [2], but the proof here is due to F. Pastijn. Howie [4] contains an extensive treatment of amalgamation of semigroups. For a thorough discussion of directed systems consult Grätzer [1].

I.13 FREE PRODUCTS

We specialize here the concepts of a free product and an amalgamated free product of algebras to the case of semigroups. The free product and the amalgamated free product have a natural place in any discussion of amalgamation. The latter is formed by dividing the former by the least congruence which identifies certain pairs of elements and can be used as a test for validity of the weak, special, or strong amalgamation property. The amalgamated free product plays a role similar to that of a free object in, say, semigroups. The principal results here show its various universal properties.

I.13.1 Definition

Let $\{S_i\}_{i\in I}$ be a family of pairwise disjoint semigroups. Define a function $\sigma\colon \bigcup_{i\in I} S_i \to I$ by letting $s\sigma = i$ if $s \in S_i$. On the set of all finite strings $s_1 s_2 \cdots s_n$ with $s_j \in \bigcup_{i\in I} S_i$ and $s_j\sigma \neq s_{j+1}\sigma$ for $j = 1, 2, \ldots, n-1$, define a multiplication by

$$(s_1 s_2 \cdots s_m)(t_1 t_2 \cdots t_n) = \begin{cases} s_1 s_2 \cdots s_m t_1 \cdots t_n \\ \qquad \text{if } s_m\sigma \neq t_1\sigma \\ s_1 s_2 \cdots (s_m t_1) \cdots t_n \\ \qquad \text{if } s_m\sigma = t_1\sigma \end{cases}$$

where in the second case, the product $s_m t_1$ is taken in $S_{s_m\sigma}$. The algebra so obtained is the *free product* of the family $\{S_i\}_{i\in I}$, denoted by $\prod^*_{i\in I} S_i$. The free product of $S_1, S_2, \ldots, S_n$ is denoted by $S_1 * S_2 * \cdots * S_n$.

For each $i \in I$, the mapping

$$\theta_i\colon s_i \to s_i \qquad (s_i \in S_i)$$

is the *natural embedding* of S_i into $\prod^*_{i\in I} S_i$.

We may safely identify the strings in $\prod^*_{i\in I} S_i$ of length 1 with the corresponding elements of $\bigcup_{i\in I} S_i$. This amounts to considering θ_i as the inclusion mapping. We will generally write $\prod^* S_i$ instead of $\prod^*_{i\in I} S_i$.

I.13.2 Lemma

The multiplication in the free product $\prod^*_{i\in I} S_i$ is associative.

Proof. The argument is long but straightforward. It consists of verifying the associative law in the following cases: for

$$a_1 \cdots a_m,\ b_1 \cdots b_n,\ c_1 \cdots c_k \in \prod^* S_i,$$

1. $a_m\sigma \neq b_1\sigma,\ b_n\sigma \neq c_1\sigma,$
2. $a_m\sigma \neq b_1\sigma,\ b_n\sigma = c_1\sigma,$
3. $a_m\sigma = b_1\sigma,\ b_n\sigma \neq c_1\sigma,$
4. $a_m\sigma = b_1\sigma,\ b_n\sigma = c_1\sigma.$

The details of this are left as an exercise.

We have thus obtained the semigroup $\Pi^*_{i\in I}S_i$, which we now use to introduce the next concept.

I.13.3 Definition

Let $\mathfrak{A} = [S_i, U; \varphi_i]_{i\in I}$ be an amalgam of semigroups. Let ρ be the congruence on Π^*S_i generated by the set

$$R = \left\{\left(u\varphi_i\theta_i,\ u\varphi_j\theta_j\right) | u \in U, i, j \in I\right\}.$$

The *free product of the amalgam* $\mathfrak{A}$ is the quotient semigroup $(\Pi^*S_i)/\rho$, to be denoted by $\Pi^*_U S_i$. If $I = \{1, 2\}$, we write $\Pi^*_U S_i = S_1 *_U S_2$.

Less formally, we speak of the *free product of semigroups S_i amalgamating U* or of the *amalgamated free product*.

Observe first that in the above notation, the composition $\mu_i = \theta_i\rho^\#\colon S_i \to \Pi^*_U S_i$ is a homomorphism and that for any $i, j \in I$, $u \in U$,

$$u\varphi_i\mu_i = u\varphi_i\theta_i\rho^\# = u\varphi_j\theta_j\rho^\# = u\varphi_j\mu_j.$$

Thus we get a homomorphism $\mu = \varphi_i\mu_i$, for any $i \in I$, for which the diagram commutes for all $i \in I$. The amalgam $\mathfrak{A}$ is thus

(i) weakly embedded in $\Pi^*_U S_i$ if all μ_i are one-to-one (9)

(ii) (strongly) embedded if in addition

$$S_i\mu_i \cap S_j\mu_j \subseteq U\mu \qquad (i, j \in I, i \neq j). \tag{10}$$

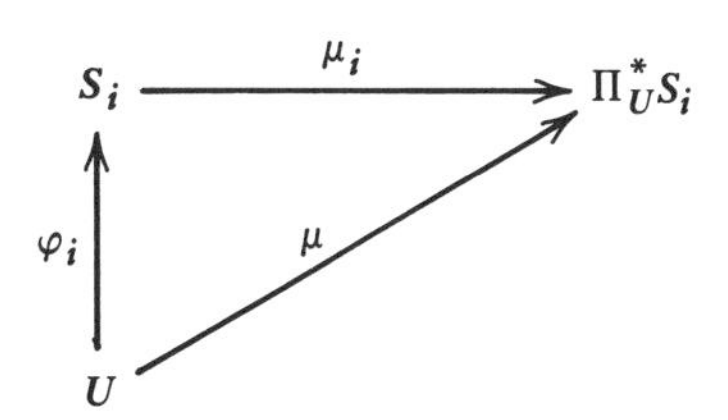

It will be convenient to introduce the following concepts.

I.13.4 Definition

The amalgam $\mathfrak{A} = [S_i, U; \varphi_i]_{i \in I}$ is *naturally strongly embedded in its free product* $\prod_U^* S_i$ if (9) and (10) above hold; and if only (9) holds, we substitute *weakly* for strongly.

The next result provides a compilation of the most important universal properties of the free product and the amalgamated free product.

I.13.5 Theorem

Let $\{S_i\}_{i \in I}$ be a family of pairwise disjoint semigroups, and let T be a semigroup.

(i) Let λ_i: $S_i \to T$ be a homomorphism for every $i \in I$, and let $F = \prod^* S_i$. Then there exists a unique homomorphism γ: $F \to T$ for which the diagram

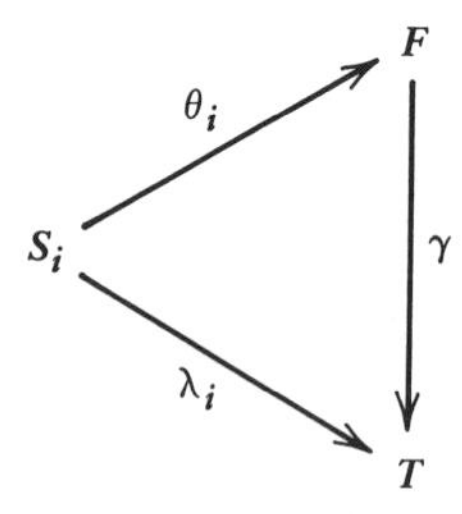

commutes for all $i \in I$.

(ii) Let $\mathfrak{A} = [S_i; U; \varphi_i]$ be an amalgam; let λ_i be as above and assume that it is one-to-one, and let $P = \prod_U^* S_i$. If $\varphi_i \lambda_i = \varphi_j \lambda_j$ for all $i, j \in I$, then μ_i is one-to-one for all $i \in I$, and there exists a unique homomorphism δ: $P \to T$ for which the diagram

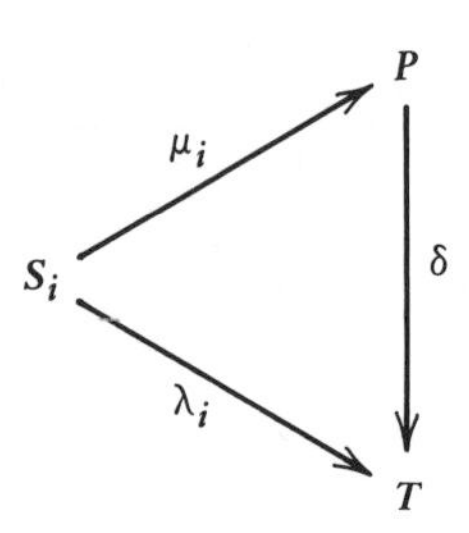

commutes for all $i \in I$. In such a case, let $\lambda = \varphi_i \lambda_i$, $\mu = \varphi_i \mu_i$.

(iii) If in addition to the above, $S_i\lambda_i \cap S_j\lambda_j \subseteq U\lambda$ for all $i, j \in I$, $i \neq j$, then also $S_i\mu_i \cap S_j\mu_j \subseteq U\mu$ for all $i, j \in I$, $i \neq j$.

Proof. (i) Every element of F can be uniquely written as $s_1s_2 \cdots s_n$ with $s_i\sigma \neq s_{i+1}\sigma$ for $i = 1, 2, \ldots, n-1$. Hence we may define γ on F by

$$\gamma: s_1s_2 \cdots s_n \to \left(s_1\lambda_{s_1\sigma}\right)\left(s_2\lambda_{s_2\sigma}\right) \cdots \left(s_n\lambda_{s_n\sigma}\right).$$

For $s = s_1s_2 \cdots s_m$ and $t = t_1t_2 \cdots t_n$, we distinguish two cases. If $s_n\sigma = t_1\sigma$, then

$$\begin{aligned}
(st)\gamma &= ((s_1s_2 \cdots s_m)(t_1t_2 \cdots t_n))\gamma \\
&= \left(s_1s_2 \cdots s_{m-1}(s_mt_1)t_2 \cdots t_n\right)\gamma \\
&= \left(s_1\lambda_{s_1\sigma}\right) \cdots (s_mt_1)\lambda_{s_m\sigma} \cdots \left(t_n\lambda_{t_n\sigma}\right) \\
&= \left(s_1\lambda_{s_1\sigma}\right) \cdots \left(s_m\lambda_{s_m\sigma}\right)\left(t_1\lambda_{t_1\sigma}\right) \cdots \left(t_n\lambda_{t_n\sigma}\right) \\
&= (s\gamma)(t\gamma),
\end{aligned}$$

and if $s_m\sigma \neq t_1\sigma$, then the equality $(st)\gamma = (s\gamma)(t\gamma)$ follows at once. Consequently, γ is a homomorphism of F into T.

Commutativity of the first diagram above follows directly from the definition of γ. The uniqueness of γ is a consequence of the fact that $\bigcup_{i \in I} S_i$ generates F, more precisely, the strings of length 1 generate the entire F. Indeed, any homomorphism of F which agrees with γ on strings of length 1 must necessarily coincide with γ.

(ii) Assume that $\varphi_i\lambda_i = \varphi_j\lambda_j$ for all $i, j \in I$. We thus have the following situation

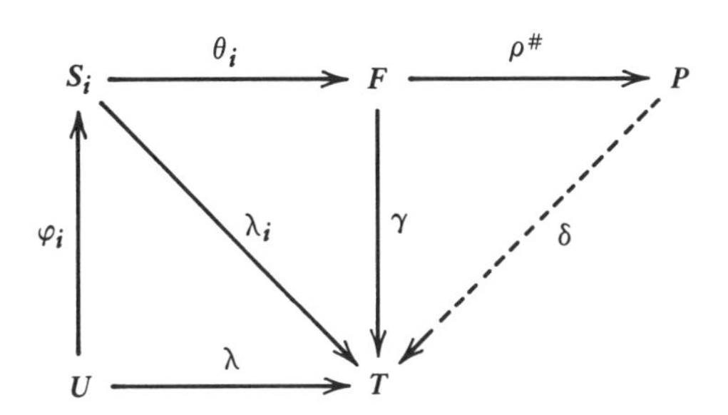

The hypothesis, together with part (i), implies that

$$\varphi_i\theta_i\gamma = \varphi_i\lambda_i = \varphi_j\lambda_j = \varphi_j\theta_j\gamma$$

for any $i, j \in I$, which shows that ρ is contained in the congruence induced on F by the homomorphism γ. It follows that there exists a unique homomorphism $\delta: P \to T$ making the right-hand part of the above diagram commutative. The same homomorphism δ thus makes the second diagram in the statement of the theorem commutative. The uniqueness of δ follows similarly as the uniqueness of γ indicated above.

The situation we arrive at can be illustrated by the commutative diagram

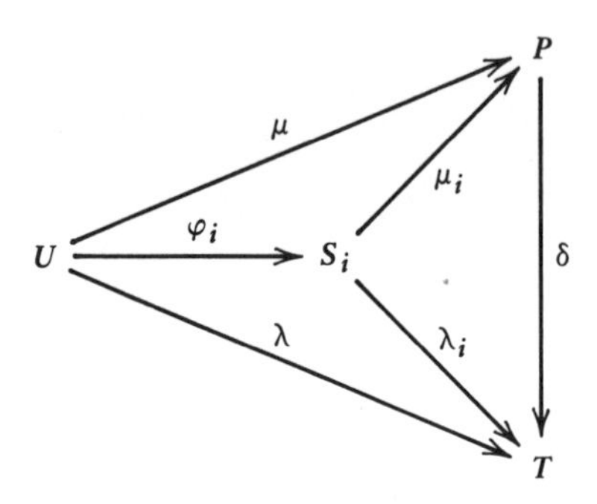

for any $i \in I$. If now $s\mu_i = t\mu_i$, then

$$s\lambda_i = s\mu_i\delta = t\mu_i\delta = t\lambda_i$$

which implies that $s = t$. Consequently, μ_i is one-to-one for all $i \in I$.

(iii) Finally, assume that also $S_i\lambda_i \cap S_j\lambda_j \subseteq U\lambda$ for all $i, j \in I$, $i \neq j$. Let $x = s_i\mu_i = s_j\mu_j$ where $s_i \in S_i$, $s_j \in S_j$, $i \neq j$. Then $x\delta = s_i\mu_i\delta = s_i\lambda_i \in S_i\lambda_i$ by part (ii), and analogously $x\delta \in S_j\lambda_j$. Since $i \neq j$, the hypothesis implies that $x\delta \in U\lambda$ and thus $x\delta = u\lambda$ for some $u \in U$. It follows that

$$s_i\lambda_i = x\delta = u\lambda = u\varphi_i\lambda_i$$

which implies $s_i = u\varphi_i$ since λ_i is one-to-one. But then

$$x = s_i\mu_i = u\varphi_i\mu_i = u\mu \in U\mu$$

which proves that $S_i\mu_i \cap S_j\mu_j \subseteq U\mu$.

I.13.6 Corollary

Let $\mathfrak{A} = [S_i, U; \varphi_i]$ be a semigroup amalgam. Then $\mathfrak{A}$ is (weakly, strongly), embeddable into a semigroup if and only if $\mathfrak{A}$ is naturally (weakly, strongly) embeddable into its free product $\Pi_U^* S_i$.

Proof. The proof of this corollary is left as an exercise.

This corollary says that the amalgamated free product $\Pi_U^* S_i$ can serve as a "universal testing object" for (weak, strong) embedding of the amalgam $\mathfrak{A}$ in the sense that the possibility or impossibility of (weak, strong) embedding can always be tested on the amalgamated free product. The reason for not doing this in every case is the complicated nature of $\Pi_U^* S_i$ and some other approach may in various cases prove more tractable.

We will need a type of converse of 13.5(i), which actually indicates that the property in 13.5(i) characterizes the free product.

I.13.7 Proposition

Let $\{S_i\}_{i \in I}$ be a family of pairwise disjoint semigroups. Let S be a semigroup with the following properties:

(i) for each $i \in I$, there is an embedding α_i: $S_i \to S$,

(ii) if T is a semigroup such that for each $i \in I$, there is a homomorphism β_i: $S_i \to T$, then there exists a unique homomorphism $\delta: S \to T$ making the diagram

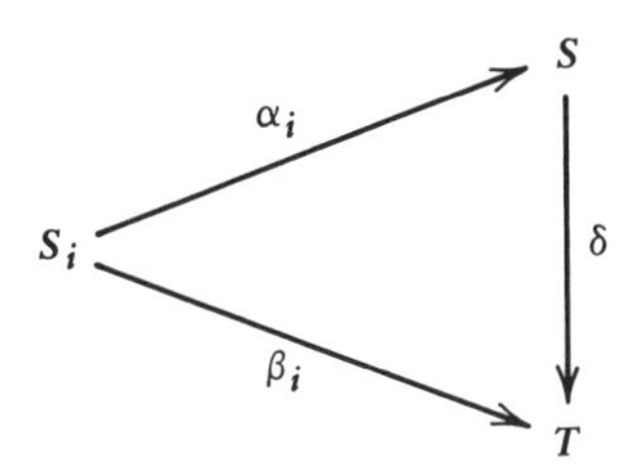

commutative for every $i \in I$.

Then S is isomorphic to $F = \Pi^* S_i$.

Proof. By 13.5(i), (F, θ_i) has the above properties of (S, α_i). Make the following specializations:

(α) in 13.5, let $T = F, \lambda_i = \theta_i$,

(β) in 13.5, let $T = S, \lambda_i = \alpha_i$,

(γ) above, let $T = S, \alpha_i = \beta_i$,

(δ) above, let $T = F, \beta_i = \theta_i$,

obtaining the unique homomorphisms $\iota_F, \gamma, \iota_S, \delta$ making the following diagram commutative:

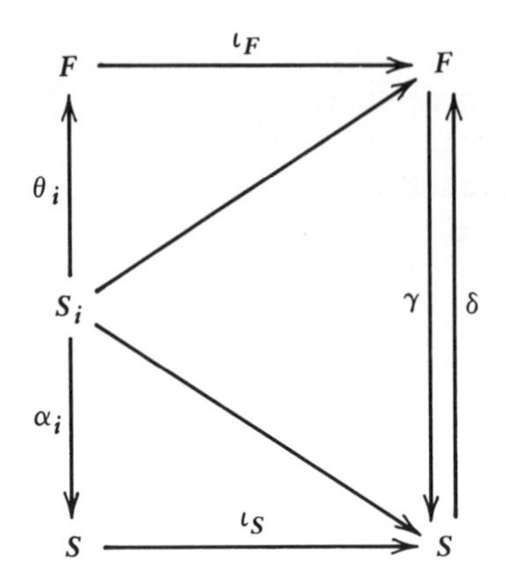

for all $i \in I$. Hence

$$\theta_i(\gamma\delta) = (\theta_i\gamma)\delta = \alpha_i\delta = \theta_i, \qquad \alpha_i(\delta\gamma) = (\alpha_i\delta)\gamma = \theta_i\gamma = \alpha_i,$$

which by the uniqueness of ι_F and ι_S yields $\gamma\delta = \iota_F$ and $\delta\gamma = \iota_S$, respectively. Hence γ and δ are mutually inverse isomorphisms of F and S.

That the class of all semigroups does not have the weak amalgamation property can be seen on the following simple instance.

I.13.8 Example

Let $S = \{a, u, v, 0\}$, $T = \{b, u, v, 0\}$, with $au = v$, $vb = u$, and all other products equal to 0. Then $S^3 = T^3 = 0$, and thus both S and T are semigroups. Let $U = \{u, v, 0\}$. Then $U = S \cap T$ and

$$(au)b = vb = u, \qquad a(ub) = a0 = 0$$

and hence the amalgam $[S, T; U]$ is not (weakly) embeddable.

Similarly, the class of all semigroups does not have the special amalgamation property, as shown by the following example.

I.13.9 Example

Let $\mathbb{Z}$ be the additive group of integers and $\mathbb{N}$ be its subsemigroup of natural numbers. Let ψ: $n \to n'$ be an isomorphism of $\mathbb{Z}$ onto its copy $\mathbb{Z}'$ disjoint from

$\mathbb{Z}$. If the amalgam $[\mathbb{Z}, \mathbb{Z}'; \mathbb{N}; \iota_{\mathbb{N}}, \psi|_{\mathbb{N}}]$ is embeddable, then since the intersection $\mathbb{Z}\lambda \cap \mathbb{Z}'\lambda'$ is nonempty, it must be a group and therefore cannot be equal to $\mathbb{N}\lambda$ $(= \mathbb{N}\lambda')$.

I.13.10 Exercises

(i) Let $\{S_i\}_{i \in I}$ be a family of pairwise disjoint semigroups. Show that $\prod^* S_i$ is isomorphic to $(\bigcup S_i)^+/\rho$ where ρ is the congruence generated by the relations valid in various S_i.

(ii) Let $\{X_i\}_{i \in I}$ be a family of pairwise disjoint nonempty sets. Show that $\prod^* X_i^+ \cong (\bigcup X_i)^+$. Deduce that for any nonempty set X, $\prod^* \{x\}^+ \cong X^+$ where x runs through all elements of X.

(iii) Let $\{X_i\}_{i \in I}$ be a family of pairwise disjoint nonempty sets. Show that $\prod^*_{\{1\}} X_i^* \cong (\bigcup X_i)^*$, where on the left-hand side the identity of each X_i^* is amalgamated.

(iv) What is an analogue of exercise (iii) in the more general situation of exercise (i)?

(v) Modify 13.9 by replacing $\mathbb{Z}$ by nonnegative integers N. Is the resulting amalgam strongly embeddable?

The material of this section is mainly folklore from group theory adapted to the semigroup situation. For a comprehensive treatment, see Howie [4]. Example 13.8 is essentially that of Kimura [1]. See also Clarke [1].

I.14 CATEGORIES

We will have several occasions to use the language of category theory. Here we give a concise list of the definitions which will be used (without reference). No results of category theory will be needed since we will be interested in the mutual relationships of categories and will not study these categories as such.

I.14.1 Definition

A *category* C consists of a class Ob C called the *objects* of C, or *C-objects*, and a class Hom C called the *morphisms* of C, or *C-morphisms*, satisfying the following conditions.

(i) With every ordered pair $a, b \in \operatorname{Ob} C$ there is associated a set $\operatorname{Hom}_C(a, b)$, or simply $\operatorname{Hom}(a, b)$, of C-morphisms in such a way that each C-morphism belongs to exactly one $\operatorname{Hom}(a, b)$. We write α: $a \to b$ as an alternative to $\alpha \in \operatorname{Hom}(a, b)$.

(ii) To every pair of morphisms $\alpha: a \to b$ and $\beta: b \to c$, there is associated a unique $\gamma: a \to c$ called the *composition* of α and β and denoted by $\alpha\beta$.
(iii) If $\alpha: a \to b$, $\beta: b \to c$, $\gamma: c \to d$, then $(\alpha\beta)\gamma = \alpha(\beta\gamma)$.
(iv) To each $a \in \operatorname{Ob} C$ there exists a C-morphism $1_a: a \to a$, called the *identity morphism*, such that for any $\beta: b \to a$ and $\gamma: a \to c$, we have $\beta 1_a = \beta$, $1_a\gamma = \gamma$.

Note that for $\alpha: a \to b$ and $\beta: c \to d$, the composition $\alpha\beta$ is defined if and only if $b = c$. Below let C be any category.

I.14.2 Definition

A *subcategory* S of C consists of a subclass of $\operatorname{Ob} C$, denoted by $\operatorname{Ob} S$, and a subclass of $\operatorname{Hom} C$, denoted by $\operatorname{Hom} S$, satisfying the following conditions:

(i) if $a \in \operatorname{Ob} S$, then $1_a \in \operatorname{Hom} S$,
(ii) if $\alpha, \beta \in \operatorname{Hom} S$ and $\alpha\beta$ is defined in C, then $\alpha\beta \in \operatorname{Hom} S$,
(iii) if $\alpha \in \operatorname{Hom} S$ and $\alpha: a \to b$, then $a, b \in \operatorname{Ob} S$.

It is clear that S is a category in its own right. For $a, b \in \operatorname{Ob} S$, we may write $\operatorname{Hom}_S(a, b)$ for the set of S-morphisms $\alpha: a \to b$, and $\operatorname{Hom}_C(a, b)$ for the set of C-morphisms $\alpha: a \to b$. Then S is a *full subcategory* of C if $\operatorname{Hom}_S(a, b) = \operatorname{Hom}_C(a, b)$ for all $a, b \in \operatorname{Ob} S$.

Note that a full subcategory of a category C is uniquely determined by its objects; this will be used to specify a full subcategory (omitting the reference to morphisms).

I.14.3 Definition

A C-morphism $\alpha: a \to b$ is an *isomorphism* if there exists a C-morphism $\beta: b \to a$ such that $\alpha\beta = 1_a$ and $\beta\alpha = 1_b$. If such β exists, it is unique and is denoted by α^{-1} and called the *inverse* of α; in such a case, the objects a and b are said to be *isomorphic*.

The concepts introduced so far were concerned essentially with a single category. The next notion compares two categories.

I.14.4 Definition

A *functor* from a category C to a category D is a pair of mappings, one from $\operatorname{Ob} C$ to $\operatorname{Ob} D$ and the other from $\operatorname{Hom} C$ to $\operatorname{Hom} D$, both denoted by the same letter, say F, satisfying the following requirements:

(i) $F1_a = 1_{Fa}$ for all $a \in \operatorname{Ob} C$,
(ii) if α: $a \to b$, then $F\alpha$: $Fa \to Fb$, and if in addition β: $b \to c$, then $F(\alpha\beta) = (F\alpha)(F\beta)$.

Furthermore, F is *full* if for any $a, b \in \operatorname{Ob} C$ and any α: $Fa \to Fb$ there exists β: $a \to b$ such that $F\beta = \alpha$; F is *representative* if for any $a \in \operatorname{Ob} D$ there exists $b \in \operatorname{Ob} C$ such that a and Fb are isomorphic. We denote by I_C the *identity functor* on C, that is I_C is the identity mapping on both $\operatorname{Ob} C$ and $\operatorname{Hom} C$.

The next concept relates two functors.

I.14.5 Definition

Let F and G be functors from a category C to a category D. A *natural transformation* τ: $F \to G$ of functors F and G is a function mapping $\operatorname{Ob} C$ into $\operatorname{Hom} D$ such that

(i) $\tau(a)$: $Fa \to Ga$ $\quad (a \in \operatorname{Ob} C, \tau(a) \in \operatorname{Hom} D)$,
(ii) if α: $a \to b$ is a C-morphism, then the diagram

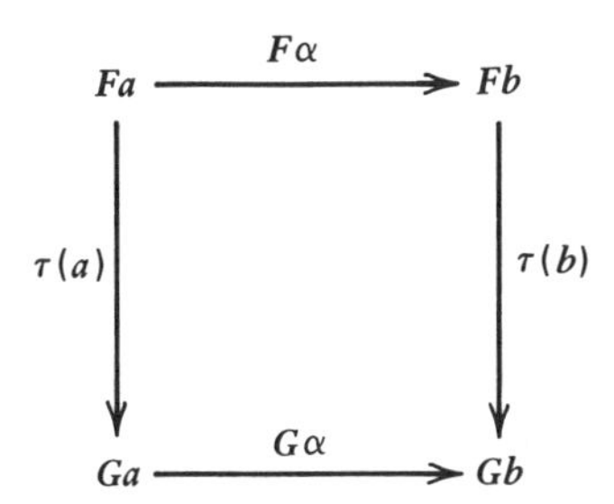

is commutative.

If also $\tau(a)$ is an isomorphism for each $a \in \operatorname{Ob} C$, then τ is a *natural equivalence* of the functors F and G.

The final concept will be used several times.

I.14.6 Definition

A quadruple (F, G, σ, τ) is an *equivalence* of the categories C and D if

(i) F is a functor from C to D,
(ii) G is a functor from D to C,

(iii) $\sigma: I_C \to GF$ is a natural equivalence,
(iv) $\tau: I_D \to FG$ is a natural equivalence.

In such a case C and D are said to be *equivalent categories*.

I.14.7 Exercises

(i) Let $\mathfrak{S}$ be given by: Ob $\mathfrak{S}$ is the class of all semigroups, Hom $\mathfrak{S}$ is the class of all homomorphisms of semigroups, and the composition of morphisms is the composition of functions (written on the right). Show that $\mathfrak{S}$ is a category.

(ii) Modify the preceding exercise by taking only monoids and monoid homomorphisms. Show that in this way we get a category, say $\mathfrak{M}$.

(iii) Show that $\mathfrak{M}$ is a subcategory of $\mathfrak{S}$ but is not a full subcategory.

(iv) Let $\mathfrak{E}$ be the category of sets and mappings between sets. For every semigroup S let FS be the set on which S is defined, and let F be the identity map on homomorphisms of semigroups. Show that F is a functor from $\mathfrak{S}$ to $\mathfrak{E}$ (F is a *forgetful functor*). Is F full or representative?

(v) Let $\mathfrak{J}$ be the category of semigroups with involution and their homomorphisms. For every semigroup with involution $(S, \cdot, ^{-1})$, let $FS = (S, \cdot)$, and let F be the identity map on Hom $\mathfrak{J}$. Show that F is a functor from $\mathfrak{J}$ to $\mathfrak{S}$. Is F full or representative?

(vi) Let Ob $\mathfrak{R}$ be the class of all equivalence relations on all nonempty sets; Hom $\mathfrak{R}$ the class of all mappings of the form: for nonempty sets X and Y with equivalence relations ρ and τ on X and Y, respectively, φ maps X onto Y in such a way that for any $x, y \in X$, $x\rho y$ implies $(x\varphi)\tau(y\varphi)$.

Let Ob $\mathfrak{P}$ be the class of all partitions of nonempty sets; Hom $\mathfrak{P}$ the class of all mappings of the form: for nonempty sets X and Y with partitions $\mathfrak{F}$ and $\mathfrak{G}$ of X and Y, respectively, φ maps X onto Y in such a way that for any $x, y \in X$, if $x, y \in F$ for some $F \in \mathfrak{F}$, then $x\varphi, y\varphi \in G$ for some $G \in \mathfrak{G}$.

Prove that $\mathfrak{R}$ and $\mathfrak{P}$ are categories and that there exists a functor of $\mathfrak{R}$ onto $\mathfrak{P}$ which is one-to-one both on Ob $\mathfrak{R}$ and Hom $\mathfrak{R}$.

(vii) Let $\mathfrak{E}$ and $\mathfrak{S}$ be as in exercises (iv) and (i). Show that there exists a functor from $\mathfrak{E}$ onto a full subcategory of $\mathfrak{S}$ distinct from $\mathfrak{S}$ and which is one-to-one both on Ob $\mathfrak{E}$ and Hom $\mathfrak{E}$.

For an extensive discussion of categories, we refer to the text of Mac Lane [1].

II

SPECIAL CLASSES

1. A semigroup S is an inverse semigroup if every element of S has a unique inverse, that is to say, the system of equations $a = axa$, $x = xax$ has a unique solution for x. An equivalent definition is that S be a regular semigroup whose idempotents commute. On every inverse semigroup one can define a partial order which plays an important role.

2. A Clifford semigroup, better known as a semilattice of groups, is a regular semigroup whose idempotents lie in its center. There is a precise structure theorem for Clifford semigroups as strong semilattices of groups. A special case of interest is a regular semigroup which is a subdirect product of a semilattice and a group.

3. A Brandt semigroup is a completely 0-simple inverse semigroup. The structure of these semigroups is explicitly described in terms of Rees matrix semigroups of a special kind. This representation of Brandt semigroups is amenable to categorical setting.

4. Primitive inverse semigroups are precisely orthogonal sums of Brandt semigroups. Regular semigroups which are subdirect products of Brandt semigroups and groups can be characterized in terms of conditions on their idempotents as well as in terms of certain functions among their $\mathcal{D}$-classes.

5. The Bruck semigroup over a monoid makes it possible to construct certain types of semigroups out of given monoids. In particular, using it, one proves that every semigroup can be embedded into a simple monoid.

6. A Reilly semigroup is a bisimple inverse semigroup whose idempotents form a chain isomorphic to negative integers. These are precisely the semigroups isomorphic to Bruck semigroups over groups. The most interesting special case of this class of semigroups is the bicyclic semigroup.

II.1 INVERSE SEMIGROUPS

We introduce here the concept of an inverse semigroup and establish some of its characterizations. We then discuss the natural partial order and Green's relations on inverse semigroups.

II.1.1 Definition

A semigroup in which every element has a unique inverse is an *inverse semigroup*. The unique universe of an element s of an inverse semigroup is denoted by s^{-1}.

In the manipulation of inverse semigroups, it is often useful to have statements equivalent to the definition of an inverse semigroup. Two of the most frequently used are as follows. First, note the obvious fact that in a regular semigroup S, we have $S^1a = Sa$ and $aS^1 = aS$ for any $a \in S$.

II.1.2 Theorem

The following conditions on a semigroup S are equivalent.

(i) S is an inverse semigroup.
(ii) S is regular and its idempotents commute.
(iii) Every $\mathfrak{L}$-class and every $\mathfrak{R}$-class of S contains exactly one idempotent.

Proof. (i) *implies* (ii). Let $e, f \in E_S$ and $a = (ef)^{-1}$. It is easy to verify that both ae and fa are inverses of ef, which by hypothesis yields that $a = ae = fa$. But then $a^2 = (ae)(fa) = a(ef)a = a$ so that $a = a^{-1} = (ef)^{-1}$. Hence $ef \in E_S$ and symmetrically $fe \in E_S$. Consequently $(ef)(fe)(ef) = ef$ and $(fe)(ef)(fe) = fe$ proving that $(ef)^{-1} = fe$. On the other hand, $(ef)^{-1} = ef$ since $ef \in E_S$. Thus $ef = fe$.

(ii) *implies* (iii). Let $e, f \in E_S$ be $\mathfrak{L}$-related. Then $e = xf$ and $f = ye$ for some $x, y \in S$, so that $e = ef = fe = f$. Similarly, $e \mathfrak{R} f$ implies $e = f$.

(iii) *implies* (i). Let x and y be inverses of an element a of S. Then $xa, ya \in E_S$ and $xa \mathfrak{L} a \mathfrak{L} ya$ and thus, by the hypothesis, $xa = ya$. Symmetrically, we get $ax = ay$. Hence $x = xax = yay = y$, as required.

It is also convenient to have a formula for the inverse of a product of two elements in an inverse semigroup. The situation is the same as in groups as we now show.

II.1.3 Lemma

For any elements a and b of an inverse semigroup S, we have

$$(ab)^{-1} = b^{-1}a^{-1}, \qquad (a^{-1})^{-1} = a.$$

Hence the operation $x \to x^{-1}$ is an involution on S.

Proof. Using commutativity of idempotents, one verifies directly that $b^{-1}a^{-1}$ is an inverse of ab, so the first formula above follows by the uniqueness of inverses. The second formula follows directly from the uniqueness of inverses.

We will use 1.2 and 1.3 without express reference. Also note that for any element a of an inverse semigroup S, we have $a^{-1}E_S a \subseteq E_S$. Every inverse semigroup admits a partial order which plays an important role in many considerations.

II.1.4 Lemma

In any inverse semigroup S, define $a \leq b$ if $a \in E_S b$. Then $\leq$ is a partial order on S.

Proof. It is obvious that $\leq$ is reflexive and transitive. Let $a \leq b$ and $b \leq a$; then $a = eb$ and $b = fa$ for some $e, f \in E_S$. Hence

$$a = eb = efa = fea = fa = b.$$

We thus introduce the following basic concept.

II.1.5 Definition

The relation $\leq$ defined on any inverse semigroup S by

$$a \leq b \Leftrightarrow a = eb \quad \text{for some } e \in E_S$$

is the *natural* (or *canonical*) *partial order* on S.

For any nonempty subset H of S, we call

$$H\omega = \{a \in S \mid a \geq h \quad \text{for some} \quad h \in H\}$$

the *closure of H in S*; if $H\omega = H$, then H is *closed in S*.

We have used ω to denote the universal relation on any set. The notation $H\omega$ conforms to the current usage and should cause no confusion as to the meaning of ω. In view of the importance of the natural partial order, we present now some alternative characterizations.

II.1.6 Lemma

Let S be an inverse semigroup and let $a, b \in S$. Then $a \leq b$ is equivalent to each of the following conditions.

(i) $a \in bE_S$.
(ii) $aa^{-1} = ab^{-1}$.
(iii) $a^{-1}a = b^{-1}a$.
(iv) $a = ab^{-1}a$.

Moreover, if $a \leq b$ and $c \in S$, then $ac \leq bc$, $ca \leq cb$, and $a^{-1} \leq b^{-1}$.

Proof. The proof of this lemma is left as an exercise.

Note that the restriction to E_S of the natural partial order on an inverse semigroup S coincides with the partial order on E_S defined in I.3.9. In the context of Green's relations on an inverse semigroup, we have the following simple statements.

II.1.7 Lemma

Let S be an inverse semigroup.

(i) $a\mathcal{L}b \Leftrightarrow a^{-1}a = b^{-1}b \qquad (a, b \in S)$.
(ii) $a\mathcal{R}b \Leftrightarrow aa^{-1} = bb^{-1} \qquad (a, b \in S)$.
(iii) $e\mathcal{D}f \Leftrightarrow e = aa^{-1}, f = a^{-1}a$ for some $a \in S \qquad (e, f \in E_S)$.
(iv) $J(e) \subseteq J(f) \Leftrightarrow e = aa^{-1}, a^{-1}a \leq f$ for some $a \in S \qquad (e, f \in E_S)$.

Proof. Note that for any $a \in S$, we have $a\mathcal{L}a^{-1}a$ and $a\mathcal{R}aa^{-1}$; using this, parts (i) and (ii) follow easily. Part (iii) follows directly from parts (i) and (ii). The details of this argument are left as an exercise.

Assume that $J(e) \subseteq J(f)$ [part (iv)] so that $e = xfy$ for some $x, y \in S$. Then

$$(ex)(ex)^{-1} = exx^{-1}e = exx^{-1}(xfy) = exfy = e,$$

$$(ex)^{-1}(ex)f = (x^{-1}ex)f = fx^{-1}ex = fx^{-1}(xfy)x$$

$$= f(x^{-1}x)f(yx) = x^{-1}(xfy)x = x^{-1}ex = (ex)^{-1}(ex),$$

and the element $a = ex$ fulfills the requirement. Conversely, if $e = aa^{-1}$ and $a^{-1}a \leq f$, then $e = a(a^{-1}a)a^{-1} = a(a^{-1}a)fa^{-1} \in SfS$ and thus $J(e) \subseteq J(f)$.

II.1.8 Corollary

Let S be an inverse semigroup. Then S is bisimple (respectively simple) if and only if for any $e, f \in E_S$ there exists $a \in S$ such that $e = aa^{-1}$ and $a^{-1}a = f$ (respectively $e = aa^{-1}$ and $a^{-1}a \leq f$).

Proof. The proof of this corollary is left as an exercise.

II.1.9 Corollary

Let S be an inverse semigroup with zero. Then S is 0-bisimple (respectively 0-simple) if and only if for any $e, f \in E_{S^*}$ there exists $a \in S$ such that $e = aa^{-1}$ and $a^{-1}a = f$ (respectively $e = aa^{-1}$ and $a^{-1}a \leq f$).

Proof. The proof of this corollary is left as an exercise.

For homomorphic images of inverse semigroups we have the following useful result.

II.1.10 Lemma

A homomorphic image of an inverse semigroup is an inverse semigroup.

Proof. Let S be an inverse semigroup and φ be a homomorphism of S onto a semigroup T. Obviously T is a regular semigroup. Let $e, f \in E_T$. By I.7.10, there exist idempotents e', f' in S such that $e'\varphi = e$ and $f'\varphi = f$. Since e' and f' commute, so do e and f, and thus T is an inverse semigroup.

II.1.11 Definition

Let A be a nonempty subset of an inverse semigroup S. Then the intersection of all inverse subsemigroups of S containing A is the *inverse subsemigroup of S generated by A*, to be denoted by $\langle A \rangle$. If $\langle A \rangle = S$, then A is a *set of generators for S* (as an inverse semigroup). In the case $A = \{s\}$, $\langle A \rangle$ is the *monogenic inverse semigroup generated by s*.

It is clear that with the above notation,

$$\langle A \rangle = \{a_1^{\varepsilon_1} a_2^{\varepsilon_2} \cdots a_n^{\varepsilon_n} | a_i \in A, \varepsilon_i = \pm 1\}.$$

We will occasionally use the following symbolism.

II.1.12 Notation

For any element a of an inverse semigroup we write $\lambda a = aa^{-1}$, $a\rho = a^{-1}a$.

Observe that $\lambda a^{-1} = a\rho$ and $a^{-1}\rho = \lambda a$.

II.1.13 Exercises

(i) Show that the following conditions on a semigroup S are equivalent.

- (α) S is an inverse semigroup.
- (β) Every principal left ideal of S has a unique right identity and every principal right ideal of S has a unique left identity.
- (γ) Every principal left and every principal right ideal of S is generated by a unique idempotent.
- (δ) Every principal factor of S is an inverse semigroup.

(ii) Show that the following conditions on a semigroup S are equivalent.

- (α) S is a group.
- (β) For every $a \in S$, there is a unique a' such that $a = aa'a$.

(γ) For every $a \in S$, there is a unique a'' such that aa'' is an idempotent.

(δ) S is an inverse semigroup and satisfies the implication $a = axa \Rightarrow x = xax$.

(iii) Show that the following conditions on a regular semigroup S are equivalent.

(α) S is an inverse semigroup.

(β) For any $a \in S$ and its inverse a', aa' and $a'a$ commute.

(γ) Every $e \in E_S$ has a unique inverse.

(δ) Every $e \in E_S$ commutes with all its inverses.

(ε) For any $e \in E_S$ and its inverse e', ee' commutes with $e'e$.

(ζ) S has an involution which fixes all idempotents of S.

(η) S satisfies the implication $a = axa = aya \Rightarrow xax = yay$.

(iv) Let V be an ideal extension of a semigroup S by a semigroup Q. Show that V is an inverse semigroup if and only if both S and Q are inverse semigroups.

(v) Show that a right cancellative inverse semigroup is a group.

(vi) Let D be a $\mathcal{D}$-class of an inverse semigroup S. Show that there is a bijection between the set of all $\mathcal{L}$-classes contained in D and the set of all $\mathcal{R}$-classes contained in D.

(vii) Let S be an inverse semigroup. Let Σ denote the set of all inverse subsemigroups of S together with the empty set ordered by inclusion. Show that Σ is a complete lattice and characterize the join of an arbitrary subset of Σ.

(viii) Show that in an inverse semigroup whose idempotents are linearly ordered, any two elements which have a common upper bound are comparable.

(ix) Characterize inverse semigroups in which no two distinct elements are comparable.

(x) Show that for any idempotents e and f in an inverse semigroup S,

$$e\mathcal{D}f \Leftrightarrow e = s^{-1}fs, \quad f = ses^{-1} \quad \text{for some} \quad s \in S.$$

Wagner [2] introduced inverse semigroups under the label of "generalized groups" using the definition that they are regular semigroups with commuting idempotents. Shortly after that A. E. Liber [1] showed that Wagner's definition is equivalent to the requirement of uniqueness of inverses. Inverse semigroups were a little later rediscovered by Preston [1] who introduced them under the name of "inverse semi-groups." The natural partial order on an inverse semigroup was introduced by Wagner [2] who proved results 1.3, 1.4, 1.6, and 1.10. For various sets of axioms for inverse semigroups, see Schein [9], [10]. Concerning the natural partial order consult Mitsch [1].

II.2 CLIFFORD SEMIGROUPS

This is the class of semigroups known under several names, the most frequent one being semilattice of groups. We start here with the concept of a semilattice of semigroups, discuss the construction of a strong semilattice of semigroups and its relation to subdirect products, and then characterize Clifford semigroups in several ways.

II.2.1 Definition

Let ρ be a congruence on a semigroup S. If S/ρ is a semilattice, then ρ is a *semilattice congruence on* S. In such a case, S is a *semilattice* $Y = S/\rho$ *of semigroups* S_α, $\alpha \in Y$, where S_α are the ρ-classes, or briefly a *semilattice of semigroups* S_α. If Y is a chain, we speak of a *chain of semigroups* S_α.

Note that if S is a semilattice Y of semigroups S_α, then $S_\alpha S_\beta \subseteq S_{\alpha\beta}$ where $\alpha\beta$ is the product of α and β in Y. In particular, $S_\alpha^2 \subseteq S_\alpha$ so S_α is a subsemigroup of S and $S_{\alpha\beta} = S_{\beta\alpha}$. A stronger concept can be found below.

II.2.2 Construction

Let Y be a semilattice. To each $\alpha \in Y$ associate a semigroup S_α and assume that $S_\alpha \cap S_\beta = \varnothing$ if $\alpha \neq \beta$. For each pair $\alpha, \beta \in Y$, $\alpha \geq \beta$, let $\varphi_{\alpha,\beta}\colon S_\alpha \to S_\beta$ be a homomorphism, and assume that the following conditions hold:

$$\varphi_{\alpha,\alpha} = \iota_{S\alpha} \qquad (\alpha \in Y),$$

$$\varphi_{\alpha,\beta}\varphi_{\beta,\gamma} = \varphi_{\alpha,\gamma} \qquad \text{if } \alpha > \beta > \gamma.$$

On the set $S = \bigcup_{\alpha \in Y} S_\alpha$ define a multiplication by

$$a * b = (a\varphi_{\alpha,\alpha\beta})(b\varphi_{\beta,\alpha\beta}) \tag{1}$$

if $a \in S_\alpha$, $b \in S_\beta$, and denote $S = [Y; S_\alpha, \varphi_{\alpha,\beta}]$.

Straightforward verification shows that the multiplication $*$ is associative, that the new multiplication coincides with the given one on each S_α, and that S is a semilattice Y of semigroups S_α. We usually denote the product in S also by juxtaposition.

II.2.3 Definition

The semigroup $[Y; S_\alpha, \varphi_{\alpha,\beta}]$ constructed in 2.2 is a *strong semilattice* Y *of semigroups* S_α *determined by the homomorphisms* $\varphi_{\alpha,\beta}$, or briefly a *strong semilattice of semigroups* S_α. More generally, this label will be attached to any semigroup which is a semilattice of semigroups S_α for which a system of functions $\varphi_{\alpha,\beta}$ exists with all the properties in 2.2 [(1) inclusive].

There is a simple relationship between strong semilattices of semigroups and subdirect products as follows.

II.2.4 Lemma

If $S = [Y; S_\alpha, \varphi_{\alpha,\beta}]$, then S is a subdirect product of semigroups S_α with a zero possibly adjoined.

Proof. For each $\alpha \in Y$, let K_α be equal to S_α with a zero 0_α adjoined if α is not the zero of Y, and $K_\alpha = S_\alpha$ otherwise. Define a function θ on S by

$$\theta\colon s \to (s_\alpha)_{\alpha \in Y}$$

where for $s \in S_\beta$, we set

$$s_\alpha = \begin{cases} s\varphi_{\beta,\alpha} & \text{if } \alpha \leq \beta \\ 0_\alpha & \text{otherwise.} \end{cases}$$

Then for $s \in S_\beta$, $t \in S_\gamma$, we obtain

$$s_\alpha t_\alpha = \begin{cases} (s\varphi_{\beta,\alpha})(t\varphi_{\gamma,\alpha}) & \text{if } \alpha \leq \beta, \gamma \\ 0_\alpha & \text{otherwise} \end{cases}$$

$$= \begin{cases} \left[(s\varphi_{\beta,\beta\gamma})(t\varphi_{\gamma,\beta\gamma})\right]\varphi_{\beta\gamma,\alpha} & \text{if } \alpha \leq \beta, \gamma \\ 0_\alpha & \text{otherwise} \end{cases}$$

$$= \begin{cases} (st)\varphi_{\beta\gamma,\alpha} & \text{if } \alpha \leq \beta\gamma \\ 0_\alpha & \text{otherwise} \end{cases}$$

$$= (st)_\alpha$$

and thus $(s\theta)(t\theta) = (st)\theta$. If $s\theta = t\theta$, then $\beta = \gamma$ and $s = s_\gamma = t_\gamma = t$, so θ is one-to-one. It is clear that θ maps S onto a subdirect product of semigroups K_α.

Note that the proof shows that one must adjoin a zero to all S_α for which α is not the zero of Y. Hence with at most one exception, one must adjoin zeros to each semigroup S_α.

We now introduce the first special class of inverse semigroups.

II.2.5 Definition

A semigroup which is a semilattice of groups is a *Clifford semigroup*.

These semigroups admit a great number of different characterizations. We limit ourselves to a few.

II.2.6 Theorem

The following conditions on a semigroup S are equivalent.

(i) S is a Clifford semigroup.
(ii) S is a strong semilattice of groups.
(iii) S is regular and is a subdirect product of groups with a zero possibly adjoined.
(iv) S is regular and its idempotents lie in its center.

Proof. (i) *implies* (ii). Let S be a semilattice Y of groups G_α with identity e_α. First let $\alpha > \beta$. Then $e_\alpha e_\beta \in G_\beta$ so there exists $x \in G_\beta$ such that $e_\alpha e_\beta x = e_\beta$ since G_β is a group. Hence $e_\alpha e_\beta = e_\alpha(e_\alpha e_\beta x) = e_\alpha e_\beta x = e_\beta$; analogously, one shows that $e_\beta e_\alpha = e_\beta$ and thus $e_\alpha > e_\beta$.

Fix $\alpha \geq \beta$ and define $\varphi_{\alpha,\beta}$ by

$$\varphi_{\alpha,\beta}: a \to ae_\beta \qquad (a \in G_\alpha).$$

For any $a, a' \in G_\alpha$, we get

$$(a\varphi_{\alpha,\beta})(a'\varphi_{\alpha,\beta}) = (ae_\beta)(a'e_\beta) = a\left[e_\beta(a'e_\beta)\right] = aa'e_\beta$$
$$= (aa')\varphi_{\alpha,\beta},$$

and $\varphi_{\alpha,\beta}$ is a homomorphism. Clearly $\varphi_{\alpha,\alpha} = \iota_{G_\alpha}$. Let $\alpha > \beta > \gamma$ and $a \in G_\alpha$. Using what we showed at the beginning of the proof, we obtain

$$a\varphi_{\alpha,\beta}\varphi_{\beta,\gamma} = (ae_\beta)e_\gamma = a(e_\beta e_\gamma) = ae_\gamma = a\varphi_{\alpha,\gamma}$$

so that $\varphi_{\alpha,\beta}\varphi_{\beta,\gamma} = \varphi_{\alpha,\gamma}$. For any $\alpha, \beta \in Y$ and $a \in G_\alpha$, $b \in G_\beta$, we have

$$ab = (ab)e_{\alpha\beta} = a(be_{\alpha\beta}) = a\left[e_{\alpha\beta}(be_{\alpha\beta})\right]$$
$$= (ae_{\alpha\beta})(be_{\alpha\beta}) = (a\varphi_{\alpha,\alpha\beta})(b\varphi_{\beta,\alpha\beta}).$$

Therefore S is a strong semilattice Y of groups G_α.

(ii) *implies* (iii). This is a direct consequence of 2.4.

(iii) *implies* (iv). Since the idempotents in a group or a group with a zero adjoined trivially lie in the center, this property is also shared by their subdirect products.

(iv) *implies* (i). Let S be a regular semigroup whose idempotents lie in its center. Let $a, b, x \in S$ be such that $a = axa$. Then $ab = a(xa)b = ab(xa) = (abx)a$ since $xa \in E_S$. This shows that $aS \subseteq Sa$, and by symmetry, we conclude that $aS = Sa$. We will use this below several times.

Now let $a, x \in S$ be such that $a = axa$. Then with $e = ax$ we obtain $a = ea = ae$, $e = ax = ya$ for some $y \in S$, and thus $a\mathcal{H}e$. Hence I.7.5 yields that all $\mathcal{H}$-classes of S are groups. Using what we proved above, it follows easily that $\mathcal{H}$ is a congruence for which $ab\mathcal{H}ba$ for all $a, b \in S$. Hence S is a Clifford semigroup.

It is convenient to introduce the following terminology.

II.2.7 Definition

If S is a Clifford semigroup isomorphic to $[Y; G_\alpha, \varphi_{\alpha,\beta}]$, then the latter is a *Clifford representation* of S.

The following result describes all homomorphisms of Clifford semigroups $[Y; G_\alpha, \varphi_{\alpha,\beta}]$.

II.2.8 Proposition

Let $S = [Y; G_\alpha, \varphi_{\alpha,\beta}]$ and $T = [Z; H_\alpha, \psi_{\alpha,\beta}]$ be Clifford semigroups. Let η: $Y \to Z$ be a homomorphism, for each $\alpha \in Y$, let χ_α: $G_\alpha \to H_{\alpha\eta}$ be a homomorphism, and assume that for any $\alpha \geq \beta$, the diagram

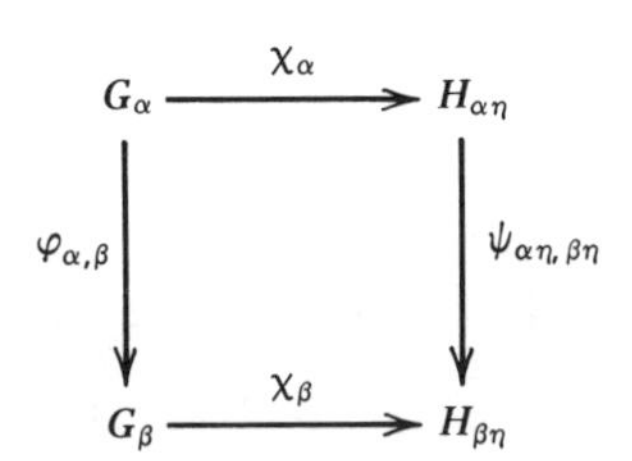

commutes. Define a function χ on S by

$$\chi\colon a \to a\chi_\alpha \qquad \text{if } a \in G_\alpha.$$

Then χ is a homomorphism of S into T. Moreover, χ is one-to-one (respectively a bijection) if and only if η and all χ_α are one-to-one (respectively bijections). Conversely, every homomorphism of S into T can be so constructed.

Proof. Let χ be as in the statement of the proposition, $a \in G_\alpha$, $b \in G_\beta$. Then

$$(a\chi)(b\chi) = (a\chi_\alpha)(b\chi_\beta) = (a\chi_\alpha\psi_{\alpha\eta,(\alpha\eta)(\beta\eta)})(b\chi_\beta\psi_{\beta\eta,(\alpha\eta)(\beta\eta)})$$

$$= (a\chi_\alpha\psi_{\alpha\eta,(\alpha\beta)\eta})(b\chi_\beta\psi_{\beta\eta,(\alpha\beta)\eta})$$

$$= (a\varphi_{\alpha,\alpha\beta}\chi_{\alpha\beta})(b\varphi_{\beta,\alpha\beta}\chi_{\alpha\beta})$$

$$= [(a\varphi_{\alpha,\alpha\beta})(b\varphi_{\beta,\alpha\beta})]\chi = (ab)\chi,$$

as required.

Conversely, let $\chi: S \to T$ be a homomorphism. Since χ preserves $\mathcal{H}$-classes, it induces a mapping $\eta: Y \to Z$ by the requirement: if $a \in G_\alpha$, then $a\chi \in H_{\alpha\eta}$. For any $a \in G_\alpha$, $b \in G_\beta$, we have $(a\chi)(b\chi) \in H_{(\alpha\eta)(\beta\eta)}$ and $(ab)\chi \in H_{(\alpha\beta)\eta}$ and thus η is a homomorphism. For every $\alpha \in Y$, let $\chi_\alpha = \chi|_{G_\alpha}$, so that χ_α: $G_\alpha \to G_{\alpha\eta}$ is a homomorphism. For any $a \in G_\alpha$ and $\alpha \geq \beta$, letting e_α be the identity of G_α, we obtain

$$a\varphi_{\alpha,\beta}\chi_\beta = (ae_\beta)\chi = (a\chi)(e_\beta\chi) = (a\chi_\alpha)e_{\beta\eta} = a\chi_\alpha\psi_{\alpha\eta,\beta\eta}$$

which proves commutativity of the above diagram. Trivially, χ is of the form as in the direct statement of the proposition.

The assertions concerning the properties of χ are obvious.

We discuss next the special case of a Clifford semigroup $S = [Y; G_\alpha, \varphi_{\alpha,\beta}]$ for which all homomorphisms $\varphi_{\alpha,\beta}$ are one-to-one. This is the case related to subdirect products of a semilattice and a group. The following auxiliary result is also of independent interest.

II.2.9 Lemma

Let $S = [Y; S_\alpha, \varphi_{\alpha,\beta}]$ and assume that all $\varphi_{\alpha,\beta}$ are one-to-one. Define a relation θ on S by: for $a \in S_\alpha$, $b \in S_\beta$,

$$a\theta b \Leftrightarrow a\varphi_{\alpha,\alpha\beta} = b\varphi_{\beta,\alpha\beta}.$$

Then θ is a congruence on S, and S is a subdirect product of Y and S/θ.

Proof. Clearly θ is reflexive and symmetric. Let $a \in S_\alpha$, $b \in S_\beta$, and $c \in S_\gamma$, assume that $a\theta b$ and $b\theta c$, and let $\delta = \alpha\beta\gamma$. Then

$$a\varphi_{\alpha,\delta} = a\varphi_{\alpha,\alpha\beta}\varphi_{\alpha\beta,\delta} = b\varphi_{\beta,\alpha\beta}\varphi_{\alpha\beta,\delta} = b\varphi_{\beta,\delta}$$

$$= b\varphi_{\beta,\beta\gamma}\varphi_{\beta\gamma,\delta} = c\varphi_{\gamma,\beta\gamma}\varphi_{\beta\gamma,\delta} = c\varphi_{\gamma,\delta}$$

which implies that $a\varphi_{\alpha,\alpha\gamma}\varphi_{\alpha\gamma,\delta} = c\varphi_{\gamma,\alpha\gamma}\varphi_{\alpha\gamma,\delta}$. It follows that $a\varphi_{\alpha,\alpha\gamma} = c\varphi_{\gamma,\alpha\gamma}$ since $\varphi_{\alpha\gamma,\delta}$ is one-to-one. Hence $a\theta c$, and thus θ is transitive.

Now assume only that $a\theta b$. Then

$$\begin{aligned}(ac)\varphi_{\alpha\gamma,\delta} &= \left[(a\varphi_{\alpha,\alpha\gamma})(c\varphi_{\gamma,\alpha\gamma})\right]\varphi_{\alpha\gamma,\delta} = (a\varphi_{\alpha,\delta})(c\varphi_{\gamma,\delta})\\ &= (a\varphi_{\alpha,\alpha\beta}\varphi_{\alpha\beta,\delta})(c\varphi_{\gamma,\delta}) = (b\varphi_{\beta,\alpha\beta}\varphi_{\alpha\beta,\delta})(c\varphi_{\gamma,\delta})\\ &= (b\varphi_{\beta,\delta})(c\varphi_{\gamma,\delta}) = (b\varphi_{\beta,\beta\gamma}\varphi_{\beta\gamma,\delta})(c\varphi_{\gamma,\beta\gamma}\varphi_{\beta\gamma,\delta})\\ &= \left[(b\varphi_{\beta,\beta\gamma})(c\varphi_{\gamma,\beta\gamma})\right]\varphi_{\beta\gamma,\delta} = (bc)\varphi_{\beta\gamma,\delta}\end{aligned}$$

and thus $ac\theta bc$. A symmetric argument shows that also $ca\theta cb$. Therefore θ is a congruence on S.

Define a mapping ψ on S by

$$\psi\colon a \to (\alpha, a\theta) \qquad \text{if } a \in S_\alpha,\ \alpha \in Y.$$

Evidently ψ is a homomorphism. If $a\psi = b\psi$ where $a \in S_\alpha$, $b \in S_\beta$, then $\alpha = \beta$ and $a\theta b$, whence $a = b$. Thus ψ is one-to-one. It is clear that the projection homomorphisms map $S\psi$ onto Y and S/θ, respectively. Therefore S is a subdirect product of Y and S/θ.

The next simple result often comes in handy.

II.2.10 Lemma

A regular semigroup with a single idempotent is a group.

Proof. Let S be such a semigroup with elements a, x, b, y such that $a = axa$, $b = byb$. The hypothesis implies that $ax = by$ whence $a = axa = b(ya)$. This shows that for any $a, b \in S$, the equation $a = bu$ is solvable in S for u; symmetrically $a = vb$ is solvable for v. Consequently, S is a group.

We can now prove the desired result.

II.2.11 Proposition

A semigroup S is regular and a subdirect product of a semilattice and a group if and only if $S \cong [Y; G_\alpha, \varphi_{\alpha,\beta}]$ where the latter is a Clifford semigroup with all $\varphi_{\alpha,\beta}$ one-to-one.

Proof

Necessity. We may suppose that $S \subseteq Y \times G$ where Y is a semilattice and G is a group. For each $\alpha \in Y$, let

$$G_\alpha = (\{\alpha\} \times G) \cap S,$$

and for any pair $\alpha, \beta \in Y$ with $\alpha \geq \beta$, define $\varphi_{\alpha,\beta}$ on G_α by

$$\varphi_{\alpha,\beta}\colon (\alpha, g) \to (\beta, g).$$

The verification that all the properties needed to have a Clifford semigroup $[Y; G_\alpha, \varphi_{\alpha,\beta}]$ are fulfilled, and that $S = [Y; G_\alpha, \varphi_{\alpha,\beta}]$ with all $\varphi_{\alpha,\beta}$ one-to-one is left as an exercise.

Sufficiency. We may take $S = [Y; G_\alpha, \varphi_{\alpha,\beta}]$, and note that by 2.9, we have that S is a subdirect product of Y and S/θ. If $e, f \in E_S$, say $e \in G_\alpha$, $f \in G_\beta$, then $e\theta(e\varphi_{\alpha,\alpha\beta}) = (f\varphi_{\beta,\alpha\beta})\theta f$ which shows that S/θ has only one idempotent, and thus by 2.10 must be a group. Therefore S is a subdirect product of a semilattice and a group.

We give below a construction of all regular semigroups which are a subdirect product of a semilattice and a group. It is convenient to introduce the following concepts.

II.2.12 Definition

Let Y be a semilattice and $\mathfrak{S}$ be a family of nonempty subsets of a set X ordered by inclusion. A function $\varphi\colon Y \to \mathfrak{S}$ is *full* if $\bigcup_{\alpha \in Y} \alpha\varphi = X$. For any group G, let $\mathfrak{L}(G)$ denote the lattice of all subgroups of G under inclusion.

II.2.13 Proposition

Let Y be a semilattice, G be a group, and η be an order inverting full function of Y into $\mathfrak{L}(G)$. Then

$$S = \{(\alpha, g) \in Y \times G \mid g \in \alpha\eta\}$$

under the induced multiplication of the direct product $Y \times G$ is a regular semigroup which is a subdirect product of Y and G, to be denoted by $[Y, G; \eta]$. Conversely, every such semigroup can be so constructed (up to an isomorphism).

Proof. Straightforward verification of the direct part is left as an exercise. Conversely, let S be a regular semigroup which is a subdirect product of a semilattice Y and a group G. We may suppose that $S \subseteq Y \times G$. For every $\alpha \in Y$, define $\alpha\eta$ by

$$\alpha\eta = \{g \in G \mid (\alpha, g) \in S\};$$

define a mapping η by

$$\eta\colon \alpha \to \alpha\eta \qquad (\alpha \in Y).$$

The verification that this η has all the requisite properties and that $S = [Y, G; \eta]$ is left as an exercise.

II.2.14 Exercises

(i) Show that the following conditions on a regular semigroup S are equivalent.

(α) S is a Clifford semigroup.

(β) In S, $a = axa$ implies $ax = xa$.

(γ) Every element of S commutes with all its inverses.

(δ) S is an inverse completely regular semigroup.

(ε) For any $a, b \in S$, $ab, ba \in E_S$ implies $ab = ba$.

(ζ) For any $x \in S$, $xS = Sx$.

(η) $\mathcal{L} = \mathcal{R}$.

(ii) Show that a semilattice of inverse semigroups is an inverse semigroup. Deduce that a semilattice of Clifford semigroups is a Clifford semigroup.

(iii) Find a property of inverse semigroups which does not carry to arbitrary semilattices of semigroups having this property.

(iv) Let S be a semilattice of monoids S_α. Show that if the identities of S_α form a subsemigroup of S, then S is a strong semilattice of semigroups S_α.

(v) Let Y be a semilattice and G be an abelian simple group. Show that there exists an inclusion preserving bijection between the set of all subdirect products of Y and G contained in $Y \times G$ and the set of all ideals of Y.

(vi) Prove that if every S_α in 2.9 is left cancellative, then θ is the least congruence on S for which S/θ is left cancellative.

(vii) Show that if every S_α in 2.9 is left reductive (respectively simple), then so is S/θ.

(viii) Show that a semilattice of commutative reductive semigroups is commutative and reductive.

(ix) Show that the following conditions on a semigroup S are equivalent.

(α) S is a chain of groups.

(β) S is completely regular and E_S is a chain.

(γ) For any $x, y \in S$, either $x \in yxSy$ or $y \in xySx$.

(x) Show that a Clifford semigroup S is a subdirect product of a semilattice and a group if and only if S satisfies the implication $xy = y \Rightarrow x^2 = x$.

(xi) Characterize the center of a Clifford semigroup $[Y; G_\alpha, \varphi_{\alpha,\beta}]$ for which all $\varphi_{\alpha,\beta}$ are epimorphisms.

(xii) Show that in a semigroup S all ideals are completely prime if and only if S is a chain of simple semigroups.

(xiii) Let n be a positive integer and S be a semigroup. Show that S satisfies the identities $x = x^{n+1}$ and $x^n y = yx^n$ if and only if S is a Clifford semigroup such that the order of any element of S divides n.

(xiv) Let $n > 1$ be an integer and S be a semigroup. Show that S satisfies the identities $x = x^n$ and $xy = yx$ if and only if S is a commutative Clifford semigroup such that the order of any element of S divides n.

(xv) For a natural number n, characterize the semigroups satisfying the identity $x = yx^n y$.

Clifford [1] introduced semilattices of groups and determined their structure in terms of what we now call strong semilattices of groups; he also characterized their isomorphisms. Part of 2.6 is due to A. E. Liber [1]. Note that strong semilattices of universal algebras are called Płonka sums in the literature outside of semigroups, see Płonka [1]. For an interesting result on semilattices of semigroups, consult Saliĭ [2].

There is a large body of literature concerning Clifford semigroups. For various aspects of general Clifford semigroups, see W. D. Burgess [1], Crawley [1], Feller–Gantos [1], Fountain–Lockley [1], Kowol–Mitsch [1], Lajos [1]–[6], Lallement–Milito [1], LaTorre [1], McMorris [1], Pastijn [1], and Rodriquez [1]–[3]. Diverse properties of commutative inverse semigroups were studied by Anderson [1], Braschear [1], Bullman–Fleming–McDowell [1], Dubikajtis–Jarek [1], Fulp [1], Gluskin [6], [7], Head–Anderson [1], Imaoka [2], Inasaridze [1], [3], Jarek [1], Kupcov [1]–[4], Lesohin [1]–[3], Manukjanc [1], Plahotnik [1], and Warne–Williams [1]. Finally, semilattices have attracted great attention; we list only works of interest from the point of view of semigroups, namely Bonzini [1], Bruns–Lakser [1], Freese–Nation [1], Grossman–Lausch [1], Hall [2], Hamilton [1], Horn–Kimura [1], Hsieh [1], Johnson–McMorris [1]–[4], Jürgensen [1], Klein–Barmen [1], [2], Papert [1], Pastijn [2], Petrich [5], Šimelfenig [1], Širjajev [4], [6], Varlet [1], and Zapletal [1], [2].

II.3 BRANDT SEMIGROUPS

We give the structure of Brandt semigroups in terms of Rees matrix semigroups of special type and discuss their homomorphisms. In addition, we consider a category whose objects are the ingredients in the construction of Brandt semigroups.

II.3.1 Definition

A completely 0-simple inverse semigroup is a *Brandt semigroup.*

In order to describe the structure of Brandt semigroups we first need some preparation.

II.3.2 Lemma

Let S be a Brandt semigroup.

(i) If e is a primitive idempotent of S and $x, y \in S$ are such that $f = xey \in E_{S^*}$, then $e = eyfxe$.
(ii) All nonzero idempotents of S are primitive.
(iii) The product of any two distinct idempotents of S is equal to zero.
(iv) For any $x, y \in S^*$, $xy \neq 0$ if and only if $x^{-1}x = yy^{-1}$.

Proof. (i) Since

$$(eyfxe)^2 = (ey)f(xey)f(xe) = eyfxe \leq e,$$

$$x(eyfxe)y = (xey)f(xey) = f \neq 0,$$

the hypothesis that e is primitive implies that $e = eyfxe$.

(ii) Let $f, g \in E_S$ be such that $0 \neq f \leq g$. Since S is 0-simple, by I.5.5, $g = xey$ for some $x, y \in S$, where e is a primitive idempotent of S. By part (i), we have $e = eygxe$. We also have $f = fg = (fx)ey$, so again by part (i), $e = eyfxe$. It follows that $eygxe = eyfxe$, which multiplied by x on the left and by y on the right yields $f = g$, as desired.

(iii) For any $e, f \in E_S$, if $ef \neq 0$, then $ef \in E_S$ and $ef \leq e$, $ef \leq f$ so by part (ii), we get $e = ef = f$.

(iv) First assume that $xy \neq 0$. Then $x(x^{-1}x)(yy^{-1})y \neq 0$ so that $(x^{-1}x)(yy^{-1}) \neq 0$ which by part (iii) gives $x^{-1}x = yy^{-1}$. Conversely, if $x^{-1}x = yy^{-1}$, then

$$x = x(x^{-1}x) = x(yy^{-1}) = (xy)y^{-1} \neq 0$$

and thus $xy \neq 0$.

II.3.3 Notation

For any semigroup S and a nonempty set I, we let

$$B(S, I) = I \times S \times I \cup \{0\}$$

where $0 \notin I \times S \times I$, with the multiplication

$$(i, g, j)(j, h, l) = (i, gh, l)$$

and all other products are equal to 0. We write $B_2 = B(S, I)$ when $|S| = 1$ and $|I| = 2$.

Note that when G is a group, $B(G, I)$ is a special case of a Rees matrix semigroup usually denoted by $\mathfrak{M}^0(I, G, I; \Delta)$ or $\mathfrak{M}^0(G; I, I; \Delta)$ and called a Brandt semigroup. Since this semigroup depends only upon the group G and the set I, the shorter notation $B(G, I)$ seems appropriate. Also notice that

$$E_{B(G,I)} = \{(i, e, i) | i \in I\} \cup 0,$$

and in $B(G, I)$,

$$(i, g, j)\mathcal{L}(k, h, l) \Leftrightarrow j = l,$$

and symmetrically for $\mathcal{R}$.

II.3.4 Lemma

Let S be a Brandt semigroup. Fix a nonzero idempotent e of S, and for each $f \in E_{S^*}$, fix an element r_f with the property that $e\mathcal{L}r_f\mathcal{R}f$. Define a mapping on S by

$$\chi: x \to \left(\lambda x, r_{\lambda x}^{-1} x r_{x\rho}, x\rho\right), \qquad 0 \to 0.$$

Then χ is an isomorphism of S onto $B(G, I)$ where $G = H_e$, $I = E_{S^*}$.

Proof. Since S is 0-simple, by 1.9, for any $e, f \in E_{S^*}$ there exists $x \in S$ such that $e = xx^{-1}$ and $x^{-1}x \leq f$. Since $x^{-1}x \neq 0$, by 3.2(ii) we get $x^{-1}x = f$ which by 1.9 implies that S is also 0-bisimple.

Let $x \in S^*$ and let $u = r_{\lambda x}^{-1}$ and $v = r_{x\rho}$. Then $e = \lambda u = v\rho$, $u\rho = \lambda x$, and $\lambda v = x\rho$. Hence

$$\lambda(uxv) = ux(vv^{-1})x^{-1}u^{-1} = uxx^{-1}u^{-1} = uu^{-1} = e$$

and similarly $(uxv)\rho = e$, so that $uxv\mathcal{H}e$. Consequently, $r_{\lambda x}^{-1} x r_{x\rho} \in H_e = G$ and thus χ maps S^* into $B(G, I)$. Let $x, y \in S$ be such that $xy \neq 0$. Then by 3.2(iv), $x\rho = \lambda y$ so that

$$\lambda(xy) = (xy)(xy)^{-1} = x(yy^{-1})x^{-1} = x(x^{-1}x)x^{-1} = xx^{-1} = \lambda x,$$

$$(xy)\rho = (xy)^{-1}(xy) = y^{-1}(x^{-1}x)y = y^{-1}(yy^{-1})y = y^{-1}y = y\rho,$$

and, in addition, $r_{\lambda y}\mathcal{R}\lambda y = yy^{-1}\mathcal{R} y$ which implies $\lambda r_{\lambda y} = \lambda y$ whence $(\lambda r_{\lambda y})y = y$. Consequently

$$\begin{aligned}(x\chi)(y\chi) &= \left(\lambda x, r_{\lambda x}^{-1} x r_{x\rho}, x\rho\right)\left(\lambda y, r_{\lambda y}^{-1} y r_{y\rho}, y\rho\right)\\ &= \left(\lambda x, r_{\lambda x}^{-1} x\left(r_{x\rho} r_{x\rho}^{-1} y\right) r_{y\rho}, y\rho\right)\\ &= \left(\lambda(xy), r_{\lambda(xy)}^{-1} xy r_{(xy)\rho}, (xy)\rho\right) = (xy)\chi.\end{aligned}$$

If $xy = 0$, then clearly $x\rho \neq \lambda y$ so $(x\chi)(y\chi) = 0 = (xy)\chi$. Therefore χ is a homomorphism.

Next let $(f, a, g) \in B(G, I)$, and let $x = r_f a r_g^{-1}$. Recalling that $r_f\rho = e = r_g\rho$, we obtain

$$\begin{aligned}\lambda x &= \left(r_f a r_g^{-1}\right)\left(r_f a r_g^{-1}\right)^{-1} = (r_f a)\left(r_g^{-1} r_g\right)\left(a^{-1} r_f^{-1}\right) = r_f a(r_g\rho)a^{-1} r_f^{-1}\\ &= r_f(aea^{-1})r_f^{-1} = r_f e r_f^{-1} = r_f(r_f\rho)r_f^{-1} = r_f r_f^{-1} = f,\end{aligned}$$

and analogously, $x\rho = g$, so that

$$r_{\lambda x}^{-1} x r_{x\rho} = r_f^{-1}\left(r_f a r_g^{-1}\right) r_g = (r_f\rho)a(r_g\rho) = eae = a$$

and thus $x\chi = (f, a, g)$. Consequently, χ maps S onto $B(G, I)$.

Assume that $x\chi = y\chi$. Then

$$\left(\lambda x, r_{\lambda x}^{-1} x r_{x\rho}, x\rho\right) = \left(\lambda y, r_{\lambda y}^{-1} y r_{y\rho}, y\rho\right)$$

which implies

$$\begin{aligned}x &= (\lambda x)x(x\rho) = r_{\lambda x}\left(r_{\lambda x}^{-1} x r_{x\rho}\right) r_{x\rho}^{-1} = r_{\lambda y}\left(r_{\lambda y}^{-1} y r_{y\rho}\right) r_{y\rho}^{-1}\\ &= (\lambda y)y(y\rho) = y\end{aligned}$$

and χ is one-to-one.

We can summarize our findings in the following form.

II.3.5 Theorem

A semigroup S is a Brandt semigroup if and only if S is isomorphic to $B(G, I)$, for some group G and nonempty set I.

Proof. The direct part is the content of 3.4. The proof of the converse consists of a straightforward verification and is left as an exercise.

It is convenient to introduce the following terminology.

II.3.6 Definition

If S is a Brandt semigroup isomorphic to $B(G, I)$, then the latter is a *Brandt representation* of S.

We now give a construction of all nontrivial homomorphisms of Brandt semigroups of the form $B(G, I)$.

II.3.7 Proposition

Let $S = B(G, I)$ and $T = B(H, J)$ be Brandt semigroups. Let $\omega\colon G \to H$ be a homomorphism, $\varphi\colon I \to J$ be a one-to-one function, and $u\colon I \to H$ be a function and write $u\colon i \to u_i$. Then the mapping θ defined by

$$\theta:(i, g, j) \to \left(i\varphi, u_i(g\omega)u_j^{-1}, j\varphi\right), \qquad 0 \to 0,$$

is a nontrivial homomorphism of S into T. Conversely, every nontrivial homomorphism of S into T can be so constructed.

Proof. A simple verification shows that the above θ is a homomorphism of S into T. Conversely, let $\theta\colon S \to T$ be a nontrivial homomorphism. Fix an element $1 \in I$ and denote the identities of both G and H by e. Since S is 0-simple and θ is nontrivial, θ must map nonzero elements of S onto nonzero elements of T. In particular, $(1, e, 1)\theta = (t, e, t)$ for some $t \in J$ since $(1, e, 1)$ is idempotent. We may thus define ω on G by the requirement that $(1, g, 1)\theta = (t, g\omega, t)$ obtaining a homomorphism. Further,

$$(i, e, 1)(1, e, 1) = (i, e, 1)$$

implies $(i, e, 1)\theta(t, e, t) = (i, e, 1)\theta$, and hence

$$(i, e, 1)\theta = (i\varphi, u_i, t)$$

for some $i\varphi \in J$, $u_i \in H$. Analogously, $(1, e, i)\theta = (t, v_i, i\psi)$ for some $v_i \in H$, $i\psi \in J$. Applying θ to the equality

$$(1, e, i)(i, e, 1) = (1, e, 1),$$

we obtain that $i\varphi = i\psi$ and $v_i u_i = e$. Thus $\varphi = \psi$ and $v_i = u_i^{-1}$. If $i\varphi = j\varphi$, then $(1, e, i\varphi)(j\varphi, e, 1) \neq 0$ so that $(1, e, i)(j, e, 1) \neq 0$ and thus $i = j$. Hence φ is one-to-one. Finally

$$\begin{aligned}(i, g, j)\theta &= [(i, e, 1)(1, g, 1)(1, e, j)]\theta \\ &= (i, e, 1)\theta(1, g, 1)\theta(1, e, j)\theta \\ &= (i\varphi, u_i, t)(t, g\omega, t)\left(t, u_j^{-1}, j\varphi\right) \\ &= \left(i\varphi, u_i(g\omega)u_j^{-1}, j\varphi\right),\end{aligned}$$

as required.

We can put the relationship of Brandt semigroups and their Brandt representations in a categorical setting as follows. This shows how strong the Brandt representation really is.

We define a category $\mathfrak{P}$ by

$$\text{Ob}\,\mathfrak{P} = \{(G, I) | G \text{ is a group and } I \text{ is a nonempty set}\}.$$

$$\text{Hom}\,\mathfrak{P} \text{ consists of triples } (\omega, u, \varphi): (G, I) \to (G', I')$$

where

$\omega: G \to G'$ is a homomorphism,

$u: I \to G'$ is a function, in notation $u: i \to u_i$,

$\varphi: I \to I'$ is a one-to-one function,

with the composition

$$(\omega, u, \varphi)(\omega', u', \varphi') = (\omega\omega', [u, \varphi, \omega', u'], \varphi\varphi')$$

where

$$[u, \varphi, \omega', u'] : i \to u'_{i\varphi}(u_i\omega') \qquad (i \in I).$$

Straightforward verification shows that $\mathfrak{P}$ satisfies the axioms for a category with $\varepsilon_{(G, I)} = (\iota_G, u, \iota_I)$, where u_i is the identity of G, being the identity morphism for any $(G, I) \in \text{Ob}\,\mathfrak{P}$.

For the second category, we let

Ob $\mathfrak{Q}$ be all Brandt semigroups,
Hom $\mathfrak{Q}$ be nontrivial homomorphisms of Brandt semigroups.

For each $(G, I) \in \text{Ob}\,\mathfrak{P}$, let $B(G, I)$ be the Brandt semigroup as before, for each $(\omega, u, \varphi) \in \text{Hom}\,\mathfrak{P}$, say $(\omega, u, \varphi): (G, I) \to (G', I')$, define a mapping on $B(G, I)$ by

$$B(\omega, u, \varphi): (i, g, j) \to (i\varphi, u_i(g\omega)u_j^{-1}, j\varphi), \qquad 0 \to 0.$$

II.3.8 Theorem

With the above notation, B is a full representative functor from $\mathfrak{P}$ to $\mathfrak{Q}$.

Proof. We know by 3.5 that for a given $(G, I) \in \text{Ob}\,\mathfrak{P}$, $B(G, I)$ is a Brandt semigroup, so $B(G, I) \in \text{Ob}\,\mathfrak{Q}$. In view of 3.7, we have that for a

$\mathfrak{P}$-morphism $(\omega, u, \varphi):(G, I) \to (G', I')$, $B(\omega, u, \varphi)$ is a nontrivial homomorphism of $B(G, I)$ into $B(G', I')$. It is clear that $B\varepsilon_{(G,I)}$ is the identical automorphism of $B(G, I)$. Assume next that we are given $\mathfrak{P}$-morphisms

$$(\omega, u, \varphi):(G, I) \to (G', I'),$$

$$(\omega', u', \varphi'):(G', I') \to (G'', I'').$$

Then for any $(i, g, j) \in B(G, I)$, we obtain

$$\begin{aligned}
&(i, g, j)[B(\omega, u, \varphi)][B(\omega', u', \varphi')]\\
&\qquad = \left(i\varphi, u_i(g\omega)u_j^{-1}, j\varphi\right)[B(\omega', u', \varphi')]\\
&\qquad = \left(i\varphi\varphi', u'_{i\varphi}\left[u_i(g\omega)u_j^{-1}\right]\omega' u'^{-1}_{j\varphi}, j\varphi\varphi'\right)\\
&\qquad = \left(i\varphi\varphi', \left[u'_{i\varphi}(u_i\omega')\right](g\omega\omega')\left[u'_{j\varphi}(u_j\omega')\right]^{-1}, j\varphi\varphi'\right)
\end{aligned}$$

which shows that B preserves the composition of morphisms. Therefore B is a functor from $\mathfrak{P}$ to $\mathfrak{Q}$.

Now B is full by 3.7 and representative by 3.5.

It follows from the definition of B that it is one-to-one on objects of $\mathfrak{P}$. The next result shows that B is not one-to-one on morphisms of $\mathfrak{P}$.

II.3.9 Proposition

Let $(\omega, u, \varphi), (\omega', u', \varphi'):(G, I) \to (G', I')$ be $\mathfrak{P}$-morphisms. Then $B(\omega, u, \varphi) = B(\omega', u', \varphi')$ if and only if $\omega' = \omega\varepsilon_c$, $u' = uc$, and $\varphi = \varphi'$ for some $c \in G'$, where $uc: i \to u_i c$ $(i \in I)$.

Proof. By definition, $B(\omega, u, \varphi) = B(\omega', u', \varphi')$ if and only if

$$\left(i\varphi, u_i(g\omega)u_j^{-1}, j\varphi\right) = \left(i\varphi', u'_i(g\omega')u'^{-1}_j, j\varphi'\right) \qquad [(i, g, j) \in B(G, I)]. \tag{2}$$

If (2) takes place, $\varphi = \varphi'$ and

$$u_i(g\omega)u_j^{-1} = u'_i(g\omega')u'^{-1}_j \qquad (i, j \in I, g \in G). \tag{3}$$

Letting g be the identity in (3), we obtain $u_i'^{-1}u_i = u_j'^{-1}u_j$ for any $i, j \in I$, so this is an element of G independent of i; we denote it by c^{-1}. From (3) we get

$$g\omega' = \left(u_i'^{-1}u_i\right)(g\omega)\left(u_j'^{-1}u_j\right)^{-1} = g\omega\varepsilon_c,$$

and thus $\omega' = \omega\varepsilon_c$. Also $u_i'^{-1}u_i = c^{-1}$ implies $u_i' = u_i c$ so that $u' = uc$.

Conversely, assume that $\omega' = \omega\varepsilon_c$, $u' = uc$, $\varphi = \varphi'$. Then for any $(i, g, j) \in B(G, I)$, we obtain

$$u_i(g\omega)u_j^{-1} = (u_i c)\left[c^{-1}(g\omega)c\right](u_j c)^{-1} = u'(g\omega')u_j'$$

which gives (3), and in conjunction with $\varphi = \varphi'$, it also yields (2), that is to say, the equality $B(\omega, u, \varphi) = B(\omega', u', \varphi')$.

II.3.10 Exercises

(i) Let $S = B(G, I)$. For a nontrivial normal subgroup N of G, define a relation ρ_N by

$$(i, g, j)\rho_N(k, h, l) \Leftrightarrow i = k, \quad gh^{-1} \in N, \ j = l, \qquad 0\rho_N 0.$$

Show that ρ_N is a proper congruence on S, and that conversely every proper congruence on S is of the form ρ_N for some nontrivial normal subgroup N of G.

(ii) Let G be a group, I and J be nonempty sets. Let

$$K = \{(j, 0, j') \mid j, j' \in J\}.$$

Show that K is an ideal of $B(B(G, I), J)$ and that

$$B(B(G, I), J)/K \cong B(G, I \times J).$$

(iii) Let $[Y; G_\alpha, \varphi_{\alpha,\beta}]$ be a Clifford semigroup, and I be a nonempty set. For every pair $\alpha, \beta \in Y$, $\alpha \geq \beta$, define $\tilde{\varphi}_{\alpha,\beta}$ on $B(G_\alpha, I)$ by

$$\tilde{\varphi}_{\alpha,\beta}: (i, g, j) \to (i, g\varphi_{\alpha,\beta}, j), \qquad 0_\alpha \to 0_\beta,$$

where 0_α and 0_β are the zeros of $B(G_\alpha, I)$ and $B(G_\beta, I)$, respectively. Let $J = \{0_\alpha \mid \alpha \in Y\}$. Show that $[Y; B(G_\alpha, I), \tilde{\varphi}_{\alpha,\beta}]$ is well defined, that J is its ideal, and that

$$\left[Y; B(G_\alpha, I), \tilde{\varphi}_{\alpha,\beta}\right]/J \cong B\left(\left[Y; G_\alpha, \varphi_{\alpha,\beta}\right], I\right).$$

(iv) On the pattern of 3.8 and the discussion preceding it, obtain an analogue of 3.8 for Clifford semigroups.

(v) Call a semigroup S a *free Brandt semigroup* if it is isomorphic to $B(G_X, X)$ for some nonempty set X. Show that the nonzero part of every Brandt semigroup is a partial homomorphic image of the nonzero part of a free Brandt semigroup, but that a Brandt semigroup need not be a homomorphic image of any free Brandt semigroup.

Brandt groupoids, which will be discussed in XIV.1, historically precede Brandt semigroups. Clifford [1] proved that Brandt groupoids with a zero adjoined, and the undefined products declared equal to zero, are precisely semigroups isomorphic to $B(G, I)$ for some group G and a nonempty set I. Munn [1] recognized that the latter are precisely inverse completely 0-simple semigroups. Also see Gelbaum–Schanuel [1] and Koževnikov [1]. Congruences on Brandt semigroups were characterized by Preston [6]. Homomorphisms of inverse semigroups onto Brandt semigroups were studied by Munn [4]. See also Barnes [1] and Hoehnke [2].

II.4 STRICT INVERSE SEMIGROUPS

These are the inverse semigroups which are subdirect products of Brandt semigroups and groups. Here we provide for them several characterizations and in XIV.4 a construction. We start with a subclass of strict inverse semigroups which can be obtained by forming of orthogonal sums of Brandt semigroups. The first lemma is of independent interest.

II.4.1 Lemma

Let S be an inverse semigroup and let $a, b \in S$. If $J_a \leq J_b$, then for every $e \in E_{J_b}$ there exists $f \in E_{J_a}$ such that $f \leq e$.

Proof. With the given hypotheses, we have $a = xey$ for some $x, y \in S$; let $f = x^{-1}xeyy^{-1}$. Then $a = xfy$ and $f = x^{-1}ay^{-1}$ so that $f \in E_{J_a}$ and clearly $f \leq e$.

II.4.2 Definition

A nontrivial inverse semigroup with zero is a *primitive inverse semigroup* if all its nonzero idempotents are primitive.

We can now prove the structure theorem for these semigroups.

II.4.3 Theorem

A semigroup S is a primitive inverse semigroup if and only if it is an orthogonal sum of Brandt semigroups.

Proof

Necessity. Let $x, a \in S$ be such that $J_x < J_a$. If $e \in E_{J_a}$, then $e > f$ for some $f \in E_{J_x}$ by 4.1, and thus $f = 0$ since e is primitive. Hence $J_x = \{0\}$ and thus $J(a) = J_a \cup \{0\}$. Now let $J_a \neq J_b$ for $a, b \in S^*$. Then $J_{ab} \leq J_a$ and $J_{ab} \leq J_b$ implies that J_{ab} must be strictly smaller than J_a or J_b, so by the above, we have $ab = 0$. This shows that S is an orthogonal sum of its nonzero ideals each of which is a Brandt semigroup.

Sufficiency. This follows directly from 3.5 and 3.2(ii).

An orthogonal sum of semigroups S_α can be easily seen to be a special subdirect product of semigroups S_α. We now turn to inverse semigroups which are arbitrary subdirect products of Brandt semigroups and groups. For the main characterization theorem we need some preparation.

II.4.4 Lemma

The following conditions on an ideal I of an inverse semigroup S are equivalent.

(i) I is a retract ideal of S.
(ii) I is a p-ideal of S.
(iii) E_I is a retract ideal of E_S.
(iv) E_I is a p-ideal of E_S.

Proof. (i) *implies* (ii). Let φ be a retraction of S onto I. For every $s \in S$, it suffices to show that

$$[s] \cap I = [s\varphi]. \tag{4}$$

If $x \in [s] \cap I$, then $x = es$ for some $e \in E_S$ and thus $x = x\varphi = (e\varphi)(s\varphi)$ which shows that $x \in [s\varphi]$. Conversely, if $x \in [s\varphi]$, then $x = e(s\varphi)$ for some $e \in E_S$ which by I.9.14 gives

$$x = x\varphi = [e(s\varphi)]\varphi = (e\varphi)(s\varphi) = (e\varphi)s$$

and thus $x \in [s] \cap I$. This proves relation (4) and shows that I is a p-ideal of S.

(ii) *implies* (i). Define φ on S by relation (4). For any $a, b \in S$, we obtain

$$\begin{aligned}[(ab)\varphi] &= [ab] \cap I = [a][b] \cap I \\ &= ([a] \cap I)([b] \cap I) = [a\varphi][b\varphi] \\ &= [(a\varphi)(b\varphi)],\end{aligned}$$

where the last equality follows from the fact that $[x] = xE_S = E_S x$ for all $x \in S$. Thus $(ab)\varphi = (a\varphi)(b\varphi)$. It is clear that $a\varphi = a$ if $a \in I$ and thus φ is a retraction of S onto I.

Applied to the semilattice E_S, the equivalence of (i) and (ii) yields the equivalence of (iii) and (iv). It is obvious that (i) implies (iii) since the retraction needed in (iii) is the one in (i) restricted to the idempotents.

(iii) *implies* (ii). It suffices to show that for any $s \in S$,

$$[s] \cap I = [s(s^{-1}s)\varphi] \tag{5}$$

where φ is a retraction of E_S onto E_I. Let $x \in [s] \cap I$. Then $x \leq s$ whence $x^{-1}x \leq s^{-1}s$ which implies $x^{-1}x \leq (s^{-1}s)\varphi$ since $(s^{-1}s)\varphi$ is the greatest element of $[s^{-1}s] \cap E_I$. It follows that

$$x = x(x^{-1}x) \leq x(s^{-1}s)\varphi \leq s(s^{-1}s)\varphi$$

which shows that $[s] \cap I \subseteq [s(s^{-1}s)\varphi]$; the opposite inclusion follows at once. This establishes relation (5) and completes the proof.

Recall that we formally write $S/\varnothing$ for S.

II.4.5 Theorem

The following conditions on an inverse semigroup S are equivalent.

(i) S is a subdirect product of Brandt semigroups and groups.

(ii) For any $e, f, g \in E_S$, $e \geq f$, $e \geq g$, $f \mathcal{D} g$ imply $f = g$.

(iii) S is completely semisimple, and for every $a \in S$, letting $P_a = \{x \in S | J_x \ngtr J_a\}$ and $Q_a = \{x \in S | J_x \ngeq J_a\}$, P_a/Q_a is a retract ideal of S/Q_a.

(iv) S is completely semisimple, and for any $\mathcal{J}$-classes $A \leq B$ of S, there exists a function $\varphi : B \to A$ such that (α) $E_B\varphi \subseteq E_A$, (β) $ab = a(b\varphi)$ if $a \in A$, $b \in B$, $ab \in A$.

Proof. (i) *implies* (ii). Let S be a subdirect product of the family S_α of Brandt semigroups and groups, $\alpha \in A$. We may suppose that $S \subseteq \prod_{\alpha \in A} S_\alpha$. Let e, f, g be as in the hypothesis of part (ii). Then $e = (e_\alpha)$, $f = (f_\alpha)$, and $g = (g_\alpha)$, and for every $\alpha \in A$, $e_\alpha \geq f_\alpha$, $e_\alpha \geq g_\alpha$, and $f_\alpha \mathcal{D} g_\alpha$. The last condition insures that $f_\alpha = 0_\alpha$ if and only if $g_\alpha = 0_\alpha$ in the case that S_α is a Brandt semigroup. This together with the conditions $e_\alpha \geq f_\alpha$ and $e_\alpha \geq g_\alpha$ implies that $f_\alpha = g_\alpha$ since either $e_\alpha = 0_\alpha$ or e_α is a primitive idempotent. In the case S_α is a group, then trivially $e_\alpha = f_\alpha = g_\alpha$. Consequently, $f_\alpha = g_\alpha$ for all $\alpha \in A$, so that $f = g$.

(ii) *implies* (iii). Let $e, f \in E_S$ be such that $e\mathcal{D}f$ and $e \geq f$. Then $e \geq e$ and $e \geq f$ with $e\mathcal{D}f$, which by the hypothesis yields $e = f$. Hence I.6.14 implies that S is completely semisimple.

Let $a \in S$. Assume that $Q_a \neq \emptyset$. For any $x, y \in S$, we have $J_{xy} \leq J_x$ and $J_{xy} \leq J_y$. Hence if $J_{xy} \geq J_a$, then $J_x \geq J_a$ and $J_y \geq J_a$. By contrapositive, we conclude that Q_a is an ideal of S. Since $P_a = J_a \cup Q_a = J(a) \cup Q_a$, it follows that P_a is always an ideal of S. We thus have that P_a/Q_a is an ideal of S/Q_a.

Since S is completely semisimple, we have $\mathcal{D} = \mathcal{J}$. The hypothesis on idempotents of S implies that E_{P_a/Q_a} is a p-ideal of E_{S/Q_a} which by 4.4 gives that P_a/Q_a is a retract ideal of S/Q_a.

(iii) *implies* (iv). Let $a, b \in S$ be such that $J_a \leq J_b$. The retraction $S/Q_a \to P_a/Q_a$ restricted to J_b trivially satisfies conditions (α) and (β) in part (iv).

(iv) *implies* (ii). Note first that $\mathcal{D} = \mathcal{J}$ in S. Let $e, f, g \in E_S$, $e \geq f$, $e \geq g$, $f\mathcal{D}g$. Then with $\varphi : J_e \to J_f$ as in part (iv), we have $e\varphi \in E_{J_a}$ and

$$f = fe = f(e\varphi), \qquad g = ge = g(e\varphi)$$

which implies that $f = e\varphi = g$ since J_f^0 is a Brandt semigroup.

(iii) *implies* (i). By hypothesis, for each $a \in S$, there is a retraction $\psi_{J_a} : S/Q_a \to P_a/Q_a$. We define a mapping χ by

$$\chi : a \to (a_{J_x})_{x \in S} \qquad (a \in S)$$

where

$$a_{J_x} = \begin{cases} a\psi_{J_x} & \text{if } J_a \geq J_x \\ 0 & \text{otherwise.} \end{cases}$$

For any $a \in S$, we have the composition $S \to S/Q_a \to P_a/Q_a$ of the homomorphism associated with the Rees quotient S/Q_a and ψ_{J_a}, which implies that χ is a homomorphism. Moreover, χ is one-to-one since ψ_{J_a} is one-to-one on J_a and elements of S not $\mathcal{J}$-related are mapped by χ onto distinct elements. It is obvious that $S\chi$ is a subdirect product of $\mathcal{J}$-classes of S with a zero possibly adjoined. Since S is completely semisimple, these components are Brandt semigroups with one possible exception which must be a group.

The functions occurring in 4.5(iv) will play an important role in XIV.4. We now establish a number of their properties.

II.4.6 Corollary

With the hypotheses and notation of 4.5(iv), write $\varphi_{BA} = \varphi : B \to A$. Then the collection of functions $\varphi = \varphi_{BA}$ also satisfies the following conditions.

(i) $ab = a(b\varphi)$ if $a \in A$, $b \in B$, $a(b\varphi) \in A$.

(ii) $ba = (b\varphi)a$ if $a \in A$, $b \in B$, and either $ba \in A$ or $(b\varphi)a \in A$.

(iii) $(bb')\varphi = (b\varphi)b' = b(b'\varphi) = (b\varphi)(b'\varphi)$ if $b, b', bb' \in B$.
(iv) $bc = (b\varphi_{BA})c = b(c\varphi_{CA}) = (b\varphi_{BA})(c\varphi_{CA})$ if $b \in B$, $c \in C$, $bc \in A$.
(v) $\varphi_{CB}\varphi_{BA} = \varphi_{CA}$ if $A \le B \le C$.

Proof. We refer to the proof of "(iii) implies (iv)" in 4.5.

(i) If $a(b\varphi) \in A$, then

$$(ab)\varphi = (a\varphi)(b\varphi) = a(b\varphi) \in A$$

which implies that $ab \in A$ and thus $ab = a(b\varphi)$.

(ii) This is symmetric to 4.5(iv) (β) and part (i).

(iii) First $bb' \in B$ implies that $(b\varphi)(b'\varphi) = (bb')\varphi \in A$ and we may apply parts (i) and (ii) to obtain the remaining equalities.

(iv) The argument here is similar to that in part (iii).

(v) This follows easily from the proof of 4.4, (ii) implies (i).

In order to facilitate some of the later considerations, we introduce the following terminology.

II.4.7 Definition

An inverse semigroup which is a subdirect product of Brandt semigroups and/or groups is a *strict inverse semigroup*.

II.4.8 Exercises

(i) Show that the following conditions on a nontrivial semigroup S with zero are equivalent.

   (α) S is a primitive inverse semigroup.
   (β) For every $a \in S^*$, there exists a unique a' such that $a = aa'a$.
   (γ) For every $a \in S^*$, there exists a unique a'' such that aa'' is a nonzero idempotent.
   (δ) S is an inverse semigroup satisfying $a = axa \neq 0 \Rightarrow x = xax$.

(ii) Show that Brandt semigroups are precisely those in the preceding exercise in which 0 is a prime ideal.

(iii) Let S be an inverse semigroup with zero and φ be a pure homomorphism of S onto a primitive inverse semigroup T. Prove the following statements.

   (α) S is categorical at zero.
   (β) T is a Brandt semigroup if and only if 0 is a prime ideal of S.

(iv) Let S be a nontrivial inverse semigroup with zero and assume that S is categorical at zero. Define a relation θ on S by

$$a\theta b \Leftrightarrow ax = bx \neq 0 \quad \text{for some } x \in S, \quad 0\theta 0.$$

Show that θ is the least pure congruence ρ on S for which S/ρ is a primitive inverse semigroup. Deduce that θ is the least congruence ρ such that S/ρ is a Brandt semigroup if and only if 0 is a prime ideal of S.

(v) Show that any primitive inverse semigroup can be embedded into a Brandt semigroup.

(vi) Characterize inverse semigroups with zero in which no two distinct nonzero elements are comparable.

(vii) Let S be an inverse semigroup. Call an element a of S *strict* if $a^{-1}Sa$ is a Clifford semigroup. Show that $a \in S$ is strict if and only if for any $b, c \in S$, $a \geq b$, $a \geq c$, and $b\mathcal{D}c$ imply $b = c$.
Denote by $\Xi(S)$ the set of all strict elements of S. Show that $\Xi(S)$ may be empty. But when $\Xi(S)$ is nonempty, prove that $\Xi(S)$ is the greatest ideal of S which is a strict inverse semigroup and that $\Xi(S/\Xi(S)) = 0$. Also show that

$$\Xi(S) = \{a \in S \mid J(a) \text{ is a strict inverse semigroup}\}.$$

(viii) Show that the following conditions on an inverse semigroup S are equivalent.

(α) S is a strict inverse semigroup.
(β) All elements of S are strict.
(γ) For every $e \in E_S$, eSe is a Clifford semigroup.
(δ) For any $a, b \in S$, $(aSa)(bSb) = abSab$.

(ix) Let S be a nontrivial inverse semigroup. Show that for all $e \in E_S$, eSe is a semilattice if and only if S is a subdirect product of Brandt semigroups with only trivial subgroups.

(x) Show that for any strict inverse semigroup S, the mapping $a \to aSa$ $(a \in S)$ is a homomorphism of S into the semigroup of all nonempty subsets of S under complex multiplication which induces $\mathcal{H}$ on S.

(xi) Show that the class of all inverse semigroups on which $\mathcal{H}$ is a congruence is closed for the formation of direct products and the taking of inverse subsemigroups. Deduce that $\mathcal{H}$ is a congruence on every strict inverse semigroup.

(xii) Prove that the function φ in 4.5(iv) has the following property: if $b \in B$, then $b\varphi$ is the unique element of A less than or equal to b. Thus φ is unique.

Various characterizations and structure theorems for primitive regular semigroups have been obtained, in chronological order, by Lallement and Petrich [1], Steinfeld [1], Venkatesan [2], and Preston [7] (this appeared also in Clifford–Preston [1]). The proof of 4.3 follows Hall [1].

Subdirect products of completely 0-simple (in particular of Brandt) semigroups were characterized by Lallement [2]; for some refinements, see Petrich [6]; 4.5(iii) was suggested by P. R. Jones. A construction for a subclass of these semigroups was provided by Lallement–Petrich [2]. Homomorphisms of inverse semigroups onto primitive inverse semigroups were studied by Lallement–Petrich [1] and Preston [7].

II.5 BRUCK SEMIGROUPS OVER MONOIDS

The construction of a Bruck semigroup over a monoid provides a means for obtaining new semigroups out of given ones. These new semigroups have certain properties not present in the given ones and will be used for describing the structure of several classes of bisimple or simple inverse semigroups having a certain configuration of idempotents. This should be compared with the construction of Rees matrix semigroups either over groups or over groups with zero. Next to the construction of Rees matrix semigroups and strong semilattices of semigroups, Bruck semigroups over monoids represent one of the basic devices for producing more complex semigroups out of given ones. We start with the construction of basic importance in our discussion.

II.5.1 Construction

Let T be a monoid, α be a homomorphism of T into its group of units and N be the set of all nonnegative integers. On the set $S = N \times T \times N$ define a multiplication by

$$(m, a, n)(p, b, q) = (m + p - r, (a\alpha^{p-r})(b\alpha^{n-r}), n + q - r)$$

where $r = \min\{n, p\}$ and α^0 is the identity mapping on T. It is clear that S is closed under this multiplication. Denote the groupoid S by $B(T, \alpha)$.

II.5.2 Lemma

$B(T, \alpha)$ is a simple monoid.

Proof. Let $(m_i, a_i, n_i) \in B(T, \alpha)$, $i = 1, 2, 3$. With $r_1 = \min\{n_1, m_2\}$ and $r_2 = \min\{n_1 + n_2 - r_1, m_3\}$, we obtain

$$\begin{aligned}
&[(m_1, a_1, n_1)(m_2, a_2, n_2)](m_3, a_3, n_3) \\
&\quad = (m_1 + m_2 - r_1, (a_1\alpha^{m_2 - r_1})(a_2\alpha^{n_1 - r_1}), n_1 + n_2 - r_1)(m_3, a_3, n_3) \\
&\quad = (m_1 + m_2 + m_3 - (r_1 + r_2), b, n_1 + n_2 + n_3 - (r_1 + r_2))
\end{aligned}$$

where

$$b = \left(a_1\alpha^{m_2+m_3-(r_1+r_2)}\right)\left(a_2\alpha^{n_1+m_3-(r_1+r_2)}\right)\left(a_3\alpha^{n_1+n_2-(r_1+r_2)}\right).$$

On the other hand, with $r_3 = \min\{n_2, m_3\}$ and $r_4 = \min\{n_1, m_2 + m_3 - r_3\}$, we similarly obtain

$$\begin{aligned}(m_1, a_1, n_1)&[(m_2, a_2, n_2)(m_3, a_3, n_3)]\\ &= (m_1, a_1, n_1)(m_2 + m_3 - r_3, (a_2\alpha^{m_3-r_3})(a_3\alpha^{n_2-r_3}), n_2 + n_3 - r_3)\\ &= (m_1 + m_2 + m_3 - (r_3 + r_4), c, n_1 + n_2 + n_3 - (r_3 + r_4))\end{aligned}$$

where

$$c = \left(a_1\alpha^{m_2+m_3-(r_3+r_4)}\right)\left(a_2\alpha^{n_1+m_3-(r_3+r_4)}\right)\left(a_3\alpha^{n_1+n_2-(r_3+r_4)}\right).$$

In order to establish associativity, it remains to prove that $r_1 + r_2 = r_3 + r_4$. To show this, we distinguish the following possibilities $n_1 \leq m_2$ or $m_2 \leq n_1$ and $n_2 \leq m_3$ or $m_3 \leq n_2$, and consider the four cases arising in this manner.

Assume that $n_1 \leq m_2$, $n_2 \leq m_3$. The hypothesis implies that $r_1 = n_1$, $r_2 = \min\{n_1 + n_2 - n_1, m_3\} = n_2$, $r_3 = n_2$, $r_4 = \min\{n_1, m_2 + m_3 - n_2\}$, and $m_2 + m_3 - n_2 \geq n_1$ so that $r_4 = n_1$. This proves that $r_1 + r_2 = r_3 + r_4$ in this case. The proof of the remaining cases is left as an exercise.

It is clear that $(0, e, 0)$ is the identity of $B(T, \alpha)$, where e is the identity of T. For any $(m, a, n), (p, b, q) \in B(T, \alpha)$ we obtain

$$\begin{aligned}(m, a, p + 1)(p, b, q)&\left(q + 1, (b\alpha)^{-1}, n\right)\\ &= \left(m, (a\alpha^0)(b\alpha), q + 1\right)\left(q + 1, (b\alpha)^{-1}, n\right)\\ &= (m, a, n)\end{aligned}$$

which in view of I.5.6 implies that $B(T, \alpha)$ is simple.

II.5.3 Definition

The monoid $B(T, \alpha)$ is the *Bruck semigroup over the monoid T with endomorphism α*, or simply a *Bruck semigroup* (*over a monoid*).

We will now explore certain properties of $B(T, \alpha)$ in relation to similar properties of T. We start with a by-product which is of general interest. First note the following simple fact.

II.5.4 Remark

With the notation as in 5.1, the mapping

$$a \to (0, a, 0) \qquad (a \in T)$$

is an embedding of T into $B(T, \alpha)$.

II.5.5 Corollary

Every semigroup can be embedded into a simple monoid.

Proof. For any semigroup S, we have the obvious embedding of S into S^1 and of the latter into $B(S^1, \alpha)$ by 5.4. The assertion now follows from 5.2.

Even though the last result is of no help in finding the structure of S, it still gives us a useful hint about simple monoids. In fact, it shows that they are in no sense "simple" except in the sense of the absence of proper ideals. For 5.5 asserts that they contain as subsemigroups all semigroups. In the following we retain the notation introduced in 5.1; we will consider regularity, Green's relations, and idempotents in T and in $B(T, \alpha)$.

II.5.6 Lemma

The element (m, a, n) is an inverse of (p, b, q) if and only if $m = q$, $n = p$, and a is an inverse of b.

Proof

Necessity. By hypothesis, we have

$$(m, a, n) = (m, a, n)(p, b, q)(m, a, n), \tag{6}$$

$$(p, b, q) = (p, b, q)(m, a, n)(p, b, q). \tag{7}$$

It follows from (6) that

$$\begin{aligned}(m, a, n) &= \left(m + p - r_1, (a\alpha^{p-r_1})(b\alpha^{n-r_1}), n + q - r_1\right)(m, a, n)\\ &= \big(2m + p - (r_1 + r_2), (a\alpha^{p+m-(r_1+r_2)})(b\alpha^{n+m-(r_1+r_2)})\\ &\quad \cdot(a\alpha^{n+q-(r_1+r_2)}), 2n + q - (r_1 + r_2)\big)\end{aligned}$$

where $r_1 = \min\{n, p\}$, $r_2 = \min\{n + q - r_1, m\}$, and thus, equating the corresponding entries, we obtain

$$m + p = r_1 + r_2 = n + q, \tag{8}$$

$$a = a(b\alpha^{n+m-(m+p)})a = a(b\alpha^{n-p})a. \tag{9}$$

Now (9) implies that $n \geq p$ which in view of (8) yields $m \geq q$. By symmetry, (7) implies that $q \geq m$ and $p \geq n$, and thus $m = q$, $n = p$. But then (9) yields $a = aba$. Again by symmetry, (7) implies that $b = bab$. Consequently, a is an inverse of b.

Sufficiency. It suffices to multiply the appropriate elements.

II.5.7 Corollary

A monoid T is a regular (respectively inverse) semigroup if and only if $B(T, \alpha)$ is a regular (respectively inverse) semigroup. Moreover, if T is an inverse semigroup, then $(m, a, n)^{-1} = (n, a^{-1}, m)$.

Proof. The proof of this corollary is left as an exercise.

II.5.8 Corollary

Every regular (respectively inverse) semigroup can be embedded into a regular (respectively inverse) simple monoid.

Proof. The proof of this corollary is left as an exercise.

II.5.9 Lemma

The following statements hold in $B(T, \alpha)$.

(i) $(m, a, n)\mathcal{L}(p, b, q) \Leftrightarrow a\mathcal{L}b,\ n = q$.
(ii) $(m, a, n)\mathcal{R}(p, b, q) \Leftrightarrow a\mathcal{R}b,\ m = p$.
(iii) $(m, a, n)\mathcal{H}(p, b, q) \Leftrightarrow m = p,\ a\mathcal{H}b,\ n = q$.
(iv) $(m, a, n)\mathcal{D}(p, b, q) \Leftrightarrow a\mathcal{D}b$.

Proof

(i) *Necessity.* By hypothesis, we have

$$(m, a, n) = (p', b', q')(p, b, q), \tag{10}$$

$$(p, b, q) = (m', a', n')(m, a, n), \tag{11}$$

for some $(p', b', q'), (m', a', n') \in B(T, \alpha)$. Equality (11) implies

$$a = (b'\alpha^{p-r})(b\alpha^{q'-r}), \tag{12}$$

$$n = q' + q - r, \tag{13}$$

where $r = \min\{q', p\}$. Hence $q' - r \geq 0$ and (13) implies that $n \geq q$. By symmetry, (11) implies that $q \geq n$, and thus $q = n$. But then (13) implies that $q' = r$, which substituted into (12) yields $a = (b'\alpha^{p-r})b$. By symmetry, (11) yields $b \in Sa$ and thus $a\mathcal{L}b$.

Sufficiency. By hypothesis, we have $a = b'b$ and $b = a'a$ for some $a', b' \in T$. Hence

$$(m, b', p)(p, b, n) = (m, b'b, n) = (m, a, n),$$

$$(p, a', m)(m, a, n) = (p, a'a, n) = (p, b, n),$$

and thus $(m, a, n)\mathcal{L}(p, b, n)$.

Part (ii) follows by symmetry; parts (iii) and (iv) follow directly from parts (i) and (ii).

II.5.10 Corollary

The sets $T/\mathcal{D}$ and $B(T, \alpha)/\mathcal{D}$ have the same cardinality. In particular, T is bisimple if and only if $B(T, \alpha)$ is bisimple.

Proof. The proof of this corollary is left as an exercise.

II.5.11 Lemma

$E_{B(T,\alpha)} = \{(m, e, m) | e \in E_T\}$ and

$$(m, e, m) \leq (n, f, n) \Leftrightarrow m = n, \quad e \leq f \quad \text{or} \quad m > n.$$

Proof. The first formula follows from 5.6 or by a simple direct calculation. Also, a straightforward verification shows that

$$(m, e, m) \leq (n, f, n) \Leftrightarrow m \geq n, \quad e \leq f\alpha^{m-n}.$$

Assume that $m \geq n$ and $e \leq f\alpha^{m-n}$. If $m = n$, then $e \leq f\alpha^0 = f$, and if $m > n$, then $e \leq f\alpha^{m-n} = 1$, the identity of T, which is automatically fulfilled. Hence the second assertion of the lemma follows.

The following concept is basic for our considerations.

II.5.12 Definition

The set N of nonnegative integers with the reverse of the usual order is an *ω-chain*, to be denoted by C_ω. Any partially ordered set order isomorphic to C_ω is also called an *ω-chain*.

The following is a consequence of 5.11; for the notation, see I.2.14.

II.5.13 Corollary

$E_{B(T,\alpha)} \cong C_\omega \circ E_T$.

Proof. According to 5.11, the mapping $(m, e, m) \to (m, e)$ is the required order isomorphism.

II.5.14 Exercises

(i) For T, α, and the set S in 5.1, define the following two multiplications on S.

(α) $(m, a, n)(p, b, q) = (m - n + t, (a\alpha^{t-n})(b\alpha^{t-p}), q - p + t)$ where $t = \max\{n, p\}$.

(β) $(m, a, n)(p, b, q) = (m + [p - n], (a\alpha^{[p-n]})(b\alpha^{[n-p]}), q + [n - p])$ where for any integer d,

$$[d] = \begin{cases} d & \text{if } d \geq 0 \\ 0 & \text{if } d < 0. \end{cases}$$

Prove that each of these multiplications coincides with the multiplication given in 5.1.

(ii) Characterize regular and completely regular elements of $B(T, \alpha)$.

(iii) Show that E_T is a subsemigroup (respectively subsemilattice) of T if and only if $E_{B(T,\alpha)}$ is a subsemigroup (respectively subsemilattice) of $B(T, \alpha)$.

(iv) Find the centralizer of the set of all idempotents of $B(T, \alpha)$.

The construction of $B(T, \alpha)$, in a slightly different form, for α the mapping of T onto its identity, was used by Bruck [1] to prove 5.5. Reilly [2] proved that every inverse semigroup can be embedded into a bisimple inverse monoid. For T a group (respectively a finite chain of groups), this construction was used by Reilly [3], see II.6 (respectively Kočin [1], see XI.4) to describe the structure of certain semigroups; Munn [9] arrived at a closely related construction. A number of statements in this section as well as an application of the construction in question were announced by Munn [12]; see also Clifford–Preston ([1], Section 8.5). Also consult Clement–Pastijn [1].

II.6 REILLY SEMIGROUPS

The only result in this section describes bisimple inverse semigroups whose idempotents form an ω-chain as Bruck semigroups over groups (up to an isomorphism). Sections XI.1–3 are devoted to certain properties of these semigroups. It is convenient to introduce the following terminology.

II.6.1 Definition

A regular semigroup S is *ω-regular* if its idempotents form an ω-chain. A bisimple ω-regular semigroup is a *Reilly semigroup*.

The next theorem represents the first structure theorem for a class of regular semigroups which are not completely semisimple. In fact, it represents a first step toward an understanding of the structure of bisimple (or simple) regular semigroups which are not completely simple.

II.6.2 Theorem

A semigroup S is a Reilly semigroup if and only if it is isomorphic to a Bruck semigroup over a group.

Proof

Necessity. Let S be a Reilly semigroup with idempotents $e_0 > e_1 > e_2 > \cdots$. The idempotents of S evidently commute so that S is an inverse semigroup by 1.2.

1. Let $e = e_0$. For any $s \in S$, by 1.2, we have that $s \mathcal{R} e_i$ for some $i > 0$. Hence

$$s = e_i s = (e e_i) s = e(e_i s) = es.$$

One shows analogously that $s = se$, which proves that e is the identity of S.

2. Let $R = R_e$ and let $a, b \in R$. Then $b \mathcal{R} e$ implies that $ab \mathcal{R} ae$. But $ae = a$ and thus $ab \mathcal{R} a$ which implies that $ab \in R$ since $a \in R$. Consequently, R is a subsemigroup of S.

3. Let $a, b \in S$ and $x \in R$ be such that $ax = bx$. Then $x \mathcal{R} e$ so that $e = xy$ for some $y \in S$. Multiplying $ax = bx$ by y on the right, we obtain $ae = be$ so that $a = b$.

4. Let $H_{ij} = R_{e_i} \cap L_{e_j}$ for $i, j = 0, 1, 2, \ldots$. Note that $H_{ij} \neq \varnothing$ since S is bisimple. Fix an element $a \in H_{01}$. Then $a \in R$; we let $a = e$ and observe that by part 2, $a^n \in R$ for $n = 0, 1, 2, \ldots$. We will prove that $a^n \in H_{0n}$ for all nonnegative integers n; it remains to show that $a^n \mathcal{L} e_n$.

First define ψ on $E = E_S$ by $e_n \psi = a^{-1} e_n a$ for all $n \geq 0$. If $e_m \psi = e_n \psi$, then $a^{-1} e_m a = a^{-1} e_n a$ which yields $e_m = e_n$ since $aa^{-1} = e$. Hence ψ is one-to-one. If $n \geq 1$, then

$$(a e_n a^{-1}) \psi = a^{-1} a e_n a^{-1} a = e_1 e_n = e_n$$

since $a^{-1} a = e_1$. Thus ψ maps E onto $E_1 = \{e_1, e_2, \ldots\}$. Since ψ obviously preserves order, it is an isomorphism of E onto E_1.

We prove next by induction that $e_n\psi = e_{n+1}$ for all $n \geq 0$. This holds for $n = 0$ since $e_0\psi = a^{-1}e_0a = e_1$. Assume the statement true for all $k \leq n$. Then $e_{n+1}\psi \geq e_{n+2}$ since ψ is one-to-one. In fact, $e_{n+1}\psi > e_{n+2}$ would contradict the conjunction of ψ being order preserving and the induction hypothesis. Consequently, $e_{n+1}\psi = e_{n+2}$ and hence this holds for all $n \geq 0$.

We have proved that $a^{-1}e_na = e_{n+1}$ for all $n \geq 0$; using this, we get

$$a^n \mathcal{L} a^{-n}a^n = a^{-(n-1)}(a^{-1}a)a^{n-1} = a^{-(n-1)}e_1a^{n-1}$$

$$= a^{-(n-2)}(a^{-1}e_1a)a^{n-2} = a^{-(n-2)}e_2e^{n-2} = \cdots = e_n,$$

as required.

5. Let $L = L_e$. For every nonnegative integer n, we have just proved that $a^n \in R \cap L_{e_n}$. By I.6.7, there exists $b \in R_{e_n} \cap L$ such that $a^nb = e$. It is verified easily that b is an inverse of a^n. Hence $b = a^{-n}$ by uniqueness of inverses. By I.6.10, we obtain that every element of S is uniquely representable in the form $a^{-m}ga^n$ with $g \in H_{0,0}$, $m, n \geq 0$.

6. Let $G = H_{0,0}$; we know by I.7.5 that G is a group. Let $g \in G$. Then $a, g \in R$ by part 2 implies that $ag \in R$. We know by part 4 that $ag \mathcal{L} e_k$ for some $k \geq 0$. If $k = 0$, then $agg^{-1}\mathcal{L} eg^{-1}$ and thus $a\mathcal{L}e$, a contradiction. Thus $k \geq 1$ and hence $ag \in Sa$, that is, $ag = xa$ for some $x \in S$.

Since $a\mathcal{R}e$, we have $e = ay$ for some $y \in S$. Hence

$$x = xe = (xa)y = a(gy), \qquad a = agg^{-1} = x(ag^{-1})$$

so that $x\mathcal{R}a$, that is, $x \in R$. Further, there exists $z \in S$ such that $ag^{-1} = za$ and $z\mathcal{R}a$, as we have just seen for x. Hence

$$x = xe = (xa)a^{-1} = (ag^{-1}a^{-1})^{-1} = (zaa^{-1})^{-1} = (ze)^{-1} = z^{-1}.$$

Now $z\mathcal{R}a$ implies $z^{-1}\mathcal{L}a^{-1}$ and thus $x\mathcal{L}a^{-1}$ which implies that $x \in L$. Consequently, $x \in G$. By part 3, x is unique and hence we may define a function α of G into itself by the requirement

$$ag = (g\alpha)a \qquad (g \in G) \tag{14}$$

If $g, h \in G$, then

$$(gh)\alpha a = a(gh) = (ag)h = [(g\alpha)a]h = (g\alpha)(ah) = (g\alpha)(h\alpha)a$$

which by part 3 implies that $(gh)\alpha = (g\alpha)(h\alpha)$. Therefore α is an endomorphism of G. We also note that by taking inverses in (14), we have $g^{-1}a^{-1} = a^{-1}(g\alpha)^{-1}$ which implies

$$ga^{-1} = a^{-1}(g\alpha) \qquad (g \in G). \tag{15}$$

A simple inductive argument shows that (14) and (15) imply

$$a^n g = (g\alpha^n)a^n, \qquad ga^{-n} = a^{-n}(g\alpha^n) \qquad (g \in G,\ n \geq 0). \tag{16}$$

7. Let $g, h \in G$, $m, n, p, q \geq 0$. Using (16), we obtain

$$\begin{aligned}
(a^{-m}ga^n)(a^{-p}ha^q) &= a^{-m}ga^{n-p}ha^q \\
&= \begin{cases} a^{-m}g(h\alpha^{n-p})a^{n-p+q} & \text{if } n-p \geq 0 \\ a^{-m+n-p}(g\alpha^{p-n})ha^q & \text{if } n-p \leq 0 \end{cases} \\
&= a^{-(m+p-r)}(g\alpha^{p-r})(h\alpha^{n-r})a^{n+q-r}
\end{aligned}$$

where $r = \min\{n, p\}$. Therefore the mapping χ defined by

$$\chi : a^{-m}ga^n \to (m, g, n) \qquad (g \in G,\ m, n \geq 0)$$

is an isomorphism of S onto $B(G, \alpha)$.

Sufficiency. This follows at once from 5.7, 5.10, and 5.13.

As a special case, we have the following important semigroup.

II.6.3 Definition

A Reilly semigroup with trivial group of units is a *bicyclic semigroup*.

According to 6.2, a bicyclic semigroup is isomorphic to a Bruck semigroup over a trivial group. In particular, any two bicyclic semigroups are isomorphic. In the construction of such a Bruck semigroup we may drop the middle entry entirely, thus arriving at the following concept.

II.6.4 Definition

The semigroup C consisting of all pairs (m, n) of nonnegative integers with multiplication

$$(m, n)(p, q) = (m + p - r, n + q - r),$$

where $r = \min\{n, p\}$, is *the bicyclic semigroup*.

II.6.5 Exercises

(i) Which Green relations on a Reilly semigroup are congruences?

(ii) Let G be a group, α and β be commuting endomorphisms of G. On $B(G, \alpha)$ define $\tilde{\beta}$ by

$$\tilde{\beta} : (m, g, n) \to (m, g\beta, n),$$

and on $B(G, \beta)$ define $\tilde{\alpha}$ by

$$\tilde{\alpha}: (m, g, n) \to (m, g\alpha, n).$$

Show that $\tilde{\alpha}$ and $\tilde{\beta}$ are endomorphisms of the respective semigroups and that

$$B(B(G, \alpha), \tilde{\beta}) \cong B(B(G, \beta), \tilde{\alpha}).$$

(iii) Show that the last semigroup in the preceding exercise is isomorphic to the following semigroup. Let C be the bicyclic semigroup, and on $S = C \times G \times C$ define a multiplication by

$$((m, n), g, (p, q))((m', n'), g', (p', q'))$$
$$= \Big((m, n)(m', n'), \big(g\alpha^{m'-r}\beta^{p'-s}\big)\big(g'\alpha^{n-r}\beta^{q-s}\big), (p, q)(p', q')\Big)$$

where $r = \min\{n, m'\}$, $s = \min\{q, p'\}$.

(iv) Let $S = [Y; G_\alpha, \varphi_{\alpha,\beta}]$ be a Clifford semigroup, and let $\theta \in \mathcal{E}(S)$ be such that $s\mathcal{H}s\theta$ for all $s \in S$. For every $\alpha \in Y$, let $\theta_\alpha = \theta|_{G_\alpha}$ so that $\theta_\alpha \in \mathcal{E}(G_\alpha)$. For any $\alpha, \beta \in Y$, $\alpha \geq \beta$, on $B(G_\alpha, \theta_\alpha)$ define $\tilde{\varphi}_{\alpha,\beta}$ by

$$\tilde{\varphi}_{\alpha,\beta}: (m, g, n) \to (m, g\varphi_{\alpha,\beta}, n).$$

Show that the functions $\tilde{\varphi}_{\alpha,\beta}$ satisfy the conditions for a strong semilattice of semigroups $B(G_\alpha, \theta_\alpha)$, and that

$$B\big([Y; G_\alpha, \varphi_{\alpha,\beta}], \theta\big) = \big[Y; B(G_\alpha, \theta_\alpha), \tilde{\varphi}_{\alpha,\beta}\big].$$

(v) Let G be a group, $\alpha \in \mathcal{E}(G)$, and I be a nonempty set. Define a mapping $\tilde{\alpha}$ on $B(G, I)$ by

$$\tilde{\alpha}: (i, g, j) \to (i, g\alpha, j), \qquad 0 \to 0.$$

Then $\alpha \in \mathcal{E}(B(G, I))$. We can define $B(B(G, I), \tilde{\alpha})$ formally as in the case of a Bruck semigroup over a monoid. Show that

$$J = \{(m, 0, n) | m, n \in N\}$$

is an ideal of $B(B(G, I), \tilde{\alpha})$ and that

$$B(B(G, I), \tilde{\alpha})/J \cong B(B(G, \alpha), I).$$

The result in this section is due to Reilly [1], [3]; an alternative proof was given by Warne [4]. Note that ω-regular semigroups are called regular ω-semigroups in the literature.

The bicyclic semigroup first appeared in print in Ljapin [1]. The relationship of the bicyclic semigroup with invertibility is due to Ljapin [5], [6]; some of these results can be found in Ljapin [7]. The bicyclic semigroup often appears as an antipode to various finiteness conditions. In this context it appears in the study of orthogonal or unitary transformations on an infinite dimensional Hilbert space, see the appendix of Halmos [1]. It is used as an example in Jacobson ([1], II.6, Example 2). In Adjan [2], it appears as an exceptional case in the study of identities on certain special classes of semigroups. Other remarkable properties of the bicyclic semigroup were found by Goralčik [1] and Justin [1]. The bicyclic semigroup appears in Perrot ([1], Chapter 6) in the context of syntactic monoids. In Clifford–Preston [1] it appears as a monoid with one defining relation, and as an inverse hull of an infinite cyclic semigroup. For certain representations of the bicyclic semigroup consult Barnes [1].

III

CONGRUENCES

Congruences play a central role in many of the structure theorems and other important considerations in the theory of inverse semigroups. The efficient handling of these congruences is a basic prerequisite for their useful application. For this reason, we have dedicated here much space to systematize the existing state of the art of congruences on inverse semigroups.

1. We start with a presentation of an arbitrary congruence ρ on an inverse semigroup S. The first such presentation of ρ is based on the fact that ρ is uniquely determined by its kernel ($\ker \rho$—the elements of S ρ-related to idempotents) and its trace ($\operatorname{tr} \rho$—the restriction of ρ to idempotents). The second presentation is based on the fact that ρ is uniquely determined by the set of ρ-classes containing idempotents.

2. It turns out that the mapping $\rho \to \operatorname{tr} \rho$ is a homomorphism of the congruence lattice of S onto the lattice of normal congruences on the semilattice E of idempotents of S. Each class of the congruence induced by this homomorphism is a complete modular lattice, whose greatest and least elements can be given a simple form.

3. Congruences on S whose trace is the equality relation on E are called idempotent separating. They form a complete modular lattice isomorphic to the lattice of all normal subsemigroups of S contained in the centralizer of E in S. These congruences are in a certain sense the nearest to congruences on a group.

4. Congruences on S whose kernel is equal to E are called idempotent pure. They form a complete lattice with greatest element the syntactic congruence on S relative to E. The mapping $\rho \to \ker \rho$ is an $\cap$-homomorphism of the congruence lattice of S onto a certain lattice of subsets of S.

5. The mapping $\rho \to \rho \vee \sigma$, where σ is the least group congruence on S, is a homomorphism of the congruence lattice of S onto the lattice of all group congruences on S. Group congruences are precisely the congruences whose trace is the universal relation on E.

6. Congruences on S whose kernel coincides with S are precisely the semilattice congruences on S. For a semilattice congruence ρ on S, we consider conditions which are equivalent to the requirement that S be a strong semilattice of ρ-classes. Clifford congruences are closely related to semilattice congruences.

7. E-unitary congruences admit several interesting characterizations. The set of all E-unitary congruences on S forms a complete $\cap$-sublattice of the congruence lattice of S, whose least element can be suitably described.

8. E-reflexive inverse semigroups arise naturally in this context as semigroups which are semilattices of E-unitary inverse semigroups. E-reflexive congruences on an inverse semigroup also form a $\cap$-sublattice of the congruence lattice of S, and their least element can be conveniently expressed.

III.1 GENERAL CONGRUENCES

A congruence ρ on an inverse semigroup S is completely determined by two ingredients: the union of ρ-classes containing idempotents and the restriction of ρ to the semilattice of idempotents of S. We characterize abstractly this pair of parameters of a congruence, and conversely, for such a pair, called a congruence pair, construct the unique congruence associated with it. Such a congruence ρ is also uniquely determined by the family of ρ-classes containing idempotents. For this characterization of ρ, we perform a similar analysis via an abstract characterization of these classes and the construction of the corresponding congruence.

These results are of basic importance for most of the material in this chapter.

The following lemma will be used often and without express reference.

III.1.1 Lemma

Let ρ be a congruence on an inverse semigroup S. For any $x, y \in S$, $x\rho y$ implies $x^{-1}\rho y^{-1}$.

Proof. In view of II.1.10, we have that S/ρ is an inverse semigroup. For any $x \in S$, it follows directly that $x^{-1}\rho$ is an inverse of $x\rho$ in S/ρ and thus $x^{-1}\rho = (x\rho)^{-1}$ by uniqueness of inverses. Now assume that $x\rho y$. Then $x^{-1}\rho = (x\rho)^{-1} = (y\rho)^{-1} = y^{-1}\rho$, that is, $x^{-1}\rho y^{-1}$.

III.1.2 Definition

For a congruence ρ on an inverse semigroup S, we define the *kernel* and the *trace* of ρ by

$$\ker \rho = \{a \in S \mid a\rho e \quad \text{for some } e \in E_S\},$$
$$\operatorname{tr} \rho = \rho|_{E_S},$$

respectively.

This associates to each congruence ρ on S the ordered pair $(\ker \rho, \operatorname{tr} \rho)$. We will introduce below a pair (K, τ) which is an abstraction of the properties of $(\ker \rho, \operatorname{tr} \rho)$ for some congruence ρ.

III.1.3 Definition

Let S be an inverse semigroup. A subset K of S is *full* if $E_S \subseteq K$; it is *self-conjugate* if $s^{-1}Ks \subseteq K$ for all $s \in S$. A full, self-conjugate inverse subsemigroup of S is a *normal subsemigroup* of S.

A congruence τ on E_S is *normal* if for any $e, f \in E_S$ and $s \in S$, $e\tau f$ implies $s^{-1}es\tau s^{-1}fs$. (Note that it suffices to require that τ be an equivalence relation with this property.)

The pair (K, τ) is a *congruence pair* for S if K is a normal subsemigroup of S, τ is a normal congruence on E_S, and these two satisfy:

(i) $ae \in K, e\tau a^{-1}a \Rightarrow a \in K \quad (a \in S, e \in E_S)$,
(ii) $k \in K \Rightarrow kk^{-1}\tau k^{-1}k$.

In such a case, define a relation $\rho_{(K,\tau)}$ on S by

$$a\rho_{(K,\tau)}b \Leftrightarrow a^{-1}a\tau b^{-1}b, \quad ab^{-1} \in K.$$

Using these concepts and notation, we have below the first characterization of congruences on inverse semigroups. But we first need some auxiliary statements.

III.1.4 Lemma

For a congruence pair (K, τ) for an inverse semigroup S, we have for any $a, b \in S$, $e \in E_S$,

(i) $aeb \in K, e\tau a^{-1}a \Rightarrow ab \in K$,
(ii) $ab \in K \Rightarrow aeb \in K$,
(iii) $ab^{-1} \in K, a^{-1}a\tau b^{-1}b \Rightarrow a^{-1}ea\tau b^{-1}eb$.

Proof. Let $a, b \in S$ and $e \in E_S$.

(i) If $aeb \in K$ and $e\tau a^{-1}a$, then

$$aeb = (ae)(bb^{-1})b = ab(b^{-1}eb), \tag{1}$$
$$(ab)^{-1}(ab) = b^{-1}(a^{-1}a)b\tau b^{-1}eb$$

which implies $ab \in K$ by 1.3(i).

(ii) If $ab \in K$, then by (1), $aeb = (ab)(b^{-1}eb) \in K$ since $b^{-1}eb \in E_S$ and K is full.

(iii) Let $ab^{-1} \in K$, $a^{-1}a\tau b^{-1}b$, and $e \in E_S$. Then

$$
\begin{aligned}
a^{-1}ea &= (a^{-1}ea)(a^{-1}a)(a^{-1}ea) \\
&\tau (a^{-1}ea)(b^{-1}b)(a^{-1}ea) \qquad \text{since } a^{-1}a\tau b^{-1}b \\
&= (a^{-1}e)(ab^{-1})(ab^{-1})^{-1}(ea) \\
&\tau (a^{-1}e)(ba^{-1})(ab^{-1})(ea) \qquad \text{using 1.3(ii)} \\
&= a^{-1}(eba^{-1})(eba^{-1})^{-1}a \\
&\tau a^{-1}(ab^{-1}e)(eba^{-1})a \qquad \text{using 1.3(ii) on } ab^{-1}e \in K \\
&\tau b^{-1}eb.
\end{aligned}
$$

We are now ready for the characterization theorem for congruences.

III.1.5 Theorem

Let S be an inverse semigroup. If (K, τ) is a congruence pair for S, then $\rho_{(K,\tau)}$ is the unique congruence ρ on S for which $\ker \rho = K$ and $\operatorname{tr} \rho = \tau$. Conversely, if ρ is a congruence on S, then $(\ker \rho, \operatorname{tr} \rho)$ is a congruence pair for S and $\rho_{(\ker \rho, \operatorname{tr} \rho)} = \rho$.

Proof. Let (K, τ) be a congruence pair for S, and let $\rho = \rho_{(K,\tau)}$. Then ρ is reflexive since K is full, and it is symmetric since τ is symmetric and K is an inverse semigroup. Let $a\rho b$ and $b\rho c$, so that $a^{-1}a\tau b^{-1}b\tau c^{-1}c$ and ab^{-1}, $bc^{-1} \in K$. Hence $a(b^{-1}b)c^{-1} \in K$ which together with $a^{-1}a\tau b^{-1}b$ by 1.4 (i) yields $ac^{-1} \in K$. Thus $a\rho c$ and ρ is transitive.

Next let $a\rho b$ and $c \in S$. Then

$$(ac)^{-1}(ac) = c^{-1}(a^{-1}a)c\tau c^{-1}(b^{-1}b)c = (bc)^{-1}(bc)$$

since $a^{-1}a\tau b^{-1}b$ and τ is a normal congruence on E_S, and

$$(ac)(bc)^{-1} = a(cc^{-1})b^{-1} \in K$$

by 1.4 (ii) since $ab^{-1} \in K$. It follows that $ac\rho bc$. Further,

$$(ca)^{-1}(ca) = a^{-1}(c^{-1}c)a\tau b^{-1}(c^{-1}c)b = (cb)^{-1}(cb)$$

by 1.4 (iii) since $a^{-1}a\tau b^{-1}b$ and $ab^{-1} \in K$; also

$$(ca)(cb)^{-1} = c(ab^{-1})c^{-1} \in K$$

since $ab^{-1} \in K$ and K is self-conjugate. Therefore ρ is a congruence on S.

If $a\rho e$ for $e \in E_S$, then $a^{-1}a\tau e$ and $ae \in K$ which by 1.3(i) yields $a \in K$. Conversely, if $a \in K$, then clearly $a\rho a^{-1}a$. Consequently, $\ker\rho = K$; and obviously $\operatorname{tr}\rho = \tau$.

Now let λ be a congruence on S such that $\ker\lambda = K$ and $\operatorname{tr}\lambda = \tau$. Assume first that $a\lambda b$. Then $a^{-1}\lambda b^{-1}$ so that $a^{-1}a\lambda b^{-1}b$; also $ab^{-1}\lambda bb^{-1}$. This shows that $a^{-1}a\tau b^{-1}b$ and $ab^{-1} \in K$, which implies that $a\rho b$. Conversely, assume that $a\rho b$. Then $a^{-1}a\lambda b^{-1}b$ and $ab^{-1}\lambda e$ for some $e \in E_S$. It follows that

$$a = a(a^{-1}a)\lambda a(b^{-1}b) = (ab^{-1})b\lambda eb,$$

$$b = b(b^{-1}b)\lambda b(a^{-1}a) = (ba^{-1})a\lambda ea.$$

Multiplying these relations by e on the left, we get $a\lambda eb\lambda ea\lambda b$. Consequently, $\rho = \lambda$ which proves uniqueness.

Conversely, let ρ be a congruence on S. A simple verification shows that $(\ker\rho, \operatorname{tr}\rho)$ is a congruence pair for S. That $\ker\rho_{(\ker\rho,\operatorname{tr}\rho)} = \ker\rho$, $\operatorname{tr}\rho_{(\ker\rho,\operatorname{tr}\rho)} = \operatorname{tr}\rho$ follows from above. Now the uniqueness just proved implies that $\rho_{(\ker\rho,\operatorname{tr}\rho)} = \rho$.

We now consider a different approach to congruences on inverse semigroups. Instead of taking the kernel and the trace of a congruence ρ on an inverse semigroup S, we consider the collection of all ρ-classes containing idempotents.

III.1.6 Notation

If ρ is a congruence on an inverse semigroup S, let

$$\mathcal{K}(\rho) = \{e\rho \mid e \in E_S\}.$$

Such collections of subsets of S can be characterized abstractly, and we thus have the following concept.

III.1.7 Definition

Let $\mathcal{K}$ be a family of pairwise disjoint inverse subsemigroups of an inverse semigroup S satisfying

(i) $E_S \subseteq \bigcup_{K\in\mathcal{K}} K$,
(ii) for each $a \in S$ and $K \in \mathcal{K}$, there exists $L \in \mathcal{K}$ such that $a^{-1}Ka \subseteq L$,
(iii) if $a, ab, bb^{-1} \in K$, $K \in \mathcal{K}$, then $b \in K$.

Then $\mathcal{K}$ is a *kernel normal system* for S. For such a family $\mathcal{K}$, we define a relation $\xi_{\mathcal{K}}$ on S by

$$a\xi_{\mathcal{K}}b \Leftrightarrow aa^{-1}, bb^{-1}, ab^{-1} \in K \quad \text{for some } K \in \mathcal{K}.$$

With this notation and these definitions, we have a second characterization of congruences on an inverse semigroup as follows.

III.1.8 Theorem

Let S be an inverse semigroup. If $\mathcal{K}$ is a kernel normal system for S, then $\xi_{\mathcal{K}}$ is the unique congruence ξ on S for which $\mathcal{K}(\xi) = \mathcal{K}$. Conversely, if ξ is a congruence on S, then $\mathcal{K}(\xi)$ is a kernel normal system for S and $\xi_{\mathcal{K}(\xi)} = \xi$.

Proof. 1. Let $\mathcal{K}$ be a kernel normal system for S, let $K = \cup_{L \in \mathcal{K}} L$ and define τ on E_S by

$$e \tau f \Leftrightarrow e, f \in L \quad \text{for some } L \text{ in } \mathcal{K}.$$

We now verify that (K, τ) is a congruence pair for S and that $\xi_{\mathcal{K}} = \rho_{(K, \tau)}$.

Since the members of $\mathcal{K}$ are pairwise disjoint and 1.7(i) holds, we obtain that τ is an equivalence relation. Condition 1.7(ii) implies that τ is a normal congruence.

We now show that K is closed under multiplication. Hence let $a \in L$ and $b \in T$ for $L, T \in \mathcal{K}$. Let $e = a^{-1}a$ and $f = bb^{-1}$. Since τ is a normal congruence, there exists $U \in \mathcal{K}$ such that $E_L E_T \subseteq E_U$. But then $eE_T e \subseteq E_U$, which by 1.7(ii) implies that $eTe \subseteq U$. Now $ebe \in eTe \subseteq U$, which implies that $(ebe)(ebe)^{-1} \in U$. Also,

$$(ebe)(ebe)^{-1}(eb) = ebeb^{-1}eb = e(beb^{-1})eb = ebeb^{-1}b$$

$$= eb(b^{-1}b)e = ebe \in U,$$

and $(eb)(eb)^{-1} = ebb^{-1} \in E_L E_L \subseteq U$. Now applying 1.7(iii), we get $eb \in U$. Since $E_T E_L = E_L E_T \subseteq E_U$ and $a^{-1} \in L$, for L is an inverse subsemigroup of S, we can similarly prove that $fa^{-1} \in U$. Thus $af = (fa^{-1})^{-1} \in U$ and we obtain

$$ab = (ae)(fb) = (af)(eb) \in UU \subseteq U$$

and therefore K is a subsemigroup of S.

Since each L in $\mathcal{K}$ is closed under taking of inverses, so is K, and is thus an inverse subsemigroup of S. Conditions 1.7(i) and (ii) insure that K is full and self-conjugate. Hence K is a normal subsemigroup of S.

We verify condition 1.3(i) next. Hence let $a \in S$, $e \in E_S$, $ae \in K$, and $e\tau a^{-1}a$. Then $ae \in L$ and $e, a^{-1}a \in T$ for some $L, T \in \mathcal{K}$. It follows that $ea^{-1} = (ae)^{-1} \in L$ and thus $(ea^{-1})(ae) = e(a^{-1}a)e \in L \cap T$ so that $L = T$. Hence $e, ea^{-1}, a^{-1}a \in L$, which by 1.7(iii) gives $a^{-1} \in L$, so that also $a \in L$. But then $a \in K$, as required. Condition 1.3(ii) follows easily: if $a \in K$, then $a \in L$ for some $L \in \mathcal{K}$, and thus $aa^{-1}, a^{-1}a \in L$ and hence $aa^{-1}\tau a^{-1}a$. Consequently, (K, τ) is a congruence pair.

2. We now check that $\rho_{(K,\tau)} = \xi_{\mathcal{K}}$. First let $a\rho_{(K,\tau)}b$, so that $a^{-1}a\tau b^{-1}b$ and $ab^{-1} \in K$. Hence $a^{-1}a, b^{-1}b \in L$ and $ab^{-1} \in T$ for some $L, T \in \mathcal{K}$. By 1.7(ii), we have $aLa^{-1} \subseteq L'$ and $bLb^{-1} \subseteq L''$ for some $L', L'' \in \mathcal{K}$. Now

$$(ab^{-1})(ba^{-1}) = a(b^{-1}b)a^{-1} \in T \cap L'$$

so that $T = L'$; also

$$(ba^{-1})(ab^{-1}) = b(a^{-1}a)b^{-1} \in T \cap L''$$

and hence $T = L''$. Since also $aa^{-1} = a(a^{-1}a)a^{-1} \in L'$ and similarly $bb^{-1} \in L''$, we deduce that $aa^{-1}, bb^{-1}, ab^{-1} \in T$ and thus $a\xi_{\mathcal{K}}b$.

Conversely, let $a\xi_{\mathcal{K}}b$. Then $aa^{-1}, bb^{-1}, ab^{-1} \in L$ for some $L \in \mathcal{K}$. We have $a^{-1}La \subseteq L'$ and $b^{-1}Lb \subseteq L''$ for some $L', L'' \in \mathcal{K}$. Now

$$a^{-1}ab^{-1}b = a^{-1}\left[(ab^{-1})(ba^{-1})\right]a = b^{-1}\left[(ba^{-1})(ab^{-1})\right]b \in L' \cap L''$$

since $ab^{-1}, ba^{-1} \in L$. It follows that $L' = L''$. Since $a^{-1}a = a^{-1}(aa^{-1})a \in L'$ and similarly $b^{-1}b \in L''$, we deduce that $a^{-1}a, b^{-1}b \in L' = L''$. We thus have $a^{-1}a\tau b^{-1}b$ and $ab^{-1} \in K$ which gives $a\rho_{(K,\tau)}b$.

3. We verify next that $\mathcal{K}(\xi_{\mathcal{K}}) = \mathcal{K}$. First let $L \in \mathcal{K}$. Then L is an inverse subsemigroup of S. In order to show that $L \in \mathcal{K}(\xi_{\mathcal{K}})$, it suffices to show that L is a $\xi_{\mathcal{K}}$-class. If $a, b \in L$, then $aa^{-1}, bb^{-1}, ab^{-1} \in L$ and thus $a\xi_{\mathcal{K}}b$. Let $a \in L$ and $a\xi_{\mathcal{K}}b$. Then $aa^{-1}, bb^{-1}, ab^{-1} \in T$ for some $T \in \mathcal{K}$. But $aa^{-1} \in L \cap T$, so that $L = T$. Since $a^{-1}\xi_{\mathcal{K}}b^{-1}$, we also have $a^{-1}a, b^{-1}b, a^{-1}b \in U$ for some $U \in \mathcal{K}$, and as above, $U = L$. Thus, in particular, $a^{-1}, a^{-1}b, bb^{-1} \in L$ and 1.7(iii) implies that $b \in L$. Consequently, L is a $\xi_{\mathcal{K}}$-class and thus $L \in \mathcal{K}(\xi_{\mathcal{K}})$.

Conversely, let $L \in \mathcal{K}(\xi_{\mathcal{K}})$. Then L contains an idempotent e, and thus $e \in T$ for some $T \in \mathcal{K}$ by 1.7(i). By the above, T also is a $\xi_{\mathcal{K}}$-class, and we must have $L = T$. Consequently, $L \in \mathcal{K}$, which completes the verification that $\mathcal{K}(\xi_{\mathcal{K}}) = \mathcal{K}$.

4. If now ρ is any congruence on S for which $\mathcal{K}(\rho) = \mathcal{K}$, then by the first part of the proof, we must have $\ker\rho = \ker\xi_{\mathcal{K}}$ and $\operatorname{tr}\rho = \operatorname{tr}\xi_{\mathcal{K}}$, and we get $\rho = \xi_{\mathcal{K}}$ by 1.5. This establishes the uniqueness of $\xi_{\mathcal{K}}$ and completes the proof of the direct part of the theorem.

5. The proof of the converse is much easier. Indeed, let ξ be a congruence on S. It is obvious that $\mathcal{K}(\xi)$ consists of a family of pairwise disjoint inverse subsemigroups of S whose union contains E_S. Let $a \in S$ and let K be a ξ-class containing an idempotent e. Then for any $k \in K$, $a^{-1}ka\xi a^{-1}ea$ so that $a^{-1}Ka \subseteq L$ where L is the ξ-class containing the idempotent $a^{-1}ea$. This verifies 1.7(ii). With the same notation, assume $a, ab, bb^{-1} \in K$. Then

$$b = (bb^{-1})b\xi eb\xi ab\xi e$$

so that $b \in K$. This verifies 1.7(iii) and completes the proof that $\mathcal{K}(\xi)$ is a kernel normal system.

6. By part 3 of this proof, we have $\mathcal{K}(\xi_{\mathcal{K}(\xi)}) = \mathcal{K}(\xi)$. Thus $\xi_{\mathcal{K}(\xi)}$ and ξ have the same kernel normal system, so by the uniqueness proved in the first part of the theorem, we obtain $\xi_{\mathcal{K}(\xi)} = \xi$.

III.1.9 Exercises

(i) Show that condition 1.3(ii) can be replaced by the requirement

$$a \in K, e \in E_S \Rightarrow a^{-1}ea\tau a^{-1}ae.$$

(ii) Let $S = [Y; G_\alpha, \varphi_{\alpha,\beta}]$ be a Clifford semigroup. For each $\alpha \in Y$, let K_α be a normal subgroup of G_α and assume that $K_\alpha \varphi_{\alpha,\beta} \subseteq K_\beta$ if $\alpha > \beta$. Let $\psi_{\alpha,\beta} = \varphi_{\alpha,\beta}|_{K_\alpha}$ if $\alpha \geq \beta$ and $K = [Y; K_\alpha, \psi_{\alpha,\beta}]$. Show that K is a normal subsemigroup of S, and that conversely, every normal subsemigroup of S is of this form.

(iii) Let the notation be as in the preceding exercise. Let ξ be a congruence on Y and assume that $K_\beta \varphi_{\alpha,\beta}^{-1} \subseteq K_\alpha$ if $\alpha > \beta$ and $\alpha \xi \beta$. Let τ be the congruence on E_S induced by ξ. Show that (K, τ) is a congruence pair for S, and, conversely, that every congruence pair for S can be so constructed.

(iv) Let I be an ideal of an inverse semigroup S. Show that every congruence ρ on I admits an extension to S of the form $\rho \cup \iota_S$.

(v) Show that a full closed inverse subsemigroup K of an inverse semigroup S is self-conjugate if and only if for any $a, b \in S$, $ab \in K$ implies $ba \in K$.

Wagner [3] was the first to prove that a congruence on an inverse semigroup is completely determined by its classes containing idempotents. A little later, Preston [1] proved a characterization theorem for congruences by means of a kernel normal system and 1.8 is due to him with a direct proof. For 20 years this was the only usable form for congruences on inverse semigroups, and still this approach has many partisans. Scheiblich [5] came up with the fresh idea of associating with each congruence on an inverse semigroup, what we now call the kernel and the trace. His set of axioms for a pair, now called a congruence pair, was improved by Green [2] [1.3(ii) is one of his axioms]. Most of the present approach to this characterization of congruences can be found in Petrich [11]. The importance of the trace was realized earlier by Reilly and Scheiblich [1]. Various results on traces can be found in Reilly [6], [8]. One-sided congruences on inverse semigroups were studied by Meakin [1]. For congruences on commutative inverse semigroups, see Jarek [1]. Also consult Mitsch [2].

Both approaches are reminiscent of the simple state of affairs for congruences on a group. Instead of the kernel in the group case, one may take the entire collection of "kernels" (hence the name kernel normal system) or one may choose to take the union of these for which one must pay the price of also taking into account the induced partition on the idempotents (namely the trace).

III.2 THE LATTICE OF CONGRUENCES

The main result here indicates that the mapping $\rho \to \operatorname{tr} \rho$ is a complete homomorphism of the congruence lattice of S onto the lattice of normal congruences on E_S, and that the congruence induced by it has all its classes complete modular lattices.

III.2.1 Notation

For any inverse semigroup S, let $\mathcal{C}(S)$ be the lattice of all congruences on S and let $\mathfrak{N}(E_S)$ be the lattice of all normal congruences on E_S both under inclusion. It is clear that both of these are complete lattices with meet equal to the intersection.

III.2.2 Corollary

Let S be an inverse semigroup. Then the join in $\mathcal{C}(E_S)$ of two normal congruences on E_S is a normal congruence on E_S.

Proof. The proof of this corollary is left as an exercise.

Hence the join in $\mathfrak{N}(E_S)$ coincides with that in $\mathcal{C}(E_S)$.

III.2.3 Proposition

Let $\mathcal{CP}(S)$ be the set of all congruence pairs for an inverse semigroup S ordered by

$$(K, \tau) \le (K', \tau') \Leftrightarrow K \subseteq K', \tau \subseteq \tau'.$$

Then the mappings

$$(K, \tau) \to \rho_{(K,\tau)}, \qquad \rho \to (\ker \rho, \operatorname{tr} \rho)$$

are mutually inverse isomorphisms of $\mathcal{CP}(S)$ and $\mathcal{C}(S)$.

Proof. The proof of this proposition is left as an exercise.

III.2.4 Notation

Let S be an inverse semigroup. For any congruence ρ on S, define two relations $\rho_{\max}$ and $\rho_{\min}$ on S by

$$a\rho_{\max}b \Leftrightarrow a^{-1}ea\rho b^{-1}eb \quad \text{for all } e \in E_S,$$

$$a\rho_{\min}b \Leftrightarrow ae = be \quad \text{for some } e \in E_S,\ e\rho a^{-1}a\rho b^{-1}b.$$

We are now ready for the main result concerning the lattice of congruences in relation to their traces.

III.2.5 Theorem

Let S be an inverse semigroup. Define a mapping tr by

$$\text{tr}: \rho \to \text{tr}\,\rho \qquad [\rho \in \mathcal{C}(S)].$$

Then tr is a complete homomorphism of $\mathcal{C}(S)$ onto $\mathfrak{N}(E_S)$. Let θ be the congruence on $\mathcal{C}(S)$ induced by tr. Then for any $\rho \in \mathcal{C}(S)$,

$$\rho\theta = [\rho_{\min}, \rho_{\max}]$$

and is a complete modular sublattice (with commuting elements) of $\mathcal{C}(S)$.

Proof. Let $\mathfrak{F}$ be a nonempty family of congruences on S. Then for any $e, f \in E_S$, we obtain

$$
\begin{aligned}
e\,\text{tr}(\cap\mathfrak{F})f &\Leftrightarrow e(\cap\mathfrak{F})f \Leftrightarrow e\rho f \quad \text{for all } \rho \in \mathfrak{F}.\\
&\Leftrightarrow e\,\text{tr}\,\rho f \quad \text{for all } \rho \in \mathfrak{F}\\
&\Leftrightarrow e \bigcap_{\rho\in\mathfrak{F}} \text{tr}\,\rho f,
\end{aligned}
$$

which proves $\text{tr}(\cap\mathfrak{F}) = \cap_{\rho\in\mathfrak{F}}\text{tr}\,\rho$. Using I.4.14, we get

$$
\begin{aligned}
e\,\text{tr}(\vee\mathfrak{F})f &\Leftrightarrow e \vee \mathfrak{F}f\\
&\Leftrightarrow e\rho_1x_1,\ x_1\rho_2x_2,\ldots,x_{n-1}\rho_nf \quad \text{for some } x_i \in S,\ \rho_i \in \mathfrak{F}\\
&\Rightarrow e\rho_1x_1x_1^{-1},\ x_1x_1^{-1}\rho_2x_2x_2^{-1},\ldots,x_{n-1}x_{n-1}^{-1}\rho_nf\\
&\Rightarrow e \bigvee_{\rho\in\mathfrak{F}} \text{tr}\,\rho f
\end{aligned}
$$

which shows that $\text{tr}(\vee\mathfrak{F}) \subseteq \vee_{\rho\in\mathfrak{F}}\text{tr}\,\rho$. The converse inclusion is even easier to establish. This proves $\operatorname{tr}(\vee \mathcal{F}) = \vee_{\rho \in \mathcal{F}} \operatorname{tr} \rho$. Consequently, tr is a complete homomorphism of $\mathcal{C}(S)$ into $\mathfrak{N}(E_S)$.

Let $\tau \in \mathfrak{N}(E_S)$ and define ρ on S by

$$a \rho b \Leftrightarrow a^{-1} e a \tau b^{-1} e b \quad \text{for all } e \in E_S.$$

It is obvious that ρ is an equivalence relation on S. Let $a \rho b$ and $c \in S$. Then for any $e \in E_S$, we have

$$(ca)^{-1} e (ca) = a^{-1}(c^{-1} e c) a \tau b^{-1}(c^{-1} e c) b = (cb)^{-1} e (cb),$$

$$(ac)^{-1} e (ac) = c^{-1}(a^{-1} e a) c \tau c^{-1}(b^{-1} e b) c = (bc)^{-1} e (bc),$$

where we have used normality of τ. It follows that ρ is a congruence on S. Further, for any $e, f \in E_S$,

$$e \rho f \Leftrightarrow e g \tau f g \quad \text{for all } g \in E_S \Leftrightarrow e \tau f$$

so that $\operatorname{tr} \rho = \tau$ which also shows that tr maps $\mathcal{C}(S)$ onto $\mathfrak{N}(E_S)$.

Let $\rho \in \mathcal{C}(S)$. The argument above shows that $\rho_{\max}$ is a congruence on S. Now let $a \rho b$. Then for any $e \in E_S$, we have $a^{-1} e a \rho b^{-1} e b$ which shows that $\rho \subseteq \rho_{\max}$. For any $e, f \in E_S$ such that $e \rho_{\max} f$, we obtain $e \rho e f \rho f$. This shows that $\operatorname{tr} \rho = \operatorname{tr} \rho_{\max}$, and since the definition of $\rho_{\max}$ depends only on idempotents, it follows that $\rho_{\max}$ is the greatest element of $\rho\theta$.

It is easy to verify that $\rho_{\min}$ is an equivalence relation. Let $ae = be$ with $e \in E_S$, $e \rho a^{-1} a \rho b^{-1} b$, and $c \in S$. Then

$$(ca)[e a^{-1}(c^{-1} c) a] = (cb)[e a^{-1}(c^{-1} c) a],$$

$$(ca)^{-1}(ca) = a^{-1}(c^{-1} c) a = (a^{-1} a)[a^{-1}(c^{-1} c) a] \rho e a^{-1}(c^{-1} c) a,$$

$$\begin{aligned}(cb)^{-1}(cb) &= b^{-1}(c^{-1} c) b \\ &= (b^{-1} b) b^{-1}(c^{-1} c) b (b^{-1} b) \rho e b^{-1}(c^{-1} c) b e \\ &= e a^{-1}(c c^{-1}) a e = e a^{-1}(c^{-1} c) a\end{aligned}$$

which shows that $ca \rho_{\min} cb$. Further,

$$(ac)(c^{-1} e c) = aec = bec = (bc)(c^{-1} e c),$$

$$(ac)^{-1}(ac) = c^{-1}(a^{-1} a) c \rho c^{-1} e c,$$

$$(bc)^{-1}(bc) = c^{-1}(b^{-1} b) c \rho c^{-1} e c,$$

and thus $ac\rho_{\min}bc$. Consequently, $\rho_{\min}$ is a congruence on S. If $ae = be$ and $e\rho a^{-1}a\rho b^{-1}b$, then $a\rho ae = be\rho b$. This proves that $\rho_{\min} \subseteq \rho$. If $e, f \in E_S$ and $e\rho f$, then $e(ef) = f(ef)$ and $ef\rho e\rho f$ so that $e\rho_{\min} f$. This shows that $\operatorname{tr} \rho = \operatorname{tr} \rho_{\min}$, and since the definition of $\rho_{\min}$ depends only on idempotents, it follows that $\rho_{\min}$ is the least element of $\rho\theta$.

It is now clear that $\rho\theta = [\rho_{\min}, \rho_{\max}]$. But any interval of a complete lattice is itself complete.

Now let $\rho, \lambda \in \mathcal{C}(S)$ be such that $\rho\theta = \lambda\theta$, and let $a\rho\lambda b$. Then $a\rho c$ and $c\lambda b$ for some $c \in S$. Hence $aa^{-1}\rho cc^{-1}$ and thus $aa^{-1}\lambda cc^{-1}$ since $\operatorname{tr} \rho = \operatorname{tr} \lambda$. Also $cc^{-1}\lambda bc^{-1}$ and thus $aa^{-1}\lambda bc^{-1}$ which yields $a\lambda bc^{-1}a$. A similar argument shows that $b\rho bc^{-1}a$ and thus $a\lambda\rho b$. Hence $\rho\lambda \subseteq \lambda\rho$ and by symmetry, we deduce that $\rho\lambda = \lambda\rho$. An application of I.2.13 now gives that $\rho\theta$ is a modular lattice.

III.2.6 Definition

We call the classes of θ in 2.5 the *trace classes* of S.

III.2.7 Corollary

For any inverse semigroup S, the mapping $\rho \to \rho\theta$ is a complete homomorphism of $\mathcal{C}(S)$ onto $\mathcal{C}(S)/\theta$ and each θ-class is a complete modular sublattice of $\mathcal{C}(S)$.

The following two lemmas will come in handy.

III.2.8 Lemma

Let ρ and ξ be congrucnces on an inverse semigroup S. If $\rho \subseteq \xi$, then $\rho_{\min} \subseteq \xi_{\min}$ and $\rho_{\max} \subseteq \xi_{\max}$.

Proof. This is immediate from 2.4.

III.2.9 Lemma

Let $\mathfrak{F}$ be a nonempty family of congruences on an inverse semigroup S. Then

$$\bigvee_{\rho \in \mathfrak{F}} \rho_{\min} = \Big(\bigvee_{\rho \in \mathfrak{F}} \rho\Big)_{\min}, \qquad \bigcap_{\rho \in \mathfrak{F}} \rho_{\max} = \Big(\bigcap_{\rho \in \mathfrak{F}} \rho\Big)_{\max}.$$

Proof. For each $\xi \in \mathfrak{F}$, we have $\xi \subseteq \bigvee_{\rho \in \mathfrak{F}} \rho$. Hence by 2.8, we get $\xi_{\min} \subseteq (\bigvee_{\rho \in \mathfrak{F}} \rho)_{\min}$ and thus $\bigvee_{\rho \in \mathfrak{F}} \rho_{\min} \subseteq (\bigvee_{\rho \in \mathfrak{F}} \rho)_{\min}$. It follows from 2.5 that $\operatorname{tr}(\bigvee_{\rho \in \mathfrak{F}} \rho_{\min}) = \operatorname{tr}(\bigvee_{\rho \in \mathfrak{F}} \rho)$, which implies that $\bigvee_{\rho \in \mathfrak{F}} \rho_{\min} \supseteq (\bigvee_{\rho \in \mathfrak{F}} \rho)_{\min}$, and the equality prevails.

For any $a, b \in S$, we have

$$a \bigcap_{\rho \in \mathcal{F}} \rho_{\max} b \Leftrightarrow a \rho_{\max} b \quad \text{for all } \rho \in \mathcal{F}$$

$$\Leftrightarrow a^{-1}ea\rho b^{-1}eb \quad \text{for all } e \in E_S, \rho \in \mathcal{F}$$

$$\Leftrightarrow a^{-1}ea \bigcap_{\rho \in \mathcal{F}} \rho b^{-1}eb \quad \text{for all } e \in E_S$$

$$\Leftrightarrow a \Big(\bigcap_{\rho \in \mathcal{F}} \rho \Big)_{\max} b$$

so that $\bigcap_{\rho \in \mathcal{F}} \rho_{\max} = (\bigcap_{\rho \in \mathcal{F}} \rho)_{\max}$.

The following example shows that in general $(\bigcap_{\rho \in \mathcal{F}} \rho)_{\min} \neq \bigcap_{\rho \in \mathcal{F}} \rho_{\min}$; that also $\bigvee_{\rho \in \mathcal{F}} \rho_{\max} \neq (\bigvee_{\rho \in \mathcal{F}} \rho)_{\max}$ in general will be shown in IX.3.13.

III.2.10 Example

Let $S = \{0, e, f, 1, a\}$ be a Clifford semigroup with idempotents

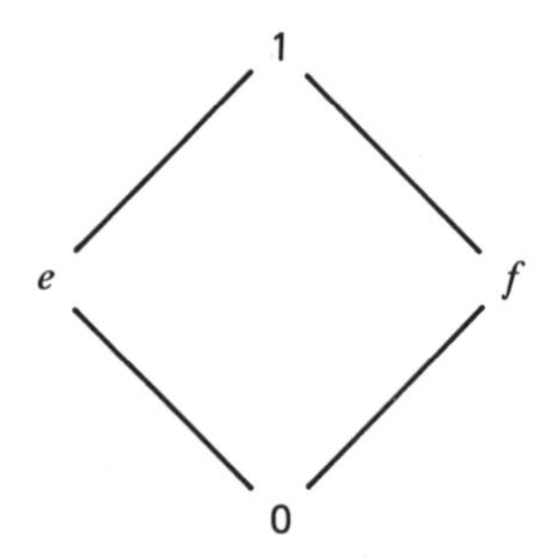

and the group of units $\{1, a\}$. Let λ and ρ be equivalence relations on S with

$$\begin{aligned} \lambda\text{-classes:} \quad & \{1, a, f\}, \{e, 0\}, \\ \rho\text{-classes:} \quad & \{1, a, e\}, \{f, 0\}. \end{aligned}$$

One checks readily that both λ and ρ are congruences with the properties: $\lambda = \lambda_{\min}$, $\rho = \rho_{\min}$, $\operatorname{tr}(\lambda \cap \rho) = \varepsilon$, $a(\lambda \cap \rho)1$. It follows that

$$(\lambda \cap \rho)_{\min} = \varepsilon \neq \lambda \cap \rho = \lambda_{\min} \cap \rho_{\min}.$$

The following relations will be useful.

III.2.11 Definition

On any inverse semigroup S, define relations $\mathcal{F}$ and $\mathcal{C}$ by

$$a\mathcal{F}b \Leftrightarrow a^{-1}b \in E_S, \qquad a\mathcal{C}b \Leftrightarrow a^{-1}b, ab^{-1} \in E_S.$$

The relation $\mathcal{C}$ is the *compatibility relation* on S, and any two $\mathcal{C}$-related elements are said to be *compatible*.

III.2.12 Proposition

For any congruence ρ on an inverse semigroup S, $\rho_{\min} = (\rho \cap \mathcal{F})^* = (\rho \cap \mathcal{C})^*$.

Proof. First let $a(\rho \cap \mathcal{F})b$. Then $e = a^{-1}b \in E_S$ and

$$ae = aa^{-1}b = (aa^{-1})(bb^{-1})b = (bb^{-1})(aa^{-1})b$$

$$= b(b^{-1}a)(a^{-1}b) = be^{-1}e = be,$$

$$e = e^{-1} = b^{-1}a\rho a^{-1}a\rho b^{-1}b.$$

Hence $\rho \cap \mathcal{F} \subseteq \rho_{\min}$ and thus

$$(\rho \cap \mathcal{C})^* \subseteq (\rho \cap \mathcal{F})^* \subseteq \rho_{\min}. \tag{2}$$

In particular, $\operatorname{tr}(\rho \cap \mathcal{C})^* \subseteq \operatorname{tr}\rho_{\min} = \operatorname{tr}\rho$.

Now let $e, f \in E_S$ be such that $e\rho f$. Then $e(\rho \cap \mathcal{C})f$ and thus $e(\rho \cap \mathcal{C})^*f$. That is to say, $\operatorname{tr}\rho \subseteq \operatorname{tr}(\rho \cap \mathcal{C})^*$, and the two traces are equal. By minimality of $\rho_{\min}$, we get $\rho_{\min} \subseteq (\rho \cap \mathcal{C})^*$ which together with (2) gives all the desired equalities.

III.2.13 Lemma

In any inverse semigroup S,

$$\mathcal{R} \cap \mathcal{F} = \mathcal{R} \cap \mathcal{C} = \mathcal{L} \cap \mathcal{C} = \mathcal{H} \cap \mathcal{F} = \mathcal{H} \cap \mathcal{C} = \varepsilon.$$

Proof. Let $a(\mathcal{R} \cap \mathcal{F})b$. Then $aa^{-1} = bb^{-1}$ and $a^{-1}b \in E_S$. Since $a^{-1}b$ is idempotent, we get

$$a^{-1}b = (a^{-1}b)(b^{-1}a) = a^{-1}(bb^{-1})a = a^{-1}(aa^{-1})a = a^{-1}a$$

and hence

$$a = a(a^{-1}a) = (aa^{-1})b = bb^{-1}b = b.$$

Thus, $\mathcal{R} \cap \mathcal{F} = \varepsilon$. Then *a fortiori* $\mathcal{R} \cap \mathcal{C} = \varepsilon$, and hence by symmetry $\mathcal{L} \cap \mathcal{C} = \varepsilon$. The assertions concerning $\mathcal{H}$ are now obvious.

III.2.14 Exercises

(i) Let S be an inverse semigroup, $\tau \in \mathcal{C}(E_S)$, and define $\hat{\tau}$ by

$$e\hat{\tau}f \Leftrightarrow a^{-1}ea\tau a^{-1}fa \quad \text{for all } a \in S.$$

Show that $\hat{\tau}$ is the greatest normal congruence on E_S contained in τ.

(ii) Show that the relations $\mathcal{C}$ and $\mathcal{F}$ on any inverse semigroup are compatible with multiplication.

(iii) For a congruence ρ on an inverse semigroup S, show

(α) $\ker \rho_{\max} = \{a \in S \mid ae\rho ea \text{ for all } e \in E_S\}$,
(β) $\ker \rho_{\min} = \{a \in S \mid ae = e \text{ for some } e \in E_S,\ e\rho a^{-1}a\}$.

(iv) Let S be an inverse semigroup, $\rho \in \mathcal{C}(S)$, $e \in E_S$. Show that $e\rho = e\rho_{\max} \cap \ker \rho$.

(v) Let S be an inverse semigroup and $\rho \in \mathcal{C}(S)$. Show that $\rho_{\min}$ is the least congruence on S containing $\operatorname{tr} \rho$.

(vi) Let ρ be a congruence on an inverse semigroup S. Show that $\rho_{\max}$ is the greatest congruence on S contained in $\rho\mathcal{L}\rho$.

(vii) In any inverse semigroup S, define a relation ξ by

$$a\xi b \Leftrightarrow ab^{-1}b = ba^{-1}a.$$

Prove that for any congruence ρ on S, $\rho_{\min} = (\rho \cap \xi)^*$.

(viii) Let I be an ideal of an inverse semigroup S and denote by ρ_I the corresponding Rees congruence. Show that for any congruence ρ on S, $\rho \vee \rho_I = \rho\rho_I\rho$.

(ix) Let I be an ideal of an inverse semigroup S. Show that the mapping $\rho \to \rho|_I$ $[\rho \in \mathcal{C}(S)]$ is a homomorphism of $\mathcal{C}(S)$ onto $\mathcal{C}(I)$.

The main result 2.5 is due to Reilly–Scheiblich [1]; this paper contains a variety of properties of congruences on regular semigroups. Example 2.10 is due to F. Pastijn. Green [2] pursued the subject of congruences on an inverse semigroup and introduced the relation $\mathcal{F}$. The compatibility relation $\mathcal{C}$ was introduced by Wagner [3]. For properties of the lattice of congruences on semilattices, consult Freese-Nation [1], Hall [2], and Papert [1].

III.3 IDEMPOTENT SEPARATING CONGRUENCES

A congruence ρ on an inverse semigroup S is said to be idempotent separating if its restriction to E_S is the equality relation, that is to say, if $\operatorname{tr} \rho = \varepsilon$. After some simple characterizations of such congruences, we establish that the lattice of idempotent separating congruences on S is isomorphic to the lattice of all normal subsemigroups of S contained in the centralizer of idempotents of S.

We then introduce the concept of an antigroup (fundamental inverse semigroup) and consider the lattice of all antigroup congruences on S.

III.3.1 Definition

A congruence ρ on a semigroup S is *idempotent separating* if for any $e, f \in E_S$, $e\rho f$ implies $e = f$.

It is clear that a congruence ρ on an inverse semigroup S is idempotent separating if and only if $\operatorname{tr}\rho = \varepsilon$, and hence the set of all idempotent separating congruences on S constitutes the θ-class of ε. Some other characterizations follow.

III.3.2 Proposition

The following conditions on a congruence ρ on an inverse semigroup S are equivalent.

(i) ρ is idempotent separating.
(ii) $\rho \subseteq \mathcal{H}$.
(iii) $\rho \subseteq \mathcal{L}$.
(iv) $\rho \cap \mathcal{C} = \varepsilon$.
(v) $\rho \cap \mathcal{F} = \varepsilon$.

Proof. (i) *implies* (ii). Let $a\rho b$. Then $aa^{-1}\rho bb^{-1}$ which by the hypothesis implies $aa^{-1} = bb^{-1}$; analogously $a^{-1}a = b^{-1}b$. Consequently, $a\mathcal{H}b$ which proves that $\rho \subseteq \mathcal{H}$.

The implications (ii) *implies* (iii) and (v) *implies* (iv) are trivial; (iii) *implies* (iv) follows from 2.13.

(iv) *implies* (i). Let $e, f \in E_S$ be ρ-related. Then $e(\rho \cap \mathcal{C})f$ and thus $e = f$.

(ii) *implies* (v). By 2.13, $\rho \cap \mathcal{F} \subseteq \mathcal{H} \cap \mathcal{F} = \varepsilon$.

III.3.3 Notation

It is customary to denote by μ, or μ_S if there is a need for emphasizing S, the greatest idempotent separating congruence on any inverse semigroup S. According to the above, μ is also the greatest congruence on S contained in $\mathcal{H}$ (or $\mathcal{L}$), and also $\mu = \varepsilon_{\max}$. Thus, by 2.4 we have

$$a\mu b \Leftrightarrow a^{-1}ea = b^{-1}eb \quad \text{for all } e \in E_S.$$

III.3.4 Notation

For any inverse semigroup S, we write $E_S\zeta$ for the *centralizer of idempotents of* S.

More elaborate notation $c_S(E_S)$ was used in I.3.11, but the present notation is quite appropriate.

III.3.5 Lemma

For any inverse semigroup S, we have $\ker \mu = E_S\zeta$.

Proof. The proof of this lemma is left as an exercise.

The situation in the next result resembles closely that in groups.

III.3.6 Theorem

Let S be an inverse semigroup with the semilattice E of idempotents. Denote by Δ the partially ordered set of all normal subsemigroups of S contained in $E\zeta$. Then Δ is a complete sublattice of the lattice of all full subsemigroups of S, and the mappings

$$K \to \rho_{(K,\varepsilon)} \qquad (K \in \Delta), \qquad \rho \to \ker \rho \qquad (\rho \in \varepsilon\theta)$$

are mutually inverse complete lattice isomorphisms of Δ and $\varepsilon\theta$.

Proof. Let $\{K_\alpha\}_{\alpha \in A}$ be a family of normal subsemigroups of S contained in $E\zeta$. It is clear that $\bigcap K_\alpha$ is a normal subsemigroup of S contained in $E\zeta$. The join $\bigvee K_\alpha$ in the lattice of all full subsemigroups of S consists of all elements of the form $x_1 x_2 \cdots x_n$ such that $x_1, x_2, \ldots, x_n \in \bigcup K_\alpha$. It is immediate that $\bigvee K_\alpha$ is a full inverse subsemigroup of S contained in $E\zeta$. For any $k = k_1 k_2 \cdots k_n$ with $k_1, k_2, \ldots, k_n \in \bigcup K_\alpha$ and $a \in S$, we obtain

$$\begin{aligned} a^{-1}ka &= a^{-1}k_1 k_2 \cdots k_n a = a^{-1}\left[(aa^{-1})k_1\right]k_2 \cdots k_n a \\ &= a^{-1}\left[k_1(aa^{-1})\right]k_2 \cdots k_n a = \left(a^{-1}k_1 a\right)a^{-1}k_2 \cdots k_n a \\ &= \cdots = \left(a^{-1}k_1 a\right)\left(a^{-1}k_2 a\right) \cdots \left(a^{-1}k_n a\right) \in \bigvee K_\alpha \end{aligned}$$

since $k_1, k_2, \ldots, k_n \in E\zeta$. Hence $\bigvee K_\alpha$ is self-conjugate and thus a normal subsemigroup of S. This shows that Δ is a complete sublattice of the lattice of all full subsemigroups of S.

It is clear that both mappings in the statement of the theorem preserve inclusion. The second one maps $\varepsilon\theta$ into Δ in view of 3.5. For any $K \in \Delta$, conditions 1.3(i) and (ii) are trivially satisfied for (K, ε). In fact,

$$K \to \rho_{(K,\varepsilon)} \to \ker \rho_{(K,\varepsilon)} = K,$$

and for any $\rho \in \varepsilon\theta$,

$$\rho \to \ker \rho \to \rho_{(\ker \rho, \varepsilon)} = \rho$$

in view of 1.5. Applying I.2.8, we deduce that both mappings are (lattice) isomorphisms. But every isomorphism of complete lattices is complete.

The strong resemblance to the group case continues in the next result.

III.3.7 Proposition

With the notation of 3.6, we have for any $K, L \in \Delta$, $K \vee L = KL = LK$ and

$$a\rho_{(K,\varepsilon)}b \Leftrightarrow a\mathcal{H}b, \quad Ka = Kb. \tag{3}$$

Proof. Let $K, L \in \Delta$, $k \in K$, $l \in L$. Then $l \in E\zeta$ and thus

$$kl = k(k^{-1}k)l = kl(k^{-1}k) = (klk^{-1})k \in LK$$

which shows that $KL \subseteq LK$. By symmetry, we conclude that $KL = LK$. It follows from this that KL is a subsemigroup of S containing both K and L. But then clearly $K \vee L = KL = LK$ where the join is that of (full) subsemigroups.

If $a\rho_{(K,\varepsilon)}b$, then $a\mathcal{H}b$ by 3.2, and thus $a = a(a^{-1}a) = (ab^{-1})b \in Kb$ so that $Ka \subseteq Kb$ and by symmetry also $Kb \subseteq Ka$. Conversely, let $a\mathcal{H}b$ and $Ka = Kb$. Then $a = kb$ for some $k \in K$ and hence $ab^{-1} = k(bb^{-1}) \in K$ and thus $a\rho_{(K,\varepsilon)}b$.

We will now have a closer look at the θ-classes.

III.3.8 Proposition

Let S be an inverse semigroup and $\xi \in \mathcal{C}(S)$. Define a function φ by

$$\varphi\colon \rho \to \rho/\xi_{\min} \qquad (\rho \in \xi\theta).$$

Then φ is a complete isomorphism of $\xi\theta$ onto the lattice of all idempotent separating congruences on $S/\xi_{\min}$.

Proof. Let $\pi = \xi_{\min}$. It is easy to verify that the mapping ψ: $\rho \to \rho/\pi$ defined for all congruences ρ containing π is a complete isomorphism onto $\mathcal{C}(S/\pi)$. Let $\rho\theta\pi$. Then $\operatorname{tr}\rho = \operatorname{tr}\pi$ which implies that $\operatorname{tr}\rho/\pi = \operatorname{tr}\pi/\pi$ where the latter is the equality relation on $E_{S/\pi}$. Hence ρ/π is idempotent separating. Conversely, let $\rho \in \mathcal{C}(S)$ be such that $\rho \supseteq \pi$ and ρ/π is idempotent separating. Then $\operatorname{tr}\rho/\pi = \operatorname{tr}\pi/\pi$, which evidently implies that $\operatorname{tr}\rho = \operatorname{tr}\pi$ and hence $\rho\theta\pi$. This shows that ψ above maps $\pi\theta$ onto the lattice of idempotent separating congruences on S/π. Consequently, $\varphi = \psi|_{\pi\theta}$ has the required properties.

The next concept will play an important role in the sequel.

III.3.9 Definition

An inverse semigroup S is an *antigroup* (or a *fundamental inverse semigroup*) if ε is the only congruence on S contained in $\mathcal{H}$ (equivalently, if $\mu = \varepsilon$).

III.3.10 Corollary

Let ρ be a congruence on an inverse semigroup S. Then $\rho = \rho_{\max}$ if and only if S/ρ is an antigroup. In particular, S/μ is an antigroup.

Proof. This follows directly from 3.8.

For the lattice of antigroup congruences, we have the following statements.

III.3.11 Theorem

Let $\mathcal{AG}(S)$ denote the set of all antigroup congruences on an inverse semigroup S ordered by inclusion. Then $\mathcal{AG}(S)$ is a complete $\cap$-subsemilattice of $\mathcal{C}(S)$ with greatest element ω and least element μ. For any nonempty subset $\mathcal{F}$ of $\mathcal{AG}(S)$, the join of $\mathcal{F}$ in $\mathcal{AG}(S)$ is given by $(\bigvee_{\rho\in\mathcal{F}}\rho)_{\max}$ [hence $\mathcal{AG}(S)$ is generally not a sublattice of $\mathcal{C}(S)$, see IX.3.13]. Moreover, $\mathcal{AG}(S) \cong \mathcal{N}(E_S)$.

Proof. If $\mathcal{F}$ is any nonempty family of congruences on S, then by 2.5, we obtain

$$\operatorname{tr}\Big(\bigvee_{\rho\in\mathcal{F}}\rho_{\max}\Big) = \bigvee_{\rho\in\mathcal{F}}\operatorname{tr}\rho_{\max} = \bigvee_{\rho\in\mathcal{F}}\operatorname{tr}\rho = \operatorname{tr}\Big(\bigvee_{\rho\in\mathcal{F}}\rho\Big) = \operatorname{tr}\Big(\bigvee_{\rho\in\mathcal{F}}\rho\Big)_{\max}$$

which proves that $\bigvee_{\rho\in\mathcal{F}}\rho_{\max}\theta(\bigvee_{\rho\in\mathcal{F}}\rho)_{\max}$. This together with 2.9 and I.2.8 implies all the assertions in the statement of the theorem including the last one since then $\rho \to \operatorname{tr}\rho$ defined for all $\rho \in \mathcal{A}(S)$ is the required (complete) isomorphism.

Kernel normal systems of idempotent separating congruences are precisely those all of whose members are groups; for this reason they are referred to as *group kernel normal systems.*

III.3.12 Exercises

(i) Let I be an ideal of an inverse semigroup S. Show that $\mu_S|_I = \mu_I$.

(ii) Let V be an inverse semigroup which is a dense extension of a semigroup S. Show that V is an antigroup if and only if S is an antigroup.

(iii) Let ρ be a congruence on an inverse semigroup S. Show that $\mu_{S/\rho} = \rho_{\max}/\rho$.

(iv) Let S be an inverse semigroup. Let $E = E_S$, $E\zeta^1 = E\zeta$, and for $n \geq 2$, let $E\zeta^n = c_S(E\zeta^{n-1})$. Show that

$$E\zeta^n = \begin{cases} E\zeta & \text{if } n \text{ is odd} \\ Z(E\zeta) & \text{if } n \text{ is even.} \end{cases}$$

(v) Let ρ be a congruence on an inverse semigroup S. Show that S/ρ is a Reilly semigroup if and only if $S/\rho_{\max}$ is a bicyclic semigroup.

(vi) Show that $\rho_{\max} = \rho \vee \mu$ for every congruence ρ on a Clifford semigroup.

(vii) Show that a subdirect product of antigroups is an antigroup.

Preston [4] was the first to study idempotent separating congruences on inverse semigroups, and 3.6 is due to him. Munn [5] showed that they coincide with the congruences contained in $\mathcal{H}$, and this result was then extended to regular semigroups by Lallement [1]. The expression for (and thus the existence of) a greatest idempotent separating congruence on an inverse semigroup is due to Howie [1]. This result excited much interest in congruences on inverse semigroups. Green [2] contains some further results on this subject.

The term "antigroups" was introduced by Wagner who studied them in [10], [11], [12]; the terminology "fundamental inverse semigroup" for the same class of semigroups belongs to Munn [10], and this latter terminology is more widely accepted; Širjajev [2] calls them "rigid inverse semigroups." In view of IV.2.5, a better name for these semigroups would be "*E-faithful.*"

III.4 IDEMPOTENT PURE CONGRUENCES

A congruence ρ on an inverse semigroup S is idempotent pure if E_S is saturated for ρ. Besides some simple characterizations and properties of idempotent pure congruences, we consider congruences on S having a fixed kernel. In particular, we prove that the mapping $\rho \to \ker \rho$ is a complete $\cap$-homomorphism of the congruence lattice of S onto a certain lattice $\mathcal{K}(S)$ of subsets of S, which are characterized abstractly. We introduce here the concept of an E-disjunctive inverse semigroup as an inverse semigroup in which the equality relation is its only idempotent pure congruence.

III.4.1 Definition

A congruence ρ on an inverse semigroup S is *idempotent pure* (or *idempotent determined*) if for any $a \in S$, $e \in E_S$, $a\rho e$ implies $a \in E_S$.

Note the obvious fact that ρ is idempotent pure if and only if $\ker \rho = E_S$. Hence idempotent pure congruences constitute least elements in their respective θ-classes.

III.4.2 Proposition

The following statements concerning a congruence ρ on an inverse semigroup S are equivalent.

(i) ρ is idempotent pure.
(ii) $\rho \cap \mathcal{L} = \varepsilon$.
(iii) $\rho \subseteq \mathcal{C}$.
(iv) $\rho \subseteq \mathcal{F}$.

Proof. (i) *implies* (ii). Let $a(\rho \cap \mathcal{L})b$. Then $ab^{-1}\rho bb^{-1}$ which gives $ab^{-1} \in E_S$ since ρ is idempotent pure. But then $ab^{-1}\mathcal{L}bb^{-1}$ implies $ab^{-1} = bb^{-1}$. Thus

$$a = a(a^{-1}a) = a(b^{-1}b) = (ab^{-1})b = (bb^{-1})b = b.$$

Consequently, $\rho \cap \mathcal{L} = \varepsilon$.

(ii) *implies* (iii). Let $a\rho b$. Then

$$a^{-1}b = (a^{-1}b)(b^{-1}b)\rho b^{-1}aa^{-1}b = (a^{-1}b)^{-1}(a^{-1}b)$$

and since also $a^{-1}b\mathcal{L}(a^{-1}b)^{-1}(a^{-1}b)$, the hypothesis yields $a^{-1}b = (a^{-1}b)^{-1}(a^{-1}b)$ and thus $a^{-1}b \in E_S$. One shows similarly that also $ab^{-1} \in E_S$, and hence $\rho \subseteq \mathcal{C}$.

(iii) *implies* (iv) trivially.

(iv) *implies* (i). Let $a \in S$, $e \in E_S$, and $a\rho e$. Then $aa^{-1}\rho e$ and thus by hypothesis $(aa^{-1})a \in E_S$, so that $a \in E_S$.

We will prove an analogue of 2.5 for the mapping $\rho \to \ker \rho$. To this end, we need some preparation.

III.4.3 Definition

Let H be a subset of and ρ be a congruence on a semigroup S. Then H is *saturated for* ρ, or ρ *saturates* H, if H is the union of some ρ-classes.

III.4.4 Proposition

For H a nonempty subset of a semigroup S, define a relation τ^H on S by

$$a\tau^H b \Leftrightarrow \left(xay \in H \Leftrightarrow xby \in H \text{ for all } x, y \in S^1\right).$$

Then τ^H is the greatest congruence on S saturating H.

Proof. The proof of this proposition is left as an exercise.

III.4.5 Definition

Let H be a nonempty subset of a semigroup S. Then τ^H is the *syntactic* (or *principal* or *Croisot*) *congruence on* S *determined by* H. Further, H is a *disjunctive subset of* S if $\tau^H = \varepsilon$. For an inverse semigroup S, we write τ instead of τ^{E_S}, or τ_S if necessary.

III.4.6 Notation

Let S be an inverse semigroup. For any congruence ρ on S, define

$$\rho^{\max} = \tau^{\ker \rho}, \qquad \rho^{\min} = (\rho \cap \mathcal{L})^*.$$

III.4.7 Definition

Let S be an inverse semigroup. A full inverse subsemigroup K of S is a *kernel in* S if it satisfies

$$ab \in K \Rightarrow aKb \subseteq K \qquad (a, b \in S).$$

If $K \neq S$, then K is said to be *proper*. We denote by $\mathcal{K}(S)$ the set of kernels in S ordered by inclusion.

It is easy to verify that $\mathcal{K}(S)$ is a lattice under inclusion. We are now ready for an analogue of 2.5 for kernels.

III.4.8 Theorem

Let S be an inverse semigroup. Define a mapping ker by

$$\ker : \rho \to \ker \rho \qquad [\rho \in \mathcal{C}(S)].$$

Then ker is a complete $\cap$-homomorphism of $\mathcal{C}(S)$ onto $\mathcal{K}(S)$. Let κ be the equivalence relation on $\mathcal{C}(S)$ induced by ker. Then for any $\rho \in \mathcal{C}(S)$,

$$\rho\kappa = \left[\rho^{\min}, \rho^{\max}\right]$$

and is a complete sublattice of $\mathcal{C}(S)$.

Proof. Let $\mathcal{F}$ be a nonempty family of congruences on S. Then for any $a \in S$, we obtain

$$a \in \ker\left(\bigcap_{\rho \in \mathcal{F}} \rho\right) \Leftrightarrow a \bigcap_{\rho \in \mathcal{F}} \rho aa^{-1}$$

$$\Leftrightarrow a\rho aa^{-1} \quad \text{for all } \rho \in \mathcal{F}$$

$$\Leftrightarrow a \in \ker \rho \quad \text{for all } \rho \in \mathcal{F}$$

$$\Leftrightarrow a \in \bigcap_{\rho \in \mathcal{F}} \ker \rho.$$

This proves that $\ker \cap_{\rho \in \mathcal{F}} \rho = \cap_{\rho \in \mathcal{F}} \ker \rho$, and hence that ker is a complete $\cap$-homomorphism on $\mathcal{C}(S)$.

Let $\rho \in \mathcal{C}(S)$, $k \in \ker \rho$, and $a, b \in S$ be such that $ab \in \ker \rho$. Then $ab\rho e$ and $k\rho f$ for some $e, f \in E_S$. Hence

$$akb\rho afb = a(fbb^{-1})b = (ab)(b^{-1}fb)\rho e(b^{-1}fb)$$

which shows that $akb \in \ker \rho$. Consequently, $\ker \rho \in \mathcal{K}(S)$ which implies that ker maps $\mathcal{C}(S)$ into $\mathcal{K}(S)$.

Now let $K \in \mathcal{K}(S)$. By 4.4, τ^K is the greatest congruence on S saturating K so that $\ker \tau^K \subseteq K$ since K is full. Let $k \in K$ and $x, y \in S$. Then $k, k^{-1} \in K$ and since $K \in \mathcal{K}(S)$, we get

$$xky \in K \Rightarrow (xk)y \in K \Rightarrow (xk)k^{-1}y \in K \Rightarrow xkk^{-1}y \in K,$$

$$xkk^{-1}y \in K \Rightarrow (xkk^{-1})ky \in K \Rightarrow x(kk^{-1}k)y \in K \Rightarrow xky \in K,$$

which proves that $k\tau^K kk^{-1}$. But then $k \in \ker \tau^K$ which yields $K = \ker \tau^K$. This shows that ker maps $\mathcal{C}(S)$ onto $\mathcal{K}(S)$. Furthermore, if $\rho \in \mathcal{C}(S)$, we get that $\tau^{\ker \rho}$ is the greatest element of the κ-class containing ρ.

Let γ be any congruence on S such that $\ker \gamma = \ker \rho$, and let $a(\rho \cap \mathcal{L})b$. Then $a\rho b$ and $a^{-1}a = b^{-1}b$. We obtain $ab^{-1}\rho bb^{-1}$ so that $ab^{-1} \in \ker \rho = \ker \gamma$, and thus

$$a = a(a^{-1}a) = a(b^{-1}b) = (ab^{-1})b\gamma(ba^{-1})(ab^{-1})b$$

$$= b(a^{-1}a)b^{-1}b = b(b^{-1}b)b^{-1}b = b.$$

We have proved that $\rho \cap \mathcal{L} \subseteq \gamma$, which gives $(\rho \cap \mathcal{L})^* \subseteq \gamma$. It follows that $\ker(\rho \cap \mathcal{L})^* \subseteq \ker \rho$. Let $a \in \ker \rho$. Then $a\rho a^{-1}a$, and since always $a\mathcal{L}a^{-1}a$, we get $a(\rho \cap \mathcal{L})a^{-1}a$. But then $a(\rho \cap \mathcal{L})^* a^{-1}a$ which says that $a \in \ker(\rho \cap \mathcal{L})^*$. This shows that $\ker(\rho \cap \mathcal{L})^* = \ker \rho$ and thus $\rho^{\min}$ is the least element of $\kappa\rho$. It follows easily that $\kappa\rho = [\rho^{\min}, \rho^{\max}]$. Since any interval of a complete lattice is a complete sublattice, $\kappa\rho$ is a complete sublattice of $\mathcal{C}(S)$.

It is clear that $\rho^{\min}$ can also be given by $(\rho \cap \mathcal{R})^*$.

III.4.9 Definition

We call the classes of κ in 4.8 the *kernel classes* of S.

It follows from 4.8 that a subset K of an inverse semigroup S is a kernel in S if and only if K is the kernel of some congruence on S. One sees directly that every kernel in S is self-conjugate and is thus a normal subsemigroup of S. The following example shows that not every normal subsemigroup of S need be a kernel in S.

III.4.10 Example

Let $S = B(1, I)$ where $I = \{1, 2, 3\}$. One verifies easily that S has no proper congruences and hence no kernels different from E_S and S. Let F be a full inverse subsemigroup of S and let $(i, 1, j) \in F$. If $(k, 1, l)^{-1}(i, 1, j)(k, 1, l) \neq 0$, then $i = k = j$ and thus

$$(k, 1, l)^{-1}(i, 1, j)(k, 1, l) = (l, 1, l) \in F.$$

Thus every full inverse subsemigroup of S is self-conjugate and is thus normal. For example,

$$F = \{(1,1,1), (1,1,2), (2,1,1), (2,1,2), (3,1,3), 0\}$$

is a normal subsemigroup of S which is not a kernel in S.

As a consequence of 4.8, we have that for congruences ρ and ξ on an inverse semigroup S, $\rho \subseteq \xi$ implies $\rho^{\min} \subseteq \xi^{\min}$. This corresponds to the first part of 2.8. The following example shows that the analogue of the second part of 2.8 does not hold in general. This example also illustrates the fact that for a nonempty family $\mathcal{F}$ of congruences on S, we may have $\vee_{\rho \in \mathcal{F}} \ker \rho \neq \ker(\vee_{\rho \in \mathcal{F}} \rho)$. The first join here is taken within $\mathcal{K}(S)$. We observe that always $\vee_{\rho \in \mathcal{F}} \ker \rho \subseteq \ker(\vee_{\rho \in \mathcal{F}} \rho)$.

III.4.11 Example

Let S be a semilattice of two groups G and H of order 2 determined by an isomorphism $\varphi: G \to H$. Let ρ be the Rees congruence on S relative to H. Then $\varepsilon \subseteq \rho$ but $\varepsilon^{\max} \not\subseteq \rho^{\max}$ since $\varepsilon^{\max} = \sigma$ and $\rho^{\max} = \rho$. Also $\sigma \vee \rho = \omega$ but

$$\ker \sigma \vee \ker \rho = E_S \vee (E_S \cup H) = E_S \cup H \neq S = \ker(\sigma \vee \rho).$$

III.4.12 Proposition

For any congruence ρ on an inverse semigroup S, we have

$$\rho = \rho_{\min} \vee \rho^{\min} = \rho_{\max} \cap \rho^{\max}.$$

Proof. Clearly,

$$\rho_{\min} \vee \rho^{\min} \subseteq \rho \subseteq \rho_{\max} \cap \rho^{\max}. \tag{4}$$

Further,

$$\ker(\rho_{\min} \vee \rho^{\min}) \supseteq \ker \rho^{\min} = \ker \rho,$$

$$\operatorname{tr}(\rho_{\min} \vee \rho^{\min}) \supseteq \operatorname{tr} \rho_{\min} = \operatorname{tr} \rho,$$

which together with (4) shows that the first inclusion in (4) is actually an equality, in the light of 1.5. Also

$$\ker(\rho_{\max} \cap \rho^{\max}) \subseteq \ker \rho^{\max} = \ker \rho$$

$$\operatorname{tr}(\rho_{\max} \cap \rho^{\max}) \subseteq \operatorname{tr} \rho_{\max} = \operatorname{tr} \rho$$

which together with (4) shows that the second inclusion in (4) is an equality, again in view of 1.5.

III.4.13 Proposition

Let S be an inverse semigroup and $\xi \in \mathcal{C}(S)$. Define a function φ by

$$\varphi : \rho \to \rho/\xi^{\min} \qquad (\rho \in \xi\kappa).$$

Then φ is a complete isomorphism of the kernel class $\xi\kappa$ onto the lattice of all idempotent pure congruences on $S/\xi^{\min}$.

Proof. The argument goes along the same lines as in the proof of 3.8. Let $\pi = \xi^{\min}$ and let $\rho\kappa\pi$. Then $\pi \subseteq \rho$. If $a \in S$ is such that $a\pi(\rho/\pi)e\pi$ for some $e \in E_S$, then $a\rho e$ so that $a \in \ker \rho = \ker \pi$ which implies that $a\pi \in E_{S/\pi}$. Hence ρ/π is idempotent pure. Conversely, let $\rho \in \mathcal{C}(S)$, $\rho \supseteq \pi$, be such that ρ/π is idempotent pure. If $a \in \ker \rho$, then $a\rho e$ for some $e \in E_S$ which implies $a\pi(\rho/\pi)e\pi$ so that $a\pi \in E_{S/\pi}$, that is $a \in \ker \pi$. Thus $\ker \rho \subseteq \ker \pi$ which gives $\ker \rho = \ker \pi$ since $\rho \supseteq \pi$. Consequently, $\rho \in \pi\kappa$. We note that $\varphi = \psi|_{\pi\kappa}$ where ψ is the complete isomorphism of the lattice of all congruences on S containing π onto $\mathcal{C}(S/\pi)$.

As a counterpart to antigroups, we introduce the following class of semigroups.

III.4.14 Definition

An inverse semigroup S is *E-disjunctive* if ε is the only congruence on S saturating E_S.

III.4.15 Remark

In view of the terminology and notation introduced earlier, the following statements are equivalent:

(α) S is E-disjunctive,
(β) E_S is a disjunctive subset of S,
(γ) ε is the only idempotent pure congruence on S,
(δ) $\tau = \varepsilon$.

III.4.16 Corollary

Let ρ be a congruence on an inverse semigroup S. Then $\rho = \rho^{\max}$ if and only if S/ρ is E-disjunctive. In particular, S/τ is E-disjunctive.

Proof. This follows from 4.13.

III.4.17 Exercises

(i) Let ρ be a congruence on an inverse semigroup S. Show that $\chi : \xi \to \xi/\rho$ is an isomorphism of the interval $[\rho, \rho^{\max}]$ onto the lattice of idempotent pure congruences on S/ρ. Deduce that $\rho = \rho^{\max}$ if and only if S/ρ is E-disjunctive.
(ii) Show that in any inverse semigroup $\mu = \mu^{\min}$, $\tau = \tau_{\min}$.
(iii) Find τ in a Brandt semigroup.
(iv) Construct an example, different from 4.10, showing that there exist normal subsemigroups which are not kernels.
(v) In any inverse semigroup S define a relation δ by $a\delta b \Leftrightarrow ab^{-1} \in E\zeta$. Show that $\mu = \delta \cap \mathcal{L}$.
(vi) Let S be a semilattice of two groups G and H whose multiplication is determined by a homomorphism $\varphi : G \to H$. Let τ be the greatest idempotent pure congruence on S. Prove that τ is the least group congruence if φ is one-to-one and is the equality relation otherwise.
(vii) Let ρ be a congruence on an inverse semigroup S. Show that $\tau_{S/\rho} = \rho^{\max}/\rho$.
(viii) Give an example of congruences ρ and ξ on an inverse semigroup S such that $\operatorname{tr}\rho = \operatorname{tr}\xi$ and $\ker\rho \vee \ker\xi \neq \ker(\rho \vee \xi)$.
(ix) Let X_α, $\alpha \in A$, be a family of nonempty subsets of an inverse semigroup S. Show that $\bigcap_{\alpha\in A}\tau^{X_\alpha} \subseteq \tau^{\bigcap_{\alpha\in A}X_\alpha}$. Deduce that for any family $\mathfrak{F}$ of congruences on S, $\bigcap_{\rho\in\mathfrak{F}}\rho^{\max} \subseteq (\bigcap_{\rho\in\mathfrak{F}}\rho)^{\max}$ and prove that $\ker(\bigcap_{\rho\in\mathfrak{F}}\rho^{\max}) = \ker(\bigcap_{\rho\in\mathfrak{F}}\rho)^{\max}$.

III.4.18 Problems

(i) For ρ a congruence on an inverse semigroup S, what can be said about $(\rho \cap \mathcal{H})^*$?

(ii) Study E-disjunctive inverse semigroups.

The main result of this section, namely 4.8, was proved by Green [2], except for the formula for $\rho^{\min}$ which was discovered by Petrich–Reilly [3]. Result 4.12 is due to Green [2]. Idempotent pure congruences were introduced under the name "idempotent determined" by Green [1]. They appear in various other papers, but prominently only in Allouch [1], [2], Green [1], O'Caroll [5], [7], and Petrich–Reilly [3].

III.5 GROUP CONGRUENCES

We first characterize group congruences on an inverse semigroup in several ways. Letting σ denote the least group congruence on an inverse semigroup S, the mapping $\rho \to \rho \vee \sigma$ is a homomorphism of $\mathcal{C}(S)$ onto the lattice of group congruences on S.

It is clear that group congruences on an inverse semigroup S are precisely those whose trace is the universal relation. Hence they form the θ-class of the universal relation ω on S, and $\omega_{\min}$ is the least group congruence.

III.5.1 Notation

The least group congruence on any inverse semigroup S is denoted by σ; we will write σ_S only if there is a danger of confusion.

We collect various information on σ in the following result.

III.5.2 Lemma

The following statements are valid in any inverse semigroup S.

(i) $$a \sigma b \Leftrightarrow ae = be \quad \text{for some } e \in E_S \qquad (a, b \in S).$$
$$\Leftrightarrow a \geq c, b \geq c \quad \text{for some } c \in S.$$

(ii) $\sigma = \mathcal{F}^* = \mathcal{C}^* = \omega_{\min}$.

Proof. The first equivalence in (i) follows from 2.4; the second equivalence obviously holds. Item (ii) follows by letting $\rho = \omega$ in 2.12.

We are now ready for characterizations of group congruences.

III.5.3 Proposition

The following statements concerning a congruence ρ on an inverse semigroup S are equivalent.

(i) ρ is a group congruence.
(ii) $\rho\mathcal{L}\rho = \omega$.
(iii) $\rho \supseteq \mathcal{F}$.
(iv) $\rho \supseteq \mathcal{C}$.

Proof. In view of 5.2(ii), items (i), (iii), and (iv) are equivalent.

(i) *implies* (ii). Let $a, b \in S$. For $e = a^{-1}ab^{-1}b$, we have

$$(ae)^{-1}(ae) = e = (be)^{-1}(be)$$

so that $ae\mathcal{L}be$. By 5.2(i), we have $a\sigma ae$, $ae\mathcal{L}be$, $be\sigma b$ and thus $a\sigma\mathcal{L}\sigma b$. Since $\rho \supseteq \sigma$, we get $\rho\mathcal{L}\rho = \omega$.

(ii) *implies* (i). Let $e, f \in E_S$. Then $e\rho\mathcal{L}\rho f$ which implies $e\rho x$, $x\mathcal{L}y$, $y\rho f$, and thus $e\rho x^{-1}x = y^{-1}y\rho f$. Consequently, ρ is a group congruence in view of II.2.10.

As preparation for the next theorem, we prove the following useful result.

III.5.4 Lemma

The following statements are true for any congruence ρ on an inverse semigroup S.

(i) $\rho \vee \sigma = \sigma\rho\sigma$.
(ii) $a(\rho \vee \sigma)b \Leftrightarrow ae\rho be$ for some $e \in E_S$.

Proof. (i) First note that $\sigma\rho\sigma \subseteq \rho \vee \sigma$. To prove the opposite inclusion, it suffices to show that $\sigma\rho\sigma$ is transitive, for then it is clearly a congruence containing both σ and ρ. Hence let $a\sigma\rho\sigma b$ and $b\sigma\rho\sigma c$. Thus

$$ae = xe, \qquad x\rho y, \qquad yf = bf,$$

$$bg = wg, \qquad w\rho z, \qquad zh = ch,$$

for some $x, y, w, z \in S$ and $e, f, g, h \in E_S$. We may evidently assume that $e = f = g = h$, which yields $ae\rho be$, $be\rho ce$, so that $ae\rho ce$. Since $a\sigma ae$ and $ce\sigma c$, we deduce that $a\sigma\rho\sigma c$, as required.

(ii) This follows from the proof of part (i).

III.5.5 Corollary

If ρ is a congruence on an inverse semigroup S, then $\ker(\rho \vee \sigma) = (\ker \rho)\omega$.

Proof. Let $a \in \ker(\rho \vee \sigma)$. Then by 5.4(ii), $af\rho ef$ for some $e, f \in E_S$ and thus $af \in \ker \rho$. But then $a \in (\ker \rho)\omega$.

Conversely, let $a \in (\ker \rho)\omega$ so that $ae \in \ker \rho$ for some $e \in E_S$ and thus $ae\rho f$ for some $f \in E_S$. But then $ae\rho fe$ which yields $a(\rho \vee \sigma)f$ and thus $a \in \ker(\rho \vee \sigma)$.

III.5.6 Theorem

Let S be an inverse semigroup. Then the mapping

$$\varphi : \rho \to \rho \vee \sigma \qquad [\rho \in \mathcal{C}(S)].$$

is a homomorphism of $\mathcal{C}(S)$ onto the lattice $\omega\theta$ of group congruences on S.

Proof. Let $\rho, \tau \in \mathcal{C}(S)$ and $a((\rho \vee \sigma) \cap (\tau \vee \sigma))b$. By 5.4(ii), we have $ae\rho be$ and $af\tau bf$ for some $e, f \in E_S$. Hence $aef(\rho \cap \tau)bef$ which again by 5.4(ii) yields $a((\rho \cap \tau) \vee \sigma)b$. Therefore, $(\rho \vee \sigma) \cap (\tau \vee \sigma) \subseteq (\rho \cap \tau) \vee \sigma$; the opposite inclusion is trivial. It is obvious that φ preserves joins, and is thus a (lattice) homomorphism of $\mathcal{C}(S)$ onto $\omega\theta$.

Note that for a normal subsemigroup K of an inverse semigroup S the pair (K, ω) is a congruence pair for S if and only if K is closed.

III.5.7 Lemma

In any inverse semigroup S,

$$\ker \sigma = E_S\omega = \{a \in S | ax = x \text{ for some } x \in S\}.$$

Proof. If $a\sigma e$ for some $e \in E_S$, then by 5.2(i), $af = ef$ for some $f \in E_S$, so $aef = ef$ and thus $a \geq ef$, that is $a \in E_S\omega$. The proof of the remaining inclusions is left as an exercise.

III.5.8 Proposition

Let S be an inverse semigroup with semilattice E of idempotents. Then S is a subdirect product of a group and an antigroup if and only if $E\omega \cap E\zeta = E$.

Proof. Let S be a subdirect product of a group G and an antigroup A and assume that $S \subseteq G \times A$. Let $(g, a) \in E\omega \cap E\zeta$. Then $(g, a) \in E\omega$ implies $(g, a)(1, e) = (1, e)$ for some $e \in E_A$, where 1 is the identity of G, so that $g = 1$. Further, $(g, a) \in E\zeta$ implies $(g, a)(1, e) = (1, e)(g, a)$ for all $e \in E_A$, and hence $ae = ea$ for all $e \in E_A$. Since A is an antigroup, we deduce from 3.5 that $a \in E_A$. Consequently, $(g, a) \in E$.

Conversely, assume that $E\omega \cap E\xi = E$. Then

$$\ker(\sigma \cap \mu) = \ker \sigma \cap \ker \mu = E$$

by 4.8, and thus $\sigma \cap \mu = \varepsilon$ and S is a subdirect product of the group S/σ and the antigroup S/μ by I.4.18.

For a factorization of any homomorphism of inverse semigroups, we first need an auxiliary statement.

III.5.9 Lemma

Let ρ be a congruence on an inverse semigroup S. Then for any $a, b \in S$, $(a\rho)\sigma(b\rho)$ in S/ρ implies $a\sigma b$ if and only if $\rho \subseteq \sigma$.

Proof.

Necessity. If $a\rho b$, then $a\rho = b\rho$ which implies $a\sigma b$. Thus $\rho \subseteq \sigma$.

Sufficiency. Let $a, b \in S$ be such that $(a\rho)\sigma(b\rho)$. Then $(a\rho)(e\rho) = (b\rho)(e\rho)$ for some $e \in E_S$, and hence $ae\rho be$. Thus, by the hypothesis $\rho \subseteq \sigma$, we get $ae\sigma be$. But then $a\sigma b$ since $a\sigma ae$, $b\sigma be$.

Note that for any congruence ρ on S, $a\sigma b$ implies $(a\rho)\sigma(b\rho)$. Hence under the hypothesis that $\rho \subseteq \sigma$, we actually have $a\sigma b$ if and only if $(a\rho)\sigma(b\rho)$. It follows that then $S/\sigma \cong (S/\rho)/\sigma$ and S and S/ρ have isomorphic maximal group homomorphic images. We may say that ρ *preserves the maximal group homomorphic images*. Since for any congruence ρ on S we have $\rho_{\min} \subseteq \rho$, we obtain the following factorization

$$S \rightarrow S/\rho_{\min} \rightarrow S/\rho \cong (S/\rho_{\min})/(\rho/\rho_{\min}).$$

Using the obvious terminology, we have the following statement.

III.5.10 Proposition

Every homomorphism of inverse semigroups can be factored into a homomorphism preserving the maximal group homomorphic images and an idempotent separating homomorphism.

Proof. It suffices to consider a congruence ρ on an inverse semigroup S. The above factorization

$$S \overset{\varphi}{\rightarrow} S/\rho_{\min} \overset{\psi}{\rightarrow} S/\rho$$

has the property that φ preserves the maximal group homomorphic images since $\rho_{min} \subseteq \sigma$ (see 5.9 and the above remarks); ψ is idempotent separating since ρ and ρ_{min} have the same trace.

III.5.11 Exercises

(i) Show that the following statements concerning congruences ρ and ξ on an inverse semigroup S are equivalent.

(α) $\operatorname{tr}\rho = \operatorname{tr}\xi$.

(β) $\rho \subseteq \xi_{max}, \quad \xi_{max}/\rho = \mu_{S/\rho}$.

(γ) $(a\rho)\mu(b\rho) \Leftrightarrow (a\xi)\mu(b\xi) \qquad (a, b \in S)$.

(δ) $(a\rho)\mathcal{H}(b\rho) \Leftrightarrow (a\xi)\mathcal{H}(b\xi) \qquad (a, b \in S)$.

(ε) $\rho \cap \xi|_{e\rho}$ and $\rho \cap \xi|_{e\xi}$ are group congruences ($e \in E_S$).

(η) $\rho/\rho \cap \xi$ and $\xi/\rho \cap \xi$ are idempotent separating congruences.

(ii) Let $\rho, \xi \in \mathcal{C}(S)$ where S is an inverse semigroup. Show that $\xi = \rho_{min}$ if and only if $\xi \subseteq \rho$ and for every $e \in E_S$, $\xi|_{e\rho}$ is the least group congruence on $e\rho$.

(iii) Show that for any congruence ρ on an inverse semigroup, $\rho \vee \sigma = (\rho\mathcal{F})^*$.

(iv) Let ρ be a congruence on an inverse semigroup S. Show that for any $a, b \in S$, $a(\rho \vee \sigma)b$ if and only if $aE_Sb^{-1} \cap \ker\rho \neq \varnothing$.

(v) Let S be an inverse semigroup. Show that the σ-classes are precisely the sets $(aE_S)\omega$ for $a \in S$, and deduce that

$$a\sigma b \Leftrightarrow (aE_S)\omega = (bE_S)\omega.$$

(vi) Show that in 5.3, part (ii) may be replaced by $\rho \vee \mathcal{L} = \omega$ where the join is taken in the lattice of equivalence relations.

(vii) Let K be a closed normal subsemigroup of an inverse semigroup S. Show that the $\rho_{(K,\omega)}$-classes are precisely the sets $(Ka)\omega$ for $a \in S$.

(viii) Let I be an ideal of an inverse semigroup S. Show that $\sigma_S|_I = \sigma_I$.

(ix) Let S be an inverse semigroup, K be a normal subsemigroup of S, and τ be a normal congruence on E_S. Show that conditions (i) and (ii) in 1.3 are equivalent to

(α) $a\tau_{max} \cap K$ is closed in $a\tau_{max} \qquad (a \in S)$,

(β) $K \subseteq \ker \tau_{max}$,

respectively, where

$$a\tau_{max}b \Leftrightarrow a^{-1}ea\tau b^{-1}eb \quad \text{for all } e \in E_S.$$

(x) Let K be a closed inverse subsemigroup of an inverse semigroup S and let $H = K\sigma^{\#}$. Prove the following statements.

(α) If K is normal in S, then H is normal in S/σ.

(β) If H is normal in S/σ and K is closed, then K is a kernel in S.

(γ) If H is normal in S/σ, K need not be normal in S.

(xi) Let S be an inverse subsemigroup of $\mathcal{I}(X)$ for some nonempty set X, not containing $\varnothing$. Show that

$$\alpha\sigma\beta \Leftrightarrow \text{there exists } \gamma \in S \text{ such that } \mathbf{d}\gamma \subseteq \mathbf{d}\alpha \cap \mathbf{d}\beta$$

$$\text{and} \quad \alpha|_{\mathbf{d}\gamma} = \gamma = \beta|_{\mathbf{d}\gamma}.$$

Wagner [3] characterized the least group congruence by the last expression in 5.2(i). The widely used formula for σ in the first part of 5.2(i) was found by Munn [4]. Both 5.6 and 5.8 can be found in Petrich [11] along with related results. The handy formula 5.4(i) for $\rho \vee \sigma$ is due to Howie [1].

III.6 SEMILATTICE CONGRUENCES AND CLIFFORD CONGRUENCES

We start with some general properties of semilattice congruences. For a semilattice congruence ρ on an inverse semigroup S, we consider statements equivalent to the condition that S be a strong semilattice of its ρ-classes. We then characterize Clifford congruences and prove that the least such coincides with $\eta_{\min}$.

First note that a congruence ρ on an inverse semigroup S is a semilattice congruence if and only if $\ker \rho = S$. Hence semilattice congruences form the κ-class of ω; and they are greatest elements in their respective θ-classes.

III.6.1 Definition

A congruence ρ on a semigroup S such that S/ρ is a Clifford semigroup is a *Clifford congruence*. For any semigroup S, η denotes the least semilattice congruence on S; if necessary, we will write η_S instead of η.

We collect some of the properties of η in the following result.

III.6.2 Lemma

In any inverse semigroup,

$$\eta = \omega^{\min} = \mathcal{L}^* = \mathcal{R}^* = \mathcal{D}^* = \mathcal{J}^*.$$

Proof. The first equality follows from the above remarks. The second and the third equalities follow directly from 4.6 and its dual. For the remaining two equalities, it suffices to prove that $\mathcal{J} \subseteq \eta$. Indeed, let $a\mathcal{J}b$. Then $a = xby$ and $b = waz$ for some $x, y, w, z \in S$. Hence $a\eta = (x\eta)(b\eta)(y\eta) \leq b\eta$, and analogously $b\eta \leq a\eta$. Thus $a\eta = b\eta$, that is, $a\eta b$ and hence $\mathcal{J} \subseteq \eta$.

III.6.3 Proposition

The following statements concerning a congruence ρ on an inverse semigroup S are equivalent.

(i) ρ is a semilattice congruence.
(ii) $\mathcal{L} \subseteq \rho$.
(iii) $\rho\mathcal{C}\rho = \omega$.
(iv) $\rho\mathcal{F}\rho = \omega$.

Proof. The equivalence of (i) and (ii) follows from 6.2.

(i) *implies* (iii). For any $a, b \in S$, we have $a\rho aa^{-1}$, $aa^{-1}\mathcal{C}bb^{-1}$, $bb^{-1}\rho b$, so that $\rho\mathcal{C}\rho = \omega$.

(iii) *implies* (iv). This is obvious.

(iv) *implies* (i). For any $a \in S$, we have $aa^{-1}\rho x$, $x\mathcal{F}y$, and $y\rho a$ for some $x, y \in S$, whence $a = (aa^{-1})a\rho x^{-1}y \in E$ and thus ρ is a semilattice congruence.

We now turn to the question: for a given congruence ρ on an inverse semigroup S, when is S a strong semilattice of ρ-classes? The inverse semigroup S in 6.4 is regarded as a partially ordered set under its natural order.

III.6.4 Theorem

Let ρ be a semilattice congruence on an inverse semigroup S. Denote by S_α the ρ-classes, where $\alpha \in Y = S/\rho$. Then the following statements are equivalent.

(i) S is a strong semilattice of ρ-classes.
(ii) For every $\alpha \in Y$, $K_\alpha = \cup_{\beta \leq \alpha} S_\beta$ is a p-ideal of S.
(iii) E_S is a strong semilattice of $\operatorname{tr}\rho$-classes.
(iv) For every $\alpha \in Y$, $F_\alpha = \cup_{\beta \leq \alpha} E_{S_\beta}$ is a p-ideal of E_S.

Proof. The argument here amounts to an easy modification of that in the proof of II.4.4 and is left as an exercise.

III.6.5 Proposition

The following statements concerning a congruence ρ on an inverse semigroup S are equivalent.

(i) ρ is a Clifford congruence.
(ii) $\rho_{\max}$ is a semilattice congruence.
(iii) $\rho_{\max} = \rho \vee \eta$.

Proof. (i) *implies* (ii). For any $a \in S$ and $e \in E_S$, we have

$$a^{-1}ea\rho a^{-1}ae = (a^{-1}a)^{-1}e(a^{-1}a)$$

and thus $a\rho_{\max}a^{-1}a$. Hence $\rho_{\max}$ is a semilattice congruence.

(ii) *implies* (iii). The hypothesis implies that $\rho_{\max} \supseteq \rho \vee \eta$. Also

$$\ker \rho_{\max} \subseteq S = \ker \eta = \ker(\rho \vee \eta).$$

$$\operatorname{tr} \rho_{\max} = \operatorname{tr} \rho \subseteq \operatorname{tr}(\rho \vee \eta),$$

and thus by 1.5, we have $\rho_{\max} = \rho \vee \eta$.

(iii) *implies* (i). The hypothesis gives $\rho_{\max} \supseteq \eta$ so that $\rho_{\max}$ is a semilattice congruence. Hence for any $a \in S$, we have $a\rho_{\max}a^{-1}a$. Thus for any $e \in E_S$,

$$a^{-1}ea\rho(a^{-1}a)^{-1}e(a^{-1}a) = a^{-1}ae,$$

which multiplied by a on the left yields $ea\rho ae$. Hence ρ is a Clifford congruence.

III.6.6 Notation

For any inverse semigroup S, let ν (or ν_S) denote the least Clifford congruence on S. Observe that ν is the intersection of all Clifford congruences on S.

III.6.7 Proposition

In any inverse semigroup S, $\nu = \eta_{\min}$.

Proof. Since $(\eta_{\min})_{\max} = \eta_{\max}$ is a semilattice congruence, by 6.5, $\eta_{\min}$ is a Clifford congruence. If ρ is any Clifford congruence on S, then by 6.5, $\eta \subseteq \rho_{\max}$ and so by 2.8,

$$\eta_{\min} \subseteq (\rho_{\max})_{\min} = \rho_{\min} \subseteq \rho.$$

Consequently $\eta_{\min} = \nu$.

III.6.8 Corollary

In any inverse semigroup S,

$$\ker \nu = \{ a \in S | ae = e \text{ for some } e \in E_S, e\eta a \}.$$

Proof. The proof of this corollary is left as an exercise.

III.6.9 Exercises

(i) Show that the lattice of all semilattice congruences on an inverse semigroup need not be modular.

(ii) Let ρ be a congruence on an inverse semigroup S. Show that $\sigma_{S/\rho} = (\rho \vee \sigma)/\rho$, $\eta_{S/\rho} = (\rho \vee \eta)/\rho$, $\nu_{S/\rho} = (\rho \vee \nu)/\rho$.

(iii) Show that for any semilattice congruence ρ on an inverse semigroup S, $\rho_{\min} \vee \mathcal{L} = \rho_{\min} \mathcal{L} \rho_{\min} = \rho$. (The join is taken in the lattice of equivalence relations on S.)

(iv) Show that the following conditions on a congruence ρ on an inverse semigroup S are equivalent.

(α) $(a\rho)\eta(b\rho)$ implies $a\eta b \qquad (a, b \in S)$.

(β) $\rho \subseteq \eta$.

(γ) $\operatorname{tr} \rho \subseteq \operatorname{tr} \eta$.

(v) Let S be any semigroup and $\mathfrak{F}$ be a family of filters on S. (A subsemigroup F of S is a *filter* if for any $a, b \in S$, $ab \in F$ implies $a, b \in F$.) For each $F \in \mathfrak{F}$ let ρ_F be any group congruence on F, and define θ_F on S by

$$a\theta_F b \Leftrightarrow a, b \in F \quad \text{and} \quad a\rho_F b \quad \text{or} \quad a, b \notin F.$$

Show that $\theta = \cap_{F \in \mathfrak{F}} \theta_F$ is a Clifford congruence on S, and that conversely, every Clifford congruence on S can be so constructed.

(vi) Let S be an inverse semigroup and $\mathfrak{F}$ be the family of all filters of S. For each $F \in \mathfrak{F}$, let σ_F be the least group congruence on F, and define θ_F as in the preceding exercise. Show that $\cap_{F \in \mathfrak{F}} \theta_F$ is the least Clifford congruence on S.

(vii) Show that a congruence ρ on an inverse semigroup is a Clifford congruence if and only if $\operatorname{tr} \rho \supseteq \operatorname{tr} \eta$.

(viii) Show that for any congruence ρ on an inverse semigroup $\rho \vee \eta = (\rho \mathcal{L})^*$.

(ix) Show that the following conditions on an inverse semigroup S are equivalent.

(α) S is a Clifford semigroup.

(β) For every $\rho \in \mathcal{C}(S)$, $\rho_{\max}$ is a semilattice congruence.

(γ) For every $\rho \in \mathcal{C}(S)$, $\rho^{\min}$ is idempotent separating.

(δ) $\mu = \eta$.

(x) Show that the following statements concerning congruences ρ and ξ on an inverse semigroup S are equivalent.

(α) $\ker \rho = \ker \xi$.

(β) $\rho \subseteq \xi^{\max}, \qquad \xi^{\max}/\rho = \tau_{S/\rho}$.

(γ) $(a\rho)\tau(b\rho) \Leftrightarrow (a\xi)\tau(b\xi) \qquad (a, b \in S)$.

(δ) $(\rho \cap \xi)|_{e\rho}$ and $(\rho \cap \xi)|_{e\xi}$ are semilattice congruences ($e \in E_S$).

(ε) $\rho/\rho \cap \xi$ and $\xi/\rho \cap \xi$ are idempotent pure congruences. [cf. III.5.11(i).]

(xi) Let S be a semigroup in which each $\mathcal{J}$-class is a subsemigroup of S. Show that $\mathcal{J} = \eta$.

That $\eta = \mathcal{D}^* = \mathcal{J}^*$ was established by Howie-Lallement [1] even for regular semigroups. p-ideals were introduced by Reilly [9] in his study of the translational hull of an inverse semigroup. A version of 6.4 was proved by O'Carroll [7], whereas 6.5 can be found in Petrich [11]. Semilattices of inverse semigroups were studied by Preston [4].

III.7 *E*-UNITARY CONGRUENCES

We first introduce E-unitary inverse semigroups and then characterize E-unitary congruences in two ways. Then we consider the lattice of all E-unitary congruences on an inverse semigroup S. This is an $\cap$-sublattice of the lattice of all congruences on S, with least element $\sigma^{\min}$. The least E-unitary Clifford congruence on S is proved to coincide with $\sigma \cap \eta$.

III.7.1 Definition

An inverse semigroup S is *E-unitary* (or *proper* or *reduced*) if E_S is a closed subset of S.

The name stems from the fact that this definition is equivalent to the requirement that E_S be a unitary subset of S; we will not need this concept. The class of E-unitary inverse semigroups can be characterized in many ways, of which we now give a sample. Notice that $\ker \sigma = E_S\omega$.

III.7.2 Proposition

The following conditions on an inverse semigroup S are equivalent.

(i) S is E-unitary.
(ii) τ is a group congruence.
(iii) σ is an idempotent pure congruence.
(iv) $\sigma \cap \mathcal{L} = \varepsilon$.
(v) S satisfies the implication $xy = x \Rightarrow y = y^2$.

Proof. The proof of this proposition is left as an exercise.

We will use $\mathcal{R}$ instead of $\mathcal{L}$ in 7.2(iv) quite frequently. The next result characterizes E-unitary congruences in terms of more familiar notions.

III.7.3 Proposition

The following statements concerning a congruence ρ on an inverse semigroup S are equivalent.

(i) ρ is E-unitary.
(ii) $\ker \rho$ is closed.
(iii) $\rho^{\max}$ is a group congruence.
(iv) $\rho^{\max} = \rho \vee \sigma$.

Proof. (i) *implies* (ii). Let $a \in (\ker \rho)\omega$. Then $a \geq b$ for some $b \in \ker \rho$, which evidently implies that $ae \in \ker \rho$ for some $e \in E_S$. We get in S/ρ, $(a\rho)(e\rho) \in E_{S/\rho}$, and since S/ρ is E-unitary, this implies $a\rho \in E_{S/\rho}$. But then $a \in \ker \rho$. We have proved that $(\ker \rho)\omega \subseteq \ker \rho$, the opposite inclusion is trivial.

(ii) *implies* (iii). By 5.5, we have $\ker(\rho \vee \sigma) = (\ker \rho)\omega = \ker \rho$ and thus $\rho \vee \sigma \subseteq \rho^{\max}$ by the maximality of the latter. But then $\rho^{\max}$ must be a group congruence.

(iii) *implies* (iv). The argument here is similar to that in the proof of 6.5 and is left as an exercise.

(iv) *implies* (i). Assume that $ae\rho e$ for some $a \in S$ and $e \in E_S$. Then $\rho \subseteq \rho^{\max}$ implies that $ae\rho^{\max}e$. Since $\rho^{\max} \supseteq \sigma$, it is a group congruence, and this gives $a\rho^{\max}e$, which in turn implies that $a \in \ker \rho^{\max} = \ker \rho$. But then $a\rho f$ for some $f \in E_S$ which proves that ρ is an E-unitary congruence.

III.7.4 Theorem

Let $\mathcal{U}(S)$ denote the set of all E-unitary congruences on an inverse semigroup S ordered by inclusion. Then $\mathcal{U}(S)$ is a complete $\cap$-subsemilattice of $\mathcal{C}(S)$ with least element $\pi = \pi_S = (\sigma \cap \mathcal{L})^*$ and greatest element ω.

Proof. All the assertions in the first sentence, except the expression for π follow directly from I.11.14 since E-unitary inverse semigroups form a quasi-variety by 7.2.

By 4.8, it suffices to prove that $\pi = \sigma^{\min}$. Since $(\sigma^{\min})^{\max} = \sigma^{\max} = \sigma$, we have by 7.3 that $\sigma^{\min}$ is an E-unitary congruence and thus $\sigma^{\min} \supseteq \pi$. Also by 7.3, $\pi^{\max}$ is a group congruence and thus $\pi \subseteq \sigma^{\min} \subseteq \sigma \subseteq \pi^{\max}$. Hence $\ker \pi \subseteq \ker \sigma \subseteq \ker \pi$ so that $\ker \pi = \ker \sigma$, which implies $\sigma^{\min} = \pi^{\min} \subseteq \pi$. Consequently $\pi = \sigma^{\min}$.

III.7.5 Corollary

In any inverse semigroup S, $\pi = \sigma^{\min}$ and the interval $[\pi, \sigma]$ is a complete sublattice of $\mathcal{U}(S)$.

Proof. The first assertion was remarked in the proof of 7.4. If $\rho \in [\sigma, \pi]$, then $\ker \rho = \ker \sigma = \ker \pi$ since $\pi = \sigma^{\min}$, and thus 7.3 gives that ρ is an E-unitary congruence.

For the intersection $\sigma \cap \eta$, we have the following simple result.

III.7.6 Proposition

For any inverse semigroup S, $\sigma \cap \eta$ is the least E-unitary Clifford congruence on S.

Proof. Since $\ker(\sigma \cap \eta) = \ker \sigma \cap \ker \eta = \ker \sigma$, by 7.3 we have that $\sigma \cap \eta$ is an E-unitary congruence. It is also the intersection of two Clifford congruences, so it is a Clifford congruence. Let ρ be an E-unitary Clifford congruence on S. By 7.3 and 6.5, we obtain $\sigma \subseteq \rho^{\max}$ and $\eta \subseteq \rho_{\max}$, which by 4.12 yields $\sigma \cap \eta \subseteq \rho^{\max} \cap \rho_{\max} = \rho$.

III.7.7 Corollary

In any inverse semigroup $\sigma \cap \eta = \nu \vee \pi$.

Proof. The proof of this corollary is left as an exercise.

For any congruence ρ on an inverse semigroup S, there exists by 7.4 the least E-unitary congruence on S containing ρ. The next result gives a characterization of such a congruence.

III.7.8 Proposition

For any congruence ρ on an inverse semigroup S, $\pi_\rho = (\sigma\rho\sigma \cap \mathcal{L}\rho\mathcal{L})^*$ is the least E-unitary congruence on S containing ρ.

Proof. The existence of the least E-unitary congruence π_ρ on S containing ρ follows from 7.4. The correspondence of congruences on S containing ρ and congruences on S/ρ shows that for any $a, b \in S$,

$$a\pi_\rho b \Leftrightarrow (a\rho)\pi_{S/\rho}(b\rho).$$

By 7.4, we have an expression for $\pi_{S/\rho}$, hence it suffices to prove

$$a(\sigma\rho\sigma \cap \mathcal{L}\rho\mathcal{L})^* b \Leftrightarrow (a\rho)(\sigma_{S/\rho} \cap \mathcal{L}_{S/\rho})^*(b\rho),$$

and for this, it is enough to show

$$a(\sigma\rho\sigma \cap \mathcal{L}\rho\mathcal{L})b \Leftrightarrow (a\rho)(\sigma_{S/\rho} \cap \mathcal{L}_{S/\rho})(b\rho).$$

It is further enough to show the equivalences

$$a(\sigma\rho\sigma)b \Leftrightarrow (a\rho)\sigma_{S/\rho}(b\rho), \tag{5}$$

$$a(\mathcal{L}\rho\mathcal{L})b \Leftrightarrow (a\rho)\mathcal{L}_{S/\rho}(b\rho). \tag{6}$$

Let $a(\sigma\rho\sigma)b$. Then $a\sigma x\rho y\sigma b$ for some $x, y \in S$, whence $ae = xe$ and $yf = bf$ for some $e, f \in E_S$. Thus $ag\rho bg$ where $g = ef$ so that $(a\rho)(g\rho) = (b\rho)(g\rho)$ with $g\rho \in E_{S/\rho}$, and thus $(a\rho)\sigma_{S/\rho}(b\rho)$.

Conversely, let $(a\rho)\sigma_{S/\rho}(b\rho)$. Then for some $g \in E_S$, we have $(a\rho)(g\rho) = (b\rho)(g\rho)$ so that $ag\rho bg$. It follows that $a\sigma(ag)\rho(bg)\sigma b$ and thus $a(\sigma\rho\sigma)b$. This verifies (5). Formula (6) is checked similarly.

The next result contains useful information concerning the congruence σ on an E-unitary inverse semigroup including a dual to 5.6.

III.7.9 Theorem

The following statements are true in any E-unitary inverse semigroup S.

(i) $a\sigma b$ if and only if $ab^{-1}a = bb^{-1}a$.

(ii) $\rho_{\min} = \rho \cap \sigma$ for any $\rho \in \mathcal{C}(S)$.

(iii) The mapping

$$\chi : \rho \to \rho \cap \sigma \qquad [\rho \in \mathcal{C}(S)]$$

is a complete lattice homomorphism of $\mathcal{C}(S)$ onto the lattice of idempotent pure congruences on S.

Proof. (i) First let $a\sigma b$. Then $ae = be$ and also $fa = fb$ for some $e, f \in E_S$. Hence $b^{-1}ae = b^{-1}be \in E_S$ and thus $b^{-1}a \in E_S$. Similarly, $fa = fb$ implies

$ab^{-1} \in E_S$. We then obtain

$$ab^{-1}a = a(b^{-1}bb^{-1})a = (ab^{-1})(bb^{-1})a = b(b^{-1}a)(b^{-1}a) = bb^{-1}a.$$

Conversely, if $ab^{-1}a = bb^{-1}a$, then $(a\sigma)(b^{-1}a)\sigma = (b\sigma)(b^{-1}a)\sigma$ so that $a\sigma b$ by cancellation.

(ii) Let $a(\rho \cap \sigma)b$. By part (i), we have $ab^{-1}a = bb^{-1}a$. It follows that $e = b^{-1}a \in E_S$ and $e = (b^{-1}a)\rho(a^{-1}a)\rho(b^{-1}b)$, and thus $a\rho_{\min}b$. Consequently, $\rho \cap \sigma \subseteq \rho_{\min}$; the opposite inclusion always holds.

(iii) By 2.9, the mapping $\rho \to \rho_{\min}$ is a complete $\vee$-homomorphism. By part (ii), $\rho_{\min} = \rho \cap \sigma$, and thus χ is a complete $\vee$-homomorphism; χ is trivially a complete $\cap$-homomorphism. Since S is E-unitary, σ is pure, and hence $\rho \cap \sigma$ is also for any $\rho \in \mathcal{C}(S)$. If ρ is idempotent pure, then $\rho \subseteq \sigma$ and hence $\rho\chi = \rho$.

III.7.10 Exercises

(i) Show that the following conditions on an inverse semigroup S are equivalent.

(α) $\sigma \cap \mathcal{L}$ is a congruence.

(β) $\sigma \cap \mathcal{R}$ is a congruence.

(γ) $\sigma \cap \mathcal{L} = \sigma \cap \mathcal{R}$.

(δ) $\sigma \cap \mathcal{L} = \sigma \cap \mu$.

(ε) There exists an idempotent separating E-unitary congruence on S.

(ii) Show that the following conditions on an inverse semigroup S are equivalent.

(α) S is E-unitary.

(β) For every $\rho \in \mathcal{C}(S)$, $\rho_{\min}$ is idempotent pure.

(γ) $\pi = \varepsilon$.

(δ) $\sigma = \tau$.

(ε) For any $a, x \in S$, $x^{-1}ax \in E_S$ implies $a \in E_S$.

(iii) Let ρ be an idempotent pure congruence on an inverse semigroup S. Show that S is E-unitary if and only if S/ρ is E-unitary.

(iv) Show that for any congruence ρ on an inverse semigroup, $\pi_\rho \vee \sigma = \rho \vee \sigma$, $\ker \pi_\rho = (\ker \rho)\omega$ and $\pi_\rho = \pi_{\rho \vee \pi}$.

(v) Let $S = [Y, \eta, G]$ be as in II.2.13. Characterize all congruences on S. For an E-unitary congruence ρ on S, construct the quotient S/ρ.

(vi) Show that for any inverse semigroup S, $\sigma/\pi = \sigma_{S/\pi}$.

(vii) Let ρ and ξ be congruences on an inverse semigroup S. Show that $\operatorname{tr}\rho \subseteq \operatorname{tr}\xi$ implies $\rho \subseteq \xi_{\max}$.

(viii) Show that in any inverse semigroup S, $\tau_{\max}$ is the greatest congruence ρ on S for which $e\rho$ is E-unitary for every $e \in E_S$.

III.7.11 Problems

(i) Investigate the classes of the congruence $\hat{\sigma}$ induced by the homomorphism $\rho \to \rho \vee \sigma$ (see 5.6). (For each $\rho \in \mathcal{C}(S)$, we have $\rho \subseteq \pi_\rho \subseteq \rho \vee \sigma$, where all three congruences are $\hat{\sigma}$-related and the last one is the greatest element of its class.) Do $\hat{\sigma}$-classes have least elements?

(ii) Characterize inverse semigroups all of whose homomorphic images are E-unitary (that is, all its kernels are closed).

(iii) If we compare 6.5 with 7.3, we see that we do not have an analogue of "$\ker\rho$ is closed" for traces. On the lattice $\mathfrak{N}(E_S)$ of normal congruences on E_S for an inverse semigroup S, we may define a closure operator by $\tau \to \tau \vee \operatorname{tr}\eta$. Then a congruence ρ on S is a Clifford congruence if and only if $\operatorname{tr}\rho$ is closed under this operator. We may go one step further and ask if there exists a partial order $\leq$ on $E_S \times E_S$ such that for any $\tau \in \mathfrak{N}(E_S)$,

$$e(\tau \vee \operatorname{tr}\eta)f \Leftrightarrow \text{there exist } e', f' \in E_S \text{ such that } (e', f') \leq (e, f)?$$

E-unitary inverse semigroups were first studied by Saitô [2] who called them proper. They were further investigated by McAlister [4], [5] and O'Carroll [1], [4] under the label of reduced inverse semigroups. The term E-unitary is now generally accepted.

The paper of Reilly–Munn [1] is entirely devoted to E-unitary congruences on an arbitrary inverse semigroup; 7.4 and a part of 7.5 is due to them, except for the form $(\sigma \cap \mathcal{L})^*$ for π which was discovered by O'Carroll [1]. The formula for π_ρ in 7.8 is due to Petrich–Reilly [3]. Theorem 7.9(iii) is due to P. R. Jones.

Chapter VII will be devoted entirely to the structure of E-unitary inverse semigroups.

III.8 *E*-REFLEXIVE CONGRUENCES

After introducing E-reflexive inverse semigroups, we provide some equivalent conditions in terms of implications as well as congruences. We then characterize E-reflexive congruences on an inverse semigroup S and prove that they form a complete $\cap$-sublattice of the lattice of all congruences on S with least element $\nu^{\min}$.

III.8.1 Definition

An inverse semigroup S is *E-reflexive* (or *strongly E-reflexive*) if for any $x, y \in S$ and $e \in E$, $xey \in E_S$ implies $yex \in E_S$.

It will be expedient to have some equivalent ways of defining E-reflexivity for the discussion below. The following lemma will be used without express mention.

III.8.2 Lemma

Let S be an inverse semigroup with semilattice E of idempotents. Then the following statements are equivalent.

(i) S is E-reflexive.
(ii) $exy \in E \Rightarrow eyx \in E \quad (e \in E^1,\ x, y \in S)$.
(iii) $exy \in E \Rightarrow eyx \in E \quad (e \in E,\ x, y \in S)$.

Proof. We will abbreviate the statement "$a \in E$ implies $b \in E$" by $a \to b$.
(i) *implies* (ii). Let $x, y \in S$. Then

$$xy = x(x^{-1}x)y \to y(x^{-1}x)x = y(y^{-1}yx^{-1}x)x \to x(y^{-1}yx^{-1}x)y$$

$$= x(x^{-1}xy^{-1}y)y = x(y^{-1}y)y \to y(y^{-1}y)x = yx$$

and if also $e \in E$, then using what we have just proved,

$$(ex)y \to y(ex) \to x(ey) \to (ey)x$$

which proves (ii).
(ii) *implies* (iii) trivially.
(iii) *implies* (ii). For $x, y \in S$ such that $xy \in E$, we obtain

$$xy \to (x^{-1}x)(xy) \to (x^{-1}x)(yx) = (x^{-1}x)(yy^{-1})(yx)$$

$$= (yy^{-1})(x^{-1}x)(yx) \to (yy^{-1})(yx)(x^{-1}x) = yx.$$

(ii) *implies* (i). For $x, y \in S$ and $e, xey \in E$, we have

$$x(ey) \to eyx \to (ex)y \to yex$$

as required.

Compare the following result with 7.2.

III.8.3 Theorem

The following conditions on an inverse semigroup S are equivalent.

(i) S is E-reflexive.
(ii) τ is a Clifford congruence.
(iii) ν is idempotent pure.
(iv) Every η-class of S is E-unitary.

Proof. (i) *implies* (ii). Let $a \in S$, $x, y \in S^1$ and $e \in E_S$. With the notation as in the proof of 8.2, we have

$$\begin{aligned} xae &\to axe \to eax \to xea, \\ aey &\to yae \to aye \to eay, \\ xaey &\to yxae \to ayxe \to xeay. \end{aligned}$$

By symmetry, we conclude that $ea\tau ae$, which proves that τ is a Clifford congruence.

(ii) *implies* (iii). By the minimality of ν, we have $\nu \subseteq \tau$. Hence $\ker \nu \subseteq \ker \tau = E_S$ so ν is idempotent pure.

(iii) *implies* (iv). We may write $S/\nu = [Y; G_\alpha, \varphi_{\alpha,\beta}]$. Let π be the natural homomorphism of S/ν onto Y. Then $\chi = \nu^{\#}\pi$ is a homomorphism of S onto the semilattice Y; for each $\alpha \in Y$, let $S_\alpha = \alpha\chi^{-1}$. It is clear that for every $\alpha \in Y$, $S_\alpha = G_\alpha \nu^{\#-1}$. Now let $a \in S_\alpha$ and $e \in E_{S_\alpha}$ be such that $ae = e$. Then $(a\nu)(e\nu) = e\nu$ in G_α where $e\nu$ is the identity of G_α, whence $a\nu = e\nu$, that is $a\nu e$. Since ν is idempotent pure, we get $a \in E_{S_\alpha}$ and S_α is E-unitary. Letting ρ be the congruence induced by χ, we obtain $\eta \subseteq \rho$ since ρ is a semilattice congruence. Since every ρ-class is E-unitary, this must also hold for every η-class.

(iv) *implies* (i). Let $x, y \in S$, $e \in E_S$ be such that $xey \in E_S$. Then

$$(x^{-1}xeyy^{-1})(yex) = x^{-1}(xey)ex \in E_S$$

where $x^{-1}xeyy^{-1}\eta yex$ so that $yex \in E_S$. Hence S is E-reflexive.

Note that 8.3 implies that E-reflexive inverse semigroups coincide with semigroups which are semilattices of E-unitary inverse semigroups. E-reflexive congruences on an arbitrary inverse semigroup admit themselves some interesting characterizations, as we now show; cf. 7.3.

III.8.4 Proposition

The following statements concerning a congruence ρ on an inverse semigroup S are equivalent.

(i) ρ is E-reflexive.
(ii) $\ker \rho \cap N$ is closed in N for every η-class N of S.
(iii) $\rho^{\max}$ is a Clifford congruence.

Proof. (i) *implies* (ii). Let N be an η-class of S and a be an element of the closure of $\ker \rho \cap N$ in N. Then $ae \in \ker \rho$ for some $e \in E_S$ such that $a\eta e$. Denote by $\bar{S}$ the semigroup S/ρ, and consider the least semilattice congruence $\bar{\eta}$ on $\bar{S}$. The relation θ defined on S by

$$a\theta b \Leftrightarrow (a\rho)\bar{\eta}(b\rho)$$

is easily seen to be a semilattice congruence on S. Hence $\eta \subseteq \theta$ and thus $a\eta e$ implies $(a\rho)\bar{\eta}(e\rho)$. By 8.3, each $\bar{\eta}$-class is an E-unitary inverse semigroup. We also have $(a\rho)(e\rho) \in E_{\bar{N}}$ since $ae \in \ker \rho$. Consequently, $a\rho \in E_{\bar{N}}$ which then gives $a \in \ker \rho$. We have shown that $\ker \rho \cap N$ is closed in N.

(ii) *implies* (i). In the light of 7.3, $\rho|_N$ is an E-unitary congruence for each η-class N of S. Let $\lambda = \rho \cap \eta$. Then S/λ is a semilattice of E-unitary inverse semigroups. It follows from 8.3 that S/λ is E-reflexive. Now let $x, y \in S$, $e \in E_S$ be such that $xey \in \ker \rho$. Then $xey \in \ker \rho \cap \ker \eta = \ker(\rho \cap \eta)$ and hence $(x\lambda)(e\lambda)(y\lambda) \in E_{S/\lambda}$, which then implies $(y\lambda)(e\lambda)(x\lambda) \in E_{S/\lambda}$. But then $yex \in \ker \lambda = \ker \rho$. Consequently, in S/ρ, $(x\rho)(e\rho)(y\rho) \in E_{S/\rho}$ implies $(y\rho)(e\rho)(x\rho) \in E_{S/\rho}$, and thus ρ is an E-reflexive congruence.

(i) *is equivalent to* (iii). According to 8.3,

$$S/\rho \text{ is } E\text{-reflexive} \Leftrightarrow \tau_{S/\rho} \text{ is a Clifford congruence}$$

$$\Leftrightarrow ((ae)\rho)\tau((ea)\rho) \qquad (a \in S, e \in E_S)$$

$$\Leftrightarrow \big[(x\rho)(ae)\rho(y\rho) \in E_{S/\rho} \Leftrightarrow (x\rho)(ea)\rho(y\rho) \in E_{S/\rho}$$
$$(x, y \in S^1, a \in S, e \in E_S)\big]$$

$$\Leftrightarrow \big[(xaey)\rho \in E_{S/\rho} \Leftrightarrow xeay \in E_{S/\rho} \ (x, y \in S^1, a \in S, e \in E_S)\big]$$

$$\Leftrightarrow \big[xaey \in \ker \rho \Leftrightarrow xeay \in \ker \rho \ (x, y \in S^1, a \in S, e \in E_S)\big]$$

$$\Leftrightarrow ae\tau^{\ker \rho}ea \qquad (a \in S, e \in E_S)$$

$$\Leftrightarrow ae\rho^{max}ea \qquad (a \in S, e \in E_S)$$

$$\Leftrightarrow \rho^{\max} \text{ is a Clifford congruence.}$$

We now turn to the set of all E-reflexive congruences on an inverse semigroup.

III.8.5 Theorem

Let $\mathfrak{R}(S)$ denote the set of all E-reflexive congruences on an inverse semigroup S ordered by inclusion. Then $\mathfrak{R}(S)$ is a complete $\cap$-subsemilattice of $\mathcal{C}(S)$ with least element $\lambda = \lambda_S = (\nu \cap \mathfrak{L})^*$ and greatest element ω.

Proof. As in the proof of 7.4, only the assertions about λ must be established. By 4.8, it suffices to prove that $\lambda = \nu^{\min}$. In a Clifford semigroup, τ is a Clifford congruence which by 8.3 gives that ν is an E-reflexive congruence. Since $(\nu^{\min})^{\max} = \nu^{\max}$, 8.4 implies that $\nu^{\min}$ is E-reflexive. Also by 8.4, $\lambda^{\max}$ is a Clifford congruence. Hence

$$\lambda \subseteq \nu^{\min} \subseteq \nu \subseteq \lambda^{\max}$$

so that $\ker \lambda \subseteq \ker \nu \subseteq \ker \lambda$ and thus $\ker \lambda = \ker \nu$. Consequently $\lambda = \nu^{\min}$.

III.8.6 Corollary

In any inverse semigroup S, $\lambda = \nu^{\min}$ and the interval $[\lambda, \nu]$ is a complete sublattice of $\mathfrak{R}(S)$.

Proof. The argument here goes along the same lines as in 7.5 and is left as an exercise.

For subdirect products of a Clifford semigroup and an E-unitary semigroup, we have the following characterization.

III.8.7 Proposition

An inverse semigroup S is a subdirect product of a Clifford semigroup and an E-unitary inverse semigroup if and only if S is E-reflexive and $\sigma \cap \mathfrak{L}$ is a congruence.

Proof. Let S be a subdirect product of a Clifford semigroup T and an E-unitary inverse semigroup K. Now both T and K are E-reflexive by 8.3. In T, $\mathfrak{L} = \eta$ and hence $\sigma \cap \mathfrak{L}$ is a congruence. By 7.2, $\sigma \cap \mathfrak{L} = \varepsilon$ in K. Thus $\sigma \cap \mathfrak{L}$ is a congruence in S.

Conversely, assume that S is E-reflexive and $\sigma \cap \mathfrak{L}$ is a congruence on S. Since S is E-reflexive, by 8.3, ν is idempotent pure. By 7.4 and the hypothesis, we obtain

$$\pi = (\sigma \cap \mathfrak{L})^* = \sigma \cap \mathfrak{L} \subseteq \mathfrak{L}$$

so π is idempotent separating. But then $\nu \cap \pi = \varepsilon$, and by I.4.18, S is a subdirect product of S/ν and S/π having all the requisite properties.

III.8.8 Review

Extremal values for the kernel and the trace, see 3.2, 4.2, 5.3, and 6.3. The congruences of the type in the left-hand column are characterized by means of $\mathfrak{F}$, $\mathfrak{L}$ and the kernel or the trace in the remaining columns.

idempotent separating	$\rho \cap \mathfrak{F} = \varepsilon$	$\rho \subseteq \mathfrak{L}$	$\operatorname{tr} \rho = \varepsilon$
idempotent pure	$\rho \subseteq \mathfrak{F}$	$\rho \cap \mathfrak{L} = \varepsilon$	$\ker \rho = E$
group	$\mathfrak{F} \subseteq \rho$	$\rho \mathfrak{L} \rho = \omega$	$\operatorname{tr} \rho = \omega$
semilattice	$\rho \mathfrak{F} \rho = \omega$	$\mathfrak{L} \subseteq \rho$	$\ker \rho = S$

In the next diagram the following procedure is indicated. On a fixed inverse semigroup S, starting with the universal relation ω, we successively apply the operation of upper and lower min thereby obtaining the sequences

$$\omega, \omega^{\min}, (\omega^{\min})_{\min}, \ldots, \qquad \omega, \omega_{\min}, (\omega_{\min})^{\min}, \ldots.$$

Each of the congruences so obtained can be expressed by means of the relations $\mathfrak{F}$ and $\mathfrak{L}$ in view of the formulae for $\rho_{\min}$ and $\rho^{\min}$. In the uppermost part of the diagram, we have the θ-class of group congruences and the κ-class of semilattice congruences. The phrase "*idem. sep. homo. im.*" should be interpreted as follows. For any congruences ρ and ξ such that $\nu \subseteq \rho \subseteq \xi \subseteq \eta$, we have that S/ξ is an idempotent separating homomorphic image of S/ρ. *Idem. pure homo. im.* has an analogous interpretation.

We may equally well start with the equality relation ε and successively perform the operation of upper and lower max thereby obtaining the sequences

$$\varepsilon, \varepsilon^{\max}, (\varepsilon^{\max})_{\max}, \ldots, \qquad \varepsilon_{\max}, (\varepsilon_{\max})^{\max}, \ldots.$$

Some of the statements above have their analogues for this part of the diagram. However, the lower part of the diagram does not seem to have so many interesting features as the upper part.

III.8.9 Diagram

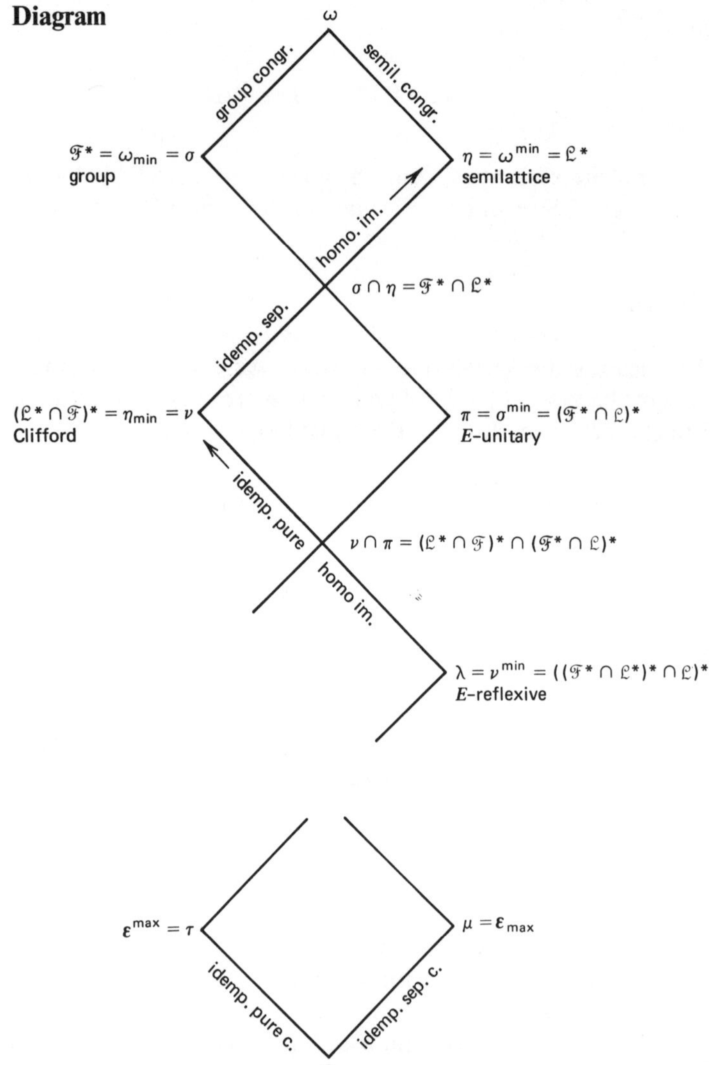

III.8.10 Coincidences

In the (α, β) position are necessary and sufficient conditions in order that the congruence in α and the congruence in β coincide. This creates many interesting classes of inverse semigroups. There is no information available for the blank spaces below the main diagonal.

	ω	σ	η	ν	π	λ	μ	τ
σ	$E\omega = S$							
η	no c. pr. ideals	$E\omega = S$ no c. pr. ideals						
ν	$\sigma = \eta = \omega$	no c. pr. ideals	$E_A\omega = A$ $\forall \eta -$ cl. A					
π	$\sigma = \eta = \omega$	tr $\pi = \omega$						
λ	$\sigma = \eta = \omega$							
μ	group	trivial	Clifford	semil.	$E\omega = E\zeta$ tr $\pi = \varepsilon$			
τ	semil.	*E*-un.	trivial	*E*-refl. tr $\tau =$ tr η	*E*-un. *E*-disj.	*E*-refl. tr $\tau =$ tr λ	*E*-disj. antig.	
ε	trivial	group	semil.	Clifford	*E*-un.	*E*-refl.	antig.	*E*-disj.

III.8.11 Exercises

(i) Show that an inverse semigroup S is *E*-reflexive if and only if it satisfies the implication

$$xy^2z = xyz \in E_S \Rightarrow zyx \in E_S \qquad (x, y, z \in S).$$

(ii) Call a nonempty subset C of a semigroup S *completely reflective* if for any $a_1, a_2, a_3 \in S^1$ and any permutation π of $\{1, 2, 3\}$, $a_1a_2a_3 \in C$ implies $a_{1\pi}a_{2\pi}a_{3\pi} \in C$. Prove that if for all $a, b, c \in S$, $abc \in C$ implies $cba \in C$ and $ab \in C$ implies $ba \in C$, then C is completely reflective.

(iii) Let S be an inverse semigroup. Show that E_S is completely reflective if and only if the least Clifford congruence ν on S is idempotent pure and S/ν is commutative.

(iv) Let ρ be a congruence on an inverse semigroup S. Show that if $\ker \rho = \ker \nu$, then ρ is *E*-reflexive.

(v) Show that neither of the conditions *E*-reflexivity and $E\omega \subseteq E\zeta$ implies the other.

(vi) Show that the statements in each column below are equivalent in any inverse semigroup.

$\nu = \varepsilon$	$\sigma = \varepsilon$	$\eta = \varepsilon$	$\pi = \varepsilon$
$\mu^{\max} = \omega$	$\mu = \omega$	$\tau = \omega$	$\tau_{\max} = \omega$
$\mu = \eta$			$\tau = \sigma$
$\eta \cap \mathcal{C} = \varepsilon$	$\mathcal{C} = \varepsilon$	$\mathcal{L} = \varepsilon$	$\sigma \cap \mathcal{L} = \varepsilon$
	$\mathcal{L} = \omega$	$\mathcal{C} = \omega$	

(vii) Prove the assertions contained in 8.10 [some were already established in 6.9(ix), 7.10(ii), and 8.11(vii)].

E-reflexive inverse semigroups were introduced by O'Carroll [8] who called them strongly *E*-reflexive. He used the term "*E*-reflexive" for the class of inverse semigroups satisfying $ab \in E \Leftrightarrow ba \in E$; this class seems to play no appreciable role. He continued the study of *E*-reflexive semigroups and congruences in O'Carroll [9], [10]. The important characterization 8.3 as well as the expression $(\nu \cap \mathcal{L})^*$ for λ are due to O'Carroll [8]. Properties of congruences obtained by starting with ω and successively forming $\rho_{\min}$ and $\rho^{\min}$, as in 8.9, were studied by Petrich–Reilly [3]. For *E*-reflexive congruences see Hardy-Tirasupa [1].

IV
REPRESENTATIONS

As elsewhere in algebra, representations serve to obtain some information about the object under study by considering homomorphisms from it into certain objects of more familiar structure. In the case of inverse semigroups, the most natural representations are those by one-to-one partial transformations of a set. This is in obvious analogy with groups and permutation groups. There is, however, an essentially new element present in the case of inverse semigroups, namely the presence of idempotents.

1. The analogue of the Cayley theorem in group theory is the Wagner representation of an inverse semigroup S by one-to-one partial transformations of a set. This representation is faithful and the domains of the partial transformations which make up the image of S are principal left ideals.

2. A different kind of representation of an inverse semigroup S is the Munn representation. It is a homomorphism of S into the semigroup of isomorphisms among principal ideals of the semilattice of idempotents of S. The congruence it induces is the greatest idempotent separating congruence on S. Semigroups in which this congruence is the equality relation are called antigroups; they play an important role in many considerations. Transitive and subtransitive representations occur naturally in the context of the Munn representation. Transitivity properties of the image of S under the Munn representation can be used to characterize properties of S such as bisimplicity, simplicity, and so on. The semilattice of idempotents of inverse semigroups S having these properties can be characterized abstractly.

3. Criteria can be given for an inverse semigroup to be congruence-free by means of concepts used in the context of the Munn representation and a rather involved condition. These conditions simplify in the case when the semigroup in question has a zero.

4. The general theory of representations of inverse semigroups by one-to-one partial transformations bifurcates into effective and noneffective ones. The

former splits into a sum of transitive ones. Each transitive representation is further equivalent to one which can be formed by means of a closed inverse subsemigroup of the inverse semigroup represented. This latter is a generalization of the representation of a group by permutations of right cosets of a subgroup.

5. With every subgroup of an inverse semigroup S there is associated a transitive representation of S. These representations enjoy some special properties. The general representation theory can be used to establish a result on extending a representation from an inverse subsemigroup U of S to the entire S. One can pass from a representation of an inverse semigroup by full transformations of a set to one by one-to-one partial transformations, and conversely.

IV.1 THE WAGNER REPRESENTATION

We start with the definition and basic properties of the symmetric inverse semigroup on a set X consisting of all one-to-one partial transformations of X with the composition of partial transformations. Then we establish the Wagner representation of an arbitrary inverse semigroup; it is a faithful representation by one-to-one partial transformations of a set.

IV.1.1 Definition

A partial transformation of a set X which is one-to-one on its domain is a *one-to-one partial transformation* of X. The set of *all one-to-one partial transformations of* X with the composition of partial transformations written on the right (respectively left) is denoted by $\mathcal{I}(X)$ [respectively $\mathcal{I}'(X)$]. Let

$$\mathcal{I}_0(X) = \{\alpha \in \mathcal{I}(X) | \operatorname{rank} \alpha \leq 1\},$$

$$\mathcal{I}_0'(X) = \{\alpha \in \mathcal{I}'(X) | \operatorname{rank} \alpha \leq 1\}.$$

Note that the empty transformation is a one-to-one partial transformation of any set X. Just as with partial transformations, the product in $\mathcal{I}(X)$ agrees with the usual composition of binary relations, whereas the product in $\mathcal{I}'(X)$ has the "opposite" multiplication. We hasten to establish some basic properties of $\mathcal{I}(X)$.

IV.1.2 Proposition

Let X be any set, $\alpha, \beta \in \mathcal{I}(X)$. Then the following statements are valid in the semigroup $\mathcal{I}(X)$.

(i) $\mathbf{d}(\alpha\beta) = (\mathbf{r}\alpha \cap \mathbf{d}\beta)\alpha^{-1}$, $\mathbf{r}(\alpha\beta) = (\mathbf{r}\alpha \cap \mathbf{d}\beta)\beta$.

(ii) $E_{\mathcal{I}(X)} = \{\iota_A | A \subseteq X\}$.

(iii) For any $A, B \subseteq X$, $\iota_A \iota_B = \iota_{A \cap B}$.

(iv) $\mathcal{I}(X)$ is an inverse semigroup with α^{-1} the usual inverse function for any $\alpha \in \mathcal{I}(X)$.

(v) $\alpha \mathcal{R} \beta \Leftrightarrow \mathbf{d}\alpha = \mathbf{d}\beta$.

(vi) $\alpha \mathcal{L} \beta \Leftrightarrow \mathbf{r}\alpha = \mathbf{r}\beta$.

(vii) $\alpha \mathcal{D} \beta \Leftrightarrow \alpha \mathcal{J} \beta \Leftrightarrow \text{rank } \alpha = \text{rank } \beta$.

(viii) $\alpha \leq \beta \Leftrightarrow \alpha \subseteq \beta$ as binary relations.

Proof. (i) This follows directly from the definition of the product $\alpha\beta$.

(ii) It is obvious that ι_A is an idempotent for any $A \subseteq X$. If $\alpha \in E_{\mathcal{I}(X)}$, then part (i) shows that $\mathbf{r}\alpha = \mathbf{d}\alpha$ since $\mathbf{d}\alpha = \mathbf{d}\alpha^2$, and for any $x \in \mathbf{d}\alpha$, $x\alpha = x\alpha^2 = (x\alpha)\alpha$, and thus $\alpha = \iota_{\mathbf{d}\alpha}$.

(iii) This is obvious.

(iv) Denoting by α^{-1} the usual inverse of a function α in $\mathcal{I}(X)$, we note that $\alpha = \alpha\alpha^{-1}\alpha$, so $\mathcal{I}(X)$ is a regular semigroup. In view of parts (ii) and (iii), its idempotents commute, and it is thus an inverse semigroup. In addition, α^{-1} is the unique inverse of α in $\mathcal{I}(X)$.

(v) If $\alpha \mathcal{R} \beta$, then $\alpha = \beta\gamma$ for some $\gamma \in \mathcal{I}(X)$ whence $\mathbf{d}\alpha \subseteq \mathbf{d}\beta$ and thus by symmetry, $\mathbf{d}\alpha = \mathbf{d}\beta$. Conversely, if $\mathbf{d}\alpha = \mathbf{d}\beta$, then $\alpha = \beta(\beta^{-1}\alpha)$ and $\beta = \alpha(\alpha^{-1}\beta)$ so that $\alpha \mathcal{R} \beta$.

(vi) This is essentially symmetric to part (v).

(vii) In any semigroup $\mathcal{D} \subseteq \mathcal{J}$. For any $\alpha, \beta \in \mathcal{I}(X)$, we clearly have

$$\text{rank}(\alpha\beta) \leq \min\{\text{rank } \alpha, \text{rank } \beta\}.$$

Hence if $\alpha \mathcal{J} \beta$, then $\alpha = \gamma\beta\delta$ for some $\gamma, \delta \in \mathcal{I}(X)$, and thus rank $\alpha \leq$ rank β, and by symmetry, rank α = rank β. Suppose that rank α = rank β, and let γ be a one-to-one mapping of $\mathbf{d}\alpha$ onto $\mathbf{r}\beta$. Then $\mathbf{d}\gamma = \mathbf{d}\alpha$ and $\mathbf{r}\gamma = \mathbf{r}\beta$, where $\gamma \in \mathcal{I}(X)$, so by parts (v) and (vi), we get $\alpha \mathcal{R} \gamma$ and $\gamma \mathcal{L} \beta$ so that $\alpha \mathcal{D} \beta$.

(viii) Assume first that $\alpha \leq \beta$. Then $\alpha = \iota_A\beta$ for some $A \subseteq X$. Hence by part (i), $\mathbf{d}\alpha = (\mathbf{r}\iota_A \cap \mathbf{d}\beta)\iota_A^{-1} = A \cap \mathbf{d}\beta$ and for any $x \in \mathbf{d}\alpha$, $x\alpha = x\iota_A\beta = x\beta$, which gives $\alpha \subseteq \beta$. Conversely, if $\alpha \subseteq \beta$, then a similar argument shows that $\alpha = \iota_{\mathbf{d}\alpha}\beta$ so that $\alpha \leq \beta$.

IV.1.3 Remark

Statements analogous to the ones above are valid for $\mathcal{I}'(X)$ and its elements. In particular, for $\mathcal{I}'(X)$:

$$\mathbf{d}(\alpha\beta) = \beta^{-1}(\mathbf{d}\alpha \cap \mathbf{r}\beta), \qquad \mathbf{r}(\alpha\beta) = \alpha(\mathbf{d}\alpha \cap \mathbf{r}\beta),$$

$$\alpha \mathcal{R} \beta \Leftrightarrow \mathbf{r}\alpha = \mathbf{r}\beta, \qquad \alpha \mathcal{L} \beta \Leftrightarrow \mathbf{d}\alpha = \mathbf{d}\beta,$$

and the rest is the same.

IV.1.4 Corollary

Let X be a set and S be an inverse subsemigroup of $\mathcal{I}(X)$. Then the natural order of S coincides with the inclusion relation of elements of S as binary relations on X.

Proof. This follows directly from 1.2(viii) in view of II.1.6.

IV.1.5 Definition

For any set X, both $\mathcal{I}(X)$ and $\mathcal{I}'(X)$ are referred to as the *symmetric inverse semigroup on X.*

Observe that the identity mapping on $\mathcal{I}(X)$ is an antiisomorphism of $\mathcal{I}(X)$ onto $\mathcal{I}'(X)$. The main result of this section follows.

IV.1.6 Theorem

Let S be an inverse semigroup. For each $a \in S$, let

$$w^a: x \to xa \qquad (x \in \mathbf{d}w^a = Sa^{-1}).$$

Then the mapping

$$w: a \to w^a \qquad (a \in S)$$

is a monomorphism of S into $\mathcal{I}(S)$.

Proof. First observe that $Sa^{-1} = Saa^{-1}$ so that $\mathbf{d}w^a$ consists of those elements of S having aa^{-1} as a right identity. If $x, y \in Sa^{-1}$ with $xa = ya$, then $xaa^{-1} = yaa^{-1}$ and thus $x = y$. Hence w maps S into $\mathcal{I}(S)$. Let $a, b \in S$. If $x \in \mathbf{d}w^{ab}$, then $x = x(ab)(ab)^{-1} = xabb^{-1}a^{-1}$, which evidently implies that $x = xaa^{-1}$ and $xa = xabb^{-1}$ so that $x \in \mathbf{d}(w^a w^b)$. Conversely, let $x \in \mathbf{d}(w^a w^b)$. Then $x = xaa^{-1}$ and $xa = xabb^{-1}$. Hence

$$x = xaa^{-1} = xabb^{-1}a^{-1} = x(ab)(ab)^{-1}$$

and thus $x \in \mathbf{d}w^{ab}$. Consequently, $\mathbf{d}(w^a w^b) = \mathbf{d}w^{ab}$, which clearly yields that $w^a w^b = w^{ab}$, and w is a homomorphism.

Assume that $w^a = w^b$. Then $Saa^{-1} = Sbb^{-1}$, which implies that each of aa^{-1} and bb^{-1} is a right identity of the other so that $aa^{-1} = bb^{-1}$. Since $aa^{-1} \in Sa^{-1}$, it follows that $aa^{-1}a = aa^{-1}b$, which implies $a = aa^{-1}b = bb^{-1}b = b$. Hence w is one-to-one and so a monomorphism of S into $\mathcal{I}(S)$.

We now introduce the concepts central in the context of this chapter.

IV.1.7 Definition

For any semigroup S and set X, a homomorphism $\varphi: S \to \mathcal{I}(X)$ [or $\mathcal{I}'(X)$] is a *representation of S by one-to-one partial transformations of X*; if φ is also one-to-one, it is *faithful*.

IV.1.8 Definition

The representation $w: S \to \mathcal{I}(S)$ in 1.6 of an inverse semigroup S is the *Wagner representation* of S. We will use the notation $w: a \to w^a$ consistently and denote the image of S under w by S^w.

As a consequence of 1.6, in the language of representations, we have the following fundamental result.

IV.1.9 Corollary

Every inverse semigroup admits a faithful representation by one-to-one partial transformations of a set.

This is a "faithful" analogue of the Cayley theorem in group theory. Hence the one-to-one partial transformations of a set bear the same relationship to inverse semigroups as do permutations to groups.

IV.1.10 Corollary

For any elements a and b of an inverse semigroup S, we have

$$a \leq b \Leftrightarrow w^a \subseteq w^b.$$

Proof. Indeed,

$$a \leq b \Leftrightarrow w^a \leq w^b \Leftrightarrow w^a \subseteq w^b$$

where the first equivalence follows from 1.6 and the second from 1.2(viii).

IV.1.11 Exercises

(i) Characterize maximal subgroups of $\mathcal{I}(X)$ for any set X.

(ii) Let α and β be partial transformations of the set $\mathbb{N}$ of positive integers defined by

$$n\alpha = n + 1 \quad \text{if } n \geq 1,$$

$$n\beta = n - 1 \quad \text{if } n > 1.$$

Show that the semigroup generated by α and β is a bicyclic subsemigroup of $\mathcal{I}(\mathbb{N})$.

(iii) In the preceding exercise extend β to a full transformation γ of $\mathbb{N}$ by letting $1\gamma = 1$. Show that the semigroup generated by α and γ is a bicyclic subsemigroup of $\mathcal{T}(\mathbb{N})$.

(iv) Show that for w^a in any inverse semigroup S, we have

$$\mathbf{d}w^a = \{s \in S \mid s\mathcal{R}sa\}.$$

(v) Let V be a vector space. Let S be the set of all linear isomorphisms between subspaces of V with the operation of composition of partial transformations. Show that S is an inverse semigroup and characterize its Green's relations.

(vi) For which inverse semigroups S is $\mathbf{d}w^a = S$ for all $a \in S$?

(vii) Find the idealizer of $\mathcal{I}_0(X)$ in $\mathcal{F}(X)$.

(viii) For a set X of n elements, prove that $\mathcal{I}(X)$ has $\Sigma_{i=0}^{n}\binom{n}{i}^2 i!$ elements.

IV.1.12 Problems

(i) For an inverse semigroup S, does the closure of S^w in $\mathcal{I}(S)$ have some interesting properties?

The Wagner representation appears for the first time in Wagner [2]. A little later, it was rediscovered by Preston [3]. It appears in the literature also under the names of Vagner–Preston, Preston–Vagner, and Preston representation. Interesting interpretations of the Wagner representation appear in Reilly [6], [14]. A generalization of the Wagner representation is due to McAlister [2]. The pleasing result 1.10 is no accident; Wagner [2] introduced the natural partial order on an arbitrary inverse semigroup by the requirement that the statement in 1.10 hold. This strips the concept of the natural partial order on an inverse semigroup of its apparent arbitrariness. Also consult Sribala [1].

IV.2 THE MUNN REPRESENTATION

This representation is a homomorphism of an arbitrary inverse semigroup into the semigroup of isomorphisms among principal ideals of its semilattice of idempotents. After constructing the Munn representation, we characterize antigroups in several ways. The second objective here is to characterize (0-)bisimple and (0-)simple inverse semigroups S by means of transitivity properties of the image S^δ of S under the Munn representation. As a corollary, we obtain which semilattices may serve as the semilattices of (0-)bisimple or (0-)simple inverse semigroups. As a further consequence, we get characterizations of (0-)bisimple and (0-)simple antigroups in terms of properties of their semilattices of idempotents.

The following concept is of basic importance for our development.

IV.2.1 Notation

For a semilattice Y, let $\Phi(Y)$ denote the set of *all isomorphisms among principal ideals of* Y under the composition of partial transformations written on the right.

We hasten to establish the main properties of $\Phi(Y)$.

IV.2.2 Lemma

For any semilattice Y, $\Phi(Y)$ is an inverse semigroup, and

$$E_{\Phi(Y)} = \{\iota_{[\alpha]} | \alpha \in Y\} \cong Y.$$

Proof. Let $\varphi, \varphi' \in \Phi(Y)$, say $\varphi: [\alpha] \to [\beta]$ and $\varphi': [\alpha'] \to [\beta']$ for some $\alpha, \beta, \alpha', \beta' \in Y$. It follows easily that $\varphi\varphi'$ maps $[(\beta\alpha')\varphi^{-1}]$ onto $[(\beta\alpha')\varphi']$ and is evidently an isomorphism. Hence $\varphi\varphi' \in \Phi(Y)$ so that $\Phi(Y)$ is closed under its product. Obviously $\Phi(Y)$ is closed under taking of inverses. Consequently $\Phi(Y)$ is an inverse subsemigroup of $\mathcal{I}(Y)$.

For any $\alpha, \beta \in Y$, we have $\iota_{[\alpha]}\iota_{[\beta]} = \iota_{[\alpha]\cap[\beta]} = \iota_{[\alpha\beta]}$ which yields the isomorphism in the statement of the lemma; the rest is obvious.

The $\Phi(Y)$ defined above is the normal hull of Y which will be treated in VI.4. We are now ready for the principal result of this section.

IV.2.3 Theorem

Let S be an inverse semigroup. For each $a \in S$, let

$$\delta^a: e \to a^{-1}ea \qquad (e \in \mathbf{d}\delta^a = [aa^{-1}]).$$

Then the mapping

$$\delta: a \to \delta^a \qquad (a \in S)$$

is a homomorphism of S into $\Phi(E_S)$ which induces the greatest idempotent separating congruence μ on S.

Proof. Let $a \in S$. It is clear that δ^a maps $[aa^{-1}]$ onto $[a^{-1}a]$. If $e\delta^a = f\delta^a$, then

$$e = (aa^{-1})e(aa^{-1}) = a(a^{-1}ea)a^{-1} = a(a^{-1}fa)a^{-1}$$

$$= (aa^{-1})f(aa^{-1}) = f$$

and thus δ^a is one-to-one. For $e, f \leq aa^{-1}$, we obtain

$$(e\delta^a)(f\delta^a) = (a^{-1}ea)(a^{-1}fa) = a^{-1}e(aa^{-1})fa = a^{-1}(ef)a = (ef)\delta^a$$

so that δ^a is a homomorphism. We have proved that $\delta^a \in \Phi(E_S)$.

Now let $a, b \in S$. If $e \in \mathbf{d}(\delta^a\delta^b)$, then $e \leq aa^{-1}$ and $a^{-1}ea \leq bb^{-1}$ which gives

$$e = (aa^{-1})e(aa^{-1}) = a(a^{-1}ea)a^{-1} \leq abb^{-1}a^{-1} = (ab)(ab)^{-1}$$

and thus $e \in \mathbf{d}\delta^{ab}$. Conversely, if $e \in \mathbf{d}\delta^{ab}$, then $e \leq abb^{-1}a^{-1}$ whence $e \leq aa^{-1}$ and $a^{-1}ea \leq a^{-1}abb^{-1}a^{-1}a \leq bb^{-1}$ and thus $e \in \mathbf{d}(\delta^a\delta^b)$. Consequently, $\mathbf{d}(\delta^a\delta^b) = \mathbf{d}\delta^{ab}$. For any e in this set, we further get

$$e\delta^a\delta^b = b^{-1}(a^{-1}ea)b = (ab)^{-1}e(ab) = e\delta^{ab}$$

which shows that $\delta^a\delta^b = \delta^{ab}$, that is to say, δ is a homomorphism of S into $\Phi(E_S)$.

Let $\bar{\delta}$ be the congruence on S induced by δ and let $a\bar{\delta}b$. Then $\mathbf{d}\delta^a = \mathbf{d}\delta^b$ and $\mathbf{r}\delta^a = \mathbf{r}\delta^b$ which implies that $aa^{-1} = bb^{-1}$ and $a^{-1}a = b^{-1}b$. Hence $a\mathcal{H}b$ and thus $\bar{\delta} \subseteq \mathcal{H}$. Let ρ be any congruence on S contained in $\mathcal{H}$, and let $a\rho b$. Then $aa^{-1}\rho bb^{-1}$ and $a^{-1}a\rho b^{-1}b$, and for any $e \in E_S$, also $a^{-1}ea\rho b^{-1}eb$. The hypothesis that $\rho \subseteq \mathcal{H}$ now implies that $aa^{-1} = bb^{-1}$, $a^{-1}a = b^{-1}b$, and $a^{-1}ea = b^{-1}eb$. Hence $\delta^a = \delta^b$ and thus $a\bar{\delta}b$. Consequently, $\rho \subseteq \bar{\delta}$ which proves that $\bar{\delta}$ is the greatest congruence on S contained in $\mathcal{H}$. But then III.3.3 yields that $\bar{\delta} = \mu$.

It is convenient to have a name for the above representation.

IV.2.4 Definition

The representation $\delta\colon S \to \Phi(E_S)$ in 2.3 of an inverse semigroup S is the *Munn representation* of S. We will use the notation $\delta\colon a \to \delta^a$ consistently and denote the image of S under δ by S^δ.

We record next several characterizations of antigroups.

IV.2.5 Theorem

The following conditions on an inverse semigroup S are equivalent.

(i) S is an antigroup.
(ii) The equality is the only idempotent separating congruence on S.
(iii) $E_S\zeta = E_S$.
(iv) The Munn representation of S is faithful.

(v) S is isomorphic to a full subsemigroup of $\Phi(E_S)$.

(vi) S is isomorphic to a semigroup of homeomorphisms among open sets of a T_0-topological space T. These open sets serve as a subbase (respectively a base) for the topology of T.

Proof. The equivalence of (i) and (ii) is a direct consequence of III.3.2; the equivalence of (ii) and (iii) follows from the fact that $\ker \mu = E_S\zeta$ and that μ is the greatest idempotent separating congruence on S; the equivalence of (iii) and (iv) stems from the fact that the Munn representation is idempotent separating with kernel $E_S\zeta$. It is obvious that (iv) implies (v).

(v) *implies* (vi). We take a full inverse subsemigroup S of $\Phi(Y)$ for some semilattice Y. The family

$$\mathfrak{B} = \{\mathbf{d}\varphi | \varphi \in S\} \tag{1}$$

consists of principal ideals of Y. By taking inverses of elements of $\mathfrak{B}$, we get

$$\mathfrak{B} = \{[\alpha] | \iota_{[\alpha]} \in S\}.$$

It is now obvious that $\mathfrak{B}$ is closed under finite intersections and hence forms a basis for a topology T on Y. For any $\alpha, \beta \in Y$, if $\alpha > \beta$, then $\alpha \notin [\beta]$ and if $\alpha \not> \beta$, then $\beta \notin [\alpha]$, which shows that T satisfies the T_0-separation axiom.

Let $\varphi \in S$. Then φ maps principal ideals of Y contained in its domain onto principal ideals contained in its range. Since some of these are members of $\mathfrak{B}$ and φ is one-to-one, we conclude that φ^{-1} is continuous. A symmetric argument shows that also φ is continuous, and is thus a homeomorphism of $\mathbf{d}\varphi$ onto $\mathbf{r}\varphi$.

(vi) *implies* (iii). Let S be a semigroup of homeomorphisms as in the statement of (vi), and assume that $\mathfrak{B}$ as in (1) generates the topology of T. Let $\varphi \in S \setminus E_S$. Then $\alpha\varphi \neq \alpha$ for some $\alpha \in \mathbf{d}\varphi$. By the T_0-property, there exists an open set V such that either $\alpha\varphi \in V$ and $\alpha \notin V$ or $\alpha\varphi \notin V$ and $\alpha \in V$. Since V is a union of members of a basis for T which consists of finite intersections of elements of $\mathfrak{B}$, we may take $V = \bigcap_{i=1}^{n} B_i$ where $B_i \in \mathfrak{B}$ for $i = 1, 2, \ldots, n$. Assume that $\alpha\varphi \in V$ and $\alpha \notin V$. Then $\alpha \notin V$ implies that $\alpha \notin B_i$ so that we can take $V = B_i$, that is $V \in \mathfrak{B}$. Hence $\iota_V \in S$. It follows that $\alpha\varphi\iota_V$ is defined and $\alpha\iota_V\varphi$ is not, and thus $\varphi\iota_V \neq \iota_V\varphi$. The other case, namely $\alpha\varphi \notin V$ and $\alpha \in V$, is treated similarly, for then $\alpha\iota_V\varphi$ is defined and $a\varphi\iota_V$ is not and again $\varphi\iota_V \neq \iota_V\varphi$. This shows that $\varphi \notin E_S\zeta$. By contrapositive, we obtain that (iii) holds.

For the next main result, we need a number of new concepts.

IV.2.6 Definition

Let Y be a semilattice and let S be an inverse subsemigroup of $\Phi(Y)$. Then S is *transitive* (respectively *subtransitive*) if for any $e, f \in Y$, there exists $\varphi \in S$ such that $e\varphi = f$ (respectively $e\varphi \leq f$).

If Y has a zero, S is *0-transitive* (respectively *0-subtransitive*) if S contains the zero of $\Phi(Y)$ and for any $e, f \in Y^*$, there exists $\varphi \in S$ such that $e\varphi = f$ (respectively $e\varphi \leq f$).

Moreover, Y is *uniform*, *subuniform*, *0-uniform*, or *0-subuniform* if $\Phi(Y)$ is transitive, subtransitive, 0-transitive, or 0-subtransitive, respectively.

IV.2.7 Theorem

An inverse semigroup S is bisimple (respectively simple, 0-bisimple, 0-simple) if and only if S^δ is transitive (respectively subtransitive, 0-transitive, 0-subtransitive).

Proof. First suppose that S is bisimple. Let $e, f \in E_S$. By II.1.8 there exists $a \in S$ such that $aa^{-1} = e$ and $a^{-1}a = f$. We have $\delta^a\delta^{a^{-1}} = \delta^e$ and $\delta^{a^{-1}}\delta^a = \delta^f$ which implies that $\mathbf{d}\delta^a = [e]$ and $\mathbf{r}\delta^a = [f]$, which evidently yields $e\delta^a = f$. Here $\delta^a \in S^\delta$ so S^δ is transitive.

Conversely, assume that S^δ is transitive. Again let $e, f \in E_S$. By hypothesis there exists $a \in S$ such that $e\delta^a = f$. It follows that $e \leq aa^{-1}$ and $a^{-1}ea = f$. Letting $b = ea$, we obtain

$$bb^{-1} = (ea)(ea)^{-1} = eaa^{-1}e = e,$$

$$b^{-1}b = (ea)^{-1}(ea) = a^{-1}ea = f,$$

which in view of II.1.8 gives that S is bisimple.

Assume next that S is simple. Let $e, f \in E_S$. By II.1.8, there exists $a \in S$ such that $aa^{-1} = e$ and $a^{-1}a \leq f$, which implies that $\mathbf{d}\delta^a = [e]$ and $\mathbf{r}\delta^a \subseteq [f]$. But then $e\delta^a \leq f$ and hence S^δ is subtransitive.

The proof that subtransitivity of S^δ implies simplicity of S goes along the same lines as the argument above for bisimplicity. Using II.1.9, an easy modification of the arguments above would complete the proof. The details of this are left as an exercise.

IV.2.8 Corollary

A semilattice Y is uniform (respectively subuniform, 0-uniform, 0-subuniform) if and only if $Y \cong E_S$ for some bisimple (respectively simple, 0-bisimple, 0-simple) inverse semigroup S.

Proof. Let Y be uniform and let $\varphi, \psi \in E_{\Phi(Y)}$. Then $\varphi = \iota_{[\alpha]}$ and $\psi = \iota_{[\beta]}$ for some $\alpha, \beta \in Y$. By the uniformity of Y, there exists an isomorphism χ of $[\alpha]$ onto $[\beta]$. It follows that $\varphi = \chi\chi^{-1}$ and $\psi = \chi^{-1}\chi$, so by II.1.8, $\Phi(Y)$ is bisimple. By 2.2, we have $Y \cong E_{\Phi(Y)}$.

Let S be bisimple and $Y \cong E_S$. Then S^δ is transitive by 2.7, and hence $\Phi(E_S)$ is transitive. That is to say, E_S is uniform and hence also Y is uniform.

The proofs of the remaining items follow along the same lines.

IV.2.9 Corollary

A semigroup S is a bisimple antigroup if and only if S is isomorphic to a full transitive inverse subsemigroup of $\Phi(Y)$ for some semilattice Y. Analogous statements are true for simple, 0-bisimple, and 0-simple antigroups.

Proof. The proof of this corollary is left as an exercise.

IV.2.10 Corollary

Let Y be a semilattice and S be a full inverse subsemigroup of $\Phi(Y)$. Then S is bisimple if and only if S is transitive. Analogous statements are true for S simple, 0-bisimple, and 0-simple.

Proof. The proof of this corollary is left as an exercise.

The opposite extreme of uniformity of a semilattice is provided by the following concept.

IV.2.11 Definition

A semilattice Y is *antiuniform* if for any $e, f \in Y$, $[e] \cong [f]$ implies $e = f$.

IV.2.12 Proposition

A semilattice Y is *antiuniform* if and only if whenever $Y \cong E_S$ for an inverse semigroup S then S is a Clifford semigroup.

Proof

Necessity. We may suppose that S is an inverse semigroup for which E_S is antiuniform. It is easy to see that then $\Phi(E_S)$ is a Clifford semigroup. Hence the Munn representation δ is an idempotent separating homomorphism of S onto a Clifford semigroup. For any $a \in S$, we have $\delta^a(\delta^a)^{-1} = (\delta^a)^{-1}\delta^a$, which implies $\delta^{aa^{-1}} = \delta^{a^{-1}a}$. Since δ is idempotent separating, we obtain $aa^{-1}\mathcal{H}a^{-1}a$ whence $aa^{-1} = a^{-1}a$. Thus S is a Clifford semigroup.

Sufficiency. In particular, $\Phi(Y)$ is a Clifford semigroup which evidently implies that Y is antiuniform.

IV.2.13 Exercises

(i) Show that a semilattice Y is uniform (respectively subuniform, 0-uniform, 0-subuniform, antiuniform) if and only if $\Phi(Y)$ is a bisimple (respectively simple, 0-bisimple, 0-simple, Clifford) semigroup.

(ii) Let X be a uniform semilattice and Y be any semilattice. Show that their ordinal product $X \circ Y$ is subuniform. Give an example showing that $X \circ Y$ need not be uniform.

(iii) Let S be an inverse semigroup. Show that S is a Clifford semigroup if and only if S^δ is a Clifford semigroup.

(iv) Prove that the following conditions on an inverse semigroup S are equivalent.

(α) S is a Reilly semigroup.

(β) S^δ is a bicyclic semigroup.

(γ) $\Phi(E_S)$ is bicyclic and S^δ is transitive.

Deduce that a semilattice Y is an ω-chain if and only if $\Phi(Y)$ is a bicyclic semigroup.

(v) Show that an inverse semigroup S is a primitive inverse (respectively Brandt) semigroup if and only if S^δ is a primitive inverse (respectively Brandt) semigroup.

(vi) Let $Y = \Sigma_{\alpha \in A} Y_\alpha$ where each Y_α is a two-element chain. Show that $\Phi(Y) \cong B(1, A)$.

(vii) Let S be an inverse semigroup. For each $a \in S$, define a relation β_a on E_S by

$$e\beta_a f \Leftrightarrow e = \lambda(eaf),\ f = (eaf)\rho$$

(see II.1.12 for notation). Show that the mapping β: $a \to \beta_a$ $(a \in S)$ is the Munn representation.

Munn [10] is the source of the Munn representation and the semigroup T_E (denoted here by $\Phi(E)$) which now bears his name. [The notation T_E is widely accepted in the literature; we chose $\Phi(E)$ for reasons which will be apparent in VI.4.1.] This representation appeared earlier under a different guise in Preston [3] [exercise (vii) above] and was nearly discovered by Howie [1]. Reilly [14] constructed an extension of the Munn representation. Uniformity was introduced by Rees [1]. Also consult Ash [2] and Reilly [12]. The characterization of antigroups in topological terms is due to Wagner [10]. Except for a different formulation and slight changes in definitions, all the remaining results are due to Munn [7], [10]. The subject of antiuniform semilattices was treated by Howie–Schein [1]; see also Schein [7] and Hickey [1], [2]. In addition, consult Sribala [1] and Širjajev [1], [2].

IV.3 CONGRUENCE-FREE INVERSE SEMIGROUPS

The semigroups in the title are characterized in the main result of this section. This result has a number of corollaries, in particular for inverse semigroups without zero. The case of inverse semigroups with zero also admits several criteria for congruence-freeness. These conditions are much more transparent than those figuring in the general case.

We start with the basic concept.

IV.3.1 Definition

A semigroup without proper congruences is *congruence-free*.

For the proof of the main characterization theorem for congruence-free inverse semigroups we will need the following auxiliary result which is also of intrinsic interest.

IV.3.2 Lemma

Let S be an inverse semigroup with semilattice E of idempotents. For a relation τ on E define τ^+ on E by

$$e\tau^+ f \Leftrightarrow e = f \quad \text{or} \quad e = s_1^{-1}u_1 s_1,$$

$$s_i^{-1}v_i s_i = s_{i+1}^{-1}u_{i+1}s_{i+1} \quad \text{for} \quad 1 \le i < n, \quad s_n^{-1}v_n s_n = f \tag{2}$$

for some $s_i \in S$ and $u_i, v_i \in E$ such that either $u_i \tau v_i$ or $v_i \tau u_i$, $i = 1, 2, \ldots, n$. Then τ^+ is the least normal congruence on E containing τ.

Proof. Clearly τ^+ is an equivalence relation. Let $e\tau^+ f$ with the above notation and $x \in S$. Then either $x^{-1}ex = x^{-1}fx$ or

$$x^{-1}ex = (s_1 x)^{-1}u_1(s_1 x),$$

$$(s_i x)^{-1}v_i(s_i x) = (s_{i+1}x)^{-1}u_{i+1}(s_{i+1}x) \quad \text{for} \quad 1 \le i < n,$$

$$(s_n x)^{-1}v_n(s_n x) - x^{-1}fx,$$

and thus $x^{-1}ex\tau^+ x^{-1}fx$. Hence τ^+ is normal and is thus also a congruence on E.

Let ξ be a normal congruence on E containing τ and let $e\tau^+ f$, $e \neq f$. In the notation of the statement of the lemma, we get $u_i \xi v_i$ and thus $s_i^{-1}v_i s_i \xi s_i^{-1}u_i s_i$ for $i = 1, 2, \ldots, n$. This together with relations (2) implies that $e\xi f$. Consequently, $\tau^+ \subseteq \xi$ which establishes the minimality of τ^+.

We are now ready for a characterization of congruence-free inverse semigroups which are not groups.

IV.3.3 Theorem

Let S be an inverse semigroup, let $E = E_S$, and assume that $|E| > 1$. Then S is congruence-free if and only if $E\zeta = E$, $E\omega = S$, and S satisfies the following condition:

(M) for any $e, f, g, h \in E$ with $e > f$ and $g > h$, there exist $t_1, t_2, \ldots, t_n \in S$ such that

$$g = t_1^{-1}et_1, \quad t_i^{-1}ft_i = t_{i+1}^{-1}et_{i+1} \quad \text{for} \quad 1 \leq i < n, \quad t_n^{-1}ft_n \leq h.$$

Proof

Necessity. If $\mu = \omega$, then S is a group, contradicting the assumption $|E| > 1$. Thus $\mu = \varepsilon$ and hence $\ker \mu = E\zeta = E$. Similarly, $\sigma \neq \varepsilon$ and thus $\sigma = \omega$ which yields $\ker \sigma = E\omega = S$.

Let $e, f, g, h \in E$ be such that $e > f$ and $g > h$. Let $\tau = \{e, f\} \times \{e, f\}$. Let τ^+ be defined on E as in 3.2; clearly $\tau^+ \neq \varepsilon$. Since τ^+ is a normal congruence on E, by III.2.5, there exists a congruence ρ on S such that $\operatorname{tr} \rho = \tau^+$. Hence $\rho \neq \varepsilon$ and thus $\rho = \omega$ in view of the hypothesis. But then $\tau^+ = \omega$ and hence, by 3.2, there exist $s_i \in S$ and $u_i, v_i \in \{e, f\}$ such that

$$g = s_1^{-1}u_1s_1, \quad s_i^{-1}v_is_i = s_{i+1}^{-1}u_{i+1}s_i \quad \text{for} \quad 1 \leq i < n, \quad s_n^{-1}v_ns_n = h. \tag{3}$$

We may suppose that $u_i \neq v_i$ since repetitions in (3) can be omitted. For $i = 1, 2, \ldots, n$, we define p_i and t_i by

$$p_i = (s_i^{-1}u_is_i)(s_{i-1}^{-1}u_{i-1}s_{i-1}) \cdots (s_1^{-1}u_1s_1), \qquad t_i = s_ip_i.$$

Note that each u_i, $s_i^{-1}u_is_i$, and p_i is an idempotent. We further obtain

$$\begin{aligned} t_i^{-1}u_it_i &= p_is_i^{-1}u_is_ip_i = p_i(s_i^{-1}u_is_i)(s_i^{-1}u_is_i) \cdots (s_1^{-1}u_1s_1) \\ &= p_i(s_i^{-1}u_is_i) \cdots (s_1^{-1}u_1s_1) = p_i \qquad (i = 1, 2, \ldots, n). \end{aligned} \tag{4}$$

In particular, $g = p_1 = t_1^{-1}u_1t_1$, and by (3) and (4), also

$$\begin{aligned} t_i^{-1}v_it_i &= p_i(s_i^{-1}v_is_i)p_i = (s_i^{-1}v_is_i)p_i = (s_{i+1}^{-1}u_{i+1}s_{i+1})p_i \\ &= p_{i+1} = t_{i+1}^{-1}u_{i+1}t_{i+1} \qquad (i = 1, 2, \ldots, n-1) \end{aligned}$$

and $t_n^{-1}v_n = (s_n^{-1}v_ns_n)p_n = hp_n \leq h$. We have shown

$$g = t_1^{-1}u_1t_1, \quad t_i^{-1}v_it_i = t_{i+1}^{-1}u_{i+1}t_{i+1} \quad \text{for} \quad 1 \leq i < n, \quad t_n^{-1}v_nt_n \leq h. \tag{5}$$

Using (4), we also have

$$t_i^{-1}u_it_i = p_i \geq p_i(s_i^{-1}v_is_i)p_i = t_i^{-1}v_it_i \qquad (i = 1, 2, \ldots, n). \tag{6}$$

In the case that $(u_i, v_i) = (f, e)$, we have by $f < e$ that $t_i^{-1}u_it_i \leq t_i^{-1}v_it_i$, which by (6) yields $t_i^{-1}u_it_i = t_i^{-1}v_it_i$. Hence we may delete all terms involving u_i and v_i whenever $(u_i, v_i) = (f, e)$. Therefore, condition (M) is satisfied.

Sufficiency. Let ρ be a congruence on S. If $\operatorname{tr}\rho = \varepsilon$, then ρ is idempotent separating, so $\rho \subseteq \mu$ and thus $\rho = \varepsilon$ in view of the hypothesis $E\zeta = E$. Hence assume that $\operatorname{tr}\rho \neq \varepsilon$. Then there exist $e, f \in E_S$ such that $e \neq f$ and $e\rho f$. Since then $e\rho(ef)\rho f$, we may suppose that $e > f$. Now let $g, h \in E_S$ be such that $g > h$. By condition (M), we have

$$g = t_1^{-1}et_1\rho t_1^{-1}ft_1 = t_2^{-1}et_2\rho \cdots \rho t_n^{-1}ft_n = t$$

so that $g\rho t$ and $t \leq h$. But then $h = hg\rho ht = t$ and thus $g\rho h$. If g and h are not comparable, then $g\rho(gh)\rho h$, which proves that $\operatorname{tr}\rho = \omega$. But then ρ is a group congruence, so $\rho \supseteq \sigma$ and thus $\rho = \omega$ in view of the hypothesis $E\omega = S$. Therefore S is congruence-free.

We can now deduce several interesting consequences of the above theorem.

IV.3.4 Corollary

Let S be a congruence-free inverse semigroup, let $E = E_S$, and assume that $|E| > 1$. Then any inverse subsemigroup T of $E_{\Phi(E)}\omega$ containing S^δ is also congruence-free.

Proof. Since $E_{\Phi(E)} \subseteq T \subseteq \Phi(E)$, we have by 2.5 that $E_T\zeta = E_T$. The hypothesis $E_{\Phi(E)} \subseteq T \subseteq E_{\Phi(E)}\omega$ implies that $E_T\omega = T$. From the first sentence, we deduce that $S \cong S^\delta$ in view of 2.5 so that the semigroup S^δ satisfies condition (M). But then the hypothesis that $S^\delta \subseteq T$ implies that T also satisfies condition (M). We may now apply 3.3 to conclude that T is congruence-free.

IV.3.5 Corollary

A semilattice E is the semilattice of a congruence-free inverse semigroup if and only if $E_{\Phi(E)}\omega$ satisfies condition (M).

Proof. The proof of this corollary is left as an exercise.

Note that $E_{\Phi(E)}\omega$ appears in the literature in the notation T_E^*. We now consider congruence-free inverse semigroups whose semilattice of idempotents has no zero.

IV.3.6 Proposition

Let S be an inverse semigroup, let $E = E_S$, and assume that E has no zero. Then S is congruence-free if and only if $E\zeta = E$, $E\omega = S$, and S satisfies condition (M) with $e = h$.

Proof

Necessity follows directly from 3.3.

Sufficiency. We show first that S is simple. Let $g, e \in E$ be such that $g > e$; it suffices to show that $J(g) = J(e)$. Since E has no zero, there exists an idempotent f such that $e > f$. We can now apply condition (M) to the sequence $g > e > f$ to obtain $t_1 \in S$ such that $g = t_1^{-1}et_1$. Letting $x = et_1$, we get $g = x^{-1}x$ and $xx^{-1} \leq e$. It follows that

$$g = x^{-1}x = x^{-1}(xx^{-1})x = x^{-1}(xx^{-1})ex \in J(e)$$

which together with $e < g$ implies $J(g) = J(e)$, as required. Therefore S is simple.

We now modify the proof of sufficiency of 3.3 and fill in only the parts to be altered. Let $e, f, g, h \in E$ be such that $e > f$ and $g > h$. Since S is simple, by II.1.8, there exists $x \in S$ such that $e = xx^{-1}$ and $x^{-1}x \leq h$. Hence

$$x^{-1}fx \leq x^{-1}ex = x^{-1}(xx^{-1})x = x^{-1}x \leq h$$

and if $x^{-1}fx = x^{-1}ex$, then

$$f = efe = x(x^{-1}fx)x^{-1} = x(x^{-1}ex)x^{-1} = e,$$

a contradiction. Hence letting $e' = x^{-1}x$ and $f' = x^{-1}fx$, we obtain the sequence $f' < e' \leq h < g$. We may thus apply condition (M) to the sequence $f' < e' < g$ to obtain

$$g = t_1^{-1}e't_1, \quad t_i^{-1}f't_i = t_{i+1}^{-1}e't_{i+1} \quad \text{for} \quad 1 \leq i < n, \quad t_n^{-1}f't_n \leq e'.$$

for some $t_1, t_2, \ldots, t_n \in S$.

Assume now that $e\rho f$ for some congruence ρ on S. Then $e'\rho f'$ and hence $t_i^{-1}e't_i\rho t_i^{-1}f't_i$ for $1 \leq i < n$ which implies that $g\rho u$, where $u = t_n^{-1}f't_n \leq e'$. It follows that

$$e' = e'g\rho e'u = u\rho g.$$

Since $e' \leq h < g$, the relation $e'\rho g$ implies $g\rho h$, as required.

IV.3.7 Corollary

Let E be a semilattice without zero. Then E is the semilattice of idempotents of a congruence-free inverse semigroup if and only if $E_{\Phi(E)}\omega$ satisfies (M) with $e = h$.

IV.3.8 Proposition

Let S be an inverse semigroup without zero, let $E = E_S$, and assume that $|E| > 1$. Then S is congruence-free if and only if $E\zeta = E$, $E\omega = S$, S is simple and satisfies condition (M) with $f = g$.

Proof. The argument here is quite similar to that in a part of the proof of 3.6 and is left as an exercise.

We have seen above that the nonexistence of a zero in an inverse semigroup plays a certain role in the conditions for it to be congruence-free. We now turn to inverse semigroups with zero with a view to obtaining somewhat simpler conditions for congruence-freeness.

IV.3.9 Lemma

Let S be an inverse semigroup with zero and the semilattice E of idempotents. Then $\operatorname{tr} \tau_S^{\{0\}} = \tau_E^{\{0\}}$ and $\mu \subseteq \tau_S^{\{0\}}$.

Proof. Let $e, f \in E$ be such that $e \tau_S^{\{0\}} f$. If $geh = 0$ with $g, h \in E^1$, then $gfh = 0$ since $e \tau_S^{\{0\}} f$, and thus $e \tau_E^{\{0\}} f$. Conversely, let $e \tau_E^{\{0\}} f$. If $xey = 0$ with $x, y \in S^1$, then $(x^{-1}x)e(yy^{-1}) = 0$ whence $(x^{-1}x)f(yy^{-1}) = 0$ since $e \tau_E^{\{0\}} f$, and thus $xey = 0$. Hence $e \tau_S^{\{0\}} f$. This proves the first assertion of the lemma. The second assertion follows from the fact that μ obviously saturates $\{0\}$.

IV.3.10 Lemma

Let S be an inverse semigroup with zero. Then S is 0-disjunctive if and only if E_S is 0-disjunctive and S is an antigroup.

Proof

Necessity. Then $\tau_S^{\{0\}} = \varepsilon$ and hence 3.9 gives that E_S is 0-disjunctive and $\mu = \varepsilon$, that is S is an antigroup.

Sufficiency. By 3.9, $\tau_S^{\{0\}}$ is idempotent separating and thus $\tau_S^{\{0\}} \subseteq \mu$. Since S is an antigroup, it follows that $\tau_S^{\{0\}} = \varepsilon$ and hence S is 0-disjunctive.

IV.3.11 Theorem

The following conditions on a nontrivial inverse semigroup S with zero are equivalent.

(i) S is congruence-free.
(ii) S is 0-simple and 0-disjunctive.
(iii) S is a 0-simple antigroup and E_S is 0-disjunctive.
(iv) S is an antigroup, E_S is 0-disjunctive, and S^δ is 0-subtransitive.

Proof. (i) *implies* (ii). Since S has no proper Rees congruences, it must be 0-simple. If S is not a semilattice, then $\tau \neq \omega$ so $\tau = \varepsilon$ and S is 0-disjunctive. If S is a semilattice, then 0-simplicity implies that S has just two elements and is thus trivially 0-disjunctive.

(ii) *implies* (i). Let ρ be a nonuniversal congruence on S. Then 0-simplicity of S implies that $\{0\}$ is a ρ-class and thus $\rho \subseteq \tau_{\{0\}}$. Now 0-disjunctiveness implies that $\rho = \varepsilon$.

Items (ii) and (iii) are equivalent by 3.10, and items (iii) and (iv) by 2.7.

IV.3.12 Corollary

Let E be a nontrivial semilattice with zero. Then E is the semilattice of idempotents of a congruence-free inverse semigroup if and only if E is 0-disjunctive and 0-subuniform. In such a case, every 0-subtransitive full inverse subsemigroup of $\Phi(E)$ is a congruence-free inverse semigroup whose semilattice of idempotents is isomorphic to E.

Proof. The proof of this corollary is left as an exercise.

IV.3.13 Exercises

(i) Show that an inverse semigroup S is congruence-free if and only if S is E-disjunctive and has neither proper completely prime ideals nor proper kernels.

(ii) Show that the following conditions on a semilattice E with zero are equivalent.

(α) E is 0-disjunctive.

(β) For any $e, f \in E$ such that $e < f$ there exists $g \in E$ such that $0 \neq g \leq f$ and $ge = 0$.

(γ) For any $e, f \in E^*$, $e \neq f$, there exists $g \in E$ such that either $eg \neq 0, fg = 0$ or $eg = 0, fg \neq 0$.

(iii) Let S be a 0-simple inverse semigroup. Show that E_S is 0-disjunctive if and only if every proper congruence on S is idempotent separating.

(iv) For any inverse semigroup S and $E = E_S$, prove

$$E_{\Phi(E)}\omega = \left\{\varphi \in \Phi(E) | \varphi|_{[\alpha]} = \iota_{[\alpha]} \text{ for some } \alpha \in E\right\}.$$

(v) Let R be the set of all real numbers with usual order. Show that $E_{\Phi(R)}\omega$ is congruence-free.

(vi) Characterize all finite congruence-free inverse semigroups.

(vii) Give an example of an inverse semigroup S satisfying: (α) S is not a semilattice, (β) S has a proper semilattice congruence, (γ) every proper congruence on S is a semilattice congruence.

(viii) Give an example of an inverse semigroup S satisfying: (α) S is not a group, (β) S has a proper group congruence, (γ) every proper congruence on S is a group congruence.

(ix) Give an example of an inverse semigroup S satisfying: (α) S is not a group, (β) S has a proper congruence, (γ) every proper congruence on S is idempotent separating.

(x) Give an example of an inverse semigroup S satisfying: (α) S has a proper congruence, (β) every proper congruence on S is idempotent pure.

(xi) Let S be an inverse semigroup and $E = E_S$. Show that S has neither proper idempotent separating nor proper group congruences if and only if S is an antigroup and $S^\delta \subseteq E_{\Phi(E)}\omega$.

The general conditions for congruence-freeness of inverse semigroups are due to Munn [14]; see also Munn [15]. Result 3.6 was suggested by N. R. Reilly. Variants of these conditions for the inverse semigroups without zero have been given by Trotter [1] and for those with zero by Baird [4]. Congruence-freeness for a special class of inverse semigroups has been investigated in some detail by Munn [14] and Reilly [10]. See also Ash [1], Gluskin [1], Howie [5], Inasaridze [2], Leemans–Pastijn [1], and Schein [11].

IV.4 REPRESENTATIONS BY ONE-TO-ONE PARTIAL TRANSFORMATIONS

An arbitrary representation of an inverse semigroup S by one-to-one partial transformations of a set is first reduced to that of an effective representation. The latter is then shown to be uniquely a sum of transitive representations. Each of these turns out to be equivalent to a representation defined by means of a closed inverse subsemigroup of S, analogous to that of coset representation in group theory. This provides an "internalization" of representations of inverse semigroups by one-to-one partial transformations.

We start with a number of new concepts.

IV.4.1 Definition

Let X be any set and let H be an inverse subsemigroup of $\mathcal{I}(X)$. The relation on X defined by

$$x\tau_H y \Leftrightarrow x\chi = y \quad \text{for some } \chi \in H$$

is the *transitivity relation* of H; the *domain* of τ_H is the set

$$X\tau_H = \bigcup_{\chi \in H} \mathbf{d}\chi.$$

If $X\tau_H = X$, then H is *effective*. If τ_H is the universal relation on X, then H is *transitive*.

Note that transitivity of H means that for any $x, y \in X$, there exists $\chi \in H$ such that $x\chi = y$ (this agrees with transitivity defined in 2.6).

The hypothesis that H is an inverse subsemigroup of $\mathcal{I}(X)$ evidently implies that τ_H is transitive and symmetric. In general τ_H need not be an equivalence relation, for there may be elements in X which are contained in the domain of no element of H. It is verified readily that

$$X\tau_H = \{x \in X | x\tau_H x\}, \tag{7}$$

and that τ_H is an equivalence relation on its domain $X\tau_H$.

IV.4.2 Definition

Let S be an inverse semigroup and X be any set. A representation φ: $S \to \mathcal{I}(X)$ is *effective* (respectively *transitive*) if $S\varphi$ is an effective (respectively transitive) inverse subsemigroup of $\mathcal{I}(X)$.

Thus, effectiveness of a representation φ: $S \to \mathcal{I}(X)$ means that "φ uses all elements of X."

The next concept concerns the construction of a new representation out of the given family of representations of an inverse semigroup. In the opposite direction, it will be used for decomposing a given representation into a sum of possibly "simpler" representations.

IV.4.3 Definition

Let S be an inverse semigroup, let $\{X_\alpha\}_{\alpha \in A}$ be a family of pairwise disjoint sets, and assume that for each $\alpha \in A$, there is given a representation φ_α: $S \to \mathcal{I}(X_\alpha)$. Let $X = \bigcup_{\alpha \in A} X_\alpha$, and for each $s \in S$, define φ^s by

$$x\varphi^s = x\varphi^s_\alpha \quad \text{if} \quad x \in \mathbf{d}\varphi^s = \bigcup_{\alpha \in A} \mathbf{d}\varphi^s_\alpha.$$

We write $\varphi = \oplus_{\alpha \in A}\varphi_\alpha$ and call it the *sum* of the family $\{\varphi_\alpha\}_{\alpha \in A}$. In the case $A = \{1, 2, \ldots, n\}$, we write $\varphi = \varphi_1 \oplus \varphi_2 \oplus \cdots \oplus \varphi_n$.

If we consider φ^s_α, with $s \in S$ and $\alpha \in A$, as a binary relation in X_α in the usual way, φ^s can be simply defined as $\bigcup_{\alpha \in A}\varphi^s_\alpha$.

Routine verification shows that the sum of representations is again a representation. The presentation of the sum of representations in terms of binary relations is sometimes the more convenient one for application. For example, the infinite commutative and distributive laws for $\oplus$ follow at once. Furthermore, if φ_α: $S \to \mathcal{I}(X_\alpha)$, ψ_α: $S \to \mathcal{I}(Y_\alpha)$, are representations for $\alpha \in A$ and the family of sets $\{X_\alpha, Y_\alpha\}_{\alpha \in A}$ is pairwise disjoint, then one verifies easily that

$$\bigoplus_{\alpha \in A} (\varphi_\alpha \oplus \psi_\alpha) = \left(\bigoplus_{\alpha \in A} \varphi_\alpha\right) \oplus \left(\bigoplus_{\alpha \in A} \psi_\alpha\right).$$

We are now ready for the basic result concerning effective representations.

IV.4.4 Proposition

An effective representation of an inverse semigroup S by one-to-one partial transformations is uniquely a sum of transitive representations of S.

Proof. Let $\varphi\colon S \to \mathcal{I}(X)$ be an effective representation, and let $H = S\varphi$. Then H is an effective inverse subsemigroup of $\mathcal{I}(X)$ which implies that its transitivity relation τ_H is an equivalence relation on X [see (7) above]. Let $\{X_\alpha\}_{\alpha \in A}$ be the collection of all equivalence classes of τ_H. For each $s \in S$ and $\alpha \in A$, we define $\varphi_\alpha^s = \varphi^s|_{X_\alpha}$. Since X is a τ_H-class, it follows that $\mathbf{r}\varphi_\alpha^s \subseteq X_\alpha$, in fact $\mathbf{r}\varphi_\alpha^s = \mathbf{r}\varphi^s \cap X_\alpha$. Thus $\varphi_\alpha^s \in \mathcal{I}(X_\alpha)$ and we have a mapping $\varphi_\alpha\colon s \to \varphi_\alpha^s$ from S into $\mathcal{I}(X_\alpha)$.

Now let $s, t \in S$, $\alpha \in A$. Then

$$\begin{aligned} \mathbf{d}\varphi^{st}|_{X_\alpha} &= \mathbf{d}(\varphi^s\varphi^t)|_{X_\alpha} = \{x \in X_\alpha | x \in \mathbf{d}\varphi^s, x\varphi^s \in \mathbf{d}\varphi^t\} \\ &= \mathbf{d}\left(\varphi^s|_{X_\alpha}\varphi^t|_{X_\alpha}\right) \end{aligned}$$

since X_α is a τ_H-class, whence it follows that φ_α is a homomorphism. For any $s \in S$, we have

$$x\varphi^s = x\varphi_\alpha^s \quad \text{if} \quad x \in \mathbf{d}\varphi_\alpha^s = \mathbf{d}\varphi^s \cap X_\alpha,$$

which shows that $\varphi = \oplus_{\alpha \in A}\varphi_\alpha$. For any $x, y \in X_\alpha$, there exists $\chi \in H$ such that $x\chi = y$ by the definition of τ_H since X_α is a τ_H-class, which shows that φ_α is transitive.

To prove uniqueness, we assume that $\varphi = \oplus_{\beta \in B}\psi_\beta$ where $\psi_\beta\colon S \to \mathcal{I}(Y_\beta)$ is a transitive representation for each $\beta \in B$ and $\bigcup_{\beta \in B} Y_\beta = X$. The transitivity classes of $\oplus_{\beta \in B}\psi_\beta$ are the sets Y_β and for each $s \in S$, we have $\psi_\beta^s = \varphi^s|_{Y_\beta}$. But then the partitions of X into the sets $\{X_\alpha\}_{\alpha \in A}$ and $\{Y_\beta\}_{\beta \in B}$ are identical, and hence the φ_α and ψ_β are pairwise equal in some order.

We will now examine transitive representations. In group theory, such representations are essentially those induced by permutations of right cosets of a subgroup. A similar idea will be useful here as well, except that we must find a suitable analogue of a subgroup and of right cosets. The obvious analogue would be to take an inverse subsemigroup H of an inverse semigroup S and $a \in S$, and call Ha a *right coset* of H in S. However, it will turn out that the right analogues are the sets $(Ha)\omega$ where H is closed and $aa^{-1} \in H$. We start with an auxiliary result which shows that closed right cosets behave very much like right cosets of a subgroup of a group and will prove quite useful later as well.

IV.4.5 Lemma

Let S be an inverse semigroup, H be its inverse subsemigroup, and $a, b \in S$ be such that $aa^{-1}, bb^{-1} \in H$. Then the following statements are equivalent.

(i) $(Ha)\omega = (Hb)\omega$.
(ii) $a \in (Hb)\omega$.
(iii) $ab^{-1} \in H\omega$.

Proof. (i) *implies* (ii). Clearly $a = (aa^{-1})a \in Ha \subseteq (Ha)\omega = (Hb)\omega$.

(ii) *implies* (iii). The hypothesis implies that $ea = hb$ for some $e \in E_S$ and $h \in H$. Hence $eab^{-1} = hbb^{-1} \in H$ so that $ab^{-1} \in H\omega$.

(iii) *implies* (i). First let $c \in (Ha)\omega$. Then $ec = ha$ for some $e \in E_S$ and $h \in H$. Hence $ecb^{-1} = hab^{-1} \in H\omega$ so that $ecb^{-1} \geq h'$ for some $h' \in H$ and thus $c \geq ecb^{-1}b \in Hb$ and $c \in (Hb)\omega$. Consequently $(Ha)\omega \subseteq (Hb)\omega$. Since $ab^{-1} \in H\omega$ implies $ba^{-1} \in H\omega$, by symmetry, we conclude that also $(Hb)\omega \subseteq (Ha)\omega$.

It is convenient to introduce the following concept.

IV.4.6 Definition

Let S be an inverse semigroup. If H is an inverse subsemigroup of S and $a \in S$ is such that $aa^{-1} \in H$, then $(Ha)\omega$ is a *right ω-coset of H*.

We are now ready for the construction of a transitive representation of an inverse semigroup induced by a closed inverse subsemigroup of it. Recall the notation $\lambda a = aa^{-1}$ in II.1.12.

IV.4.7 Lemma

Let H be a closed inverse subsemigroup of an inverse semigroup S. For every $s \in S$, let

$$\varphi_H^s\colon (Ha)\omega \to (Has)\omega \qquad ((Ha)\omega \in \mathbf{d}\varphi_H^s = \{(Hb)\omega \mid \lambda(bs) \in H\}).$$

Then the mapping $\varphi_H\colon S \to \mathcal{I}(\mathcal{X})$,

$$\varphi_H\colon s \to \varphi_H^s \qquad (s \in S)$$

is a transitive representation of S, where $\mathcal{X}$ is the set of all right ω-cosets of H in S.

Proof. First note that if $\lambda(as) \in H$, then $\lambda a \geq \lambda(as)$ implies that $\lambda a \in H$ and thus $(Ha)\omega$ is a right ω-coset. Let $(Ha)\omega, (Hb)\omega, (Has)\omega, (Hbs)\omega \in \mathcal{X}$ so that λa, λb, $\lambda(as)$, $\lambda(bs) \in H$. Note that

$$(as)(bs)^{-1} = ass^{-1}b^{-1} = ass^{-1}a^{-1}ab^{-1} = \lambda(as)(ab^{-1}), \tag{8}$$

and that by hypothesis $H\omega = H$ and $\lambda(as) \in H$. If $(Ha)\omega = (Hb)\omega$, then by 4.5, $ab^{-1} \in H$ and thus $\lambda(as)(ab^{-1}) \in H$, whence by (8), $(as)(bs)^{-1} \in H$ which again by 4.5 implies $(Has)\omega = (Hbs)\omega$. Conversely, if $(Has)\omega = (Hbs)\omega$, then by 4.5, $(as)(bs)^{-1} \in H$ so by (8), $\lambda(as)(ab^{-1}) \in H$ whence $ab^{-1} \in H\omega = H$ and hence, again by 4.5, $(Ha)\omega = (Hb)\omega$. Consequently, φ_H^s is a one-to-one partial transformation of $\mathcal{X}$ so that φ maps S into $\mathcal{I}(\mathcal{X})$.

Now let $s, t \in S$; we must show that $\varphi_H^{st} = \varphi_H^s\varphi_H^t$ and for this, it is evidently enough to prove that the domains of these functions are equal. Let $(Ha)\omega$ be a right ω-coset of H. Using $\lambda(as) \geq \lambda(ast)$ and the closure of H, we obtain

$$\begin{aligned}(Ha)\omega \in \mathbf{d}\varphi_H^{st} &\Leftrightarrow \lambda(ast) \in H \Leftrightarrow \lambda(as), \lambda(ast) \in H \\ &\Leftrightarrow (Ha)\omega \in \mathbf{d}\varphi_H^s, \quad (Has)\omega \in \mathbf{d}\varphi_H^t \\ &\Leftrightarrow (Ha)\omega \in \mathbf{d}(\varphi_H^s\varphi_H^t).\end{aligned}$$

We conclude that φ_H is a homomorphism.

Now let $(Ha)\omega, (Hb)\omega$ be right ω-cosets of H. Then $\lambda(aa^{-1}b) = aa^{-1}bb^{-1}aa^{-1} \in H$ and thus $(Ha)\omega \in \mathbf{d}\varphi_H^{a^{-1}b}$. Further, $(aa^{-1}b)b^{-1} = (\lambda a)(\lambda b) \in H$ which by 4.5 implies that $(Ha)\omega\varphi_H^{a^{-1}b} = (Haa^{-1}b)\omega = (Hb)\omega$. Therefore, φ_H is a transitive representation.

We now turn to a kind of converse of 4.7. If φ: $S \to \mathcal{I}(X)$ is a representation, it is often necessary to replace the set X by some other set of the same cardinality as X. This situation arises in particular in considering sums of representations where the corresponding sets are required to be pairwise disjoint. This calls for the following concept.

IV.4.8 Definition

Let φ: $S \to \mathcal{I}(X)$ and ψ: $S \to \mathcal{I}(Y)$ be representations of a semigroup S. Then φ and ψ are *equivalent* if there exists a bijection θ: $X \to Y$ such that

$$(x\varphi^s)\theta = (x\theta)\psi^s \quad \text{if} \quad x\theta \in \mathbf{d}\psi^s = (\mathbf{d}\varphi^s)\theta \qquad (s \in S).$$

In terms of binary relations,

$$\psi^s = \{(x\theta, x'\theta) \in Y \times Y | (x, x') \in \varphi^s\} \qquad (s \in S).$$

Equivalent representations may be considered as "equal" for they differ only in the labeling of elements of the representation set. The sum $\varphi \oplus \psi$ of two representations $\varphi: S \to \mathcal{I}(X)$ and $\psi: S \to \mathcal{I}(Y)$ is not defined unless X and Y are disjoint. In order to get around this, we can construct representations $\varphi': S \to \mathcal{I}(X')$ and $\psi': S \to \mathcal{I}(Y')$, where φ is equivalent to $\varphi'$and ψ is equivalent to ψ', $X' \cap Y' = \varnothing$, and then form the sum $\varphi' \oplus \psi'$. This procedure can of course be extended to any number of summands and is quite satisfactory in applications.

IV.4.9 Lemma

Let S be an inverse semigroup, X be a set, and $\psi: S \to \mathcal{I}(X)$ be a transitive representation. Fix an element z in X and let

$$H = \{s \in S | z\psi^s = z\}.$$

Then H is a closed inverse subsemigroup of S and ψ is equivalent to φ_H.

Proof. First H is nonempty since ψ is transitive. Since ψ is a homomorphism, it follows that H is an inverse subsemigroup of S. Let $t \in H\omega$ so that $te \in H$ for some $e \in E_S$. Then

$$(z\psi^t)\psi^e = z\psi^{te} = z = z\psi^t(\psi^e)^2 = (z\psi^{te})\psi^e = z\psi^e$$

whence $z\psi^t = z$ since ψ^e is one-to-one. Consequently, $t \in H$ and H is closed.

We will show that the one-to-one correspondence $\theta: X \to \mathcal{X}$ needed for establishing that ψ and φ_H are equivalent is given by

$$\theta: x \to (Ha)\omega \quad \text{if} \quad z\psi^a = x. \tag{9}$$

If $z\psi^a = x = z\psi^b$, then

$$z = x\psi^{b^{-1}} = z\psi^a\psi^{b^{-1}} = z\psi^{ab^{-1}}$$

which shows that $ab^{-1} \in H$, so by 4.5, $(Ha)\omega = (Hb)\omega$, and thus θ is single valued. If $x\theta = y\theta = (Ha)\omega$, then $x = z\psi^a = y$, and θ is one-to-one. If $(Ha)\omega \in \mathcal{X}$, then $aa^{-1} \in H$, so $z\psi^a\psi^{a^{-1}} = z$ and putting $x = z\psi^a$, we obtain $x\theta = (Ha)\omega$, which shows that θ maps X onto $\mathcal{X}$. Consequently, θ is a bijection.

Let $x \in \mathbf{d}\psi^s$ and $(x\psi^s)\theta = (Ha)\omega$, so that $z\psi^a = x\psi^s$. Also let $x\theta = (Hb)\omega$, so that $z\psi^b = x$ and $(x\theta)\varphi_H^s = (Hbs)\omega$, where $x\theta \in \mathbf{d}\varphi_H^s$ since $\lambda(bs) \geq \lambda(bsa^{-1}) \in H$. Hence $z\psi^{bs} = z\psi^b\psi^s = x\psi^s = z\psi^a$ and thus $z\psi^{(bs)a^{-1}} = z$. But then $(bs)a^{-1} \in H$ and 4.5 implies that $(Hbs)\omega = (Ha)\omega$, which yields

$(x\theta)\varphi_H^s = (x\psi^s)\theta$. Now let $(Ha)\omega \in \mathbf{d}\varphi_H^s$ so that $\lambda(as) \in H$. Then $z\psi^{\lambda(as)} = z$, and letting $x = z\psi^a$, we get $x \in \mathbf{d}\psi^s$ and $x\theta = (Ha)\omega$ since $z\psi^a = x$. Therefore, θ has all the requisite properties.

From 4.4, 4.7, and 4.9, we deduce the desired result.

IV.4.10 Theorem

Every effective representation of an inverse semigroup S is uniquely a sum of transitive representations ψ_α for $\alpha \in A$, each of which is equivalent to φ_{H_α} for some closed inverse subsemigroup H_α of S.

For a better understanding of the representations φ_H, it is useful to have more information concerning their properties and their relationship with arbitrary transitive representations.

IV.4.11 Proposition

Let H be a closed inverse subsemigroup of an inverse semigroup S.

(i) For any $h \in H$, $(Hh)\omega = H$ and thus H is a right ω-coset of H.
(ii) $H = \{s \in S | H\varphi_H^s = H\}$.
(iii) If $\psi: S \to \mathcal{I}(X)$ is a transitive representation equivalent to φ_H, then there exists $z \in X$ such that $H = \{s \in S | z\psi^s = z\}$.

Proof. (i) Let $h \in H$. Then $(Hh)\omega \subseteq H\omega = H$ since H is closed. If also $k \in H$, then $kh^{-1} \in H$ and 4.5 implies $k \in (Hh)\omega$ so that $H \subseteq (Hh)\omega$.

(ii) If $s \in H$, then part (i) directly implies that $H\varphi_H^s = H$. If $H\varphi_H^s = H$, then $H \in \mathbf{d}\varphi_H^s$ so $ss^{-1} \in H$ and thus

$$s = (ss^{-1})s \in Hs \subseteq (Hs)\omega = H\varphi_H^s = H,$$

as required.

(iii) Let $\psi: S \to \mathcal{I}(X)$ be a transitive representation equivalent to φ_H, and let θ be the corresponding bijection of the set $\mathcal{X}$ of right ω-cosets of H onto X. By part (i), we have $H \in \mathcal{X}$, so we may set $z = H\theta$. By the definition of equivalence, we get for any $s \in S$,

$$ss^{-1} \in H \Leftrightarrow H \in \mathbf{d}\varphi_H^s \Leftrightarrow z \in \mathbf{d}\psi^s$$

and if this is the case,

$$(H\varphi_H^s)\theta = (H\theta)\psi^s = z\psi^s.$$

Using part (ii) and the fact that θ is one-to-one, we deduce for any $s \in S$,

$$s \in H \Leftrightarrow H\varphi_H^s = H \Leftrightarrow z\psi^s = z,$$

as required.

IV.4.12 Corollary

Let H be a closed inverse subsemigroup of an inverse semigroup S, and let ψ: $S \to \mathcal{I}(X)$ be a transitive representation. Then φ_H and ψ are equivalent if and only if there exists $z \in X$ such that $H = \{s \in S | z\psi^s = z\}$.

Proof. This is a direct consequence of 4.9 and 4.11(iii).

The following result provides a criterion for equivalence of transitive representations of the form φ_H and φ_K directly in terms of H and K.

IV.4.13 Proposition

Let H and K be closed inverse subsemigroups of an inverse semigroup S. Then φ_H and φ_K are equivalent if and only if there exists $a \in S$ such that $a^{-1}Ha \subseteq K$ and $aKa^{-1} \subseteq H$. If this is the case, then $aa^{-1} \in H$ and $a^{-1}a \in K$.

Proof

Necessity. Let θ be the bijection of the set $\mathcal{X}_K$ of all right ω-cosets of K onto the set $\mathcal{X}_H$ of all right ω-cosets of H in the definition of equivalence of φ_K and φ_H. Since $K \in \mathcal{X}_K$, there exists $a \in S$ such that $K\theta = (Ha)\omega$ where $(Ha)\omega \in \mathcal{X}_H$. As in the proof of 4.11(iii), we obtain

$$K = \{s \in S | (Ha)\omega\varphi_H^s = (Ha)\omega\},$$

whence using 4.5, we get

$$K = \{s \in S | \lambda(as), asa^{-1} \in H\}.$$

It then follows that $aKa^{-1} \subseteq H$.

Now let $h \in H$. Then $(Ha)\omega \in \mathbf{d}\varphi_H^{a^{-1}ha}$ since $\lambda(aa^{-1}ha) = aa^{-1}haa^{-1}h^{-1}aa^{-1} \in H$; furthermore, $(Ha)\omega\varphi_H^{a^{-1}ha} = (Ha)\omega$ by 4.5 since $aa^{-1}haa^{-1} \in H$. Hence $(K\theta)\varphi_H^{a^{-1}ha} = K\theta$ and applying θ^{-1} to this formula, we get $K\varphi_K^{a^{-1}ha} = K$. But then 4.11(ii) yields $a^{-1}ha \in K$. Consequently $a^{-1}Ha \subseteq K$.

Sufficiency. For any $e \in E_H$, we obtain

$$aa^{-1} \geq a(a^{-1}ea)a^{-1} \in aKa^{-1} \subseteq H$$

and thus $aa^{-1} \in H$ since H is closed. Hence $(Ha)\omega$ is a right ω-coset of H. In view of 4.9, in order to show that φ_H and φ_K are equivalent, it suffices to show that

$$K = \{s \in S | (Ha)\omega\varphi_H^s = (Ha)\omega\};$$

equivalently, using 4.5, that

$$K = \{ s \in S | \lambda(as), asa^{-1} \in H \}. \tag{10}$$

If $s \in K$, then $a(ss^{-1})a^{-1}, asa^{-1} \in aKa^{-1} \subseteq H$, which gives one inclusion in (10). If $asa^{-1} \in H$, then

$$s \geq a^{-1}(asa^{-1})a \in a^{-1}Ha \subseteq K$$

and thus $s \in K$ since K is closed, giving the other inclusion in (10).

We have seen above that $aa^{-1} \in H$; analogously $a^{-1}a \in K$.

We now illustrate some of the above theory on a bicyclic semigroup.

IV.4.14 Example

As usual, let C denote the bicyclic semigroup. We consider the representation $\varphi = \varphi_H$ where $H = \{(0,0)\}$. For any $(m, n) \in C$, we have

$$\begin{aligned} \mathbf{d}\varphi^{(m,n)} &= \{ (p,q)\omega | \lambda(p,q) = \lambda((p,q)(m,n)) = (0,0) \} \\ &= \{ (p,q)\omega | (p,p) = (p+m-r, p+m-r) = (0,0) \} \\ &= \{ (0,q)\omega | q \geq m \} \\ &= \{ \{(0,m)\}, \{(0,m+1)\}, \{(0,m+2)\}, \ldots \}, \end{aligned} \tag{11}$$

where $r = \min\{q, m\}$,

$$\{(0,q)\}\varphi^{(m,n)} = \{(0,q)(m,n)\} = \{(0, n-m+q)\}, \qquad q \geq m \tag{12}$$

so that $\mathbf{d}\varphi^{(m,n)} \subseteq R_{(0,0)}$ and the effect of $\varphi^{(m,n)}$ is a shift for $n - m$. Further, if $\varphi^{(m,n)} = \varphi^{(p,q)}$, then $\mathbf{d}\varphi^{(m,n)} = \mathbf{d}\varphi^{(p,q)}$ which by (11) implies that $m = p$ and by (12) that $n - m = q - p$, so that $n = q$. Hence φ is faithful and thus the bicyclic semigroup is isomorphic to a transitive semigroup of one-to-one partial transformations on the set $\mathcal{X}$ of all right ω-cosets of H. Explicitly,

$$\mathcal{X} = \{ \{(0,0)\}, \{(0,1)\}, \{(0,2)\}, \ldots \}$$

which can be identified with $N = \{0, 1, 2, \ldots\}$. In fact, φ is equivalent to the representation ψ: $C \to \mathcal{I}(N)$ where for every $(m, n) \in C$,

$$\mathbf{d}\psi^{(m,n)} = \{ m, m+1, m+2, \ldots \},$$

$$q\psi^{(m,n)} = n - m + q \qquad \text{if} \quad q \geq m.$$

We may interpret N as the semilattice of idempotents of C, with the opposite of the usual order of N, in which case $\psi = \delta$ is the Munn representation. Indeed, using the mapping θ: $m \to e_m$, we obtain that ψ is actually equivalent to the Munn representation of C. This shows again that ψ is one-to-one.

Note that in C

$$(m, n) \geq (p, q) \Leftrightarrow p - m = q - n \geq 0.$$

Now fix $m > 0$, and let

$$H = \{(p, q) \in C | m \text{ divides } p - q\}.$$

Straightforward verification shows that H is the closed inverse subsemigroup of C generated by the element $(0, m)$. Let $\varphi = \varphi_H$. Since $E_S \subseteq H$, the domain of each φ^s is equal to $\mathcal{X}$, the set of all right ω-cosets of H in C.

For any $(p, q), (u, v) \in C$,

$$\varphi^{(p,q)} = \varphi^{(u,v)} \Leftrightarrow (p, q)(v, u) \in H \Leftrightarrow (p + v - r, q + u - r) \in H$$

$$\Leftrightarrow m \text{ divides } (p - q) - (u - v),$$

where $r = \min\{q, v\}$, and thus $S\varphi \cong \mathbb{Z}/(m)$.

IV.4.15 Exercises

(i) Are the Wagner and the Munn representations effective? Find their transitivity relations.

(ii) Show that no semilattice having more than two elements admits a faithful transitive representation.

(iii) Show that two equivalent representations of an inverse semigroup S induce the same congruence on S, but that the converse does not hold in general.

(iv) Give an example of two distinct closed inverse subsemigroups H and K of an inverse semigroup S such that φ_H and φ_K are equivalent.

(v) Let S be an inverse semigroup, H a nonempty subset of S, and s an element of S. Show that $((H\omega)s)\omega = (Hs)\omega$.

(vi) Let S be an inverse semigroup and let $E = E_S$. Prove the following statements.

(α) $S\varphi_{E\omega}$ is a simply transitive group of permutations on S/σ. (A group G of permutations on a set X is *simply transitive* if for any $x, y \in X$, there exists a unique $\alpha \in G$ such that $x\alpha = y$.)

(β) The mapping $\varphi^s_{E\omega} \to s\sigma$ is an isomorphism of $S\varphi_{E\omega}$ onto S/σ.

(γ) $\varphi_{E\omega}$ induces σ on S.

(vii) Let H and K be closed inverse subsemigroups of an inverse semigroup S. Show that $\varphi_H = \varphi_K$ if and only if there exist $a, b \in S$ such that

$$K = (Ha)\omega, \qquad H = (Kb)\omega,$$

$$aE_K^1 a^{-1} \subseteq E_H, \qquad bE_H^1 b^{-1} \subseteq E_K.$$

(viii) Let $S = B(G, I)$ be a Brandt semigroup.

(α) Characterize all closed inverse subsemigroups of S.
(β) Characterize all transitive representations of S.
(γ) Find the number of inequivalent transitive representations of S in terms of the number of conjugacy classes of G.

(ix) Let S be an inverse semigroup. Show that the following conditions on a closed inverse subsemigroup H of S are equivalent.

(α) H is full.
(β) $\varphi_H: S \to \mathcal{S}(\mathcal{X})$.
(γ) $S\varphi_H$ is a group.

If this is the case, show that $\ker \varphi_H$ is the greatest normal subsemigroup of S contained in H and

$$\ker \varphi_H = \{s \in S \mid a^{-1}sa \in H \quad \text{for all} \quad a \in S\}.$$

(x) Show that for any idempotent e of an inverse semigroup S,

$$\{z \in S \mid e\delta^z = e\} = H_e\omega = \{z \in S \mid ez \in H_e\}.$$

(xi) Show that the Wagner representation of an inverse semigroup S is transitive if and only if S is a group.

(xii) Let ρ be an E-unitary congruence on an inverse semigroup S. Show that $S\varphi_{\ker \rho} \cong (S/\rho)/\sigma$.

IV.4.16 Problems

(i) In groups, Brandt semigroups, and bicylic semigroups every congruence is induced by a transitive representation by one-to-one partial transformation of a set. Find more classes of inverse semigroups with this property.

(ii) Let ρ be a congruence on an inverse semigroup S. What are necessary and sufficient conditions on the kernel and the trace of ρ in order that there exists a transitive representation of S which induces ρ? (Note that every congruence on S is induced by some effective representation of S, and that every group congruence on S is induced by a transitive representation of S.)

(iii) What is a generalization of 4.15(ix) to the situation of φ_H for the general case; for example, what is the relationship of H and $\ker \varphi_H$?

All the results in this section are due to Schein [5]. Result 4.4 was also proved by Ponizovskiĭ [4].

IV.5 SUPPLEMENTS

This section contains results of three kinds: (i) those that supplement the results in IV.4, (ii) an application of the results in IV.4, and (iii) results on representations of general interest.

We have seen in 4.10 that transitive representations, and thus all effective representations, of an inverse semigroup S reduce to representations of the form φ_H for some closed inverse subsemigroup H of S. Further, 4.13 provides a criterion for equivalence of such representations. It is now clear that all transitive representations of a general inverse semigroup S are not easily available. We construct below a class of transitive representations for an arbitrary inverse semigroup S; these are equivalent to the representations φ_H where H has a group ideal. When S satisfies the minimal condition on idempotents, all transitive representations of S can be so found. The construction of these representations and the criterion for their equivalence are more suitable for manipulation than the general case. We start with an auxiliary result, parts of which are of independent interest.

IV.5.1 Lemma

Let S be an inverse semigroup, G be a subgroup of S with identity e, $H = G\omega$, and $R = R_e$. Then the following statements hold.

(i) H is a closed inverse subsemigroup of S.

(ii) $He = eH = H \cap R = G$.

(iii) $H = \{a \in S | ea, ae \in G\} = i_S(G)$.

(iv) Let $a \in S$ and recall the notation in 4.7; then

$$(Ha)\omega \in \mathcal{X} \Leftrightarrow aa^{-1} \geq e \Leftrightarrow ea \in R \Leftrightarrow Ga \subseteq R.$$

(v) If (iv) takes place, then

$$e((Ha)\omega) = Ga = (Ha)\omega \cap R.$$

(vi) $Ga \cap Gb \neq \varnothing \Rightarrow Ga = Gb \quad (a, b \in S)$.

(vii) $Ga \cap R \neq \varnothing \Rightarrow Ga \subseteq R \quad (a \in S)$.

(viii) $Gab \subseteq R \Rightarrow Ga \subseteq R \quad (a, b \in S)$.

Proof. (i) If $a, b \in H$, then $fa, bg \in G$ for some $f, g \in E_S$, so that $f(ab)g \in G$ and thus $ab \in H$; also $fa \in G$ implies $a^{-1}f \in G$ and thus $a^{-1} \in H$. Hence H is an inverse subsemigroup of S and is obviously closed.

(ii) Let $a \in H$, so that $af \in G$ for some $f \in E_S$. Then

$$fa^{-1}af = (af)^{-1}(af) = e$$

so that $f \geq e$. Thus $ae = a(fe) = (af)e = af \in G$ which shows that $He \subseteq G$. Conversely, if $b \in G$, then $b = be \in He$, and thus also $G \subseteq He$. Hence $He = G$; a symmetric argument shows that also $eH = G$.

Next let $a \in H \cap R$. Then $ea \in G$ since $eH = G$ and $a = ea$ since $a \mathcal{R} e$, so that $a \in G$. This shows that $H \cap R \subseteq G$; the opposite inclusion is trivial.

(iii) Let A and B denote the second and the third sets in part (iii), respectively. If $a \in H$, then $ea \in eH = G$ by part (ii); analogously $ae \in G$, and thus $H \subseteq A$. Let $a \in A$ and $g \in G$. Then $ag = a(eg) = (ae)g \in G$ and symmetrically $ga \in G$. Thus G is an ideal of H and hence $A \subseteq B$. Let $b \in B$. Then $be \in G$ and thus $b \in H$. Consequently $B \subseteq H$.

(iv) If $(Ha)\omega \in \mathfrak{X}$, then $aa^{-1} \in H$ and thus $aa^{-1} \geq e$. If $aa^{-1} \geq e$, then $e = eaa^{-1}$ whence $ea \mathcal{R} e$ so that $ea \in R$. If $ea \in R$ and $g \in G$, then $ga = gea \mathcal{R} ge = g$ and hence $Ga \subseteq R$. If $Ga \subseteq R$, then $ea \in R$ whence $e = (ea)(ea)^{-1} = eaa^{-1}$ so that $aa^{-1} \in H$ and $(Ha)\omega \in \mathfrak{X}$.

(v) Assume that (iv) takes place. Let $x \in e((Ha)\omega)$; then $x = ey$ where $fy \in Ha$ for some $f \in E_S$, and thus $fy = ha$ with $h \in H$ and hence $eh \in G$ by part (ii). Thus

$$fx = fey = efy = eha \in Ga$$

where

$$f \geq f(xx^{-1})f = (fx)(fx)^{-1} = eh(aa^{-1})h^{-1}e \geq eheh^{-1}$$

$$= (eh)(eh)^{-1} = e$$

so that $x = ex = (ef)x = e(fx) \in eGa = Ga$. Consequently $e(Ha) \subseteq Ga$. The inclusion $Ga \subseteq (Ha)\omega \cap R$ follows from part (iv). If $x \in (Ha)\omega \cap R$, then $x = ex \in e(Ha)\omega$, so that $(Ha)\omega \cap R \subseteq e(Ha)\omega$.

(vi) Let $c \in Ga \cap Gb$ and $d \in Ga$. Then $c = ga = hb$ and $d = xa$ for some $g, h, x \in G$. Hence

$$d = xa = xea = xg^{-1}(ga) = xg^{-1}(hb) = (xg^{-1}h)b \in Gb$$

so that $Ga \subseteq Gb$; the opposite inclusion follows by symmetry.

(vii) Let $g \in G$ be such that $ga \in R$. Then $ga \mathcal{R} e$ which gives

$$ea = g^{-1}ga \mathcal{R} g^{-1}e = g^{-1} \mathcal{R} e.$$

Hence $ea \in R$ and part (iv) yields $Ga \subseteq R$.

(viii) Let $g \in G$ be such that $gab \in R$. Then $gab \mathcal{R} g$ which implies that $ga \mathcal{R} g$ whence $ea \mathcal{R} e$. Thus $ea \in R$ and part (iv) gives $Ga \subseteq R$.

We can now establish the desired result.

IV.5.2 Theorem

Let S be an inverse semigroup, G be a subgroup of S with identity e, and $R = R_e$. For every $s \in S$, let

$$\psi_G^s\colon Ga \to Gas \quad \text{if} \quad Ga \in \mathbf{d}\psi_G^s = \{Gb \mid Gbs \subseteq R\},$$

and define

$$\psi_G\colon s \to \psi_G^s \qquad (s \in S).$$

Also let $\mathcal{Y} = \{Ga \mid Ga \subseteq R\}$, $H = G\omega$, and recall the notation in 4.7. Then ψ_G: $S \to \mathcal{I}(\mathcal{Y})$ is a transitive representation of S equivalent to φ_H.

Proof. We first define

$$\theta\colon (Ha)\omega \to Ga \qquad [(Ha)\omega \in \mathcal{X}].$$

If $(Ha)\omega = (Hb)\omega \in \mathcal{X}$, then by 5.1(v),

$$Ga = e((Ha)\omega) = e((Hb)\omega) = Gb$$

and thus θ is single valued. By 5.1(iv), we have that $(Ha)\omega \in \mathcal{X}$ if and only if $Ga \in \mathcal{Y}$. Hence θ maps $\mathcal{X}$ onto $\mathcal{Y}$. Let $Ga = Gb \in \mathcal{Y}$. Then $ea = gb$ for some $g \in G$, and by 5.1(iv), $bb^{-1} \geq e$. Hence

$$eab^{-1} = gbb^{-1} = (ge)bb^{-1} = g(ebb^{-1}) = ge = g \in G$$

which implies $ab^{-1} \in H$. Now 4.5 gives $(Ha)\omega = (Hb)\omega$ which proves that θ is one-to-one. Therefore θ is a bijection of $\mathcal{X}$ onto $\mathcal{Y}$.

For any $s \in S$, we have by 5.1(iv)

$$(\mathbf{d}\varphi_H^s)\theta = \{(Ha)\omega \mid \lambda(as) \in H\}\theta = \{Ga \mid Gas \subseteq R\} = \mathbf{d}\psi_G^s$$

and if $(Ha)\omega \in \mathbf{d}\varphi_H^s$, then

$$((Ha)\omega\varphi_H^s)\theta = ((Has)\omega)\theta = Gas = (Ga)s = ((Ha)\omega)\theta\psi_G^s.$$

Let $s \in S$. Then $\mathbf{r}\psi_G^s \in \mathcal{Y}$ by the definition of $\mathbf{d}\psi_G^s$, and $\mathbf{d}\psi_G^s \in \mathcal{Y}$ by 5.1(viii). Further, ψ_G^s is obviously single valued, and is thus a partial transformation of $\mathcal{Y}$. Let $Gas = Gbs \in \mathbf{r}\psi_G^s$. Applying θ^{-1}, we obtain $(Has)\omega = (Hbs)\omega \in \mathbf{d}\varphi_H^s$ which implies $(Ha)\omega\varphi_H^s = (Hb)\omega\varphi_H^s$. Since φ_H^s is one-to-one, it follows that $(Ha)\omega = (Hb)\omega$; now applying θ, we obtain $Ga = Gb$. Consequently, ψ_G^s is one-to-one, and thus ψ_G: $S \to \mathcal{I}(\mathcal{Y})$. Next let $s, t \in S$. Then, using 5.1(viii), we get

$$Ga \in \mathbf{d}(\psi_G^s\psi_G^t) \Leftrightarrow Gas \subseteq R,\ Gast \subseteq R$$

$$\Leftrightarrow Gast \subseteq R \Leftrightarrow Ga \in \mathbf{d}\psi_G^{st},$$

and if this is the case, clearly $(Ga)\psi_G^s\psi_G^t = (Ga)\psi_G^{st}$. Therefore, ψ_G is a homomorphism.

We have now established that ψ_G: $S \to \mathcal{I}(\mathcal{Y})$ is a representation, which together with the first part of the proof will yield that ψ_G is equivalent to φ_H. Since the latter is a transitive representation, so is the former.

For the representations ψ_G, we have the following equivalence criterion.

IV.5.3 Proposition

Let S be an inverse semigroup, G and M be subgroups of S with identities e and f, respectively. Then ψ_G and ψ_M are equivalent representations if and only if there exists $x \in S$ such that $x^{-1}Gx = M$ and $xMx^{-1} = G$. If this is the case, x can be chosen to satisfy $e\mathcal{R}x\mathcal{L}f$ and thus $e\mathcal{D}f$.

Proof. Let $H = G\omega$ and $K = M\omega$.

Necessity. By 5.2, we have that φ_H and φ_K are equivalent, and thus by 4.13, there exists $a \in S$ such that $a^{-1}Ha \subseteq K$ and $aKa^{-1} \subseteq H$. Let $x = eaf$. If $g \in G$, then

$$x^{-1}gx = fa^{-1}(ege)af = f(a^{-1}ga)f \in fKf = M$$

in view of 5.1(ii) and thus $x^{-1}Gx \subseteq M$. Further, $x^{-1}x = f(a^{-1}ea)f = f$ since $a^{-1}ea \in K$ and thus $a^{-1}ea \geq f$. If $m \in M$, then

$$m = fmf = x^{-1}(xmx^{-1})x = x^{-1}(eafmfa^{-1}e)x$$

$$= x^{-1}e(ama^{-1})ex \in x^{-1}Gx$$

since $ama^{-1} \in H$ and thus by 5.1(ii), $e(ama^{-1})e \in G$; hence $M \subseteq x^{-1}Gx$. We have proved that $x^{-1}Gx = M$; one shows symmetrically that also $xMx^{-1} = G$.

Sufficiency. Assume that $x^{-1}Gx = M$ and $xMx^{-1} = G$, and let $y = exf$. For any $h \in H$, we get $ehe \in G$ and thus

$$y^{-1}hy = fx^{-1}(ehe)xf \in fx^{-1}Gxf = fMf = M$$

and hence $y^{-1}Hy \subseteq K$. A symmetric argument shows that also $yKy^{-1} \subseteq H$. Now 4.13 yields that φ_H and φ_K are equivalent. But then by 5.2, we also have that ψ_G and ψ_M are equivalent.

Furthermore, $yy^{-1} = e(xfx^{-1})e = e$ and similarly $y^{-1}y = f$, and hence $e\mathcal{R}y\mathcal{L}f$. For any $g \in G$, we obtain

$$y^{-1}gy = fx^{-1}egexf = f(x^{-1}gx)f = x^{-1}gx$$

since $x^{-1}gx \in M$. Consequently $y^{-1}Gy = M$, and analogously $yMy^{-1} = G$.

We now investigate briefly how extensive are closed inverse subsemigroups of an inverse semigroup S of the form $G\omega$ for a subgroup G of S.

IV.5.4 Lemma

Let S be an inverse semigroup. If e is the zero of E_S, then H_e is an ideal of S.

Proof. Let $a \in H_e$ and $b \in S$. Then

$$(ab)(ab)^{-1} = ae(bb^{-1})a^{-1} = aea^{-1} = aa^{-1} = e,$$

$$(ab)^{-1}(ab) = b^{-1}(a^{-1}a)b = b^{-1}eb \geq e \tag{13}$$

where the inequality takes place since e is the least element of E_S. Hence

$$e = (bb^{-1})e(bb^{-1}) = b(b^{-1}eb)b^{-1} \geq beb^{-1} \geq e$$

so that $e = beb^{-1}$. This also holds if we write b^{-1} instead of b, which gives $e = b^{-1}eb$. Thus by (13), we obtain $(ab)^{-1}(ab) = e$. Consequently, $ab \in H_e$ and H_e is a right ideal of S. We conclude by symmetry that H_e is also a left ideal of S.

IV.5.5 Proposition

Let H be a closed inverse subsemigroup of an inverse semigroup S. Then the following statements are equivalent.

(i) $H = G\omega$ for some subgroup G of S.
(ii) H has a group ideal.
(iii) E_H has a zero.

Proof. (i) *implies* (ii). This follows from 5.1(iii).
(ii) *implies* (iii) trivially.
(iii) *implies* (i). This follows from 5.4.

Not all closed inverse subsemigroups of an inverse semigroup S need be of the form $G\omega$ for a subgroup G of S; a sufficient condition is that S satisfies the minimal condition on descending chains of idempotents. For the case when 5.2 is applicable, it gives more precise information concerning the structure of the representation and 5.3 for the criterion for equivalence than that available in 4.10 and 4.13 for the general case.

Notice that in 5.2, $\mathscr{Y} = \{Ga \mid a \in R\}$ and that the sets Ga are pairwise disjoint, as in the case of groups.

The next result asserts that every representation of an inverse subsemigroup U of an inverse semigroup S by one-to-one partial transformations of a set Y

can be "extended" to a representation of the entire semigroup S by one-to-one partial transformations of a set $Y \cup Z$ for some set Z disjoint from Y.

IV.5.6 Notation

Let S be a semigroup, X a nonempty set, and $\varphi: S \to \mathcal{I}(X)$ be a representation, say $\varphi: s \to \varphi^s$. If Y is a nonempty subset of X such that φ^s maps Y into itself, we write

$$\varphi_Y: s \to \varphi^s|_Y \qquad (s \in S).$$

It is clear that $\varphi_Y: S \to \mathcal{I}(X)$ is a representation.

IV.5.7 Theorem

Let U be an inverse subsemigroup of an inverse semigroup S and let $\varphi: U \to \mathcal{I}(Y)$ be a representation of U. Then there exists a set Z disjoint from Y and a representation $\psi: S \to \mathcal{I}(Y \cup Z)$ of S such that

$$\psi|_U = \varphi \oplus (\psi|_U)_Z.$$

Proof. The argument will be divided into three steps: first we consider $\varphi = \varphi_H$ for a closed inverse subsemigroup H of U, secondly we study the case when φ is effective, and we conclude with the general case.

Step 1. Let H be a closed inverse subsemigroup of U. By this we mean that H is closed in U, and it need not be closed in S. We will have to make a distinction between closure in U and S, so for the former we will write $A\omega_U$ and for the latter $A\omega_S$ for any subset A of U.

Let $\mathcal{Y}$ be the set of all right ω-cosets of H in U. We set $K = H\omega_S$ and let $\mathcal{W}$ be the set of all right ω-cosets of K in S. We thus arrive at the two representations

$$\varphi_H: U \to \mathcal{I}(\mathcal{Y}), \qquad \varphi_K: S \to \mathcal{I}(\mathcal{W}).$$

Note that

$$K \cap U = H, \tag{14}$$

for if $u \in K \cap U$, then $ue \in H$ for some $e \in E_S$, and thus $ue = u[u^{-1}(ue)]$ where $u^{-1}(ue) \in E_U$ so that $u \in H$; this proves that $K \cap U \subseteq H$; the opposite inclusion is trivial. We now define

$$\xi: (Ha)\omega_U \to (Ka)\omega_S \qquad [(Ha)\omega_U \in \mathcal{Y}].$$

Let $(Ha)\omega_U, (Hb)\omega_U \in \mathcal{Y}$; then

$$\begin{aligned} (Ha)\omega_U = (Hb)\omega_U &\Leftrightarrow ab^{-1} \in H && \text{by 4.5} \\ &\Leftrightarrow ab^{-1} \in K && \text{by (14)} \\ &\Leftrightarrow (Ka)\omega_S = (Kb)\omega_S && \text{by 4.5} \end{aligned}$$

which proves that ξ is single valued and is one-to-one. Letting $\mathcal{Z} = \mathcal{W} \setminus \mathcal{Y}\xi$, we obtain $\mathcal{W}$ as a disjoint union of $\mathcal{Y}\xi$, which is in a one-to-one correspondence with $\mathcal{Y}$, and $\mathcal{Z}$.

We now consider $\varphi_K|_U$. Let $u \in U$ and $s \in S$, and assume that $(Ks)\omega_S, (Ksu)\omega_S \in \mathcal{W}$. We will show that

$$(Ks)\omega_S \in \mathcal{Y}\xi \Leftrightarrow (Ksu)\omega_S \in \mathcal{Y}\xi. \tag{15}$$

Suppose first that $(Ks)\omega_S \in \mathcal{Y}\xi$. Then $(Ks)\omega_S = (Kv)\omega_S$ for some $v \in H$ such that $(Hv)\omega_U \in \mathcal{Y}$. Hence $(Ksu)\omega_S = (Kvu)\omega_S$ where $(Hvu)\omega_U \in \mathcal{Y}$ since $\lambda(vu) \in K \cap U = H$ by (14). Thus $(Ksu)\omega_S \in \mathcal{Y}\xi$. Conversely, assume that $(Ksu)\omega_S \in \mathcal{Y}\xi$. Then $(Ksu)\omega_S = (Kt)\omega_S$ for some $t \in U$ such that $(Ht)\omega_U \in \mathcal{Y}$. Let $p = tu^{-1}$. Then $\lambda p = tu^{-1}ut^{-1} \geq \lambda t$ where $\lambda t \in H$ so $\lambda p \in H$ since H is closed in U. Hence $(Hp)\omega_U \in \mathcal{Y}$. Further, $sp^{-1} = s(ut^{-1}) = (su)t^{-1} \in K$ by 4.5 since $(Ksu)\omega_S = (Kt)\omega_S$, which again by 4.5 implies $(Ks)\omega_S = (Kp)\omega_S \in \mathcal{Y}\xi$. This proves relation (15) which indicates that $\varphi_K|_U$ maps each of the sets $\mathcal{Y}\xi$ and $\mathcal{Z}$ into itself, and thus

$$\varphi_K|_U = (\varphi_K|_U)_{\mathcal{Y}\xi} \oplus (\varphi_K|_U)_{\mathcal{Z}}. \tag{16}$$

In order to show that $(\varphi_K|_U)_{\mathcal{Y}\xi}$ is equivalent to φ_H, we let $(Ha)\omega_U \in \mathcal{Y}$ and $u \in U$. Then, in view of (14),

$$(Ha)\omega_U \in \mathbf{d}\varphi_H^u \Leftrightarrow \lambda(au) \in H \Leftrightarrow \lambda(au) \in K \Leftrightarrow (Ka)\omega_S \in \mathbf{d}(\varphi_K^u|_U),$$

and for $(Ha)\omega_U \in \mathbf{d}\varphi_H^u$, we get

$$((Ha)\omega_U\varphi_H^u)\xi = ((Hau)\omega_U)\xi = (Kau)\omega_S = (Ka)\omega_S(\varphi_K^u|_U).$$

Therefore, the representations $\varphi_K|_U$ and φ_H are equivalent.

Step 2. Now let $\varphi\colon U \to \mathcal{I}(Y)$ be an effective representation. By 4.10, $\varphi = \oplus_{\alpha \in A}\varphi_\alpha$ where $\varphi_\alpha\colon U \to \mathcal{I}(Y_\alpha)$ is a transitive representation equivalent to φ_{H_α} for some closed inverse subsemigroup of U, and the sets Y_α form a partition of Y. Let $\mathcal{Y}_\alpha$ be the set of all right ω-cosets of H_α in Y_α, and let θ_α: $Y_\alpha \to \mathcal{Y}_\alpha$ be the bijection which figures in the equivalence of φ_α and φ_{H_α}. By Step 1 above, there is a bijection $\xi_\alpha\colon \mathcal{Y}_\alpha \to \mathcal{Y}_\alpha\xi_\alpha$ with $\mathcal{W}_\alpha = \mathcal{Y}_\alpha\xi_\alpha \cup \mathcal{Z}_\alpha$ the set of all right ω-cosets of $K_\alpha = H_\alpha\omega_S$ in S and $\mathcal{Y}_\alpha\xi_\alpha \cap \mathcal{Z}_\alpha = \varnothing$. Choose a system of sets Z_α for $\alpha \in A$ satisfying

$$Y_\alpha \cap Z_\alpha = \varnothing, \qquad (Y_\alpha \cup Z_\alpha) \cap (Y_\beta \cup Z_\beta) = \varnothing \quad \text{if} \quad \alpha \neq \beta, \tag{17}$$

and for each $\alpha \in A$, let χ_α be a bijection of Z_α and $\mathcal{Z}_\alpha$. Define λ_α on $X_\alpha = Y_\alpha \cup Z_\alpha$ by

$$\lambda_\alpha\colon x \to \begin{cases} x\theta_\alpha\xi_\alpha & \text{if } x \in Y_\alpha \\ x\chi_\alpha & \text{if } x \in Z_\alpha, \end{cases}$$

so that λ_α is a bijection of X_α onto $\mathcal{Y}_\alpha \xi_\alpha \cup \mathcal{Z}_\alpha$. For every $s \in S$, define

$$\psi_\alpha^s = \{(w, w') \in X_\alpha \times X_\alpha | (w\lambda_\alpha, w'\lambda_\alpha) \in \varphi_{K_\alpha}^s\}.$$

Then ψ_α: $S \to \mathcal{I}(X_\alpha)$ is a representation equivalent to φ_{K_α}, and

$$\psi_\alpha|_U = (\psi_\alpha|_U)_{Y_\alpha} \oplus (\psi_\alpha|_U)_{Z_\alpha} = \varphi_\alpha \oplus (\psi_\alpha|_U)_{Z_\alpha}. \tag{18}$$

Now let $Z = \cup_{\alpha \in A} Z_\alpha$, so that $Y \cap Z = \varnothing$, and let $\psi = \oplus_{\alpha \in A} \psi_\alpha$ [note that the sets X_α are pairwise disjoint, see (17)]. Then by (18),

$$\begin{aligned}
\psi|_U &= \Big(\bigoplus_{\alpha \in A} \psi_\alpha\Big)|_U = \bigoplus_{\alpha \in A} \psi_\alpha|_U = \bigoplus_{\alpha \in A} \big(\varphi_\alpha \oplus (\psi_\alpha|_U)_{Z_\alpha}\big) \\
&= \Big(\bigoplus_{\alpha \in A} \varphi_\alpha\Big) \oplus \Big(\bigoplus_{\alpha \in A} (\psi_\alpha|_U)_{Z_\alpha}\Big) = \Big(\bigoplus_{\alpha \in A} \varphi_\alpha\Big) \oplus \Big(\bigoplus_{\alpha \in A} \psi_\alpha|_U\Big)_Z \\
&= \varphi \oplus (\psi|_U)_Z.
\end{aligned}$$

This is the desired form for the special case of the theorem for an effective representation.

Step 3. Finally, let φ: $U \to \mathcal{I}(Y)$ be any representation. Letting $Y' = \cup_{u \in U} \mathbf{d}\varphi^u$, we obtain an effective representation φ: $U \to \mathcal{I}(Y')$. The result obtained in Step 2 above may be applied to the latter representation obtaining the required ψ: $S \to \mathcal{I}(Y' \cup Z)$. Here Z can be chosen such that $Y \cap Z = \varnothing$, so we may consider ψ as a representation of S in $\mathcal{I}(Y \cup Z)$. Therefore, ψ has all the required properties.

The transition from a representation by partial transformations to one by full transformations can be effected as follows.

IV.5.8 Lemma

Let φ: $S \to \mathcal{F}(X)$ be a representation of a semigroup S by partial transformations of a set X. Let 0 be an element not contained in X. Letting φ: $s \to \varphi^s$, we define φ_0^s on $X_0 = X \cup \{0\}$ by

$$x\varphi_0^s = \begin{cases} x\varphi^s & \text{if } x \in \mathbf{d}\varphi^s \\ 0 & \text{otherwise.} \end{cases}$$

Then the mapping φ_0 on S defined by φ_0: $s \to \varphi_0^s$ is a representation φ_0: $S \to \mathcal{T}(X_0)$.

Proof. Since φ is a representation, for any $a, b \in S$, we have $x\varphi_0^a\varphi_0^b \neq 0$ if and only if $x\varphi_0^{ab} \neq 0$. If this is the case, then

$$x\varphi_0^a\varphi_0^b = x\varphi^a\varphi^b = x\varphi^{ab} = x\varphi_0^{ab},$$

and otherwise $x\varphi_0^a\varphi_0^b = x\varphi_0^{ab} = 0$. Consequently, φ_0 is a homomorphism.

The next result provides a passage from a representation of an inverse semigroup by transformations of a set to a representation by one-to-one partial transformations. This is a kind of converse of the preceding result but only for inverse semigroups.

IV.5.9 Theorem

Let S be an inverse semigroup, X be a set, and $\varphi: S \to \mathcal{T}(X)$ be a representation, say $\varphi: s \to \varphi^s$. For every $s \in S$, let $\overline{\varphi^s} = \varphi|_{\mathbf{r}\varphi^{s^{-1}}}$, and define

$$\overline{\varphi}: s \to \overline{\varphi^s} \qquad (s \in S).$$

Then $\overline{\varphi}: S \to \mathcal{I}(X)$ is a representation such that φ and $\overline{\varphi}$ induce the same congruence on S.

Proof. First note that for any $s \in S$, $\mathbf{d}\overline{\varphi^s} = \mathbf{r}\varphi^{s^{-1}}$. Let $x, y \in \mathbf{d}\overline{\varphi^s}$ be such that $x\overline{\varphi^s} = y\overline{\varphi^s}$. Then $x, y \in \mathbf{r}\varphi^s$ and thus $x = u\varphi^{s^{-1}}$ and $y = v\varphi^{s^{-1}}$ for some $u, v \in \mathbf{r}\varphi^s$. It follows that

$$x\overline{\varphi^s} = x\varphi^s = u\varphi^{s^{-1}}\varphi^s = u\varphi^{s^{-1}s} = u$$

and analogously $y\overline{\varphi^s} = v$, which implies that $x = y$. Consequently, $\overline{\varphi^s}$ is one-to-one so that $\overline{\varphi^s} \in \mathcal{I}(X)$.

We verify next that $\overline{\varphi}$ is a homomorphism. Let $x \in \mathbf{d}\overline{\varphi^s}$ and $x\overline{\varphi^s} \in \mathbf{d}\overline{\varphi^t}$. Then $x \in \mathbf{r}\varphi^{s^{-1}}$ and $x\varphi^s \in \mathbf{r}\varphi^{t^{-1}}$ so that $x = u\varphi^{s^{-1}}$ and $x\varphi^s = v\varphi^{t^{-1}}$ for some $x, y \in X$. Hence

$$x = u\varphi^{s^{-1}} = \left(u\varphi^{s^{-1}}\right)\varphi^s\varphi^{s^{-1}} = \left(x\varphi^s\right)\varphi^{s^{-1}} = v\varphi^{t^{-1}}\varphi^{s^{-1}} = v\varphi^{(st)^{-1}}$$

and thus $x \in \mathbf{d}\overline{\varphi^{st}}$. Conversely, assume $x \in \mathbf{d}\overline{\varphi^{st}}$. Then $x \in \mathbf{r}\varphi^{(st)^{-1}}$ so that $x = w\varphi^{(st)^{-1}}$ for some $w \in X$. It follows that $x = (w\varphi^{t^{-1}})\varphi^{s^{-1}}$ whence $x \in \mathbf{r}\varphi^{s^{-1}}$ and

$$x\varphi^s = w\varphi^{t^{-1}}\varphi^{s^{-1}s} = x\varphi^{t^{-1}}\varphi^{tt^{-1}}\varphi^{s^{-1}s} = \left(x\varphi^{t^{-1}}\varphi^{s^{-1}s}\varphi^t\right)\varphi^{t^{-1}},$$

that is to say, $x\varphi^s \in \mathbf{r}\varphi^{t^{-1}}$. But then $x \in \mathbf{d}\overline{\varphi^s}$ and $x\overline{\varphi^s} \in \mathbf{d}\overline{\varphi^t}$ which implies that $x \in \mathbf{d}(\overline{\varphi^s}\overline{\varphi^t})$. We have proved that $\mathbf{d}(\overline{\varphi^s}\overline{\varphi^t}) = \mathbf{d}\overline{\varphi^{st}}$ which evidently implies that $\overline{\varphi^s}\overline{\varphi^t} = \overline{\varphi^{st}}$. Therefore, $\overline{\varphi}$ is a homomorphism.

Suppose next that $\varphi^s = \varphi^t$. By taking inverses, we get $\varphi^{s^{-1}} = \varphi^{t^{-1}}$ and thus $\mathbf{r}\varphi^{s^{-1}} = \mathbf{r}\varphi^{t^{-1}}$ which gives $\overline{\varphi^s} = \overline{\varphi^t}$. Conversely, assume that $\overline{\varphi^s} = \overline{\varphi^t}$. Again by taking inverses, we get $\overline{\varphi^{s^{-1}}} = \overline{\varphi^{t^{-1}}}$, and thus for any $x \in \mathbf{d}\varphi^s$, we have

$$x\varphi^s = \left(x\varphi^s\varphi^{s^{-1}}\right)\overline{\varphi^s} = \left(x\varphi^s\varphi^{s^{-1}}\right)\overline{\varphi^t} = x\varphi^{ss^{-1}}\varphi^{tt^{-1}}\varphi^t$$

$$= x\varphi^{tt^{-1}}\varphi^{ss^{-1}}\varphi^t = \left(x\varphi^t\right)\overline{\varphi^{t^{-1}}}\,\overline{\varphi^t}\,\overline{\varphi^{t^{-1}}}\,\overline{\varphi^t} = x\varphi^t\overline{\varphi^{t^{-1}}}\,\overline{\varphi^t} = x\varphi^t$$

which shows that $\varphi^s = \varphi^t$. Consequently, φ and $\overline{\varphi}$ induce the same congruence on S.

IV.5.10 Exercises

(i) Let H be a subgroup of a group G. Show that

$$\ker \psi_H = \{h \in H \mid g^{-1}hg \in H \text{ for all } g \in G\},$$

and that $\ker \psi_H$ is the greatest normal subgroup of G contained in H. Also show that every transitive representation of G is equivalent to one so constructed.

(ii) Let $S = B(G, I)$ be a Brandt semigroup, let H be a subgroup of G, $i \in I$, and

$$H_i = \{(i, h, i) \in S \mid h \in H\}.$$

Show that $\varphi_{H_i} = \psi_{H_i}$ and

$$\ker \psi_{H_i} = \{(j, s, j) \in S \mid g^{-1}sg \in H \text{ for all } g \in G\} \cup \{0\}$$

$$= \{(j, s, j) \in S \mid s \in \ker \psi_H \text{ within } G\} \cup \{0\}.$$

Also show that every transitive representation of S by one-to-one partial transformations of a set with more than one element is equivalent to one so constructed.

(iii) Let $S = B(G, I)$ be a Brandt semigroup and let k be the number of classes of conjugate subgroups of G. Show that S has exactly $k + 1$ inequivalent transitive representations by one-to-one partial transformations.

(iv) Let X be a set with (a finite number) n elements. Show that $\mathcal{I}(X)$ has exactly $\Sigma_{i=1}^{n} k_i$ inequivalent transitive representations by one-to-one partial transformations, where k_i is the number of classes of conjugate subgroups of the symmetric group $\mathfrak{S}_i$ on i letters.

(v) Show that for any inverse semigroup S, the Wagner representation is equivalent to the sum $\oplus_{e \in E_S} \psi_{\{e\}}$.

(vi) Show that for any inverse semigroup S and for a set Δ of idempotents of S taken one from each $\mathcal{D}$-class of S, the Munn representation is equivalent to the sum $\oplus_{e \in \Delta} \psi_{H_e}$.

(vii) Let S be an inverse semigroup and ρ be the right regular representation of S. Show that $\bar{\rho}$ defined in 5.9 is the Wagner representation.

(viii) Let K be a normal subsemigroup of an inverse semigroup S. Define a function $\gamma = \gamma(S : K)$ by

$$\gamma : a \to \gamma^a \qquad (a \in S)$$

where γ^a is defined by

$$\gamma^a : k \to a^{-1}ka \qquad (k \in K).$$

Show that $\gamma : S \to \mathfrak{T}(K)$ is a representation of S and that $\overline{\gamma(S : E_S)}$ is the Munn representation of S.

A variant of 5.2 and 5.3 and parts of 5.1, together with a converse, can be found in Ponizovskiĭ [4]; see also Petrich [4] and Reilly [1]. Result 5.7 is due to Hall [4]; the proof here follows Howie [4]. Lemmas 5.8 and 5.9 are due to Wagner [3]; see also Preston [10].

V

THE TRANSLATIONAL HULL

1. The Wagner representation of an inverse semigroup S can be extended to an embedding of the translational hull of S into the symmetric inverse semigroup on S. As a consequence, we get that the translational hull of an inverse semigroup is again an inverse semigroup. Several lemmas are proved here which come in very handy in the consideration of the translational hull of an inverse semigroup.

2. There are two hulls intimately related to the translational hull of an inverse semigroup S. The first of these hulls consists of all one-to-one partial right translations whose domains are left ideals under composition. The second one is defined by means of certain subsets of S under the usual multiplication of complexes. These two hulls are proved to be isomorphic and each of them contains a copy of the translational hull of S.

3. The translational hull of a Clifford semigroup S is again a Clifford semigroup. If S is given the Clifford representation, then the translational hull of S admits such a representation in terms of retract ideals of the semilattice of idempotents of S and inverse limits of maximal subgroups of S. An inverse semigroup which is a subdirect product of a semilattice and a group represents an interesting special case.

4. If a Brandt semigroup S is given in the form $B(G, I)$, then its translational hull can be expressed in terms of the wreath product of G and $\mathcal{I}(I)$. This description is sufficiently explicit to make it possible to construct all ideal extensions of one Brandt semigroup by another. The construction is notationally complex but quite transparent and depends on a few independent parameters. All this can be extended to a construction of all ideal extensions of one primitive inverse semigroup by another.

V.1 THE TRANSLATIONAL HULL OF AN INVERSE SEMIGROUP

We prove here that the translational hull of an arbitrary inverse semigroup S is isomorphic to the idealizer of S^w in $\mathscr{I}(S)$. Hence the Wagner representation provides a dense embedding of any inverse semigroup. As a consequence we also have that the translational hull of an inverse semigroup is again an inverse semigroup.

The following lemma is not surprising.

V.1.1 Lemma

Every inverse semigroup is reductive.

Proof. Let S be an inverse semigroup and $a, b \in S$ be such that $ax = bx$ for all $x \in S$. Then

$$a = aa^{-1}a = (aa^{-1})^{-1}a = (ba^{-1})^{-1}a = ab^{-1}a,$$

$$b^{-1} = b^{-1}(bb^{-1}) = b^{-1}ab^{-1},$$

which by the uniqueness of inverses yields $a^{-1} = b^{-1}$ and thus $a = b$. A symmetric argument shows that $xa = xb$ for all $x \in S$ implies $a = b$.

V.1.2 Corollary

Let S be an inverse semigroup. Then the homomorphisms $\pi : S \to \Omega(S)$, and

$$(\lambda, \rho) \to \lambda, \qquad (\lambda, \rho) \to \rho \qquad [(\lambda, \rho) \in \Omega(S)]$$

are all monomorphisms.

Proof. This is a direct consequence of 1.1, I.8.6, and I.8.13.

Hence for an inverse semigroup S, we have

$$S \cong \Pi(S), \qquad \Omega(S) \cong \tilde{\Lambda}(S) \cong \tilde{\mathrm{P}}(S).$$

It is natural to try to "extend" the Wagner representation to the translational hull of the semigroup [identifying S and $\Pi(S)$]. Since $\Pi(S)$ is an ideal of $\Omega(S)$, such an extension would map $\Omega(S)$ into the idealizer of S^w in $\mathscr{I}(S)$. Moreover, the Wagner representation maps each element a of S onto a truncated part of the inner right translation ρ_a. Hence an extension can be performed on the projection $\tilde{\mathrm{P}}(S)$ of $\Omega(S)$ in the semigroup $\mathrm{P}(S)$ of right translations. The basic idea behind the proof below that the Wagner representation can be "extended" to the translational hull is the same one as in IV.5.9.

V.1.3 Theorem

Let S be an inverse semigroup. The mapping χ defined by

$$\chi:(\lambda,\rho)\to\rho|_{A_\rho}\qquad[(\lambda,\rho)\in\Omega(S)]$$

where $A_\rho=\{x\in S|x\mathcal{R}(x\rho)\}$, is an isomorphism of $\Omega(S)$ onto the idealizer I of S^w in $\mathcal{I}(S)$. Moreover, χ is the only homomorphism of $\Omega(S)$ into $\mathcal{I}(S)$ making the diagram

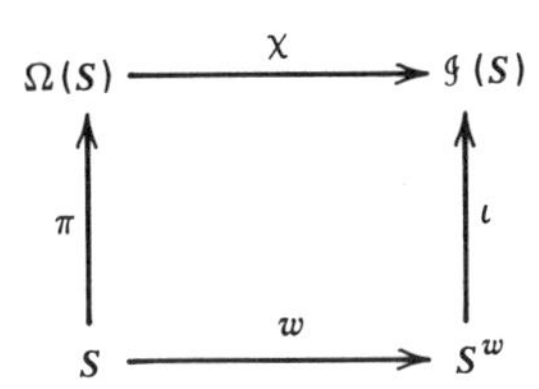

commute (ι is the inclusion mapping).

Proof. 1. Let $(\lambda,\rho)\in\Omega(S)$, $x,y\in A_\rho$ and assume that $x\rho=y\rho$. Then

$$\left[\lambda(x\rho)^{-1}x\right]\left[\lambda(x\rho)^{-1}x\right]=\lambda(x\rho)^{-1}(x\rho)(x\rho)^{-1}x=\lambda(x\rho)^{-1}x$$

so that $\lambda(x\rho)^{-1}x\in E_S$, which gives

$$x=(xx^{-1})x=(x\rho)(x\rho)^{-1}x=(y\rho)(x\rho)^{-1}x=y\left[\lambda(x\rho)^{-1}x\right]\leq y.$$

Symmetrically, we also have $y\leq x$ and thus $x=y$. Hence χ maps $\Omega(S)$ into $\mathcal{I}(S)$.

2. Let $(\lambda,\rho),(\sigma,\tau)\in\Omega(S)$. Then

$$\mathbf{d}\big((\rho\tau)|_{A_{\rho\tau}}\big)=\{x\in S|x\mathcal{R}(x\rho\tau)\},$$

$$\mathbf{d}\big(\rho|_{A_\rho}\tau|_{A_\tau}\big)=\{x\in S|x\mathcal{R}(x\rho),(x\rho)\mathcal{R}(x\rho\tau)\}.$$

If $x\mathcal{R}(x\rho\tau)$, then

$$x\mathcal{R}xx^{-1}=(x\rho\tau)(x\rho\tau)^{-1}=(x\rho)\left[\sigma(x\rho\tau)^{-1}\right]\mathcal{R}(x\rho),$$

and thus the two domains are equal. It is now clear that $(\rho\tau)|_{A_{\rho\tau}}=\rho|_{A_\rho}\tau|_{A_\tau}$. Consequently, χ is a homomorphism.

3. With the same notation, assume that $\rho|_{A_\rho} = \tau|_{A_\tau}$. For any $x \in S$, we have

$$x\rho = (x\rho)(x\rho)^{-1}(x\rho)(x\rho)^{-1}(x\rho) = \left[(x\rho)(x\rho)^{-1}x\right]\lambda\left[(x\rho)^{-1}(x\rho)\right]$$

which gives

$$(x\rho)(x\rho)^{-1}x\mathcal{R}(x\rho) = \left[(x\rho)(x\rho)^{-1}x\right]\rho$$

so that $(x\rho)(x\rho)^{-1}x \in A_\rho$. Further,

$$x\rho = (x\rho)(x\rho)^{-1}(x\rho) = \left[(x\rho)(x\rho)^{-1}x\right]\rho$$

$$= \left[(x\rho)(x\rho)^{-1}x\right]\tau = (x\rho)(x\rho)^{-1}(x\tau) \leq x\tau.$$

Symmetrically, we have $x\tau \leq x\rho$ so that $x\rho = x\tau$. This shows that $\rho = \tau$, and thus 1.2 also gives $\lambda = \sigma$. Therefore, χ is one-to-one.

4. For any $s \in S$, we have

$$A_{\rho_s} = \{x \in S | x\mathcal{R}xs\} = \{x \in S | x = xss^{-1}\} = Ss^{-1}$$

so that $\rho_s|_{A_{\rho_s}} = w^s$. Hence $\pi_s\chi = w^s$, and the above diagram commutes.

5. Since $\Pi(S)$ is an ideal of $\Omega(S)$, we have that $\Pi(S)\chi$ is an ideal of $\Omega(S)\chi$. We have just seen that $\Pi(S)\chi = S^w$ which then implies that $\Omega(S)\chi \subseteq I$, where $I = i_{\mathcal{I}(S)}(S^w)$.

Let $\alpha \in I$. Then for any $x \in S$, we have $w^x\alpha, \alpha w^x \in S^w$ and thus we may define a pair (λ, ρ) of functions by the requirement

$$w^x\alpha = w^{x\rho}, \qquad \alpha w^x = w^{\lambda x} \qquad (x \in S).$$

Easy verification shows that $(\lambda, \rho) \in \Omega(S)$.

Now let $x \in A_\rho$. Since $x\mathcal{R}(x\rho)$, we have

$$xx^{-1} = (x\rho)(x\rho)^{-1} \in \mathbf{d}w^{x\rho}$$

whence $x = (xx^{-1})w^x \in \mathbf{d}\alpha$. Conversely, let $x \in \mathbf{d}\alpha$. Then

$$xx^{-1} \in \mathbf{d}(w^x\alpha) = \mathbf{d}w^{x\rho} = S(x\rho)^{-1}$$

so that

$$xx^{-1} = (xx^{-1})(x\rho)(x\rho)^{-1} = (x\rho)(x\rho)^{-1},$$

and thus $x \in A_\rho$. If this is the case, then

$$x\alpha = (xx^{-1})w^x\alpha = (xx^{-1})w^{x\rho} = (xx^{-1})(x\rho) = x\rho$$

and we get $\rho|_{A_\rho} = \alpha$. Consequently, $(\lambda, \rho)\chi = \alpha$, which shows that χ maps $\Omega(S)$ onto I.

6. Let $\chi' : \Omega(S) \to \mathcal{I}(S)$ be a homomorphism which makes the above diagram commute. As above, we get that $\Omega(S)\chi' \subseteq I$. For any $x \in S$ and $(\lambda, \rho) \in \Omega(S)$, we obtain

$$\begin{aligned}[(\lambda, \rho)\chi'\chi^{-1}]\pi_x &= [(\lambda, \rho)\chi'\chi^{-1}][\pi_x\chi'\chi^{-1}] = [(\lambda, \rho)\pi_x]\chi'\chi^{-1} \\ &= (\lambda, \rho)\pi_x\end{aligned}$$

since $(\lambda, \rho)\pi_x \in \Pi(S)$; analogously, we get $\pi_x[(\lambda, \rho)\chi'\chi^{-1}] = \pi_x(\lambda, \rho)$. Since this holds for all $x \in S$, by I.8.7, we have that $(\lambda, \rho)\chi'\chi^{-1} = (\lambda, \rho)$ and thus $\chi' = \chi$. This establishes uniqueness of χ.

As important consequences of 1.3, we have the following results.

V.1.4 Corollary

If S is an inverse semigroup, so is $\Omega(S)$.

Proof. It is obvious that the idealizer of an inverse subsemigroup of an inverse semigroup is itself an inverse semigroup. Hence the corollary follows immediately from 1.3.

V.1.5 Corollary

For any inverse semigroup S, the Wagner representation of S is a dense embedding of S into $\mathcal{I}(S)$.

Proof. This follows directly from 1.3.

The next two lemmas are very useful in the study of the translational hull of an inverse semigroup.

V.1.6 Lemma

In any inverse semigroup S, we have

$$\lambda e = e\rho \in E_S \qquad [e \in E_S, (\lambda, \rho) \in E_{\Omega(S)}].$$

Proof. Indeed,

$$\begin{aligned}[(\lambda e)(\lambda e)^{-1}]\rho[(\lambda e)(\lambda e)^{-1}]\rho &= [(\lambda e)(\lambda e)^{-1}(\lambda^2 e)(\lambda e)^{-1}]\rho \\ &= [(\lambda e)(\lambda e)^{-1}]\rho \in E_S\end{aligned}$$

which implies

$$\lambda e = (\lambda e)(\lambda e)^{-1}(\lambda e) = \left[(\lambda e)(\lambda e)^{-1}\right]\rho e$$
$$= e\left[(\lambda e)(\lambda e)^{-1}\right]\rho e = e(\lambda e),$$

and symmetrically $e\rho = (e\rho)e$, which gives $\lambda e = e\rho$. It now follows easily that $\lambda e \in E_S$.

V.1.7 Lemma

In any inverse semigroup S,

$$(\lambda a)^{-1} = a^{-1}\rho^{-1}, \qquad (a\rho)^{-1} = \lambda^{-1}a^{-1} \qquad [a \in S, (\lambda, \rho) \in \Omega(S)].$$

Proof. Indeed, using 1.6, we obtain

$$(\lambda a)(a^{-1}\rho^{-1})(\lambda a) = \lambda\left[(aa^{-1})\rho^{-1}\rho\right]a = \lambda\lambda^{-1}\lambda(aa^{-1})a = \lambda a,$$
$$(a^{-1}\rho^{-1})(\lambda a)(a^{-1}\rho^{-1}) = a^{-1}\left[\lambda^{-1}\lambda(aa^{-1})\right]\rho^{-1} = a^{-1}\left[(aa^{-1})\rho^{-1}\rho\rho^{-1}\right]$$
$$= a^{-1}\rho^{-1}$$

which implies $a^{-1}\rho^{-1} = (\lambda a)^{-1}$. The second formula is proved similarly.

Note that the assertions both in 1.6 and 1.7 are reminiscent of the state of affairs in an inverse semigroup; indeed, 1.6 looks like commuting of idempotents and 1.7 resembles the formula for the inverse of the product of two elements.

The set A_ρ in 1.3 admits an interesting interpretation as follows.

V.1.8 Proposition

Let S be an inverse semigroup and $(\lambda, \rho) \in \Omega(S)$. Then $A_\rho = S\rho^{-1}$.

Proof. If $x \in A_\rho$, then using 1.6 and 1.7, we get

$$x = (x\rho)(x\rho)^{-1}x = (x\rho)(\lambda^{-1}x^{-1})x = x(\lambda\lambda^{-1}x^{-1}x)$$
$$= x\left(x^{-1}x\rho\rho^{-1}\right) = x\rho\rho^{-1} \in S\rho^{-1},$$

and thus $A_\rho \subseteq S\rho^{-1}$. Conversely, if $x \in S\rho^{-1}$, then using 1.7, we obtain

$$xx^{-1} = x\left(x\rho\rho^{-1}\right)^{-1} = x\left[\lambda(x\rho)^{-1}\right] = (x\rho)(x\rho)^{-1}$$

so that $x \in A_\rho$. Consequently, $S\rho^{-1} \subseteq A_\rho$ and the equality prevails.

From 1.3 and 1.8 we see that the mapping $\rho \to \rho|_{A_\rho}$, which is part of χ, is precisely the mapping in IV.5.9, so that $(\lambda, \rho)\chi = \bar{\rho}$.

V.1.9 Exercises

(i) Let S be an inverse semigroup and $(\lambda, \rho) \in \Omega(S)$. Show that the following statements are equivalent.

(α) $\lambda^2 = \lambda$.

(β) $\rho^2 = \rho$.

(γ) $\lambda E_S \subseteq E_S$.

(δ) $E_S \rho \subseteq E_S$.

Deduce that $E_{\Omega(S)} \cong \Omega(E_S)$. Also prove that $E_{\Omega(S)}\zeta \cong \Omega(E_S\zeta)$.

(ii) Let S be an inverse semigroup. Show that any automorphism of $\Omega(S)$ which leaves all idempotents of $\Pi(S)$ fixed also leaves all idempotents of $\Omega(S)$ fixed.

(iii) Let S be an inverse semigroup with zero. Show that 0 is a prime ideal of S if and only if π_0 is a prime ideal of $\Omega(S)$.

That the translational hull of an inverse semigroup is an inverse semigroup was first proved directly by Ponizovskiĭ [5]. A somewhat modified version of his proof appeared in Petrich [8]. Gould [1] devised a simple proof for this result. The handy formula in 1.6 was discovered by Ault [3]; this paper and Reilly [9] contain further information concerning the translational hull of an inverse semigroup. The paper Schein [17] contains many results related to the subject of this section.

V.2 ONE-TO-ONE PARTIAL RIGHT TRANSLATIONS

The semigroup of all mappings in the title of this section of an inverse semigroup S is proved here to be isomorphic to the semigroup of all permissible subsets of S under complex multiplication. Several interesting subsemigroups of the latter are then identified, including a copy of the translational hull of S.

We start with partial translations.

V.2.1 Definition

A *one-to-one partial right translation* of a semigroup S is a one-to-one partial transformation ρ which satisfies: $\mathbf{d}\rho$ is a left ideal of S and

$$x(y\rho) = (xy)\rho \qquad (x \in S,\ y \in \mathbf{d}\rho).$$

V.2.2 Lemma

For any inverse semigroup S, the set $\hat{S}$ of all one-to-one partial right translations is an inverse subsemigroup of $\mathcal{I}(S)$.

Proof. Let $\rho \in \hat{S}$ and $x \in S$, $y \in \mathbf{d}\rho^{-1}$. Then $y \in \mathbf{r}\rho$, so letting $z = y\rho^{-1}$, we obtain $xy = x(z\rho) = (xz)\rho$ and thus $xy \in \mathbf{r}\rho = \mathbf{d}\rho^{-1}$. Hence $\mathbf{d}\rho^{-1}$ is a left ideal of S. Further, $x(y\rho^{-1}) = xz = (xy)\rho^{-1}$ which shows that $\rho^{-1} \in \hat{S}$.

Now let $\rho, \sigma \in \hat{S}$. We have just seen that $\mathbf{r}\rho$ is a left ideal of S and hence $\mathbf{r}\rho \cap \mathbf{d}\sigma \neq \varnothing$ since S is an inverse semigroup. This implies that $\rho\sigma \neq \varnothing$. Further, for any $x \in S$ and $y \in \mathbf{d}(\rho\sigma)$, we have $y\rho \in \mathbf{d}\sigma$ so that $(xy)\rho \in \mathbf{r}\rho \cap \mathbf{d}\sigma$ and thus $xy \in \mathbf{d}(\rho\sigma)$. It follows that $\mathbf{d}(\rho\sigma)$ is a left ideal, whence it is clear that $\rho\sigma \in \hat{S}$.

V.2.3. Notation

For any semigroup S, let $\hat{S}$ denote the *semigroup of all one-to-one partial right translations on* S.

In view of IV.1.4, the natural order relation on the inverse semigroup $\hat{S}$ coincides with the inclusion relation of elements of S as binary relations on S.

We now turn to complexes.

V.2.4 Definition

A nonempty subset H of an inverse semigroup S is *permissible* if

(i) $a \leq b, b \in H \Rightarrow a \in H \quad (a \in S)$,

(ii) $a, b \in H \Rightarrow a^{-1}b, ab^{-1} \in E_S$.

According to the terminology introduced in I.2.4 and III.2.11, permissible subsets are precisely the order ideals in which any two elements are compatible. It is convenient to introduce the following notions.

V.2.5 Definition

Let S be an inverse semigroup. For any nonempty subsets H and K of S,

$$HK = \{hk \mid h \in H, k \in K\} \text{ is the } \textit{complex product} \text{ of } H \text{ and } K,$$

$$H^{-1} = \{h^{-1} \mid h \in H\} \text{ is the } \textit{inverse} \text{ of } H.$$

Condition (i) in 2.4 can be expressed as $HE_S \subseteq H$ (or $E_SH \subseteq H$, or $HE_S = H$ or $E_SH = H$). Condition (ii) in 2.4 can be written as $H^{-1}H, HH^{-1} \subseteq E_S$.

V.2.6 Lemma

For any inverse semigroup S, the set $C(S)$ of all permissible subsets of S is an inverse semigroup under complex multiplication in which H^{-1} is the inverse of H and the natural order coincides with inclusion.

Proof. Let $H, K \in C(S)$. Then $(HK)E_S = H(KE_S) = HK$ and HK is an order ideal of S. Further,

$$(HK)^{-1}(HK) = (K^{-1}H^{-1})(HK) = K^{-1}(H^{-1}H)K \subseteq K^{-1}E_SK$$

$$= K^{-1}(E_SK) = K^{-1}K \subseteq E_S$$

and analogously $(HK)(HK)^{-1} \subseteq E_S$ which proves that HK is permissible. Moreover, $H^{-1}E_S = (E_SH)^{-1} = H^{-1}$ and H^{-1} is an order ideal of S. Since $(H^{-1})^{-1} = H$, we deduce that H^{-1} is permissible. Finally, $HH^{-1}H \subseteq E_SH \subseteq H$, and for any $h \in H$, $h = hh^{-1}h \in HH^{-1}H$, so that $H = HH^{-1}H$. Thus $C(S)$ is regular.

Let $H^2 = H$ and let $a \in H$. Then $a = kg$ for some $k, g \in H$ since $H^2 = H$. Hence $a = (kga^{-1})a = ha$ with $h = kga^{-1}$. Since $k \in H$ and $ga^{-1} \in E_S$, we have $h = k(ga^{-1}) \in HE_S \subseteq H$. Hence $aa^{-1} = haa^{-1} \in H$ and thus $aa^{-1} \in H$. Now $aa^{-1}, a \in H$ implies $(aa^{-1})^{-1}a \in E_S$ so that $a \in E_S$. It follows that $H \subseteq E_S$. Consequently, any two idempotents of $C(S)$ commute.

Let $A, B \in C(S)$. Assume first that $A \leq B$. Then $A = BK$ for some $K \in E_{C(S)}$. By the above, we have $A = BK \subseteq BE_S = B$. Conversely, suppose that $A \subseteq B$. Then

$$A = AA^{-1}A \subseteq BA^{-1}A = BK$$

where $K = A^{-1}A \in E_{C(S)}$. Let $a \in BK$ so that $a = bc^{-1}d$ with $b \in B$, $c, d \in A$. Hence $b, c \in B$ and thus $bc^{-1} \in E_S$ which yields $a = (bc^{-1})d \in E_SA \subseteq A$. Consequently, $A = BK$ and $A \leq B$.

V.2.7 Notation

For any inverse semigroup S, let $C(S)$ denote the (inverse) semigroup of permissible subsets of S under complex multiplication.

The relationship of $\hat{S}$ and $C(S)$ is the subject of the next result.

V.2.8 Theorem

Let S be an inverse semigroup. Define a function Φ by

$$\Phi : \rho \to \Phi_\rho = E_{\mathbf{d}\rho}\rho \qquad (\rho \in \hat{S}).$$

Then Φ is an isomorphism of $\hat{S}$ onto $C(S)$ and

$$\Phi : w^a \to [a] \qquad (a \in S).$$

Proof. 1. Let $\rho \in \hat{S}$; we will show first that $\Phi_\rho \in C(S)$. Let $e \in E_{\mathbf{d}\rho}$, $a \le e\rho$. Then $a = f(e\rho) = (fe)\rho$ for some $f \in E_S$. Hence $fe \in E_{\mathbf{d}\rho}$ and thus $a \in \Phi_\rho$, which proves that Φ_ρ is an order ideal of S.

Next let $e, f \in E_{\mathbf{d}\rho}$. Then

$$e\rho = (e\rho)(e\rho)^{-1}(e\rho) = \left[(e\rho)(e\rho)^{-1}e\right]\rho$$

implies

$$e = (e\rho)(e\rho)^{-1}e = e(e\rho)(e\rho)^{-1} = (e\rho)(e\rho)^{-1} \tag{1}$$

since ρ is one-to-one. It follows that

$$(e\rho)^{-1} = (e\rho)^{-1}(e\rho)(e\rho)^{-1} = (e\rho)^{-1}e \tag{2}$$

whence

$$(e\rho)^{-1}(f\rho) = (e\rho)^{-1}e(f\rho) = (e\rho)^{-1}(ef)\rho = (e\rho)^{-1}f(e\rho) \in E_S.$$

Further,

$$(e\rho)^{-1}(e\rho)(f\rho)^{-1}(f\rho) = (f\rho)^{-1}(f\rho)(e\rho)^{-1}(e\rho)$$

implies

$$(e\rho)^{-1}(e\rho)(f\rho)^{-1}f = (f\rho)^{-1}(f\rho)(e\rho)^{-1}e$$

since ρ is one-to-one, whence by (2) applied to f and e,

$$(e\rho)^{-1}(e\rho)(f\rho)^{-1} = (f\rho)^{-1}(f\rho)(e\rho)^{-1}$$

so that

$$(e\rho)(f\rho)^{-1} = (e\rho)\left[(e\rho)^{-1}(e\rho)(f\rho)^{-1}\right] = (e\rho)(f\rho)^{-1}(f\rho)(e\rho)^{-1} \in E_S.$$

Consequently, $\Phi_\rho \in C(S)$.

2. In order to show that Φ is a homomorphism, we let $\rho, \sigma \in \hat{S}$. If $e \in E_{\mathbf{d}\rho}$ and $f \in E_{\mathbf{d}\sigma}$, then

$$\begin{aligned}(e\rho)(f\sigma) &= \left[(e\rho)(e\rho)^{-1}(e\rho)f\right]\sigma = \left[(e\rho)f(e\rho)^{-1}(e\rho)\right]\sigma \\ &= \left[(e\rho)f(e\rho)^{-1}e\right]\rho\sigma \in E_{\mathbf{d}(\rho\sigma)}\end{aligned}$$

which proves that $\Phi_\rho\Phi_\sigma \subseteq \Phi_{\rho\sigma}$.

Conversely, let $e \in E_{\mathbf{d}(\rho\sigma)}$ and $a = e\rho\sigma$. Then $e \in \mathbf{d}\rho$, $e\rho \in \mathbf{d}\sigma$, and thus $(e\rho)^{-1}(e\rho) \in \mathbf{d}\sigma$, so that

$$a = (e\rho)\left[(e\rho)^{-1}(e\rho)\right]\sigma \in (E_{\mathbf{d}\rho}\rho)(E_{\mathbf{d}\sigma}\sigma)$$

which proves the inclusion $\Phi_{\rho\sigma} \subseteq \Phi_\rho\Phi_\sigma$. Hence Φ is a homomorphism.

3. To prove that Φ is one-to-one, we let $\Phi_\rho = \Phi_\sigma$ with $\rho, \sigma \in \hat{S}$. Let $a \in \mathbf{d}\rho$; then $e = a^{-1}a \in \mathbf{d}\rho$ and hence $e\rho \in \Phi_\rho = \Phi_\sigma$. It follows that $e\rho = f\sigma$ for some $f \in E_{\mathbf{d}\sigma}$. Hence by (1),

$$e = (e\rho)(e\rho)^{-1} = (f\sigma)(f\sigma)^{-1} = f$$

so that $e \in \mathbf{d}\sigma$, and thus $a = ae \in \mathbf{d}\sigma$. Further,

$$a\rho = a(e\rho) = a(e\sigma) = a\sigma.$$

By symmetry, we conclude that $\mathbf{d}\rho = \mathbf{d}\sigma$ and thus $\rho = \sigma$. Consequently, Φ is one-to-one.

4. To establish that Φ maps $\hat{S}$ onto $C(S)$, we let $H \in C(S)$. We first claim

$$aa^{-1} = bb^{-1} \quad \text{or} \quad a^{-1}a = b^{-1}b \quad \Rightarrow \quad a = b \qquad (a, b \in H). \qquad (3)$$

Indeed, assume that $aa^{-1} = bb^{-1}$ with $a, b \in H$. Then $b^{-1}a \in E_S$ and

$$b^{-1}a = (b^{-1}a)(b^{-1}a)^{-1} = b^{-1}(aa^{-1})b = b^{-1}bb^{-1}b = b^{-1}b$$

which implies

$$a = (aa^{-1})a = b(b^{-1}a) = bb^{-1}b = b.$$

The proof of the second part of (3) is symmetric.

We now define a partial transformation ρ by

$$\mathbf{d}\rho = \{a \in S \mid a^{-1}a = hh^{-1} \text{ for some } h \in H\},$$

$$a\rho = ah \quad \text{if} \quad a^{-1}a = hh^{-1}, \quad h \in H.$$

Relation (3) guarantees that ρ is single valued. Note that $\mathbf{d}\rho$ is not empty since $H^{-1} \subseteq \mathbf{d}\rho$.

We prove next that $\rho \in \hat{S}$. So let $x \in S$ and $a \in \mathbf{d}\rho$ with $a^{-1}a = hh^{-1}$ where $h \in H$, and set $u = (xa)^{-1}(xa)$. Then $uh \in H$ since H is an order ideal and

$$(xa)^{-1}(xa) = u = ua^{-1}a = uhh^{-1} = (uh)(uh)^{-1}$$

so that $xa \in \mathbf{d}\rho$. Hence $\mathbf{d}\rho$ is a left ideal of S. Further,

$$(xa)\rho = (xa)(uh) = (xa)h = x(ah) = x(a\rho).$$

Now suppose that $x^{-1}x = hh^{-1}$, $y^{-1}y = kk^{-1}$ with $h, k \in H$, and $x\rho = y\rho$. Then $xh = yk$ and

$$h^{-1}h = h^{-1}(hh^{-1})h = (h^{-1}x^{-1})(xh) = (xh)^{-1}(xh)$$
$$= (yk)^{-1}(yk) = k^{-1}(y^{-1}y)k = k^{-1}(kk^{-1})k = k^{-1}k$$

and relation (3) yields $h = k$. But then

$$x = x(x^{-1}x) = xhh^{-1} = y(kk^{-1}) = y(y^{-1}y) = y$$

and ρ is one-to-one. Consequently, $\rho \in \hat{S}$.

Finally,

$$\Phi_\rho = E_{\mathbf{d}\rho}\rho = \{e\rho | e \in E_S, e = hh^{-1} \text{ for some } h \in H\}$$
$$= \{h | h \in H\} = H$$

which proves that Φ is onto.

Thus Φ is an isomorphism of $\hat{S}$ onto $C(S)$.

For any $a \in S$, we obtain

$$\Phi_{w^a} = E_{\mathbf{d}w^a}w^a = E_{Sa^{-1}a} = \{x \in S | x = ya^{-1}a,\ ya^{-1} \in E_S\} = [a]$$

which completes the proof.

We can review the mappings in 1.3 and here by the following diagram.

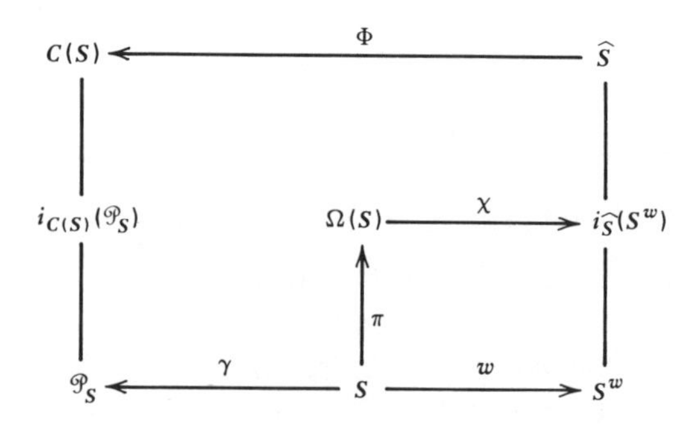

where

$$\mathcal{P}_S = \{[a] | a \in S\}, \qquad \gamma : a \to [a] \qquad (a \in S),$$

and 1.3 implies that $i_{\mathcal{I}(S)}(S^w) = i_{\hat{S}}(S^w)$.

V.2.9 Corollary

For any inverse semigroup S, $\chi\Phi$ is an isomorphism of $\Omega(S)$ onto $i_{C(S)}(\mathcal{P}_S)$ which maps $\Pi(S)$ onto $\mathcal{P}_S$. Thus γ is a dense embedding of S into $C(S)$.

Proof. This follows easily from 1.3 and 2.8; see the above diagram.

We will next identify the image of $\Omega(S)$ in $C(S)$, that is the idealizer of $\mathcal{P}_S$ in $C(S)$.

V.2.10 Proposition

Let S be an inverse semigroup, denote the set of all retract ideals of E_S by $\mathcal{R}_{E_S}$, and let $H \in C(S)$. Then $H \in i_{C(S)}(\mathcal{P}_S)$ if and only if $HH^{-1}, H^{-1}H \in \mathcal{R}_{E_S}$.

Proof. First note that $[a]H = aH$ and $H[a] = Ha$ for all $a \in S$. It is easily verified that

$$eHH^{-1} = [e] \cap HH^{-1} \qquad (e \in E_S). \tag{4}$$

Now assume that $H \in i_{C(S)}(\mathcal{P}_S)$. Then for every $a \in S$, there exists a (unique) $a' \in S$ such that $aH = [a']$. Hence for any $e \in E_S$, using (4) we get

$$[e] \cap HH^{-1} = (eH)H^{-1} = e'H^{-1} = [e'']$$

for some $e'' \in S$ since $H^{-1} \in i_{C(S)}(\mathcal{P}_S)$. Since HH^{-1} is clearly an ideal of E_S, we have $HH^{-1} \in \mathcal{R}_{E_S}$. The same type of argument shows that $H^{-1}H \in \mathcal{R}_{E_S}$.

Conversely, assume that $HH^{-1}, H^{-1}H \in \mathcal{R}_{E_S}$. Let $a \in S$ and $e = a^{-1}a$. Then $[e] \cap HH^{-1} = [f]$ for some $f \in E_S$. By (4), $eHH^{-1} = [f]$ and hence $aHH^{-1} = aeHH^{-1} = a[f] = [af]$. Thus there exist $h, k \in H$ such that $af = ahk^{-1}$. We will show that $aH = [afk]$. Let $x \in aH$ so that $x = at$ for some $t \in H$, and $e = a^{-1}a$. Then $ett^{-1} \in eHH^{-1} = [f]$ so $ett^{-1} \leq f$. Hence

$$\begin{aligned} x = a(ett^{-1})t \leq (af)t = ahk^{-1}t = (ahk^{-1})kk^{-1}t \\ = (afk)(k^{-1}t) \leq afk \end{aligned}$$

since $k^{-1}t \in E_S$; therefore $x \in [afk]$. Conversely, let $x \in [afk]$. Then $x \leq a(fk) \in aH$ since $f \in E_S$ and $k \in H$. Thus $x \in aH$, which proves that $aH = [afk]$. A symmetric argument shows that $Ha = [b]$ for some $b \in S$. Consequently, $H \in i_{C(S)}(\mathcal{P}_S)$.

V.2.11 Exercises

(i) Characterize the elements of $C(S)$ where $S = [Y; G_\alpha, \varphi_{\alpha,\beta}]$ is a Clifford semigroup.

(ii) Let H be a nonempty subset of an inverse semigroup S. Show that if $HH^{-1} \subseteq E_S$, then H is contained in a σ-class of S. Also prove the converse in the case S is E-unitary.

(iii) Find explicitly an isomorphism of C onto $\hat{C}$.

V.2.12 Problems

(i) Construct $\hat{S}$ for a Reilly semigroup $S = B(G, \alpha)$.

Schein [17] defined $C(S)$ pointing out that permissible subsets appear already in Wagner [3] under a different guise. In the same paper, Schein establishes many results related to $\Omega(S)$ and $C(S)$ for an inverse semigroup. The semigroup $\hat{S}$ was introduced by McAlister [8], [10] who established its relationship 2.8 to $C(S)$. McAlister's papers treat a variety of topics connected with $\hat{S}$.

V.3 THE TRANSLATIONAL HULL OF A CLIFFORD SEMIGROUP

We consider first translations on a semilattice and characterize its translational hull in terms of retract ideals. We then construct the translational hull of a Clifford semigroup by finding the Clifford representation for it. For the special case of an inverse semigroup which is a subdirect product of a semilattice and a group, we construct its translational hull explicitly.

V.3.1 Lemma

Let Y be a semilattice. Then right translations of Y coincide with I-endomorphisms of Y where I is an ideal of Y.

Proof. For $\rho \in \mathrm{P}(Y)$ and $\alpha \in Y$, we obtain

$$\alpha\rho = (\alpha\rho)^2 = (\alpha\rho)(\alpha\rho) = [(\alpha\rho)\alpha]\rho = \alpha\rho^2,$$

and hence for any $\alpha, \beta \in Y$,

$$(\alpha\beta)\rho = (\alpha\beta)\rho^2 = [\alpha(\beta\rho)]\rho = (\alpha\rho)(\beta\rho).$$

It follows that ρ is an endomorphism of Y which fixes every element of $Y\rho$. It is obvious that $Y\rho$ is an ideal of Y.

Conversely, let ρ be an I-endomorphism of Y for some ideal I. Then for any $\alpha, \beta \in Y$, we have

$$\alpha(\beta\rho) = [\alpha(\beta\rho)]\rho = (\alpha\rho)(\beta\rho^2) = (\alpha\rho)(\beta\rho) = (\alpha\beta)\rho$$

since $\alpha(\beta\rho) \in I$ is left fixed by ρ. Hence $\rho \in \mathrm{P}(Y)$.

In the light of II.4.4, we note that in a semilattice, order p-ideals coincide with retract ideals.

V.3.2 Theorem

Let Y be a semilattice and $\mathfrak{R}_Y$ be the set of all retract ideals of Y with the operation of intersection. Then

$$\sigma : \rho \to Y\rho \qquad [\rho \in \mathrm{P}(Y)],$$

$$\tau : I \to \rho \qquad (I \in \mathfrak{R}_Y),$$

where $[\alpha\rho] = [\alpha] \cap I$ for all $\alpha \in Y$, are mutually inverse isomorphisms of $\mathrm{P}(Y)$ and $\mathfrak{R}_Y$.

Proof. If $\rho \in \mathrm{P}(Y)$, then a simple verification shows that $[\alpha\rho] = [\alpha] \cap Y\rho$ for all $\alpha \in Y$, so that $Y\rho \in \mathfrak{R}_Y$. Next let $\rho, \rho' \in \mathrm{P}(Y)$ and $\alpha \in Y$. Then

$$\alpha\rho\rho' = [\alpha(\alpha\rho)]\rho' = (\alpha\rho)(\alpha\rho') = (\alpha\rho')(\alpha\rho) = [(\alpha\rho')\alpha]\rho$$

which implies $Y\rho\rho' \subseteq Y\rho \cap Y\rho'$. Conversely, if $\alpha = \beta\rho = \gamma\rho'$, then

$$\alpha = (\beta\rho)(\gamma\rho') = [(\beta\rho)\gamma]\rho' = [\gamma(\beta\rho)]\rho' = (\gamma\beta)\rho\rho'$$

which shows that $Y\rho \cap Y\rho' \subseteq Y\rho\rho'$. Consequently, $Y\rho\rho' = Y\rho \cap Y\rho'$ and σ is a homomorphism.

Now let $I \in \mathfrak{R}_Y$ and define ρ as in the statement of the theorem. For $\alpha, \beta, \gamma \in Y$, we obtain

$$\begin{aligned} \gamma \leq \alpha(\beta\rho) &\Leftrightarrow \gamma \leq \alpha, \quad \gamma \leq \beta\rho \\ &\Leftrightarrow \gamma \leq \alpha, \quad \gamma \leq \beta, \quad \gamma \in I \\ &\Leftrightarrow \gamma \leq \alpha\beta, \quad \gamma \in I \\ &\Leftrightarrow \gamma \leq (\alpha\beta)\rho \end{aligned}$$

which shows that $\alpha(\beta\rho) = (\alpha\beta)\rho$. If also $\tau : I' \to \rho'$, then for any $\alpha, \beta \in Y$, we have

$$\beta \leq \alpha\rho\rho' \Leftrightarrow \beta \leq \alpha\rho, \quad \beta \in I' \Leftrightarrow \beta \leq \alpha, \quad \beta \in I \cap I'$$

which shows that τ is a homomorphism.

A straightforward argument shows that $\sigma\tau$ and $\tau\sigma$ are the identity mappings on the respective sets.

V.3.3 Corollary

If Y is a semilattice, so is $\Omega(Y)$.

Proof. It follows from 3.2 that $\mathrm{P}(Y) \cong \mathfrak{R}_Y$. Every right translation ρ of Y, when written on the left, is a left translation λ of Y and the two are linked.

Hence the projection of $\Omega(Y)$ into $P(Y)$ maps $\Omega(S)$ onto $P(S)$. This projection is one-to-one by I.8.13 since Y is obviously reductive. Hence $\Omega(S) \cong P(S)$.

For a Clifford semigroup $S = [Y; G_\alpha, \varphi_{\alpha,\beta}]$, we will construct the semigroup $P(S)$ of right translations by means of a device which has its origin in group theory, namely a generalization of inverse limit of groups. For a discussion of inverse limits, consult Grätzer [1].

V.3.4 Notation

Let $S = [Y; G_\alpha, \varphi_{\alpha,\beta}]$ be a Clifford semigroup. We denote by

(i) $\mathcal{I}_Y$ the *set of all ideals of* Y,
(ii) $\mathcal{R}_Y$ the *set of all retract ideals of* Y,
(iii) $\mathcal{P}_Y$ the *set of all principal ideals of* Y,

with the operation of intersection.

For each $I \in \mathcal{I}_Y$, we write

$$\text{inv lim}\{G_\alpha\}_{\alpha\in I} = \left\{(g_\alpha) \in \prod_{\alpha\in I} G_\alpha \,|\, g_\alpha\varphi_{\alpha,\beta} = g_\beta \text{ if } \alpha > \beta\right\},$$

with the multiplication it inherits from the direct product $\prod_{\alpha\in I} G_\alpha$. We next set

$$\text{Inv lim}\{G_\alpha\}_{\alpha\in Y} = \bigcup_{I\in\mathcal{I}_Y} \text{inv lim}\{G_\alpha\}_{\alpha\in I}$$

with the multiplication

$$(g_\alpha)_{\alpha\in I}\cdot(h_\beta)_{\beta\in J} = (g_\gamma h_\gamma)_{\gamma\in I\cap J}.$$

V.3.5 Lemma

For each $I \in \mathcal{I}_Y$, inv $\lim\{G_\alpha\}_{\alpha\in I}$ is a group and

$$\text{Inv lim}\{G_\alpha\}_{\alpha\in Y} \cong [\mathcal{I}_Y; \text{inv lim}\{G_\alpha\}_{\alpha\in I}, \Phi_{I,J}]$$

where, for any $I \supseteq J$,

$$(g_\alpha)_{\alpha\in I}\Phi_{I,J} = (g_\alpha)_{\alpha\in J}.$$

Proof. The first assertion is easy to verify. The second assertion follows from the fact that the very definition of Inv $\lim\{G_\alpha\}_{\alpha\in Y}$ indicates that it is the semilattice $\mathcal{I}_Y$ of groups inv $\lim\{G_\alpha\}_{\alpha\in I}$; its multiplication is evidently determined by the transitive system of homomorphisms $\Phi_{I,J}$. The details of this argument are left as an exercise.

We are now ready for the main construction theorem for the right translations of a Clifford semigroup.

V.3.6 Theorem

Let $S = [Y; G_\alpha, \varphi_{\alpha,\beta}]$ be a Clifford semigroup and set $A = \text{Inv lim}\{G_\alpha\}_{\alpha \in Y}$. Define a function ψ by

$$\psi : \rho \to (e_\alpha \rho)_{\alpha \in I} \qquad [\rho \in \mathrm{P}(S)] \tag{5}$$

where e_α is the identity of G_α and $I = \{\alpha \in Y | S\rho \cap G_\alpha \neq \varnothing\}$. Then ψ is an embedding of $\mathrm{P}(S)$ into A and

$$\Delta(S)\psi = \{(g_\alpha)_{\alpha \in I} \in A | I \in \mathcal{P}_Y\}, \tag{6}$$

$$\mathrm{P}(S)\psi = \{(g_\alpha)_{\alpha \in I} \in A | I \in \mathcal{R}_Y\} = i_A(\Delta(S)\psi). \tag{7}$$

Proof. Let $\rho \in \mathrm{P}(S)$. For $x, y \in G_\alpha$, $\alpha \in Y$, we have $x\rho \in G_\gamma$ and $y\rho \in G_\delta$, which implies $x\rho = xy^{-1}(y\rho)$ so that $\gamma \leq \delta$; by symmetry, we deduce that $\gamma = \delta$. Letting $g \to \bar{g}$ denote the canonical homomorphism of S onto Y, we can thus define a function $\bar{\rho}$ by

$$\bar{g}\bar{\rho} = \overline{g\rho} \qquad (g \in S).$$

It follows that

$$(\bar{x}\bar{y})\bar{\rho} = \overline{xy}\bar{\rho} = \overline{(xy)\rho} = \overline{x(y\rho)} = \bar{x}(\bar{y}\bar{\rho}),$$

which shows that $\bar{\rho} \in \mathrm{P}(Y)$. According to 3.2, $\bar{\rho}$ is completely determined by the retract ideal $I = Y\bar{\rho}$. It is clear that this I coincides with that defined in the statement of the theorem. The restriction $\bar{\rho}|_I$ is the identity transformation, which implies that $e_\alpha \rho \in G_\alpha$ for all $\alpha \in I$, and if $\alpha > \beta$, then

$$e_\beta \rho = (e_\beta e_\alpha)\rho = e_\beta(e_\alpha \rho) = e_\alpha \rho \varphi_{\alpha,\beta}.$$

Consequently, ψ maps $\mathrm{P}(S)$ into A.

Let $\rho, \sigma \in \mathrm{P}(S)$, and let

$$I = \{\alpha \in Y | S\rho \cap G_\alpha \neq \varnothing\},$$

$$J = \{\alpha \in Y | S\sigma \cap G_\alpha \neq \varnothing\}.$$

With the notation introduced above, we have $\bar{\rho}, \bar{\sigma} \in P(Y)$ and $Y\bar{\rho}\bar{\sigma} = Y\bar{\rho} \cap Y\bar{\sigma} = I \cap J$. Evidently, $\overline{\rho\sigma} = \bar{\rho}\bar{\sigma}$, and so $Y\overline{\rho\sigma} = I \cap J$. Hence

$$I \cap J = \{\alpha \in Y | S\rho\sigma \cap G_\alpha \neq \varnothing\}.$$

We now put $\rho\psi = (g_\alpha)_{\alpha \in I}$ and $\sigma\psi = (h_\alpha)_{\alpha \in J}$. For any $\alpha \in I \cap J$, we then have

$$e_\alpha \rho\sigma = (e_\alpha \rho)\sigma = g_\alpha \sigma = (g_\alpha e_\alpha)\sigma = g_\alpha(e_\alpha \sigma) = g_\alpha h_\alpha$$

which evidently implies that ψ is a homomorphism.

Next assume that $\rho\psi = \sigma\psi = (g_\alpha)_{\alpha \in I}$ and let $h \in G_\alpha$. According to 3.2, we have $[\alpha\bar{\rho}] = [\alpha] \cap I = [\alpha\bar{\sigma}]$ for all $\alpha \in I$, so that

$$hp = (he_\alpha)\rho = h(e_\alpha \rho) = h(e_{\alpha\bar{\rho}}\rho)$$

$$= h(e_{\alpha\bar{\sigma}}\sigma) = h(e_\alpha \sigma) = (he_\alpha)\sigma = h\sigma.$$

Hence ψ is one-to-one. This proves the first assertion of the theorem.

For any $g_\alpha \in G_\alpha$, it follows at once that $Y\bar{\rho}_{g_\alpha} = [\alpha]$ which implies that $\rho_{g_\alpha}\psi$ is of the form $(g_\beta)_{\beta \in [\alpha]}$. Conversely, let $(g_\beta)_{\beta \in [\alpha]} \in A$. Then for any $\beta \leq \alpha$, we obtain

$$g_\beta = g_\alpha \varphi_{\alpha, \beta} = e_\beta g_\alpha = e_\beta(e_\alpha g_\alpha) = e_\beta(e_\alpha \rho_{g_\alpha})$$

$$= (e_\beta e_\alpha)\rho_{g_\alpha} = e_\beta \rho_{g_\alpha},$$

that is to say $\rho_{g_\alpha}\psi = (g_\beta)_{\beta \in [\alpha]}$. This proves (6).

We have seen above that for any $\rho \in P(S)$ and $\rho\psi = (g_\alpha)_{\alpha \in I}$, we actually have $I \in \mathfrak{R}_Y$. This shows that the first set in (7) is contained in the second. Since $\mathfrak{P}_Y$ is an ideal of $\mathfrak{R}_Y$, it follows easily that the second semigroup in (6) is an ideal of the second semigroup in (7). This implies that the latter is contained in the third semigroup in (7).

Finally, let $(g_\gamma)_{\gamma \in I} \in i_A(\Delta(S)\psi)$. For any $\alpha \in Y$, the product $(g_\gamma)_{\gamma \in I} \cdot (e_\beta)_{\beta \in [\alpha]}$ is contained in $\Delta(S)\psi$ and hence $I \cap [\alpha]$ is a principal ideal of Y and thus $I \in \mathfrak{R}_Y$. We now define a function ρ on S by

$$h\rho = hg_\gamma \quad \text{if} \quad h \in G_\alpha \quad \text{and} \quad I \cap [\alpha] = [\gamma]. \tag{8}$$

It is clear that the function $\bar{\rho}$, defined above for $\rho \in P(S)$, can be defined here as well, and that it gives $I \cap [\alpha] = [\alpha\bar{\rho}]$. It follows that $h\rho = hg_{\alpha\bar{\rho}}$ if $h \in G_\alpha$. For any $a \in G_\alpha$, $b \in G_\beta$, we obtain

$$a(b\rho) = a(bg_{\beta\bar{\rho}}) = (a\varphi_{\alpha, \alpha(\beta\bar{\rho})})(b\varphi_{\beta, \alpha(\beta\bar{\rho})})(g_{\beta\bar{\rho}}\varphi_{\beta\bar{\rho}, \alpha(\beta\bar{\rho})})$$

$$= [(a\varphi_{\alpha, \alpha\beta})(b\varphi_{\beta, \alpha\beta})]\varphi_{\alpha\beta, (\alpha\beta)\bar{\rho}}(g_{(\alpha\beta)\bar{\rho}}) = (ab)\rho$$

and $\rho \in \mathrm{P}(S)$. It follows from (8) that $\rho\psi = (g_\gamma)_{\gamma \in I}$ and hence $i_A(\Delta(S)\psi) \subseteq \mathrm{P}(S)\psi$, which completes the proof of (7).

We can now easily characterize the translational hull of a Clifford semigroup and obtain a dense embedding of the semigroup itself.

V.3.7 Theorem

Let $S = [Y; G_\alpha, \varphi_{\alpha,\beta}]$ be a Clifford semigroup, and let $A = \text{Inv}\lim\{G_\alpha\}_{\alpha \in Y}$. The function χ defined by

$$\chi: g \to (g\varphi_{\alpha,\beta})_{\beta \leq \alpha} \qquad (g \in G_\alpha, \alpha \in Y)$$

is a dense embedding of S into A. Moreover,

$$\Omega(S) \cong \Lambda(S) \cong \mathrm{P}(S) \cong [\mathfrak{R}_Y; \text{inv}\lim\{G_\alpha\}_{\alpha \in I}, \Phi_{I,J}], \tag{9}$$

$$S\chi = \{(g_\alpha)_{\alpha \in I} \in A \,|\, I \in \mathcal{P}_Y\} \tag{10}$$

(see 3.4 for notation).

Proof. We show first that for every right translation ρ of S there exists a left translation λ of S linked to ρ. Let $\rho \in \mathrm{P}(S)$, so by 3.6, $\rho\psi = (g_\gamma)_{\gamma \in I}$ where ρ can be expressed as in (8). We now define λ on S by

$$\lambda h = g_\gamma h \quad \text{if} \quad h \in G_\alpha \quad \text{and} \quad I \cap [\alpha] = [\gamma].$$

The proof that λ is a left translation is symmetric to the argument in 3.6 that ρ is a right translation. For any $a \in G_\alpha$, $b \in G_\beta$, we have $I \cap [\alpha] = [\gamma]$, $I \cap [\beta] = [\delta]$, and

$$\begin{aligned}[\alpha\delta] &= [\alpha] \cap [\delta] = [\alpha] \cap (I \cap [\beta]) \\ &= ([\alpha] \cap I) \cap [\beta] = [\gamma] \cap [\beta] = [\gamma\beta].\end{aligned}$$

It follows that $\alpha\delta = \gamma\beta$ which yields $g_\delta\varphi_{\delta,\alpha\delta} = g_\gamma\varphi_{\gamma,\beta\gamma}$ since $(g_\theta)_{\theta \in I} \in A$, and thus

$$\begin{aligned}a(\lambda b) &= a(g_\delta b) = (a\varphi_{\alpha,\alpha\delta})(g_\delta\varphi_{\delta,\alpha\delta})(b\varphi_{\beta,\alpha\delta}) \\ &= (a\varphi_{\alpha,\beta\gamma})(g_\gamma\varphi_{\gamma,\beta\gamma})(b\varphi_{\beta,\beta\gamma}) = (ag_\gamma)b = (a\rho)b\end{aligned}$$

and λ and ρ are linked.

By a symmetric argument, using the left–right dual of 3.6 we conclude that for every left translation λ of S there exists a linked right translation ρ of S. By 1.2, the projection homomorphisms $(\lambda, \rho) \to \lambda$ and $(\lambda, \rho) \to \rho$ are one-to-one.

By the above, they are also onto, which establishes the first two isomorphisms in (9); the third follows easily from 3.5.

For any $g \in G_\alpha$, using the notation of 3.6, we have

$$\rho_g\psi = (e_\beta\rho_g)_{\beta\leq\alpha} = (e_\beta g)_{\beta\leq\alpha} = (g\varphi_{\alpha,\beta})_{\beta\leq\alpha}.$$

It follows that the function χ in the statement of the theorem is the composition of the canonical isomorphism $g \to \rho_g$ of S onto $\Delta(S)$ and ψ restricted to $\Delta(S)$. We have already proved that the projection homomorphism of $\Omega(S)$ onto $\mathrm{P}(S)$ is an isomorphism. This implies that $\Delta(S)$ is a densely embedded ideal of $\mathrm{P}(S)$ according to I.9.18. In view of 3.6, $S\chi = \Delta(S)\psi$ is then a densely embedded ideal of $i_A(S\chi) = \mathrm{P}(S)\psi$, which, by definition, means that χ is a dense embedding of S into A. Finally, (10) follows from these remarks and 3.6.

V.3.8 Corollary

If S is a Clifford semigroup, so is $\Omega(S)$.

We will now characterize the translational hull of an inverse semigroup which is a subdirect product of a semilattice and a group.

V.3.9 Theorem

With the notation in II.2.13, let $S = [Y, G; \eta]$. Then

$$\Omega(S) \cong \left\{(I, c) \in \mathcal{R}_Y \times G \,|\, c \in \bigcap_{\alpha\in I} \alpha\eta\right\}$$

and the latter semigroup is a subdirect product of $\mathcal{R}_Y$ and G.

Proof. In the light of 3.6, we may restrict our attention to left translations of S. Hence let $\lambda \in \Lambda(S)$. Note that $E_S = Y \times \{1\}$, where 1 is the identity of G. We may define the functions σ and τ by the requirement that

$$\lambda(\alpha, 1) = (\sigma\alpha, \tau\alpha) \qquad (\alpha \in Y).$$

Then

$$[\lambda(\alpha,1)](\beta,1) = (\sigma\alpha, \tau\alpha)(\beta,1) = ((\sigma\alpha)\beta, \tau\alpha),$$

$$\lambda[(\alpha,1)(\beta,1)] = (\sigma(\alpha\beta), \tau(\alpha\beta)),$$

whence we get $(\sigma\alpha)\beta = \sigma(\alpha\beta)$ and $\tau\alpha = \tau(\alpha\beta)$. But then $\sigma \in \Lambda(Y)$ and $\tau\alpha = \tau(\alpha\beta) = \tau\beta$ is a constant which we denote by c_λ. We also know by 3.2 that σ is determined uniquely by its range, so we write $I_\lambda = \mathbf{r}\sigma$.

For any element (α, g) of S, we then have

$$\lambda(\alpha, g) = [\lambda(\alpha,1)](\alpha, g) = (\sigma\alpha, c_\lambda)(\alpha, g) = (\sigma\alpha, c_\lambda g).$$

This shows that λ is uniquely determined by the pair (I_λ, c_λ). Since $\lambda(\alpha, 1) = (\sigma\alpha, c_\lambda) \in S$, we must have $c_\lambda \in (\sigma\alpha)\eta$. This holds for all $\alpha \in Y$, and hence $c_\lambda \in \cap_{\alpha \in I_\lambda} \alpha\eta$. We have thus arrived at the function

$$\chi : \lambda \to (I_\lambda, c_\lambda) \in \mathcal{R}_Y \times G$$

where $c_\lambda \in \cap_{\alpha \in I_\lambda} \alpha\eta$.

Now let also $\lambda' \in \Lambda(S)$, and denote by σ' the left translation of Y with range $I_{\lambda'}$. Then for $(\alpha, g) \in S$, we get

$$\lambda\lambda'(\alpha, g) = \lambda(\sigma'\alpha, c_{\lambda'}g) = (\sigma\sigma'\alpha, c_\lambda c_{\lambda'} g).$$

It follows from 3.2 that $\mathbf{r}(\sigma\sigma') = I_\lambda \cap I_{\lambda'}$, and thus χ defined above is a homomorphism. We have seen above that (I_λ, c_λ) determines λ uniquely, and hence χ is one-to-one. Now let $(I, c) \in \mathcal{R}_Y \times G$ with $c \in \cap_{\alpha \in I} \alpha\eta$. Let σ be the left translation on Y with range I, and define λ by

$$\lambda(\alpha, g) = (\sigma\alpha, cg) \qquad [(\alpha, g) \in S].$$

Let $(\alpha, g) \in S$, so that $g \in \alpha\eta$. By hypothesis $c \in (\sigma\alpha)\eta$ and since $\sigma\alpha \leq \alpha$, we have $\alpha\eta \subseteq (\sigma\alpha)\eta$. Hence $cg \in (\sigma\alpha)\eta$ and thus $(\sigma\alpha, cg) \in S$. Consequently, λ maps S into itself. Routine verification shows that λ is a left translation of S and clearly $\lambda\chi = (I, c)$.

For any $I \in \mathcal{R}_Y$, we have $\lambda\chi = (I, 1)$ where $\lambda(\alpha, g) = (\sigma\alpha, g)$ and $\sigma \in \Lambda(Y)$ with $\mathbf{r}\sigma = I$. Next let $g \in G$. Since $\eta : Y \to \mathcal{L}(G)$ is full, there exists $\alpha \in Y$ such that $g \in \alpha\eta$. We obtain $\cap_{\beta \leq \alpha} \beta\eta = \alpha\eta$ since η inverts the order. Hence the pair $([\alpha], g)$ represents a left translation of S since $g \in \cap_{\beta \in [\alpha]} \beta\eta$. This proves that the semigroup $\Lambda(S)\chi$ is a subdirect product of $\mathcal{R}_Y$ and G.

V.3.10 Exercises

(i) If S is a Clifford semigroup, show directly that $\Omega(S)$ is also. Show that if S is a semilattice, so is $\Omega(S)$.

(ii) Characterize the least commutative congruence on a Clifford semigroup $[Y; G_\alpha, \varphi_{\alpha,\beta}]$.

(iii) For a Clifford semigroup $S = [Y; G_\alpha, \varphi_{\alpha,\beta}]$, prove that $\hat{S} \cong \mathrm{Inv\,lim}\{G_\alpha\}_{\alpha \in Y}$.

(iv) In the setting of 3.9 and the notation of 3.6, prove the following statements.

(α) A is isomorphic to a subdirect product of $\mathcal{I}_Y$ and G given by the function $\bar{\eta} : \mathcal{I}_Y \to \mathcal{L}(G)$ where $\bar{\eta} : I \to \cap_{\alpha \in I} \alpha\eta$.

(β) $P(S)$ is isomorphic to a subdirect product of $\mathcal{R}_Y$ and G.

(γ) S can be densely embedded into $\mathcal{I}_Y \times G$.

(v) If Y is a semilattice and G is a group, show that $\Omega(Y \times G) \cong \mathcal{R}_Y \times G$.

(vi) Show that for every subsemigroup S of a group G, the inclusion mapping $S \to G$ is a dense embedding.

V.3.11 Problems

(i) If a semigroup S is embeddable into an inverse semigroup, is $\Omega(S)$ also embeddable into an inverse semigroup? [According to 3.10(vi), if S is embeddable into a group, so is $\Omega(S)$.]

The material on semilattices can be found in Petrich [5], even though various bits and pieces have their origin elsewhere. The translational hull of a Clifford semigroup was constructed by Petrich [7] and Schein [17]. The results on the translational hull of a subdirect product of a semilattice and a group were deduced from the general case in both papers. The construction of the translational hull of a Clifford semigroup was generalized by Petrich [9] to strong semilattices of weakly reductive semigroups.

V.4 IDEAL EXTENSIONS OF ONE BRANDT SEMIGROUP BY ANOTHER

In the light of the general considerations in I.9 concerning ideal extensions, in order to treat the subject in the title of this section, we must first have a good understanding of the structure of the translational hull of a Brandt semigroup. To this end, the following concept will be quite useful.

V.4.1 Definition

Let I be a nonempty set and G be a group. We denote by G^I the *set of all functions* (*written on the left*) *from subsets of I into G* with the "pointwise" multiplication:

$$(\varphi \cdot \varphi')i = (\varphi i)(\varphi' i) \qquad [i \in \mathbf{d}(\varphi \cdot \varphi') = \mathbf{d}\varphi \cap \mathbf{d}\varphi'].$$

For any $\alpha \in \mathcal{F}'(I)$ and $\varphi \in G^I$, we define a mapping φ^α by

$$\varphi^\alpha i = \varphi \alpha i \qquad (i \in \mathbf{d}\alpha,\ \alpha i \in \mathbf{d}\varphi).$$

Note that $\varphi^\alpha \in G^I$, and that the empty function $\varnothing$ is an element of G^I (with $\mathbf{d}\varnothing = \varnothing$).

Next let P be any subsemigroup of $\mathcal{F}'(I)$. The *left wreath product* of P and G, denoted by $P \operatorname{wl} G$, is the set

$$\{(\alpha, \varphi) \in P \times G^I \mid \mathbf{d}\alpha = \mathbf{d}\varphi\}$$

with the multiplication

$$(\alpha, \varphi)(\alpha', \varphi') = (\alpha\alpha', \varphi^{\alpha'} \cdot \varphi').$$

It will follow from the next result that $P \operatorname{wl} G$ is actually a semigroup.

V.4.2 Lemma

Let $S = B(G, I)$ be a Brandt semigroup. Then the function

$$\mathfrak{a} : \lambda \to (\alpha, \varphi) \qquad [\lambda \in \Lambda(S)] \tag{11}$$

where

$$\lambda(i, 1, i) = (\alpha i, \varphi i, i) \quad \text{if} \quad \lambda(i, 1, i) \neq 0, \tag{12}$$

is an isomorphism of $\Lambda(S)$ onto $\mathcal{F}'(I)\,\mathrm{wl}\,G$.

Proof. First note that for any $\lambda \in \Lambda(S)$, $(i, g, j) \in S$, we have

$$\lambda(i, g, j) = \lambda[(i, 1, i)(i, g, j)] = [\lambda(i, 1, i)](i, g, j) \tag{13}$$

is of the form $(\ ,\ , j)$ if $\lambda(i, g, j) \neq 0$. Hence (12) defines $\alpha \in \mathcal{F}'(I)$ and $\varphi \in G^I$ such that $\mathbf{d}\alpha = \mathbf{d}\varphi = \{i \in I \mid \lambda(i, 1, i) \neq 0\}$. It follows that $\mathfrak{a}$ maps $\Lambda(S)$ into $\mathcal{F}'(I)\,\mathrm{wl}\,G$. Further, from (13) we obtain

$$\lambda(i, g, j) = \begin{cases} (\alpha i, \varphi i, i)(i, g, j) & \text{if } i \in \mathbf{d}\alpha \\ 0 & \text{if } i \notin \mathbf{d}\alpha \end{cases}$$

$$= \begin{cases} (\alpha i, (\varphi i) g, j) & \text{if } i \in \mathbf{d}\alpha \\ 0 & \text{if } i \notin \mathbf{d}\alpha. \end{cases} \tag{14}$$

Now let $\lambda, \lambda' \in \Lambda(S)$ with $\mathfrak{a}\lambda = (\alpha, \varphi)$, $\mathfrak{a}\lambda' = (\alpha', \varphi')$. Using (14), we obtain

$$\lambda\lambda'(i, 1, i) = \begin{cases} \lambda(\alpha' i, \varphi' i, i) & \text{if } i \in \mathbf{d}\alpha' \\ \lambda 0 & \text{if } i \notin \mathbf{d}\alpha' \end{cases}$$

$$= \begin{cases} (\alpha\alpha' i, (\varphi\alpha' i)(\varphi' i), i) & \text{if } i \in \mathbf{d}\alpha', \alpha' i \in \mathbf{d}\alpha \\ 0 & \text{otherwise} \end{cases}$$

which shows that $\mathfrak{a}(\lambda\lambda') = (\alpha\alpha', \varphi^{\alpha'} \cdot \varphi')$, so that $\mathfrak{a}$ is a homomorphism.

For any $(\alpha, \varphi) \in \mathcal{F}'(I)\,\mathrm{wl}\,G$, we may define λ by (14), and verify easily that λ is a left translation of S for which $\mathfrak{a}\lambda = (\alpha, \varphi)$. This shows that $\mathfrak{a}$ maps $\Lambda(S)$ onto $\mathcal{F}'(I)\,\mathrm{wl}\,G$. It also follows from (14) that $\mathfrak{a}$ is one-to-one, and thus an isomorphism.

We now introduce the notation and concepts which represent the left–right dual of those in 4.1 in order to describe right translations of $B(G, I)$.

V.4.3 Definition

Let I be a nonempty set and G be a group. We denote by ${}^I G$ the *set of all functions (written on the right) from subsets of I into G* with the "pointwise" multiplication:

$$i(\psi \cdot \psi') = (i\psi)(i\psi') \qquad \left[i \in \mathbf{d}(\psi \cdot \psi') = \mathbf{d}\psi \cap \mathbf{d}\psi'\right].$$

For any $\beta \in \mathcal{F}(I)$ and $\psi \in {}^I G$, we define a mapping ${}^\beta\psi$ by

$$i^\beta\psi = i\beta\psi \qquad (i \in \mathbf{d}\beta, i\beta \in \mathbf{d}\psi).$$

Note that ${}^\beta\psi \in {}^I G$, and that the empty function $\varnothing$ is an element of ${}^I G$ (with $\mathbf{d}\varnothing = \varnothing$).

Next let Q be any subsemigroup of $\mathcal{F}(I)$. The *right wreath product* of G and Q, denoted by $G \operatorname{wr} Q$, is the set

$$\{(\psi, \beta) \in {}^I G \times Q | \mathbf{d}\psi = \mathbf{d}\beta\}$$

with the multiplication

$$(\psi, \beta)(\psi', \beta') = (\psi \cdot {}^\beta\psi', \beta\beta'). \tag{15}$$

As in 4.2 for left translations, one proves the following result.

V.4.4 Lemma

Let $S = B(G, I)$ be a Brandt semigroup. Then the function

$$\mathfrak{b} : \rho \to (\psi, \beta) \qquad [\rho \in \mathrm{P}(S)]$$

where

$$(i, 1, i)\rho = (i, i\psi, i\beta) \qquad \text{if} \quad (i, 1, i)\rho \neq 0,$$

is an isomorphism of $\mathrm{P}(S)$ onto $G \operatorname{wr} \mathcal{F}(I)$.

The proof is analogous to that of 4.2 and is left as an exercise. For reference, we only state: for $(\psi, \beta) \in G \operatorname{wr} \mathcal{F}(I)$, the corresponding right translation ρ is given by

$$(i, g, j)\rho = \begin{cases} (i, g(j\psi), j\beta) & \text{if} \quad j \in \mathbf{d}\beta \\ 0 & \text{if} \quad j \notin \mathbf{d}\beta. \end{cases} \tag{16}$$

We need next a criterion for linking of left and right translations.

V.4.5 Lemma

Let $S = B(G, I)$, $\lambda \in \Lambda(S)$, $\mathfrak{a}\lambda = (\alpha, \varphi)$, $\rho \in \mathrm{P}(S)$, $\rho\mathfrak{b} = (\psi, \beta)$. Then $(\lambda, \rho) \in \Omega(S)$ if and only if both α and β are one-to-one, are mutually inverse, and $(\alpha j)\psi = \varphi j$ for all $j \in \mathbf{d}\alpha$.

Proof. The argument amounts to the following calculation:

$$(i,1,i)[\lambda(j,1,j)] = \begin{cases} (i, \varphi j, j) & \text{if } j \in \mathbf{d}\alpha, \alpha j = i \\ 0 & \text{otherwise,} \end{cases}$$

$$[(i,1,i)\rho](j,1,j) = \begin{cases} (i, i\psi, j) & \text{if } i \in \mathbf{d}\beta, i\beta = j \\ 0 & \text{otherwise,} \end{cases}$$

and thus $(\lambda, \rho) \in \Omega(S)$ if and only if

$$i \in \mathbf{d}\beta, \quad i\beta = j \Leftrightarrow j \in \mathbf{d}\alpha, \quad \alpha j = i \Rightarrow i\psi = \varphi j$$

which is equivalent to the claim of the lemma.

We now deduce the desired characterization of the translational hull of a Brandt semigroup.

V.4.6 Theorem

For any Brandt semigroup $S = B(G, I)$, we have

$$\Omega(S) \cong \mathcal{I}'(I) \,\mathrm{wl}\, G \cong G \,\mathrm{wr}\, \mathcal{I}(I).$$

Proof. This follows easily from 4.2–4.5; the details of this argument are left as an exercise.

In order to construct all ideal extensions of one Brandt semigroup $S = B(G, I)$ by another Brandt semigroup we will use the isomorphic copy $G \,\mathrm{wr}\, \mathcal{I}(I)$ of $\Omega(S)$. From the above results, an isomorphism of $\Omega(S)$ onto $G \,\mathrm{wr}\, \mathcal{I}(I)$ is given by: $(\lambda, \rho) \to (\psi, \beta)$, where

$$\mathbf{d}\beta = \mathbf{d}\psi = \{ j \in I | (j,1,j)\rho \neq 0 \}, \tag{17}$$

$$(i,1,j)\rho = (i, j\psi, j\beta) \qquad (j \in \mathbf{d}\beta), \tag{18}$$

with 1 the identity of G. One verifies easily that the inner right translations of S correspond to the elements of the wreath product of rank at most one, that is for $\rho\mathfrak{b} = (\psi, \beta)$,

$$\rho \in \Delta(S) \Leftrightarrow \mathrm{rank}(\psi, \beta) = \mathrm{rank}\, \beta \leq 1.$$

From (14), 4.5, and (16), we deduce

$$\lambda(i, g, j) = \begin{cases} (i\beta^{-1}, (i\beta^{-1}\psi)g, j) & \text{if } i \in \mathbf{r}\beta \\ 0 & \text{otherwise.} \end{cases}$$

It is now easy to reformulate the extension construction in I.9.5 for the special case we are interested in at present. We extend the notion of an *extension function* for S by Q to include functions mapping Q^* into an isomorphic copy of $\Omega(S)$ and satisfying the appropriate conditions.

V.4.7 Theorem

Let $S = B(G, I)$ be a Brandt semigroup and Q be a semigroup with zero disjoint from S. Let $\varphi : Q^* \to G \operatorname{wr} \mathcal{I}(I)$, denoted by $\varphi : q \to (\psi_q, \beta_q)$, be a partial homomorphism such that rank $\beta_q \beta_r \leq 1$ if $qr = 0$ in Q. On $V = S \cup Q^*$ define a multiplication $*$ by: for $q, r \in Q^*$, $(i, g, j) \in S$,

$$(i, g, j) * q = \big(i, g(j\psi_q), j\beta_q\big) \quad \text{if} \quad j \in \mathbf{d}\beta_q,$$

$$q * (i, g, j) = \big(i\beta_q^{-1}, (i\beta_q^{-1}\psi_q)g, j\big) \quad \text{if} \quad i \in \mathbf{r}\beta_q,$$

and if $qr = 0$ in Q,

$$q * r = \big(k\beta_q^{-1}, (k\beta_q^{-1}\psi_q)(k\psi_r), k\beta_r\big) \quad \text{if} \quad k = \mathbf{r}\beta_q \cap \mathbf{d}\beta_r,$$

$$a * b = ab \quad \text{if} \quad a, b \in S, \quad \text{or} \quad a, b \in Q^* \quad \text{and} \quad ab \neq 0,$$

and all other products are equal to 0. Then V is an ideal extension of S by Q. Conversely, every ideal extension of S by Q can be so constructed.

Proof. It follows from the discussion preceding the theorem that φ is an extension function. It is easy to verify that the multiplication here agrees with that in I.9.5 relative to the given extension function. We consider only the case $qr = 0$ in Q; the remaining cases follow directly from the remarks preceding the theorem.

Hence let $q, r \in Q^*$ with $qr = 0$. Then rank $\beta_q\beta_r \leq 1$. If rank $\beta_q\beta_r = 0$, then $(q\varphi)(r\varphi) = \varnothing$ and thus $q * r = 0$. Assume that rank $\beta_q\beta_r = 1$. Then $\mathbf{r}\beta_q \cap \mathbf{d}\beta_r = k \in I$. From (17) and (18), it follows for any $(i, g, j) \in S$,

$$\rho_{(i,g,j)}\mathfrak{b} = (\psi, \beta) \quad \text{where} \quad i\psi = g, i\beta = j, \mathbf{d}\beta = i.$$

In our case, writing $q * r = (i, g, j)$, we obtain

$$i = \mathbf{d}(\beta_q\beta_r) = k\beta_q^{-1}, \qquad g = \big(k\beta_q^{-1}\psi_q\big)\big(k\beta_q^{-1}\beta_q\psi_r\big), \qquad j = k\beta_r$$

which gives the expression for $q * r$ in the statement of the theorem.

We can now establish the desired result by particularizing Q to be a Brandt semigroup and explicitly computing the extension functions figuring in 4.7.

V.4.8 Theorem

Let $S = B(G, I)$ and $Q = B(H, X)$ be two disjoint Brandt semigroups. Let ν be a cardinal number with $\nu \leq |I|$, and $\mathcal{P}_\nu$ be the family of all subsets of I of cardinality ν. Fix $P_0 \in \mathcal{P}_\nu$, and let the following parameters be given:

(i) let $\pi : X \to \mathcal{P}_\nu$ be any function such that $|x\pi \cap y\pi| \leq 1$ if $x \neq y$;
(ii) let $\theta : H \to G \,\mathrm{wr}\, \mathcal{S}(P_0)$ be a homomorphism with $h\theta = (\sigma_h, \tau_h)$;
(iii) for each $x \in X$, let $\xi_x : P_0 \to x\pi$ be a one-to-one correspondence, and $\eta_x : P_0 \to G$ be any function.

Then $\varphi : Q^* \to G \,\mathrm{wr}\, \mathcal{I}(I)$, defined by: for $q = (x, h, y) \in Q^*$, $q\varphi = (\psi_q, \beta_q)$ with

$$i\psi_q = \left(i\xi_x^{-1}\eta_x\right)^{-1}\left(i\xi_x^{-1}\sigma_h\right)\left(i\xi_x^{-1}\tau_h\eta_y\right) \quad \text{for all} \quad i \in \mathbf{d}\psi_q = \mathbf{d}\beta_q = x\pi, \tag{19}$$

$$\beta_q = \xi_x^{-1}\tau_h\xi_y, \tag{20}$$

is an extension function.

Conversely, every extension function relative to S and Q can be so obtained.

Proof. For the direct part, we must show that φ as defined above is an extension function. Hence let $q = (x, g, y)$ and $r = (w, h, z)$ be any elements of Q^*.

Assume first that $qr \neq 0$. Then $y = w$ and thus

$$\beta_q\beta_r = \xi_x^{-1}\tau_g\xi_y\xi_y^{-1}\tau_h\xi_z = \xi_x^{-1}\tau_g\tau_h\xi_z = \xi_x^{-1}\tau_{gh}\xi_z = \beta_{qr},$$

and for $i \in x\pi = \mathbf{d}\beta_q = \mathbf{d}(\beta_q\beta_r)$,

$$\begin{aligned}
i\left(\psi_q \cdot {}^{\beta_q}\psi_r\right) &= (i\psi_q)(i\beta_q\psi_r) \\
&= \left[\left(i\xi_x^{-1}\eta_x\right)^{-1}\left(i\xi_x^{-1}\sigma_g\right)\left(i\xi_x^{-1}\tau_g\eta_y\right)\right]\left[\left(i\xi_x^{-1}\tau_g\xi_y\right)\xi_y^{-1}\eta_y\right]^{-1} \\
&\quad \cdot\left[\left(i\xi_x^{-1}\tau_g\xi_y\right)\xi_y^{-1}\sigma_h\right]\left[\left(i\xi_x^{-1}\tau_g\xi_y\right)\xi_y^{-1}\tau_h\eta_z\right] \\
&= (i\xi_x\eta_x)^{-1}\left(i\xi_x^{-1}\sigma_g\right)\left[\left(i\xi_x^{-1}\tau_g\eta_y\right)\left(i\xi_x^{-1}\tau_g\eta_y\right)^{-1}\right] \\
&\quad \cdot\left(i\xi_x^{-1}\tau_g\sigma_h\right)\left(i\xi_x^{-1}\tau_g\tau_h\eta_z\right) \\
&= (i\xi_x\eta_x)^{-1}\left(i\xi_x^{-1}\sigma_{gh}\right)\left(i\xi_x^{-1}\tau_{gh}\eta_z\right) \\
&= i\psi_{qr}
\end{aligned}$$

and φ is a partial homomorphism.

Suppose next that $qr = 0$. Then $y \neq w$ which by part (i) of the hypothesis implies that $|y\pi \cap w\pi| \leq 1$. Since $\mathbf{r}\beta_q = \mathbf{r}\xi_y = y\pi$ and $\mathbf{d}\beta_r = \mathbf{d}\xi_w = w\pi$, we obtain rank $\beta_q\beta_r \leq 1$ which shows that φ satisfies the conditions in 4.7. Consequently, φ is an extension function for S by Q.

Conversely, let φ be an extension function for S by Q. We may suppose that φ maps Q^* directly into $G \operatorname{wr} \mathcal{I}(I)$, so let $q\varphi = (\psi_q, \beta_q)$ for every $q \in Q^*$. Then the mapping $q \to \beta_q$ is a partial homomorphism of Q^* into $\mathcal{I}(I)$. Since Q is 0-simple, Q^* consists of a single $\mathcal{J}$-class. It is easy to see that a partial homomorphism preserves $\mathcal{J}$-classes. It follows that φ maps Q^* into a single $\mathcal{J}$-class of $G \operatorname{wr} \mathcal{I}(I)$. Hence by IV.1.2(vii), there exists a cardinal number ν such that $\nu \leq |I|$ and rank $\beta_q = \nu$ for all $q \in Q^*$. Note that $|\mathbf{d}\beta_q| = |\mathbf{r}\beta_q| = \nu$.

Let $\mathcal{P}_\nu$ be as in the statement of the theorem, and define $\pi : X \to \mathcal{P}_\nu$ by

$$x\pi = \mathbf{d}\beta_{(x,1,x)} \qquad (x \in X).$$

Since φ satisfies the conditions of 4.7, if $x \neq y$, then rank $\beta_{(x,1,x)}\beta_{(y,1,y)} \leq 1$. Further, $\beta_{(x,1,x)}$ being an idempotent, we have $\mathbf{d}\beta_{(x,1,x)} = \mathbf{r}\beta_{(x,1,x)} = x\pi$, so that $|x\pi \cap y\pi| \leq 1$. Hence condition (i) in the statement of the theorem is satisfied.

Fix $x_0 \in X$ and let $P_0 = x_0\pi$. Define a function τ on H, $\tau : h \to \tau_h$, by

$$\tau_h = \beta_{(x_0,h,x_0)} \qquad (h \in H).$$

Then

$$\mathbf{d}\beta_{(x_0,h,x_0)} = \mathbf{r}\beta_{(x_0,h,x_0)} = \mathbf{d}\beta_{(x_0,1,x_0)} = \mathbf{r}\beta_{(x_0,1,x_0)} = x_0\pi = P_0$$

whence it follows that $\tau : H \to \mathcal{S}(P_0)$ and it is evidently a homomorphism.

For each $h \in H$, define σ_h by

$$\sigma_h = \psi_{(x_0,h,x_0)}.$$

Then $\sigma_h : P_0 \to G$, and for any $i \in P_0$, we obtain by the wreath product multiplication,

$$\begin{aligned}(i\sigma_g)(i\tau_g\sigma_h) &= \left(i\psi_{(x_0,g,x_0)}\right)\left(i\beta_{(x_0,g,x_0)}\psi_{(x_0,h,x_0)}\right)\\ &= i\psi_{(x_0,g,x_0)(x_0,h,x_0)} = i\psi_{(x_0,gh,x_0)} = i\sigma_{gh}.\end{aligned}$$

It follows that $\theta : H \to G \operatorname{wr} \mathcal{S}(P_0)$, defined by $h\theta = (\sigma_h, \tau_h)$, is a homomorphism. This verifies (ii).

Next, for each $x \in X$, let

$$\xi_x = \beta_{(x_0,1,x)}, \qquad \eta_x = \psi_{(x_0,1,x)}.$$

It follows at once that (iii) is satisfied. For any $(x, h, y) \in Q^*$, we obtain

$$\begin{aligned}\beta_{(x,h,y)} &= \beta_{(x,1,x_0)(x_0,h,x_0)(x_0,1,y)} \\ &= \beta_{(x,1,x_0)}\beta_{(x_0,h,x_0)}\beta_{(x_0,1,y)} = \xi_x^{-1}\tau_h\xi_y\end{aligned}$$

which verifies (20). In order to check (19), we let $(x, h, y) \in Q^*$, $i \in x\pi$, $j = i\xi_x^{-1}$, and compute

$$\begin{aligned}&\left(i\xi_x^{-1}\eta_x\right)^{-1}\left(i\xi_x^{-1}\sigma_h\right)\left(i\xi_x^{-1}\tau_h\eta_y\right) \\ &= \left(j\psi_{(x_0,1,x)}\right)^{-1}\left(j\psi_{(x_0,h,x_0)}\right)\left(j\beta_{(x_0,h,x_0)}\psi_{(x_0,1,y)}\right) \\ &= \left(j\psi_{(x_0,1,x)}\right)^{-1}\left[\left(j\psi_{(x_0,1,x)}\right)\left(j\beta_{(x_0,1,x)}\psi_{(x,h,x_0)}\right)\right]\left(j\beta_{(x_0,h,x_0)}\psi_{(x_0,1,y)}\right) \\ &= \left(j\beta_{(x_0,1,x)}\psi_{(x,h,x_0)}\right)\left(j\beta_{(x_0,h,x_0)}\psi_{(x_0,1,y)}\right) \\ &= \left(i\psi_{(x,h,x_0)}\right)\left(i\beta_{(x,h,x_0)}\psi_{(x_0,1,y)}\right) \\ &= i\psi_{(x,h,y)}\end{aligned}$$

as required.

We say that the extension function φ in 4.8, and hence the ideal extension V in 4.7, is *determined by the parameters* $\{\nu, \pi, \theta; \xi_x, \eta_x\}$. Note that these parameters are *independent*, that is, there are no relations that two or more of them must satisfy. Here ν is a cardinal number, π is a function satisfying a simple condition, θ is a group homomorphism, ξ_x is a one-to-one correspondence, and η_x is just a function. The problem of finding all ideal extensions is thus reduced to functions among sets and group homomorphisms.

If we drop the requirement in 4.8 that $|x\pi \cap y\pi| \leq 1$, then φ constructed in 4.8 is a partial homomorphism of Q^* into $G \operatorname{wr} \mathcal{I}(I)$, and conversely, all such partial homomorphisms can be so constructed.

The above result 4.8 combined with 4.7 gives all ideal extensions of one Brandt semigroup by another. We can now easily extend this to a construction of all ideal extensions of one primitive inverse semigroup by another, since primitive inverse semigroups coincide with orthogonal sums of Brandt semigroups by II.4.3. Again we only need to construct all extension functions.

V.4.9 Theorem

Let $S = \Sigma_{j \in J} S_j$ and $Q = \Sigma_{\alpha \in A} Q_\alpha$ be orthogonal sums of Brandt semigroups

$$S_j = B(G_j, I_j), \qquad Q_\alpha = B(H_\alpha, X_\alpha),$$

where all the sets S, Q, I_j, X_α are pairwise disjoint. For every $\alpha \in A, j \in J$, let

$$\varphi^{\alpha j} : Q_\alpha^* \to G_j \operatorname{wr} \mathcal{I}(I_j)$$

be an extension function for S_j by Q_α determined by the parameters

$$\left\{ \nu^{\alpha j}, \pi^{\alpha j}, \theta^{\alpha j}; \xi_x^{\alpha j}, \eta_x^{\alpha j} \right\}.$$

Assume that for any $\alpha, \beta \in A$, $x \in X_\alpha$, $y \in X_\beta$, $x \neq y$,

(i) $|x\pi^{\alpha j} \cap y\pi^{\beta j}| \leq 1$ for all $j \in J$,
(ii) there exists at most one $j \in J$ for which $x\pi^{\alpha j} \cap y\pi^{\beta j} \neq \varnothing$.

Then $\varphi : Q^* \to \prod_{j \in J} G_j \operatorname{wr} \mathcal{I}(I_j)$ defined by

$$\varphi : q \to \left(q\varphi^{\alpha j} \right)_{j \in J} \qquad \text{if} \quad q \in Q_\alpha^*$$

is an extension function.

Conversely, every extension of S by Q gives rise to such an extension function.

Proof. First note that by I.8.11, we have

$$\Omega\left(\sum_{j \in J} S_j \right) \cong \prod_{j \in J} \Omega(S_j)$$

and by the proof of it, that to the inner part of $\Omega(\Sigma_{j \in J} S_j)$ correspond elements $(\omega_j) \in \prod_{j \in J} \Omega(S_j)$ with $\omega_k = \pi_a$ for some $a \in S_k$ and $\omega_j = \pi_0$ if $j \neq k$. We can thus speak of an extension function $\varphi : Q^* \to \prod_{j \in J} G_j \operatorname{wr} \mathcal{I}(I_j)$ since

$$\Omega(S) \cong \prod_{j \in J} G_j \operatorname{wr} \mathcal{I}(I_j).$$

Observe that for $\alpha = \beta$, condition (i) is already included in the hypothesis that $\varphi^{\alpha j}$ is an extension function. Let φ be defined as in the statement of the theorem and let $q, r \in Q^*$. If $qr \neq 0$, then $q, r \in Q_\alpha^*$ for some $\alpha \in A$, and hence $(q\varphi^{\alpha j})(r\varphi^{\alpha j}) = (qr)\varphi^{\alpha j}$ for all $j \in J$ which evidently implies that $(q\varphi)(r\varphi) = (qr)\varphi$. Assume that $qr = 0$. Writing $q = (x, g, y)$, $r = (w, h, z)$, we must have $y \neq w$ since $qr = 0$ and the index sets X_α are pairwise disjoint. We deduce from 4.8 that $\mathbf{r}(q\varphi^{\alpha j}) = y\pi^{\alpha j}$ and $\mathbf{d}(r\varphi^{\beta j}) = w\pi^{\beta j}$ if $q \in Q_\alpha^*$, $r \in Q_\beta^*$. Conditions (i) and (ii) assure that $y\pi^{\alpha j} \cap w\pi^{\beta j}$ may be nonempty only for one j in which case this set contains exactly one element. This means that $(q\varphi^{\alpha j})(r\varphi^{\beta j})$ is nonzero for at most one $j \in J$ and for that j, $\operatorname{rank}(q\varphi^{\alpha j})(r\varphi^{\beta j}) \leq 1$. According to the remarks made at the beginning of this proof, $(q\varphi)(r\varphi)$ is contained in the part of $\prod_{j \in J} G_j \operatorname{wr} \mathcal{I}(I_j)$ which corresponds to the inner part of $\Omega(S)$. Consequently, φ is an extension function.

The proof of the converse goes along the same lines, in the reverse order; the details are left as an exercise.

V.4.10 Exercises

(i) Let $S = B(G, I)$ be a Brandt semigroup. Let $M(G, I)$ be the set of all $I \times I$ matrices over G^0 with at most one nonzero entry, under the usual row by column multiplication of matrices setting equal to 0 any sums of zeros. Show that the mapping

$$(i, g, j) \to (g)_{ij}$$

where $(g)_{ij}$ is the $I \times I$ matrix with g in the (i, j)th position and 0 elsewhere, is an isomorphism of S onto $M(G, I)$.

(ii) Let S be as in exercise (i). Let $L(I, G)$ be the set of all $I \times I$ matrices over G^0 with at most one nonzero entry in each column, with matrix multiplication (indicated above). Show that the mapping

$$(\alpha, \varphi) \to (g_{ij})$$

where (g_{ij}) is the $I \times I$ matrix with $g_{ij} = \varphi_j$ if $j \in \mathbf{d}\alpha$, $\alpha j = i$ and $g_{ij} = 0$ otherwise, is an isomorphism of $\mathcal{F}'(I) \operatorname{wl} G$ onto $L(I, G)$.

(iii) In an analogy to exercise (ii), define $R(G, I)$ and devise an isomorphism of $G \operatorname{wr} \mathcal{F}(I)$ onto $R(G, I)$.

(iv) Let S be as in exercise (i). Let $T(G, I)$ be the set of all $I \times I$ matrices over G^0 with at most one nonzero entry in each row and each column under the usual multiplication. Show that $\Omega(S) \cong T(G, I)$.

(v) For the semigroup V constructed in 4.7, using the extension function in 4.8, show that every idempotent in Q^* covers exactly ν idempotents in S^*. Also show that

(α) V is a retract extension of S if and only if $\nu \leq 1$,
(β) V is an orthogonal sum of S and Q if and only if $\nu = 0$,
(γ) V is a pure extension of S if and only if $\nu > 1$.

(vi) Prove the statements corresponding to exercise (v) for the case of an ideal extension of one primitive inverse semigroup by another using the extension function in 4.9.

(vii) Let the situation be as in exercise (v) and assume additionally that $|I| = n$ is finite. Prove that every ideal extension of S by Q is strict (and thus a retract extension) if and only if $|X| > \binom{n}{2}$.

(viii) For a nonempty set I and a group G, let $T = G \operatorname{wr} \mathcal{I}(I)$. Show that a nonzero element (ψ, β) of T is idempotent if and only if $\mu\psi = 1$ and $\mu\beta = \mu$ for all $\mu \in \mathbf{d}\beta$. If this is the case, prove that

$$H_{(\psi, \beta)} = \{(\xi, \eta) \in T \mid \mathbf{d}\xi = \mathbf{r}\xi = \mathbf{d}\beta, \eta \in \mathcal{S}(\mathbf{d}\beta)\} \cong G \operatorname{wr} \mathcal{S}(\mathbf{d}\beta).$$

(ix) For a nonempty set I, show that $\mathcal{I}_0(I)$ is a densely embedded ideal of $\mathcal{I}(I)$ contained in all nonzero ideals of $\mathcal{I}(I)$.

(x) Let Y be a nontrivial semilattice with the property that every principal ideal of Y is a chain with at most three elements. Show that $\Phi(Y)$ has $\mathcal{H}$ trivial and is either a Brandt semigroup or a retract extension of one Brandt semigroup by another. Conversely, let S be a semigroup having all these properties. Show that E_S has the property of Y above and that $S \cong \Phi(E_S)$.

The first result concerning an ideal extension of a Brandt semigroup by a semigroup with zero was established by Warne [5]. Lallement and Petrich [4] explicitly constructed all ideal extensions of a Brandt semigroup having a finite number of idempotents by an arbitrary Brandt semigroup. The general case, as well as the extensions of primitive inverse semigroups is due to Ault [1], [2]. These papers contain further information on the subject. The extensions treated here have also been constructed by Schein [17]. For a special case, see Hamilton–Tamura [1]. Concerning the wreath product, consult Hoehnke [5].

VI

THE CONJUGATE HULL

The conjugate hull of an inverse semigroup S is the inverse semigroup of all isomorphisms among subsemigroups of S of the form $\lambda S \rho$ for some idempotent bitranslation (λ, ρ) of S. It arose in the study of normal extensions of inverse semigroups. In view of its interesting properties it merits a study in its own right. This is enhanced by its appearance in the consideration of conjugate extensions, of which only embryonic beginnings are now available.

1. A rather long verification shows that the conjugate hull $\Psi(S)$ of an inverse semigroup S is closed under the composition of functions. The sets $\lambda S \rho$ with $(\lambda, \rho) \in E_{\Omega(S)}$ which figure in the definition of $\Psi(S)$ can be characterized in several ways, the most interesting being that they are subsemigroups of S which are full closure of retract ideals of E_S.

2. An inverse semigroup S is a conjugate extension of an inverse semigroup K if K is a self-conjugate subsemigroup of S. Every such extension gives rise to a canonical homomorphism $\theta(S : K)$ of S into $\Psi(K)$. The kernel of the congruence on S induced by $\theta_S = \theta(S : S)$ is the metacenter $M(S)$ of S. The image $\Theta(S)$ of S under θ_S is a self-conjugate subsemigroup of $\Psi(S)$. In the case $M(S) = E_S$, that is, S has idempotent metacenter, $\Psi(S)$ is a maximal essential conjugate extension of $\Theta(S)$.

3. There is a homomorphism of $\Omega(S)$ into $\Psi(S)$ which has some interesting features. In particular, when S has idempotent metacenter, then the canonical isomorphism θ_S is a dense embedding of S into $\Psi(S)$. There is also a homomorphism of $\Omega(S)$ into $\Psi(E_S)$ which induces the greatest idempotent separating congruence on $\Omega(S)$.

4. The normal hull $\Phi(S)$ consists of isomorphisms among subsemigroups of S of the form eSe for some idempotent e of S. This is actually the full closure of $\Theta(S)$ in $\Psi(S)$. Many interesting properties of $\Phi(S)$ as well as its application to normal extensions make it a worthwhile object of study.

5. The conjugate hull of an inverse semigroup which is a subdirect product of a semilattice and a group can be characterized explicitly. The hulls discussed

here also represent a valuable source of examples of inverse semigroups having some special features.

6. As in the Schreier theory of group extensions, one may consider normal extensions of inverse semigroups. Indeed, for a given triple (K, π, Q) of which K and Q are inverse semigroups and π is a homomorphism of K onto E_Q, one searches for an inverse semigroup S and a congruence ρ on S such that, roughly speaking, $\ker \rho = K$, $\operatorname{tr} \rho = \operatorname{tr} \pi$, and $S/\rho \cong Q$. The general solution of this problem is quite involved. For the case where K has idempotent metacenter, one gets a general overview of all solutions to this extension problem. A more precise description of the solutions is possible when in addition Q is an antigroup.

7. In the case that K is a semilattice, and thus ρ is required to be idempotent pure, a more elaborate solution has been devised. It consists of a construction which generalizes the semidirect product of two groups. This construction simplifies considerably in the case Q is a group and gives a description of all E-unitary inverse semigroups.

8. Specializing the last-mentioned construction to the case when K is a semilattice and Q is E-disjunctive, one obtains an equivalence of the category of such triples, and suitable morphisms, and the category of inverse semigroups and homomorphisms which preserve the greatest idempotent pure congruences.

VI.1 THE CONSTRUCTION

In order to study normal extensions of inverse semigroups, we will need yet another hull, namely the normal hull of an inverse semigroup. We start, more generally, with conjugate extensions, for which we need the conjugate hull. For its definition we need some preparation. We then establish two characterizations of certain semigroups figuring in its definition.

VI.1.1 Lemma

Let S be an inverse semigroup. For any $(\lambda, \rho) \in E_{\Omega(S)}$, define a function σ by

$$\sigma : a \to \lambda a \rho \qquad (a \in S). \tag{1}$$

Then $\sigma|_{E_S}$ is a right translation of E_S and

$$a\sigma = (aa^{-1})\sigma a(a^{-1}a)\sigma \qquad (a \in S). \tag{2}$$

Conversely, if σ is a function on S such that $\sigma|_{E_S}$ is a right translation on E_S and (2) holds, then there exists a unique $(\lambda, \rho) \in E_{\Omega(S)}$ such that (1) holds.

Proof. First let $(\lambda, \rho) \in E_{\Omega(S)}$ and define σ by (1). Then for any $e \in E_S$, we have $e\sigma = \lambda e = e\rho$ by V.1.6, so $\sigma|_{E_S}$ is a right translation, and for any $a \in S$,

$$(aa^{-1})\sigma a(a^{-1}a)\sigma = \lambda(aa^{-1})a(a^{-1}a)\rho = \lambda a\rho = a\sigma.$$

Conversely, let σ be a function with the properties enunciated in the statement of the lemma. Define λ and ρ by

$$\lambda a = (aa^{-1})\sigma a, \qquad a\rho = a(a^{-1}a)\sigma \qquad (a \in S).$$

Then

$$\begin{aligned}(\lambda a)b &= (aa^{-1})\sigma ab = \left[(aa^{-1})\sigma(ab)(ab)^{-1}\right](ab)\\ &= \left[(ab)(ab)^{-1}(aa^{-1})\sigma\right](ab) \qquad \text{since } (aa^{-1})\sigma \in E_S\\ &= \left[(ab)(ab)^{-1}(aa^{-1})\right]\sigma(ab)\\ &= \left[(aa^{-1})(ab)(ab)^{-1}\right]\sigma(ab) = \left[(ab)(ab)^{-1}\right]\sigma(ab) = \lambda(ab),\end{aligned}$$

symmetrically $a(b\rho) = (ab)\rho$,

$$\begin{aligned}a(\lambda b) &= a\left[(bb^{-1})\sigma b\right] = a\left[(a^{-1}a)(bb^{-1})\sigma\right]b\\ &= a(a^{-1}a)\sigma(bb^{-1})b = (a\rho)b,\end{aligned}$$

$$\begin{aligned}\lambda^2 a &= \lambda(\lambda a) = \lambda\left[(aa^{-1})\sigma a\right] = \lambda\left[(aa^{-1})\sigma\right]a = (aa^{-1})\sigma^2(aa^{-1})\sigma a\\ &= \left[(aa^{-1})\sigma\right]^2 a = (aa^{-1})\sigma a = \lambda a,\end{aligned}$$

symmetrically $a\rho^2 = a\rho$,

$$\lambda a\rho = \left[(aa^{-1})\sigma a\right]\rho = (aa^{-1})\sigma(a\rho) = (aa^{-1})\sigma a(a^{-1}a)\sigma = a\sigma$$

which establishes all the assertions of the converse except for uniqueness. If $\lambda e\rho = \lambda' e\rho'$ for some $(\lambda, \rho), (\lambda', \rho') \in E_{\Omega(S)}$ and all $e \in E$, then for every $a \in S$,

$$\lambda a = \lambda(aa^{-1})a = \left[\lambda(aa^{-1})\rho\right]a = \left[\lambda'(aa^{-1})\rho'\right]a = \lambda'(aa^{-1})a = \lambda' a$$

so $\lambda = \lambda'$ and symmetrically $\rho = \rho'$.

VI.1.2 Lemma

Let σ, σ', τ be functions on an inverse semigroup S satisfying the conditions in 1.1. Let φ be an isomorphism of $S\sigma$ onto $S\sigma'$ and assume that $S\tau \subseteq S\sigma$. Define a function τ' by

$$\tau' : x \to x\sigma'\varphi^{-1}\tau\varphi \qquad (x \in S).$$

Then τ' satisfies the conditions in 1.1 and $S\tau\varphi = S\tau'$.

Proof. First note that τ' is well defined since $S\sigma' = \mathbf{r}\varphi$ and $S\tau \subseteq S\sigma = \mathbf{d}\varphi$. For any $e, f \in E_S$, we obtain

$$\begin{aligned}(ef)\tau' &= (ef)\sigma'\varphi^{-1}\tau\varphi = [(e\sigma')(f\sigma')]\varphi^{-1}\tau\varphi \\ &= [(e\sigma'\varphi^{-1})(f\sigma'\varphi^{-1})]\tau\varphi = [(e\sigma'\varphi^{-1})(f\sigma'\varphi^{-1}\tau)]\varphi \\ &= (e\sigma')(f\sigma'\varphi^{-1}\tau\varphi) = e(f\sigma'\varphi^{-1}\tau\varphi\sigma') \\ &= e(f\sigma'\varphi^{-1}\tau\varphi) = e(f\tau')\end{aligned}$$

and hence $\tau'|_{E_S} \in \mathrm{P}(E_S)$. Let (λ', ρ') correspond to σ' and (α, β) to τ as in 1.1. Then for any $x \in S$, we get

$$\begin{aligned}(xx^{-1})\tau'x(x^{-1}x)\tau' &= [(xx^{-1})\sigma'\varphi^{-1}\tau\varphi]x[(x^{-1}x)\sigma'\varphi^{-1}\tau\varphi] \\ &= [(xx^{-1})\sigma'\varphi^{-1}\tau\varphi]\rho'x\lambda'[(x^{-1}x)\sigma'\varphi^{-1}\tau\varphi] \\ &= [(xx^{-1})\sigma'\varphi^{-1}\tau\varphi](x\sigma')[(x^{-1}x)\sigma'\varphi^{-1}\tau\varphi] \\ &= \{[(xx^{-1})\sigma'\varphi^{-1}\tau](x\sigma'\varphi^{-1})[(x^{-1}x)\sigma'\varphi^{-1}\tau]\}\varphi \\ &= \{[(xx^{-1})\sigma'\varphi^{-1}](x\sigma'\varphi^{-1})[(x^{-1}x)\sigma'\varphi^{-1}]\}\tau\varphi \\ &= \{[(xx^{-1})\sigma'](x\sigma')[(x^{-1}x)\sigma']\}\varphi^{-1}\tau\varphi \\ &= x\sigma'\varphi^{-1}\tau\varphi = x\tau'\end{aligned}$$

and τ' satisfies the conditions in 1.1. Further, for any $x \in S\tau$, we have

$$x\varphi\tau' = (x\varphi)\sigma'\varphi^{-1}\tau\varphi = x\varphi\varphi^{-1}\tau\varphi = x\tau\varphi = x\varphi$$

so that $S\tau\varphi \subseteq S\tau'$. Conversely, $S\tau' = S\sigma'\varphi^{-1}\tau\varphi \subseteq S\tau\varphi$. Consequently, $S\tau\varphi = S\tau'$.

We may now introduce the desired concept.

VI.1.3 Definition

For any inverse semigroup S, the set consisting of isomorphisms among subsemigroups of S of the form $\lambda S\rho$ with $(\lambda, \rho) \in E_{\Omega(S)}$ and with composition of these isomorphisms as right operators is the *conjugate hull* $\Psi(S)$ of S.

We hasten to prove that $\Psi(S)$ is closed under its operation.

VI.1.4 Proposition

For any inverse semigroup S, $\Psi(S)$ is a subsemigroup of $\mathcal{I}(S)$.

Proof. Let $\varphi, \psi \in \Psi(S)$ with

$$\varphi : \lambda S\rho \to \lambda' S\rho', \qquad \psi : \alpha S\beta \to \alpha' S\beta'$$

in the obvious notation. Then $\mathbf{d}(\varphi\psi) = (\mathbf{r}\varphi \cap \mathbf{d}\psi)\varphi^{-1}$ and

$$\mathbf{r}\varphi \cap \mathbf{d}\psi = \lambda' S\rho' \cap \alpha S\beta = \lambda'\alpha S\beta\rho'$$

since (λ', ρ') and (α, β) are idempotents, and thus

$$\mathbf{d}(\varphi\psi) = \left(\lambda'\alpha S\beta\rho'\right)\varphi^{-1}.$$

Now letting

$$\theta : a \to \lambda' a\rho', \qquad \theta' : a \to \lambda a\rho, \qquad \tau : a \to \lambda'\alpha a\beta\rho' \qquad (a \in S)$$

and using φ^{-1}, we deduce from 1.2 that $\mathbf{d}(\varphi\psi) = S\tau'$ for some τ' satisfying the conditions in 1.2. Consequently, 1.1 implies that $\mathbf{d}(\varphi\psi)$ is of the form $\xi S\eta$ for some $(\xi, \eta) \in E_{\Omega(S)}$. One proves similarly that $\mathbf{r}(\varphi\psi) = (\lambda'\alpha S\beta\rho')\psi$ has such a form.

The definition of the conjugate hull of an inverse semigroup S is based upon the sets $\lambda S\rho$ with $(\lambda, \rho) \in E_{\Omega(S)}$. Hence these sets deserve special attention, and we now proceed to give two characterizations of these sets. To this end, we need a new concept.

VI.1.5 Definition

A subsemigroup B of a semigroup S is a *biideal* of S if $BSB \subseteq B$.

Recall from II.1.5 that an inverse semigroup is endowed with its natural partial order. This partial order in a semilattice coincides with its usual semilattice order. Recall also the concepts of an order ideal and a p-ideal introduced in I.2.4. By II.4.4, in a semilattice, order p-ideals coincide with retract ideals.

VI.1.6 Theorem

Let B be a subset of an inverse semigroup S. Then $B = \lambda S\rho$ for some $(\lambda, \rho) \in E_{\Omega(S)}$ if and only if B is an inverse semigroup which is a biideal and an order p-ideal of S.

Proof

Necessity. Let $(\lambda, \rho) \in E_{\Omega(S)}$ and $B = \lambda S\rho$. Clearly B is a subsemigroup of S. If $x \in B$, then $x = \lambda x = x\rho$ and thus using V.1.7,

$$x^{-1} = (\lambda x)^{-1} = x^{-1}\rho = (x\rho)^{-1} = \lambda x^{-1}$$

so that $x^{-1} \in B$. Hence B is an inverse semigroup. Let

$$(\lambda x\rho)y(\lambda z\rho) = \lambda[(x\rho)y(\lambda z)]\rho \in B$$

so that B is a biideal of S. If $y \leq x$ and $x \in B$, then $y = ex$ for some $e \in E_S$ and thus

$$y = ex = e(\lambda x\rho) = (e\rho)(x\rho) = (\lambda e)(x\rho) = \lambda(ex)\rho = \lambda y\rho$$

and hence $y \in B$. Consequently, B is an order ideal of S. Further,

$$\lambda x\rho = \lambda[(xx^{-1})x(x^{-1}x)]\rho = [\lambda(xx^{-1})]x[(x^{-1}x)\rho] \leq x$$

since $\lambda(xx^{-1}), (x^{-1}x)\rho \in E_S$ so that $[\lambda x\rho] \subseteq B \cap [x]$. Conversely, if $y \in B$ and $y \leq x$, then $y = ex$ for some $e \in E_S$, which implies

$$y = \lambda y\rho = \lambda(ex)\rho = (\lambda e)(x\rho) = (e\rho)(x\rho) = e(\lambda x\rho) \leq \lambda x\rho$$

so that $B \cap [x] \subseteq [\lambda x\rho]$. Hence B is a p-ideal of S.

Sufficiency. We will verify the conditions of 1.1. Let B have the requisite properties. We define a function σ by the requirement

$$[x\sigma] = B \cap [x] \qquad (x \in S).$$

If $e \in E_S$, then $[e] = E_S e$ and thus $e\sigma \in E_S$. For $e, f \in E_S$ we obtain

$$[e(f\sigma)] = [e][f\sigma] = [e](B \cap [f]) = [e]B \cap [ef], \tag{3}$$

$$[(ef)\sigma] = B \cap [ef]. \tag{4}$$

Now let $g \in [e]B \cap [ef]$. Then $g = tb$ for some $t \leq e$ and $b \in B$. Hence

$$g = tb = (bb^{-1})tb = b(b^{-1}t)b \in BSB \subseteq B,$$

which proves that $[e]B \cap [ef] \subseteq B \cap [ef]$. Conversely, let $g \in B \cap [ef]$. Then $g = eg \in [e]B$ so that $B \cap [ef] \subseteq [e]B \cap [ef]$. Relations (3) and (4) now yield $\sigma|_{E_S} \in \mathrm{P}(E_S)$.

Let $y \leq (xx^{-1})\sigma x(x^{-1}x)\sigma$. Using $aE_S = E_S a$ for any $a \in S$, we obtain

$$y \in (xx^{-1})\sigma x(x^{-1}x)\sigma E_S = \left[(xx^{-1})\sigma E_S\right]x\left[(x^{-1}x)\sigma E_S\right]$$

$$= \left(xx^{-1}E_S \cap B\right)x\left(x^{-1}xE_S \cap B\right)$$

which implies $y \in BSB \subseteq B$ and $y \in E_S x E_S = xE_S$. It follows that $y \in [x] \cap B = [x\sigma]$ and hence $y \leq x\sigma$. We have proved that $(xx^{-1})\sigma x(x^{-1}x)\sigma \leq x\sigma$.

In order to prove the opposite inequality we first let $y \in [xx^{-1}]x[x^{-1}x] \cap B$. Then $y = uxv$ where $u = exx^{-1}$ and $v = x^{-1}xf$ for some $e, f \in E_S$. Hence $y = exf$. Noting that B is an inverse subsemigroup of S, we get

$$yy^{-1} = (exf)(exf)^{-1} = exfx^{-1}e \in [xx^{-1}] \cap B,$$

$$y^{-1}y = (exf)^{-1}(exf) = fx^{-1}exf \in [x^{-1}x] \cap B,$$

$$(exfx^{-1}e)x(fx^{-1}exf) = (exf)(exf)^{-1}(exf)(exf)^{-1}(exf) = exf = y$$

which shows that $y \in ([xx^{-1}] \cap B)x([x^{-1}x] \cap B)$. Using this, we obtain

$$[x\sigma] = [x] \cap B = \left[(xx^{-1})x(x^{-1}x)\right] \cap B$$

$$= [xx^{-1}]x[x^{-1}x] \cap B \subseteq \left([xx^{-1}] \cap B\right)x\left([x^{-1}x] \cap B\right)$$

$$= \left[(xx^{-1})\sigma\right]x\left[(x^{-1}x)\sigma\right] = \left[(xx^{-1})\sigma x(x^{-1}x)\sigma\right]$$

which implies $x\sigma \leq (xx^{-1})\sigma x(x^{-1}x)\sigma$.

We have proved that the conditions in 1.1 are satisfied, which gives that $B = \lambda S\rho$ for some $(\lambda, \rho) \in E_{\Omega(S)}$.

For the second characterization we need further concepts.

VI.1.7 Definition

Let K be an inverse subsemigroup of an inverse semigroup S. An inverse subsemigroup F of S containing K is the *full closure* of K in S if F is the greatest inverse subsemigroup of S containing K as a full subsemigroup. If K coincides with its full closure, it is *fully closed*.

VI.1.8 Lemma

Let K be an inverse subsemigroup of an inverse semigroup S, and assume that E_K is an ideal of E_S. Then

$$F = \{s \in S \mid ss^{-1}, s^{-1}s \in K\}$$

is the full closure of K in S.

Proof. Let $s, t \in F$. Then $(st)(st)^{-1} = (ss^{-1})[(st)(st)^{-1}] \in E_K$ since $ss^{-1} \in E_K$ and E_K is an ideal of E_S. Similarly, $(st)^{-1}(st) \in E_K$ and thus $st \in F$. Hence F is a subsemigroup of S, which then easily implies that F is the full closure of K in S.

The desired characterization can now be proved.

VI.1.9 Theorem

Let B be a subset of an inverse semigroup S. Then $B = \lambda S\rho$ for some $(\lambda, \rho) \in E_{\Omega(S)}$ if and only if E_B is a retract ideal of E_S and B is the full closure of E_B.

Proof

Necessity. Let $(\lambda, \rho) \in E_{\Omega(S)}$, $B = \lambda S\rho$ and $\lambda' = \lambda|_{E_S}$, $\rho' = \rho|_{E_S}$. Using V.1.6, it follows easily that $(\lambda', \rho') \in \Omega(E_S)$ and $E_B = \lambda' E_S \rho'$. By V.3.2, E_B is a retract ideal of E_S. It follows from V.1.7 that B is an inverse subsemigroup of S, so that

$$B \subseteq \{s \in S \mid ss^{-1}, s^{-1}s \in E_B\}. \tag{5}$$

In order to prove the opposite inclusion, we let $ss^{-1}, s^{-1}s \in E_B$. Then

$$s = (ss^{-1})s = [\lambda'(ss^{-1})]s = [\lambda(ss^{-1})]s = (\lambda s)s^{-1}s = \lambda s$$

and analogously $s = s\rho$ so that $s \in \lambda S\rho = B$. This establishes the other inclusion in (5). We can now apply 1.8, since E_B is an ideal of E_S, to get that B is the full closure of E_B.

Sufficiency. According to V.3.2, there exists $(\alpha, \beta) \in \Omega(E_S)$ such that $E_B = \alpha E_S \beta$. Define λ and ρ by

$$\lambda s = [\alpha(ss^{-1})]s, \qquad s\rho = s[(s^{-1}s)\beta] \qquad (s \in S).$$

We then obtain for any $s, t \in S$,

$$(\lambda s)t = [\alpha(ss^{-1})]st = [\alpha(ss^{-1})](st)(st)^{-1}(st)$$

$$= \alpha[(ss^{-1})(st)(st)^{-1}](st) = \alpha[(st)(st)^{-1}](st) = \lambda(st)$$

and analogously $s(t\rho) = (st)\rho$. Further,

$$s(\lambda t) = s[\alpha(tt^{-1})]t = s(s^{-1}s)[\alpha(tt^{-1})]t$$

$$= s[(s^{-1}s)\beta](tt^{-1})t = (s\rho)t$$

and hence $(\lambda, \rho) \in \Omega(S)$. Also

$$\lambda(\lambda s) = \lambda[\alpha(ss^{-1})s] = \lambda[\alpha(ss^{-1})]s$$

$$= \alpha^2(ss^{-1})s = \alpha(ss^{-1})s = \lambda s$$

and analogously $(s\rho)\rho = s\rho$, so that $(\lambda, \rho) \in E_{\Omega(S)}$.

Using 1.8, we finally obtain for any $s \in S$,

$$s \in B \Leftrightarrow ss^{-1}, s^{-1}s \in E_B \Leftrightarrow ss^{-1} = \alpha(ss^{-1}),\ s^{-1}s = (s^{-1}s)\beta$$

$$\Leftrightarrow s = \lambda s = s\rho \Leftrightarrow s \in \lambda S\rho$$

which proves that $B = \lambda S\rho$.

VI.1.10 Exercises

(i) Let S be an inverse semigroup and B be a subset of S. Show that the following statements are equivalent.

- (α) $B = \lambda S\rho$ for some $(\lambda, \rho) \in E_{\Omega(S)}$.
- (β) B is a fully closed inverse subsemigroup of S and E_B is a retract ideal of E_S.
- (γ) $B = YSY$ for some retract ideal Y of E_S.

(ii) Let K be an inverse subsemigroup of an inverse semigroup S. Show that

$$K^c = \{s \in S \mid ss^{-1}, s^{-1}s \in E_K,\ sE_Ks^{-1}, s^{-1}E_Ks \subseteq E_K\}$$

is the full closure of K in S.

(iii) Show that for a semilattice Y, $\Psi(Y)$ is simple if and only if every (principal) ideal of Y contains an isomorphic copy of Y.

The conjugate hull was introduced and investigated by Petrich [13]. The material here is taken from that paper except 1.8, which was remarked by McAlister [8], and 1.9. In the case when X is a semilattice, $\Psi(X)$ appears in Reilly [9] in the notation U_X. If a semilattice E has an identity, then $\Psi(E)$ coincides with the semigroup T_E introduced by Munn [10].

VI.2 CONJUGATE EXTENSIONS

After introducing conjugate extensions, we prove that if S is a conjugate extension of K, there is a canonical homomorphism of S into $\Psi(K)$. The next result says that for $K = S$, $\Psi(S)$ is a conjugate extension of the image of S under this homomorphism. We then consider essential conjugate extensions of S and establish some of their properties under a certain restriction on S.

VI.2.1 Definition

Let K be an inverse subsemigroup of an inverse semigroup S. Then S is a *conjugate extension* of K if K is self-conjugate in S.

VI.2.2 Theorem

Let an inverse semigroup S be a conjugate extension of an inverse semigroup K. For each $a \in S$, let

$$\theta^a : k \to a^{-1}ka \qquad (k \in \mathbf{d}\theta^a = aKa^{-1}).$$

Then the mapping

$$\theta = \theta(S : K) : a \to \theta^a \qquad (a \in S)$$

is a homomorphism of S into $\Psi(K)$.

Proof. Let $a \in S$. Define

$$\sigma : k \to aa^{-1}kaa^{-1} \qquad (k \in K).$$

It follows immediately that σ satisfies the conditions in 1.1 and $K\sigma = aKa^{-1}$. Also $a^{-1}(aKa^{-1})a = a^{-1}Ka$ has the property that $K\sigma' = a^{-1}Ka$ for some σ' satisfying the conditions in 1.1. Let $x, y \in aKa^{-1}$ and assume that $a^{-1}xa = a^{-1}ya$. Then $aa^{-1}xaa^{-1} = aa^{-1}yaa^{-1}$ which yields $x = y$ since $x, y \in aKa^{-1}$. Further,

$$(x\theta^a)(y\theta^a) = (a^{-1}xa)(a^{-1}ya) = a^{-1}x(aa^{-1})ya = a^{-1}xya = (xy)\theta^a$$

since $x = x(aa^{-1})$. Consequently, $\theta^a \in \Psi(K)$.

Now let $a, b \in S$. Then

$$\mathbf{d}(\theta^a\theta^b) = (\mathbf{r}\theta^a \cap \mathbf{d}\theta^b)(\theta^a)^{-1} = a(a^{-1}Ka \cap bKb^{-1})a^{-1}$$

$$= aa^{-1}Kaa^{-1} \cap abKb^{-1}a^{-1} = (ab)K(ab)^{-1} = \mathbf{d}\theta^{ab}$$

and for any $x \in \mathbf{d}\theta^{ab}$, we have

$$x\theta^a\theta^b = (a^{-1}xa)\theta^b = b^{-1}(a^{-1}xa)b = (ab)^{-1}x(ab) = x\theta^{ab}.$$

Consequently, $\theta^a\theta^b = \theta^{ab}$ and θ is a homomorphism.

The following concepts are of basic importance in the study of conjugate extensions.

VI.2.3 Definition

With the notation in 2.2, the mapping $\theta(S : K)$ is the *canonical homomorphism* of S into $\Psi(K)$.

For any homomorphism φ of inverse semigroups, we denote by $\ker\varphi$ and $\operatorname{tr}\varphi$ the kernel and the trace, respectively, of the congruence induced by φ (see III.1.2 for the definitions).

We write θ_S for $\theta(S : S)$. The semigroup

$$\Theta(S) = \{\theta_S^s \mid s \in S\}$$

is the *inner part* of $\Psi(S)$, and

$$M(S) = \ker\theta_S$$

the *metacenter* of S. If $M(S) = E_S$, we say that S has *idempotent metacenter*.

The next lemma provides an explicit expression for the metacenter. Recall that $Z(S)$ denotes the center of S.

VI.2.4 Lemma

For any inverse semigroup S, θ_S is idempotent separating and has kernel

$$M(S) = \{a \in S \mid aa^{-1}xa = axa^{-1}a \quad \text{for all} \quad x \in S\}$$

$$= \bigcup_{e \in E_S} (Z(eSe) \cap H_e).$$

Proof. For any $a, b \in S$, we obtain

$$\theta_S^a = \theta_S^b \Leftrightarrow a^{-1}xa = b^{-1}xb \text{ if } x \in aSa^{-1} = bSb^{-1}$$

$$\Leftrightarrow a^{-1}xa = b^{-1}xb \text{ if } x \in aSa^{-1} \text{ and } aa^{-1} = bb^{-1},$$

$$a^{-1}a = b^{-1}b$$

$$\Leftrightarrow a^{-1}xa = b^{-1}xb \text{ if } x \in aSa^{-1} \text{ and } a\mathcal{H}b$$

which implies that θ is idempotent separating. Further, for $e \in E_S$,

$$\theta_S^a = \theta_S^e \Leftrightarrow a^{-1}xa = x \text{ if } x \in eSe \text{ and } a\mathcal{H}e$$

$$\Leftrightarrow xa = ax \text{ if } x \in eSe \text{ and } a\mathcal{H}e \tag{6}$$

which shows that $M(S)$ is equal to the third set in the statement of the lemma.

Now assume that $aa^{-1}xa = axa^{-1}a$ for all $x \in S$. For $x = a^{-1}a$, we get

$$aa^{-1}(a^{-1}a)a = a(a^{-1}a)a^{-1}a = a$$

so that $a \in Sa^2$. One shows similarly that $a \in a^2S$, which implies that $a \in H_e$ for some $e \in E_S$. For $x \in eSe$, we obtain

$$xa = exa = aa^{-1}xa = axa^{-1}a = axe = ax$$

and hence $a \in Z(eSe) \cap H_e$.

Finally, let $\theta_S^a = \theta_S^e$ where $e \in E_S$. Then for any $x \in S$, by (6) we have $(exe)a = a(exe)$ which implies $aa^{-1}xa = axa^{-1}a$ since $e = aa^{-1} = a^{-1}a$ again by (6). Hence a is contained in the second set in the statement of the lemma.

It is the expression

$$M(S) = \{a \in S \mid aa^{-1}xa = axa^{-1}a \quad \text{for all} \quad x \in S\}$$

that will be used frequently for the metacenter of S. After some preparation, we will show that $\Psi(S)$ is a conjugate extension of $\Theta(S)$.

VI.2.5 Lemma

Let S be an inverse semigroup, $a \in S$ and $(\lambda, \rho) \in E_{\Omega(S)}$. Then $\lambda a^{-1}\rho = (\lambda a \rho)^{-1}$.

Proof. This follows easily from V.1.7.

VI.2.6 Lemma

Let S be an inverse semigroup, $a \in S$ and $\psi \in \Psi(S)$ with $\mathbf{d}\psi = \lambda S\rho$, $(\lambda, \rho) \in E_{\Omega(S)}$. Then $\psi^{-1}\theta^{a}\psi = \theta^{(\lambda a\rho)\psi}$.

Proof. For $x \in \mathbf{d}(\psi^{-1}\theta^{a}\psi)$, we obtain

$$x(\lambda a\rho)\psi[(\lambda a\rho)\psi]^{-1} = \left[(x\psi^{-1})(\lambda a\rho)(\lambda a\rho)^{-1}\right]\psi$$

$$= \left[(x\psi^{-1}\rho)(a\rho)(\lambda a\rho)^{-1}\right]\psi$$

$$= \left[(x\psi^{-1})(a\rho)(\lambda a\rho)^{-1}\right]\psi \qquad \text{since } x\psi^{-1} \in \lambda S\rho$$

$$= \left[(x\psi^{-1})(a\rho)(\lambda a^{-1}\rho)\right]\psi \qquad \text{by 2.5}$$

$$= \left[(x\psi^{-1})a(\lambda a^{-1}\rho)\right]\psi \qquad \text{since } (\lambda, \rho) \in E_{\Omega(S)}$$

$$= \left\{a\left[a^{-1}(x\psi^{-1})a\right]\rho(a^{-1}\rho)\right\}\psi \qquad \text{since } x\psi^{-1} \in aSa^{-1}$$

$$= \left[aa^{-1}(x\psi^{-1})a(a^{-1}\rho)\right]\psi \qquad \text{since } a^{-1}(x\psi^{-1})a \in \lambda S\rho$$

$$= \left[(x\psi^{-1})(aa^{-1})\rho\right]\psi \qquad \text{since } x\psi^{-1} \in aSa^{-1}$$

$$= \left\{\left[(x\psi^{-1})aa^{-1}\right]\rho\right\}\psi$$

$$= (x\psi^{-1})\rho\psi = x\psi^{-1}\psi = x$$

and an analogous argument shows that $(\lambda a\rho)\psi[(\lambda a\rho)\psi]^{-1}x = x$. Consequently, $x \in \mathbf{d}\theta^{(\lambda a\rho)\psi}$.

Conversely, let $x \in \mathbf{d}\theta^{(\lambda a\rho)\psi}$ so that

$$x = (\lambda a\rho)\psi[(\lambda a\rho)\psi]^{-1}x(\lambda a\rho)\psi[(\lambda a\rho)\psi]^{-1}.$$

Since $(\lambda a\rho)\psi \in \mathbf{r}\psi$, it follows that $x \in \mathbf{r}\psi$. Further,

$$x\psi^{-1} = (\lambda a\rho)(\lambda a\rho)^{-1}(x\psi^{-1}) = \left[\lambda(aa^{-1})\right](a\rho)(\lambda a\rho)^{-1}(x\psi^{-1})$$

$$= (aa^{-1})\rho(a\rho)(\lambda a\rho)^{-1}(x\psi^{-1}) = aa^{-1}(\lambda a\rho)(\lambda a\rho)^{-1}(x\psi^{-1})$$

$$= aa^{-1}(x\psi^{-1})$$

and symmetrically $x\psi^{-1} = (x\psi^{-1})aa^{-1}$ so that $x\psi^{-1} \in aSa^{-1}$. Finally,

$$\lambda\left[a^{-1}(x\psi^{-1})a\right] = (\lambda a^{-1})(\lambda a\rho)(\lambda a\rho)^{-1}aa^{-1}(x\psi^{-1})a$$

$$= \lambda\left[a^{-1}(\lambda a\rho)(\lambda a\rho)^{-1}a\right]a^{-1}(x\psi^{-1})a$$

$$= a^{-1}(\lambda a\rho)(\lambda a\rho)^{-1}(a\rho)a^{-1}(x\psi^{-1})a \qquad \text{by V.1.6}$$

$$= a^{-1}\left[(\lambda a\rho)(\lambda a^{-1}\rho)(\lambda a\rho)\right]a^{-1}(x\psi^{-1})a \qquad \text{by 2.5}$$

$$= a^{-1}(\lambda a\rho)a^{-1}(x\psi^{-1})a \qquad \text{by 2.5}$$

$$= a^{-1}(\lambda a\rho)(\lambda a^{-1})\left[\lambda(x\psi^{-1})\right]a \qquad \text{since } x\psi^{-1} \in \mathbf{d}\psi = \lambda S\rho$$

$$= a^{-1}(\lambda a\rho)(\lambda a^{-1}\rho)(x\psi^{-1})a$$

$$= a^{-1}\left[(\lambda a\rho)(\lambda a\rho)^{-1}(x\psi^{-1})\right]a \qquad \text{by 2.5}$$

$$= a^{-1}(x\psi^{-1})a$$

and symmetrically $[a^{-1}(x\psi^{-1})a]\rho = a^{-1}(x\psi^{-1})a$, so that $x \in \mathbf{d}(\psi^{-1}\theta^a\psi)$.

We have proved that $\mathbf{d}(\psi^{-1}\theta^a\psi) = \mathbf{d}\theta^{(\lambda a\rho)\psi}$. For any x in this set, we get

$$x(\psi^{-1}\theta^a\psi) = \left[a^{-1}(x\psi^{-1})a\right]\psi = \left\{\lambda\left\{a^{-1}\left[\lambda(x\psi^{-1})\rho\right]a\right\}\rho\right\}\psi$$

$$= \left[(\lambda a^{-1}\rho)(x\psi^{-1})(\lambda a\rho)\right]\psi = \left[(\lambda a\rho)^{-1}(x\psi^{-1})(\lambda a\rho)\right]\psi$$

$$= \left[(\lambda a\rho)\psi\right]^{-1}x\left[(\lambda a\rho)\psi\right] = x\theta^{(\lambda a\rho)\psi}$$

which completes the proof.

VI.2.7 Corollary

For any inverse semigroup S, $\Psi(S)$ is a conjugate extension of $\Theta(S)$.

We now introduce a standard universal algebraic concept. Note that the same type of concept was called in I.9.17 a dense extension (for historical reasons).

VI.2.8 Definition

Let K be a subsemigroup of a semigroup S having some property P, possibly empty, relative to the entire semigroup S; call S a *P-extension* of K. Then S is an *essential P-extension* of K if the equality relation on S is the only congruence on S whose restriction to K is the equality relation on K. If in addition for any essential P-extension S' of K containing S we have $S = S'$, then S is a *maximal essential P-extension* of K.

We first prove the following simple result.

VI.2.9 Proposition

Let an inverse semigroup S be a conjugate extension of an inverse semigroup K. Then $\theta(S:K)$ is one-to-one if and only if $M(K) = E_K$ and S is an essential extension of K.

Proof. Let $\theta = \theta(S:K)$.

Necessity. Since $\theta|_K = \theta_K$, it follows that $M(K) = E_K$. Let ξ be a congruence on S whose restriction to K is the equality relation. Assume that $a\xi b$. Then $a^{-1}\xi b^{-1}$ and hence for any $k \in K$, we have $a^{-1}ka\xi b^{-1}kb$, which by the hypothesis on ξ implies $a^{-1}ka = b^{-1}kb$. Similarly, $aka^{-1} = bkb^{-1}$ which implies $aKa^{-1} = bKb^{-1}$. But then $\theta^a = \theta^b$ and hence $a = b$ since θ is assumed to be one-to-one. Consequently, S is an essential extension of K.

Sufficiency. Let ξ be the congruence on S induced by θ. Then $\xi|_K$ is the equality relation since $M(K) = E_K$. Since the extension is essential, it follows that ξ is the equality relation. But then θ is one-to-one.

We now come to the main result of this section.

VI.2.10 Theorem

If S is an inverse semigroup with idempotent metacenter, then $\Psi(S)$ is a maximal essential conjugate extension of $\Theta(S)$.

Proof. Let ξ be a congruence on $\Psi(S)$ whose restriction to $\Theta(K)$ is the equality relation, and let $\varphi\xi\psi$. Then, as in the proof of 2.9, we obtain

$$\varphi^{-1}\theta^a\varphi = \psi^{-1}\theta^a\psi, \qquad \varphi\theta^a\varphi^{-1} = \psi\theta^a\psi^{-1} \qquad (a \in S).$$

According to 2.6, this yields

$$\theta^{(\lambda a\rho)\varphi} = \theta^{(\alpha a\beta)\psi}, \qquad \theta^{(\lambda' a\rho')\varphi^{-1}} = \theta^{(\alpha' a\beta')\psi^{-1}} \qquad (a \in S),$$

where $\varphi: \lambda S\rho \to \lambda' S\rho'$ and $\psi: \alpha S\beta \to \alpha' S\beta'$. The hypothesis $M(S) = E_S$ implies

$$(\lambda a\rho)\varphi = (\alpha a\beta)\psi, \qquad (\lambda' a\rho')\varphi^{-1} = (\alpha' a\beta')\psi^{-1} \qquad (a \in S). \tag{7}$$

Let $x \in \lambda S\rho$. Then $x\varphi \in \lambda' S\rho'$ so by (7),

$$x = (x\varphi)\varphi^{-1} = [\alpha'(x\varphi)\beta']\psi^{-1} = \alpha\{[\alpha'(x\varphi)\beta']\psi^{-1}\}\beta = \alpha x\beta$$

so that $x \in \alpha S\beta$. Hence $\lambda S\rho \subseteq \alpha S\beta$ and by symmetry $\lambda S\rho = \alpha S\beta$. The first part of (7) now shows that $\varphi = \psi$. Consequently, $\Psi(S)$ is an essential extension of $\Theta(S)$.

Recall that by 2.7, $\Psi(S)$ is a conjugate extension of $\Theta(S)$. The hypothesis that $M(S) = E_S$ implies $M(\Theta(S)) = E_{\Theta(S)}$ since $S \cong \Theta(S)$. We have shown above that $\Psi(S)$ is an essential extension of $\Theta(S)$. Hence by 2.9, $\theta = \theta(\Psi(S) : \Theta(S))$ is one-to-one.

Now let T be an essential conjugate extension of $\Theta(S)$ containing $\Psi(S)$ and let $a \in T$. Then $\theta^a \in \Psi(\Theta(K))$ according to 2.2. Now θ^a_S induces an element φ^a of $\Psi(S)$, explicitly

$$\mathbf{d}\varphi^a = \{k \in S | \theta^k = aa^{-1}\theta^k aa^{-1}\},$$

$$\varphi^a : k \to k' \quad \text{if} \quad \theta^a : \theta^k \to a^{-1}\theta^k a = \theta^{k'}.$$

It follows that $\theta^{\varphi^a} = \theta^a$ where φ^a, $a \in T$. Since θ is one-to-one, we must have $\varphi^a = a$. But this means that $a \in \Psi(S)$ so that $T = \Psi(S)$. Therefore, $\Psi(S)$ is a maximal essential conjugate extension of $\Theta(S)$.

The converse of 2.10 will be proved in XIII.5.7. As in the case of ideal extensions, it is convenient to introduce the following concept.

VI.2.11 Definition

Let an inverse semigroup S be a conjugate extension of an inverse semigroup K. Then the image $T(S : K)$ of S in $\Psi(K)$ under the homomorphism $\theta(S : K)$ is the *type* of the extension.

Now 2.10 has the following interesting consequences.

VI.2.12 Corollary

Let S be an inverse semigroup with idempotent metacenter. Then S has a maximal essential conjugate extension. Every essential conjugate extension of S is isomorphic to its type in $\Psi(S)$, and every maximal such is isomorphic to $\Psi(S)$. Every inverse subsemigroup of $\Psi(S)$ containing $\Theta(S)$ is the type of some essential conjugate extension of S.

Proof. The proof of this corollary is left as an exercise.

VI.2.13 Exercises

(i) Let S be an inverse semigroup. Show that for $\theta = \theta(S : E_S)$ and $a \in S$,

$$\mathbf{d}\theta^a = \left\{ e \in E_S | e \leq aa^{-1} \right\}$$

$$= \left\{ e \in E_S | e = eafa^{-1}e,\, f = fa^{-1}eaf \text{ for some } f \in E_S \right\}.$$

(ii) Find the center and the metacenter of (α) a Brandt semigroup $B(G, I)$ and (β) the bicyclic semigroup.

(iii) Let K be an inverse semigroup with idempotent metacenter. Let S be a conjugate extension of K such that $T(S : K) = \Theta(K)$. Show that K is a retract of S.

(iv) Let e and f be idempotents of an inverse semigroup S. Show that $e \mathcal{D} f$ if and only if $e\theta_S^x = f$ for some $x \in S$.

(v) Let S be an inverse semigroup. Show that $M(S) = Z(S)$ if and only if S is a Clifford semigroup. Also show that $M(S) = S$ if and only if S is commutative.

(vi) Let S be an inverse semigroup with idempotent metacenter. Show that every inverse subsemigroup of $\Psi(S)$ containing $\Theta(S)$ has idempotent metacenter.

(vii) For conjugate extensions, define strict and pure extensions in analogy to I.9.8 using $\Psi(K)$ and $\Theta(K)$. In this setting, prove analogues of I.9.9, I.9.10, I.9.12, and I.9.18.

(viii) Let K be an inverse semigroup with idempotent metacenter and C be a maximal essential conjugate extension of K. If S is a self-conjugate inverse subsemigroup of C which contains K, show that C is a maximal essential conjugate extension of S.

(ix) Let S and S' be conjugate extensions of an inverse semigroup K. Show that if S and S' are K-isomorphic, then they have the same type in $\Psi(K)$.
Conversely, if K has idempotent metacenter, S and S' are essential extensions of K, and $T(S : K) = T(S' : K)$, show that S and S' are K-isomorphic.

(x) Let S be a conjugate extension of an inverse semigroup K. Show that $\mu_S|_K = \mu_K$. Deduce that if S is an antigroup, then so is K, and that the converse holds if S is an essential extension of K. Also deduce that the conjugate hull of an antigroup is an antigroup.

(xi) Give an example showing that for an inverse semigroup S, $\Psi(S)$ need not be an essential conjugate extension of $\Theta(S)$.

(xii) State and prove the analogue of 2.12 for semilattices.

(xiii) Show that for any normal subsemigroup K of an inverse semigroup S,

$$\ker \theta(S : K) = \bigcup_{e \in E_S} [Z(eKe) \cap G_e].$$

VI.2.14 Problems

(i) (P. R. Jones) Let S be an inverse semigroup. For which self-conjugate inverse subsemigroups K of S is $\theta(S : K)$ idempotent separating?

(ii) What are necessary and sufficient conditions on an inverse semigroup S in order that $M(S) = E_S\zeta$? [$M(S) \subseteq E_S\zeta$ always holds.]

(iii) For an inverse semigroup S, what kind of "hull" is the closure of $\Theta(S)$ in $\Psi(S)$?

Conjugate extensions were introduced and investigated in Petrich [13]. The metacenter was characterized in Petrich [15]. The material of this section is taken from these two papers. The functions θ^s and the metacenter appear in Preston [5] and Melnik [1]. Consult also Gluskin [6], [7].

VI.3 RELATIONSHIP BETWEEN $\Omega(S)$ AND $\Psi(S)$

For any inverse semigroup S, we construct here homomorphisms of $\Omega(S)$ into $\Psi(S)$ and also into $\Psi(E_S)$. The image of $\Omega(S)$ under the first one has several interesting features. In the case that S has idempotent metacenter, the canonical isomorphism θ_S turns out to be a dense embedding of S into $\Psi(S)$. The second homomorphism induces the greatest idempotent separating congruence on $\Omega(S)$.

VI.3.1 Lemma

Let κ be an isomorphism of an inverse semigroup S onto a semigroup T. For each $\psi \in \Psi(S)$, let

$$\bar{\psi} : t \to t\kappa^{-1}\psi\kappa \qquad [t \in \mathbf{d}\bar{\psi} = (\mathbf{d}\psi)\kappa].$$

Then the mapping

$$\bar{\kappa} : \psi \to \bar{\psi} \qquad [\psi \in \Psi(S)]$$

is an isomorphism of $\Psi(S)$ onto $\Psi(T)$, said to be *induced* by κ.

Proof. The proof of this lemma is left as an exercise.

VI.3.2 Theorem

Let S be an inverse semigroup and K be an inverse subsemigroup of S such that $K\pi$ is self-conjugate in $\Omega(S)$, where $\pi : S \to \Omega(S)$. With $\nu_K = \overline{\pi|_K}$ (see 3.1) the diagram

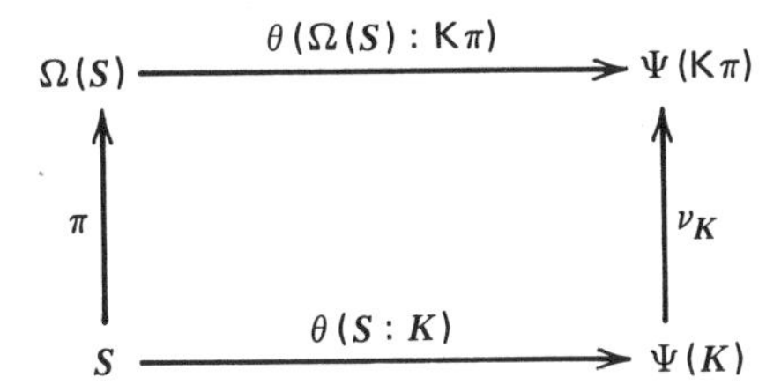

commutes. If K is also full in S, then both $\theta(S: K)$ and $\theta(\Omega(S): K\pi)$ are idempotent separating.

Proof. We write ν for ν_K. Note that for any $\psi \in \Psi(S)$,

$$\mathbf{d}(\psi\nu) = (\mathbf{d}\psi)\pi, \qquad \psi\nu : s \to s\pi^{-1}\psi\pi \text{ if } s \in \mathbf{d}(\psi\nu).$$

Let $\theta_1 = \theta(S: K)$ and $\theta_2 = \theta(\Omega(S): K\pi)$. Then for any $a \in S$, we get

$$a \xrightarrow{\pi} \pi_a \xrightarrow{\theta_2} \theta_2^{\pi_a}, \qquad a \xrightarrow{\theta_1} \theta_1^a\nu,$$

where $\mathbf{d}\theta_2^{\pi_a} = \pi_a(K\pi)\pi_a^{-1} = (aKa^{-1})\pi = \mathbf{d}(\theta_1^a\nu)$ and for any $k \in aKa^{-1}$,

$$\pi_k\theta_2^{\pi_a} = \pi_a^{-1}\pi_k\pi_a = \pi_{a^{-1}ka} = \pi_{k\theta_1^a} = k\theta_1^a\nu$$

which establishes the commutativity of the above diagram.

Now suppose that K is full in S. First let $e, f \in E_S$ be such that $\theta_1^e = \theta_1^f$. Then $eKe = fKf$ which implies that $e = ef$ since $e \in E_S \subseteq K$, and analogously $f = fe$ so that $e = f$. Consequently, $\theta_1 = \theta(S: K)$ is idempotent separating.

Next let $(\lambda, \rho), (\alpha, \beta) \in E_{\Omega(S)}$ be such that $\theta_2^{(\lambda,\rho)} = \theta_2^{(\alpha,\beta)}$. Hence $(\lambda, \rho)(K\pi)(\lambda, \rho) = (\alpha, \beta)(K\pi)(\alpha, \beta)$ so that $\lambda K\rho = \alpha K\beta$. For $e \in E_S$, there exists $k \in K$ such that $\lambda e\rho = \alpha k\beta$. It follows that $\lambda e = \alpha\lambda e$, and since this holds for an arbitrary idempotent of S, we obtain $\lambda = \alpha\lambda$. By symmetry, $\alpha = \lambda\alpha$ and hence $\lambda = \alpha$, and again by symmetry $\rho = \beta$. Hence $(\lambda, \rho) = (\alpha, \beta)$ and $\theta_2 = \theta(\Omega(S): K\pi)$ is idempotent separating.

The case considered in 3.3 below corresponds in the above theorem to $K = S$. In 3.9, we will be considering the case $K = E_S$.

VI.3.3 Theorem

Let S be an inverse semigroup. Then $\theta = \theta(\Omega(S):\Pi(S))$ is an idempotent separating homomorphism of $\Omega(S)$ onto a full inverse subsemigroup of $\Psi(\Pi(S))$ with kernel $M(\Omega(S))$.

Proof. First, θ is a homomorphism by 2.2 and is idempotent separating by 3.2.

In order to see that the image $\Omega(S)\theta$ is full in $\Psi(\Pi(S))$, we let $\psi \in E_{\Psi(\Pi(S))}$. Then $\psi = \iota_{\alpha\Pi(S)\beta}$ for some $(\alpha, \beta) \in E_{\Omega(\Pi(S))}$. Let $(\lambda, \rho) = (\alpha, \beta)\overline{\pi}^{-1}$ where the upper bar has the meaning given to it in I.8.10; hence $(\lambda, \rho) \in E_{\Omega(S)}$. Note that

$$\theta^{(\lambda,\rho)}: \pi_s \to (\lambda, \rho)\pi_s(\lambda, \rho) \text{ if } \pi_s \in (\lambda, \rho)\Pi(S)(\lambda, \rho).$$

A simple argument shows that then

$$\theta^{(\lambda,\rho)}: \pi_s \to \pi_{\lambda s \rho} \text{ if } \pi_s \in \{\pi_{\lambda t \rho} | t \in S\},$$

whence $\theta^{(\lambda,\rho)} = \iota_{\{\pi_{\lambda s\rho}|s\in S\}}$. On the other hand, $(\lambda, \rho)\overline{\pi} = (\alpha, \beta)$ so that $\overline{\lambda} = \alpha$, $\overline{\rho} = \beta$ (for the notation, see again I.8.10). For any $s \in S$, we get

$$\overline{\lambda}\pi_s = [\lambda(\pi_s\pi^{-1})]\pi = \pi_{\lambda s} = \alpha\pi_s$$

and analogously $\pi_{s\rho} = \pi_s\beta$. Consequently,

$$\alpha\Pi(S)\beta = \{\pi_{\lambda s\rho} | s \in S\} = \mathbf{d}\theta^{(\lambda,s)}$$

which finally gives that $\psi = \iota_{\alpha\Pi(S)\beta} = \theta^{(\lambda,\rho)}$, and $\Omega(S)\theta$ is full in $\Psi(\Pi(S))$.

It remains to show that $\ker\theta = M(\Omega(S))$. Let $\omega = (\lambda, \rho) \in \Omega(S)$. Then, for some $\varepsilon = (\alpha, \beta) \in E_{\Omega(S)}$,

$$\begin{aligned} \omega \in \ker\theta &\Leftrightarrow \theta^{\omega} = \theta^{\varepsilon} \\ &\Leftrightarrow \omega^{-1}\pi_s\omega = \varepsilon\pi_s\varepsilon \text{ if } \pi_s \in \omega\Pi(S)\omega^{-1} = \varepsilon\Pi(S)\varepsilon \\ &\Leftrightarrow \pi_{\lambda^{-1}s\rho} = \pi_{\alpha s\beta} \text{ if } \pi_s = \{\pi_{\lambda t\rho^{-1}} | t \in S\} = \{\pi_{\alpha t\beta} | t \in S\} \\ &\Leftrightarrow \lambda^{-1}s\rho = s \text{ if } s \in \lambda S\rho^{-1} = \alpha S\beta. \end{aligned} \tag{8}$$

We now verify that the last part of (8) is equivalent to

$$s = \lambda\lambda^{-1}s\rho\rho^{-1} \quad \text{implies} \quad s = \lambda^{-1}s\rho. \tag{9}$$

First, (8) trivially implies (9). Conversely, assume that (9) holds. Take $\alpha = \lambda\lambda^{-1}$, $\beta = \rho\rho^{-1}$. Obviously

$$\lambda S\rho^{-1} = \lambda\lambda^{-1}(\lambda S\rho^{-1})\rho\rho^{-1} \subseteq \alpha S\beta = \lambda(\lambda^{-1}S\rho)\rho^{-1} \subseteq \lambda S\rho^{-1}$$

and the equality holds throughout. If $s \in \lambda S \rho^{-1}$, then $s = \lambda\lambda^{-1}s\rho\rho^{-1}$ so (9) gives $s = \lambda^{-1}s\rho$. Thus (8) holds. From (8) and (9), we deduce that

$$\omega \in \ker\theta \Leftrightarrow s = \lambda^{-1}s\rho \text{ if } s = \lambda\lambda^{-1}s\rho\rho^{-1} \quad \text{for any} \quad s \in S. \tag{10}$$

Assume that the (last) condition in (10) holds. Since θ is idempotent separating, we have $\ker\theta \subseteq E_{\Omega(S)}\zeta$ by III.3.5 which implies that $\omega\omega^{-1} = \omega^{-1}\omega$ since $E_{\Omega(S)}\zeta$ is a Clifford semigroup. Hence for any $a \in S$, $b = \lambda\lambda^{-1}a\rho\rho^{-1}$ has the property $b = \lambda\lambda^{-1}b\rho\rho^{-1}$ so that $\lambda^{-1}b\rho = b$. It follows that

$$\lambda^{-1}(\lambda\lambda^{-1}a\rho\rho^{-1})\rho = \lambda\lambda^{-1}a\rho\rho^{-1}$$

which implies $\lambda^{-1}a\rho = \lambda\lambda^{-1}a\rho\rho^{-1}$. Consequently,

$$\lambda(\lambda^{-1}a)\rho = \lambda^2\lambda^{-1}a\rho\rho^{-1} = \lambda\lambda^{-1}\lambda a\rho^{-1}\rho = \lambda(a\rho^{-1})\rho.$$

This can be written in $\Omega(S)$ as $\omega\omega^{-1}\pi_a\omega = \omega\pi_a\omega^{-1}\omega$ for all $a \in S$. Now let $\tau \in \Omega(S)$ and $a \in S$. Then

$$\begin{aligned}
(\omega\omega^{-1}\tau\omega)\pi_a &= \omega\omega^{-1}\tau\omega(\omega^{-1}\omega)\pi_{aa^{-1}}\pi_a = \omega\omega^{-1}\tau\omega\pi_{aa^{-1}}\omega^{-1}\omega\pi_a \\
&= \left(\omega\omega^{-1}\pi_{\tau\omega aa^{-1}\omega^{-1}}\omega\right)\pi_a = \left(\omega\pi_{\tau\omega aa^{-1}\omega^{-1}}\omega^{-1}\omega\right)\pi_a \\
&= \omega\tau\left(\omega\pi_{aa^{-1}}\omega^{-1}\omega\right)\omega^{-1}\pi_a = \omega\tau\left(\omega\omega^{-1}\pi_{aa^{-1}}\omega\right)\omega^{-1}\pi_a \\
&= \omega\tau\omega\omega^{-1}\left(\pi_{aa^{-1}}\omega\omega^{-1}\right)\pi_a = \omega\tau\omega\omega^{-1}\pi_a = (\omega\tau\omega^{-1}\omega)\pi_a
\end{aligned}$$

which implies $(\omega\omega^{-1}\tau\omega)a = (\omega\tau\omega^{-1}\omega)a$, and analogously $a(\omega\omega^{-1}\tau\omega) = a(\omega\tau\omega^{-1}\omega)$. This means that $\omega\omega^{-1}\tau\omega = \omega\tau\omega^{-1}\omega$ so that $\omega \in M(\Omega(S))$.

Conversely, let $\omega \in M(\Omega(S))$. Then for any $a \in S$, we have $\lambda(\lambda^{-1}a)\rho = \lambda(a\rho^{-1})\rho$ and $\omega\omega^{-1} = \omega^{-1}\omega$ since $M(\Omega(S)) \subseteq E_{\Omega(S)}\zeta$. For any $a \in S$ such that $a = \lambda\lambda^{-1}a\rho\rho^{-1}$, we get

$$\lambda^{-1}a\rho = \lambda^{-1}\left[\lambda(\lambda^{-1}a)\rho\right] = \lambda^{-1}\left[\lambda(a\rho^{-1})\rho\right] = \lambda\lambda^{-1}a\rho\rho^{-1} = a$$

which in view of (10) yields $\omega \in \ker\theta$.

The above theorem has interesting consequences. To state some of them, it is convenient to introduce new notation.

VI.3.4 Notation

Let S be an inverse semigroup. For each $(\lambda, \rho) \in \Omega(S)$, let

$$\tau_{(\lambda,\rho)} : s \to \lambda^{-1}s\rho \qquad \left(s \in \mathbf{d}\tau_{(\lambda,\rho)} = \lambda S_\rho^{-1}\right).$$

Define

$$\tau : (\lambda, \rho) \to \tau_{(\lambda, \rho)} \qquad [(\lambda, \rho) \in \Omega(S)],$$
$$T(S) = \{\tau_{(\lambda, \rho)} | (\lambda, \rho) \in \Omega(S)\}.$$

VI.3.5 Corollary

Let S be an inverse semigroup.

(i) The following diagram commutes:

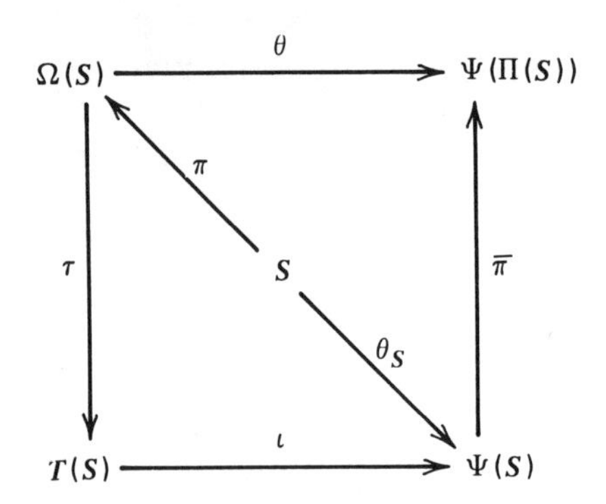

with the notation of 3.1 for $\bar{\pi}$, 3.3 for θ, 3.4 for τ; ι is the inclusion mapping.

(ii) τ is a homomorphism of $\Omega(S)$ onto $T(S)$.

(iii) $T(S)$ is a full inverse subsemigroup of $\Psi(S)$.

(iv) θ, $\theta_{\Omega(S)}$, and τ induce the same congruence on $\Omega(S)$.

(v) $E_{\Psi(S)} \cong E_{\Omega(S)}$.

Proof. (i) For the commutativity of the square diagram, we take $(\lambda, \rho) \in \Omega(S)$ and compute

$$\theta^{(\lambda, \rho)} : \pi_s \to (\lambda, \rho)^{-1} \pi_s (\lambda, \rho) \text{ if } \pi_s \in (\lambda, \rho)\Pi(S)(\lambda, \rho)^{-1},$$

that is,

$$\theta^{(\lambda, \rho)} : \pi_s \to \pi_{\lambda^{-1} s \rho} \text{ if } \pi_s \in \{\pi_{\lambda t \rho^{-1}} | t \in S\}.$$

On the other hand, $\bar{\pi} : \tau_{(\lambda, \rho)} \to \overline{\tau_{(\lambda, \rho)}}$, where

$$\overline{\tau_{(\lambda, \rho)}} : \pi_s \to \pi_s \pi^{-1} \tau_{(\lambda, \rho)} \pi = \pi_{\lambda^{-1} s \rho} \text{ if } \pi_s \in \{\pi_{\lambda t \rho^{-1}} | t \in S\}.$$

It follows that $\tau\bar{\pi} = \theta$.

Commutativity of the lower triangle in the diagram is obvious. For the upper triangle, we observe that

$$s \xrightarrow{\theta_S} \theta_S^s \xrightarrow{\overline{\pi}} \overline{\theta_S^s}, \qquad s \xrightarrow{\pi} \pi_s \xrightarrow{\theta} \theta^{\pi_s}.$$

An argument of the type already used above shows that $\theta^{\pi_s} = \overline{\theta_S^s}$.

(ii) Since by part (i), $\tau = \theta\overline{\pi}^{-1}$, τ is a homomorphism, and it trivially maps $\Omega(S)$ onto $T(S)$.

(iii) Again by part (i), we have $T(S)\overline{\pi} = \Omega(S)\theta$, where the latter is full in $\Psi(\Pi(S))$ by 3.3. But then $T(S)$ is full in $\Psi(S)$ since $\overline{\pi}$ is an isomorphism.

(iv) According to 2.4, $\theta_{\Omega(S)}$ is idempotent separating and has kernel equal to $M(\Omega(S))$. By 3.3, θ shares these properties, so $\theta_{\Omega(S)}$ and θ induce the same congruence on $\Omega(S)$. By part (i), θ and τ must induce the same congruence on $\Omega(S)$ since $\overline{\pi}$ is an isomorphism.

(v) This follows from 3.3 since θ is idempotent separating, $\Omega(S)\theta$ is full in $\Psi(\Pi(S))$, and $\overline{\pi}$ is an isomorphism.

Note that according to V.1.9(i), we also have $E_{\Omega(S)} \cong \Omega(E_S)$. In order to derive some consequences of 3.5 we first prove an auxiliary result.

VI.3.6 Lemma

Let I be an ideal of an inverse semigroup S. Then $M(S) \cap I = M(I)$.

Proof. We will use the form of $M(S)$ in 2.4. Let $a \in M(I)$ and $x \in S$. Note that since $M(I) \subseteq E_I\zeta$, we have $aa^{-1} = a^{-1}a$. Thus

$$aa^{-1}xa = aa^{-1}(aa^{-1}x)a = a(aa^{-1}x)a^{-1}a = (aa^{-1}a)xa^{-1}a = axa^{-1}a$$

which proves that $a \in M(S)$. Thus $M(I) \subseteq M(S) \cap I$, the opposite inclusion is trivial.

We now deduce several consequences of 3.5.

VI.3.7 Corollary

For any inverse semigroup S, $M(\Omega(S)) \cap \Pi(S) = M(\Pi(S))$, and $M(S) = E_S$ if and only if $M(\Omega(S)) = E_{\Omega(S)}$.

Proof. The first assertion follows from 3.6.

Assume that $M(S) = E_S$. Then $M(\Pi(S)) = E_{\Pi(S)}$ which by 3.5(iv) implies that $\tau|_{\Pi(S)}$ is one-to-one. By I.9.18, $\Omega(S)$ is a dense ideal extension of $\Pi(S)$, and hence τ must be one-to-one. But then $M(\Omega(S)) = E_{\Omega(S)}$. The converse of the second assertion of the corollary follows from its first assertion.

Note that 3.5(ii) implies that $T(S)$ is an inverse subsemigroup of $\Psi(S)$ and that $\Theta(S)$ is an ideal of $T(S)$. The case of S with idempotent metacenter is treated in the next result.

VI.3.8 Theorem

Let S be an inverse semigroup with idempotent metacenter. Then θ_S is a dense embedding of S into $\Psi(S)$.

Proof. According to 3.7, $M(\Omega(S)) = E_{\Omega(S)}$, so that by 3.5(iv), τ is an isomorphism of $\Omega(S)$ onto $T(S)$. One verifies easily that τ maps $\Pi(S)$ onto $\Theta(S)$. Since by I.9.18, $\Pi(S)$ is a densely embedded ideal of $\Omega(S)$, it follows that $\Theta(S)$ is a densely embedded ideal of $T(S)$. To complete the proof of the theorem, it remains to show that the idealizer $I = i_{\Psi(S)}(\Theta(S))$ of $\Theta(S)$ in $\Psi(S)$ is contained in $T(S)$ (the opposite inclusion is now obvious).

Hence let $\psi \in I$ and write $\theta = \theta_S$. Then for every $s \in S$, $\psi\theta^s, \theta^s\psi \in \Theta(S)$, and since by hypothesis θ is one-to-one, we are able to define functions λ and ρ on S by the formulae

$$\psi\theta^s = \theta^{\lambda s}, \qquad \theta^s\psi = \theta^{s\rho}.$$

It follows easily that $(\lambda, \rho) \in \Omega(S)$. Using V.1.7, we obtain for any $x \in S$,

$$\begin{aligned}\theta^{\lambda^{-1}x} &= \theta^{(x^{-1}\rho)^{-1}} = \left(\theta^{x^{-1}\rho}\right)^{-1} = \left(\theta^{x^{-1}}\psi\right)^{-1} \\ &= \psi^{-1}\left(\theta^{x^{-1}}\right)^{-1} = \psi^{-1}\theta^x\end{aligned}$$

and thus

$$\theta^{\lambda\lambda^{-1}x} = \psi\theta^{\lambda^{-1}x} = \psi\psi^{-1}\theta^x = \iota_{\lambda' S\rho'}\theta^x$$

where $\mathbf{d}\psi = \lambda' S\rho'$ with $(\lambda', \rho') \in E_{\Omega(S)}$. Hence

$$\mathbf{d}\theta^{\lambda\lambda^{-1}x} = (\lambda' S\rho') \cap xSx^{-1}, \tag{11}$$

and on the other hand,

$$\mathbf{d}\theta^{\lambda' x} = (\lambda' x)S(\lambda' x)^{-1}. \tag{12}$$

If y is an element in the set in (11), then $y = \lambda' s\rho' = xtx^{-1}$ for some $s, t \in S$, and thus by V.1.7,

$$y = (\lambda' x)t(x^{-1}\rho') = (\lambda' x)t(\lambda' x)^{-1}$$

and y is in the set in (12). Conversely, if y is in the set in (12), then for some $s \in S$, using V.1.7 and V.1.6,

$$\begin{aligned}y &= (\lambda' x)s(\lambda' x)^{-1} = \lambda'(xsx^{-1})\rho' \in \lambda' S\rho', \\ y &= (\lambda' x)s(x^{-1}\rho') = (\lambda' xx^{-1})xsx^{-1}(xx^{-1}\rho') \\ &= (xx^{-1}\rho')xsx^{-1}(\lambda' xx^{-1}) \in xSx^{-1}\end{aligned}$$

which proves that y is also in the set in (11). Hence $\mathbf{d}\theta^{\lambda\lambda^{-1}x} = \mathbf{d}\theta^{\lambda' x}$.

Now let $y \in \mathbf{d}\theta^{\lambda' x}$. Then $y \in \lambda' S \rho'$ whence $y = \lambda' y \rho'$. By the above and V.1.7, we get

$$y\theta^{\lambda\lambda^{-1}x} = y\left(\iota_{\lambda' S\rho'}\theta^{x}\right) = y\theta^{x} = x^{-1}yx$$

$$= x^{-1}(\lambda' y \rho')x = (x^{-1}\rho')y(\lambda' x)$$

$$= (\lambda' x)^{-1}y(\lambda' x) = y\theta^{\lambda' x}.$$

We have proved that $\theta^{\lambda\lambda^{-1}x} = \theta^{\lambda' x}$, and since θ is one-to-one, it follows that $\lambda\lambda^{-1}x = \lambda' x$. Since $x \in S$ is arbitrary, we must have $\lambda\lambda^{-1} = \lambda'$. An analogous argument can be used to show that also $\rho\rho^{-1} = \rho'$.

We now have

$$\mathbf{d}\tau_{(\lambda,\rho)} = \lambda S \rho^{-1} = \lambda\lambda^{-1}S\rho\rho^{-1} = \lambda' S \rho' = \mathbf{d}\psi.$$

Further, for $x \in \lambda S \rho^{-1}$, we have by the above and 2.6,

$$\theta^{\lambda^{-1}x\rho} = \psi^{-1}\theta^{x}\psi = \theta^{x\psi}$$

and thus

$$x\tau_{(\lambda,\rho)} = \lambda^{-1}x\rho = x\psi.$$

This shows that $\psi \in T(S)$ and hence $T(S)$ is the idealizer of $\Theta(S)$ in $\Psi(S)$.

We now construct a homomorphism of $\Omega(S)$ into $\Psi(E_S)$.

VI.3.9 Theorem

Let S be an inverse semigroup. For each $(\lambda, \rho) \in \Omega(S)$, let

$$\xi_{(\lambda,\rho)} : e \to \lambda^{-1}e\rho \qquad \left(e \in \mathbf{d}\xi_{(\lambda,\rho)} = \lambda E_S \rho^{-1}\right).$$

Then the mapping

$$\xi : (\lambda, \rho) \to \xi_{(\lambda,\rho)} \qquad [(\lambda, \rho) \in \Omega(S)]$$

is a homomorphism of $\Omega(S)$ into $\Psi(E_S)$ which induces the greatest idempotent separating congruence on $\Omega(S)$.

Proof. The first assertion of the theorem will follow from 3.2 if we show that $\xi = \theta\nu_{E_S}^{-1}$ where $\theta = \theta(\Omega(S) : E_{\Pi(S)})$. Indeed, let $(\lambda, \rho) \in \Omega(S)$. Then

$$\mathbf{d}\theta^{(\lambda,\rho)} = (\lambda, \rho)E_{\Pi(S)}(\lambda, \rho)^{-1} = \left\{(\lambda, \rho)\pi_e(\lambda, \rho)^{-1} \mid e \in E_S\right\}$$

$$= \left\{\pi_{\lambda e \rho^{-1}} \mid e \in E_S\right\},$$

$$\pi_e\theta^{(\lambda,\rho)} = (\lambda, \rho)^{-1}\pi_e(\lambda, \rho) = \pi_{\lambda^{-1}e\rho} \text{ if } \pi_e \in \mathbf{d}\theta^{(\lambda,\rho)},$$

and letting $\psi = \theta^{(\lambda,\rho)}\nu_{E_S}^{-1}$, we obtain from above,

$$\psi : e \to \lambda^{-1}e\rho \text{ if } e \in \mathbf{d}\psi = \lambda E_S \rho^{-1}$$

which proves that $\xi = \theta\nu_{E_S}^{-1}$. Now 3.2 implies that ξ is an idempotent separating homomorphism of $\Omega(S)$ into $\Psi(E_S)$. Let $\omega = (\lambda, \rho) \in E_{\Omega(S)}\zeta$. Then $\omega\varepsilon = \varepsilon\omega$ for all $\varepsilon \in E_{\Omega(S)}$ and in particular, $\omega\pi_e = \pi_e\omega$ for all $e \in E_S$. The latter is equivalent to $\lambda e = e\rho$ for all $e \in E_S$. It follows that for any $e \in E_S$,

$$\lambda^{-1}e = (e\rho)^{-1} = (\lambda e)^{-1} = e\rho^{-1}.$$

Using what we just proved, we get $\mathbf{d}\xi_\omega = \lambda\lambda^{-1}E_S$ and $\xi_\omega : e \to \lambda\lambda^{-1}e$ if $e \in \lambda\lambda^{-1}E_S$, which evidently implies that ξ_ω is the identity mapping on its domain. Hence $\xi_\omega \in E_{\Psi(S)}$ which yields that $E_{\Omega(S)}\zeta \subseteq \ker\xi$. Since ξ is idempotent separating and $E_{\Omega(S)}\zeta$ is the kernel of the greatest idempotent separating congruence on $\Omega(S)$, we must have $E_{\Omega(S)}\zeta = \ker\xi$.

The connection between the two homomorphisms τ and ξ can be stated as follows.

VI.3.10 Remark

It follows easily from 3.4 and 3.9 that

$$\mathbf{d}\xi_{(\lambda,\rho)} = \mathbf{d}\tau_{(\lambda,\rho)} \cap E_S,$$

$$\xi_{(\lambda,\rho)} = \tau_{(\lambda,\rho)}|_{\mathbf{d}\xi_{(\lambda,\rho)}},$$

$$\xi\nu_{E_S} = \theta\big(\Omega(S) : E_{\Pi(S)}\big).$$

VI.3.11 Exercises

(i) Let S be an inverse semigroup and let $E = E_S$. Show that for K equal to E, $M(S)$, $Z(E\zeta)$, $E\zeta$, and $E\omega$, we have $(\mathbf{d}\psi \cap K)\psi \subseteq K$ for all $\psi \in \Psi(S)$. Also show that

$$E \subseteq M(S) \subseteq Z(E\zeta) \subseteq E\zeta$$

and that $Z(E\zeta)$ is the centralizer of $E\zeta$ in S.

(ii) For a set X, let $S = \mathcal{I}(X)$ and $E = E_S$. Find $E\zeta$, $Z(E\zeta)$, $M(S)$, and $Z(S)$.

(iii) Show that the following statements concerning an inverse semigroup S are equivalent.

(α) ξ is one-to-one.

(β) $\xi|_{\Pi(S)}$ is one-to-one.

(γ) S is an antigroup.
(δ) $\Omega(S)$ is an antigroup.
(ε) S is an essential conjugate extension of E_S.

(iv) Show that for any semilattice Y, $T(Y) = E_{\Psi(Y)}$.
(v) Show that for any inverse semigroup S,

$$\Theta(E_S) = E_{\Phi(S)} \subseteq \Theta(S) \cap E_{\Phi(S)}\zeta = \Theta(E_S\zeta).$$

(vi) Let S be an inverse semigroup. If K is a self-conjugate inverse subsemigroup of S, show that $i_S(K)$ is also.
(vii) Show that if S is an inverse semigroup with idempotent metacenter, then $T(S)$ is a self-conjugate subsemigroup of $\Omega(S)$.

VI.3.12 Problems

(i) Is $T(S)$ self-conjugate in $\Psi(S)$?

Except for 3.8 and parts of 3.3. and 3.5, most of the material is taken from Petrich [13].

VI.4 THE NORMAL HULL

We introduce here the concept of a normal hull and establish a number of its properties. In addition, we construct a homomorphism of $\Psi(S)$ into $\mathcal{C}(S/\sigma)$ which provides another interesting self-conjugate inverse subsemigroup of $\Psi(S)$.

VI.4.1 Definition

For an inverse semigroup S,

$$\Phi(S) = \{\psi \in \Psi(S) | \mathbf{d}\psi = eSe \text{ for some } e \in E_S\}$$

together with the operation in $\Psi(S)$ is the *normal hull* of S.

Note that $eSe = \lambda_e S \rho_e$, and that for $\psi \in \Psi(S)$ with domain eSe we get

$$\mathbf{r}\psi = (e\psi)S(e\psi) = \lambda_{e\psi} S \rho_{e\psi}.$$

Hence the elements ψ of $\Phi(S)$ are precisely those elements of $\Psi(S)$ for which $\mathbf{d}\psi = \lambda S \rho$, $\mathbf{r}\psi = \lambda' S \rho'$ with $(\lambda, \rho), (\lambda', \rho') \in \Pi(S)$. For the case when S is a semilattice, the normal hull was introduced in IV.2.1. For the Munn representation of an inverse semigroup S we have $\delta = \theta(S : E_S)$ so that $S^\delta = T(S : E_S)$.

We will now establish several properties of the normal hull. See I.9.2 and 3.9 for the notation.

VI.4.2 Theorem

The following statements are true for any inverse semigroup S.

(i) $\Phi(S)$ is an ideal of $\Psi(S)$.
(ii) The mapping $\tau = \tau(\Psi(S) : \Phi(S))$ is an idempotent separating homomorphism of $\Psi(S)$ onto a full inverse subsemigroup of $\Omega(\Phi(S))$.
(iii) $\Phi(S)$ is the full closure of $\Theta(S)$ in $\Psi(S)$.
(iv) $\Phi(S) \cap T(S) = \Theta(S)$.
(v) $\Pi(S) = \Phi(E_S)\xi^{-1}$ (see 3.9).

Proof. (i) Let $\psi \in \Psi(S)$ with $\psi : \lambda S\rho \to \alpha S\beta$, where $(\lambda, \rho), (\alpha, \beta) \in E_{\Omega(S)}$; and $\varphi \in \Phi(S)$ with $\varphi : eSe \to fSf$, where $e, f \in E_S$. Then

$$\mathbf{d}(\psi\varphi) = (\alpha S\beta \cap eSe)\psi^{-1}, \qquad \mathbf{r}(\psi\varphi) = (\alpha S\beta \cap eSe)\varphi. \tag{13}$$

Let $B = \alpha S\beta$. Then by 1.9, E_B is a retract ideal of E_S and B is its full closure in S. Thus $B \cap eSe$ is an inverse subsemigroup of S, and

$$E_{B \cap eSe} = E_B \cap [e] = [\bar{e}]$$

for some $\bar{e} \in E_S$. If $a \in B \cap eSe$, then $aa^{-1}, a^{-1}a \leq \bar{e}$ so that

$$a = (aa^{-1})a(a^{-1}a) = \bar{e}(aa^{-1})a(a^{-1}a)\bar{e} = \bar{e}a\bar{e} \in \bar{e}S\bar{e}.$$

Conversely, let $a \in \bar{e}S\bar{e}$. Then $a \in eSe$ since $\bar{e} \leq e$, and $aa^{-1}, a^{-1}a \leq \bar{e}$, so $aa^{-1}, a^{-1}a \in B$. But then $a \in B$ since B is the full closure of E_B in S. Consequently, $B \cap eSe = \bar{e}S\bar{e}$ which in view of (13) yields that $\psi\varphi \in \Phi(S)$.

A similar proof shows that also $\varphi\psi \in \Phi(S)$ and therefore $\Phi(S)$ is an ideal of $\Psi(S)$.

(ii) Let $\psi, \psi' \in E_{\Psi(S)}$ be such that $\psi\tau = \psi'\tau$, and let $e \in E_{\mathbf{d}\psi}$. By hypothesis, $\theta^e\psi = \theta^e\psi'$, so that

$$e = e\theta^e = e\theta^e\psi = e\theta^e\psi' = e\psi'$$

which shows that $e \in E_{\mathbf{d}\psi'}$. Hence $E_{\mathbf{d}\psi} \subseteq E_{\mathbf{d}\psi'}$ and by symmetry, we get $E_{\mathbf{d}\psi} = E_{\mathbf{d}\psi'}$. By 1.9, $\mathbf{d}\psi$ is the full closure of $E_{\mathbf{d}\psi}$, which then yields $\mathbf{d}\psi = \mathbf{d}\psi'$. Since both ψ and ψ' are idempotents, we obtain $\psi = \psi'$, so that τ is idempotent separating.

We show next that $\Psi(S)\tau$ is full in $\Omega(\Phi(S))$. To this end, we first illustrate the arguments below by a diagram.

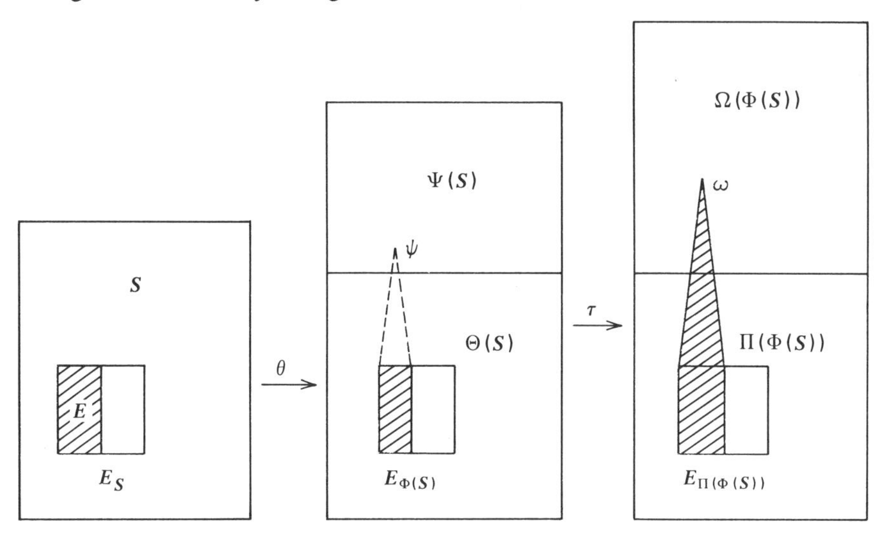

Let $\omega \in E_{\Omega(\Phi(S))}$ and let

$$E = \{e \in E_S | \pi_{\theta^e} \leq \omega\}.$$

It follows at once that $E\theta\tau$ is a retract ideal of $E_{\Pi(\Phi(S))}$, and since $\theta\tau$ restricted to E_S is an isomorphism of E_S onto $E_{\Pi(\Phi(S))}$, we obtain that E is a retract ideal of E_S. Let B be the full closure of E in S. In the light of 1.9, we have that $\psi = \iota_B$ is an idempotent of $\Psi(S)$.

Let $\varphi \in \Phi(S)$, say $\mathbf{d}\varphi = eSe$, $\mathbf{r}\varphi = fSf$, $e, f \in E_S$. Then $\theta^e \mathcal{R} \varphi \mathcal{L} \theta^f$ in $\Phi(S)$. Since $E \in \mathfrak{R}_{E_S}$, there exist $\bar{e}, \bar{f} \in E_S$ such that

$$[\bar{e}] = [e] \cap E, \qquad [\bar{f}] = [f] \cap E.$$

On the one hand,

$$\psi\varphi = \psi(\theta^e\varphi) = (\psi\theta^e)\varphi = \theta^{\bar{e}}\varphi$$

and similarly $\varphi\psi = \varphi\theta^{\bar{f}}$, and on the other hand,

$$\omega\pi_\varphi = \omega(\pi_{\theta^e}\pi_\varphi) = (\omega\pi_{\theta^e})\pi_\varphi = \pi_{\theta^{\bar{e}}}\pi_\varphi = \pi_{\theta^{\bar{e}}\varphi}$$

and similarly $\pi_\varphi\omega = \pi_{\varphi\theta}\bar{f}$. It now follows that $\psi\tau = \omega$, as desired.

(iii) Let $\varphi \in E_{\Phi(S)}$ with $\mathbf{d}\varphi = \mathbf{r}\varphi = eSe$ where $e \in E_S$. Then $\varphi = \iota_{eSe} = \theta^e$ which shows that $\Theta(S)$ is full in $\Phi(S)$. Conversely, let $\psi \in \Psi(S)$ be such that $\psi\psi^{-1} = \theta^e$ and $\psi^{-1}\psi = \theta^f$ for some $e, f \in E_S$. Then

$$\mathbf{d}\psi = \mathbf{d}(\psi\psi^{-1}) = \mathbf{d}\theta^e = eSe$$

and thus $\psi \in \Phi(S)$. Hence $\Phi(S)$ is the full closure of $\Theta(S)$ in $\Psi(S)$.

(iv) Let $\omega = (\lambda, \rho) \in \Omega(S)$ be such that $\tau_\omega \in \Phi(S)$. Then $\mathbf{d}\tau_\omega = \lambda S\rho^{-1} = eSe$ and $\mathbf{r}\tau_\omega = \lambda^{-1}S\rho = fSf$ for some $e, f \in E_S$. Letting $a = e\rho$, we obtain

$$aa^{-1} = (e\rho)(e\rho)^{-1} = (e\rho)(\lambda^{-1}e) = (e\rho\rho^{-1})e = e\rho\rho^{-1} = e,$$

$$a^{-1}a = (e\rho)^{-1}(e\rho) = (\lambda^{-1}e)(e\rho) = \lambda^{-1}e\rho = f$$

since $e \in \lambda S\rho^{-1}$ implies $e = e\rho\rho^{-1}$; and for $x \in eSe$,

$$x\tau_\omega = \lambda^{-1}x\rho = \lambda^{-1}(exe)\rho = (\lambda^{-1}e)x(e\rho) = (e\rho)^{-1}x(e\rho) = a^{-1}ea.$$

Consequently, $\tau_\omega = \theta^a \in \Theta(S)$. This proves that $\Phi(S) \cap T(S) \subseteq \Theta(S)$, the opposite inclusion is obvious.

(v) For any $a \in S$, we get $\mathbf{d}\xi_{\pi_a} = aE_Sa^{-1}$ and $\mathbf{r}\xi_{\pi_a} = a^{-1}E_Sa$ which shows that $\xi_{\pi_a} \in \Phi(E_S)$. Conversely, let $\omega = (\lambda, \rho) \in \Omega(S)$ be such that $\xi_\omega \in \Phi(E_S)$. Then $\mathbf{d}\xi_\omega = \lambda E_S\rho^{-1} = eE_S$ and $\mathbf{r}\xi_\omega = \lambda^{-1}E_S\rho = fE_S$ for some $e, f \in E_S$. In particular, $\lambda^{-1}e\rho = f$ whence

$$e\rho = e\rho\rho^{-1}\rho = \lambda(\lambda^{-1}e\rho) = \lambda f = a.$$

For any $x \in S$, we obtain

$$\begin{aligned} x\rho &= x\left[(x^{-1}x)\rho\rho^{-1}\right]\rho = x\left[\lambda\lambda^{-1}(x^{-1}x)\right]\rho \\ &= (x\rho)\left[\lambda^{-1}(x^{-1}x)\rho\right]f = x\left[\lambda\lambda^{-1}(x^{-1}x)\right]\rho f \\ &= x\left[(x^{-1}x)\rho\rho^{-1}\right]\rho f = (x\rho)f = x(\lambda f) = xa \end{aligned}$$

and analogously $\lambda x = ax$, which gives $\omega = \pi_a \in \Pi(S)$.

In the case of idempotent metacenter, we have the following statement.

VI.4.3 Proposition

Let S be an inverse semigroup with idempotent metacenter. Then $\tau = \tau(\Psi(S) : \Phi(S))$ is an isomorphism of $\Psi(S)$ onto the greatest conjugate extension of $\pi(\Theta(S))$ in $\Omega(\Phi(S))$.

Proof. Since $\Phi(S)$ is weakly reductive, $\tau|_{\Phi(S)} = \pi$ is one-to-one and thus $\tau|_{\Theta(S)}$ is one-to-one. Since S has idempotent metacenter, 2.10 gives that $\Psi(S)$ is an essential conjugate extension of $\Theta(S)$. But then τ itself must be one-to-one. Since τ is an isomorphism, we have that $\tau(\Psi(S))$ is a maximal essential extension of $\tau(\Theta(S)) = \pi(\Theta(S))$.

For $\omega_1, \omega_2 \in \Omega(\Phi(S))$ such that

$$\omega_i^{-1}\pi(\Theta(S))\omega_i \subseteq \pi(\Theta(S)), \qquad i = 1, 2,$$

we have

$$(\omega_1\omega_2)^{-1}\pi(\Theta(S))(\omega_1\omega_2) \subseteq \pi(\Theta(S)).$$

It follows that there exists a greatest conjugate extension $\overline{S}$ of $\pi(\Theta(S))$ in

$\Omega(\Phi(S))$. We evidently have

$$\pi(\Theta(S)) \subseteq \Pi(\Phi(S)) \subseteq \tau(\Psi(S)) \subseteq \bar{S}$$

and must show that the last inclusion is actually an equality. Let ρ be a congruence on $\bar{S}$ such that $\rho|_{\pi(\Theta(S))} = \varepsilon$. Since $\tau(\Psi(S))$ is an essential conjugate extension of $\pi(\Theta(S))$, we obtain $\rho|_{\tau(\Psi(S))} = \varepsilon$. But then also $\rho|_{\Pi(\Phi(S))} = \varepsilon$. Using I.9.18, we conclude that $\bar{S}$ is a dense ideal extension of $\Pi(\Phi(S))$, so that we must have $\rho = \varepsilon$. Consequently, $\bar{S}$ is an essential conjugate extension of $\pi(\Theta(S))$ which contains $\tau(\Psi(S))$. We have seen above that $\tau(\Psi(S))$ is a maximal essential conjugate extension of $\pi(\Theta(S))$, which evidently implies that $\bar{S} = \tau(\Psi(S))$.

VI.4.4 Corollary

For any semilattice Y, we have $\Psi(Y) \cong \Omega(\Phi(Y))$.

Proof. Let $\omega \in \Omega(\Phi(Y))$ and $\alpha \in Y$. Then $\omega^{-1}\pi_{\theta^\alpha}\omega \in \Pi(\Phi(Y))$ since $\Pi(\Phi(Y))$ is an ideal of $\Omega(\Phi(Y))$. But $\omega^{-1}\pi_{\theta^\alpha}\omega$ is an idempotent, which implies that $\omega^{-1}\pi_{\theta^\alpha}\omega \in \pi(\Theta(Y))$ since $\pi(\Theta(Y))$ is full in $\Pi(\Phi(Y))$. Consequently, $\pi(\Theta(Y))$ is self-conjugate in $\Omega(\Theta(Y))$. The assertion now follows from 3.2.

We will need the following simple result.

VI.4.5 Lemma

Let I be an ideal of an inverse semigroup S. Then $E_S\zeta \cap I = E_I\zeta$.

Proof. Let $a \in E_I\zeta$ and $e \in E_S$. Then

$$\begin{aligned} ae &= a[(a^{-1}a)e] = [(a^{-1}a)e]a = [e(a^{-1}a)]a \\ &= e[(a^{-1}a)a] = e[a(a^{-1}a)] = ea \end{aligned}$$

which proves that $E_I\zeta \subseteq E_S\zeta \cap I$. The opposite inclusion is trivial.

We now construct a homomorphism of $\Psi(S)$ into $\mathfrak{A}(S/\sigma)$, the automorphism group of the greatest group homomorphic image of S.

VI.4.6 Proposition

Let S be an inverse semigroup. For each $\psi \in \Psi(S)$, let

$$\tilde{\psi} : a\sigma \to a\psi\sigma \qquad (a \in \mathbf{d}\tilde{\psi} = \lambda S\rho),$$

where $\mathbf{d}\psi = \lambda S\rho$, $(\lambda, \rho) \in E_{\Omega(S)}$. Then the mapping

$$\chi : \psi \to \tilde{\psi} \qquad [\psi \in \Psi(S)]$$

is a homomorphism of $\Psi(S)$ into $\mathfrak{A}(S/\sigma)$. Moreover,

$$\chi : \theta_S^a \to \varepsilon_{a\sigma} \qquad (a \in S).$$

Proof. First let $(\lambda, \rho) \in E_{\Omega(S)}$. Then $\lambda S\rho$ is an inverse semigroup and it thus contains idempotents. Now let $e \in E_{\lambda S\rho}$ and $a \in S$. Then $a\sigma eae$ and $\lambda(eae)\rho = eae$ which shows that every σ-class intersects $\lambda S\rho$.

Now let $\psi \in \Psi(S)$ with $\mathbf{d}\psi = \lambda S\rho$, $(\lambda, \rho) \in E_{\Omega(S)}$. It follows from above that ψ is defined on all of S/σ. Assume that $x\sigma = y\sigma$ where $x, y \in \lambda S\rho$. Then $xe = ye$ for some $e \in E_S$ by III.5.2(i), so that $(xe)\rho = y(e\rho)$ which in view of V.1.6 yields $x(\lambda e\rho) = y(\lambda e\rho)$. But then

$$(x\psi)(\lambda e\rho)\psi = (y\psi)(\lambda e\rho)\psi$$

which proves that $x\psi\sigma = y\psi\sigma$ since $(\lambda e\rho)\psi \in E_S$. This shows that $\tilde{\psi}$ is single valued.

Continuing with the same notation, we obtain

$$(x\sigma)\tilde{\psi}(y\sigma)\tilde{\psi} = (x\psi\sigma)(y\psi\sigma) = (xy)\psi\sigma = (xy)\sigma\tilde{\psi}$$

and $\tilde{\psi}$ is a homomorphism of S/σ. Assume next that $(x\sigma)\tilde{\psi} = (y\sigma)\tilde{\psi}$. Then $x\psi\sigma = y\psi\sigma$ and thus $(x\psi)e = (y\psi)e$ for some $e \in E_S$. We have $\mathbf{r}\psi = \alpha S\beta$ for some $(\alpha, \beta) \in E_{\Omega(S)}$. It follows that $(x\psi)(\alpha e\beta) = (y\psi)(\alpha e\beta)$, and letting $f = (\alpha e\beta)\psi^{-1}$, we get

$$(xf)\psi = (x\psi)(f\psi) = (x\psi)(\alpha e\beta) = (y\psi)(\alpha e\beta) = (y\psi)(f\psi) = (yf)\psi.$$

But then $xf = yf$ which shows that $x\sigma = y\sigma$. Consequently, $\tilde{\psi}$ is one-to-one. For any $a \in S$ there is $b \in \alpha S\beta$ such that $a\sigma = b\sigma$ (see above). Letting $x = b\psi^{-1}$, we obtain

$$(x\sigma)\tilde{\psi} = x\psi\sigma = b\sigma = a\sigma$$

which proves that $\tilde{\psi}$ maps S/σ onto itself. Therefore, $\tilde{\psi} \in \mathcal{Q}(S/\sigma)$.

Now let $\psi, \psi' \in \Psi(S)$ with

$$\psi\colon \lambda S\rho \to \alpha S\beta, \qquad \psi'\colon \lambda' S\rho' \to \alpha' S\beta'.$$

For any $x \in \mathbf{d}(\psi\psi') = (\alpha S\beta \cap \lambda' S\rho')\psi^{-1}$, we get

$$(x\sigma)\tilde{\psi}\widetilde{\psi'} = (x\psi)\sigma\widetilde{\psi'} = (x\psi\psi')\sigma = (x\sigma)\widetilde{\psi\psi'}$$

which shows that $\chi\colon \psi \to \tilde{\psi}$ is a homomorphism.

For any $a \in S$, we have for $x \in aSa^{-1}$,

$$(x\sigma)\widetilde{\theta^a} = (x\theta^a)\sigma = (a^{-1}xa)\sigma = (a\sigma)^{-1}(x\sigma)(a\sigma) = (a\sigma)\varepsilon_{a\sigma}$$

which establishes the last assertion of the proposition.

This proposition makes it possible to introduce the following symbolism.

VI.4.7 Notation

For any inverse semigroup S, let

$$I(S) = \{\psi \in \Psi(S) | \tilde{\psi} \in \mathcal{I}\mathcal{A}(S/\sigma)\}.$$

The next diagram gives a summary of the semigroups and homomorphisms considered in this and the preceding sections.

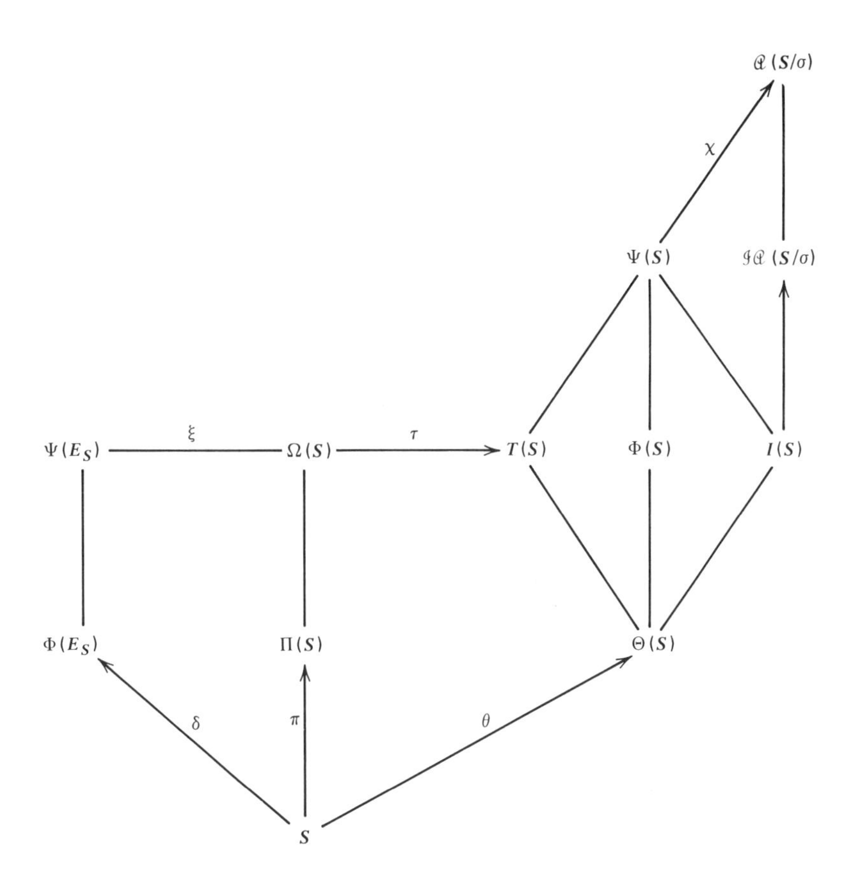

VI.4.9 Exercises

(i) Show that every isomorphism of an inverse semigroup S onto a semigroup S' induces an isomorphism of $\Psi(S)$ onto $\Psi(S')$ which maps $\Theta(S)$, $\Phi(S)$, $T(S)$, and $I(S)$ onto the corresponding subsemigroup of $\Psi(S')$.

(ii) For any inverse semigroup S, prove that $I(S)$ is a normal subsemigroup of $\Psi(S)$.

(iii) Find a suitable isomorphic copy of the normal hull of a Brandt semigroup.

(iv) Show that the following conditions on an inverse semigroup S are equivalent.

 (α) S has an identity.
 (β) $\Pi(S) = \Omega(S)$.
 (γ) $\Phi(S) = \Psi(S)$.

(v) Construct an example showing that τ in 4.3 need not be one-to-one.

The normal hull of an inverse semigroup was introduced in Petrich [15]. Results 4.2(i), (ii), 4.3, and 4.4 are due to Pastijn [6]; statements 4.2(iii)–(v) can be found in Petrich [13].

VI.5 EXAMPLES OF CONJUGATE HULLS

We illustrate the concepts studied in this chapter by characterizing the conjugate hull and some of its subsemigroups on the example of an inverse semigroup which is a subdirect product of a semilattice and a group. Two further examples illustrate the conjugate hull.

VI.5.1 Theorem

Let $S = [Y, G; \eta]$ be as in II.2.13. Then

$$\Psi(S) \cong \{(\sigma, \tau) \in \Psi(Y) \times \mathcal{Q}(G) | \alpha \in \mathbf{d}\sigma, g \in \alpha\eta \Rightarrow g\tau \in \alpha\sigma\eta\}$$

under coordinatewise multiplication.

Proof. Let $\psi \in \Psi(S)$ with ψ: $\lambda S\rho \to \lambda' S\rho'$, where (λ, ρ), $(\lambda', \rho') \in E_{\Omega(S)}$. According to V.3.9, we have the correspondence $(\lambda, \rho) \to (I, c)$, which by idempotency of (λ, ρ) yields $c = 1$. It then follows easily that

$$\lambda S\rho = (I \times G) \cap S$$

and analogously

$$\lambda' S\rho' = (I' \times G) \cap S$$

with $I, I' \in \mathfrak{R}_Y$.

We now define functions σ and τ on $\mathbf{d}\psi$ by the formula

$$(\alpha, g)\psi = ((\alpha, g)\sigma, (\alpha, g)\tau). \tag{14}$$

Then

$$
\begin{aligned}
(\alpha,1)\psi &= \left[(\alpha, g)(\alpha, g)^{-1}\right]\psi = (\alpha, g)\psi[(\alpha, g)\psi]^{-1} \\
&= ((\alpha, g)\sigma, (\alpha, g)\tau)((\alpha, g)\sigma, (\alpha, g)\tau)^{-1} \\
&= ((\alpha, g)\sigma, 1)
\end{aligned}
$$

and thus $(\alpha, 1)\sigma = (\alpha, g)\sigma$. Hence we may write $\alpha\sigma = (\alpha, g)\sigma$. Now let (α, g), $(\beta, g) \in \mathbf{d}\psi$. Then

$$
\begin{aligned}
(\alpha\beta,1)\psi &= \left[(\alpha, g)(\beta, g)^{-1}\right]\psi = (\alpha, g)\psi[(\beta, g)\psi]^{-1} \\
&= (\alpha\sigma, (\alpha, g)\tau)(\beta\sigma, (\beta, g)\tau)^{-1} \\
&= \left((\alpha\beta)\sigma, (\alpha, g)\tau[(\beta, g)\tau]^{-1}\right)
\end{aligned}
$$

which implies that $(\alpha, g)\tau[(\beta, g)\tau]^{-1} = 1$, that is, $(\alpha, g)\tau = (\beta, g)\tau$. Hence we may write $g\tau = (\alpha, g)\tau$. Now (14) becomes $(\alpha, g)\psi = (\alpha\sigma, g\tau)$. In particular, $(\alpha, 1)\psi = (\alpha\sigma, 1)$ so that σ must be an isomorphism of I onto I'. Further, the projection of $(I \times G) \cap S$ in G is

$$\{g \in G | g \in \alpha\eta \text{ for some } \alpha \in I\} = \bigcup_{\alpha \in I} \alpha\eta.$$

For any $g \in G$, there is $\alpha \in Y$ such that $g \in \alpha\eta$. For such an α there is $\beta \in I$ such that $\alpha \geq \beta$. Hence $g \in \alpha\eta \subseteq \beta\eta$. This shows that $\bigcup_{\alpha \in I}\alpha\eta = S$. Hence $\tau \in \mathcal{Q}(G)$. In addition, $(\alpha, g) \in \mathbf{d}\psi$ implies $(\alpha\sigma, g\tau) \in \mathbf{r}\psi$, which yields

$$\alpha \in I, \quad g \in \alpha\eta \Rightarrow g\tau \in \alpha\sigma\eta. \tag{15}$$

We may thus define a function

$$\kappa: \psi \to (\sigma_\psi, \tau_\psi) \qquad [\psi \in \Psi(S)]$$

where σ_ψ and τ_ψ stand for σ and τ above, respectively. Since σ_ψ and τ_ψ determine ψ uniquely, the function κ is one-to-one.

Conversely, let $I, I' \in \mathcal{R}_Y$, σ be an isomorphism of I onto I', $\tau \in \mathcal{Q}(G)$, and assume that relation (15) holds. Define ψ by

$$(\alpha, g)\psi = (\alpha\sigma, g\tau) \qquad [(\alpha, g) \in (I \times G) \cap S]. \tag{16}$$

Routine verification shows that ψ is an isomorphism of $(I \times G) \cap S$ onto $(I' \times G) \cap S$. In view of V.3.2, we have that all subsets of S of the form $\lambda S\rho$ for some $(\lambda, \rho) \in E_{\Omega(S)}$ coincide with the subsets $(I \times G) \cap S$ for $I \in \mathcal{R}_Y$. It follows that ψ defined in (16) is in the conjugate hull of S. It is trivial that $\psi\kappa = (\sigma, \tau)$. Consequently, κ is a one-to-one correspondence of $\Psi(S)$ and the set stated in the theorem.

To prove that κ is a homomorphism, we let $\varphi, \psi \in \Psi(S)$. Then we have the following situation

$$(I \times G) \cap S \xrightarrow{\varphi} (I' \times G) \cap S$$

$$(J \times G) \cap S \xrightarrow{\psi} (J' \times G) \cap S$$

which illustrates the fact that $\sigma_\varphi \sigma_\psi$ is the product of partial transformations on Y and that $\tau_\varphi \tau_\psi$ is the product of automorphisms of G. Therefore, κ is a homomorphism, and hence the required isomorphism.

IV.5.2 Proposition

Let everything be as in 5.1 and denote by H the set on the right-hand side in 5.1. Then the following statements are valid.

(i) $\Phi(S) \cong \{(\sigma, \tau) \in H | \sigma \in \Phi(Y)\}$.
(ii) $T(S) \cong \{(\iota_I, \varepsilon_c) | I \in \mathfrak{R}_Y, c \in \bigcap_{\alpha \in \mathbf{d}\sigma} \alpha\eta\}$.
(iii) $I(S) \cong \{(\sigma, \tau) \in H | \tau \in \mathcal{IA}(G)\}$.
(iv) $\Theta(S) \cong \{(\iota_{[\alpha]}, \varepsilon_c) | (\alpha, c) \in S\}$.

Proof. (i) Let $\psi \in \Psi(S)$, $\psi\kappa = (\sigma, \tau)$, $I = \mathbf{r}\sigma$, and $\mathbf{d}\psi = \lambda S \rho$ where $(\lambda, \rho) \in E_{\Omega(S)}$. We have seen in the proof of 5.1 that $\lambda S \rho = (I \times G) \cap S$. It follows that $\psi \in \Phi(S)$ if and only if I is a principal ideal of Y which is evidently equivalent to $\sigma \in \Phi(Y)$.

(ii) Let $(\lambda, \rho) \in \Omega(S)$ and consider $\tau_{(\lambda, \rho)} \in T(S)$. According to V.3.9, we have the correspondence $(\lambda, \rho) \to (I, c)$ with $I \in \mathfrak{R}_Y$ and $c \in \bigcap_{\alpha \in I} \alpha\eta$, where for any $(\alpha, g) \in S$,

$$\left. \begin{aligned} \lambda(\alpha, g) &= (\bar{\alpha}, cg) \\ (\alpha, g)\rho &= (\bar{\alpha}, gc) \end{aligned} \right\} \tag{17}$$

with $[\bar{\alpha}] = I \cap [\alpha]$. In this correspondence we have $(\lambda, \rho)^{-1} \to (I, c^{-1})$. Assume that

$$(\alpha, g) \in \mathbf{d}\tau_{(\lambda, \rho)} = (I \times G) \cap S.$$

Then $\alpha \in I$, and

$$(\alpha, g)\tau_{(\lambda, \rho)} = \lambda^{-1}(\alpha, g)\rho = (\alpha, c^{-1}gc).$$

It follows that

$$\tau_{(\lambda, \rho)}\kappa = (\iota_I, \varepsilon_c), \qquad c \in \bigcap_{\alpha \in I} \alpha\eta.$$

Conversely, let $I \in \mathfrak{R}_Y$ and $c \in \bigcap_{\alpha \in I} \alpha\eta$. Defining λ and ρ by (17), we obtain $(\iota_I, \varepsilon_c) = \tau_{(\lambda, \rho)}\kappa$.

(iii) For $\psi \in \Psi(S)$, we have defined $\tilde{\psi}$ in 4.6 by

$$\tilde{\psi}: a\sigma \to a\psi\sigma$$

where σ is the least group congruence on S. In our case, for $(\alpha, g) \in S$,

$$\tilde{\psi}: (\alpha, g)\sigma \to (\alpha, g)\psi\sigma = (\alpha\sigma_\psi, g\tau_\psi)\sigma.$$

Since S is a subdirect product of a semilattice and a group, this means that $\tilde{\psi}$ induces the automorphism τ_ψ of G. Part (iii) follows from this and 5.1.

(iv) This is a consequence of parts (i) and (ii) and 4.2(iv).

We now consider some examples. For a three-element nonchain semilattice we give below the conjugate hull. In this example $\Theta(Y)$, $\Phi(Y)$, $T(Y)$, and $\Psi(Y)$ are all distinct, but $I(Y) = \Psi(Y)$ which is the case for any semilattice Y.

VI.5.3 Example

Let $Y = \{a, b, c\}$ be a semilattice with $ab = c$. Then with the following notation and multiplication we obtain $\Psi(Y)$.

$$\varepsilon_a = \begin{pmatrix} a & c \\ a & c \end{pmatrix}, \quad \varepsilon_b = \begin{pmatrix} b & c \\ b & c \end{pmatrix}, \quad \varepsilon_c = \begin{pmatrix} c \\ c \end{pmatrix},$$

$$\alpha = \begin{pmatrix} a & c \\ b & c \end{pmatrix}, \quad \beta = \begin{pmatrix} b & c \\ a & c \end{pmatrix},$$

$$\varepsilon = \begin{pmatrix} a & b & c \\ a & b & c \end{pmatrix}, \quad \gamma = \begin{pmatrix} a & b & c \\ b & a & c \end{pmatrix}.$$

	ε_a	ε_b	ε_c	α	β	γ	ε
ε_a	ε_a	ε_c	ε_c	α	ε_c	α	ε_a
ε_b	ε_c	ε_b	ε_c	ε_c	β	β	ε_b
ε_c	ε_c	ε_c	ε_c	ε_c	ε_c	ε_c	ε_c
α	ε_c	α	ε_c	ε_c	ε_a	ε_a	α
β	β	ε_c	ε_c	ε_b	ε_c	ε_b	β
γ	β	α	ε_c	ε_b	ε_a	ε	γ
ε	ε_a	ε_b	ε_c	α	β	γ	ε

$$\Theta(Y) = \{\varepsilon_a, \varepsilon_b, \varepsilon_c\},$$

$$\Phi(Y) = \Theta(Y) \cup \{\alpha, \beta\},$$

$$T(Y) = \Theta(Y) \cup \{\varepsilon\},$$

$$\Psi(Y) = \Theta(Y) \cup \{\alpha, \beta, \gamma, \varepsilon\} = I(Y).$$

We have shown by the example III.4.11 that we might have $\ker(\vee_{\rho\in\mathcal{F}}\rho) \neq \vee_{\rho\in\mathcal{F}}\ker\rho$ for a collection $\mathcal{F}$ of congruences on an inverse semigroup. The following is a second example for this, now using the normal hull.

VI.5.4 Example

Let Y be the semilattice pictured by

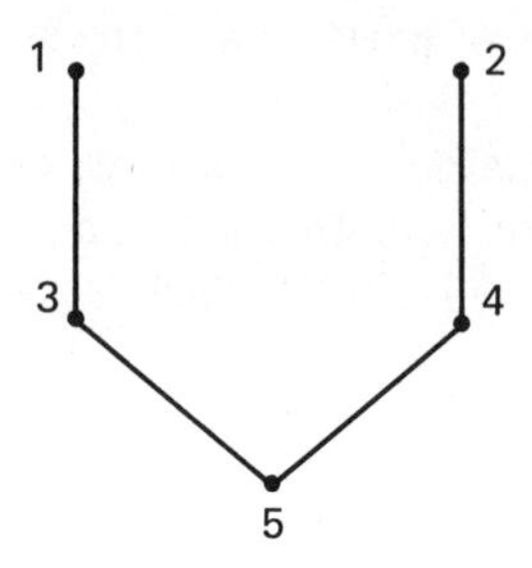

and let $S = \Phi(Y)$. Then the elements of S can be put into a convenient table as follows:

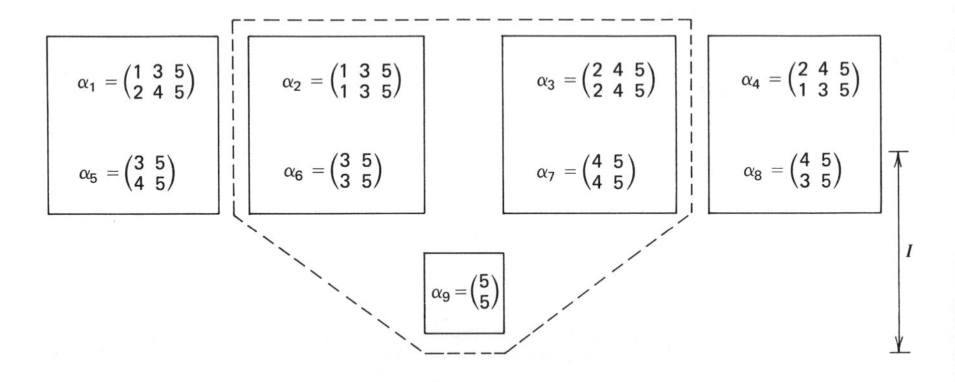

The elements within the dashed line are the idempotents of S. Let ρ be the Rees congruence on S relative to the ideal $I = \{\alpha_5, \alpha_6, \ldots, \alpha_9\}$. Let ξ be the equivalence relation on S whose classes are $\{\alpha_i, \alpha_{i+4}\}$, $i = 1, 2, 3, 4$, and $\{\alpha_9\}$ (indicated on the picture by boxes). It is readily verified that ξ is a congruence. Also

$$\ker\rho = E_S \cup \{\alpha_5, \alpha_8\}, \qquad \ker\xi = E_S$$

so $\ker\rho \vee \ker\xi = E_S \cup \{\alpha_5, \alpha_8\}$. On the other hand, clearly $\rho \vee \xi = \omega$ so that $\ker(\rho \vee \xi) = S$. Consequently

$$\ker\rho \vee \ker\xi \neq \ker(\rho \vee \xi).$$

VI.5.5 Exercises

(i) Construct the conjugate hull of a Brandt semigroup and identify in it $\Phi(S)$, $T(S)$, $I(S)$, and $\Theta(S)$.

The last example is due to N. R. Reilly.

VI.6 NORMAL EXTENSIONS

We first single out normal extensions among conjugate extensions of an inverse semigroup. Then we define the extension problem for normal extensions analogous to the one in group theory solved by the Schreier extensions. The added feature here is the part concerning a partition of idempotents. In the case that an inverse semigroup K has idempotent metacenter, we give a general solution for the extension problem and a more specific solution when the quotient is restricted to be an antigroup.

VI.6.1 Definition

An inverse semigroup S is a *normal extension* of an inverse semigroup K if K is a normal subsemigroup of S.

The next lemma restricts the type of a normal extension.

VI.6.2 Lemma

If S is a normal extension of K, then the canonical homomorphism $\theta(S:K)$ maps S into $\Phi(K)$.

Proof. For any $a \in S$, we have $aKa^{-1} = aa^{-1}Kaa^{-1}$ where $aa^{-1} \in K$ since K is full in S. Hence $\mathbf{d}\theta^a = \lambda_e K \rho_e$ where $e = aa^{-1}$ and thus $\theta^a \in \Phi(K)$.

Let S be an inverse semigroup. If $\varphi \in E_{\Phi(S)}$, then $\varphi = \iota_{eSe}$ for some $e \in E_S$ which implies that $\varphi = \theta_S^e$. Hence $\Theta(S)$ is full in $\Phi(S)$, and is self-conjugate in $\Phi(S)$ by 2.7. Hence $\Phi(S)$ is a normal extension of $\Theta(S)$.

Let ρ be a congruence on an inverse semigroup S. We encounter here the following parameters:

$$K = \ker \rho, \qquad \tau = \operatorname{tr} \rho, \qquad Q = S/\rho. \tag{18}$$

One may ask whether it is possible to start with (K, τ, Q), evidently satisfying certain conditions relating them, and construct all inverse semigroups S and all congruences ρ on S so that (18) may be fulfilled. The triple (K, τ, Q) consists of an inverse semigroup K, a (necessarily normal) congruence τ on E_K, and an inverse semigroup Q. It is also necessary that there exist a homomorphism π of

K onto E_Q which induces τ. It will turn out more convenient (and is equivalent) to consider the triple (K, π, Q) and then try to give all solutions to the problem just discussed, which we may call the (*normal*) *extension problem*. We must make provisions for isomorphisms of the inverse semigroups under consideration, so we proceed formally as follows.

VI.6.3 Definition

We call (K, π, Q) a (*normal*) *extension triple* if K and Q are inverse semigroups and π is a homomorphism of K onto E_Q. A triple (ξ, S, η) is a *normal extension of K by Q along π* if S is an inverse semigroup, ξ is an isomorphism of K into S, η is a homomorphism of S onto Q, $\mathbf{r}\xi = \ker \eta$, and $\xi\eta = \pi$; that is, the diagram

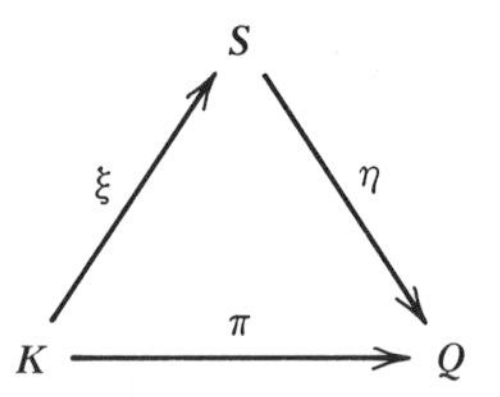

commutes. We then say that the triple (ξ, S, η) is a *solution* (*of the normal extension problem*) *for the triple* (K, π, Q).

For a given extension triple (K, π, Q), we will consider certain solutions of the extension problem as essentially identical so we now postulate when two such extensions will be termed equivalent.

VI.6.4 Definition

The solutions (ξ, S, η) and (ξ', S', η') of the extension problem for the extension triple (K, π, Q) are *equivalent* if there exists an isomorphism ψ of S onto S' such that $\xi\psi = \xi'$; that is, the diagram

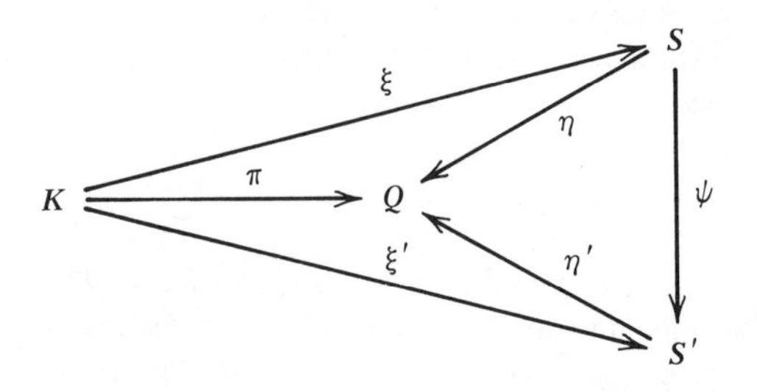

commutes.

Equivalence of solutions of the extension problem for a given extension triple is evidently an equivalence relation on the class of all such solutions. We are interested in "all" solutions of the extension problem, but all we really want is at least one solution in each equivalence class, which we state formally as follows.

VI.6.5 Definition

A class $\mathcal{C}$ of solutions for the extension triple (K, π, Q) is a *complete class of solutions* (of the extension problem) *for the triple* (K, π, Q) if every solution for this triple is equivalent to some solution in $\mathcal{C}$.

VI.6.6 Remark

Let (ξ, S, η) be a solution for the extension triple (K, π, Q). It will often be convenient to identify K with $\mathbf{r}\xi$, which amounts to the case of ξ being the identity map on K. In such a case, let ρ be the congruence on S induced by η; we get

$$K = \ker \rho, \qquad \operatorname{tr} \pi = \operatorname{tr} \rho, \qquad S/\rho \cong Q.$$

If we denote this extension by (S, ρ), we obtain that two such extensions (S, ρ) and (S', ρ') are equivalent if and only if there exists a K-isomorphism ψ of S onto S' for which $a\rho b \Leftrightarrow (a\psi)\rho'(b\psi)$ for all $a, b \in S$.

Let (K, π, Q) be an extension triple. The next theorem provides a complete *set* of solutions for this triple under the restriction that K have idempotent metacenter. If in addition Q is an antigroup, then 6.9 supplies a complete set of inequivalent solutions. The subject of the next section is a more explicit solution for the case when K is a semilattice.

VI.6.7 Theorem

Let (K, π, Q) be an extension triple and assume that K has idempotent metacenter. Define a function ξ by

$$\xi\colon k \to \left(\theta_K^k, k\pi\right) \qquad (k \in K).$$

Let S be any inverse subsemigroup of $\Phi(K) \times Q$ such that

$$S \cap \left(\Phi(K) \times E_Q\right) = K\xi \tag{19}$$

and the projection η of S into Q is onto. Then the collection of all such (ξ, S, η) is a complete set of solutions for the extension problem for (K, π, Q).

Proof. Let the notation be as in the statement of the theorem. First observe that ξ is a monomorphism since K has trivial metacenter, and that η is a homomorphism of S onto Q.

Let $k \in K$. Then

$$k\xi\eta = (\theta_K^k, k\pi)\eta = k\pi \in E_Q$$

which shows that $\xi\eta = \pi$ and that $\mathbf{r}\xi \subseteq \ker \eta$. Next let $s \in \ker \eta$. Then $s = (\varphi, q) \in \Phi(K) \times Q$ and $s\eta = q \in E_Q$. Hence using the hypothesis, we get

$$s = (\varphi, q) \in S \cap (\Phi(K) \times E_Q) = K\xi$$

which shows that $s \in \mathbf{r}\xi$. Thus $\mathbf{r}\xi = \ker \eta$, and (ξ, S, η) is a solution for (K, π, Q).

Conversely, let (ξ, S, η) be a solution for the triple (K, π, Q). We may suppose that ξ is the identity mapping. Then as in 6.6, we may consider the pair (S, ρ) where ρ is the congruence induced by η. We thus have

$$K = \ker \rho, \qquad \tau = \operatorname{tr} \rho, \qquad S/\rho \cong Q.$$

Set $\theta = \theta(S : K)$ and define a mapping ψ by

$$\psi : a \to (\theta^a, a\eta) \qquad (a \in S).$$

Letting $\bar{\theta}$ be the congruence induced by θ, we obtain

$$\operatorname{tr}(\bar{\theta} \cap \rho) = \operatorname{tr} \bar{\theta} \cap \operatorname{tr} \rho \subseteq \operatorname{tr} \bar{\theta} = \varepsilon$$

since K is a full subsemigroup of S and has idempotent metacenter. Further,

$$\ker(\bar{\theta} \cap \rho) = \ker \bar{\theta} \cap \ker \rho = \ker \bar{\theta} \cap K = E_K$$

since $\bar{\theta}|_K = \varepsilon$. Hence ψ is one-to-one and is thus an isomorphism of S onto $S' = S\psi$. It is obvious that the projection η' of S' into Q is onto.

Let $\psi' = \psi|_K$. For any $k \in K$, we have $k\eta = k\pi$ since ξ is the identity mapping. Hence

$$k\xi' = k\psi = (\theta^k, k\eta) = (\theta_K^k, k\pi)$$

and ξ' is of prescribed form.

Let $(\varphi, q) \in S\psi \cap (\Phi(K) \times E_Q)$. Then $(\varphi, q) \in S\psi$ yields $(\varphi, q) = (\theta^a, a\eta)$ for some $a \in S$, and thus $(\varphi, q) \in \Phi(K) \times E_Q$ gives $a\eta \in E_Q$. Since η induces ρ, it follows that $a \in \ker \rho = K$. Furthermore, $(\varphi, q) = a\xi' \in K\xi'$. Conversely, let $k \in K$. Then

$$k\xi' = (\theta^k, k\eta) \in S' \cap (\Phi(K) \times E_Q)$$

since $k\xi' = k\psi$, η induces ρ and $K = \ker \rho$. This verifies relation (19).

It follows that the triple (ξ', S', η') has the form as in the statement of the theorem. For any $a \in S$, we have

$$a\psi\eta' = (\theta^a, a\eta)\eta' = a\eta$$

so that $\psi\eta' = \eta$, and the two extensions (ι_K, S, η) and (ξ', S', η') are equivalent.

For the next result we will need an auxiliary statement which is also of independent interest.

VI.6.8 Lemma

Let ρ be an antigroup congruence on an inverse semigroup S. Then S is an essential normal extension of $\ker\rho$.

Proof. Let ξ be a congruence on S such that $\xi|_{\ker\rho} = \varepsilon$. Then ξ is idempotent separating and thus $\xi \subseteq \mu$. Since ρ is an antigroup congruence, III.3.11 implies that $\mu \subseteq \rho$. Thus $\xi \subseteq \rho$ which evidently implies that $\ker\xi = E_S$. But then $\xi = \varepsilon$ and S is an essential normal extension of $\ker\rho$.

VI.6.9 Theorem

Let (K, π, Q) be an extension triple and assume that K has idempotent metacenter and that Q is an antigroup. Let S be an inverse subsemigroup of $\Phi(K)$ containing $\Theta(K)$, and let η be a homomorphism of S onto Q with kernel $\Theta(K)$ and such that $\theta_K\eta = \pi$. Then the collection of all such (θ_K, S, η) is a complete set of inequivalent extensions of K by Q along π.

Proof. If (θ_K, S, η) is as constructed in the statement of the theorem, it follows at once that it is a solution for the triple (K, π, Q).

For the converse, let (ξ', S', η') be a solution for the triple (K, π, Q). Letting $\theta = \theta(S' : K\xi')$, we deduce from 6.8 and 2.9 that θ is one-to-one. We also have an isomorphism $\overline{\xi'}$ of $\Phi(K)$ onto $\Phi(K\xi)$ according to 3.1. Now letting

$$S = S'\theta\overline{\xi'}^{-1}, \qquad \psi = \left(\overline{\xi'}|_S\right)\theta^{-1},$$

$$\xi = \xi'\psi^{-1}, \qquad \eta = \psi\eta',$$

we get the following commutative diagram

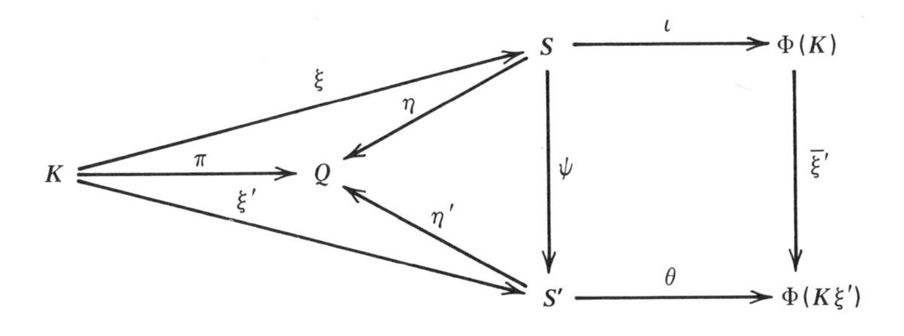

where ι is the inclusion mapping. Thus the extensions $(\overline{\xi'}, S', \eta')$ and (ξ, S, η) are equivalent. For any $k \in K$, using the definition of $\overline{\xi'}$ in 3.1, we obtain

$$k\xi = k\xi'\theta\overline{\xi'}^{-1} = \theta^{k\xi'}\overline{\xi'}^{-1} = \theta^{k\xi'\xi'^{-1}} = \theta^{k}$$

so that $\xi = \theta_K$, as required.

Now assume that two extensions (θ_K, S, η) and (θ_K, S', η') as in the statement of the theorem are equivalent, and let ψ be the corresponding isomorphism of S onto S'. Let $\sigma \in S$ and $k \in \mathbf{d}(\sigma\psi)$. Using 2.6, we obtain

$$\begin{aligned}\theta^{k(\sigma\psi)} &= (\sigma\psi)^{-1}\theta_K^k(\sigma\psi) = (\sigma\psi)^{-1}(\theta_K^k\psi)(\sigma\psi)\\ &= (\sigma^{-1}\theta_K^k\sigma)\psi = \theta_K^{k\sigma}\psi = \theta^{k\sigma}\end{aligned}$$

so that $k(\sigma\psi) = k\sigma$ since θ_K is one-to-one. Since $k \in \mathbf{d}(\sigma\psi) = \mathbf{d}\sigma$ is arbitrary, we deduce that $\sigma\psi = \sigma$, which implies that $S = S'$ and $\eta = \eta'$. Thus the two triples (θ_K, S, η) and (θ_K, S', η') are equal.

Under the conditions of 6.9, we may take $K \subseteq S$ [identifying K and $\Theta(K)$] and thus all solutions (ι_K, S, η) are obtained by taking an inverse subsemigroup of $\Phi(K)$ containing $\Theta(K)$ and a homomorphism of S onto Q with kernel $\Theta(K)$. Different semigroups S give inequivalent extensions.

VI.6.10 Exercises

(i) Let $K = B(1, I)$. Characterize all inverse semigroups having K as a normal subsemigroup. Use this to find all solutions of the extension problem for the triple (K, π, Q) with Q and π arbitrary.

(ii) Let K be a normal subsemigroup of an inverse semigroup S. Show that S is an essential extension of K if and only if for any normal subsemigroup N of S contained in $E_S\zeta$, $K \cap N = E_S$ implies $N = E_S$ (cf. essential extensions of rings and groups).

(iii) For any inverse semigroup S, find the kernel and the trace of the congruence induced by $\theta(S\colon E_S\zeta)$.

(iv) Call a normal extension S of K *strict* if $T(S:K) = \Theta(K)$. Show that Clifford semigroups coincide with strict normal extensions of semilattices.

(v) Find a complete solution for the extension triple (K, π, Y) where K and π are arbitrary and Y is a semilattice.

(vi) Let K be an inverse subsemigroup of an inverse semigroup S. Find the expressions for the greatest inverse subsemigroup of S which is a conjugate (respectively normal) extension of K.

(vii) For an inverse semigroup S, prove directly that $\Phi(S)$ is a subsemigroup of $\mathcal{I}(S)$.

(viii) Let S be an inverse semigroup, and let $a \in S$, $\varphi \in \Phi(S)$. Prove directly that $\varphi^{-1}\theta_S^a\varphi = \theta_S^{(eae)\varphi}$ where $\mathbf{d}\varphi = eSe$.

(ix) Let $S = G \cup H$ be a semilattice of groups G and H with multiplication given by a homomorphism $\varphi\colon G \to H$. Let S be an essential normal extension of an inverse semigroup K. Show that

(α) H is an essential extension of $K \cap H$,

(β) G need not be an essential extension of $K \cap G$.

VI.6.11 Problems

(i) Let an inverse semigroup S be a normal extension of an inverse semigroup K. Is it true that $\theta(S : K)$ is the greatest congruence on S contained in the relation ρ defined by

$$a\rho b \Leftrightarrow aK = bK, \quad Ka = Kb?$$

(ii) Find a suitable variant of the wreath product of two inverse semigroups K and Q with the property that every normal extension of K by Q (with some trace) can be embedded into it. (This is possible when both K and Q are groups. A special case has been solved by Houghton [1].)

The material in this section is taken from Petrich [15]. Idempotent separating extensions [those corresponding to the case (K, π, Q) with π one-to-one on idempotents] were studied by Coudron [1], D'Alarcao [1], and Širjajev [1], providing a somewhat complicated solution. The solution of the extension problem in full generality, including many special cases, was obtained by Allouch [2]. Further special cases are discussed by Allouch [1], [3], Green [1], O'Carroll [5], see also Edwards [1], Houghton [1], Lausch [1], Sribala [2], and Širjajev [3].

VI.7 NORMAL EXTENSIONS OF A SEMILATTICE

We have seen in the preceding section that normal extensions of an inverse semigroup with idempotent metacenter can be characterized in a relatively explicit fashion. If we also consider such extensions by an antigroup, then we have seen that a much more concrete characterization is possible. We study here normal extensions of a semilattice (its metacenter is trivially idempotent) by an arbitrary inverse semigroup, and for such extensions obtain a construction which is considerably more explicit than the one arrived at in the preceding section for inverse semigroups with idempotent metacenter.

We start with several new concepts which are needed in the construction of the sought extensions.

VI.7.1 Notation

For a semilattice Y, we denote by $\Sigma(Y)$ the *set of all isomorphisms among ideals of* Y, written as operators on the left and composed as such.

Simple verification shows that the operation in $\Sigma(Y)$ is closed under the composition of functions, which shows that $\Sigma(Y)$ is an inverse subsemigroup of $\mathcal{I}'(Y)$.

VI.7.2 Definition

Let S and T be inverse semigroups. A mapping $\psi\colon S \to T$ is a *prehomomorphism* if

(i) $(\psi a)(\psi b) \leq \psi(ab) \quad (a, b \in S)$,
(ii) $\psi a^{-1} = (\psi a)^{-1} \quad (a \in S)$.

The above concept will prove to be useful in various situations. Here we need a special case which we formulate as follows.

VI.7.3 Definition

Let (Y, π, P) be an extension triple, where Y is a semilattice and P is an arbitrary inverse semigroup. A *prehomomorphism* $\psi\colon P \to \Sigma(Y)$, in notation $\psi\colon p \to \psi_p$, *is compatible with* π if

(i) $\psi_e = \iota_{(\pi^{-1}e)Y} \quad (e \in E_P)$,
(ii) $\pi\psi_p \subseteq \theta\wp^{-1}\pi \quad (p \in P)$,
(iii) $\Gamma_p = \mathbf{r}\psi_p \cap \pi^{-1}(pp^{-1}) \neq \varnothing$ and $\mathbf{r}\psi_p = \Gamma_p Y \quad (p \in P)$.

For convenience let $\Delta_p = \psi_p^{-1}\Gamma_p$ for every $p \in P$. In order to conform with the rest of the notation, we have written $\theta\wp$ and π as left operators. The operation in Y will be denoted by $\wedge$, and $p\alpha$ will stand for $\psi_p\alpha$.

VI.7.4 Lemma

Let the notation be as in 7.3 and $p, q \in P$.

(i) $\Delta_p = \mathbf{d}\psi_p \cap \pi^{-1}(p^{-1}p) = \Gamma_{p^{-1}}$.
(ii) $\psi_p(\Delta_p\Gamma_q) \subseteq \Gamma_{pq}$.

Proof. (i) Let $\alpha \in \Delta_p$. Then $\alpha = \psi_p^{-1}\beta$ for some $\beta \in \mathbf{r}\psi_p$, and thus $\alpha \in \mathbf{d}\psi_p$. Using the fact that ψ is a prehomomorphism, we get

$$\alpha = \psi_p^{-1}\psi_p\alpha = \psi_{p^{-1}}\psi_p\alpha = \psi_{p^{-1}}\beta.$$

Since $\beta = \psi_p\alpha \in \Gamma_p$, by 7.3(iii), we have $\beta \in \pi^{-1}(pp^{-1})$ so that $\pi\beta = pp^{-1}$. Now using 7.3(ii), we obtain

$$\pi\alpha = \pi(\psi_{p^{-1}}\beta) = \theta_\beta^p\pi\beta = p^{-1}(\pi\beta)p = p^{-1}(pp^{-1})p = p^{-1}p$$

and thus $\Delta_p \subseteq \mathbf{d}\psi_p \cap \pi^{-1}(p^{-1}p)$. Conversely, let $\alpha \in \mathbf{d}\psi_p \cap \pi^{-1}(p^{-1}p)$. Then $\psi_p\alpha \in \mathbf{r}\psi_p$ and $\pi\alpha = p^{-1}p$. By 7.3(ii), we get

$$\pi\psi_p\alpha = \theta_\beta^{p^{-1}}\pi\alpha = p(p^{-1}p)p^{-1} = pp^{-1}$$

and thus $\alpha \in \Delta_p$. Clearly $\Delta_p = \Gamma_{p^{-1}}$.

(ii) Let $\alpha \in \Delta_p$ and $\beta \in \Gamma_q$. Then $\beta = \psi_q\gamma$ for some $\gamma \in \Delta_q$. Thus

$$\psi_p(\alpha \wedge \beta) = \psi_p(\alpha \wedge \psi_q\gamma) = \psi_p\psi_q\delta = \psi_{pq}\eta$$

for some $\delta \in \mathbf{d}\psi_q$ since $\alpha \wedge \psi_q\gamma \in \mathbf{r}\psi_q$, and for some $\eta \in \mathbf{d}\psi_{pq}$ in view of 7.2(i). Hence $\psi_p(\alpha \wedge \beta) \in \mathbf{r}\psi_{pq}$. Further, by 7.3(ii) and part (i),

$$\begin{aligned}\pi\psi_p(\alpha \wedge \beta) &= \theta_\beta^{p^{-1}}\pi(\alpha \wedge \beta) = \theta_\beta^{p^{-1}}((\pi\alpha)(\pi\beta)) = p(p^{-1}p)(qq^{-1})p^{-1} \\ &= (pq)(pq)^{-1}\end{aligned}$$

and thus $\alpha \wedge \beta \in \Gamma_{pq}$.

The stage is now set for the desired device.

VI.7.5 Construction

Let (Y, π, P) be an extension triple, Y be a semilattice, $\psi\colon P \to \Sigma(Y)$ be a prehomomorphism compatible with π. On

$$S = \{(\alpha, p) \in Y \times P \mid \alpha \in \Gamma_p\}$$

define a multiplication by

$$(\alpha, p)(\beta, q) = (p(p^{-1}\alpha \wedge \beta), pq).$$

We denote this system by $S = Q(Y, \pi, P; \psi)$.

VI.7.6 Lemma

The multiplication in 7.5 is closed and regular.

Proof. Let $\alpha \in \Gamma_p$ and $\beta \in \Gamma_q$. Then $p^{-1}\alpha \wedge \beta \in \Delta_p\Gamma_q$ and by 7.4(ii), we have $p(p^{-1}\alpha \wedge \beta) \in \psi_p(\Delta_p\Gamma_q) \subseteq \Gamma_{pq}$. Hence the product $(\alpha, p)(\beta, q)$ is again an element of S. Further, by 7.4(i), $p^{-1}\alpha \in \Delta_p = \Gamma_{p^{-1}}$ so that $(p^{-1}\alpha, p^{-1}) \in S$, and

$$\begin{aligned}(\alpha, p)(p^{-1}\alpha, p^{-1})(\alpha, p) &= (p(p^{-1}\alpha \wedge p^{-1}\alpha), pp^{-1})(\alpha, p)\\ &= (\alpha, pp^{-1})(\alpha, p)\\ &= (pp^{-1}(pp^{-1}\alpha \wedge \alpha), pp^{-1}p) = (\alpha, p).\end{aligned}$$

Instead of verifying associativity of the multiplication in 7.5, we prove the following statement.

VI.7.7 Lemma

With the notation of 7.5, for every $(\alpha, p) \in S$, let

$$\varphi_{(\alpha, p)}: \beta \to \psi_p^{-1}\beta \qquad (\beta \in \mathbf{d}\varphi_{(\alpha, p)} = [\alpha]),$$

with $\varphi_{(\alpha, p)}$ written on the right. Then

$$\varphi: (\alpha, p) \to (\varphi_{(\alpha, p)}, p) \qquad [(\alpha, p) \in S]$$

is an isomorphism of S onto an inverse subsemigroup of $\Phi(Y) \times P$.

Proof. It is clear that for any $(\alpha, p) \in S$, $\varphi_{(\alpha, p)} \in \Phi(Y)$ so that φ maps S into $\Phi(Y) \times P$. Let $(\alpha, p), (\beta, q) \in S$. Then

$$\begin{aligned}\gamma \in \mathbf{d}(\varphi_{(\alpha, p)}\varphi_{(\beta, q)}) &\Leftrightarrow \gamma \le \alpha, \ p^{-1}\gamma \le \beta\\ &\Leftrightarrow p^{-1}\gamma \le p^{-1}\alpha \wedge \beta\\ &\Leftrightarrow \gamma \le p(p^{-1}\alpha \wedge \beta)\\ &\Leftrightarrow \gamma \in \mathbf{d}\varphi_{(p(p^{-1}\alpha \wedge \beta), pq)} = \mathbf{d}\varphi_{(\alpha, p)(\beta, q)}\end{aligned}$$

and if this is the case, using 7.2, we get

$$\begin{aligned}\gamma\varphi_{(\alpha, p)}\varphi_{(\beta, q)} &= \psi_q^{-1}\psi_p^{-1}\gamma = \psi_{q^{-1}}\psi_{p^{-1}}\gamma = \psi_{q^{-1}p^{-1}}\gamma = \psi_{(pq)^{-1}}\gamma\\ &= \psi_{pq}^{-1}\gamma\end{aligned}$$

which proves that $\varphi_{(\alpha, p)}\varphi_{(\beta, q)} = \varphi_{(\alpha, p)(\beta, q)}$. It follows that φ is a homomorphism.

If $(\varphi_{(\alpha, p)}, p) = (\varphi_{(\beta, q)}, q)$, then $\varphi_{(\alpha, p)} = \varphi_{(\beta, q)}$ and $p = q$ so also $\alpha = \beta$. Thus φ is an isomorphism into $\Phi(Y) \times P$. Since S is regular, so is $S\varphi$ and thus $S\varphi$ is an inverse subsemigroup of $\Phi(Y) \times P$.

We are now ready for the first half of the desired result.

VI.7.8 Theorem

Let $S = Q(Y, \pi, P; \psi)$ and define

$$\xi\colon \alpha \to (\alpha, \pi\alpha) \qquad (\alpha \in Y),$$
$$\eta\colon (\alpha, p) \to p \qquad [(\alpha, p) \in S].$$

Then (ξ, S, η) is a normal extension of Y by P along π.

Proof. It follows from 7.7 that S is an inverse semigroup. Let $(\alpha, p) \in E_S$. Then $p \in E_P$ and thus $\alpha \in \Gamma_p = \mathbf{r}\psi_p \cap \pi^{-1}p$ implies that $\pi\alpha = p$. Hence $(\alpha, p) = (\alpha, \pi\alpha)$ whence we deduce that ξ is an isomorphism of Y onto E_S.

By 7.3(iii), η maps S onto P, and it is obviously a homomorphism. If for $(\alpha, p) \in S$ we have $p \in E_P$, then by the above, $(\alpha, p) \in E_S$ which shows that the congruence on S induced by η is pure. Clearly, $\xi\eta = \pi$. Therefore, (ξ, S, η) is a normal extension of Y by P along π.

In order to prove the converse, we first analyze pure congruences on an inverse semigroup.

VI.7.9 Lemma

Let ρ be a pure congruence on an inverse semigroup S. For $s\rho t$ and $e, f \in E_S$, the following statements hold.

(i) $s(t^{-1}te)s^{-1} = t(s^{-1}se)t^{-1}$.
(ii) $s(t^{-1}tef)s^{-1} = (ses^{-1})(tft^{-1})$.

Proof. (i) First $s\rho t$ implies $st^{-1}\rho tt^{-1}$ so that $st^{-1} \in \ker\rho = E_S$. Similarly, $tes^{-1}\rho tet^{-1}$ so that $tes^{-1} \in \ker\rho = E_S$. Using this, we obtain

$$s(t^{-1}te)s^{-1} = (st^{-1})(tes^{-1}) = (tes^{-1})(st^{-1}) = t(es^{-1}s)t^{-1}$$
$$= t(s^{-1}se)t^{-1}.$$

(ii) As above, we deduce that $tefs^{-1}, t^{-1}s \in E_S$, whence we obtain

$$s(t^{-1}tef)s^{-1} = st^{-1}(tefs^{-1})^{-1} = st^{-1}seft^{-1} = s(t^{-1}s)^{-1}eft^{-1}$$
$$= s(s^{-1}t)eft^{-1} = (ses^{-1})(tft^{-1}).$$

VI.7.10 Lemma

Let ρ be a pure congruence on an inverse semigroup S. For every $p \in P = S/\rho$, let

$$\psi_p\colon s^{-1}se \to ses^{-1} \qquad (s\rho = p, e \in E = E_S).$$

Then

$$\psi : p \to \psi_p \qquad (p \in P)$$

is a prehomomorphism of P into $\Sigma(E)$ compatible with $\pi = \rho^{\#}|_E$.

Proof. 1. *ψ_p is single valued.* Let $s\rho = t\rho = p$, $e, f \in E$ be such that $s^{-1}se = t^{-1}tf$. Let $x = s^{-1}se = t^{-1}tf$ so that $s^{-1}sx = x = t^{-1}tx$. Using 7.9(i), we obtain

$$\begin{aligned} ses^{-1} &= s(s^{-1}se)s = sxs^{-1} = s(t^{-1}tx)s^{-1} = t(s^{-1}sx)t^{-1} \\ &= txt^{-1} = t(t^{-1}tf)t^{-1} = tft^{-1}. \end{aligned}$$

2. *ψ_p is a homomorphism.* Continuing the notation and using 7.9(ii), we obtain

$$\begin{aligned} \psi_p(s^{-1}se)\psi_p(t^{-1}tf) &= (ses^{-1})(tft^{-1}) = s(t^{-1}tef)s^{-1} \\ &= \psi_p(s^{-1}s(t^{-1}tef)) = \psi_p((s^{-1}se)(t^{-1}tf)). \end{aligned}$$

3. *ψ_p is one-to-one.* Assume that $\psi_p(s^{-1}se) = \psi_p(t^{-1}tf)$, so that $ses^{-1} = tft^{-1}$. Since $s\rho t$, we have $s^{-1}t, t^{-1}s \in E$. Hence

$$\begin{aligned} s^{-1}se &= s^{-1}(ses^{-1})se = s^{-1}(tft^{-1})se = (s^{-1}t)e(t^{-1}s)f \\ &= (t^{-1}s)e(s^{-1}t)f = t^{-1}(ses^{-1})tf = t^{-1}(tft^{-1})tf = t^{-1}tf. \end{aligned}$$

4. $\mathbf{d}\psi_p$ is clearly an ideal of E. Since $(ses^{-1})f = s(es^{-1}fs)s^{-1}$, also $\mathbf{r}\psi_p$ is an ideal of E.

We have proved that $\psi_p \in \Sigma(E)$.

5. $\psi_p\psi_q \le \psi_{pq}$. Let $s\rho = p$, $t\rho = q$, and $e \in E$ be such that

$$s^{-1}se = tft^{-1} \in \mathbf{d}\psi_p \cap \mathbf{r}\psi_q.$$

Then

$$x = t^{-1}tf = t^{-1}(tft^{-1})tf = t^{-1}(s^{-1}se)tf = (st)^{-1}(st)[(t^{-1}et)f]$$

and thus

$$\begin{aligned} \psi_p\psi_q x &= \psi_p\psi_q(t^{-1}tf) = \psi_p(tft^{-1}) = \psi_p(s^{-1}se) = ses^{-1} \\ &= s(s^{-1}se)es^{-1} = s(tft^{-1})es^{-1} = (st)t^{-1}(tft^{-1})es^{-1} \\ &= (st)(t^{-1}et)ft^{-1}s^{-1} = (st)[(t^{-1}et)f](st)^{-1} = \psi_{pq}x. \end{aligned}$$

6. $\psi_p^{-1} = \psi_{p^{-1}}$. Noting that

$$\mathbf{d}\psi_{p^{-1}} = \{t^{-1}te \mid t\rho = p^{-1}, e \in E\} = \{ss^{-1}e \mid s\rho = p, e \in E\},$$

we easily deduce that $\mathbf{d}\psi_{p^{-1}} = \mathbf{r}\psi_p$. For $s\rho = p$ and $e \in E$, we have

$$\psi_{p^{-1}}(ss^{-1}e) = s^{-1}es = \psi_p^{-1}(ss^{-1}e)$$

whence $\psi_{p^{-1}} = \psi_p^{-1}$.

We have shown that $\psi\colon P \to \Sigma(E)$ is a prehomomorphism.

7. 7.3(i) *holds* since idempotents of S commute and ρ is idempotent pure.

8. 7.3(ii) *holds*. Let $s\rho = p$ and $e \in E$. Then

$$\begin{aligned}\pi\psi_p(s^{-1}se) &= \pi(ses^{-1}) = (s\rho)(e\rho)(s\rho)^{-1} = p(\pi e)p^{-1}\\ &= p(p^{-1}p)(\pi e)p^{-1} = p(s\rho)^{-1}(s\rho)(e\rho)p^{-1}\\ &= \theta_p^{p^{-1}}\pi(s^{-1}se)\end{aligned}$$

and thus $\pi\psi_p \subseteq \theta_p^{p^{-1}}\pi$.

9. 7.3(iii) *holds*. Let $\Gamma_p = \mathbf{r}\psi_p \cap \pi(pp^{-1})$ and $s\rho = p$. Then

$$ss^{-1} = s(s^{-1}s)s^{-1} \in \mathbf{r}\psi_p \cap \pi^{-1}(pp^{-1})$$

so that $\Gamma_p \neq \varnothing$ and $\Gamma_p E \subseteq \mathbf{r}\psi_p$. Since $ses^{-1} = (ss^{-1})(ses^{-1}) \in \Gamma_p E$, we obtain $\mathbf{r}\psi_p = \Gamma_p E$.

Therefore ψ is compatible with π.

The desired converse can now be established.

VI.7.11 Theorem

Let (ξ, S, η) be a normal extension of a semilattice Y by an inverse semigroup P along π. Let ρ be the congruence on S induced by η, $\pi_S = \rho^{\#}|_{E_S}$, and ψ be as constructed in 7.10.

Let $S' = Q(E_S, \pi_S, S/\rho; \psi)$ and define

$$\xi'\colon \alpha \to (\xi\alpha, \pi_S\xi\alpha) \qquad (\alpha \in Y),$$

$$\eta'\colon (e, p) \to p \qquad [(e, p) \in S'].$$

Then (ξ, S, η) and (ξ', S', η') are equivalent normal extensions of Y by P along π.

Proof. It is verified easily that (ξ', S', η') is a normal extension of Y by P along π. Define

$$\chi\colon s \to (ss^{-1}, s\rho) \qquad (s \in S).$$

Then for any $s, t \in S$, we have

$$(\chi s)(\chi t) = (ss^{-1}, s\rho)(tt^{-1}, t\rho) = \left((s\rho)\left((s\rho)^{-1}ss^{-1} \wedge tt^{-1}\right), (s\rho)(t\rho)\right)$$

$$= \left(\psi_{s\rho}\left(\psi_{s\rho}^{-1}(ss^{-1}) \wedge tt^{-1}\right), (st)\rho\right) = \left(\psi_{s\rho}(s^{-1}stt^{-1}), (st)\rho\right)$$

$$= \left(stt^{-1}s^{-1}, (st)\rho\right) = \left((st)(st)^{-1}, (st)\rho\right) = \chi(st)$$

and χ is a homomorphism. Assume that $\chi s = \chi t$. Then $ss^{-1} = tt^{-1}$ and $s\rho = t\rho$ which implies $s(\mathcal{R} \cap \rho)t$. Since ρ is idempotent pure, III.4.2 implies that $s = t$. Hence χ is one-to-one.

Let $(e, s\rho) \in S'$. Then $e \in \Gamma_{s\rho} = \mathbf{r}\psi_{s\rho} \cap \pi_S^{-1}((s\rho)(s\rho)^{-1})$ which gives $\pi e = (ss^{-1})\rho$, that is, $e\rho ss^{-1}$ and thus $es\rho s$. Since $e \in \mathbf{r}\psi_{s\rho}$, we have $e = sfs^{-1}$ for some $f \in E_S$. We obtain

$$\chi(sf) = \left((sf)(sf)^{-1}, (sf)\rho\right) = \left(sfs^{-1}, (sfs^{-1}s)\rho\right) = \left(e, (es)\rho\right) = (e, s\rho)$$

which shows that χ maps S onto S'.

It follows easily that $\chi\xi = \xi'$ and $\chi\eta' = \eta$ which completes the proof that (ξ, S, η) and (ξ', S', η') are equivalent extensions of Y by P along π.

The following result provides a complete solution for the extension problem for a semilattice.

VI.7.12 Theorem

Let (Y, π, P) be an extension triple where Y is a semilattice. Then the collection of all (ξ, S, η), where $S = Q(Y, \pi, P; \psi)$ and

$$\xi\colon \alpha \to (\alpha, \pi\alpha) \qquad (\alpha \in Y),$$

$$\eta\colon (\alpha, p) \to p \qquad [(\alpha, p) \in S],$$

is a complete set of inequivalent normal extensions of Y by P along π.

Proof. By 7.8 such triples (ξ, S, η) are normal extensions of Y by P along π, and by 7.11, any normal extension of Y by P along π is equivalent to one as in the statement of the theorem. It remains to show that any two different such extensions are inequivalent. Hence let (ξ, S, η) and (ξ', S', η') be equivalent, say

$$S = Q(Y, \pi, P; \psi), \qquad S' = Q(Y, \pi, P; \psi')$$

with the corresponding isomorphism $\chi\colon S \to S'$.

By hypothesis, $\chi\xi = \xi'$ which implies $\chi(\alpha, \pi\alpha) = (\alpha, \pi\alpha)$, that is, χ fixes idempotents of S and hence S and S' have the same idempotents. Also $\eta'\chi = \eta$ which implies that for any $\alpha \in \Gamma_p$, $\chi(\alpha, p) = (\alpha_p, p)$ for some $\alpha_p \in \Gamma'_p$. Further,

$$\chi(\alpha, p) = \chi\big((\alpha, pp^{-1})(\alpha, p)\big) = \big(\alpha, pp^{-1}\big)\chi(\alpha, p) = \big(\alpha \wedge \alpha_p, p\big)$$

which together with $\chi(\alpha, p) = (\alpha_p, p)$ gives $\alpha_p = \alpha \wedge \alpha_p$ and thus $\alpha \geq \alpha_p$. We can analogously define α^p by the requirement $\chi^{-1}(\alpha, p) = (\alpha^p, p)$ for any $\alpha \in \Gamma'_p$, and obtain $\alpha \geq \alpha^p$. For any $\alpha \in \Gamma_p$, we then obtain $\alpha \geq \alpha_p \geq (\alpha_p)^p = \alpha$ which yields $\alpha = \alpha_p$. Consequently, χ is the identity mapping, and thus $S = S'$ as semigroups. It follows that $\Gamma_p = \Gamma'_p$ and thus $\mathbf{r}\psi_p = \Gamma_p Y = \Gamma'_p Y = \mathbf{r}\psi'_p$. Also $(\psi_p^{-1}\alpha, p^{-1}) = (\alpha, p)^{-1} = (\psi_p'^{-1}\alpha, p^{-1})$ which yields $\psi_p'^{-1} = \psi_p^{-1}$. Therefore $\psi = \psi'$.

As an example, we now compute all extensions of an arbitrary semilattice Y by the Brandt semigroup B_2 (see II.3.3). For this, according to the results of this section, for a given extension triple (Y, π, B_2), it suffices to find all prehomomorphisms ψ of B_2 into $\Sigma(Y)$ compatible with π.

VI.7.13 Theorem

Let (Y, π, B_2) be an extension triple where Y is a semilattice. Denote the nonzero elements of B_2 by (i, j) with $1 \leq i, j \leq 2$ and let

$$A = \pi^{-1}(1,1), \qquad B = \pi^{-1}(2,2), \qquad C = \pi^{-1}0.$$

Let D be an ideal of B and R be an ideal of A such that $DY \cap C = RY \cap C$. Let θ be an isomorphism of D onto R such that $\theta\alpha \wedge \beta = \alpha \wedge \beta$ for all $\alpha \in D$, $\beta \in DY \cap C$.

Define ψ: $(i, j) \to \psi_{ij}$, $0 \to \psi_0$ on B_2 by

$$\psi_{11} = \iota_{AY}, \qquad \psi_{12} = \theta \cup \iota_{DY\cap C},$$

$$\psi_{21} = \psi_{12}^{-1}, \qquad \psi_{22} = \iota_{BY}, \qquad \psi_0 = \iota_C.$$

Then ψ is a prehomomorphism of B_2 into $\Sigma(Y)$ compatible with π. Conversely, every such can be so constructed.

Proof

Direct part. Note first that ψ_{12} is a one-to-one correspondence of $DY = D \cup (DY \cap C)$ onto $RY = R \cup (RY \cap C)$. The hypotheses on θ and $DY \cap C = RY \cap C$ insure that ψ_{12} is an isomorphism. Consequently, $\psi_{12} \in \Sigma(Y)$ which then implies that ψ maps B_2 into $\Sigma(Y)$. Clearly, $(\psi a)^{-1} = \psi a^{-1}$ for all $a \in B_2$. The property $(\psi a)(\psi b) \leq \psi(ab)$ is straightforward to verify for all 25 possibilities. For example,

$$\psi_{12}\psi_{11} = \psi_{12}\iota_{AY} = \iota_{DY\cap C} \leq \iota_C = \psi_0,$$

$$\psi_{12}\psi_{21} = \psi_{12}\psi_{12}^{-1} = \iota_{RY} \leq \iota_{AY} = \psi_{11},$$

the verification of the remaining cases is just as simple.

Condition 7.3(i) holds obviously. For condition 7.3(ii), we check only the cases $p = (1,1)$ and $p = (1,2)$. Indeed,

$$\pi\psi_{11} = \pi\iota_{AY} = \{((1,1),\alpha)|\alpha \in A\} \cup \{(0,\alpha)|\alpha \in AY \cap C\},$$

$$\theta_{B_2}^{(1,1)}\pi = \{((1,1),\alpha)|\alpha \in A\} \cup \{(0,\alpha)|\alpha \in C\}$$

and the desired inclusion takes place; similarly,

$$\pi\psi_{12} = \{((1,1),\alpha)|\alpha \in D\} \cup \{(0,\alpha)|\alpha \in DY \cap C\},$$

$$\theta_{B_2}^{(2,1)}\pi = \{((1,1),\alpha)|\alpha \in B\} \cup \{(0,\alpha)|\alpha \in C\}$$

and again the desired inclusion holds. The remaining cases are verified similarly. For condition 7.3(iii), we compute

$$\Gamma_{11} = \mathbf{r}\psi_{11} \cap \pi^{-1}(1,1) = AY \cap A = A, \qquad \mathbf{r}\psi_{11} = AY = \Gamma_{11}Y,$$

$$\Gamma_{12} = \mathbf{r}\psi_{12} \cap \pi^{-1}(1,1) = RU \cap A = R, \qquad \mathbf{r}\psi_{12} = RY = \Gamma_{11}Y,$$

and the remaining cases follow just as easily.

Therefore ψ is compatible with π.

Converse. Let ψ be a prehomomorphism of B_2 into $\Sigma(Y)$ compatible with π, and let A, B, C, (i, j), and 0 have the same meaning as above, and let ψ: $(i, j) \to \psi_{ij}$, $0 \to \psi_0$. Define $D = \Gamma_{21}$ and $R = \Gamma_{12}$, so that

$$\mathbf{D} = \mathbf{r}\psi_{21} \cap \pi^{-1}((2,1)(1,2)) = \mathbf{d}\psi_{12} \cap \pi^{-1}((2,2)) = \mathbf{d}\psi_{12} \cap B,$$

$$\mathbf{d}\psi_{12} = \mathbf{r}\psi_{21} = \Gamma_{21}Y = DY$$

and similarly,

$$R = \mathbf{r}\psi_{12} \cap A, \qquad \mathbf{r}\psi_{12} = RY.$$

Since $\mathbf{d}\psi_{12}$ is an ideal of Y, we have that D is an ideal of B; analogously R is an ideal of A. Further, by 7.3(i), $\psi_{11} = \iota_{AY}$, $\psi_{22} = \iota_{BY}$, $\psi_0 = \iota_C$. For any $\alpha \in \mathbf{d}\psi_{12} \cap C$, we have

$$\psi_{12}\alpha = \psi_{12}\iota_C\alpha = \iota_C\alpha = \alpha$$

since $\psi_{12}\iota_C = \psi_{12}\psi_0 \leq \psi_0 = \iota_C$. It follows that $\psi_{12}|_{DY\cap C} = \iota_{DY\cap C}$ which also yields $DY \cap C = RY \cap C$. But then $\theta = \psi_{12}|_D$ is an isomorphism of D onto R. For any $\alpha \in D$ and $\beta \in DY \cap C$, we have $\alpha \wedge \beta \in DY \cap C$ and thus

$$\theta\alpha \wedge \beta = \psi_{12}\alpha \wedge \psi_{12}\beta = \psi_{12}(\alpha\beta) = \alpha \wedge \beta.$$

In particular, $\psi_{21} = \psi_{12}^{-1}$ and $\psi_{12} = \theta \cup \iota_{DY\cap C}$, which completes the verification that ψ is of the form prescribed in the statement of the theorem.

We can illustrate the situation in the above theorem as follows.

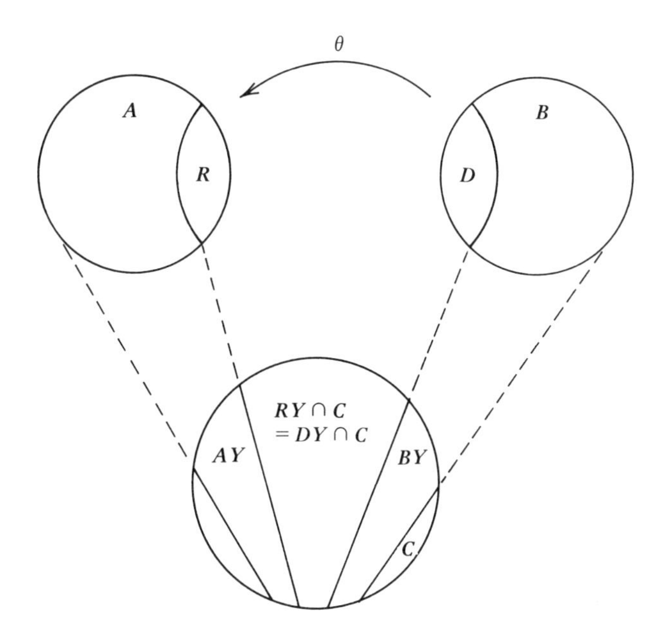

With the ψ in the theorem, we obtain

$$S = A \times \{(1,1)\} \cup R \times \{(1,2)\} \cup D \times \{(2,1)\} \cup B \times \{(2,2)\} \cup C \times \{0\}$$

which can be illustrated by

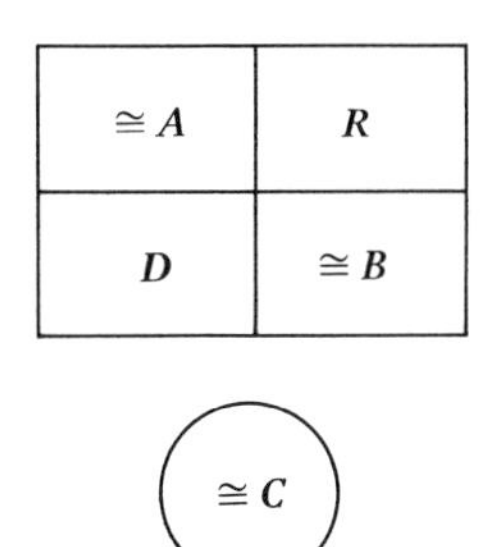

The multiplication in S can be arranged in a 5×5 table using the formula for multiplication and the given parameters. For example,

$$(\alpha, (1,1))(\beta, (1,1)) = (\alpha \wedge \beta, (1,1)),$$

$$(\alpha, (1,2))(\beta, (1,1)) = \left(\theta(\theta^{-1}\alpha \wedge \beta), 0\right),$$

and so on.

VI.7.14 Exercises

(i) Let Y and Z be semilattices. Let $\theta\colon \alpha \to I_\alpha$ be a mapping from Z into the set of all ideals of Y such that $I_\alpha \cap I_\beta \subseteq I_{\alpha\beta}$. Show that θ induces a prehomomorphism of Z into $\Sigma(Y)$, and that conversely, every prehomomorphism of Z into $\Sigma(Y)$ can be so obtained.

(ii) Let Y be a semilattice. Show that $\Sigma(Y)$ is a Clifford semigroup if and only if no two distinct ideals of Y are isomorphic. In such a case, construct the Clifford representation for $\Sigma(Y)$.

(iii) Let Y be an orthogonal sum of a three-element and a two-element chain. Find explicitly all normal extensions of Y by B_2.

(iv) Call a semigroup S *group-congruence-free* if ω is the only group congruence on S. Prove the following statements for a semilattice E.

(α) The type of a group-congruence-free normal extension of E is contained in $E_{\Phi(E)}\omega$.

(β) The type of a maximal essential group-congruence-free normal extension of E coincides with $E_{\Phi(E)}\omega$.

(γ) Every full inverse subsemigroup of $E_{\Phi(E)}\omega$ is the type of some essential group-congruence-free normal extension of E.

All the results here, except for the last theorem, are due to Allouch [2]; see also Allouch [1]. Normal extensions of semilattices were also studied by Green [1] and O'Carroll [5], [7], [8].

VI.8 A CATEGORY OF INVERSE SEMIGROUPS

In the system $(Y, \pi, P; \psi)$ constructed in the preceding section, we may take Y to be a semilattice and P an E-disjunctive inverse semigroup, and equip the class of all such objects with morphisms, thereby obtaining a category. We will prove below that this category is equivalent to the category of inverse semigroups and homomorphisms compatible with the greatest idempotent pure congruence. As special cases, we consider P to be a group or a Clifford semigroup, which corresponds to E-unitary and E-reflexive inverse semigroups, respectively.

Recall from III.4.5 that τ has been used to denote the greatest idempotent pure congruence on any inverse semigroup S, and that it coincides with the syntactic congruence on S relative to the set of all idempotents of S. Also recall from III.4.14 that in the case $\tau = \varepsilon$, S is called E-disjunctive.

VI.8.1 Notation

Let $\mathfrak{D}$ be the category whose objects are quadruples $(Y, \pi, P; \psi)$ satisfying the following requirements:

(i) Y is a semilattice (whose operation is denoted by $\wedge$),
(ii) π is a homomorphism of Y onto E_P,
(iii) P is an E-disjunctive inverse semigroup,
(iv) $\psi: p \to \psi_p$ is a prehomomorphism of P into $\Sigma(Y)$ compatible with π.

A morphism in $\mathfrak{D}$ is a pair (θ, ω): $(Y, \pi, P; \psi) \to (Y', \pi', P'; \psi')$ such that $\theta: Y \to Y'$ and $\omega: P \to P'$ are homomorphisms satisfying

$$(\psi_p \alpha)\theta = \psi'_{p\omega}(\alpha\theta) \qquad (\theta \in \mathbf{d}\psi_p).$$

Notice that ψ_p acts on the left and θ on the right.

Routine verification shows that $\mathfrak{D}$ is indeed a category. We continue the earlier practice of writing $p\alpha$ instead of $\psi_p\alpha$, so the last equation above takes on the simple form

$$(p\alpha)\theta = (p\omega)(\alpha\theta) \qquad (\alpha \in \mathbf{d}\psi_p).$$

The following concept will be convenient.

VI.8.2 Definition

Let ρ stand for any of the congruences considered in Chapter III on any inverse semigroup and having some extremal property, for instance, σ, η, ν, τ, and so on. A homomorphism φ of an inverse semigroup S into an inverse semigroup T is *ρ-preserving* if for any $a, b \in S$, $a\rho b$ implies $(a\varphi)\rho(b\varphi)$.

VI.8.3 Notation

Let $\mathfrak{S}$ be the category whose objects are inverse semigroups and whose morphisms are τ-preserving homomorphisms of inverse semigroups.

In order to show that the categories $\mathfrak{D}$ and $\mathfrak{S}$ are equivalent, we first devise a functor from $\mathfrak{D}$ to $\mathfrak{S}$.

VI.8.4 Lemma

For every $(Y, \pi, P; \psi) \in \operatorname{Ob} \mathfrak{D}$ let $Q(Y, \pi, P; \psi)$ be the semigroup constructed in 7.5. For a $\mathfrak{D}$-morphism $(\theta, \omega) : (Y, \pi, P; \psi) \to (Y', \pi', P'; \psi')$, let

$$Q(\theta, \omega): (\alpha, p) \to (\alpha\theta, p\omega) \qquad [(\alpha, p) \in Q(Y, \pi, P; \psi)].$$

Then Q is a functor from $\mathfrak{D}$ to $\mathfrak{S}$.

Proof. We have seen in 7.7 that $Q(Y, \pi, P; \psi)$ is an inverse semigroup if P is an inverse semigroup. Let (θ, ω) be as in the statement of the lemma, and let $\varphi = Q(\theta, \omega)$. For any $(\alpha, p), (\beta, q) \in S = Q(Y, \pi, P; \psi)$, we obtain

$$\begin{aligned} (\alpha, p)\varphi(\beta, q)\varphi &= (\alpha\theta, p\omega)(\beta\theta, q\omega) \\ &= \big((p\omega)\big((p\omega)^{-1}(\alpha\theta) \wedge \beta\theta\big), (p\omega)(q\omega)\big) \\ &= \big((p(p^{-1}\alpha \wedge \beta))\theta, (pq)\omega\big) = ((\alpha, p)(\beta, q))\varphi \end{aligned}$$

and φ is a homomorphism.

We have seen in 7.8 that the congruence ρ defined on S by

$$(\alpha, p)\rho(\beta, q) \Leftrightarrow p = q$$

is idempotent pure. Hence $\rho \subseteq \tau$ which yields that τ/ρ is a congruence on S/ρ. Assume that $((\alpha, p)\rho)(\tau/\rho)((\beta, \pi\beta)\rho)$ and recall that $(\beta, \pi\beta) \in E_S$. Then $(\alpha, p)\tau(\beta, \pi\beta)$ and thus $(\alpha, p) \in E_S$ since τ is idempotent pure. It follows that τ/ρ is idempotent pure and thus $\tau/\rho = \varepsilon$ since $S/\rho \cong P$ and P is assumed to be E-disjunctive. But then $\rho = \tau$.

Hence $(\alpha, p)\tau(\beta, q)$ in S implies $p = q$ and thus

$$(\alpha, p)\varphi = (\alpha\theta, p\omega)\tau(\beta\theta, p\omega) = (\beta, p)\varphi.$$

The same type of statement concerning τ is valid for $S' = Q(Y', \pi', P'; \psi')$. Consequently, φ is τ-preserving and thus is an $\mathfrak{S}$-morphism.

The remaining axioms for a functor follow without difficulty.

For a functor from $\mathfrak{S}$ to $\mathfrak{D}$, we have the following construction.

VI.8.5 Lemma

For every inverse semigroup S, let $T(S) = (E_S, \pi, S/\tau; \psi)$ where $\pi = \tau^{\#}|_{E_S}$ and ψ is as defined in 7.10 relative to $\rho = \tau$. For an $\mathfrak{S}$-morphism $\varphi: S \to S'$, let $T(\varphi) = (\varphi|_{E_S}, \omega)$, where ω is defined by

$$\omega: s\tau \to s\varphi\tau \qquad (s \in S).$$

Then T is a functor from $\mathfrak{S}$ to $\mathfrak{D}$.

Proof. By III.4.16, S/τ is E-disjunctive, which together with 7.10 gives that $T(S) = (E_S, \pi, S/\tau; \psi) \in \operatorname{Ob} \mathfrak{D}$.

Next let $\varphi \in \operatorname{Hom} \mathfrak{S}$, say $\varphi: S \to S'$ for inverse semigroups S and S'. Let $\theta = \varphi|_{E_S}$ and ω be defined as in the statement of the lemma. Note that the hypothesis that φ is τ-preserving implies that ω is single valued. It follows easily that ω is a homomorphism of S/τ into S'/τ. Let $p \in S/\tau$ and $e \in \mathbf{d}\psi_p$. Then $e = s^{-1}sf$ for some $s\tau = p$ and $f \in E_S$. Now writing pe for $\psi_p e$, we have

$$(pe)\theta = (sfs^{-1})\varphi = (s\varphi)(f\varphi)(s\varphi)^{-1} = (p\omega)\left[(s\varphi)^{-1}(s\varphi)(f\varphi)\right]$$

$$= (p\omega)(s^{-1}sf)\theta = (p\omega)(e\theta)$$

which shows that $T(\varphi) = (\varphi|_{E_S}, \omega) \in \operatorname{Hom} \mathfrak{D}$.

The remaining axioms for a functor can be easily checked.

VI.8.6 Lemma

For every $(Y, \pi, P; \psi) \in \operatorname{Ob} \mathfrak{D}$, define $\zeta(Y, \pi, P; \psi) = (\xi, \omega)$, where

$$\xi: \alpha \to (a, \pi\alpha) \qquad (\alpha \in Y),$$

$$\omega: p \to (\quad, p)\tau \qquad (p \in P),$$

(the entry in the empty space is of no importance here). Then ζ is a natural equivalence of the functors $I_{\mathfrak{D}}$ and TQ.

Proof. Let $(Y, \pi, P; \psi) \in \operatorname{Ob} \mathfrak{D}$. Then $TQ(Y, \pi, P; \psi) = (Y', \pi', P', \psi')$, where

$$S = Q(Y, \pi, P; \psi),$$

$$Y' = \{(\alpha, \pi\alpha) \in S | \alpha \in Y\},$$

$$\pi': (\alpha, \pi\alpha) \to (\alpha, \pi\alpha)\tau,$$

$$P' = S/\tau,$$

and, we claim, that for all $p \in P$,

$$\psi'_{(\ ,p)\tau}: (\alpha, \pi\alpha) \to (p\alpha, \pi(p\alpha)) \qquad (\alpha \in \mathbf{d}\psi_p). \tag{20}$$

In order to prove the claim, we first note that by 7.8, the mapping $\eta: S \to P$, defined by $\eta: (\alpha, p) \to p$, is a homomorphism of S onto P which induces an idempotent pure congruence on S. Since P is E-disjunctive, it follows from III.4.15 that η induces the congruence τ on S. Consequently,

$$(\alpha, p)\tau(\beta, q) \Leftrightarrow p = q. \tag{21}$$

Let $p \in P$ and consider $\mathbf{d}\psi'_{(\ ,p)\tau}$. Let $(\alpha, \pi\alpha) \in \mathbf{d}\psi'_{(\ ,p)\tau}$. It follows from 7.10 that there exists $(\delta, p) \in S$ such that

$$(\alpha, \pi\alpha) = (\delta, p)^{-1}(\delta, p)(\alpha, \pi\alpha)$$

which easily implies that

$$(\alpha, \pi\alpha) = \left(p^{-1}\delta \wedge \alpha, p^{-1}p(\pi\alpha)\right).$$

The last equality indicates that $\alpha \in \mathbf{d}\psi_p$. Conversely, let $\alpha \in \mathbf{d}\psi_p$. Since $\mathbf{d}\psi_p = \Delta_p Y$, there exists $\gamma \in \Delta_p$ such that $\alpha \leq \gamma$. Now $\gamma \in \Delta_p$ implies $p\gamma \in \Gamma_p$ and hence $(p\gamma, p) \in S$. Using 7.4, we obtain $\gamma \in \mathbf{d}\psi_p \cap \pi^{-1}(p^{-1}p)$, whence

$$\begin{aligned}(p\gamma, p)^{-1}(p\gamma, p)(\alpha, \pi\alpha) &= (\gamma, p^{-1})(p\gamma, p)(\alpha, \pi\alpha)\\ &= \left(p^{-1}(p\gamma \wedge p\gamma), p^{-1}p\right)(\alpha, \pi\alpha)\\ &= ((\pi\gamma)\gamma, \pi\gamma)(\alpha, \pi\alpha)\\ &= ((\pi\gamma)(\gamma \wedge \alpha), (\pi\gamma)(\pi\alpha))\\ &= (\alpha, \pi\alpha)\end{aligned}$$

so that $(\alpha, \pi\alpha) \in \mathbf{d}\psi'_{(\ ,p)\tau}$. We have proved that

$$\mathbf{d}\psi'_{(\ ,p)\tau} = \left\{(\alpha, \pi\alpha) | \alpha \in \mathbf{d}\psi_p\right\}.$$

Now let $\alpha \in \mathbf{d}\psi_p$ and let γ be as above. Then by 7.10 and 7.3, we obtain

$$\begin{aligned}\psi'_{(\ ,p)\tau}(\alpha, \pi\alpha) &= (p\gamma, p)(\alpha, \pi\alpha)(p\gamma, p)^{-1}\\ &= \left(p(p^{-1}p\gamma \wedge \alpha), p(\pi\alpha)\right)(\gamma, p^{-1})\\ &= (p\alpha, p(\pi\alpha))(\gamma, p^{-1})\\ &= \left(p(\pi\alpha)\left((\pi\alpha)^{-1}p^{-1}p\alpha \wedge \gamma\right), p(\pi\alpha)p^{-1}\right)\\ &= (p\alpha, \pi(p\alpha)).\end{aligned}$$

This concludes the proof of relation (20).

It follows that $\psi'_{p\omega}(\alpha\xi) = (p\alpha)\xi$ for all $p \in P$ and $\alpha \in Y$, which shows that $(\xi, \omega) \in \operatorname{Hom} \mathfrak{D}$. Relation (21) implies that ω defined in the statement of this lemma is an isomorphism of P onto S/τ. Recall from 7.8 that ξ is an isomorphism of Y onto E_S. It then follows that (ξ, ω) is an isomorphism of $(Y, \pi, P; \psi)$ onto $(Y', \pi', P'; \psi')$.

It remains to show commutativity of the diagram

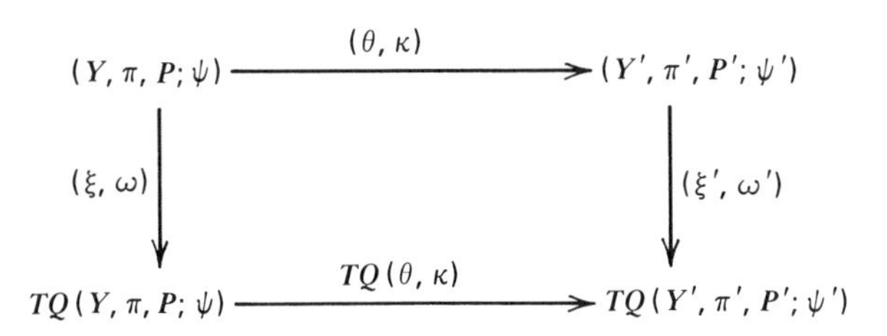

for a given $(\theta, \kappa) \in \operatorname{Hom} \mathfrak{D}$. Letting $TQ(\theta, \kappa) = (\theta', \kappa')$, for any $p \in P$ we have

$$p\kappa\omega' = (\quad, p\kappa)\tau = (\quad, p)\tau\kappa' = p\omega\kappa'$$

and for any $\alpha \in Y$,

$$\alpha\theta\xi' = (\alpha\theta, \pi(\alpha\theta)) = (\alpha, \pi\alpha)\theta' = \alpha\xi\theta'$$

which establishes commutativity of the diagram.

VI.8.7 Lemma

For every $S \in \operatorname{Ob} \mathfrak{S}$, define $\chi = \chi(S)$ by

$$\chi \colon s \to (ss^{-1}, s\tau) \qquad (s \in S).$$

Then χ is a natural equivalence of the functors $I_{\mathfrak{S}}$ and QT.

Proof. For any inverse semigroup S, the case at hand corresponds to $\rho = \tau$ in 7.11, and we have that χ is an isomorphism of S onto $QT(S)$.

Let $\varphi \in \operatorname{Hom} \mathfrak{S}$, say $\varphi \colon S \to S'$. For any $s \in S$, we have

$$\begin{aligned}(s\varphi)[\chi(S')] &= \left((s\varphi)(s\varphi)^{-1}, s\varphi\tau\right) = (ss^{-1}, s\tau)[QT(\varphi)] \\ &= s[\chi(S)][QT(\varphi)]\end{aligned}$$

which proves commutativity of the diagram

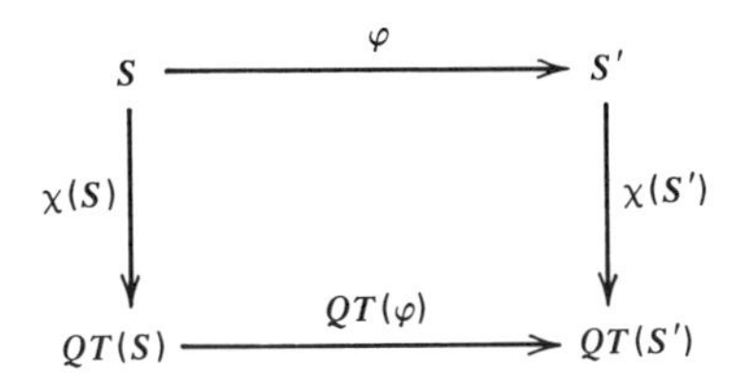

Now collecting 8.4–8.7, we deduce the desired result.

VI.8.8 Theorem

The quadruple (Q, T, ζ, χ) is an equivalence of the categories $\mathfrak{D}$ and $\mathfrak{S}$.

The above equivalence of $\mathfrak{D}$ and $\mathfrak{S}$ may be restricted to some subcategories of particular interest. We consider below two such subcategories. For this, we need some new concepts.

VI.8.9 Definition

A prehomomorphism φ of an inverse monoid S into an inverse monoid T is a *monoid prehomomorphism* if φ maps the identity of S onto the identity of T.

VI.8.10 Definition

A triple $(Y, G; \psi)$ is a *unitary triple* if Y is a semilattice, G is a group, and ψ: $G \to \Sigma(Y)$ is a monoid prehomomorphism. For two unitary triples, (θ, ω): $(Y, G, \psi) \to (Y', G', \psi)$ is a morphism if θ: $Y \to Y'$ and ω: $G \to G'$ are homomorphisms such that for all $g \in G$,

$$(g\alpha)\theta = (g\omega)(\alpha\theta) \qquad (\alpha \in \mathbf{d}\psi_g)$$

in the notation used earlier.

Easy verification shows that unitary triples and their morphisms form a category.

VI.8.11 Notation

Let $\mathfrak{U}$ denote the category of unitary triples and their morphisms. Denote by $\mathfrak{E}$ the category of E-unitary inverse semigroups and their homomorphisms.

VI.8.12 Theorem

The categories $\mathfrak{U}$ and $\mathfrak{E}$ are equivalent.

Proof. We will verify that for the needed equivalence quadruple, we may take the suitable restriction of the quadruple in 8.8. First consider $(Y, \pi, G; \psi) \in \mathrm{Ob}\,\mathfrak{D}$ where G is a group. Since G has only one idempotent, π maps Y onto the identity of G and the reference to π may safely be omitted. We thus arrive at the triple $(Y, G; \psi)$.

Since $\pi^{-1}1 = Y$, condition 7.3(i) becomes $\psi_1 = \iota_Y$, that is ψ is a monoid prehomomorphism. Condition 7.3(ii) is easily seen to be automatically satisfied. Again $\pi^{-1}1 = Y$ implies in condition 7.3(iii) that $\Gamma_p = \mathbf{r}\psi_p$. Hence $S = Q(Y, \pi, G; \psi) = Q(Y, G; \psi)$ is of the form

$$S = \{(\alpha, g) \in Y \times G | \alpha \in \mathbf{r}\psi_g\}$$

with the same multiplication as in 7.5. It follows at once that S is an E-unitary inverse semigroup. Hence for $(Y, G; \psi) \in \operatorname{Ob} \mathfrak{U}$, we get $Q(Y, G; \psi) \in \operatorname{Ob} \mathfrak{E}$. Since we did not change morphisms, Q maps $\operatorname{Hom} \mathfrak{U}$ into $\operatorname{Hom} \mathfrak{E}$.

Conversely, let $S \in \operatorname{Ob} \mathfrak{E}$. In view of III.7.2, the least group congruence σ on S is idempotent pure and thus $\sigma = \tau$ since we always have $\tau \subseteq \sigma$. Hence we have $T(S) = (E_S, \pi, S/\sigma; \psi)$ and we may again omit all reference to π. In view of the first part of the proof, we deduce $T(S) = (E_S, S/\sigma; \psi) \in \operatorname{Ob} \mathfrak{U}$.

Next let $\varphi \in \operatorname{Hom} \mathfrak{E}$, say $\varphi \colon S \to S'$ is a homomorphism of E-unitary inverse semigroups. If $a\sigma b$ in S, then $ae = be$ for some $e \in E_S$ and thus $(a\varphi)(e\varphi) = (b\varphi)(e\varphi)$ so that $(a\varphi)\sigma(b\varphi)$. Hence φ is σ-preserving. But in an E-unitary inverse semigroup $\tau = \sigma$, and thus φ is τ-preserving. Hence $\varphi \in \operatorname{Hom} \mathfrak{S}$ which gives that $T(\varphi) \in \operatorname{Hom} \mathfrak{D}$ and thus also $T(\varphi) \in \operatorname{Hom} \mathfrak{U}$.

The equivalence of $\mathfrak{U}$ and $\mathfrak{E}$ now follows from 8.8.

We may restrict P in $(Y, \pi, P; \psi)$ to be a Clifford semigroup, thus obtaining a subcategory of $\mathfrak{D}$, and restrict S in $\operatorname{Ob} \mathfrak{S}$ to be E-reflexive. In view of III.8.3 and 8.8 here, we obtain the equivalence of the resulting categories.

That not all homomorphisms of inverse semigroups are τ-preserving can be seen in the following simple instance.

VI.8.13 Example

Let G be a group with at least three elements, $Y_2 = \{0, 1\}$ be a chain, $S = Y_2 \times G$, and $T = G^0$. Define a mapping χ by

$$\chi\colon (1, g) \to g, \qquad \chi\colon (0, g) \to 0 \qquad (g \in G).$$

Clearly, χ is a homomorphism of S onto T. Let $g \in G$, $g \neq 1_G$. It can be easily verified that $(1, g)\tau(0, g)$. Also, for any $h \in G$, $h \neq g^{-1}$, we have

$$(1, g)\chi h = gh \neq 1_G, \qquad (0, g)\chi h = 0h = 0$$

which shows that $(1, g)\chi$ and $(0, g)\chi$ are not τ-related.

It will be convenient to have the following concept.

VI.8.14 Definition

If S is an inverse semigroup isomorphic to $Q(Y, \pi, P; \psi)$, where Y is a semilattice and P is E-disjunctive, we call the latter a *Q-representation* of S.

VI.8.15 Exercises

(i) Let $(Y_i, \pi_i, P_i; \psi_i) \in \operatorname{Ob} \mathfrak{D}$ for $i \in I$. Show that

$$\prod_{i \in I} Q(Y_i, \pi_i, P_i; \Psi_i) \cong Q\Big(\prod_{i \in I} Y_i, \pi, \prod_{i \in I} P_i; \psi\Big)$$

for suitable π and ψ.

(ii) Let $S = [Y; G_\alpha, \varphi_{\alpha,\beta}]$ be a Clifford semigroup. Show that S is E-disjunctive if and only if it satisfies: if $\alpha, \beta \in Y$ are such that $\ker \varphi_{\gamma,\alpha\gamma} = \ker \varphi_{\gamma,\beta\gamma}$ for all $\gamma \in Y$, then $\alpha = \beta$. Also show that the latter condition holds if for no $\alpha > \beta$, $\varphi_{\alpha,\beta}$ is one-to-one.

(iii) Show that the following classes of inverse semigroups are E-disjunctive: groups, Brandt semigroups, non-E-unitary Reilly semigroups. Find some more classes of E-disjunctive inverse semigroups.

(iv) Find all normal extensions of an arbitrary semilattice by a two-element group.

(v) Construct the Q-representation of the bicyclic semigroup.

Theorem 8.12 has in Petrich [14] a direct proof. The main part of it, namely the Q-representation of an E-unitary inverse semigroup, stems from Petrich–Reilly [1]; for an alternative form see the earlier paper by Žitomirskiĭ [8]. For other categories of inverse semigroups, consult Batbedat [1], McAlister [8], [10], and Širjajev [5].

VII

E-UNITARY INVERSE SEMIGROUPS

1. Besides the representation of the form $Q(Y, G; \psi)$, E-unitary inverse semigroups admit the important P-representation, in symbols $P(Y, G; X)$. Here we have an equivalence of the categories of the McAlister triples $(Y, G; X)$ and their morphisms and E-unitary inverse semigroups and their homomorphisms.

2. Congruences on P-semigroups $P(Y, G; X)$ admit an elegant characterization in terms of a congruence τ on Y and a collection of subgroups of G indexed by Y/τ, satisfying relatively simple conditions. These conditions simplify considerably for some special kinds of congruences, such as idempotent separating ones, and so on.

3. The translational hull of $S = P(Y, G; X)$ is isomorphic to $P(\mathcal{R}_Y, G; G\mathcal{R}_Y)$ where G acts on $\mathcal{R}_Y$ in a natural way. Similarly, $\hat{S}$ is isomorphic to $P(\mathcal{I}_Y, G; G\mathcal{I}_Y)$ under the same action. In particular, we see that both $\Omega(S)$ and $\hat{S}$ are again E-unitary inverse semigroups.

4. All E-unitary covers of an inverse semigroup S over a group G can be constructed by means of a full prehomomorphism of G into $C(S)$. Every inverse semigroup admits an E-unitary cover, which may be further restricted. The theory may be illustrated by calculation of all E-unitary covers of a Brandt semigroup.

5. There is a whole spectrum of special cases of E-unitary inverse semigroups arising from the restrictions imposed on their σ-classes, or the parameters figuring in $P(Y; G; X)$ or $Q(Y, G; \psi)$. In this way, we obtain several classes of E-unitary inverse semigroups which have some interesting features and for which alternative constructions may be devised.

6. One of these classes is the class of F-inverse semigroups. Analogous to McAlister triples for E-unitary inverse semigroups, one may define here

F-pairs and their morphisms. The category so arising is equivalent to the category of *F*-inverse semigroups and their homomorphisms preserving maximal elements of σ-classes. *E*-unitary covers have their analogue for *F*-inverse semigroups.

VII.1 *P*-SEMIGROUPS

We introduce McAlister triples and construct *P*-semigroups $P(Y, G; X)$. This is used to devise an equivalence between the category $\mathfrak{M}$ of McAlister triples and their morphisms and *E*-unitary inverse semigroups and their homomorphisms. We thus arrive at a new construction of *E*-unitary inverse semigroups related to the one in VI.8. In particular, the categories $\mathfrak{M}$, $\mathfrak{E}$, and $\mathfrak{U}$ are equivalent.

VII.1.1 Definition

Let G be a group and X be a partially ordered set. Then *G acts on X by order automorphisms on the left* if there exists a function $G \times X \to X$, denoted by $(g, \alpha) \to g\alpha$ such that

(i) $\alpha \to g\alpha \quad (\alpha \in X)$ is an automorphism of X for each $g \in G$,
(ii) $(gh)\alpha = g(h\alpha) \quad (g, h \in G, \alpha \in X)$.

If Y is a nonempty subset of X, we write

$$gY = \{g\alpha | \alpha \in Y\} \qquad (g \in G),$$

$$GY = \{g\alpha | g \in G, \alpha \in Y\}.$$

Note that if we write $\varphi_g \alpha$ instead of $g\alpha$, then the above is equivalent to the function $\varphi : g \to \varphi_g$ being a homomorphism of G into the group of order automorphisms of X. In view of (ii) above, we may write $gh\alpha$.

VII.1.2 Lemma

With the setup of 1.1, let $g \in G$, $\alpha, \beta \in X$ and assume that $\alpha \wedge \beta$ exists. Then $g(\alpha \wedge \beta) = g\alpha \wedge g\beta$.

Proof. The proof of this lemma is left as an exercise.

VII.1.3 Lemma

Let G be a group acting on a partially ordered set X by order automorphisms on the left. Let Y be a subsemilattice and an ideal of X such that $GY = X$.

Then the following conditions are equivalent.

(i) $gY \cap Y \neq \varnothing$ for all $g \in G$.
(ii) Y is an essential ideal of X.
(iii) X is lower directed.

Proof. (i) *implies* (ii). Let $\alpha \in X$. By the condition $GY = X$ there exist $g \in G$ and $\beta \in Y$ such that $\alpha = g\beta$. Hence by (i), there exist $\gamma, \delta \in Y$ such that $g\gamma = \delta$. It follows that

$$g(\beta \wedge \gamma) \leq g\gamma = \delta \in Y$$

and thus $g(\beta \wedge \gamma) \in Y$. Further, $g(\beta \wedge \gamma) \leq g\beta = \alpha$ and hence $g(\beta \wedge \gamma)$ is the required element.

(ii) *implies* (iii). Let $\alpha, \beta \in X$. By (ii), there exist $\gamma, \delta \in Y$ such that $\gamma \leq \alpha$ and $\delta \leq \beta$. But then $\gamma \wedge \delta \leq \alpha$ and $\gamma \wedge \delta \leq \beta$ so X is lower directed.

(iii) *implies* (i). Let $g \in G$ and $\alpha \in Y$. By (iii), there exists $\beta \in X$ such that $\beta \leq \alpha$ and $\beta \leq g^{-1}\alpha$. Now $\beta \leq \alpha$ and $g\beta \leq \alpha$ imply $\beta, g\beta \in Y$ so that $gY \cap Y \neq \varnothing$.

We now arrive at the concept which is of fundamental importance for our discussion.

VII.1.4 Definition

Let G be a group acting by order automorphisms on a partially ordered set X on the left. Let Y be a subsemilattice and an essential ideal of X such that $GY = X$. We call $(Y, G; X)$ a *McAlister triple*.

According to 1.3, in this definition, we can substitute "essential" by "lower directed" for X or by the condition $gY \cap Y \neq \varnothing$ for all $g \in G$. Both "essential" and "lower directed" make the statement of the definition somewhat shorter, but for the practical manipulation, the third condition is usually more easily applied. In fact, this condition means: for every g there exists $\alpha \in Y$ such that $g\alpha \in Y$. We will now provide these triples with appropriate morphisms.

VII.1.5 Definition

Let $(Y, G; X)$ and $(Y', G'; X')$ be McAlister triples. An ordered pair (θ, ω) is a *morphism of* $(Y, G; X)$ *into* $(Y', G'; X')$ if

(i) θ is an order preserving mapping of X into X',
(ii) $\theta|_Y$ is a (semilattice) homomorphism of Y into Y',
(iii) ω is a homomorphism of G into G',
(iv) $(g\alpha)\theta = (g\omega)(\alpha\theta)$ for all $g \in G$, $\alpha \in X$.

The axioms for a category for McAlister triples and their morphisms can be easily verified.

VII.1.6 Notation

Let $\mathfrak{M}$ be the category of McAlister triples and their morphisms.

We will show that the categories $\mathfrak{M}$ and $\mathfrak{E}$ (see VI.8.11) are equivalent. For this, we first need a functor from $\mathfrak{M}$ to $\mathfrak{E}$, which we now define.

VII.1.7 Definition

For each $(Y, G; X) \in \text{Ob}\,\mathfrak{M}$, let $P(Y, G; X)$ be the set

$$\{(\alpha, g) \in Y \times G | g^{-1}\alpha \in Y\}$$

with multiplication

$$(\alpha, g)(\beta, h) = (\alpha \wedge g\beta, gh).$$

For each $(\theta, \omega) \in \text{Hom}\,\mathfrak{M}$, say $(\theta, \omega) : (Y, G; X) \to (Y', G'; X')$, let $P(\theta, \omega)$ be the mapping

$$P(\theta, \omega) : (\alpha, g) \to (\alpha\theta, g\omega) \qquad [(\alpha, g) \in P(Y, G; X)].$$

VII.1.8 Lemma

With the notation just introduced, P is a functor from $\mathfrak{M}$ to $\mathfrak{E}$.

Proof. Let $(Y, G; X) \in \text{Ob}\,\mathfrak{M}$, and let $S = P(Y, G; X)$.

In order to see that S is closed under this multiplication, let $(\alpha, g), (\beta, h) \in S$. Then $g^{-1}\alpha \in Y$ and thus $g^{-1}\alpha \wedge \beta \in Y$; let $\gamma = g(g^{-1}\alpha \wedge \beta)$. Then $g^{-1}\gamma \leq g^{-1}\alpha$ so that $\gamma \leq \alpha$. Also $g^{-1}\gamma \leq \beta$ and thus $\gamma \leq g\beta$. If $\delta \leq \alpha$ and $\delta \leq g\beta$, then $g^{-1}\delta \leq g^{-1}\alpha \wedge \beta$ and hence $\delta \leq \gamma$. Consequently, $\gamma = \alpha \wedge g\beta$, that is, $\alpha \wedge g\beta$ exists. Further,

$$(gh)^{-1}(\alpha \wedge g\beta) = h^{-1}g^{-1}(\alpha \wedge g\beta) \leq h^{-1}g^{-1}g\beta = h^{-1}\beta \in Y$$

so that $(gh)^{-1}(\alpha \wedge g\beta) \in Y$. Hence $(\alpha \wedge g\beta, gh) \in S$ and S is closed under our multiplication.

For any elements of S, we get

$$[(\alpha, g)(\beta, h)](\gamma, t) = (\alpha \wedge g\beta, gh)(\gamma, t)$$
$$= ((\alpha \wedge g\beta) \wedge gh\gamma, ght),$$
$$(a, g)[(\beta, h)(\gamma, t)] = (\alpha, g)(\beta \wedge h\gamma, ht)$$
$$= (\alpha \wedge g(\beta \wedge h\gamma), ght)$$

which in view of I.2.2 and 1.2 implies associativity of this multiplication.

If $(\alpha, g) \in S$, then clearly $(g^{-1}\alpha, g^{-1}) \in S$ and by straightforward multiplication, we see that $(g^{-1}\alpha, g^{-1})$ is an inverse of (α, g). Hence S is regular. An equally simple argument shows that

$$E_S = \{(\alpha, 1) | \alpha \in Y\}$$

with the product $(\alpha, 1)(\beta, 1) = (\alpha \wedge \beta, 1)$ so that idempotents commute. Consequently, S is an inverse semigroup with

$$(\alpha, g)^{-1} = (g^{-1}\alpha, g^{-1}) \qquad [(\alpha, g) \in S].$$

If $(\alpha, g), (\beta, 1) \in S$ are such that $(\alpha, g)(\beta, 1) = (\beta, 1)$, then the product of the second components immediately yields $g = 1$, that is, $(\alpha, g) \in E_S$. Therefore, S is an E-unitary inverse semigroup, that is $P(Y, G; X) \in \mathrm{Ob}\, \mathfrak{E}$.

Next let $(\theta, \omega) \in \mathrm{Hom}\, \mathfrak{M}$, say $(\theta, \omega) : (Y, G; X) \to (Y', G', X')$. Then for any $(\alpha, g), (\beta, h) \in P(Y, G; X)$, first

$$(g\omega)^{-1}(\alpha\theta) = (g^{-1}\omega)(\alpha\theta) = (g^{-1}\alpha)\theta \in Y'$$

so that $(\alpha\theta, g\omega) \in P(Y', G'; X')$. Secondly, letting $\varphi = P(\theta, \omega)$, we obtain

$$(\alpha, g)\varphi(\beta, h)\varphi = (\alpha\theta, g\omega)(\beta\theta, h\omega) = (\alpha\theta \wedge (g\omega)(\beta\theta), (g\omega)(h\omega))$$
$$= \big((g\omega)\big((g\omega)^{-1}(\alpha\theta) \wedge (\beta\theta)\big), (gh)\omega\big)$$
$$= \big((g\omega)\big((g^{-1}\alpha)\theta \wedge (\beta\theta)\big), (gh)\omega\big)$$
$$= \big((g\omega)(g^{-1}\alpha \wedge \beta)\theta, (gh)\omega\big) = \big(g(g^{-1}\alpha \wedge \beta)\theta, (gh)\omega\big)$$
$$= ((\alpha \wedge g\beta)\theta, (gh)\omega) = (\alpha \wedge g\beta, gh)\varphi$$
$$= [(\alpha, g)(\beta, h)]\varphi.$$

Consequently, $P(\theta, \omega)$ is a homomorphism of $P(Y, G; X)$ into $P(Y', G'; X')$.

The remaining axioms for a functor are clearly satisfied.

We now devise a functor from $\mathfrak{E}$ to $\mathfrak{M}$.

VII.1.9 Definition

Let $S \in \operatorname{Ob} \mathfrak{E}$, $G = S/\sigma$, $E = E_S$. For each $g \in G$ and $e \in E$, let

$$ge = \{(s^{-1}es, g(s\sigma)) | s \in S\}.$$

Further let

$$\mathscr{X} = \{ge | g \in G, e \in E\}, \qquad \mathscr{Y} = \{1e | e \in E\}$$

with $\mathscr{X}$ ordered by inclusion, and let G act on $\mathscr{X}$ by

$$g(he) = (gh)e \qquad (g, h \in G, e \in E).$$

Let $R(S) = (\mathscr{Y}, G; \mathscr{X})$ with this action.

Next let $\varphi \in \operatorname{Hom} \mathfrak{E}$, say $\varphi: S \to S'$ is a homomorphism. Define ω and θ by

$$(s\sigma)\omega = (s\varphi)\sigma \qquad (s \in S),$$

$$(ge)\theta = (g\omega)(e\varphi) \qquad (g \in G, e \in E),$$

and let $R(\varphi) = (\theta, \omega)$.

VII.1.10 Lemma

With the notation just introduced, R is a functor from $\mathfrak{E}$ to $\mathfrak{M}$.

Proof. Let $S \in \operatorname{Ob} \mathfrak{E}$, $\mathscr{G} = S/\sigma$, $E = E_S$. Let $g, h, k \in \mathscr{G}$ and $e, f \in E$. Assume that $ge \subseteq hf$. Since $(e, g) \in ge$, we get $(e, g) \in hf$ which implies that $(e, g) = (s^{-1}fs, h(s\sigma))$ for some $s \in S$. Hence $e = s^{-1}fs$, $g = h(s\sigma)$. Conversely, suppose that $e = s^{-1}fs$ and $g = h(s\sigma)$ for some $s \in S$. Then for any $t \in S$, we have

$$\begin{aligned}(t^{-1}et, g(t\sigma)) &= (t^{-1}s^{-1}fst, h(s\sigma)(t\sigma)) \\ &= ((st)^{-1}f(st), h(st)\sigma) \in hf\end{aligned}$$

so that $ge \subseteq hf$. We have proved

$$ge \subseteq hf \Leftrightarrow e = s^{-1}fs, \quad g = h(s\sigma) \quad \text{for some } s \in S. \tag{1}$$

It follows that $ge \subseteq hf$ implies $(kg)e \subseteq (kh)f$, so by the definition of the action of $\mathcal{G}$ on $\mathcal{X}$, we obtain $k(ge) \subseteq k(hf)$. Conversely, if $k(ge) \subseteq k(hf)$, then applying k^{-1}, we obtain $ge \subseteq hf$. Consequently, $\mathcal{G}$ acts by order automorphisms on $\mathcal{X}$ on the left.

Define a function ψ by

$$\psi : e \to 1e \qquad (e \in E).$$

Assume that $e \leq f$ in E. Then $e = efe$ and $1 = 1(e\sigma)$ and hence $1e \subseteq 1f$. Conversely, suppose that $1e \subseteq 1f$. By (1), there exists $s \in S$ such that $e = s^{-1}fs$ and $1 = s\sigma$. Hence $s \in E$ since S is E-unitary and thus $e = sf \leq f$. Consequently, ψ is an isomorphism of E onto $\mathcal{Y}$. In particular, $\mathcal{Y}$ is a subsemilattice of $\mathcal{X}$.

In order to prove that $\mathcal{Y}$ is an ideal of $\mathcal{X}$, we assume that $ge \subseteq 1f$. By (1), $e = s^{-1}fs$ and $g = s\sigma$ for some $s \in S$. Let $p = fss^{-1}$. For $u = fs$, we get $e = u^{-1}pu$ and $g = u\sigma$ which by (1) yields $ge \subseteq 1p$. For $v = s^{-1}$, we have $p = vev^{-1}$ and $1 = g(v\sigma)$, so again by (1), $1p \subseteq ge$. Consequently, $ge = 1p$ and $\mathcal{Y}$ is an ideal of $\mathcal{X}$.

By the very definitions, we have $\mathcal{G}\mathcal{Y} = \mathcal{X}$. Let $g \in \mathcal{G}$ and let $t\sigma = g$. Then

$$\begin{aligned} g(t^{-1}t) &= \{(s^{-1}t^{-1}ts, g(s\sigma)) | s \in S\} \\ &= \{((ts)^{-1}(tt^{-1})(ts), (ts)\sigma) | s \in S\} = 1(tt^{-1}), \end{aligned}$$

where the last equality is obtained as just above. Consequently, $g(t^{-1}t) = 1(tt^{-1})$ which proves $g\mathcal{Y} \cap \mathcal{Y} \neq \varnothing$.

Therefore, $(\mathcal{Y}, \mathcal{G}; \mathcal{X}) \in \operatorname{Ob} \mathfrak{M}$.

Now let $\varphi \in \operatorname{Hom} \mathfrak{E}$, say $\varphi : S \to S'$ for E-unitary inverse semigroups S and S'. Let $s, t \in S$ be such that $s\sigma t$. Then $se = te$ for some $e \in E = E_S$, so that $(s\varphi)(e\varphi) = (t\varphi)(e\varphi)$ which yields $(s\varphi)\sigma(t\varphi)$. This shows that ω defined by

$$(s\sigma)\omega = (s\varphi)\sigma \qquad (s \in S)$$

is single valued. It is evident that ω is a homomorphism of $\mathcal{G} = S/\sigma$ into $\mathcal{G}' = S'/\sigma$.

Next let $g, h \in \mathcal{G}$, $e, f \in E$, and assume that $ge \subseteq hf$. By (1), we have $e = s^{-1}fs$ and $g = h(s\sigma)$ for some $s \in S$. Hence $e\varphi = (s\varphi)^{-1}(f\varphi)(s\varphi)$ and $g\omega = (h\omega)(s\sigma)\omega = (h\omega)(s\varphi)\sigma$ which again by (1) means that $(g\omega)(e\varphi) \subseteq (h\omega)(f\varphi)$. Now considering the equality $ge = hf$, we conclude that the function θ defined by

$$(ge)\theta = (g\omega)(e\varphi) \qquad (g \in G, e \in E)$$

is single valued. We have also shown that θ is order preserving and maps $\mathcal{X}$ into $\mathcal{X}'$ where $R(S) = (\mathcal{Y}, \mathcal{G}; \mathcal{X})$ and $R(S') = (\mathcal{Y}', \mathcal{G}'; \mathcal{X}')$.

It is clear that $\mathcal{Y}\theta \subseteq \mathcal{Y}'$. For any $e, f \in E$, we have

$$(1e \cap 1f)\theta = (1(ef))\theta = 1((ef)\varphi) = 1((e\varphi)(f\varphi)) = 1(e\varphi) \cap 1(f\varphi)$$

since ψ above is an isomorphism of E_S onto $\mathcal{Y}$, analogously $E_{S'}$ maps onto $\mathcal{Y}'$. Hence $\theta|_{\mathcal{Y}}$ is a homomorphism.

For any $g, h \in \mathcal{G}$, $e \in E$, we further obtain

$$[g(he)]\theta = [(gh)e]\theta = (gh)\omega(e\varphi) = [(g\omega)(h\omega)](e\varphi)$$

$$= (g\omega)[(h\omega)(e\varphi)] = (g\omega)(he)\theta$$

which concludes the proof that $R(\varphi) = (\theta, \omega)$ is a morphism of $R(S)$ into $R(S')$. Thus $R(\varphi) \in \operatorname{Hom} \mathfrak{M}$.

The remaining axioms for a functor can be verified without difficulty. Therefore, R is a functor from $\mathfrak{E}$ to $\mathfrak{M}$.

VII.1.11 Lemma

For every $(Y, G; X) \in \operatorname{Ob} \mathfrak{M}$, let $\tau(Y, G; X) = (\theta, \omega)$ where

$$\theta : g\alpha \to [(\ \ , g)\sigma](\alpha, 1) \qquad (g \in G, \alpha \in Y),$$

$$\omega : g \to (\ \ , g)\sigma \qquad (g \in G),$$

where the variables in the empty spaces are of no interest here. Then τ is a natural equivalence of the functors $I_{\mathfrak{M}}$ and RP.

Proof. Let $(Y, G; X) \in \operatorname{Ob} \mathfrak{M}$. Since in $S = P(Y, G; X)$, we have $(\alpha, g)\sigma(\beta, h)$ if and only if $g = h$, it follows at once that ω is an isomorphism of G onto $\mathcal{G} = S/\sigma$.

Next let $g, h \in G$ and $\alpha, \beta \in Y$. Assume that $g\alpha \le h\beta$. Then $h^{-1}g\alpha \le \beta$ and thus $h^{-1}g\alpha \in Y$, so $s = (h^{-1}g\alpha, h^{-1}g)$ is an element of S. Let $e = (\alpha, 1)$, $f = (\beta, 1)$, and calculate

$$s^{-1}fs = (\alpha, g^{-1}h)(\beta, 1)(h^{-1}g\alpha, h^{-1}g)$$

$$= (\alpha \wedge g^{-1}h\beta, g^{-1}h)(h^{-1}g\alpha, h^{-1}g)$$

$$= (\alpha \wedge g^{-1}h\beta, 1) = (\alpha, 1) = e$$

and $(\ \ , h)\sigma(s\sigma) = (\ \ , hh^{-1}g)\sigma = (\ \ , g)\sigma$. It now follows from (1) that

$$[(\ \ , g)\sigma](\alpha, 1) \subseteq [(\ \ , h)\sigma](\beta, 1),$$

that is $(g\alpha)\theta \subseteq (h\beta)\theta$. This also shows that θ is single valued.

Conversely, assume that $(g\alpha)\theta \subseteq (h\beta)\theta$. Then by (1), there exists $(\gamma, k) \in S$ such that

$$(\alpha, 1) = (\gamma, k)^{-1}(\beta, 1)(\gamma, k)$$

and $g = hk$. Hence

$$\begin{aligned}(\alpha, 1) &= (\gamma, h^{-1}g)^{-1}(\beta, 1)(\gamma, h^{-1}g)\\ &= (g^{-1}h\gamma, g^{-1}h)(\beta \wedge \gamma, h^{-1}g)\\ &= (g^{-1}h(\beta \wedge \gamma), h^{-1}g)\end{aligned}$$

which yields $\alpha = g^{-1}h(\beta \wedge \gamma)$. But then $g\alpha = h(\beta \wedge \gamma) \leq h\beta$.

We have proved that θ is an order isomorphism of partially ordered sets X and

$$\mathfrak{X} = \{[(\ \ , g)\sigma](\alpha, 1) | g \in G, \alpha \in Y\}.$$

It is clear that θ maps Y onto $\mathcal{Y} = \{1(\alpha, 1) | \alpha \in Y\}$ where we set $[(\ \ ,1)\sigma] \cdot(\alpha, 1) = 1(\alpha, 1)$.

Let $g, h \in G$, $\alpha \in Y$; then

$$\begin{aligned}[g(h\alpha)]\theta = [(gh)\alpha]\theta &= [(\ \ , gh)\sigma](\alpha, 1)\\ &= [(\ \ , g)\sigma][(\ \ , h)\sigma](\alpha, 1)\\ &= (g\omega)(h\alpha)\theta.\end{aligned}$$

We finally note that $RP(Y, G; X) = (\mathcal{Y}, \mathcal{G}; \mathfrak{X})$ as constructed above, and hence (θ, ω) is an isomorphism of $(Y, G; X)$ onto $RP(Y, G; X)$.

It remains to establish commutativity of the diagram

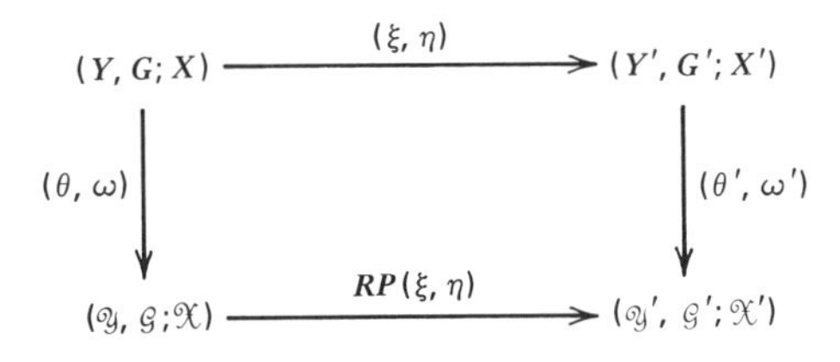

where (ξ, η) is a morphism of $(Y, G; X)$ into $(Y', G'; X')$ and the rest of the notation is obvious. Let $RP(\xi, \eta) = (\xi', \eta')$. For any $g \in G$, we have

$$g\eta\omega' = (\ \ , g\eta)\sigma = (\ \ , g)\sigma\eta' = g\omega\eta'$$

and if also $\alpha \in Y$, then

$$(g\alpha)\xi\theta' = [(g\eta)(\alpha\xi)]\theta' = (g\eta\omega')[1(\alpha\xi, 1)] = (\ \ , g\eta)\sigma[1(\alpha, 1)]\xi'$$

$$= \{[(\ \ , g)\sigma][1(\alpha, 1)]\}\xi' = (g\alpha)\theta\xi'.$$

VII.1.12 Lemma

For every $S \in \text{Ob}\,\mathfrak{E}$, let $\chi = \chi(S)$ be defined by

$$\chi : s \to (1(ss^{-1}), s\sigma) \qquad (s \in S).$$

Then χ is a natural equivalence of the functors $I_{\mathfrak{E}}$ and PR.

Proof. Let $S \in \text{Ob}\,\mathfrak{E}$ and let $(\mathcal{Y}, \mathcal{G}; \mathcal{X}) = R(S)$, $T = P(\mathcal{Y}, \mathcal{G}; \mathcal{X})$, χ be as defined above. If $s \in S$, then

$$\begin{aligned}(s\sigma)^{-1}[1(ss^{-1})] &= \{(t^{-1}ss^{-1}t, (s^{-1}t)\sigma) | t \in S\} \\ &= \{((s^{-1}t)^{-1}ss^{-1}(s^{-1}t), (s^{-1}t)\sigma) | t \in S\} \\ &= 1(s^{-1}s) \in \mathcal{Y}\end{aligned}$$

where the last equality obtains similarly as in the proof of 1.10. Hence χ maps S into T.

Next let $(1e, g) \in T$. Then $g^{-1}(1e) \in \mathcal{Y}$ so that $g^{-1}(1e) = 1(tt^{-1})$ for some $t \in S$ such that $e = t^{-1}t$ and $t\sigma = g$ as in the proof of 1.10. But then $t\chi = (1(tt^{-1}), t\sigma) = (1e, g)$ and thus χ maps S onto T.

Next assume that for $s, t \in S$, we have $(1(ss^{-1}), s\sigma) = (1(tt^{-1}), t\sigma)$. Then $1(ss^{-1}) = 1(tt^{-1})$ and hence $ss^{-1} = tt^{-1}$ as we have seen in the proof of 1.10. Thus $s\mathcal{R}t$ and $s\sigma t$ so that $s = t$ since S is E-unitary. Consequently, χ is one-to-one.

In order to show that χ is a homomorphism, we first prove

$$1(ss^{-1}) \cap (s\sigma)[1(tt^{-1})] = 1[(st)(st)^{-1}]. \tag{2}$$

Let $(e, g) \in 1(ss^{-1}) \cap (s\sigma)[1(tt^{-1})]$. Then

$$(e, g) = (u^{-1}ss^{-1}u, u\sigma) = (v^{-1}tt^{-1}v, (sv)\sigma)$$

so that $e = u^{-1}ss^{-1}u = v^{-1}tt^{-1}v$ and $g = u\sigma = (sv)\sigma$. We deduce that

$(s^{-1}uv^{-1})\sigma = 1$ and thus $s^{-1}uv \in E_S$ since S is E-unitary. It follows that

$$\begin{aligned}(e, g) = (e^2, g) &= \left(u^{-1}s(s^{-1}uv^{-1})tt^{-1}v, u\sigma\right)\\ &= \left(u^{-1}stt^{-1}(s^{-1}uv^{-1})v, u\sigma\right)\\ &= \left((v^{-1}v)u^{-1}stt^{-1}(s^{-1}uv^{-1})v, u\sigma\right)\\ &= \left((uv^{-1}v)^{-1}(st)(st)^{-1}(uv^{-1}v), (uv^{-1}v)\sigma\right)\\ &\in 1\left[(st)(st)^{-1}\right].\end{aligned}$$

Conversely, let $(e, g) \in 1[(st)(st)^{-1}]$, so that $(e, g) = (u^{-1}stt^{-1}s^{-1}u^{-1}, u\sigma)$ for some $u \in S$, and thus $e = u^{-1}stt^{-1}s^{-1}u$, $g = u\sigma$. It follows that

$$e = (s^{-1}u)^{-1}(tt^{-1})(s^{-1}u), \qquad g = (s\sigma)(s^{-1}u)\sigma$$

and hence $(e, g) \in (s\sigma)[1(tt^{-1})]$. Since $(st)(st)^{-1} = stt^{-1}s^{-1} \leq ss^{-1}$, by the proof of 1.10, we have $1[(st)(st)^{-1}] \subseteq 1(ss^{-1})$ which yields $(e, g) \in 1(ss^{-1})$.

We have proved (2). It now follows

$$\begin{aligned}(s\chi)(t\chi) &= \left(1(ss^{-1}), s\sigma\right)\left(1(tt^{-1}), t\sigma\right)\\ &= \left(1(ss^{-1}) \cap (s\sigma)\left[1(tt^{-1})\right], (s\sigma)(t\sigma)\right)\\ &= \left(1\left[(st)(st)^{-1}\right], (st)\sigma\right) = (st)\chi\end{aligned}$$

and χ is indeed an isomorphism of S onto T.

Letting φ be a homomorphism of S into an E-unitary inverse semigroup S', we must show commutativity of the diagram

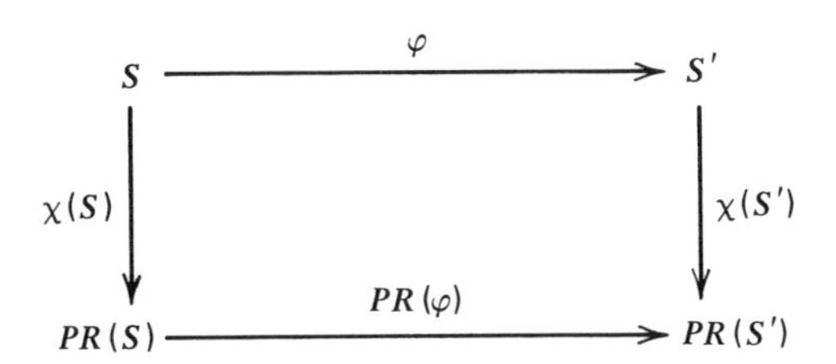

Indeed, for any $s \in S$, we have

$$\begin{aligned}s\varphi\chi(S') &= \left(1\left[(s\varphi)(s\varphi)^{-1}\right], s\varphi\sigma\right) = \left(1\left[(ss^{-1})\varphi\right], s\varphi\sigma\right)\\ &= \left(1(ss^{-1}), s\sigma\right)[PR(\varphi)] = s[\chi(S)][PR(\varphi)].\end{aligned}$$

Now collecting 1.8 and 1.10–1.12, we deduce the desired result.

VII.1.13 Theorem

The quadruple (P, R, τ, χ) is an equivalence of the categories $\mathfrak{M}$ and $\mathfrak{E}$.

We may restate a portion of the above result as follows.

VII.1.14 Corollary

If $(Y, G; X)$ is a McAlister triple, then $P(Y, G; X)$ is an E-unitary inverse semigroup. Conversely, every E-unitary inverse semigroup can be so constructed (up to an isomorphism).

Proof. This is actually a weakening of 1.13.

For ease of reference, we introduce the following concepts.

VII.1.15 Definition

The semigroups $P(Y, G; X)$ are *P-semigroups*. The representation of an E-unitary inverse semigroup S as a P-semigroup is its *P-representation*.

VII.1.16 Exercises

(i) Show that $P(Y, G; X)$ is bisimple if and only if G acts transitively on Y.

(ii) Show that $P(Y, G; X)$ is simple if and only if for any $\alpha, \beta \in Y$, there exists $g \in G$ such that $g\alpha \geq \beta$.

(iii) Characterize right ideals and ideals of $P(Y, G; X)$ in terms of ideals of Y.

(iv) Let $S = P(Y, G; X)$, $a = (\alpha, g) \in S$, $b = (\beta, h) \in S$. Prove the following statements.

(α) $a\mathfrak{L}b \Leftrightarrow g^{-1}\alpha = h^{-1}\beta$.

(β) $a\mathfrak{R}b \Leftrightarrow \alpha = \beta$.

(γ) $a\sigma b \Leftrightarrow g = h$.

(δ) $a \leq b \Leftrightarrow \alpha \leq \beta, g = h$.

(v) Let $(Y, G; X)$ be a McAlister triple and $S = P(Y, G; X)$. Let F be a filter of Y and N be a subgroup of G such that

$$g \in N, \alpha \in F, g\alpha \in Y \Rightarrow g\alpha \in F.$$

Show that $H = S \cap (F \times N)$ is a closed inverse subsemigroup of S, and that conversely, every closed inverse subsemigroup of S is of this form for unique F and N.

(vi) With the notation of the preceding exercise, write $H \sim (F, N)$. Let also $H' \sim (F', N')$ in S. Show that φ_H and $\varphi_{H'}$ are equivalent representations if and only if there exists $(\alpha, g) \in S$ such that $gF = F'$, $gNg^{-1} = N'$ and $N'\alpha \subseteq F'$. For the notation, see IV.4.7 and observe that

$$N'\alpha = \{n'\alpha \mid n' \in N'\}.$$

(vii) Let $(Y, G; X)$ be a McAlister triple and $S = P(Y, G; X)$. Let $\alpha \in Y$ and N be a subgroup of G. Show that

$$K = \{(\alpha, n) \in S \mid n \in N, n\alpha = \alpha\}$$

is a subgroup of S, and that conversely, every subgroup of S is of this form for unique α and N.

(viii) With the notation of the preceding exercise, write $K \sim (\alpha, N)$. Let also $K' \sim (\alpha', N')$ in S. Show that ψ_K and $\psi_{K'}$ are equivalent if and only if there exists $g \in G$ such that $\alpha = g\alpha'$ and $g^{-1}Ng = N'$.

(ix) Construct the P-representation of the bicyclic semigroup C. Use it to find all transitive representations of C.

VII.1.17 Problems

(i) Find the structure of inverse semigroups S for which $E_S\omega \subseteq E_S\zeta$.

The construction of P-semigroups and their relationship to E-unitary inverse semigroups was established by McAlister [4], [5]. Alternative proofs of McAlister's result were devised by Munn [16] and Schein [20]. The class of E-unitary inverse semigroups was first studied by Saitô [2]. The categorical setting we have presented can be found in Petrich [14]. Further properties of P-semigroups are contained in McAlister [4], [5], [8], [10], McAlister–McFadden [1], O'Carroll [1], [4], [6], and Yoshida [1].

McAlister's work was evidently influenced by Scheiblich's construction of a free inverse semigroup on a set, a variant of which will be presented in the next chapter. His work has in its turn inspired much research activity in the field of structure theorems for various classes of inverse semigroups.

VII.2 CONGRUENCES ON *P*-SEMIGROUPS

We present here a construction of congruences on a P-semigroup. From it we deduce simpler conditions for various special congruences such as idempotent separating, idempotent pure, and E-unitary congruences.

VII.2.1 Theorem

Let $S = P(Y, G; X)$. Let τ be a congruence on Y satisfying

(i) if $\alpha\tau\beta$ and $g\alpha, g\beta \in Y$, then $(g\alpha)\tau(g\beta)$.

For every $A \in Y/\tau$, let K_A be a subgroup of G, and assume

(ii) if $g \in K_A$, then there exists $\alpha \in A$ such that $g\alpha \in A$,

(iii) if $\alpha \leq \beta$, $g^{-1}\alpha \in Y$, then $g^{-1}K_{\beta\tau}g \subseteq K_{(g^{-1}\alpha)\tau}$.

Define a relation $\rho = \rho_{(\tau, K_{\alpha\tau})}$ on S by

$$(\alpha, g)\rho(\beta, h) \Leftrightarrow \alpha\tau\beta, \quad gh^{-1} \in K_\alpha.$$

Then ρ is a congruence on S. Conversely, every congruence on S can be constructed in this way.

Proof

Direct part. Let ρ be as defined in the statement of the theorem. Clearly ρ is an equivalence relation. Let $(\alpha, g)\rho(\beta, h)$ and $(\gamma, k) \in S$. Then $\alpha\tau\beta$ and $gh^{-1} \in K_{\alpha\tau}$.

In order to show that ρ is a left congruence, we first establish two claims. Note that for $g = 1$ in (iii) we obtain

$$\alpha \leq \beta \Rightarrow K_{\beta\tau} \subseteq K_{\alpha\tau}.$$

We also claim that

$$t \in G, \quad \delta, t^{-1}\delta \in Y \Rightarrow t^{-1}K_{\delta\tau}t = K_{(t^{-1}\delta)\tau}.$$

Indeed, with $\delta \leq \delta$ and $t^{-1}\delta \in Y$, condition (iii) gives $t^{-1}K_{\delta\tau}t \subseteq K_{(t^{-1}\delta)\tau}$, and with $t^{-1}\delta \leq t^{-1}\delta$ and $t(t^{-1}\delta) \in Y$, condition (iii) yields $tK_{(t^{-1}\delta)\tau}t^{-1} \subseteq K_{t\tau}$, which proves the claim.

Going back to our hypothesis, we note that

$$k(k^{-1}\gamma \wedge \alpha) = \gamma \wedge k\alpha \in Y, \qquad k^{-1}\gamma \wedge \alpha \in Y$$

so the above claims imply that

$$kg(kh)^{-1} = k(gh^{-1})k^{-1} \in kK_{\alpha\tau}k^{-1} \subseteq kK_{(k^{-1}\gamma \wedge \alpha)\tau}k^{-1} = K_{(\gamma \wedge k\alpha)\tau}.$$

On the other hand, $\alpha\tau\beta$ implies that $(k^{-1}\gamma \wedge \alpha)\tau(k^{-1}\gamma \wedge \beta)$ and

$$k(k^{-1}\gamma \wedge \alpha), k(k^{-1}\gamma \wedge \beta) \in Y$$

and condition (i) gives that $k(k^{-1}\gamma \wedge \alpha)\tau k(k^{-1}\gamma \wedge \beta)$. But then

$$(\gamma \wedge k\alpha)\tau(\gamma \wedge k\beta)$$

which finally yields

$$(\gamma, k)(\alpha, g) = (\gamma \wedge k\alpha, kg)\rho(\gamma \wedge k\beta, kh) = (\gamma, k)(\beta, h)$$

and ρ is a left congruence.

In order to show that ρ is a right congruence, we first prove

$$\delta, \eta \in Y, \quad t \in K_{\delta\tau} \Rightarrow (\delta \wedge \eta)\tau(\delta \wedge t\eta).$$

By the first claim above, we have $t \in K_{\delta\tau} \subseteq K_{(\delta \wedge \eta)\tau}$ which by condition (ii) implies the existence of $\xi, \xi' \in Y$ such that $\xi\tau\delta\tau(t\xi)$ and $\xi'\tau(\delta \wedge \eta)\tau(t\xi')$. It follows that $\xi'\tau(\delta \wedge \eta)\tau(\xi \wedge \eta)$ whence

$$(\delta \wedge \eta)\tau(t\xi')\tau(t\xi \wedge t\eta)\tau(\delta \wedge t\eta)$$

since $t\xi \wedge t\eta \leq t\xi \in Y$. This establishes the third claim.

Going back to our hypothesis and using the first claim, we get

$$gk(hk)^{-1} = gh^{-1} \in K_{\alpha\tau} \subseteq K_{(\alpha \wedge g\gamma)\tau}.$$

Now using the third claim, we get

$$\alpha \wedge g\gamma = (\alpha \wedge ghh^{-1}\gamma)\tau(\alpha \wedge h\gamma)\tau(\alpha \wedge \beta \wedge h\gamma)\tau(\beta \wedge h\gamma)$$

since $gh^{-1} \in K_{\alpha\tau}$ and $\alpha\tau\beta$. Consequently

$$(\alpha, g)(\gamma, k) = (\alpha \wedge g\gamma, gk)\rho(\beta \wedge h\gamma, hk) = (\beta, h)(\gamma, k)$$

and ρ is also a right congruence.

Converse. Let ρ be a congruence on S. Since the mapping $(\alpha, a) \to \alpha$ is an isomorphism of E_S onto Y, τ defined by

$$\alpha\tau\beta \Leftrightarrow (\alpha, 1)\rho(\beta, 1)$$

is a congruence on Y.

We now claim that with $t \in G$, $\delta, \eta \in Y$ and $t^{-1}\delta, t^{-1}\eta \in Y$, we have

$$\delta\tau\eta \Leftrightarrow (\delta, t)\rho(\eta, t). \tag{3}$$

We assume first that $\delta\tau\eta$ so that $(\delta, 1)\rho(\eta, 1)$. It follows that

$$\begin{aligned}(\delta, t) &= (\delta, 1)(\delta, t)\rho(\eta, 1)(\delta, t) = (\eta \wedge \delta, t)\\ &= (\delta, 1)(\eta, t)\rho(\eta, 1)(\eta, t) = (\eta, t).\end{aligned}$$

Conversely, if $(\delta, t)\rho(\eta, t)$, then taking inverses, we get

$$(\delta, 1) = (\delta, t)(\delta, t)^{-1}\rho(\eta, t)(\eta, t)^{-1} = (\eta, 1).$$

In order to establish condition (i), we assume that $\alpha\tau\beta$ and $g\alpha, g\beta \in Y$. By relation (3), we get $(\alpha, g^{-1})\rho(\beta, g^{-1})$. Taking inverses, we have

$$(g\alpha, g)\rho(g\beta, g)$$

which again by (3) gives $(g\alpha)\tau(g\beta)$.

For every $\alpha \in Y$, we now define $K_{\alpha\tau}$ to be the projection of $(\alpha, 1)\rho$ in G, explicitly

$$K_{\alpha\tau} = \{g \in G \mid (\beta, g)\rho(\alpha, 1) \text{ for some } \beta \in Y\}.$$

Since $(\alpha, 1)\rho$ is an inverse subsemigroup of S, it follows that $K_{\alpha\tau}$ is a subgroup of G.

To verify condition (ii), we suppose that $g \in K_{\alpha\tau}$. By the definition of $K_{\alpha\tau}$, there exists $\beta \in Y$ such that $(\beta, g)\rho(\alpha, 1)$. Then

$$(\beta, 1)\rho(g^{-1}\beta, 1)\rho(\alpha, 1)$$

so that for $\delta = g^{-1}\beta$, we have $\delta\tau(g\delta)\tau\alpha$, as required.

For condition (iii), assume that $\alpha \leq \beta$ and $g^{-1}\alpha \in Y$. Let $h \in K_{\beta\tau}$ so that $(\gamma, h)\rho(\beta, 1)$ for some $\gamma \in Y$. Then $h^{-1}\gamma \in Y$; thus $h^{-1}(\alpha \wedge \gamma) \in Y$ and

$$(\alpha \wedge \gamma, h) = (\alpha, 1)(\gamma, h)\rho(\alpha, 1)(\beta, 1) = (\alpha, 1).$$

Multiplying on the left by $(\alpha, g)^{-1}$ and on the right by (α, g), we obtain

$$(g^{-1}\alpha, g^{-1})(\alpha \wedge \gamma, h)(\alpha, g)\rho(g^{-1}\alpha, g^{-1})(\alpha, 1)(\alpha, g)$$

which gives

$$(g^{-1}\alpha \wedge g^{-1}(\alpha \wedge \gamma) \wedge g^{-1}h\alpha, g^{-1}hg)\rho(g^{-1}\alpha, 1)$$

yielding $g^{-1}hg \in K_{(g^{-1}\alpha)\tau}$.

Now that we have verified conditions (i), (ii), and (iii), it remains to show that ρ coincides with the relation $\rho_{(\tau, K_{\alpha\tau})}$.

Assume that $(\alpha, g)\rho(\beta, h)$. Then $(g^{-1}\alpha, g^{-1})\rho(h^{-1}\beta, h^{-1})$ which then yields $(\alpha, 1)\rho(\beta, 1)$ so that $\alpha\tau\beta$. Moreover,

$$(\alpha, g)(h^{-1}\beta, h^{-1}) = (\alpha \wedge gh^{-1}\beta, gh^{-1})\rho(\alpha, 1)$$

which yields $gh^{-1} \in K_{\alpha\tau}$.

Conversely, suppose that $(\alpha, g), (\beta, h) \in S$, $\alpha\tau\beta$ and $gh^{-1} \in K_{\alpha\tau}$. We have $(\delta, gh^{-1})\rho(\alpha, 1)\rho(\beta, 1)$ for some $\delta \in Y$, which implies

$$(\delta \wedge gh^{-1}\beta, gh^{-1}) = (\delta, gh^{-1})(\beta, 1)\rho(\beta, 1).$$

Multiplying on the right by (β, h) yields $(\delta \wedge gh^{-1}\beta, g)\rho(\beta, h)$. It follows that

$$(\delta \wedge gh^{-1}\beta, 1)\rho(\beta, 1)\rho(\alpha, 1)$$

in the usual way, and since $g^{-1}\alpha \in Y$ and $g^{-1}(\delta \wedge gh^{-1}\beta) \leq h^{-1}\beta \in Y$, relation (3) gives $(\delta \wedge gh^{-1}\beta, g)\rho(\alpha, g)$. This together with the $(\delta \wedge gh^{-1}\beta, g)\rho(\beta, h)$ above finally yields that $(\alpha, g)\rho(\beta, h)$.

In the above proof, we established directly that $\rho_{(\tau, K_{\alpha\tau})}$ is a congruence and conversely, that every congruence on S is of this form. Alternatively, we may have passed via congruence pairs in both directions. The parameters and the form of $\rho_{(\tau, K_{\alpha\tau})}$ are actually closest to a kernel normal system and this approach may also be successfully used.

Note that conditions (i), (ii), and (iii) in 2.1 are such that we can transcribe them easily to the case of $Q(Y, G; \psi)$.

We now deduce simpler conditions for various types of special congruences on an inverse semigroup.

VII.2.2 Corollary

Let $S = P(Y, G; X)$. For every $\alpha \in Y$, let K_α be a subgroup of G and assume

(i) if $g \in K_\alpha$, then $g\alpha = \alpha$,
(ii) if $\alpha \leq \beta$, $g^{-1}\alpha \in Y$, then $g^{-1}K_\beta g \subseteq K_{g^{-1}\alpha}$.

Define a relation $\rho = \rho_{K_\alpha}$ on S by

$$(\alpha, g)\rho(\beta, h) \Leftrightarrow \alpha = \beta, \quad gh^{-1} \in K_\alpha.$$

Then ρ is an idempotent separating congruence on S. Conversely, every idempotent separating congruence on S can be constructed this way.

Proof. This is an obvious specialization of the result in 2.1 to the case when $\tau = \varepsilon$.

For the next specialization, we need an auxiliary statement of independent interest.

VII.2.3 Lemma

For $S = P(Y, G; X)$, we have

$$E_S\zeta = \{(\alpha, g) \in S \mid g\beta = \beta \text{ for all } \beta \leq \alpha\}.$$

Proof. Let $(\alpha, g) \in E_S\zeta$. Then for any $\beta \in Y$,

$$(\alpha, g)(\beta, 1) = (\beta, 1)(\alpha, g)$$

so that $\alpha \wedge g\beta = \alpha \wedge \beta$. Letting $\beta = g^{-1}\alpha$, we get $\alpha \leq g^{-1}\alpha$ so that $g\alpha \leq \alpha$. Now let $\beta \leq \alpha$. Then $\alpha \wedge g\beta = \beta$ whence $\beta \leq g\beta$. In particular $\alpha \leq g\alpha$ and thus $\alpha = g\alpha$. Further, the equation $\alpha \wedge g\beta = \beta$ implies $g^{-1}\alpha \wedge \beta = g^{-1}\beta$ which implies $\alpha \wedge \beta = g^{-1}\beta$. But then $\beta = g^{-1}\beta$ and $g\beta = \beta$.

Conversely, assume that $g\beta = \beta$ for all $\beta \leq \alpha$. Then for any $\beta \in Y$, we get

$$\alpha \wedge \beta = g(\alpha \wedge \beta) = g\alpha \wedge g\beta = \alpha \wedge g\beta$$

which evidently implies that (α, g) commutes with $(\beta, 1)$.

VII.2.4 Corollary

In $S = P(Y, G: X)$, we have

$$(\alpha, g)\mu(\beta, h) \Leftrightarrow \alpha = \beta, \quad g\gamma = h\gamma \quad \text{for all } \gamma \leq g^{-1}\alpha.$$

Proof. Since $\ker \mu = E_S\zeta$, using 2.3, we get

$$\begin{aligned}(\alpha, g)\mu(\beta, h) &\Leftrightarrow (\alpha, g)(\alpha, g)^{-1} = (\beta, h)(\beta, h)^{-1}, \quad (\alpha, g)^{-1}(\beta, h) \in E_S\zeta \\ &\Leftrightarrow \alpha = \beta, \quad (g^{-1}\alpha, g^{-1}h) \in E_S\zeta \\ &\Leftrightarrow \alpha = \beta, \quad g^{-1}h\gamma = \gamma \quad \text{for all } \gamma \leq g^{-1}\alpha \\ &\Leftrightarrow \alpha = \beta, \quad g\gamma = h\gamma \quad \text{for all } \gamma \leq g^{-1}\alpha.\end{aligned}$$

VII.2.5 Corollary

Let $S = P(Y, G; X)$. Let τ be a congruence on Y satisfying 2.1(i). Define a relation $\rho = \rho_\tau$ on S by

$$(\alpha, g)\rho(\beta, h) \Leftrightarrow \alpha\tau\beta, \quad g = h.$$

Then ρ is an idempotent pure congruence on S. Conversely, every idempotent pure congruence on S can be constructed in this way.

Proof. It suffices to observe that in S,

$$\ker \rho = \{(\alpha, g) \in S \mid g \in K_{\alpha\tau}\}.$$

Hence $\ker \rho = E_S$ if and only if $K_{\alpha\tau} = 1$ for all $\alpha \in Y$.

VII.2.6 Corollary

Let $S = P(Y, G; X)$. For a normal subgroup K of G, define a relation $\rho = \rho_K$ on S by

$$(\alpha, g)\rho(\beta, h) \Leftrightarrow gh^{-1} \in K.$$

Then ρ is a group congruence on S. Conversely, every group congruence on S can be constructed in this way.

Proof. A simple verification shows that ρ_K is a group congruence on S. Conversely, let ρ be a group congruence on S. Let K be the projection of $\ker \rho$ in G. Then K is a normal subgroup of G since $\ker \rho$ is a normal subsemigroup of S. Also, $\rho = \rho_{(w, K)}$ by 2.1 and thus $\rho = \rho_K$.

The corresponding result for semilattice congruences is obtained by specifying that the kernel must be equal to the entire semigroup S with the resulting simplifications. In fact, we may consider any of the special congruences we encountered in Chapter III on a P-semigroup, thereby obtaining deeper insight into their nature. For example, we may study E-unitary, E-reflexive, antigroup congruences on a P-semigroup and observe the specific form their parameters take in terms of Y and G. In particular, if ρ is an E-unitary congruence on a P-semigroup S, we may construct the P-representation of the quotient semigroup S/ρ. One can perform a similar analysis with the Q-representation.

VII.2.7 Exercises

(i) Let $S = P(Y, G; X)$. Show that a relation ρ on E_S is a normal congruence if and only if the induced relation on Y is a congruence τ satisfying condition 2.1(i). Also give an expression for $\rho_{\min}$ and $\rho_{\max}$ directly in terms of τ.

(ii) Let $S = P(Y, G; X)$. Let τ be an equivalence relation on X such that $\tau|_Y$ is a congruence, let π be a congruence on G, and assume that $g\pi h$, $\alpha\tau\beta$ imply $(g\alpha)\tau(h\beta)$. Define a relation ρ on S by

$$(\alpha, g)\rho(\beta, h) \Leftrightarrow \alpha\tau\beta,\ g\pi h.$$

Show that ρ is an E-unitary congruence on S, and that, conversely, every E-unitary congruence on S can be so constructed.

(iii) State and prove the analogue of 2.1 for congruences on $Q(Y, G; \psi)$.

(iv) State and prove an analogue of 2.5 for idempotent pure congruences ρ on $Q(Y, G; \psi)$ and express $Q(Y, G; \psi)/\rho$ in the form $Q(Y', G'; \psi')$.

The form for congruences on a P-semigroup we have given is, in a different notation, due to Jones [2]; the argument presented here is due to S. A. Rankin. Alternative characterizations for general or special congruences on P-semigroups were given by McAlister [5], [9] and Petrich [11]. Idempotent pure congruences on P-semigroups were studied by Reilly-Munn [1]. See also Reilly [10].

VII.3 THE TRANSLATIONAL HULL OF A *P*-SEMIGROUP

For a P-semigroup $S = P(Y, G; X)$, we provide the P-representations of $\hat{S}$ in the form $P(\mathcal{I}_Y, G; G\mathcal{I}_Y)$ and of $\Omega(S)$ in the form $P(\mathcal{R}_Y, G; G\mathcal{R}_Y)$.

Recall the definition and properties of $\hat{S}$ discussed in V.2.

VII.3.1 Lemma

Let S be an inverse semigroup and $\rho \in \hat{S}$. If $x \in \mathbf{d}\rho$, then $x\mathcal{R}(x\rho)$.

Proof. Let $x \in \mathbf{d}\rho$, $\rho \in \hat{S}$. Then

$$x\rho = (xx^{-1}x)\rho = x(x^{-1}x)\rho \in xS,$$

$$x\rho = (x\rho)(x\rho)^{-1}(x\rho) = \left[(x\rho)(x\rho)^{-1}x\right]\rho,$$

and the latter implies $x = (x\rho)(x\rho)^{-1}x \in (x\rho)S$. Consequently, $x\mathcal{R}(x\rho)$.

Now let $S = P(Y, G; X)$ and $\rho \in \hat{S}$. If $e, f \in E_S \cap \mathbf{d}\rho$, then

$$e(f\rho) = (ef)\rho = (fe)\rho = f(e\rho)$$

which implies that $(f\rho)\sigma(e\rho)$. This together with 3.1 gives that for any $(\alpha, 1) \in \mathbf{d}\rho$,

$$(\alpha, 1)\rho = (\alpha, g_\rho) \tag{4}$$

where g_ρ is independent of α. This makes it possible to define the function ψ below.

VII.3.2 Theorem

Let $S = P(Y, G; X)$ and define a function ψ by

$$\psi : \rho \to (I_\rho, g_\rho) \qquad (\rho \in \hat{S}),$$

where g_ρ is defined by (4) and

$$I_\rho = \{\alpha \in Y | (\alpha, 1) \in \mathbf{d}\rho\}.$$

Then ψ is an isomorphism of $\hat{S}$ onto $P(\mathcal{I}_Y, G; G\mathcal{I}_Y)$.

Proof. Let $\rho \in \hat{S}$. Then $I_\rho \in \mathcal{I}_Y$ since $\mathbf{d}\rho$ is a left ideal of S. If $\alpha \in I_\rho$, then $(\alpha, 1) \in \mathbf{d}\rho$ which yields $(\alpha, g_\rho) \in S$ and thus $g_\rho^{-1}\alpha \in Y$. Hence $g_\rho^{-1}I_\rho \subseteq Y$ and it follows easily that $g_\rho^{-1}I_\rho$ is in fact an ideal of Y. Consequently, $g_\rho^{-1}I_\rho \in \mathcal{I}_Y$ which implies that the function ψ maps $\hat{S}$ into $P(\mathcal{I}_Y, G; G\mathcal{I}_Y)$.

Next let $\rho, \sigma \in \hat{S}$. Then for any $\alpha \in Y$, we have

$$\begin{aligned}(\alpha, 1) \in \mathbf{d}(\rho\sigma) &\Leftrightarrow (\alpha, 1) \in \mathbf{d}\rho, \quad (\alpha, 1)\rho \in \mathbf{d}\sigma \\ &\Leftrightarrow \alpha \in I_\rho, \quad (\alpha, g_\rho) \in \mathbf{d}\sigma \\ &\Leftrightarrow \alpha \in I_\rho, \quad (\alpha, g_\rho)^{-1}(\alpha, g_\rho) \in \mathbf{d}\sigma \\ &\Leftrightarrow \alpha \in I_\rho, \quad \left(g_\rho^{-1}\alpha, 1\right) \in \mathbf{d}\sigma \\ &\Leftrightarrow \alpha \in I_\rho, \quad g_\rho^{-1}\alpha \in I_\sigma\end{aligned}$$

which proves that $I_{\rho\sigma} = I_\rho \cap g_\rho I_\sigma$. For $(\alpha, 1) \in \mathbf{d}(\rho\sigma)$, we also have

$$\begin{aligned}(\alpha, 1)\rho\sigma = (\alpha, g_\rho)\sigma &= (\alpha, g_\rho)\left[\left(g_\rho^{-1}\alpha, 1\right)\sigma\right] \\ &= (\alpha, g_\rho)\left(g_\rho^{-1}\alpha, g_\sigma\right) = (\alpha, g_\rho g_\sigma)\end{aligned}$$

so that $g_{\rho\sigma} = g_\rho g_\sigma$. It follows that ψ is a homomorphism of $\hat{S}$ into $P(\mathcal{I}_Y, G; G\mathcal{I}_Y)$.

Now let $(\alpha, g) \in \mathbf{d}\rho$. Then

$$(\alpha, g)\rho = (\alpha, g)\left[(g^{-1}\alpha, 1)\rho\right] = (\alpha, g)\left(g^{-1}\alpha, g_\rho\right) = (\alpha, g_\rho).$$

Hence if $\rho\psi = \sigma\psi$, then $I_\rho = I_\sigma$, $g_\rho = g_\sigma$, whence it follows that $\rho = \sigma$. Consequently, ψ is one-to-one.

Let $(I, g) \in P(\mathcal{I}_Y, G; G\mathcal{I}_Y)$ and define ρ by

$$\mathbf{d}\rho = \left\{(\alpha, h) \in S \mid h^{-1}\alpha \in I\right\},$$

$$(\alpha, h)\rho = (\alpha, hg) \qquad \text{if } (\alpha, h) \in \mathbf{d}\rho.$$

The verification that $\rho \in \hat{S}$ and $\rho\psi = (I, g)$ is left as an exercise.

In order to get a similar representation for $\Omega(S)$, we need an auxiliary statement.

VII.3.3 Lemma

Let S and ψ be as in 3.2 and w be the Wagner representation of S. Then

$$w\psi : (\alpha, g) \to ([\alpha], g) \qquad [(\alpha, g) \in S]$$

and is a monomorphism of S into $P(\mathcal{I}_Y, G; G\mathcal{I}_Y)$.

Proof. The proof of this lemma is left as an exercise.

VII.3.4 Theorem

For $S = P(Y, G; X)$, $\Omega(S)$ is isomorphic to $P(\mathcal{R}_Y, G; G\mathcal{R}_Y)$.

Proof. It follows from V.1.3 that $i_{\hat{S}}(S^w) \cong \Omega(S)$. Thus in order to prove the theorem, it suffices to show that the idealizer A of $S^w\psi$ in $P(\mathcal{I}_Y, G; G\mathcal{I}_Y)$ coincides with $P(\mathcal{R}_Y, G; G\mathcal{R}_Y)$, where ψ is as in 3.2. We will use 3.3 repeatedly without express reference.

Let $(I, g) \in A$. Then $(I, 1) \in A$ since A is an inverse subsemigroup of T. Hence $(I, 1)([\alpha], 1) \in S^w\psi$ for all $\alpha \in Y$, and thus $I \cap [\alpha] \in \mathcal{P}_Y$ for all $\alpha \in Y$, so that $I \in \mathcal{R}_Y$. Further, for all $\alpha \in Y$, we have

$$(I, g)([\alpha], 1) = (I \cap g[\alpha], g) \in S^w\psi,$$

and thus $I \cap g[\alpha] = [\beta]$ with $(\beta, g) \in S$. Hence $g^{-1}\beta \in Y$, say $g^{-1}\beta = \alpha'$ so that $\beta = g\alpha'$. It follows that $\alpha' = g^{-1}\beta \le g^{-1}(g\alpha) = \alpha$ and $I \cap g[\alpha] = g[\alpha']$ which implies $g^{-1}I \cap [\alpha] = [\alpha']$. Since $(I, g) \in P(\mathcal{I}_Y, G; G\mathcal{I}_Y)$, $g^{-1}I$ is an ideal of Y, which together with the property established shows that $g^{-1}Y \in \mathcal{R}_Y$. Therefore, $A \subseteq P(\mathcal{R}_Y, G; G\mathcal{R}_Y)$.

Conversely, let $(I, g) \in P(\mathcal{R}_Y, G; G\mathcal{R}_Y)$ and $([\alpha], h) \in S^w\psi$. If $g^{-1}I \cap [\alpha] = [\alpha']$, then $([g\alpha'], g) \in S^w\psi$ and

$$\begin{aligned}(I, g)([\alpha], h) &= (I, g)([\alpha], 1)([\alpha], h)\\ &= (I \cap g[\alpha], g)([\alpha], h)\\ &= ([g\alpha'], g)([\alpha], h) \in S^w\psi.\end{aligned}$$

Further, if $g^{-1}I \cap [g^{-1}h^{-1}\alpha] = [\beta]$, then $I \cap [h^{-1}\alpha] = [\alpha'']$ with $\alpha'' = g\beta$, so that $([\alpha''], g) \in S^w\psi$, and

$$\begin{aligned}([\alpha], h)(I, g) &= ([\alpha], h)([h^{-1}\alpha], 1)(I, g)\\ &= ([\alpha], h)([h^{-1}\alpha] \cap I, g)\\ &= ([\alpha], h)([\alpha''], g) \in S^w\psi.\end{aligned}$$

Consequently, $(I, g) \in A$, which proves that A is the idealizer of $S^w\psi$ in $P(\mathcal{I}_Y, G; G\mathcal{I}_Y)$.

VII.3.5 Remark

Note that with the isomorphisms χ in V.1.3 and ψ in 3.2, we have that $\chi(\psi|_{i_{\hat{S}}(S^w)})$ is an isomorphism of $\Omega(S)$ onto $P(\mathcal{R}_Y, G; G\mathcal{R}_Y)$.

Notice the similarity of the *P*-representation for $\hat{S}$ and $\Omega(S)$ when $S = P(Y, G; X)$.

VII.3.6 Exercises

(i) Prove the following statements for an *E*-unitary inverse semigroup *S*.

(α) $E_{C(S)} = C(E_S) = \mathcal{I}_{E_S}$.

(β) Every element of $C(S)$ is contained in some σ-class of *S*.

(γ) For any $A, B \in C(S)$, $A\sigma_{C(S)}B$ if and only if *A* and *B* are contained in the same σ-class of *S*. Hence $S/\sigma \cong C(S)/\sigma$.

(δ) For any $a \in S$, $a\sigma$ is the greatest element of its σ-class in $C(S)$.

(ε) The mapping $s \to [s]$ is an embedding of *S* into $C(S)$.

(ii) Let *V* be an inverse semigroup and a dense ideal extension of a semigroup *S*. Show that if *S* is *E*-unitary, so is *V*.

(iii) Show that an inverse semigroup *S* is *E*-reflexive if and only if *S* is a subdirect product of *E*-unitary inverse semigroups with a zero possibly adjoined.

VII.3.7 Problems

(i) Is the translational hull of an *E*-reflexive inverse semigroup *E*-reflexive?

All the results in this section are due to McAlister [8].

VII.4 *E*-UNITARY COVERS

We provide a construction of all *E*-unitary covers for an arbitrary inverse semigroup over a group. It is then demonstrated that every inverse semigroup has an *E*-unitary cover over some group. This result can be considerably sharpened. As an example, we construct all *E*-unitary covers of a Brandt semigroup.

VII.4.1 Definition

Let *S* be an inverse semigroup. An inverse semigroup *T* is an *E-unitary cover for S over a group G* if *T* is *E*-unitary, $T/\sigma \cong G$, and there is an idempotent separating homomorphism of *T* onto *S*. A function $\theta : G \to C(S)$, say $\theta : g \to \theta_g$, is *full* if $\cup_{g \in G} \theta_g = S$.

VII.4.2 Theorem

Let S be an inverse semigroup and G be a group. If $\theta: G \to C(S)$ is a full prehomomorphism, then

$$T = \{(s, g) \in S \times G | s \in \theta_g\}$$

is an E-unitary cover for S over G. Conversely, every E-unitary cover for S over G can be so constructed.

Proof. Recall that by V.2.6 the natural order in $C(S)$ coincides with inclusion.

Direct part. Let T be as in the statement of the theorem. Since θ is a prehomomorphism, T is closed under multiplication and taking of inverses. Let $e \in E_S$. Since θ is full, there exists $g \in G$ such that $e \in \theta_g$. Hence $e = ee \in \theta_g\theta_g^{-1} = \theta_g\theta_{g^{-1}} \subseteq \theta_1$ and thus $(e, 1) \in T$. It follows that $E_T = \{(e, 1) | e \in E_S\}$. Let $(s, g) \in T$ and $e \in E_S$ be such that $(s, g)(e, 1) = (e, 1)$. Then $g = 1$ so $s \in \theta_1$. Hence

$$s = ss^{-1}s \in \theta_1\theta_1^{-1}\theta_1 = \theta_1\theta_1\theta_1 \subseteq \theta_1\theta_1 \subseteq E_S$$

since θ_1 is permissible, which gives $(s, 1) \in E_T$. Consequently, T is an E-unitary inverse semigroup.

Fullness of θ implies that the projection mapping $(s, g) \to s$ maps T onto S, and is thus an idempotent separating homomorphism of T onto S. Next let $(s, g), (t, h) \in T$. If $(s, g)\sigma(t, h)$, then

$$(s, g)(e, 1) = (t, h)(e, 1)$$

for some $e \in E_S$ and thus $se = te$ and $g = h$. Conversely, let $g = h$. Then $s, t \in \theta_g$ and thus $st^{-1} = f \in E_S$ since θ_g is permissible. Hence $s(t^{-1}t) = ft = t(t^{-1}ft)$ so that $se = te$ for $e = t^{-1}ft$, and thus $(s, g)\sigma(t, h)$. Consequently, $(s, g)\sigma(t, h)$ if and only if $g = h$ which implies $T/\sigma \cong G$. Therefore, T is an E-unitary cover for S over G.

Converse. Let φ be an idempotent separating homomorphism of an E-unitary inverse semigroup T onto S and let $G = T/\sigma$. For each $g \in G$, let

$$\theta_g = \{s \in S | s = t\varphi, t\sigma = g \text{ for some } t \in T\}.$$

We will show that the function $\theta: g \to \theta_g$ is the required one.

Let $g \in G$. Since $\sigma: T \to G$ is onto, $\theta_g \neq \varnothing$.

Next let $s \in \theta_g$ and $e \in E_S$. Then $s = t\varphi$, $t\sigma = g$, and $e = f\varphi$ for some $t \in T, f \in E_T$. Hence $se = (tf)\varphi$ and $(tf)\sigma = g$ so that $se \in \theta_g$.

Let $s, t \in \theta_g$. There exist $u, v \in T$ such that $u\varphi = s$, $v\varphi = t$, $u\sigma = v\sigma = g$. It follows that $(uv^{-1})\sigma = (u\sigma)(v\sigma)^{-1} = gg^{-1} = 1$ and thus $uv^{-1} \in E_T$ since T is

E-unitary. But then $st^{-1} = (uv^{-1})\varphi$ yields that $st^{-1} \in E_S$; one shows analogously that $s^{-1}t \in E_S$.

Consequently, θ maps G into $C(S)$.

Let $s \in \theta_g$ and $t \in \theta_h$. Then $s = u\varphi$, $u\sigma = g$, $t = v\varphi$, $v\sigma = h$ for some $u, v \in T$. Hence $st = (uv)\varphi$, $(uv)\sigma = gh$, which proves that $\theta_g\theta_h \subseteq \theta_{gh}$.

Note that $s \in \theta_g$ is equivalent to $s = t\varphi$, $t\sigma = g$ for some $t \in T$, and this is equivalent to $s^{-1} = t^{-1}\varphi$, $t^{-1}\sigma = g^{-1}$ for some $t^{-1} \in T$, which is finally equivalent to $s^{-1} \in \theta_{g^{-1}}$. It follows that $\theta_g^{-1} = \theta_{g^{-1}}$.

Fullness of θ follows from the fact that φ maps T onto S.

Now define

$$T' = \{(s, g) \in S \times G | s \in \theta_g\},$$

$$\psi : t \to (t\varphi, t\sigma) \qquad (t \in T).$$

It is clear that ψ is a homomorphism of T onto T'. Assume that $(t\varphi, t\sigma) = (u\varphi, u\sigma)$ for some $t, u \in T$. Then $t\varphi = u\varphi$ implies $t\mathcal{H}u$ since φ is idempotent separating. We thus have $t(\mathcal{H} \cap \sigma)u$ which gives $t = u$ since T is *E*-unitary. Consequently, ψ is an isomorphism of T onto T'.

VII.4.3 Remark

For the *E*-unitary inverse semigroup T constructed in 4.2 we can easily give the *Q*-representation. For the mapping

$$(s, g) \to (ss^{-1}, g)$$

provides an isomorphism of T onto $Q(E_S, G; \psi)$ where

$$\mathbf{d}\psi_g = \{s^{-1}s | s \in \theta_g\}, \qquad \psi_g(s^{-1}s) = ss^{-1}.$$

Note that conversely, if $S = Q(Y, G; \psi)$, then the mapping $\theta : g \to \theta_g$, where

$$\theta_g = (Y \times \{g\}) \cap S = \mathbf{r}\psi_g \times \{g\},$$

is a function $G \to C(S)$ satisfying the conditions in 4.2. A verification of these statements is left as an exercise.

We show next that the hypotheses of 4.2 are realizable for any inverse semigroup S and for some group G.

Let S be an inverse semigroup. Let S' be a set of the same cardinality as S disjoint from S, and let

$$X = \begin{cases} S & \text{if } S \text{ is finite} \\ S \cup S' & \text{if } S \text{ is infinite.} \end{cases}$$

Denote by $w : s \to w^s$ the Wagner representation of S, and set

$$G = \{ g \in \mathcal{S}(X) | w^s \subseteq g \text{ for some } s \in S \}.$$

Define a function $\theta : g \to \theta_g$ where for each $g \in G$,

$$\theta_g = \{ s \in S | w^s \subseteq g \}.$$

VII.4.4 Lemma

The function θ satisfies the conditions in 4.2 and

$$T = \{ (s, g) \in S \times G | s \in \theta_g \}$$

is an E-unitary cover for S over G.

Proof. Let $g \in G$. By the very construction, $\theta_g \neq \varnothing$. Let $s \in \theta_g$ and $e \in E_S$. Then w^e is an idempotent, so $w^e \subseteq \iota_X$ which implies $w^{se} = w^s w^e \subseteq \theta_g \theta_1 \subseteq \theta_g$ and shows that $se \in \theta_g$. Thus θ_g is an order ideal of S. If $s, t \in \theta_g$, then $st^{-1} \in \theta_g \theta_g^{-1} = \theta_g \theta_{g^{-1}} \subseteq \theta_1$ which implies that $st^{-1} \in E_S$ since $\theta_1 = E_S$. One shows analogously that $s^{-1}t \in E_S$. Consequently, θ maps G into $C(S)$.

Let $s \in \theta_g$ and $t \in \theta_h$. Then $w^s \subseteq g$ and $w^t \subseteq h$ whence $w^{st} = w^s w^t \subseteq gh$ and thus $st \in \theta_{gh}$, which verifies that $\theta_g \theta_h \subseteq \theta_{gh}$. Also, $s \in \theta_g$ is equivalent to $w^s \subseteq g$ and this to $w^{s^{-1}} \subseteq g^{-1}$ and this to $s^{-1} \in \theta_{g^{-1}}$. Hence $\theta_g^{-1} = \theta_{g^{-1}}$.

Let $a \in S$. Then $w^a \in \mathcal{I}(X)$. If S is finite, then

$$|S \setminus Sa^{-1}| = |S \setminus Sa|. \tag{5}$$

If S is infinite, then

$$|X \setminus Sa^{-1}| = |S \setminus Sa^{-1}| + |S'| = |S|$$

and analogously $|X \setminus Sa| = |S|$, so again (5) holds. Now letting h be any one-to-one mapping of $X \setminus Sa^{-1}$ onto $X \setminus Sa$, we get that $g = h \cup w^a$ is a permutation of X such that $w^a \subseteq g$. In particular, $g \in G$ and $a \in \theta_g$. This verifies that θ is full.

By 4.2, T is an E-unitary cover of S over G.

A less restrictive concept concerning E-unitary covers is the following.

VII.4.5 Definition

An E-unitary inverse semigroup T is an *E-unitary cover* for an inverse semigroup S if there exists an idempotent separating homomorphism of T onto S.

Now 4.4 has the following interesting consequence.

VII.4.6 Corollary

Every inverse semigroup has an E-unitary cover.

We can sharpen this result considerably. To this end, we first prove an auxiliary result.

VII.4.7 Lemma

Let φ be an idempotent separating homomorphism of an inverse semigroup S onto a semigroup T. Then for all $a, b \in S$, $a\mu b \Leftrightarrow (a\varphi)\mu(b\varphi)$.

Proof. The direct implication is obvious. Let $(a\varphi)\mu(b\varphi)$. Then $(a\varphi)^{-1}(e\varphi)(a\varphi) = (b\varphi)^{-1}(e\varphi)(b\varphi)$ for all $e \in E_S$, so that $(a^{-1}ea)\varphi = (b^{-1}eb)\varphi$ for all $e \in E_S$. Since φ is idempotent separating, we must have $a^{-1}ea = b^{-1}eb$ for all $e \in E_S$ and hence $a\mu b$.

The following result will prove very useful later.

VII.4.8 Theorem

For every inverse semigroup S there exists a group G, a (inverse semigroup) subdirect product T of S/μ and G, and an idempotent separating homomorphism φ of T onto S.

Proof. Let S be an inverse semigroup. By 4.6, S has an E-unitary cover T. That T is E-unitary means that $\ker \sigma = E_T$. Hence

$$\ker(\mu \cap \sigma) \subseteq \ker \sigma = E_T,$$

$$\operatorname{tr}(\mu \cap \sigma) \subseteq \operatorname{tr} \mu = \varepsilon,$$

so that $\mu \cap \sigma = \varepsilon$, and thus T is a subdirect product of the antigroup T/μ and the group $G = T/\sigma$. By 4.7, we have that $T/\mu \cong S/\mu$, and the assertion follows.

In order to illustrate the theory, we now construct all E-unitary covers for a Brandt semigroup $B(H, I)$.

VII.4.9 Lemma

Let K be a group, $Q = B(H, I)$ be a Brandt semigroup disjoint from K, $u : i \to u_i$ be a function mapping I into K, and τ be a homomorphism of H into K. Then

$$\psi : (i, h, j) \to u_i^{-1}(h\tau)u_j \qquad [(i, h, j) \in Q^*]$$

is a partial homomorphism of Q^* into K. Conversely, every partial homomorphism of Q^* into K can be so constructed.

Proof. It is obvious that ψ in the statement of the lemma is a partial homomorphism. Conversely, let $\psi: Q^* \to K$ be a partial homomorphism. Fix an element, say k, in I. Define τ and $u: i \to u_i$ by the requirements:

$$h\tau = (k, h, k)\psi, \qquad u_i = (k, 1, i)\psi \qquad (h \in H, i \in I).$$

Then

$$(i, g, j)\psi = (k, 1, i)^{-1}\psi(k, g, k)\psi(k, 1, j)\psi = u_i^{-1}(h\tau)u_j$$

as required.

VII.4.10 Lemma

With the notation of 4.9, let $T = K \cup Q^*$ with the multiplication

$$(i, g, j)(p, h, q) = u_i^{-1}(g\tau)u_j u_p^{-1}(h\tau)u_q \quad \text{if } j \neq p,$$

$$(i, g, j)k = u_i^{-1}(g\tau)u_j k, \qquad k(i, g, j) = ku_i^{-1}(g\tau)u_j,$$

if $(i, g, j), (p, h, q) \in Q^*$, $k \in K$, and the remaining products are as in K and Q^*. Then T is an ideal extension of K by Q. Conversely, every ideal extension of K by Q can be so constructed.

Proof. If K is an ideal of any semigroup S, then K is a retract of S by I.9.16. By I.9.14, every ideal extension of K is determined by a partial homomorphism. The lemma now follows (see I.9.13).

VII.4.11 Lemma

The inverse semigroup T constructed in 4.10 is E-unitary if and only if both $u: i \to u_i$ and τ are one-to-one and

$$\{u_i u_j^{-1} | i, j \in I\} \cap H\tau = \{1\}. \tag{6}$$

Proof

Necessity. Assume that $u_i = u_j$. Then

$$(i, 1, j)1 = u_i^{-1}u_j = u_i^{-1}u_i = 1$$

so by the hypothesis, we must have $i = j$. Next suppose that $h\tau = 1$. Then

$$(i, h, i)1 = u_i^{-1}(h\tau)u_i = u_i^{-1}u_i = 1$$

and again by the hypothesis, $h = 1$. Now assume that $u_i u_j^{-1} = h\tau$. Then

$$(i, h, j)1 = u_i^{-1}(h\tau)u_j = u_i^{-1}(u_i u_j^{-1})u_j = 1$$

and the hypothesis yields $i = j$ and $h = 1$, which gives $u_i^{-1}u_j = 1$.

Sufficiency. We consider several cases. If

$$(i, h, j)(p, 1, p) = (p, 1, p),$$

then $i = p = j$ and $h = 1$, so $(i, h, j) \in E_T$. If $(i, h, j)1 = 1$, then $u_i^{-1}(h\tau)u_j = 1$ so that $h\tau = u_i u_j^{-1}$. The hypothesis implies $h\tau = u_i u_j^{-1} = 1$ and thus $h = 1$, $u_i = u_j$. But then also $i = j$, and we have that $(i, h, j) \in E_T$. The case $k \cdot 1 = 1$ is trivial for $k \in K$, and no other case may occur.

VII.4.12 Lemma

Let $S = B(G, I)$ and T be as in 4.10. Let $v : i \to v_i$ be a function mapping I into G, ω be a homomorphism of H onto G, and ξ be a permutation of I. Then

$$\varphi : (i, g, j) \to \left(i\xi, v_i^{-1}(g\omega)v_j, j\xi\right) \qquad [(i, g, j) \in Q^*]$$

$$k \to 0 \qquad [k \in K]$$

is an idempotent separating homomorphism of T onto S. Conversely, every idempotent separating homomorphism of T onto S can be so constructed.

Proof. The verification of the direct part is straightforward and is left as an exercise. Conversely, let φ be an idempotent separating homomorphism of T onto S. Since K is the only nonzero proper ideal of T, and S is 0-simple, we must have $0\varphi^{-1} = K$. Thus φ induces a homomorphism χ of T/K onto S which coincides with φ on $(T/K)^* = T \setminus K$. Then χ maps T/K onto S and is idempotent separating. By the specialization of II.3.7 to the present setting, we see that χ maps (i, g, j) onto $(i\xi, v_i^{-1}(g\omega)v_j, j\xi)$ with all the parameters as in the statement of the lemma.

The next lemma is of general interest.

VII.4.13 Lemma

Let φ be an idempotent separating homomorphism of an inverse semigroup S onto a semigroup T. Then for any $a, b \in S$ and any Green's relation $\mathcal{K}$, $a\mathcal{K}b$ if and only if $(a\varphi)\mathcal{K}(b\varphi)$.

Proof. The direct part is true for any semigroups and any homomorphisms and can be checked readily. Assume that $(a\varphi)\mathcal{L}(b\varphi)$. Then $(a\varphi)^{-1}(a\varphi) =$

$(b\varphi)^{-1}(b\varphi)$ and thus $(a^{-1}a)\varphi = (b^{-1}b)\varphi$. By the hypothesis, it follows that $a^{-1}a = b^{-1}b$, so that $a\mathcal{L}b$. The case $\mathcal{K} = \mathcal{R}$ is symmetric, and the cases $\mathcal{K} = \mathcal{H}$ and $\mathcal{K} = \mathcal{D}$ follow from these. Let $(a\varphi)\mathcal{J}(b\varphi)$. Then $a\varphi = (x\varphi)(b\varphi)(y\varphi)$ for some $x, y \in S$. Hence $a\varphi = (xby)\varphi$ and the hypothesis yields $a\mathcal{H}xby$. But then $a = uxby$ for some $u \in S$. Consequently, $a \in SbS$, and symmetrically $b \in SaS$ so that $a\mathcal{J}b$.

We are finally ready for a description of all E-unitary covers for a Brandt semigroup.

VII.4.14 Theorem

Let $S = B(G, I)$ be a Brandt semigroup.

Let K be a group, $Q = B(H, I)$, $u : I \to K$, and $\tau : H \to K$ be one-to-one functions, and τ be a homomorphism such that (6) holds. Let $T = K \cup Q^*$ with multiplication

$$(i, g, j)(p, h, q) = u_i^{-1}(g\tau)u_j u_p^{-1}(h\tau)u_q \quad \text{if } j \neq p,$$

$$(i, g, j)k = u_i^{-1}(g\tau)u_j k, \qquad k(i, g, j) = ku_i^{-1}(g\tau)u_j,$$

if $(i, g, j), (p, h, q) \in Q^*$, $k \in K$, and the remaining products are as in Q^* and K. Then T is an inverse semigroup.

Let $v : I \to G$ be a function and ω be a homomorphism of H onto G. Then φ defined on T by

$$\varphi : (i, g, j) \to \left(i, v_i^{-1}(g\omega)v_j, j\right),$$

$$k \to 0 \quad \text{if } k \in K,$$

is an idempotent separating homomorphism of T onto S.

The semigroup T is an E-unitary cover for S over K. Conversely, every E-unitary cover for S over some group K can be so constructed.

Proof. The direct part follows directly from 4.12. Conversely, let T be an E-unitary cover of S over K and let φ be an idempotent separating homomorphism of T onto S. In view of 4.13, the $\mathcal{D}$-classes of T are $0\varphi^{-1}$ and $T \setminus 0\varphi^{-1}$ since S is 0-bisimple. Further, $K = 0\varphi^{-1}$ contains exactly one idempotent since φ is idempotent separating, and thus must be a group. Now φ induces an idempotent separating homomorphism χ of T/K onto S which coincides with φ on $T \setminus K$. Let e and f be two distinct nonzero idempotents of T/K. Then $e\varphi$ and $f\varphi$ are two distinct nonzero idempotents of S. Since S is a Brandt semigroup, we must have $(e\chi)(f\chi) = 0$. But then $(ef)\chi = 0$ and it follows that $ef = 0$ in T/K. Now T/K is a 0-bisimple inverse semigroup in which the product of any two distinct idempotents is equal to zero. It follows from II.4.3 that T/K must be a Brandt semigroup.

We thus have that T is an ideal extension of the group K by the Brandt semigroup $Q = T/K$. All such T were constructed in 4.10 and in 4.11 necessary and sufficient conditions on the parameters in the construction of the ideal extension were established. This construction and the restrictions on the parameters are precisely those in the statement of the present theorem. All idempotent separating homomorphisms of T onto S were given in 4.12. In this reference, for the present purposes, we may take ξ to be the identity mapping as we may relabel the index set of the Brandt semigroup Q. Since K is an ideal of T, it is easy to see that $\sigma_T|_K = \sigma_K$. As K is itself a group, σ_K is the equality relation, so $T/\sigma \cong K/\sigma \cong K$. This agrees with the hypothesis that T is an E-unitary cover for S over K.

VII.4.15 Remark

It is easy to verify that in the above notation

$$(i, h, j)\sigma k \Leftrightarrow u_i^{-1}(h\tau)u_j = k$$

and for the construction of T by means of a function $\theta: K \to C(S)$, we may use

$$\theta_k = \left\{\left(i, v_i^{-1}(h\omega)v_j, j\right) \in S | u_i^{-1}(h\tau)u_j = k\right\} \cup 0.$$

Condition (6) means that Iu is a cross section of some right cosets of $H\tau$ in K. In view of this, we can construct an E-unitary cover for S of the least possible cardinality as follows.

Let $S = B(G, I)$ be as above. We may suppose that I is provided with a group structure (since groups of all cardinalities exist). In the notation of 4.14, we let $H = G$, $v_i = 1$ for all $i \in I$, $\omega = \iota_G$, $K = G \times I$, where this is the direct product of the groups G and I, $u_i = (1, i)$ for all $i \in I$, and $g\tau = (g, 1)$ for all $g \in G$. We thus arrive at the following construction.

For $S = B(G, I)$, let $T = S^* \cup (G \times I)$ with multiplication

$$(i, g, j)(p, h, q) = \left(gh, i^{-1}jp^{-1}q\right) \quad \text{if } j \neq p,$$

$$(i, g, j)(h, k) = \left(gh, i^{-1}jk\right),$$

$$(h, k)(i, g, j) = \left(hg, ki^{-1}j\right),$$

for all $g, h \in G$, $i, j, k, p, q \in I$. The mapping

$$(i, g, j) \to (i, g, j),$$

$$G \times I \to 0,$$

is an idempotent separating homomorphism of T onto S. Hence T is an E-unitary cover for S over $G \times I$.

VII.4.16 Exercises

(i) Describe all E-unitary covers of a semilattice Y over a group G.
(ii) Find all E-unitary covers of a semigroup which is a semilattice of two groups.

VII.4.17 Problems

(i) Characterize all E-unitary covers of $B(G, \alpha)$.

The concept of an E-unitary cover was introduced by McAlister–Reilly [1] and 4.2 and 4.4 are due to them. However, 4.6 was first proved by McAlister [4]. Another proof can be found in Reilly–Munn [1]. Result 4.8 is due to Petrich [11]. Further results on E-unitary covers can be found in Batbedat [1], McAlister [8], and Petrich–Reilly [4].

VII.5 SPECIAL CASES

For E-unitary inverse semigroups represented either as $P(Y, G; X)$ or as $Q(Y, G; \psi)$, we consider various restrictions on the semigroup itself and on the parameters figuring in these representations. For example, a restriction on S itself may be expressed by HH^{-1} being a retract ideal of E_S for all σ-classes H of S; on $P(Y, G; X)$ that X be a semilattice; on $Q(Y, G; \psi)$ that $\psi : G \to \Phi(Y)$. For each such special case, several equivalent conditions are provided. The restrictions are made stronger step by step and they terminate with the direct product of a semilattice and a group. Various cases considered are illustrated by a diagram at the end of the section.

In order to simplify the notation, we will often write E instead of E_S.

VII.5.1 Proposition

The following conditions on an inverse semigroup S are equivalent.

(i) S is E-unitary.
(ii) For every σ-class H, $HH^{-1} \in \mathcal{I}_E$.
(iii) For every σ-class H and every $e \in E$, $HeH^{-1} \in \mathcal{I}_E$.

Proof. (i) *implies* (ii). Let $a\sigma b$. Then $ab^{-1}\sigma bb^{-1}$ so that $ab^{-1} \in E$ since E constitutes a σ-class by the hypothesis that S is E-unitary. Thus $HH^{-1} \subseteq E$ and then clearly $HH^{-1} \in \mathcal{I}_E$.

(ii) *implies* (iii). Let H be a σ-class of S, $s, t \in H$, and $e \in E$. Then by the hypothesis $st^{-1} \in E$ whence

$$set^{-1} = se(t^{-1}t)t^{-1} = s(t^{-1}t)et^{-1}$$

$$= (st^{-1})(tet^{-1}) \in E$$

so that $HeH^{-1} \subseteq E$. But then also $HeH^{-1} \in \mathcal{I}_E$.

(iii) *implies* (i). Let $ae, e \in E$. Then $a\sigma ae\sigma e$ implies $a\sigma a^{-1}a$ and so

$$a = a(a^{-1}a)a^{-1}a \in (a\sigma)(a^{-1}a)(a\sigma)^{-1} \in HeH^{-1} \subseteq E$$

and hence S is E-unitary.

VII.5.2 Lemma

In an E-unitary inverse semigroup S for any σ-class H and $e \in E$, we have

(i) $HeH^{-1} = \{heh^{-1} | h \in H\}$,
(ii) $HH^{-1} = \{hh^{-1} | h \in H\}$.

Proof. Let H be a σ-class of S and $s, t \in H$, $e \in E$. Then

$$(set^{-1})\sigma = (s\sigma)(e\sigma)(t\sigma)^{-1} = (s\sigma)(t\sigma)^{-1} = HH^{-1} \subseteq E$$

in view of 5.1, and thus $set^{-1} \in E$ again since S is E-unitary. Hence

$$set^{-1} = (set^{-1})(set^{-1})^{-1} = set^{-1}tes^{-1} = (st^{-1}t)e(t^{-1}ts^{-1})$$

$$= (st^{-1}t)e(st^{-1}t)^{-1}$$

where $st^{-1}t \in H$. Consequently, $HeH^{-1} \subseteq \{heh^{-1} | h \in H\}$ and the opposite inclusion is trivial. This proves (i); the proof of (ii) is analogous.

VII.5.3 Lemma

Let $S = P(Y, G; X) \cong Q(Y, G; \psi)$, and for $g \in G$, let H be the corresponding σ-class of S, $e = (\alpha, 1) \in E$.

(i) $H = \{(\beta, g) | g^{-1}\beta \in Y\} = \{(\beta, g) | \beta \in \mathbf{r}\psi_g\}$.
(ii) $HH^{-1} = \{(\beta, 1) | g^{-1}\beta \in Y\} = \{(\beta, 1) | \beta \in \mathbf{r}\psi_g\}$.
(iii) $HeH^{-1} = \{(\beta, 1) | \beta \in [g\alpha] \cap Y\}$.
(iv) $HH^{-1} = \cup_{e \in E} HeH^{-1}$.

Proof. Items (i) and (ii) are obvious. By 5.2, we get

$$HeH^{-1} = \{(\gamma, g)(\alpha, 1)(\gamma, g)^{-1} | g^{-1}\gamma \in Y\}$$

$$= \{(\gamma \wedge g\alpha, 1) | g^{-1}\gamma \in Y\}. \tag{7}$$

Letting $\beta = \gamma \wedge g\alpha$, we obtain $\beta \in [g\alpha] \cap Y$. Conversely, if $\beta \in [g\alpha] \cap Y$, then $\beta \leq g\alpha$ so that $g^{-1}\beta \in Y$ and $\beta = \beta \wedge g\alpha$. It follows that the last expression in (7) coincides with the right-hand side of (iii).

A straightforward verification shows that (iv) also holds.

VII.5.4 Proposition

The following conditions on $S = P(Y, G; X)$ are equivalent.

(i) For all σ-classes H and all $e \in E$, $HeH^{-1} \in \mathcal{R}_E$.
(ii) For all $\alpha \in X$, $[\alpha] \cap Y \in \mathcal{R}_Y$.
(iii) X is a semilattice.
(iv) For any σ-class H and any $e, f \in E$, $H \cap eSf$ has a greatest element.

Proof. The equivalence of (i) and (ii) follows directly from 5.3.

(ii) *implies* (iii). Let $g, h \in G$ and $\alpha, \beta \in Y$. Then by the hypothesis, we have

$$[h^{-1}g\alpha] \cap [\beta] = ([h^{-1}g\alpha] \cap Y) \cap [\beta] = [\gamma]$$

for some $\gamma \in Y$ and thus $h^{-1}g\alpha \wedge \beta = \gamma$ whence $g\alpha \wedge h\beta = h\gamma$. Since $GY = X$, this shows that X is a semilattice.

(iii) *implies* (iv). Let H be a σ-class and $e, f \in E_S$. Then $H = \{(\gamma, h) | h^{-1}\gamma \in Y\}$ for a fixed $h \in G$. For $e = (\alpha, 1)$ and $f = (\beta, 1)$, by hypothesis $\alpha \wedge h\beta$ exists, and $h^{-1}(\alpha \wedge h\beta) = h^{-1}\alpha \wedge \beta \in Y$. Hence $(\alpha \wedge h\beta, h) \in S$, and thus $(\alpha \wedge h\beta, h) \in H$. The elements of eSf are of the form $(\alpha, 1)(\gamma, g)(\beta, 1) = (\alpha \wedge \gamma \wedge g\beta, g)$ with $g^{-1}\gamma \in Y$, those which are also in H are thus of the form $(\alpha \wedge \gamma \wedge h\beta, h)$. Since

$$(\alpha \wedge \gamma \wedge h\beta, h) = (\gamma, 1)(\alpha \wedge h\beta, h)$$

we deduce that $(\alpha \wedge h\beta, h)$ is the greatest element of $H \cap eSf$.

(iv) *implies* (i). Let H be a σ-class and $e, f \in E$, and consider $HeH^{-1} \cap [f]$. Let a be the greatest element of $H \cap fSe$. For $b \in HeH^{-1} \cap [f]$, we have by 5.2, $b = heh^{-1} = gf$ for some $h \in H$ and $g \in E$. Hence

$$b = fheh^{-1}f = (fhe)(fhe)^{-1} \leq aa^{-1}$$

which shows that $HeH^{-1} \cap [f] \subseteq [aa^{-1}]$. Conversely, let $x \leq aa^{-1}$. Then

$x = taa^{-1}$ for some $t \in E$. Using the fact that $a \in H \cap fSe$, we get

$$x = t(fae)(fae)^{-1} = (tfa)e(a^{-1}f) \in HeH^{-1} \cap [f],$$

which shows that $[aa^{-1}] \subseteq HeH^{-1} \cap [f]$, and the equality prevails. This proves that $HeH^{-1} \cap [f] \in \mathcal{P}_E$ so that $HeH^{-1} \in \mathcal{R}_E$.

VII.5.5 Proposition

The following conditions on $S \cong P(Y, G; X) \cong Q(Y, G; \psi)$ are equivalent.

(i) For all σ-classes H, $HH^{-1} \in \mathcal{R}_E$.
(ii) For all $g \in G$, $gY \cap Y \in \mathcal{R}_Y$.
(iii) For all σ-classes H and all $e \in E$, $HeH^{-1} \in \mathcal{P}_E$.
(iv) For all $\alpha \in X$, $[\alpha] \cap Y \in \mathcal{P}_Y$.
(v) X is a semilattice and $Y \in \mathcal{R}_X$.
(vi) $\psi : G \to \Psi(Y)$.

Proof. The equivalence of (i) and (ii), and of (iii) and (iv) follows directly from 5.3.

(ii) *implies* (iv). In view of the hypothesis $GY = X$, it suffices to prove that $[g\alpha] \cap Y \in \mathcal{P}_Y$ for all $g \in G$, $\alpha \in Y$. First

$$g^{-1}Y \cap [\alpha] = (g^{-1}Y \cap Y) \cap [\alpha] = [\beta]$$

for some $\beta \in Y$ by the hypothesis. Thus

$$[g\alpha] \cap Y = g([\alpha] \cap g^{-1}Y) = g[\beta] = [g\beta]$$

so that $[g\alpha] \cap Y \in \mathcal{P}_Y$.

(iv) *implies* (v). It follows from 5.4 that X is a semilattice. The very condition $[\alpha] \cap Y \in \mathcal{P}_Y$ implies $[\alpha] \cap Y \in \mathcal{P}_X$ for all $\alpha \in X$ which implies that $Y \in \mathcal{R}_X$.

(v) *implies* (ii). Let $\alpha \in Y$, $g \in G$. Using the hypothesis, we get

$$[\alpha] \cap (gY \cap Y) = [\alpha] \cap gY = g([g^{-1}\alpha] \cap Y) = g[\beta] = [g\beta]$$

for some $\beta \in Y$ and hence $gY \cap Y \in \mathcal{R}_Y$.

The equivalence of (i) and (vi) follows directly from 5.3.

We give below an alternative construction for the semigroups appearing in the above proposition, so for convenience we give them a name.

VII.5.6 Definition

An *R-inverse semigroup* is an inverse semigroup in which $HH^{-1} \in \mathcal{R}_E$ for every σ-class H.

Note that by 5.1 an R-inverse semigroup is E-unitary so that R-inverse semigroups are precisely those appearing in 5.5.

VII.5.7 Construction

Let G be a group and Y a semilattice. Let G act on Y on the left and assume that the action satisfies the following axioms:

(F1) $1\alpha = \alpha$,

(F2) $g(\alpha \wedge \beta) = g\alpha \wedge g\beta$,

(R3) $[g(h\alpha)] = [(gh)\alpha] \cap gY$,

for all $g, h \in G$, $\alpha, \beta \in Y$. Let

$$S = \{(\alpha, g) \in Y \times G | \alpha \in gY\}$$

with the multiplication

$$(\alpha, g)(\beta, h) = (\alpha \wedge g\beta, gh).$$

Denote this structure by $R(Y, G)$.

VII.5.8 Lemma

With the above notation, $S = R(Y, G)$ is an R-inverse semigroup.

Proof. We start with closure of the operation. First let $g \in G$ and $\alpha \in Y$. Then by (F1) and (R3), we have

$$[g\alpha] = [g(1\alpha)] = [(g1)\alpha] \cap gY = [g\alpha] \cap gY,$$

which implies that gY is an ideal of Y for all $g \in G$. Now let $(\alpha, g), (\beta, h) \in S$. Then $\alpha \in gY$ and $\beta \in hY$ so that $\beta = h\gamma$ for some $\gamma \in Y$. Using (R3) and the fact that gY and $(gh)Y$ are ideals of Y, we obtain

$$\begin{aligned} [\alpha \wedge g\beta] &= [\alpha] \cap [g(h\gamma)] \\ &= [\alpha] \cap [(gh)\gamma] \cap gY \\ &= [\alpha] \cap [(gh)\gamma] \subseteq (gh)Y \end{aligned}$$

which yields that $\alpha \wedge g\beta \in (gh)Y$. Consequently, $(\alpha \wedge g\beta, gh) \in S$ and S is closed under this operation.

We will need below the formula

$$g[\beta] = [g\beta] \qquad (g \in G, \beta \in Y). \tag{8}$$

Indeed, by (F2), we get $g[\beta] \subseteq [g\beta]$. Conversely, let $\gamma \in [g\beta]$. Then $\gamma = g\delta$ for some $\delta \in Y$ since, as we have seen above, $[g\beta] \subseteq gY$. Hence by (F2),

$$\gamma = g\delta = g\beta \wedge g\delta = g(\beta \wedge \delta) \in g[\beta]$$

which completes the proof of relation (8).

Next we verify associativity. On the one hand,

$$\begin{aligned}[(\alpha, g)(\beta, h)](\gamma, p) &= (\alpha \wedge g\beta, gh)(\gamma, p) \\ &= (\alpha \wedge g\beta \wedge (gh)\gamma, ghp)\end{aligned}$$

and on the other hand,

$$\begin{aligned}(\alpha, g)[(\beta, h)(\gamma, p)] &= (\alpha, g)(\beta \wedge h\gamma, hp) \\ &= (\alpha \wedge g\beta \wedge g(h\gamma), ghp),\end{aligned}$$

and using (8),

$$\begin{aligned}[\alpha \wedge g\beta \wedge (gh)\gamma] &= [\alpha] \cap [g\beta] \cap [(gh)\gamma] \\ &= [\alpha] \cap g[\beta] \cap [g(h\gamma)] \cap gY \\ &= [\alpha] \cap g[\beta] \cap [g(h\gamma)] \\ &= [\alpha \wedge g\beta \wedge g(h\gamma)]\end{aligned}$$

which establishes associativity.

Let $(\alpha, g) \in S$. Then $g^{-1}\alpha \in g^{-1}Y$ implies $(g^{-1}\alpha, g^{-1}) \in S$, and

$$\begin{aligned}(\alpha, g)(g^{-1}\alpha, g^{-1})(\alpha, g) &= (\alpha \wedge g(g^{-1}\alpha), 1)(\alpha, g) \\ &= (\alpha \wedge g(g^{-1}\alpha), g)\end{aligned}$$

with

$$[\alpha \wedge g(g^{-1}\alpha)] = [\alpha] \cap [g(g^{-1}\alpha)] = [\alpha] \cap [\alpha] \cap gY = [\alpha] \tag{9}$$

since $\alpha \in gY$. It follows that

$$(\alpha, g)(g^{-1}\alpha, g^{-1})(\alpha, g) = (\alpha, g),$$

and hence S is regular.

Routine verification shows that

$$E_S = \{(\alpha, 1) | \alpha \in Y\},$$

that idempotents commute, that S is E-unitary, and that

$$(\alpha, g)\sigma(\beta, h) \Leftrightarrow g = h.$$

Let H be the σ-class corresponding to $g \in G$. Then by 5.2(ii) and (9), we obtain

$$\begin{aligned} HH^{-1} &= \{(\alpha, g)(\alpha, g)^{-1} | \alpha \in gY\} \\ &= \{(\alpha \wedge g(g^{-1}\alpha), 1) | \alpha \in gY\} \\ &= \{(\alpha, 1) | \alpha \in gY\}. \end{aligned}$$

Hence

$$\begin{aligned} HH^{-1} \cap [(\gamma, 1)] &= \{(\alpha, 1) | \alpha \in [\gamma] \cap gY\} \\ &= \{(\alpha, 1) | \alpha \in [g(g^{-1}\gamma)]\} \\ &= [(g(g^{-1}\gamma), 1)] \end{aligned}$$

and S is an R-inverse semigroup.

VII.5.9 Lemma

Let $S = P(Y, G; X)$ be an R-inverse semigroup. Let G act on Y as follows: $g \cdot \alpha$ is given by

$$[g \cdot \alpha] = [g\alpha] \cap Y.$$

Then $R(Y, G)$ is equal to S.

Proof. First note that the hypothesis that S is an R-inverse semigroup implies by 5.5 that X is a semilattice and $Y \in \mathfrak{R}_X$. Hence the definition of $g \cdot \alpha$ makes sense.

We now verify the axioms for $g \cdot \alpha$. First

$$[1 \cdot \alpha] = [1\alpha] \cap Y = [\alpha]$$

so that $1 \cdot \alpha = \alpha$; secondly,

$$\begin{aligned} [g \cdot (\alpha \wedge \beta)] &= [g(\alpha \wedge \beta)] \cap Y = [g\alpha \wedge g\beta] \cap Y \\ &= ([g\alpha] \cap Y) \cap ([g\beta] \cap Y) = [g \cdot \alpha] \cap [g \cdot \beta] \\ &= [g \cdot \alpha \wedge g \cdot \beta] \end{aligned}$$

which yields $g \cdot (\alpha \wedge \beta) = g \cdot \alpha \wedge g \cdot \beta$; thirdly,

$$\begin{aligned}[g \cdot (h \cdot \alpha)] &= [g(h \cdot \alpha)] \cap Y = g([h\alpha] \cap Y) \cap Y \\ &= g[h\alpha] \cap gY \cap Y = ([(gh)\alpha] \cap Y) \cap (gY \cap Y) \\ &= [(gh) \cdot \alpha] \cap g \cdot Y.\end{aligned}$$

Furthermore,

$$(\alpha, g) \in S \Leftrightarrow g^{-1}\alpha \in Y \Leftrightarrow \alpha \in gY \cap Y \Leftrightarrow \alpha \in g \cdot Y$$

and thus $S = R(Y, G)$ as sets. Since

$$\begin{aligned}[\alpha \wedge g\beta] &= [\alpha] \cap [g\beta] = [\alpha] \cap ([g\beta] \cap Y) \\ &= [\alpha] \cap [g \cdot \beta] = [\alpha \wedge g \cdot \beta]\end{aligned}$$

they also have the same multiplication.

We can summarize the essential parts of the two preceding lemmas as follows.

VII.5.10 Theorem

Let a group G act on a semilattice Y obeying the axioms (F1), (F2), and (R3). Then $R(Y, G)$ is an R-inverse semigroup. Conversely, every R-inverse semigroup is isomorphic to some $R(Y, G)$.

Proof. The direct part is the content of 5.8. Conversely, let S be an R-inverse semigroup. We have remarked after 5.6 that S is then E-unitary. By 1.14, S is isomorphic to a P-semigroup $P(Y, G; X)$ which is equal to $R(Y, G)$ by 5.9.

VII.5.11 Proposition

The following conditions on $S \cong P(Y, G; X) \cong Q(Y, G; \psi)$ are equivalent.

(i) For all σ-classes H, $HH^{-1} \in \mathcal{P}_E$.
(ii) For all $g \in G$, $gY \cap Y \in \mathcal{P}_Y$.
(iii) X is a semilattice and $Y \in \mathcal{P}_X$.
(iv) $\psi : G \to \Phi(Y)$.

Proof. The equivalence of (i) and (ii) follows directly from 5.3.

(ii) *implies* (iii). It follows from 5.4 that X is a semilattice. For $g = 1$, we obtain $1Y \cap Y = Y \in \mathcal{P}_Y$.

(iii) *implies* (ii). By hypothesis, Y has an identity element, say ε. Hence

$$gY \cap Y = g[\varepsilon] \cap [\varepsilon] = [g\varepsilon] \cap [\varepsilon] \in \mathcal{P}_Y$$

since X is a semilattice and Y is an ideal of X.

The equivalence of (i) and (iv) follows directly from 5.3(ii).

We will have an alternative construction for the semigroups appearing in the above proposition. We first prove the following statement.

VII.5.12 Lemma

Let H be a σ-class of an inverse semigroup S.

(i) If H has a greatest element h, then $HH^{-1} = Ehh^{-1}$.
(ii) If $HH^{-1} = Ee$ for some $e \in E$ and $H^{-1}H \subseteq E$, then H has a greatest element h and $hh^{-1} = e$.

Proof. (i) If h is the greatest element of H, then $H = Eh$ and

$$HH^{-1} = (Eh)(Eh)^{-1} = (Eh)(h^{-1}E) = Ehh^{-1}.$$

(ii) Let $HH^{-1} = Ee$, so that $e = hh^{-1}$ for some $h \in H$ in view of 5.2(ii). For any $a \in H$, we then have $aa^{-1} \leq e = hh^{-1}$. Further $h^{-1}a \in H^{-1}H \subseteq E$ by the hypothesis and thus

$$a = aa^{-1}a \leq (hh^{-1})a = h(h^{-1}a) \leq h.$$

Hence h is the greatest element of H.

VII.5.13 Definition

An *F-inverse semigroup* is an inverse semigroup in which every σ-class has a greatest element.

VII.5.14 Proposition

F-inverse semigroups are precisely those appearing in 5.11.

Proof. Let S be an F-inverse semigroup. Then for any σ-class H of S, we have $HH^{-1} = Ehh^{-1}$ by 5.12, where h is the greatest element of H. Hence $HH^{-1} \subseteq E$ and thus by 5.1, S is E-unitary. Since also $HH^{-1} \in \mathcal{P}_E$, S is a semigroup satisfying the conditions of 5.11.

Conversely, if S satisfies the conditions of 5.11, then any σ-class H of S satisfies the condition of 5.12(ii), and thus S must be an F-inverse semigroup.

VII.5.15 Construction

Let Y in 5.7 have an identity element ε. For any $g \in G$, $\alpha \in Y$, by (8), we have $[g\varepsilon] = g[\varepsilon] = gY$, so that

$$[g(h\alpha)] = [(gh)\alpha] \cap [g\varepsilon] \qquad (g, h \in G, \alpha \in Y)$$

and we may substitute axiom (R3) by

$$\text{(F3)} \quad g(h\alpha) = (gh)\alpha \wedge g\varepsilon \qquad (g, h \in G, \alpha \in Y).$$

We denote the semigroup $R(Y, G)$ by $F(Y, G)$ in this case and note that in it $HH^{-1} \in \mathcal{P}_E$ for all σ-classes H.

VII.5.16 Theorem

Let a group G act on a semilattice Y with identity ε obeying the axioms (F1), (F2), and (F3). Then $F(Y, G)$ is an F-inverse semigroup. Conversely, every F-inverse semigroup is isomorphic to some $F(Y, G)$.

Proof. The direct part follows from 5.15, 5.8, and 5.14. For the converse, taking into account 5.14, we have that an F-inverse semigroup is an R-inverse semigroup in which the σ-class E, and thus Y, has an identity element. The assertion now follows from 5.11.

VII.5.17 Definition

An isomorphism of an F-inverse semigroup S onto a semigroup of the form $F(Y, G)$ is an *F-representation* of S.

VII.5.18 Proposition

The following conditions on $S \cong P(Y, G; X) \cong Q(Y, G; \psi)$ are equivalent.

(i) S is a Clifford semigroup.
(ii) The action of G on X is trivial.
(iii) $\psi : G \to E_{\Sigma(Y)}$.

Proof. The semigroup S is a Clifford semigroup if and only if $xx^{-1} = x^{-1}x$ for all $x \in S$; this is equivalent to $g^{-1}\alpha = \alpha$ for all $\alpha \in \mathbf{d}\psi_g$ and all $g \in G$, which is in turn equivalent to $\psi_g \in E_{\Sigma(Y)}$ for all $g \in G$. Thus (i) and (iii) are equivalent, and analogously (i) and (ii) are equivalent.

Note that the condition in 5.18 is equivalent to S being a subdirect product of Y and G since the multiplication is componentwise.

We will now make a little digression of independent interest which helps the treatment of another special case.

VII.5.19 Notation

Let S be an inverse semigroup. For every $g \in S/\sigma$, define a relation φ_g on E_S by

$$e\varphi_g f \Leftrightarrow e = ss^{-1}, \quad f = s^{-1}s, \quad s\sigma = g \quad \text{for some } s \in S.$$

Note that φ_g is a relation which coincides with the inverse of the function ψ_g in the case S is E-unitary.

VII.5.20 Lemma

For any inverse semigroup S, we have

(i) $\varphi_g\varphi_h \subseteq \varphi_{gh} \quad (g, h \in S/\sigma)$,
(ii) $\varphi_g^{-1} = \varphi_{g^{-1}} \quad (g \in S/\sigma)$.

Proof. (i) Let $e\varphi_g f$, $f\varphi_h q$. Then

$$e = ss^{-1}, \qquad f = s^{-1}s = tt^{-1},$$

$$q = t^{-1}t, \qquad s\sigma = g, \qquad t\sigma = h,$$

for some $s, t \in S$. We obtain

$$(st)(st)^{-1} = s(tt^{-1})s^{-1} = sfs^{-1} = s(s^{-1}s)s^{-1} = ss^{-1} = e,$$

$$(st)^{-1}(st) = t^{-1}(s^{-1}s)t = t^{-1}ft = t^{-1}(tt^{-1})t = t^{-1}t = q,$$

$$(st)\sigma = (s\sigma)(t\sigma) = gh,$$

which implies $e\varphi_{gh}q$, and proves $\varphi_g\varphi_h \subseteq \varphi_{gh}$.
(ii) We calculate

$$e\varphi_{g^{-1}}f \Leftrightarrow e = ss^{-1}, \quad f = s^{-1}s, \quad s\sigma = g^{-1} \quad \text{for some } s \in S$$

$$\Leftrightarrow f = s^{-1}s, \quad e = ss^{-1}, \quad s^{-1}\sigma = g \quad \text{for some } s \in S$$

$$\Leftrightarrow f\varphi_g e$$

so that $\varphi_{g^{-1}} = \varphi_g^{-1}$.

The following concept is of interest for arbitrary semigroups.

VII.5.21 Definition

A congruence ρ on a semigroup S is *perfect* if for any $a, b \in S$, $(a\rho)(b\rho) = (ab)\rho$ where the product and the equality are the product and the equality of sets, respectively.

Note that in a group every congruence is perfect. For φ_g as in 5.19, let $\mathbf{d}\varphi_g$ and $\mathbf{r}\varphi_g$ be the first and the second projections of φ_g, respectively.

VII.5.22 Proposition

Let S be an inverse semigroup and let $G = S/\sigma$. With the notation of 5.19, the following statements are equivalent.

(i) $\varphi_g\varphi_h = \varphi_{gh} \quad (g, h \in G)$.
(ii) $\mathbf{d}\varphi_g = E \quad (g \in G)$.
(iii) $\mathbf{r}\varphi_g = E \quad (g \in G)$.
(iv) $\mathcal{L}\sigma = \omega$.
(v) σ is perfect.

Proof. (i) *implies* (ii). Since $\varphi_g\varphi_{g^{-1}} = \varphi_1 \supseteq \varepsilon$, we must have $\mathbf{d}\varphi_g = E$.

(ii) *implies* (iii). By 5.20, we have $\varphi_g^{-1} = \varphi_{g^{-1}}$ so that $\mathbf{r}\varphi_g = \mathbf{d}\varphi_{g^{-1}} = E$.

(iii) *implies* (iv). Let $s, t \in S$ and let $t\sigma = g$. By hypothesis $s^{-1}s \in \mathbf{r}\varphi_g$ so there exists $u \in S$ such that $s^{-1}s = u^{-1}u$ and $u\sigma = g$. Thus $s\mathcal{L}u$ and $u\sigma t$ which yields $s\mathcal{L}\sigma t$.

(iv) *implies* (v). Let $a, b \in S$ and observe that $(a\sigma)(b\sigma) \subseteq (ab)\sigma$ by the definition of a congruence. Let $c\sigma ab$. By the hypothesis $c\mathcal{L}\sigma b$, so that $c\mathcal{L}d$ and $d\sigma b$ for some $d \in S$. Letting $e = cd^{-1}$, we get

$$c = cc^{-1}c = cd^{-1}d = ed \tag{10}$$

whence $ed\sigma ab$, which implies $ed\sigma ad$. But then $e\sigma a$ since S/σ is cancellative. By (10), we have that $c = ed$ where $e\sigma a$ and $b\sigma d$.

(v) *implies* (i). Let $e\varphi_{gh}f$. Then $e = ss^{-1}, f = s^{-1}s$, $s\sigma = gh$ for some $s \in S$. By perfectness of σ, there exist $t, u \in S$ such that $s = tu$, $t\sigma = g$, $u\sigma = h$. We compute

$$(et)(et)^{-1} = tt^{-1}e = tt^{-1}ss^{-1} = tt^{-1}tus^{-1} = ss^{-1} = e,$$

$$\begin{aligned}(et)^{-1}(et) &= t^{-1}et = t^{-1}ss^{-1}t = (t^{-1}t)(uu^{-1})(t^{-1}t)\\ &= (uu^{-1})(t^{-1}t)(uu^{-1}) = us^{-1}su^{-1} = ufu^{-1}\\ &= (uf)(uf)^{-1},\end{aligned}$$

$$(uf)^{-1}(uf) = u^{-1}uf = u^{-1}us^{-1}s = u^{-1}uu^{-1}t^{-1}s = s^{-1}s = f,$$

$$(et)\sigma = t\sigma = g, \qquad (uf)\sigma = u\sigma = h,$$

hence letting $q = t^{-1}et$, we obtain $e\varphi_g q$, $q\varphi_h f$. Consequently, $\varphi_{gh} \subseteq \varphi_g\varphi_h$ and the equality holds by 5.20.

The following concept is standard.

VII.5.23 Definition

Let Y be a semilattice, G be a group, and let G act on Y by automorphisms written on the left. Then S defined on the Cartesian product $Y \times G$ with the multiplication

$$(\alpha, g)(\beta, h) = (\alpha \wedge g\beta, gh)$$

is a *semidirect product* of Y and G. Any semigroup isomorphic to such an S is also a semidirect product of Y and G.

We can now return to E-unitary inverse semigroups.

VII.5.24 Proposition

The following conditions on $S \cong P(Y, G; X) \cong Q(Y, G; \psi)$ are equivalent.

(i) σ is perfect.
(ii) For all σ-classes H, $HH^{-1} = E$.
(iii) $X = Y$.
(iv) $\psi: G \to \mathcal{Q}(Y)$.
(v) S is a semidirect product of Y and G.

Proof. (i) *implies* (ii) trivially.

(ii) *implies* (iii). Let $g \in G$ and $\alpha \in Y$. By hypothesis, and using 5.2(ii), there exists $\beta \in Y$ such that $g\beta \in Y$ and $(\alpha, 1) = (\beta, g^{-1})(\beta, g^{-1})^{-1}$. But then $\alpha = \beta$ so that $g\alpha \in Y$. Hence $X = GY = Y$, as required.

(iii) *implies* (iv). Note that for any $g \in G$,

$$\mathbf{d}\psi_g = gY \cap Y = gX \cap X = X = Y$$

and analogously $\mathbf{r}\psi_g = Y$, so that $\psi_g \in \mathcal{Q}(Y)$.

(iv) *implies* (v) trivially.

(v) *implies* (i). Note that

$$(\alpha, g)\sigma(\beta, h) \Leftrightarrow g = h.$$

Given $\alpha \in Y$ and $g, h \in G$, we obtain $(\alpha, gh) = (\alpha, g)(g^{-1}\alpha, h)$ which shows that σ is perfect.

VII.5.25 Corollary

The semigroup $Q(Y, G; \psi)$ is a direct product of Y and G if and only if $\psi : G \to \{\iota_Y\}$.

Proof. This is a direct consequence of 5.18 and 5.24.

Hence the conjunction of the subdirect product and the semidirect product amounts to the direct product. We now illustrate the various special cases discussed with the following diagram.

VII.5.26 Diagram

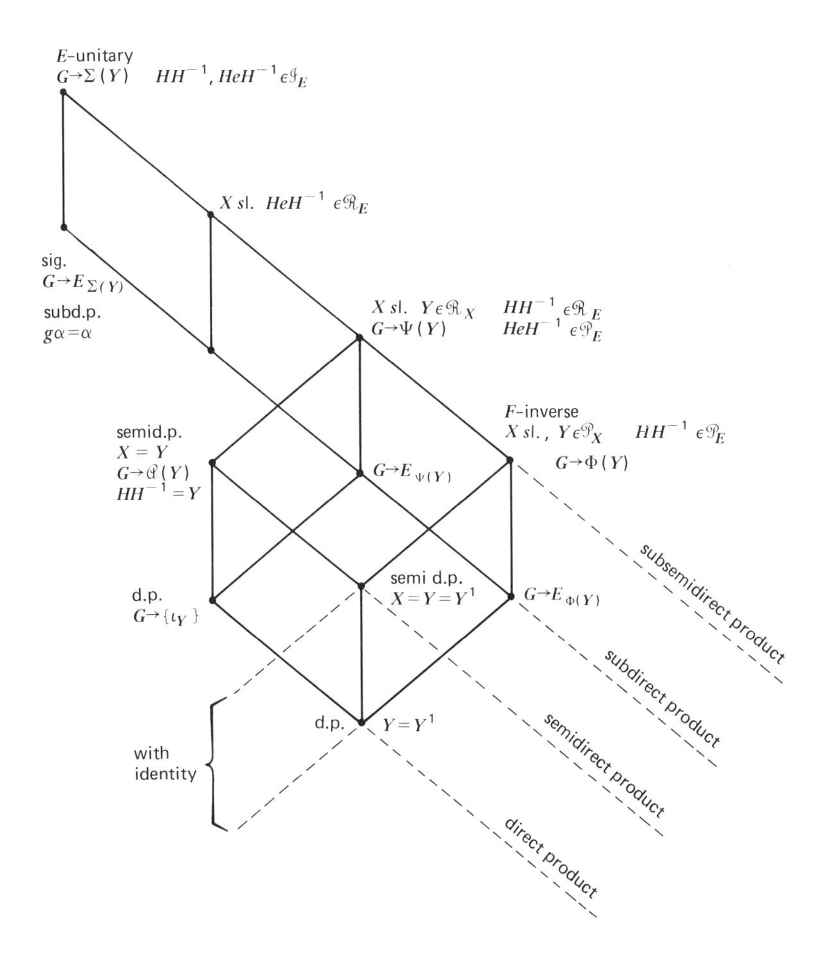

VII.5.26 Diagram.

VII.5.27 Exercises

(i) Let $S \cong P(Y, G, X) \cong Q(Y, G; \psi)$. Show that the following conditions are equivalent.

(α) For all σ-classes H, $HH^{-1} = H^{-1}H$.

(β) For all $g \in G$, $\alpha \in Y$, $g^{-1}\alpha \in Y$ if and only if $g\alpha \in Y$.

(γ) For all $g \in G$, ψ_g is contained in some subgroup of $\Sigma(Y)$.

(ii) Let $S = P(Y, G; X)$. Show that an element (α, g) of S is the greatest element in its σ-class if and only if $gY \cap Y = [\alpha]$. Deduce that S is an F-inverse semigroup if and only if $gY \cap Y \in \mathcal{P}_Y$ for all $g \in G$.

(iii) For $S = P(Y, G; X)$, prove that $C(S) \cong F(\mathcal{I}_Y, G)$ with the action $g \cdot I = gI \cap Y$.

(iv) Show that an inverse semigroup S is E-unitary if and only if every σ-class of S is permissible.

(v) Let $S = F(Y, G)$. Show that axiom (F3) can be substituted by the axiom

$$\text{(F4)} \quad g(h\alpha) = (gh)\alpha \qquad (g, h \in G, \alpha \in Y)$$

if and only if the greatest elements of the various σ-classes form a subsemigroup.

(vi) Let $S = Q(Y, G; \psi)$ and assume that for every $g \in G$, there exists $\bar{g} \in Y$ such that $\mathbf{d}\psi_g = [\bar{g}]$. Define an action of G on Y by

$$g \cdot \alpha = g(\alpha \wedge \bar{g}) \qquad (g \in G, \alpha \in Y).$$

Show that this action satisfies (F1), (F2), and (F3) and that $S = F(Y, G)$.

(vii) The preceding exercise provides a transition from $Q(Y, G; \psi)$ to $F(Y, G)$. Describe a transition from $F(Y, G)$ to both $P(Y, G; X)$ and $Q(Y, G; \psi)$.

(viii) Let S be an F-inverse semigroup, and for each $g \in S/\sigma$, denote by m_g the greatest element of the σ-class g. Show that the mapping $g \to m_g$ is a prehomomorphism.

(ix) Show that a semigroup S is a semidirect product of a semilattice and a group if and only if it is E-unitary and satisfies any of the following conditions.

(α) For every $a \in S$, $e \in E_S$, there exists $b \in S$ such that $e = bb^{-1}$, $a^{-1}b \in E_S$.

(β) For every $a \in S$, $e \in E_S$, there exists $b \in S$ such that $e = bb^{-1}$, $aa^{-1}b = ea$.

VII.5.28 Problems

(i) What are the conditions on the function $\psi: G \to \Sigma(Y)$ equivalent to the conditions in 5.4?

For $S = P(Y, G; X)$, the case when X is a semilattice was studied in some detail by McAlister [11] (he calls S in such a case "an E-unitary inverse semigroup over a semilattice"). F-inverse semigroups were introduced by McFadden–O'Carroll [1] who gave a construction of them along the lines described here. Their system of axioms was simplified by Batbedat [1]. The material concerning the relations φ_g in connection with perfect congruences and semidirect products is due to Fortunatov [1]; see also McAlister [8]. Various special cases of $Q(Y, G; \psi)$ were investigated by Petrich–Reilly [1]. For further results concerning perfect congruences consult Hamilton–Tamura [1].

VII.6 F-INVERSE SEMIGROUPS

We perform an analysis for F-inverse semigroups analogous to that in VII.1 for E-unitary inverse semigroups. To this end, we introduce the category of F-pairs and their morphisms and prove that it is equivalent to the category of F-inverse semigroups and their homomorphisms which preserve greatest elements of σ-classes. We also introduce F-inverse covers for inverse monoids and provide a construction for them.

For the sake of completeness and convenience, we give below complete proofs for several statements which have been, usually indirectly, established in the preceding section.

VII.6.1 Definition

Let G be a group and Y be a semilattice with identities 1 and ε, respectively. Let G act on Y satisfying the axioms

(F1) $1\alpha = \alpha$,
(F2) $g(\alpha \wedge \beta) = g\alpha \wedge g\beta$,
(F3) $g(h\alpha) = (gh)\alpha \wedge g\varepsilon$,

for all $g, h \in G$, $\alpha, \beta \in Y$. We call such a pair (Y, G) an *F-pair*.

Let (Y, G) and (Y', G') be F-pairs. An ordered pair (θ, ω) is a *morphism* of (Y, G) to (Y', G') if $\theta: Y \to Y'$ and $\omega: G \to G'$ are monoid homomorphisms satisfying

$$(g\alpha)\theta = (g\omega)(\alpha\theta) \qquad (g \in G, \alpha \in Y).$$

It is straightforward to verify that F-pairs and their morphisms form a category.

VII.6.2 Notation

Let $\mathfrak{T}$ be the category of F-pairs and their morphisms. Also let $\mathfrak{F}$ be the category of F-inverse semigroups and their homomorphisms which map the greatest element of a σ-class onto the greatest element of a σ-class.

We will show that these two categories are equivalent. We first devise a functor from $\mathfrak{T}$ to $\mathfrak{F}$ as follows.

VII.6.3 Notation

For any $(Y, G) \in \text{Ob}\,\mathfrak{T}$, let $F(Y, G)$ be the set

$$\{(\alpha, g) \in Y \times G | \alpha \leq g\varepsilon\}$$

together with the multiplication

$$(\alpha, g)(\beta, h) = (\alpha \wedge g\beta, gh).$$

For any $(\theta, \omega) \in \text{Hom}\,\mathfrak{T}$, say $(\theta, \omega) : (Y, G) \to (Y', G')$, define a mapping

$$F(\theta, \omega) : (\alpha, g) \to (\alpha\theta, g\omega) \qquad [(\alpha, g) \in F(Y, G)].$$

VII.6.4 Lemma

With the notation just introduced, F is a functor from $\mathfrak{T}$ to $\mathfrak{F}$.

Proof. Let $(Y, G) \in \text{Ob}\,\mathfrak{T}$ and $S = F(Y, G)$.
Let $(\alpha, g), (\beta, h) \in S$. Then $\beta \leq h\varepsilon$ and thus by (F2) and (F3),

$$\alpha \wedge g\beta \leq g\beta \leq g(h\varepsilon) \leq (gh)\varepsilon$$

so that $(\alpha, g)(\beta, h) \in S$. Hence S is closed under this operation.
For any elements of S, we obtain

$$\begin{aligned}
[(\alpha, g)(\beta, h)](\gamma, p) &= (\alpha \wedge g\beta, gh)(\gamma, p) \\
&= (\alpha \wedge g\beta \wedge (gh)\gamma, ghp) \\
&= (\alpha \wedge g\beta \wedge (gh)\gamma \wedge g\varepsilon, ghp) \\
&= (\alpha \wedge g\beta \wedge g(h\gamma), ghp) \\
&= (\alpha \wedge g(\beta \wedge h\gamma), ghp) \\
&= (\alpha, g)(\beta \wedge h\gamma, hp) \\
&= (\alpha, g)[(\beta, h)(\gamma, p)].
\end{aligned}$$

Further, $(\alpha, g) \in S$ implies $\alpha \leq g\varepsilon$ which gives

$$g^{-1}\alpha \leq g^{-1}(g\varepsilon) = (g^{-1}g)\varepsilon \wedge g^{-1}\varepsilon = g^{-1}\varepsilon$$

so that $(g^{-1}\alpha, g^{-1}) \in S$. We now have

$$\begin{aligned}(\alpha, g)(g^{-1}\alpha, g^{-1})(\alpha, g) &= (\alpha \wedge g(g^{-1}\alpha), 1)(\alpha, g)\\ &= (\alpha \wedge (gg^{-1})\alpha \wedge g\varepsilon, 1)(\alpha, g)\\ &= (\alpha, 1)(\alpha, g) = (\alpha, g)\end{aligned}$$

which shows that S is regular. It follows at once that

$$E_S = \{(\alpha, 1) | \alpha \in Y\}$$

and that idempotents commute. Consequently, S is an inverse semigroup. Furthermore,

$$\begin{aligned}(\alpha, g)\sigma(\beta, h) &\Leftrightarrow (\gamma, 1)(\alpha, g) = (\gamma, 1)(\beta, h)\\ &\Leftrightarrow (\gamma \wedge \alpha, g) = (\gamma \wedge \beta, h)\end{aligned}$$

for some $\gamma \in Y$, which evidently implies

$$(\alpha, g)\sigma(\beta, h) \Leftrightarrow g = h. \tag{11}$$

Also,

$$\begin{aligned}(\alpha, g) \leq (\beta, h) &\Leftrightarrow (\alpha, g) = (\gamma, 1)(\beta, h)\\ &\Leftrightarrow \alpha \leq \beta, \quad g = h.\end{aligned} \tag{12}$$

It follows from (11) and (12) that $(g\varepsilon, g)$ is the greatest element of its σ-class. Therefore, $S \in \mathrm{Ob}\,\mathfrak{F}$.

Now let $(\theta, \omega) \in \mathrm{Hom}\,\mathfrak{T}$, say $(\theta, \omega) : (Y, G) \to (Y', G')$. Letting $\zeta = F(\theta, \omega)$, we obtain for any elements of $F(Y, G)$,

$$\begin{aligned}(\alpha, g)\zeta(\beta, h)\zeta &= (\alpha\theta, g\omega)(\beta\theta, h\omega)\\ &= (\alpha\theta \wedge (g\omega)(\beta\theta), (g\omega)(h\omega))\\ &= (\alpha\theta \wedge (g\beta)\theta, (gh)\omega)\\ &= ((\alpha \wedge g\beta)\theta, (gh)\omega)\\ &= [(\alpha, g)(\beta, h)]\zeta.\end{aligned}$$

We also have $(g\varepsilon, g)\zeta = ((g\varepsilon)\theta, g\omega) = ((g\omega)\varepsilon', g\omega)$, which shows that ζ maps the greatest element of a σ-class onto the greatest element of a σ-class. Thus $\zeta \in \operatorname{Hom} \mathfrak{F}$. The remaining axioms for a functor follow without difficulty.

We now need a functor in the opposite direction.

VII.6.5 Notation

Let $S \in \operatorname{Ob} \mathfrak{F}$, $Y = E_S$, $G = S/\sigma$. Denote by m_g the greatest element of the σ-class g, where $g \in G$. Define an action of G on Y by $g\alpha = m_g \alpha m_g^{-1}$, and let $K(S) = (Y, G)$ stand for the pair Y and G with this action.

Next let $\varphi \in \operatorname{Hom} \mathfrak{F}$, say $\varphi : S \to S'$. Let $\theta = \varphi|_{E_S}$ and ω be defined by

$$(s\sigma)\omega = s\varphi\sigma \qquad (s \in S).$$

Denote the pair (θ, ω) by $K(\varphi)$.

VII.6.6 Lemma

With the notation just introduced, K is a functor from $\mathfrak{F}$ to $\mathfrak{T}$.

Proof. Let $S \in \operatorname{Ob} \mathfrak{F}$ and $K(S) = (Y, G)$. We verify axioms (F1), (F2), and (F3). Let 1 and ε be the identities of G and Y, respectively; $g, h \in G$, $\alpha, \beta \in Y$. Then

$$1\alpha = m_1 \alpha m_1^{-1} = \varepsilon \wedge \alpha \wedge \varepsilon = \alpha,$$

$$\begin{aligned} g(\alpha \wedge \beta) &= m_g(\alpha \wedge \beta)m_g^{-1} = (m_g\alpha\beta)(m_g^{-1}m_g)m_g^{-1} \\ &= m_g\alpha(m_g^{-1}m_g)\beta m_g^{-1} = (m_g\alpha m_g^{-1})(m_g\beta m_g^{-1}) \\ &= g\alpha \wedge g\beta. \end{aligned}$$

Noting that $m_g^{-1} = m_{g^{-1}}$, we obtain

$$\begin{aligned} g(h\alpha) &= m_g(m_h\alpha m_h^{-1})m_g^{-1} = (m_g m_h)\alpha(m_{h^{-1}}m_{g^{-1}})(m_g m_g^{-1}) \\ &\le (m_{gh}\alpha m_{h^{-1}g^{-1}})(m_g\varepsilon m_g^{-1}) \\ &= (m_{gh}\alpha m_{gh}^{-1})(m_g\varepsilon m_g^{-1}) = (gh)\alpha \wedge g\varepsilon \\ &\le m_g(m_{g^{-1}}m_{gh})\alpha(m_{h^{-1}g^{-1}}m_g)m_g^{-1} \\ &\le m_g(m_h\alpha m_h^{-1})m_g^{-1} = g(h\alpha) \end{aligned}$$

which yields $g(h\alpha) = (gh)\alpha \wedge g\varepsilon$. Consequently, $K(S) = (Y, G) \in \operatorname{Ob} \mathfrak{T}$.

Next let $\varphi \in \operatorname{Hom} \mathfrak{F}$ with the notation of 6.5. It follows at once that ω is a homomorphism of G into G', where $K(S) = (Y, G)$, $K(S') = (Y', G')$. Further, for any $s \in S$, $\alpha \in Y$,

$$((s\sigma)\alpha)\theta = (m_{s\sigma}\alpha m_{s\sigma}^{-1})\varphi = (m_{s\sigma}\varphi)(\alpha\varphi)(m_{s\sigma}^{-1}\varphi)$$
$$= m_{s\varphi\sigma}(\alpha\theta)m_{s\varphi\sigma}^{-1} = (s\varphi\sigma)(\alpha\theta) = (s\sigma)\omega(\alpha\theta)$$

which proves that $K(\varphi) = (\theta, \omega) \in \operatorname{Hom} \mathfrak{T}$.

The remaining axioms for a functor are verified readily.

VII.6.7 Lemma

For any $(Y, G) \in \operatorname{Ob} \mathfrak{T}$, let $\tau(Y, G) = (\xi, \eta)$, where $S = F(Y, G)$ and

$$\xi : \alpha \to (\alpha, 1) \qquad (\alpha \in Y),$$
$$\eta : g \to (\ \ , g)\sigma \qquad (g \in G).$$

Then τ is a natural equivalence of the functors $I_{\mathfrak{T}}$ and KF.

Proof. First note that ξ maps Y onto E_S in view of axiom (F1), so that ξ is an isomorphism of Y onto E_S. It follows from the proof of 6.4 that η is an isomorphism of G onto S/σ. For any $g \in G$, $\alpha \in Y$, we obtain

$$(g\eta)(\alpha\xi) = (\ \ , g)\sigma(\alpha, 1) = m_{(\ \ ,g)\sigma}(\alpha, 1)m_{(\ \ ,g)\sigma}^{-1}$$
$$= (g\varepsilon, g)(\alpha, 1)(g^{-1}\varepsilon, g^{-1}) = (g\alpha, 1) = (g\alpha)\xi$$

and hence (ξ, η) is an isomorphism of (Y, G) onto $KF(Y, G)$.

It remains to verify commutativity of the diagram

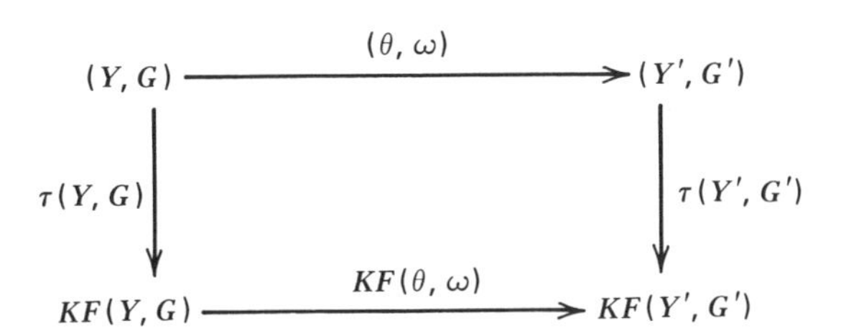

Let $\tau(Y, G) = (\xi, \eta)$, $\tau(Y', G') = (\xi', \eta')$, $KF(\theta, \omega) = (\theta', \omega')$. Then for any $\alpha \in Y$, $g \in G$, we have

$$\alpha\theta\xi' = (\alpha\theta, 1) = (\alpha, 1)\theta' = \alpha\xi\theta',$$
$$g\omega\eta' = (\ \ , g\omega)\sigma = (\ \ , g)\sigma\omega' = g\eta\omega',$$

which verifies commutativity of the diagram.

VII.6.8 Lemma

For any $S \in \operatorname{Ob} \mathfrak{F}$, let

$$\lambda(S): s \to (ss^{-1}, s\sigma) \qquad (s \in S).$$

Then λ is a natural equivalence of the functors $I_{\mathfrak{F}}$ and FK.

Proof. Let $s, t \in S$ where $S \in \operatorname{Ob} \mathfrak{F}$. Then $s\sigma m_{s\sigma}$ implies that $s^{-1}m_{s\sigma} \in E_S$ since S is in particular E-unitary. Also, $s \leq m_{s\sigma}$ gives $ss^{-1} \leq m_{s\sigma}m_{s\sigma}^{-1}$. Using this, we get with $\lambda = \lambda(S)$,

$$\begin{aligned}(s\lambda)(t\lambda) &= (ss^{-1}, s\sigma)(tt^{-1}, t\sigma)\\ &= (ss^{-1} \wedge (s\sigma)(tt^{-1}), (s\sigma)(t\sigma))\\ &= \left(s(s^{-1}m_{s\sigma})tt^{-1}m_{s\sigma}^{-1}, (st)\sigma\right)\\ &= \left(stt^{-1}s^{-1}m_{s\sigma}m_{s\sigma}^{-1}, (st)\sigma\right)\\ &= \left(st(st)^{-1}, (st)\sigma\right) = (st)\lambda.\end{aligned}$$

If $s\lambda = t\lambda$, then $s(\mathcal{R} \cap \sigma)t$ so that $s = t$ since $\mathcal{R} \cap \sigma = \varepsilon$ by III.7.2. Hence λ is one-to-one. Let $(e, g) \in FK(S)$. Then $e \in E_S$, $g \in S/\sigma$ and $e \leq m_g m_g^{-1}$. Letting $s = em_g$, we obtain $s\lambda = (e, g)$ which shows that λ maps S onto $FK(S)$. Consequently, λ is an isomorphism of these semigroups, and thus clearly maps the greatest elements of σ-classes of S onto the greatest elements of σ-classes of $FK(S)$. Therefore, λ is an isomorphism of objects of $\mathfrak{F}$.

Commutativity of the diagram

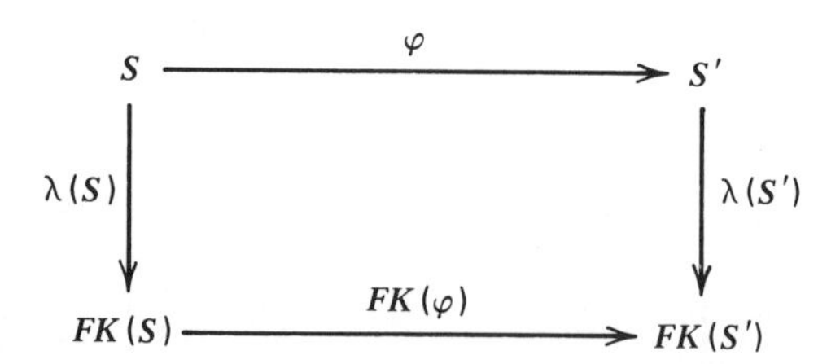

is obvious. Thus λ is a natural equivalence of the functors $I_{\mathfrak{F}}$ and FK.

Collecting 6.4 and 6.6–6.8, we deduce the desired result.

VII.6.9 Theorem

The quadruple (F, K, τ, λ) is an equivalence of the categories $\mathfrak{T}$ and $\mathfrak{F}$.

We discuss next a natural analogue of an E-unitary cover of an inverse semigroup for the case of F-inverse semigroups and inverse monoids.

VII.6.10 Definition

If M is an inverse monoid and T is an E-unitary cover for M over a group G, and T is an F-inverse semigroup, we call T an *F-inverse cover for M over G*.

These covers are constructed in the next result.

VII.6.11 Theorem

Let M be an inverse monoid and G be a group. If $\theta : G \to M$ is a prehomomorphism such that for every $m \in M$ there exists $g \in G$ for which $m \leq g\theta$, then

$$T = \{(m, g) \in M \times G | m \leq g\theta\}$$

is an F-inverse cover for M over G. Conversely, every F-inverse cover for M over G can be so constructed.

Proof

Direct part. We can verify directly that all the assertions made are true. It is shorter to apply 4.2 to the present case. For let θ be as given. For each $g \in G$, let $\theta_g = [g\theta]$. It is easy to verify that $[g\theta] \in C(M)$ and the conditions in 4.2 clearly follow from the hypotheses imposed here on θ. Hence T is an E-unitary cover for M over G. We have seen in the proof of 6.4 that in T,

$$(s, g)\sigma(t, h) \Leftrightarrow g = h.$$

Hence σ-classes of T are of the form

$$A_g = \{(s, g) | s \in \theta_g\} = \{(s, g) | s \leq g\theta\}$$

for each $g \in G$. Since $(s, g) \leq (t, g)$ if and only if $s \leq t$, it follows that $g\theta$ is the greatest element of A_g. Consequently, T is an F-inverse semigroup.

Converse. Let T be an F-inverse cover for M over G. Then 4.2 provides the function $\theta : G \to C(S)$. From the expression for σ-classes in T discussed above, it follows that each θ_g must be principal, that is, have a greatest element, which we denote by $g\theta$. Conditions in 4.2 insure that $\theta : g \to g\theta$ be a prehomomorphism satisfying the additional hypothesis. It is evident that the expression for T in the present theorem coincides with that in 4.2 if we take into account that $\theta_g = [g\theta]$.

Note that the prehomomorphism θ in 4.2 maps G into $C(S)$ whereas in 6.11 it maps G into M.

VII.6.12 Exercises

(i) Show that if S is an F-inverse semigroup, so is $\Omega(S)$.

(ii) Which σ-classes of the bicyclic semigroup have a greatest element?

(iii) Let ρ be an idempotent pure congruence on an F-inverse semigroup S. Show that S/ρ is an F-inverse semigroup.

(iv) Show that an inverse semigroup S is E-unitary if and only if $C(S)$ is an F-inverse semigroup.

(v) Let Y be a semilattice. Prove that Y is inversely well ordered if and only if all ideals of Y are principal. Also show that if this is the case, all E-unitary inverse semigroups S for which $E_S \cong Y$ are F-inverse.

VII.6.13 Problems

(i) Does the last property in 6.12(v) characterize the inversely well ordered semilattices?

For further results concerning the subject of this section consult Batbedat [1].

VIII

FREE INVERSE SEMIGROUPS

The semigroups in the title represent one of the most interesting and important classes of inverse semigroups. They merit special attention because of the great richness of their structural properties as well as for their impact on the study of varieties of inverse semigroups. The discovery of the various phenomena exhibited by free inverse semigroups seems far from completion.

1. A free inverse semigroup I_X on a nonempty set X can be described as a quotient of a free semigroup with involution by the least inverse semigroup congruence. A more concrete and usable way of describing it is as a P-semigroup $P(E, G_X; \mathfrak{X})$ where G_X is the free group on X, E is a semilattice, and $\mathfrak{X}$ is a partially ordered set both obtained from G_X.

2. By taking the greatest semilattice and Clifford homomorphic images of I_X one obtains a free semilattice and a free Clifford semigroup on X. Both of these admit simple constructions: the former as finite nonempty subsets of X under union, the latter as a subdirect product of the former and the free group on X.

3. The free inverse semigroup on X admits an isomorphic copy whose elements are birooted word trees with a natural multiplication. These elements form a kind of graph with two distinguished vertices.

4. Necessary and sufficient conditions on a subset X of an arbitrary inverse semigroup S can be given in order that X generates a free inverse subsemigroup. These conditions simplify considerably when S itself is free or X has only one element.

5. An inverse semigroup S is said to have the basis property if for any inverse subsemigroup T of S, any two bases of T have the same cardinality. A stronger related property is called the strong basis property. Every completely semisimple combinatorial inverse semigroup, and in particular a free inverse semigroup, has the strong basis property.

6. Sufficient conditions can be given on an inverse semigroup S in order that every inverse subsemigroup has a property somewhat stronger than having a basis. In particular, these conditions are fulfilled in a free inverse semigroup.

VIII.1 THE CONSTRUCTION

We construct here a free inverse monoid, and thus also a free inverse semigroup, in two ways: as a quotient of a free monoid with involution divided by a certain congruence and as a P-semigroup. The former is interesting generically, but it is the latter that has a simple usable form. From this, we derive a short proof that every inverse semigroup has an E-unitary cover of a restricted nature.

We take up again the construction in I.10.5. Let X be a nonempty set, φ be a bijection of X onto some set X' disjoint from X, and let $Y = X \cup X'$. Define a unary operation $^{-1}$ on Y^* by

$$x^{-1} = \begin{cases} x\varphi & \text{if } x \in X \\ x\varphi^{-1} & \text{if } x \in X' \end{cases} \qquad \varnothing^{-1} = \varnothing,$$

and

$$(x_1 x_2 \cdots x_n)^{-1} = x_n^{-1} x_{n-1}^{-1} \cdots x_1^{-1}.$$

Let $Z = (Y^*, \cdot, ^{-1})$ and recall I.10.6 which asserts that Z is a free monoid with involution on X.

We are now ready for the first construction of a free inverse monoid.

VIII.1.1 Theorem

Let ρ be the congruence on Z generated by the relation

$$\{(uu^{-1}u, u) | u \in Z\} \cup \{(uu^{-1}vv^{-1}, vv^{-1}uu^{-1}) | u, v \in Z\}.$$

Then $(Z/\rho, \rho^{\#}|_X)$ is a free inverse monoid on X.

Proof. It follows from the definition of ρ that Z/ρ is regular. In order to show that idempotents of Z/ρ commute, again by the definition of ρ, it suffices to show that every idempotent of Z/ρ is the product of the elements of the form $(uu^{-1})\rho$ for some $u \in Y^*$. If $e\rho$ is an idempotent of Z/ρ, then

$$\begin{aligned} e\rho &= (e\rho)^2 = (ee^{-1}e)\rho(ee^{-1}e)\rho = (e\rho)(e^{-1}e)\rho(ee^{-1})\rho(e\rho) \\ &= (e\rho)(ee^{-1})\rho(e^{-1}e)\rho(e\rho) = (e\rho)^2(e^{-1}\rho)^2(e\rho)^2 \\ &= (e\rho)(e^{-1}\rho)^2(e\rho) = (ee^{-1})\rho(e^{-1}e)\rho. \end{aligned}$$

Hence Z/ρ is an inverse semigroup. Since X generates Z as a monoid with involution, $X\rho^{\#}$ generates Z/ρ as an inverse monoid.

Let M be an inverse monoid and $\psi\colon X\rho^{\#} \to M$ be any mapping. Then $\varphi = \rho^{\#}\psi$ maps X into M and hence can be extended to Y by letting $x^{-1}\varphi = (x\rho^{\#}\psi)^{-1}$. Now Y^* is a free monoid on Y, so φ extends uniquely to a homomorphism of Y^* into M, again denoted by φ. Since M is an inverse monoid, by minimality of ρ, we conclude that $\rho \subseteq \overline{\varphi}$, where $\overline{\varphi}$ is the congruence on Y^* induced by φ. Hence there is a unique homomorphism $\chi\colon Z/\rho \to M$ such that $\varphi^{\#}\chi = \varphi$. Since $X\varphi^{\#}$ generates Z/ρ as an inverse monoid, χ is the unique homomorphism with this property.

We are now ready for the second construction of a free inverse monoid.

VIII.1.2 Construction

In order to avoid ambiguity, we will denote by 1 the identity of both Z and G_X. For any word $w = x_1 x_2 \cdots x_n$, we let

$$\hat{w} = \{1, x_1, x_1x_2, \ldots, x_1x_2 \cdots x_n\}.$$

Let $\mathcal{P}(W)$ denote the set of all subsets of a set W. Let

$$E = \{A \in \mathcal{P}(G_X) | 0 < |A| < \infty, w \in A \Rightarrow \hat{w} \subseteq A\},$$

and

$$M_X = \{(A, g) \in E \times G_X | g \in A\}$$

with multiplication

$$(A, g)(B, h) = (A \cup gB, gh)$$

where both gB and gh are products computed in G_X.

We put the above construction on familiar grounds in the following statement.

VIII.1.3 Proposition

Let the notation be as in 1.2. Let

$$\mathcal{X} = \{gA | g \in G_X, A \in E\}.$$

and define an action of G on χ by

$$g(hA) = (gh)A \qquad (g, h \in G_X, A \in E).$$

Let a partial order be defined on χ by $A \leq B$ if and only if $A \supseteq B$. Then $(E, G_X; \mathcal{X})$ is a McAlister triple and $M_X = P(E, G_X; \mathcal{X})$.

Proof. Clearly, for $A, B \in E$, $A \cup B \in E$ and is the greatest lower bound of A and B so that E is a semilattice. Let $g \in G_X$, say $g = y_1 y_2 \cdots y_n$ with $y_i \in Y$. Then $\widehat{g^{-1}} \in E$ and

$$g\widehat{g^{-1}} = y_1 y_2 \cdots y_n \{ y_n^{-1}, y_n^{-1} y_{n-1}^{-1}, \ldots, y_n^{-1} y_{n-1}^{-1} \cdots y_1^{-1} \}$$

$$= \{ y_1 y_2 \cdots y_{n-1}, y_1 y_2 \cdots y_{n-2}, \ldots, y_1, 1 \} \in E$$

so that $gE \cap E \neq \varnothing$. Trivially, $G_X E = \mathcal{X}$ and G_X evidently acts on $\mathcal{X}$ by order automorphisms.

Let $g \in G_X$ and $A \in E$. We verify next that

$$g^{-1}A \in E \Leftrightarrow g \in A. \tag{1}$$

Indeed, if $g^{-1}A \in E$, then $g^{-1}A = B$ for some $B \in E$, so $A = gB$ and $g \in A$ since $1 \in B$. Conversely, let $g \in A$ and let $w = x_1 x_2 \cdots x_m \in g^{-1}A$ where $x_i \in Y$. Then

$$gw = (y_1 \cdots y_n)(x_1 \cdots x_m) \in A.$$

In reduced form

$$gw = \begin{cases} y_1 \cdots y_n x_1 \cdots x_m & \text{if } x_1 \neq y_n^{-1} \\ y_1 \cdots y_{n-i} x_{i+1} \cdots x_m & \text{if } x_j = y_{n-j+1}^{-1} \text{ for} \\ & j \leq i < n, m, \ \ x_{i+1} \neq y_{n-i}^{-1} \\ y_1 \cdots y_{n-m} & \text{if } x_j = y_{n-j+1}^{-1} \text{ for all} \\ & j \leq m < n \\ x_{n+1} \cdots x_m & \text{if } x_j = y_{n-j+1}^{-1} \text{ for all} \\ & j \leq n < m \\ 1 & \text{otherwise.} \end{cases}$$

Let $1 \leq k \leq m$. In the first case, we get $y_1 \cdots y_n x_1 \cdots x_k \in A$ and thus $x_1 x_2 \cdots x_k \in g^{-1}A$. In the second case, for $k > i$, we have $y_1 \cdots y_{n-i} x_{i+1} \cdots x_k \in A$ and hence $x_{i+1} \cdots x_k \in y_{n-i}^{-1} \cdots y_1^{-1}A$ whence

$$x_1 \cdots x_k \in (x_1 \cdots x_i)(y_{n-i}^{-1} \cdots y_1^{-1})A = g^{-1}A;$$

if $k \leq i$, then

$$x_1 \cdots x_k = y_n^{-1} \cdots y_{n-k+1}^{-1} = g^{-1} y_1 y_2 \cdots y_{n-k} \in g^{-1}A$$

since $g \in A$. In the third case, we have

$$x_1 x_2 \cdots x_k = y_n^{-1} \cdots y_{n-k+1}^{-1} = g^{-1} y_1 \cdots y_{n-k} \in g^{-1} A$$

again since $g \in A$. In the fourth case, for $k > n$, we get $x_{n+1} \cdots x_k \in A$ and thus

$$x_1 \cdots x_k = (x_1 \cdots x_n) x_{n+1} \cdots x_k = g^{-1} x_{n+1} \cdots x_k \in g^{-1} A;$$

and if $k \leq n$, then

$$x_1 \cdots x_k = y_n^{-1} \cdots y_{n-k+1}^{-1} = g^{-1} y_1 \cdots y_{n-k} \in g^{-1} A$$

since $g \in A$. The fifth case follows similarly. Since $|g^{-1}A| = |A|$, we deduce that $g^{-1}A \in E$. This proves relation (1).

Now let $g \in G_X$ and $A, B \in E$ be such that $g^{-1}A \in \mathcal{X}$, $B \subseteq g^{-1}A$. Since $1 \in B$, we get $1 = g^{-1}a$ for some $a \in A$, and thus $g \in A$. By (1), we obtain that $g^{-1}A \in E$. Hence E is an ideal of $\mathcal{X}$ and $(E, G_X; \mathcal{X})$ is a McAlister triple.

Relation (1) also implies that $(A, g) \in M_X$ if and only if $(A, g) \in P(E, G_X; \mathcal{X})$ and since the operations in M_X and $P(E, G_X; \mathcal{X})$ obviously agree, we obtain $M_X = P(E, G_X; \mathcal{X})$.

It will be convenient to have the following terminology.

VIII.1.4 Definition

The congruence ρ in 1.1 is the *Wagner congruence* on Z. Recalling that $r(w)$ is the reduced form of the word w in Z, for any nonempty subset A of Z, we let

$$r(A) = \{r(w) | w \in A\}.$$

We are finally ready for the main result.

VIII.1.5 Theorem

Define a mapping χ by

$$\chi : w \to (r(\hat{w}), r(w)) \qquad (w \in Z).$$

Then χ is a homomorphism of Z onto M_X which induces the Wagner congruence on Z.

Proof. Let $u \in r(\hat{w})$ where $w = x_1 x_2 \cdots x_n$. Then $u = r(x_1 x_2 \cdots x_i)$ for some $1 \leq i \leq n$, so $u = x_{j_1} x_{j_2} \cdots x_{j_m}$ for some $1 \leq j_1 \leq \cdots \leq j_m \leq i$. Now $x_{j_1} x_{j_2} \cdots x_{j_k} = r(x_1 x_2 \cdots x_{j_k}) \in r(\hat{w})$ for any $1 \leq k \leq m$. Consequently, $r(\hat{w}) \in E$ and χ maps Z into M_X.

If also $v = y_1 y_2 \cdots y_n$, then

$$\begin{aligned}(w\chi)(v\chi) &= (r(\hat{w}), r(w))(r(\hat{v}), r(v))\\ &= (r(\hat{w}) \cup r(w)r(\hat{v}), r(wv))\\ &= (r(\hat{w}) \cup r(\{w, wy_1, \dots, r(wv)\}), r(wv))\\ &= (r(\widehat{wv}), r(wv)) = (wv)\chi,\end{aligned}$$

so that χ is a homomorphism.

Let $(A, g) \in M_X$. Then $A = \{w_1, w_2, \dots, w_m\}$ for some $w_i \in G_X$, and since $g \in A$, we may take $g = w_m$. Letting

$$w = (w_1 w_1^{-1})(w_2 w_2^{-1}) \cdots (w_{m-1} w_{m-1}^{-1}) w_m,$$

we claim that $w\chi = (A, g)$. Indeed, $w_i = r(w_1 w_1^{-1} w_2 w_2^{-1} \cdots w_i)$ so that

$$A \subseteq r(\hat{w}). \tag{2}$$

For the opposite inclusion, if $w = x_1 x_2 \cdots x_n$ then $r(x_1 x_2 \cdots x_i)$ must be contained in $\hat{w}_j$ for some w_j and hence is in A since $A \in E$. This gives the opposite inclusion in (2). It is clear that $r(w) = w_m$ which shows that $w\chi = (A, g)$. Consequently, χ maps Z onto M_X.

By the minimality of the Wagner congruence ρ on Z, we must have $\rho \subseteq \tau$, where τ is the congruence on Z induced by χ, since M_X is an inverse semigroup by 1.3. For $w = x_1 x_2 \cdots x_n$, letting

$$\begin{aligned}\overline{w} = (x_1 x_1^{-1})&\left[r(x_1 x_2) r(x_1 x_2)^{-1}\right]\\ &\cdots\left[r(x_1 \cdots x_{n-1}) r(x_1 \cdots x_{n-1})^{-1}\right] r(w),\end{aligned}$$

we claim that $w\rho\overline{w}$. The argument is by induction on n. This is trivial for $n = 1$. First let $v = x_1 x_2 \cdots x_{n-1}$ and suppose that $r(v)x_n \neq r(vx_n)$. In reduced form $r(u) = y_1 y_2 \cdots y_k$ for some variables $y_1, y_2, \dots, y_k$, so that the hypothesis implies that $x_n = y_k^{-1}$. We obtain

$$\begin{aligned}r(v)r(v)^{-1}r(vx_n) &= (y_1 \cdots y_k)(y_1 \cdots y_k)^{-1}(y_1 \cdots y_{k-1})\\ &\rho(y_1 \cdots y_{k-1})(y_k y_k^{-1})(y_1 \cdots y_{k-1})^{-1}(y_1 \cdots y_{k-1})\\ &\rho(y_1 \cdots y_{k-1})(y_1 \cdots y_{k-1})^{-1}(y_1 \cdots y_{k-1})(y_k y_k^{-1})\\ &\rho y_1 \cdots y_k y_k^{-1} = r(v)x_n.\end{aligned} \tag{3}$$

Now assume that the above claim is true for all $k < n$. Using (3), we obtain

$$x_1x_2 \cdots x_n \rho \left(x_1x_1^{-1}\right) \cdots \left[r(x_1 \cdots x_{n-2}) r(x_1 \cdots x_{n-2})^{-1}\right]$$
$$\cdot r(x_1 \cdots x_{n-1}) x_n$$
$$\rho \left(x_1x_1^{-1}\right) \cdots \left[r(x_1 \cdots x_{n-1}) r(x_1 \cdots x_{n-1})^{-1}\right] r(x_1 \cdots x_{n-1}) x_n$$
$$\rho \left(x_1x_1^{-1}\right) \cdots \left[r(x_1 \cdots x_{n-1}) r(x_1 \cdots x_{n-1})^{-1}\right] r(x_1 \cdots x_n).$$

This shows that $w \rho \overline{w}$. If $w\chi = u\chi$, then $\overline{w} \rho \overline{u}$ and thus $w \rho u$. Consequently, $\rho = \tau$.

VIII.1.6 Corollary

The pair (M_X, φ), where

$$\varphi : x \to (\hat{x}, x) \qquad (x \in X),$$

is a free inverse monoid on X.

Proof. This is a direct consequence of 1.1 and 1.5.

For the nonmonoid case, we first introduce the following.

VIII.1.7 Notation

For any nonempty set X, let

$$I_X = M_X \backslash \{(\{1\}, 1)\}$$

with the induced multiplication.

VIII.1.8 Proposition

With the notation of 1.6 and 1.7, (I_X, φ) is a free inverse semigroup on X.

Proof. Let S be an inverse semigroup and $\psi : X \to S$ be a function. Let $\overline{S}$ be the semigroup obtained from S by the adjunction of an identity 1. Then we have a unique monoid homomorphism $\overline{\psi} : M_X \to \overline{S}$ making the diagram

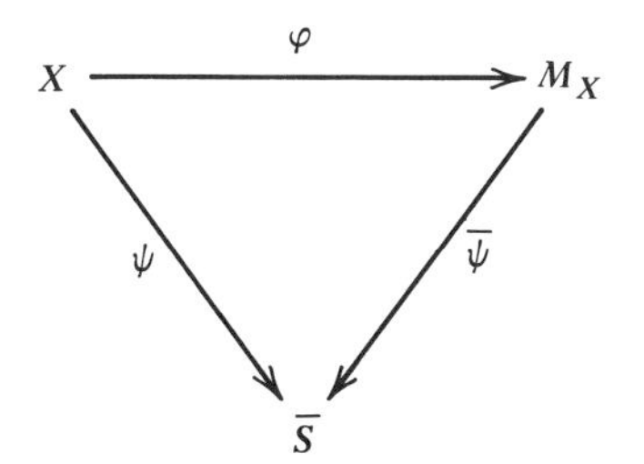

commutative. Since every element of M_X different from its identity is a product of elements in $X\varphi$ and their inverses, the commutativity of the diagram implies that $\bar{\psi}$ maps only the identity of M_X onto 1, the identity of $\bar{S}$. It follows that $\chi = \bar{\psi}|_{I_X}$ is a homomorphism of I_X into S for which the resulting diagram commutes. Since $X\varphi$ generates I_X, χ is unique with these properties (alternatively, if θ is another such, extending θ to map the identity of M_X onto 1, we would be able to conclude that $\theta = \bar{\psi}$).

We will generally work with I_X and call it *the free inverse semigroup on* X.

VIII.1.9 Corollary

Every inverse semigroup is homomorphic image of a free inverse semigroup.

Proof. For any inverse semigroup S, we may take for X any set of generators of S, and according to 1.8 obtain S as a homomorphic image of I_X.

We are now in a position to give a different proof of VII.4.6.

VIII.1.10 Theorem

Every inverse semigroup has an E-unitary cover.

Proof. Let S be an inverse semigroup. By 1.9, S is a homomorphic image of a free inverse semigroup F, so we have $F/\rho \cong S$ for some congruence ρ on F. As in the proof of III.5.10, we factor

$$F \overset{\varphi}{\to} F/\rho_{\min} \overset{\psi}{\to} F/\rho.$$

Let $a \in F$, $e \in E_F$ be such that $(a\varphi)(e\varphi) = e\varphi$. Then $ae\rho_{\min}e$ whence $aef = ef$ for some $f \in E_F$. Since F is E-unitary, we get $a \in E_F$ and thus $a\varphi \in E_{F/\rho_{\min}}$. Consequently, $F/\rho_{\min}$ is E-unitary. Since $\rho_{\min}$ and ρ have the same trace, it follows that the natural homomorphism $\psi\colon F/\rho_{\min} \to F/\rho$ is idempotent separating. Therefore, S is an idempotent separating homomorphic image of the E-unitary inverse semigroup $F/\rho_{\min}$.

We can sharpen the above result, towards which end we first prove the following statement.

VIII.1.11 Lemma

Let $S = P(Y, G; X)$ where X is a semilattice and let ρ be a congruence on S contained in σ. Then S/ρ can be represented as $P(Y', G'; X')$ where X' is a semilattice.

Proof. Let $\overline{S} = S/\rho$ and for any $x \in S$, $H \subseteq S$, write $\bar{x} = x\rho$, $\overline{H} = \{h\rho | h \in H\}$. Let $a, b \in S$ be such that $\bar{a}\sigma\bar{b}$. Then $\bar{a}\bar{e} = \bar{b}\bar{e}$ for some idempotent e in S. Hence $ae\rho be$ which by the hypothesis yields $ae\sigma be$. But then $a\sigma b$. The converse always holds, namely $a\sigma b$ implies $\bar{a}\sigma\bar{b}$.

Since E_S is a σ-class of S, it now follows that $E_{\overline{S}}$ is a σ-class, so that $\overline{S}$ is E-unitary. We let H be a σ-class of S and $e, f \in E_S$. By the above, $\overline{H}$ is a σ-class of $\overline{S}$. By VII.5.4, there exists $a \in S$ such that $H \cap eSf = [a]$. In view of VII.5.4, it suffices to prove that $\overline{H} \cap \bar{e}\overline{S}\bar{f} = [\bar{a}]$.

Let $x \in S$. Let $\bar{x} \in \bar{e}\overline{S}\bar{f} \cap \overline{H}$, then $x\rho esf$ for some $s \in S$ and $x \in H$ by the above. Hence $x\sigma(esf)\sigma s$ so that $s \in H$ and we get $esf \in H \cap eSf = [a]$. Thus $esf = ta$ for some $t \in E_S$, and we obtain $x\rho ta$ which implies $\bar{x} \leq \bar{a}$. Conversely, let $\bar{x} \leq \bar{a}$ so that $x\rho ta$ for some $t \in E_S$. Then $x\sigma a$ and thus $x \in H$, and also $x\rho exf$ since $a \in H \cap eSf$. Thus $\bar{x} = \overline{exf} \in \overline{H} \cap \bar{e}\overline{S}\bar{f}$, as required.

VIII.1.12 Theorem

Every inverse semigroup is an idempotent separating homomorphic image of a P-semigroup $P(Y, G; X)$ where X is a semilattice.

Proof. For any $(A, g) \in M_X$, a simple verification shows that $(\hat{g}, g)$ is the greatest element of the σ-class containing (A, g). Hence M_X is an F-inverse semigroup. By VII.5.14 and VII.5.11, we deduce that $\mathcal{X}$ is a semilattice. Since I_X equals M_X with $(\{1\}, 1)$ removed, in the representation of M_X, I_X corresponds to $E' = E \setminus \{1\}$ and $\mathcal{X}' = \mathcal{X} \setminus G_X$. In VII.5.4, condition (ii) is still fulfilled in this new setting, and thus by VII.5.4, $\mathcal{X}'$ is a semilattice. We now go into the proof of 1.10 and note that $\rho_{\min} \subseteq \sigma$ for any congruence ρ on any inverse semigroup. Applying 1.11 we reach the desired conclusion.

We terminate this section with some properties of I_X. The following terminology is widely used.

VIII.1.13 Definition

A semigroup in which $\mathcal{H} = \varepsilon$ is *combinatorial.*

VIII.1.14 Proposition

Every free inverse semigroup is combinatorial.

Proof. We consider I_X. Let $(A, g)\mathcal{H}(B, h)$. Then $A = B$ and $g^{-1}A = h^{-1}B$. Hence $g^{-1}A = h^{-1}A$ which implies $tA = A$ with $t = gh^{-1}$. Since $1 \in A$, this yields $t \in A$. But then $t^n \in A$ for all $n \geq 1$, which by finiteness of A gives $t = 1$. Hence $g = h$ and thus $(A, g) = (B, h)$.

VIII.1.15 Proposition

If $s = (A, g) \in I_X$, then $|L_s| = |R_s| = |A|$, $|D_s| = |J_s| = |A|^2$.

Proof. For $(A, g), (B, h) \in I_X$, $(A, g)\mathcal{R}(B, h)$ if and only if $A = B$. Since $g \in A$, we conclude that $|R_s| = |A|$. In any inverse semigroup S, the mapping $s \to s^{-1}$ is a permutation on S which maps any $\mathcal{L}$-class of S onto an $\mathcal{R}$-class of S contained in the same $\mathcal{D}$-class. Thus $|L_s| = |A|$. But then $|D_s| = |A|^2$ since $\mathcal{H} = \varepsilon$ by 1.14. It follows that each $\mathcal{D}$-class of I_X is finite, and hence every principal factor is a Brandt semigroup or a group in view of I.6.14. In particular, $\mathcal{D} = \mathcal{J}$ and hence $|D_s| = |J_s|$.

VIII.1.16 Corollary

For any set X, I_X is completely semisimple with all $\mathcal{D}$-classes finite.

VIII.1.17 Exercises

(i) Prove that a free inverse semigroup satisfies the ascending chain condition for principal right ideals.

(ii) Characterize $\mathcal{D}$-equivalence in a free inverse semigroup.

(iii) Find the Q-representation of I_X for any nonempty set X.

(iv) For any nonempty set X, show that M_X is an F-inverse semigroup and find its F-representation.

Wagner [8] was the first to prove the existence of a free inverse semigroup and 1.1 is due to him. The first concrete construction of free inverse semigroups is that of Scheiblich [4]. Alternative constructions were then offered by McAlister–McFadden [1], Munn [13] (see VIII.3), Preston [8], and Schein [21]. Further results were obtained by Kleiman [1] and O'Carroll [3]. The presentation of 1.5 is essentially that of Petrich [16] containing basic ideas of Scheiblich's and Schein's approaches. A P-representation of a free inverse semigroup was established in McAlister–McFadden [1]. The proof of 1.10 stems from Reilly–Munn [1]; see also O'Carroll [6]. An interesting universal property for inverse semigroups was found by McAlister [1]. Jones [7] established a property of free inverse semigroups in terms of their inverse subsemigroups. Free inverse semigroups with commuting generators were investigated by McAlister–McFadden [2]. Also, see Rosenblat [1], Saitô [4], and Važenin [1]. Reilly [11] is an extensive survey of the results on free universe semigroups.

VIII.2 FREE CLIFFORD SEMIGROUPS

We consider here free semilattices (with identity) and free Clifford monoids on a nonempty set X. A freeness-type interpretation is given to E_{I_X}. We then conclude with some other properties of I_X.

The notation established in the preceding section will be used here as well. We start with a free semilattice with identity.

VIII.2.1 Proposition

Let η be the congruence on X^* generated by the relation

$$\{(u^2, u) | u \in X^*\} \cup \{(uv, vu) | u, v \in X^*\}.$$

Then $(X^*/\eta, \eta^{\#}|_X)$ is a free semilattice with identity on X.

Proof. It is obvious that X^*/η is a semilattice. The proof of freeness goes along the same lines as the argument in the last paragraph of the proof of 1.1.

In order to get a concrete copy of the free semilattice with identity, we introduce the following symbols.

VIII.2.2 Notation

Let Y_X denote the set of *all finite subsets of X with the operation of set theoretical union*. Then Y_X is a monoid and a semilattice.

For each element $w = x_1 x_2 \cdots x_n$ of Z define the *content* $c(w)$ of w by

$$c(w) = \{|x_1|, |x_2|, \ldots, |x_n|\}$$

where

$$|x| = \begin{cases} x & \text{if } x \in X \\ x^{-1} & \text{if } x \in X', \end{cases} \qquad |\varnothing| = \varnothing.$$

It follows that $c(w) \in Y_X$. Also let

$$|w| = |x_1||x_2| \cdots |x_n|.$$

VIII.2.3 Proposition

The mapping $c: w \to c(w)$ is a homomorphism of Z onto Y_X which induces the least semilattice congruence η on Z. The pair (Y_X, ι_X) is a free semilattice with identity on X.

Proof. Clearly the mapping c is multiplicative. For $x \in X$ and $y \in X'$, we obtain

$$c(x^{-1}) = \{x\} = c(x) = c(x)^{-1},$$

$$c(y^{-1}) = \{y^{-1}\} = c(y) = c(y)^{-1},$$

which implies $c(w^{-1}) = c(w)^{-1}$ for all $w \in Z$. Consequently, c is a homomorphism of Z onto Y_X. Letting τ be the congruence on Z induced by c, we get $\eta \subseteq \tau$ by the minimality of η. For any $w \in Z$, $c(w) = c(|w|)$. Hence for any $w, u \in Z$, $c(w) = c(u)$ implies $w\eta(|w|)\eta(|u|)\eta u$which proves that $\tau \subseteq \eta$.
The second assertion of the proposition now follows from 2.1.

One can easily prove directly that Y_X is a free semilattice with identity on X.

VIII.2.4 Definition

For any subset A of G_X, define the *content* of A by $c(A) = \cup_{w \in A} c(w)$.

VIII.2.5 Corollary

The mapping $(A, g) \to c(A)$ is a homomorphism of M_X onto Y_X which induces the least semilattice congruence on M_X.

Proof. It suffices to observe that $c(x_1 \cdots x_n) = c(x_1 r(x_1 x_2) \cdots r(x_1 \cdots x_n))$ and apply 2.3.

The semilattice of idempotents of a free inverse semigroup admits an interesting interpretation. To obtain it, we first prove an auxiliary statement.

VIII.2.6 Lemma

Let G_X be the free group on a set X. For any $u, v \in G_X$, define $u \leq v$ if and only if $u \in \hat{v}$. Then $\leq$ is a partial order on G_X.

Proof. The proof of this lemma is left as an exercise.

The convexity below refers to the partial ordering just introduced. The following result lends an interesting interpretation to the elements of E in 1.2.

VIII.2.7 Remark

Let A be a nonempty subset of G_X. Then A is convex and $1 \in A$ if and only if for any $w \in G_X$, $w \in A$ implies $\hat{w} \subseteq A$.

Proof. The proof is left as an exercise.

A variant of freeness is provided by the following concept.

VIII.2.8 Definition

Let P be a partially ordered set. The pair (Y, φ) is a *free semilattice over* P if Y is a semilattice, $\varphi : P \to Y$ is an order preserving map such that for any order preserving map ψ of P into a semilattice X there exists a unique homomorphism $\bar{\psi}$ of Y into X making the diagram

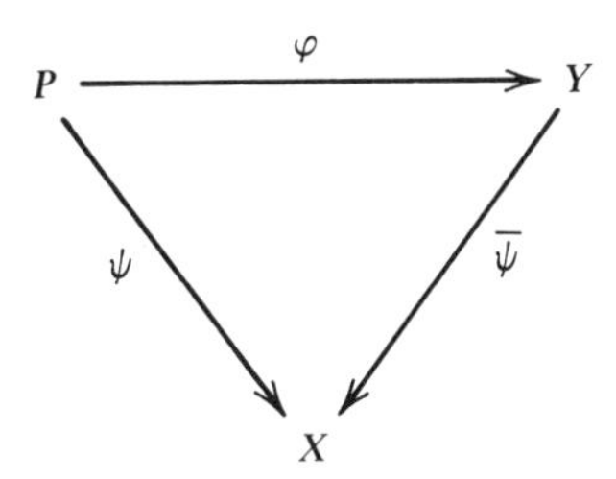

commutative.

We are now all set for the desired result.

VIII.2.9 Theorem

Let X be a nonempty set, P_X be the set G_X together with the partial order opposite to the one defined in 2.6. Then the pair $(E_{I_X}, w \to (r(\hat{w}), 1))$ is a free semilattice over P_X.

Proof. Denote the order in P_X by $\preccurlyeq$ and the mapping $w \to (r(\hat{w}), 1)$ by φ. Assume that $w \preccurlyeq u$ in P_X. Then $u \leq w$ so that $u \in r(\hat{w})$. Hence $r(\hat{u}) \subseteq r(\hat{w})$ which implies that $w\varphi \leq u\varphi$.

Next let $(A,1) \in E_{I_X}$. Then $A = \{1, w_1, w_2, \ldots, w_n\}$ and one verifies easily that

$$(A,1) = (\hat{w}_1, 1)(\hat{w}_2, 1) \cdots (\hat{w}_n, 1).$$

This shows that $P_X\varphi$ generates E_{I_X}.

Now let Y be a semilattice and $\psi: P_X \to Y$ be an order preserving map. Define a mapping $\bar{\psi}$ on E_{I_X} by

$$\bar{\psi}: (A,1) \to (w_1\psi)(w_2\psi) \cdots (w_n\psi) \quad \text{if} \quad A = \{1, w_1, w_2, \ldots, w_n\}.$$

For any $(A,1), (B,1) \in E_{I_X}$, say

$$A = \{1, w_1, w_2, \ldots, w_n\}, \qquad B = \{1, u_1, u_2, \ldots, u_m\},$$

we obtain

$$\begin{aligned}(A,1)\bar{\psi}(B,1)\bar{\psi} &= (w_1\psi) \cdots (w_n\psi)(u_1\psi) \cdots (u_m\psi) \\ &= (\{1, w_1, \ldots, w_n, u_1, \ldots, u_m\}, 1)\bar{\psi} \\ &= (A \cup B, 1)\bar{\psi} = ((A,1)(B,1))\bar{\psi}.\end{aligned}$$

Consequently, $\bar{\psi}$ is a homomorphism of E_{I_X} into Y, and it evidently makes the requisite diagram commutative. Uniqueness of $\bar{\psi}$ for a given ψ is a consequence of the fact that $P_X \varphi$ generates E_{I_X}.

We now turn to free Clifford semigroups.

VIII.2.10 Proposition

Let ν be the congruence on Z generated by the relation

$$\{(uu^{-1}u, u) | u \in Z\} \cup \{(uvv^{-1}, vv^{-1}u) | u, v \in Z\}.$$

Then $(Z/\nu, \nu^{\#}|_X)$ is a free Clifford monoid on X.

Proof. The argument here goes along the same lines as in 1.1 and is left as an exercise.

VIII.2.11 Notation

For a nonempty set X, let

$$C_X = \{(A, w) \in Y_X \times G_X | c(w) \subseteq A\}$$

(with componentwise multiplication).

VIII.2.12 Proposition

The mapping $\psi: w \to (c(w), r(w))$ is a homomorphism of Z onto C_X which induces the least Clifford congruence ν on Z. The pair $(C_X, x \to (\{x\}, x))$ is a free Clifford monoid on X.

Proof. It is easy to see that ψ is a homomorphism of Z into C_X. If $(A, u) \in C_X$, where $u = x_1^{\varepsilon_1} x_2^{\varepsilon_2} \cdots x_m^{\varepsilon_m}$, $\varepsilon_i = \pm 1$, $x_i \in X$, and $A = \{x_1, x_2, \ldots, x_m, \ldots, x_n\}$, then $(u(x_{m+1}x_{m+1}^{-1}) \cdots (x_n x_n^{-1}))\psi = (A, u)$. Hence ψ maps Z onto C_X.

Since C_X is a Clifford monoid, the congruence τ induced by ψ satisfies $\nu \subseteq \tau$. For $w = w(y_1, \ldots, y_n) \in Z$, we have

$$w\nu(y_1 y_1^{-1})(y_2 y_2^{-1}) \cdots (y_n y_n^{-1}) r(w) \tag{4}$$

Hence if $w \tau u$, then $c(w) = c(u)$, $r(w) = r(u)$ which by (4) implies $w \nu u$. Consequently, $\nu = \tau$ and ψ induces ν.

The last assertion of the proposition now follows from 2.10.

VIII.2.13 Corollary

The mapping $(A, g) \to (c(A), g)$ is a homomorphism of M_X onto C_X which induces the least Clifford congruence on M_X.

Proof. This follows easily from 2.12.

VIII.2.14 Corollary

The intersection of the least semilattice congruence and the least group congruence equals the least Clifford congruence both in Z and M_X.

Proof. The proof of this corollary is left as an exercise.

VIII.2.15 Corollary

For any nonempty set X, C_X is a subdirect product of Y_X and G_X.

Proof. The proof is left as an exercise.

We can represent the monoid C_X as a Clifford semigroup as follows:

$$C_X \cong [Y_X; G_A, i_{A,B}]$$

where $i_{A,B}: G_A \to G_B$ is the inclusion homomorphism for $A \subseteq B$.

VIII.2.16 Exercises

(i) Show directly that for a nonempty set X, the set of all finite subsets of X under union is a free semilattice with identity on X.

(ii) Let X be a nonempty set. Prove directly that $[Y_X; G_A, i_{A,B}]$, where $i_{A,B}: G_A \to G_B$ for $A \subseteq B$ is the inclusion homomorphism, is a free Clifford semigroup on X.

The form of a free semilattice on a set is folklore. The interpretation of E_{I_X} as a free object is due to Schein [21]. S. A. Liber [1] announced the form of the free Clifford semigroup. See also Petrich [16].

VIII.3 THE GRAPH REPRESENTATION

We present here a remarkable isomorphic copy of I_X for any nonempty set X in terms of birooted word trees. The latter are special kinds of labeled graphs with two distinguished vertices. Certain properties of the free inverse semigroup can be easily read off from this representation.

VIII.3.1 Definition

A *graph* is a finite nonempty set of elements called *vertices* together with a set of unordered pairs of distinct vertices called *edges*. The set of all vertices of a graph G is denoted by $V(G)$. Let $|G|$ be the cardinality of $V(G)$. If the vertices v_1, v_2 form an edge of the graph, we say that they are *adjacent* and that they are *joined by an edge*. A graph H is a *subgraph* of G if all vertices and edges of H are also vertices and edges of G, respectively. A vertex is *extreme* if it belongs to exactly one edge.

A *walk* in G is a sequence $(\gamma_0, \gamma_1, \ldots, \gamma_n)$ of vertices of G such that γ_{i-1}, γ_i are adjacent for $i = 1, 2, \ldots, n$. This is a walk of *length* n, and we refer to it also as to a (γ_0, γ_n)-*walk*. A *path* is a walk $(\gamma_0, \gamma_1, \ldots, \gamma_n)$ in which all vertices γ_i are distinct; we also talk of a (γ_0, γ_n)-*path* in this case. The graph G is *connected* if every pair of vertices of G is joined by a path. A (α, α)-walk is said to be *closed*. A *cycle* is a closed walk all of whose vertices are distinct and with at least three vertices. A *tree* is a connected graph without cycles. A tree in which a vertex is distinguished is a *rooted tree*.

A walk W *spans* G, or is a *spanning walk*, if all vertices of G occur among vertices of W. In a tree T, for any $\alpha, \beta \in V(T)$, there is a unique (α, β)-path; we denote it by $\Pi(\alpha, \beta)$.

An edge with vertices α, β is *oriented* if we consider the edge together with (α, β) as an ordered pair; we write $\alpha \to \beta$ and denote the edge by $\alpha\beta$. An edge is *labeled* if a symbol is associated to it.

We thus arrive at the following concept basic for our discussion.

VIII.3.2 Definition

A *word tree* T on a nonempty set X is a tree with at least one edge satisfying

(i) each edge is oriented and labeled by an element of X,

(ii) T has no subgraph of the form $\circ \xrightarrow[x]{} \circ \xleftarrow[x]{} \circ$ or $\circ \xleftarrow[x]{} \circ \xrightarrow[x]{} \circ$.

We now extend the set of labels from X to $Y = X \cup X'$ in the notation of I.10.5 and make the convention that ${}^{\alpha}\circ \xrightarrow[x^{-1}]{} \circ^{\beta}$ means the same as ${}^{\alpha}\circ \xleftarrow[x]{} \circ^{\beta}$.

The next notion serves to compare word trees.

VIII.3.3 Definition

Let T and T' be word trees on X. An *isomorphism* of T onto T' is a bijection of $V(T)$ onto $V(T')$ which preserves adjacency, orientation of edges, and labeling of edges. In such a case, we write $T \cong T'$. (It is clear that isomorphism is an equivalence relation on the class of all word trees on X.) Let $\mathfrak{T}_X$ be a *cross section of the isomorphism classes of word trees on* X.

We now describe composition of word trees in $\mathfrak{T}_X$ furnished with a root. Informally, we proceed as follows. Let (T, α) and (T', α') be two rooted word trees. Paste T' on T by identifying all pairs (β, β'), where $\beta \in V(T)$ and $\beta' \in V(T')$, for which the (α, β)-path is isomorphic to the (α', β')-path. The resulting graph is the composition of (T, α) and (T', α'). Formally, we have the following procedure.

VIII.3.4 Construction

Let $T, T' \in \mathfrak{T}_X$ and $\alpha \in V(T)$, $\alpha' \in V(T')$. Let $\gamma \in V(T')$ be an extreme vertex and let $\Pi(\alpha', \gamma) = (\alpha' = \gamma_0, \gamma_1, \dots, \gamma_n = \gamma)$ in T'. There exists $\delta_m \in V(T)$ such that $\Pi(\alpha, \delta_m) = (\alpha = \delta_0, \delta_1, \dots, \delta_m)$ in T is isomorphic to $(\alpha' = \gamma_0, \gamma_1, \dots, \gamma_m)$ and m is the greatest integer with this property. Note that $\Pi(\alpha, \delta_m)$ is unique by 3.2(ii). We now identify $\gamma_i = \delta_i$ for $i = 0, 1, 2, \dots, m$. If $m < n$, we attach the graph $(\gamma_m, \gamma_{m+1}, \dots, \gamma_n)$ to the vertex $\gamma_m = \delta_m$. Doing this for all extreme vertices $\gamma \in V(T')$, we evidently obtain a word tree on X. Let $T(\alpha, \alpha')T'$ denote its representative in $\mathfrak{T}_X$. It is a notational convenience to identify the vertices of T and T' with the corresponding vertices of $T(\alpha, \alpha')T'$.

The final basic concept is the following.

VIII.3.5 Definition

A triple (α, T, β) is a *birooted word tree on* X if $T \in \mathfrak{T}_X$ and $\alpha, \beta \in V(T)$. Let B_X denote the set of *all birooted word trees on* X together with the multiplication

$$(\alpha, T, \beta)(\alpha', T', \beta') = (\alpha, T(\beta, \alpha')T', \beta').$$

We will prove below that there is an isomorphism of I_X onto B_X. Hence there is no need to establish any properties of this multiplication for the moment, but to devise a function from I_X to B_X. We now approach that task.

VIII.3.6 Construction

Let $(A, g) \in I_X$. First let $w = x \in A$ be of length 1. Form an edge (α, γ) labeled by x. Keep α fixed and apply the same procedure obtaining edges of the form $(\alpha, \delta), \dots$. Assume that we have assigned a path to each word in A of length less than k, and let $w = x_1 x_2 \cdots x_k \in A$. There exists a unique path $(\alpha = \gamma_0, \gamma_1, \dots, \gamma_{k-1})$ labeled $x_1, x_2, \dots, x_{k-1}$ in the graph already constructed. Attach an edge (γ_{k-1}, γ_k) labeled by x_k. Do this for all words in A of length k.

We thus inductively construct a word tree; denote by T its representative in $\mathfrak{T}_X$, and let β be the vertex of T for which the (α, β)-path is the one labeled by $x_1, x_2, \dots, x_n$ where $g = x_1 x_2 \cdots x_n$ in reduced form. We thus arrive at a birooted word tree (α, T, β) on X and write $(A, g)\tau = (\alpha, T, \beta)$.

The following example illustrates the constructions in 1.2, 3.4, and 3.6.

VIII.3.7 Example

Let $X = \{a, b, c\}$ and

$$w = aaa^{-1}a^{-1}a^{-1}abb^{-1}ab^{-1}bcaa^{-1}cc^{-1}.$$

Direct inspection shows that

$$A = r(\hat{w}) = \{1, a, a^2, a^{-1}, b, ab^{-1}, ac, aca, ac^2\},$$

$$g = r(w) = ac.$$

Thus in the representation I_X, to w corresponds the element (A, g). Note that (A, g) can also be obtained by multiplying out in I_X the expression for w where we have first performed the substitution

$$a \to (\{1, a\}, a), \qquad b \to (\{1, b\}, b), \qquad c \to (\{1, c\}, c).$$

The birooted word tree (α, T, β) corresponding to (A, g), according to the above procedure, is the following

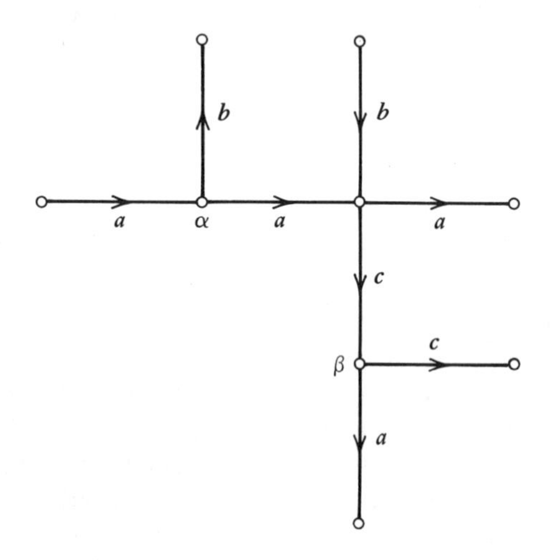

In other words, we successively form edges labeled $a, a, a^{-1}, b, \ldots,$ starting from an arbitrary point α and let β be the vertex with (α, β)-path labeled ac.

Now take another word, say

$$w' = cc^{-1}aa^{-1}abb^{-1}a^{-1}c^{-1}a^{-1}bbb^{-1}.$$

For it, we have the pair

$$A' = \{1, c, a, ab, c^{-1}, c^{-1}a^{-1}, c^{-1}a^{-1}b, c^{-1}a^{-1}b^2\},$$

$$g' = c^{-1}a^{-1}b,$$

and the birooted word tree

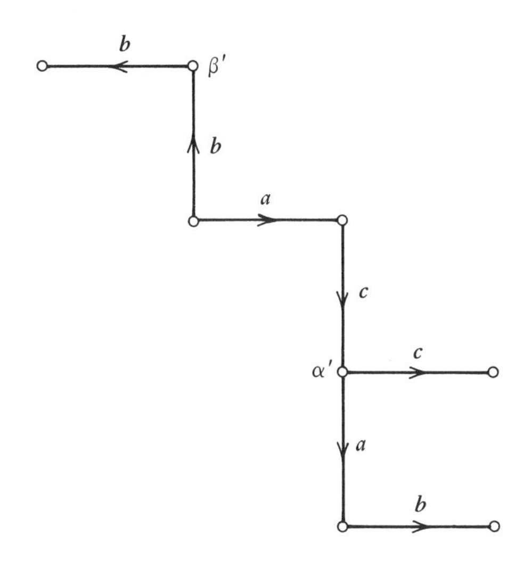

For the product, we have

$$ww' = aaa^{-1}a^{-1}a^{-1}abb^{-1}ab^{-1}bcaa^{-1}cc^{-1}abb^{-1}a^{-1}c^{-1}a^{-1}bbb^{-1}$$

$$(A, g)(A', g') = (A \cup gA', gg') = (B, b)$$

where

$$B = \{1, a, a^2, a^{-1}, b, ab^{-1}, ac, aca, ac^2, ac^2a,$$

$$ac^2ab, aca^{-1}, aca^{-1}b, aca^{-1}b^2\}.$$

We now paste the second graph over the first by identifying β and α' and other corresponding vertices.

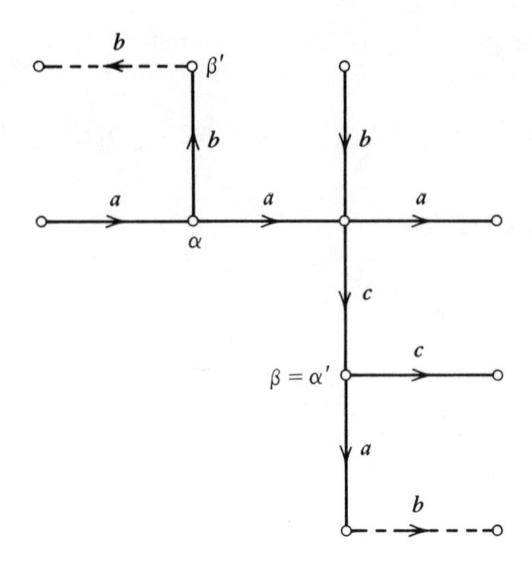

We have actually added to the first graph the two arrows indicated by a broken line. The product of the two birooted word trees is (α, T'', β'), where T'' is represented by the third graph.

We now prove the desired result.

VIII.3.8 Theorem

The function τ defined in 3.6 is an isomorphism of I_X onto B_X.

Proof. Note that for $(A, g) \in I_X$, $(A, g)\tau$ is the unique birooted word tree (α, T, β) on X whose set of all (α, γ)-paths bears the labels of words in A and the (α, β)-path bears the label of g. Hence τ maps I_X into B_X. Let $(\alpha, T, \beta) \in B_X$. Let A be the set of all words which label the (α, γ)-paths as γ runs over $V(T)$, and let g be the word which labels $\Pi(\alpha, \beta)$. It is clear that with every $w \in A$, all the left factors of w are contained in A, and that $g \in A$. Thus $(A, g) \in I_X$, and according to the remark made at the outset of the proof, we must have $(A, g)\tau = (\alpha, T, \beta)$. Consequently, τ maps I_X onto B_X. Since the constructed associations are unique, we get as a consequence that τ is one-to-one.

Recall that for $(A, g), (A', g') \in I_X$, we have

$$(A, g)(A', g') = (A \cup gA', gg').$$

Let $(A, g)\tau = (\alpha, T, \beta)$ and $(A', g')\tau = (\alpha', T', \beta')$. The word tree corresponding to $A \cup gA'$ is obtained by applying the procedure in 3.6. This amounts to taking the union of T and T' and performing the identifications described in 3.4 (in particular, β and α' are identified). In other words, we obtain the word tree $T(\beta, \alpha')T'$. Note that elimination of loops in the graph corresponds to cancellation of the corresponding products of letters. It follows easily that $\Pi(\alpha, \beta')$ can be obtained by following $\Pi(\alpha, \beta)$ by $\Pi(\alpha', \beta')$ since we have identified β and α' and then eliminated all loops (if any). This shows that gg' labels $\Pi(\alpha, \beta')$. Consequently, τ is a homomorphism.

We may combine the homomorphism χ in 1.5 of Z onto M_X with the isomorphism τ in 3.6 extended to map the identity of M_X onto the triple $(\alpha, \{\alpha\}, \alpha)$ for a fixed vertex α to obtain a homomorphism $\chi\tau$ of Z onto B_X so extended which induces the Wagner congruence on Z. Indeed,

$$\chi\tau : x_1 x_2 \cdots x_n \to (\alpha, T, \beta)$$

if and only if $T \in \mathfrak{T}_X$ and there exists a spanning walk $(\alpha = \gamma_0, \gamma_1, \ldots, \gamma_n = \beta)$ on T whose edges are labeled $x_1, x_2, \ldots, x_n$.

For any word tree T on X, denote

$$l(T) = \{|x| \mid x \text{ labels an edge of } T\}.$$

VIII.3.9 Proposition

The following statements are valid in B_X (cf. 1.14).

(i) $(\alpha, T, \beta)^{-1} = (\beta, T, \alpha)$.
(ii) $(\alpha, T, \beta)\mathcal{R}(\alpha', T', \beta') \Leftrightarrow T = T', \alpha = \alpha'$.
(iii) $(\alpha, T, \beta)\mathcal{L}(\alpha', T', \beta') \Leftrightarrow T = T', \beta = \beta'$.
(iv) $(\alpha, T, \beta)\mathcal{H}(\alpha', T', \beta') \Leftrightarrow (\alpha, T, \beta) = (\alpha', T', \beta')$.
(v) $(\alpha, T, \beta)\mathcal{D}(\alpha', T', \beta') \Leftrightarrow T = T'$.
(vi) $(\alpha, T, \beta)\sigma(\alpha', T', \beta') \Leftrightarrow \Pi(\alpha, \beta) \cong \Pi(\alpha', \beta')$.
(vii) $(\alpha, T, \beta)\eta(\alpha', T', \beta') \Leftrightarrow l(T) = l(T')$.
(viii) $|L_{(\alpha, T, \beta)}| = |R_{(\alpha, T, \beta)}| = |T|$.
(ix) $|D_{(\alpha, T, \beta)}| = |T|^2$.
(x) $(\alpha, T, \beta)^2 = (\alpha, T, \beta) \Leftrightarrow \alpha = \beta$.

Proof. (i) This follows from the remarks preceding the proposition, and can be verified easily.

(ii) First note

$$(A, g)\mathcal{R}(A', g') \Leftrightarrow A = A'.$$

By the construction of $(A, g)\tau = (\alpha, T, \beta)$, we see that A and (α, T) mutually determine each other uniquely, so that $A = A'$ if and only if $\alpha = \alpha'$, $T = T'$ where $(A', g')\tau = (\alpha', T', \beta')$.

(iii) This follows from (i) and (ii) since for any elements of an inverse semigroup, $a\mathcal{L}b \Leftrightarrow a^{-1}\mathcal{R}b^{-1}$.

(iv) This follows directly from (ii) and (iii), and confirms what we have seen in 1.14 for I_X.

(v) For any elements of B_X, we have

$$(\alpha, T, \beta)\mathcal{D}(\alpha', T', \beta') \Leftrightarrow (\alpha, T, \beta)\mathcal{R}(\alpha'', T'', \beta''),$$

$$(\alpha'', T'', \beta'')\mathcal{L}(\alpha', T', \beta') \quad \text{for some } (\alpha'', T'', \beta'')$$

$$\Leftrightarrow T = T'.$$

(vi) This follows directly from $(A, g)\sigma(B, h) \Leftrightarrow g = h$ in view of the isomorphism τ.

(vii) This follows from 2.5 since for $(A, g)\tau = (\alpha, T, \beta)$, we evidently have $c(A) = l(T)$.

(viii) This is a consequence of (ii) and (iii) since there are $|T|$ choices for α and $|T|$ choices for β.

(ix) This follows easily from (viii).

(x) This follows from (i).

Munn [13] introduced birooted word trees and proved that they are in a one-to-one correspondence with elements of a free inverse monoid on the same set. In Petrich [16] the multiplication exhibited here was introduced for birooted word trees making the one-to-one correspondence an isomorphism. See also Mitchell [1].

VIII.4 FREE GENERATORS

Motivated by the well-known result in group theory that a subgroup of a free group is itself free, it is natural to ask the same type of question for inverse semigroups. The answer turns out to be negative, so one asks for necessary and sufficient conditions on a subset of the free inverse semigroup in order that it generates a free inverse semigroup. The main result here answers the more general question of when a subset of an arbitrary inverse semigroup generates a free inverse semigroup. The condition is necessarily quite complicated, but it is usable and simplifies considerably in various special cases.

VIII.4.1 Notation

Let S be any inverse semigroup. If K is a nonempty subset of S, we denote by $\langle K \rangle$ the inverse subsemigroup of S generated by K, and write $K^{-1} = \{k^{-1} | k \in K\}$. For elements $a_1, a_2, \ldots, a_n \in S$, we write $\prod_{i=1}^{n} a_i = a_1 a_2 \cdots a_n$. For any element $u = (A, g)$ of I_X, we put $\Delta(u) = A$.

VIII.4.2 Definition

Let K be a nonempty subset of an inverse semigroup S, and denote by $\iota: K \to S$ the inclusion mapping. If $(\langle K \rangle, \iota)$ is a free inverse semigroup on K, we say that K is *a set of free generators for* $\langle K \rangle$.

The next result characterizes all sets of free generators of a free inverse semigroup (I_X, φ).

VIII.4.3 Proposition

A subset K of I_X is a set of free generators for I_X if and only if $K \subseteq X\varphi \cup (X\varphi)^{-1}$ and for each $x \in X$, $|K \cap \{x\varphi, (x\varphi)^{-1}\}| = 1$.

Proof

Necessity. Let $x \in X$. Then

$$x\varphi = (\hat{x}, x) = w_1 w_2 \cdots w_n$$

for some $w_1, w_2, \ldots, w_n \in K \cup K^{-1}$. Hence $\Delta(w_1) \subseteq \hat{x} = \{1, x\}$ so that $\Delta(w_1) = \hat{x}$. Thus either $w_1 = (\hat{x}, x)$ or $w_1 = (\hat{x}, 1)$. In the latter case, w_1 is an idempotent which is impossible since K is a set of free generators. Hence $x\varphi = w_1 \in K \cup K^{-1}$ and thus $X\varphi \cup (X\varphi)^{-1} \subseteq K \cup K^{-1}$. Since both K and $X\varphi$ are sets of free generators for I_X, we must have $X\varphi \cup (X\varphi)^{-1} = K \cup K^{-1}$. It follows that $K \subseteq X\varphi \cup (X\varphi)^{-1}$ and that for any $x \in X$, $\{x\varphi, (x\varphi)^{-1}\} \cap K \neq \varnothing$. Since K is a set of free generators, we cannot have $x\varphi, (x\varphi)^{-1} \in K$, which implies that $|K \cap \{x\varphi, (x\varphi)^{-1}\}| = 1$.

Sufficiency is clear.

We are now ready to establish the main criterion on a subset K of an arbitrary inverse semigroup to be a set of free generators for $\langle K \rangle$. Recall the notation $\lambda x = xx^{-1}$ in II.1.12.

VIII.4.4 Theorem

Let K be a nonempty subset of an inverse semigroup S. Then K is a set of free generators for $\langle K \rangle$ if and only if

(i) $K \cap K^{-1} = \varnothing$,

(ii) if $y \in K \cup K^{-1}$ and $\lambda y \geq e_1 e_2 \cdots e_n$ where $e_i = \lambda(y_{i1} y_{i2} \cdots y_{in_i})$ for some $y_{ij} \in K \cup K^{-1}$ such that $y_{ij} \neq y_{i(j+1)}^{-1}$ for $j = 1, 2, \ldots, n_i - 1$, $i = 1, 2, \ldots, n$, then $y = y_{i1}$ for some i.

Proof. Let $\kappa: X \to K$ be a one-to-one correspondence of some set X onto K. As usual, we denote by (I_X, φ) the free inverse semigroup on X. Then κ

determines a unique homomorphism θ of I_X onto $\langle K \rangle$ which makes the diagram

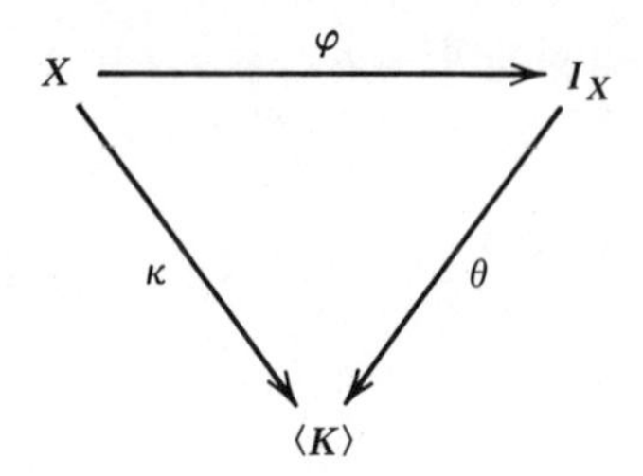

commute. Then K is a set of free generators for $\langle K \rangle$ if and only if θ is one-to-one.

Necessity. Let K be a set of free generators for $\langle K \rangle$. Then θ is one-to-one, and we may identify I_X and $\langle K \rangle$ thereby setting θ to be the identity mapping. Then K satisfies the conditions of 4.3 and thus condition (i) is satisfied.

Let the hypothesis of condition (ii) be satisfied, and let

$$y = (\hat{z}, z), \qquad y_{ij} = (\hat{z}_{ij}, z_{ij}) \qquad (1 \leq j \leq n_i, 1 \leq i \leq n).$$

The hypothesis $\lambda y \geq e_1 e_2 \cdots e_n$ implies

$$\Delta(y) = \hat{z} = \Delta(\lambda y) \subseteq \Delta(e_1 e_2 \cdots e_n) = \bigcup_{i=1}^{n} \Delta(e_i)$$

$$= \bigcup_{i=1}^{n} \Delta(y_{i1} y_{i2} \cdots y_{in_i}).$$

Consequently, $z \in \Delta(y_{i1} y_{i2} \cdots y_{in_i})$ for some i. Since $y_{ij} \neq y_{i(j+1)}^{-1}$, we have $z_{ij} \neq z_{i(j+1)}$ for $j = 1, 2, \ldots, n_i - 1$. Hence all the elements $z_{i1}, z_{i1}z_{i2}, \ldots, z_{i1}z_{i2} \cdots z_{in_i}$ are in reduced form, and thus

$$\Delta(y_{i1} y_{i2} \cdots y_{in_i}) = \Delta(y_{i1}) \cup z_{i1}\Delta(y_{i2}) \cup z_{i1}z_{i2}\Delta(y_{i3}) \cup \cdots$$

$$= \{1, z_{i1}, z_{i1}z_{i2}, z_{i1}z_{i2}z_{i3}, \ldots, z_{i1}z_{i2} \cdots z_{in_i}\}. \tag{5}$$

Since $y \in K \cup K^{-1}$, by 4.3, we must have $z \in X \cup X^{-1}$. But $z \in \Delta(y)$, so by the above, z is an element of the set (5). The only element in (5) which is an element of $X \cup X^{-1}$ is z_{i1}. Hence $z = z_{i1}$ and thus $y = y_{i1}$, as required.

Sufficiency. By the remarks at the beginning of the proof, it suffices to show that θ is one-to-one. We have seen in 1.14 that I_X is combinatorial. Thus it suffices to show that the congruence $\bar{\theta}$ induced by θ is contained in $\mathcal{H}$, which in view of III.3.2, is equivalent to $\bar{\theta}$ being idempotent separating. Our proof reduces to showing that θ restricted to idempotents of I_X is one-to-one.

By contradiction, we assume the existence of distinct idempotents e, f in I_X for which $e\theta = f\theta$. Then $\Delta(e) \neq \Delta(f)$ and we may suppose that $z_1 z_2 \cdots z_n \in \Delta(e) \setminus \Delta(f)$, where $n \geq 1$, $z_1, z_2, \ldots, z_n \in X \cup X^{-1}$ and $z_t \neq z_{t+1}^{-1}$ for $t = 1, 2, \ldots, n-1$. Let $u_i = (\hat{z}_i, z_i)$ for $i = 1, 2, \ldots, n$. Since $z_1 z_2 \cdots z_n \in \Delta(e)$, we obtain

$$\begin{aligned}\Delta(\lambda(u_1 u_2 \cdots u_n)) &= \Delta(u_1 u_2 \cdots u_n) \\ &= \{1, z_1, z_1 z_2, \ldots, z_1 z_2 \cdots z_n\} \subseteq \Delta(e)\end{aligned}$$

and hence $e \leq \lambda(u_1 u_2 \cdots u_n)$. It follows that

$$f\theta = e\theta \leq \lambda((u_1\theta)(u_2\theta) \cdots (u_n\theta)). \tag{6}$$

Let $\Delta(f) = \{g_1, g_2, \ldots, g_r\}$ for $r \geq 1$, where for $i = 1, 2, \ldots, r$, $g_i = v_{i1} v_{i2} \cdots v_{ik_i}$ for some $v_{ij} \in X \cup X^{-1}$ with $v_{ij} \neq v_{i(j+1)}^{-1}$ for $j = 1, 2, \ldots, k_i - 1$. Let $u_{ij} = (\hat{v}_{ij}, v_{ij})$ for $i = 1, 2, \ldots, r$ and $j = 1, 2, \ldots, k_i$. For $i = 1, 2, \ldots, r$, let $f_i = (\hat{g}_i, 1)$. Then

$$\Delta(f) = \bigcup_{i=1}^{r} \hat{g}_i, \qquad f = f_1 f_2 \cdots f_r, \qquad f_i = \lambda(u_{i1} u_{i2} \cdots u_{in_i}) \tag{7}$$

for $i = 1, 2, \ldots, r$. We will show by induction that for $s = 1, 2, \ldots, n$, there exists an i such that $u_1 = u_{i1}, u_2 = u_{i2}, \ldots, u_s = u_{is}$.

First consider the case $s = 1$. Since $\lambda(u_1\theta) \geq \lambda((u_1\theta) \cdots (u_n\theta))$, by (6) we have that $\lambda(u_1\theta) \geq f\theta$. Let $u_i\theta = y_i$ for $i = 1, 2, \ldots, n$ and $u_{ij}\theta = y_{ij}$ for $i = 1, 2, \ldots, r$ and $j = 1, 2, \ldots, k_i$. Then $\lambda y_i \geq \prod_{i=1}^{r} \lambda(y_{i1} \cdots y_{ik_i})$ and condition (ii) in the statement of the theorem yields $y_1 = y_{i1}$ for some i. Condition (i) insures that θ is one-to-one on $X\varphi \cup (X\varphi)^{-1}$, which implies $u_1 = u_{i1}$.

Assume next that $u_1 = u_{i1}, u_2 = u_{i2}, \ldots, u_{s-1} = u_{i(s-1)}$ for $i = 1, 2, \ldots, k$, but not for $i = k+1, \ldots, r$. Then $y_1 = y_{i1}, y_2 = y_{i2}, \ldots, y_{s-1} = y_{i(s-1)}$ for $i = 1, 2, \ldots, k$ but not for $i = k+1, \ldots, r$. For convenience, we will sometimes write $y = y_1 y_2 \cdots y_{s-1}$.

By (6), we have $\lambda(y y_s) \geq \lambda(y_1 y_2 \cdots y_n) \geq f\theta$ which gives

$$\lambda y_s \geq y^{-1} y (\lambda y_s) y^{-1} y = y^{-1} \lambda(y y_s) y \geq y^{-1}(f\theta) y = \prod_{i=1}^{r} y^{-1}(f_i\theta) y. \tag{8}$$

If $i \leq k$, then by (7) and the hypothesis, we obtain

$$\begin{aligned}y^{-1}(f_i\theta) y &= y^{-1}(\lambda(u_{i1} u_{i2} \cdots u_{in_i}))\theta y = y^{-1} \lambda(y_{i1} y_{i2} \cdots y_{in_i}) y \\ &= y^{-1} \lambda(y y_{is} \cdots y_{in_i}) y = y^{-1} y \lambda(y_{is} \cdots y_{in_i}).\end{aligned} \tag{9}$$

If $i > k$, then for some integer $p = p(i)$ such that $0 \leq p < s - 1$, we have

$$y^{-1}(f_i\theta)y = y^{-1}(\lambda(u_{i1} \cdots u_{in_i}))\theta y = y^{-1}\lambda(y_{i1} \cdots y_{in_i})y$$

$$= y^{-1}\lambda(y_1 \cdots y_p y_{i(p+1)} \cdots y_{in_i})y$$

$$= y^{-1}(y_1 \cdots y_p)\lambda(y_{i(p+1)} \cdots y_{in_i})$$

$$\cdot[(y_1 \cdots y_p)^{-1}(y_1 \cdots y_p)](y_{p+1} \cdots y_{s-1})$$

$$= y^{-1}(y_1 \cdots y_p)[(y_1 \cdots y_p)^{-1}(y_1 \cdots y_p)]$$

$$\cdot\lambda(y_{i(p+1)} \cdots y_{in_i})(y_{p+1} \cdots y_{s-1})$$

$$= y^{-1}(y_1 \cdots y_p)\lambda(y_{i(p+1)} \cdots y_{in_i})$$

$$\cdot[\lambda(y_{p+1} \cdots y_{s-1})(y_{p+1} \cdots y_{s-1})]$$

$$= y^{-1}[(y_1 \cdots y_p)\lambda(y_{p+1} \cdots y_{s-1})]$$

$$\cdot\lambda(y_{i(p+1)} \cdots y_{in_i})(y_{p+1} \cdots y_{s-1})$$

$$= y^{-1}y(y_{s-1}^{-1} \cdots y_{p+1}^{-1})\lambda(y_{i(p+1)} \cdots y_{in_i})$$

$$\cdot(y_{p+1} \cdots y_{s-1}), \tag{10}$$

where $y_{p+1} \neq y_{i(p+1)}$. Letting

$$e_i = \lambda(y_{is} \cdots y_{in_i}) \quad \text{for} \quad i = 1, 2, \ldots, k,$$

$$e_i = \lambda(y_{s-1}^{-1} \cdots y_{p+1}^{-1}y_{i(p+1)} \cdots y_{in_i}) \quad \text{for} \quad i = k + 1, \ldots, r,$$

$$e_{r+1} = \lambda(y^{-1}) = \lambda(y_{s-1}^{-1} \cdots y_1^{-1}),$$

we obtain from (8), (9), and (10) that $\lambda y_s \geq e_1 e_2 \cdots e_{r+1}$; applying condition (ii) in the statement of the theorem we obtain either $y_s = y_{is}$ for some $1 \leq i \leq k$ or $y_s = y_{s-1}^{-1}$. The latter alternative implies $u_s = u_{s-1}^{-1}$ contradicting the hypothesis that $z_s \neq z_{s-1}^{-1}$. Consequently, for some i, we have $y_1 = y_{i1}$, $y_2 = y_{i2}, \ldots, y_s = y_{is}$ and thus $u_1 = u_{i1}$, $u_2 = u_{i2}, \ldots, u_s = u_{is}$.

By induction we conclude that $u_1 = u_{i1}$, $u_2 = u_{i2}, \ldots, u_n = u_{in}$ for some i which implies that $\lambda(u_1 u_2 \cdots u_n) \geq f_i \geq f$. But then $z_1 z_2 \cdots z_n \in \Delta(f)$ contradicting the choice of $z_1 z_2 \cdots z_n$. Therefore θ is one-to-one on idempotents and hence an isomorphism.

The conditions in 4.4 simplify considerably in certain special cases. One of these is the case when the inverse semigroup S is itself free.

VIII.4.5 Corollary

A nonempty subset K of I_X is a set of free generators for $\langle K \rangle$ if and only if

(i) $K \cap K^{-1} = \varnothing$,
(ii) if $y \in K \cup K^{-1}$ and $\Delta(y) \subseteq \cup_{i=1}^{n} \Delta(y_{i1} y_{i2} \cdots y_{in_i})$ for some $y_{ij} \in K \cup K^{-1}$ such that $y_{ij} \neq y_{i(j+1)}^{-1}$ for $j = 1, 2, \ldots, n_i - 1$, $i = 1, 2, \ldots, n$, then $y = y_{i1}$ for some i.

Proof. The proof of this corollary is left as an exercise.

Another special case of interest is when $|K| = 1$. In this case, we have the following simple criterion for an element of an inverse semigroup to generate a free monogenic inverse subsemigroup.

VIII.4.6 Corollary

Let S be an inverse semigroup and let $a \in S$. Then $\langle a \rangle$ is a free inverse semigroup on a single generator if and only if $aa^{-1} \ngeq a^{-n}a^{n}$ and $a^{-1}a \ngeq a^{n}a^{-n}$ for all positive integers n.

Proof. The proof is left as an exercise.

We can apply the last result to obtain further information about free inverse semigroups as follows.

VIII.4.7 Corollary

An nonidempotent of a free inverse semigroup I_X generates a free inverse subsemigroup of I_X.

Proof. Let $a \in I_X$ and suppose that for some positive integer n, $aa^{-1} \geq a^{-n}a^{n}$, that is $aa^{-1}a^{-n}a^{n} = a^{-n}a^{n}$, or equivalently, $aa^{-(n+1)} = a^{-n}$. Now

$$a^{2}a^{-(n+2)} = a(aa^{-(n+1)})a^{-1} = aa^{-n}a^{-1} = aa^{-(n+1)} = a^{-n},$$

and a simple induction yields $a^{n}a^{-2n} = a^{-n}$, whence $a^{n}a^{-n} \geq a^{-n}a^{n}$. Since $a^{n} \mathcal{J} a^{-n}$ and I_X is completely semisimple by 1.16, we obtain $a^{n}a^{-n} = a^{-n}a^{n}$, that is, a^{n} belongs to a subgroup of I_X. But I_X is combinatorial by 1.14, so a^{n} is an idempotent. It follows that $(a\sigma)^{n} = 1$ in the free group I_X/σ, whence $a\sigma = 1$. Since I_X is E-unitary, it follows that a is idempotent.

This argument used on a^{-1} instead of a shows that if $a^{-1}a \geq a^{n}a^{-n}$, then a^{-1} is idempotent. The desired conclusion now follows from 4.6.

VIII.4.8 Exercises

(i) Let S be a Clifford semigroup and let A be a nonempty subset of S. Show that if $\langle A, \iota \rangle$ is a free Clifford semigroup, then

(α) $\mathcal{H}|_A = \varepsilon$,

(β) for every maximal subgroup G of S, either $\langle A \rangle \cap G = \varnothing$ or $\langle A \rangle \cap G$ is a free group,

(γ) $\langle A \rangle / \mathcal{H}$ is a free semilattice.

(Note that if S is free, (β) holds automatically.)

(ii) Show that 4.6 ceases to be valid if we replace the conditions $aa^{-1} \ngeq a^{-n}a^n$ and $a^{-1}a \ngeq a^n a^{-n}$ for all n by a finite subcollection.

VIII.4.9 Problems

(i) Establish an analogue of 4.4 for the following classes: semilattices, groups, Clifford semigroups.

All results in this section are due to Reilly [7]; this reference contains further relevant material. An alternative proof of result 4.4 was devised by O'Carroll [3].

VIII.5 THE BASIS PROPERTY

In analogy with the situation in vector spaces, we say that a minimal generating set of an inverse semigroup S is a basis for S, and if any two such subsets of S have the same cardinality, S is said to have the basis property. We consider here a somewhat more general condition, and prove that completely semisimple combinatorial inverse, and thus in particular free inverse, semigroups have this property. This further enlarges the list of properties of free inverse semigroups established in the earlier sections of this chapter. We will conclude this chapter in the next section where we will prove that inverse subsemigroups of a free inverse semigroup indeed possess bases. In the entire discussion it is convenient to regard the empty subset of an inverse semigroup as an inverse subsemigroup.

We start with a number of auxiliary statements.

VIII.5.1 Lemma

In any semigroup S, for $a, c \in S^1$, $b \in S$, we have $J_{abc} \leq J_b$.

Proof. Trivially, $J(abc) = S^1abcS^1 \subseteq S^1bS^1 = J(b)$.

We will use this result often and without express reference, mostly when w is a word containing a letter b or its inverse to conclude that $J_w \leq J_b$.

VIII.5.2 Notation

Let S be an inverse semigroup. If A and B are nonempty subsets of S, we write $\langle A, B\rangle$ for $\langle A \cup B\rangle$. For $x, y \in S$, the notation $\langle A, x, y\rangle$ means $\langle A \cup \{x, y\}\rangle$.

VIII.5.3 Lemma

Let T be an inverse subsemigroup of an inverse semigroup S. If $a, b \in S$, $J_a \nleq J_b$ and $a \in \langle T, b\rangle$, then $a \in T$.

Proof. This follows directly from 5.1.

VIII.5.4 Lemma

The following statements concerning elements a, b, c of a completely semisimple combinatorial inverse semigroup S are true.

(i) If $a = ba^{-1}c$ and $a \mathcal{J} b$, then $a = b$.
(ii) If $aba \mathcal{J} a$, then $aba = a$.
(iii) If $aba^{-1} \mathcal{J} a$, then $aba^{-1} = aa^{-1}$.

Proof. (i) Let $a = ba^{-1}c$ and $a\mathcal{J}b$. Then $a = b(a^{-1}a)(a^{-1}c)$ where all the elements $a, b, a^{-1}a, a^{-1}c$ are $\mathcal{J}$-related and thus can be taken as elements of a combinatorial Brandt semigroup. We thus arrive at an equation of the form

$$(i,1,j) = (k,1,l)(j,1,j)(m,1,p)$$

which evidently gives $i = k$ and $j = l$. Consequently, $a = b$.

(ii) Let $aba\mathcal{J}a$. Then $aba = a(baa^{-1})a$ where a, baa^{-1}, and $a(baa^{-1})a$ are all $\mathcal{J}$-related and thus can be taken to be elements of a combinatorial Brandt semigroup B. We thus arrive at an equation $(i,1,j)(k,1,l)(i,1,j) \neq 0$ in B, which evidently yields $j = k$ and $l = i$. But then $a = a(baa^{-1})a = aba$ as required.

(iii) The argument here is similar to the preceding one and is left as an exercise.

The next lemma represents the first step in an inductive process making up the proof of the principal result of this section.

VIII.5.5 Lemma

Let S be a completely semisimple combinatorial inverse semigroup, T be an inverse subsemigroup of S, and $x, y, b \in S$. If $\langle T, x, y\rangle = \langle T, b\rangle$, then either $b \in \langle T, x\rangle$ or $b \in \langle T, y\rangle$.

Proof. If $J_b \nleq J_x$, then by 5.3, we have $b \in \langle T, y \rangle$. Similarly, if $J_b \nleq J_y$, then $b \in \langle T, x \rangle$. Further, if $J_x \nleq J_b$, then $x \in T$, and thus $b \in \langle T, y \rangle$. Similarly, $J_y \nleq J_b$ implies $b \in \langle T, x \rangle$. There remains the case $J_x = J_y = J_b$.

Hence suppose that $x \mathcal{J} y \mathcal{J} b$. The conclusion certainly holds if $b \in T$, so we assume that $b \notin T$. Since $b \in \langle T, x, y \rangle$, b can be written as a product of elements from T and from $X = \{x, x^{-1}, y, y^{-1}\}$. Without loss of generality, we may suppose that of all elements of X appearing in the expression for b it is x that appears *first* (by renaming or taking inverses, if necessary).

Thus $b = vxw$ for some $v \in T^1$ and $w \in \langle T, x, y \rangle$. Similarly, $x \in \langle T, b \rangle$ implies that x is a product of elements from T and from $B = \{b, b^{-1}\}$. We may also assume that $x \notin T$, for otherwise $b \in \langle T, y \rangle$. Here we distinguish two cases: the *last* appearance of an element from B is either (I) equal to b or (II) equal to b^{-1}.

Case I. $x = pbq$ for some $p \in \langle T, b \rangle^1$, $q \in T^1$. Then

$$b = vxw = vxx^{-1}xw = vx(pbq)^{-1}xq = (vxq^{-1})b^{-1}(p^{-1}xq)$$

and $vxq^{-1} = v(pbq)q^{-1} \in J(b)$ so that $b \mathcal{J} vxq^{-1}$. Hence by 5.4(i) we deduce that $b = vxq^{-1} \in \langle T, x \rangle$.

Case II. $x = pb^{-1}q$ for some $p \in \langle T, b \rangle$, $q \in T^1$. Now consider the *first* appearance of elements of B in x. By repeated applications of 5.4(ii) and (iii), we can reduce this case to two subcases as follows:

$$\begin{aligned} &(\text{A}) \quad x = pb^{-1}q \quad \text{for some} \quad p, q \in T^1, \\ &(\text{B}) \quad x = pbb^{-1}q \quad \text{for some} \quad p, q \in T^1, \end{aligned}$$

For (A), we now have

$$b = vxw = vxx^{-1}xw = vxx^{-1}(pb^{-1}q)w = (vxx^{-1}p)b^{-1}(qw)$$

and $vxx^{-1}p = vpb^{-1}qx^{-1}p \in J(b^{-1})$ so that $b \mathcal{J} vxx^{-1}p$. Hence by 5.4(i), we obtain that $b = vxx^{-1}p \in \langle T, x \rangle$.

For (B), we must again consider two subcases. This time consider the *last* appearance of x or x^{-1} in the expression for b. As before, by repeated applications of 5.4(ii) and (iii), we arrive at the following two cases:

$$\begin{aligned} &(\text{B1}) \quad b = vxw \qquad \text{for some} \quad v \in T^1 \quad \text{and} \quad w \in \langle T, y \rangle^1, \\ &(\text{B2}) \quad b = vxx^{-1}w \quad \text{for some} \quad v \in T^1, w \in \langle T, y \rangle^1. \end{aligned}$$

Note that if $w \in T^1$, in either case we have $a \in \langle T, x \rangle$. Hence suppose that $w \notin T^1$. Then $w \mathcal{J} b$ since $b \in J(w)$ and $w \in J(y) = J(b)$.

If (B1) takes place, we have

$$b = vxw = v(pbb^{-1}q)w = (vpb)b^{-1}(qw)$$

and $b \mathfrak{J} qw$, so by the dual of 5.4(i), we have $b = qw \in \langle T, y \rangle$.

If (B2) takes place, we have

$$b = vxx^{-1}w = vx(q^{-1}bb^{-1}p^{-1})w = (vxq^{-1}b)b^{-1}(p^{-1}w)$$

and $b \mathfrak{J} p^{-1}w$, so by the dual of 5.4(i), we have $b = p^{-1}w \in \langle T, y \rangle$.

We have exhausted all the cases.

We are now ready for the principal result of this section.

VIII.5.6 Theorem

Let S be a completely semisimple combinatorial inverse semigroup. Let T be an inverse subsemigroup of S, A and B be subsets of S such that $|A| > |B|$, and $\langle T, A \rangle = \langle T, B \rangle$. Then there exists $a \in A$ such that $a \in \langle T, A \setminus a \rangle$.

Proof. We consider first the case when A and B are finite, and will use an inductive argument on $k = |B|$. In the case $k = 0$, we have $\langle T, A \rangle = T$ and thus $A \subseteq T$, so the desired conclusion holds trivially.

Case $k = 1$. Let $B = \{b\}$ and $A = \{a_1 a_2, \ldots, a_r\}$, $r \geq 2$.

First let $r = 2$. Then $\langle T, a_1, a_2 \rangle = \langle T, b \rangle$ and by 5.5, either $b \in \langle T, a_1 \rangle$ or $b \in \langle T, a_2 \rangle$. In the first case

$$a_2 \in \langle T, a_1, a_2 \rangle = \langle T, b \rangle = \langle T, a_1 \rangle$$

and similarly, in the second case, $a_1 \in \langle T, a_2 \rangle$, as required.

Assume next that $r \geq 3$, and set $V = \langle T, a_3, \ldots, a_r \rangle$. Then

$$\langle V, b \rangle = \langle T, a_3, \ldots, a_r, b \rangle \subseteq \langle T, A \rangle = \langle T, b \rangle \subseteq \langle V, b \rangle$$

since $b \in \langle T, A \rangle$, and thus

$$\langle V, a_1, a_2 \rangle = \langle T, A \rangle = \langle V, b \rangle.$$

Hence 5.5 yields that either $b \in \langle V, a_1 \rangle$ or $b \in \langle V, a_2 \rangle$. In the first case

$$a_2 \in \langle T, A \rangle = \langle V, a_1 \rangle = \langle T, A \setminus a_2 \rangle$$

and analogously, in the second case, $a_1 \in \langle T, A \setminus a_1 \rangle$.

Case $k = n$. Assume the result true for $k \leq n - 1$, and let $k = n$. Set $B = \{b_1, b_2, \ldots, b_n\}$ and $A = \{a_1, a_2, \ldots, a_{n+r}\}$, $r \geq 1$. Hence we have

$$\langle T, a_1, a_2, \ldots, a_{n+r} \rangle = \langle T, b_1, b_2, \ldots, b_n \rangle$$

which implies

$$\langle\langle T, b_2, \ldots, b_n\rangle, a_1, \ldots, a_{n+r}\rangle = \langle\langle T, b_2, \ldots, b_n\rangle, b_1\rangle.$$

The case $k = 1$ above, taken $(n + r - 1)$ times implies the existence of a_i for some $1 \leq i \leq n + r$ such that

$$\langle\langle T, b_2, \ldots, b_n\rangle, a_i\rangle = \langle\langle T, b_2, \ldots, b_n\rangle, b_1\rangle.$$

We can rewrite this equation in the form

$$\langle\langle T, a_i\rangle, b_2, \ldots, b_n\rangle = \langle\langle T, a_i\rangle, A \setminus a_i\rangle.$$

Letting $U = \langle T, a_i\rangle$, we have $\langle U, B \setminus b_1\rangle = \langle U, A \setminus a_i\rangle$; we can apply the induction hypothesis: there exists $a_j \in A \setminus a_i$ such that

$$a_j \in \langle U, A \setminus \{a_i, a_j\}\rangle = \langle U, A \setminus a_j\rangle = \langle T, A \setminus a_j\rangle.$$

This takes care of the case when both A and B are finite. Now assume that $B = \{b_1, b_2, \ldots, b_n\}$ is finite and A is infinite. Then for each b_i, we have $b_i \in \langle T, A_i\rangle$ for some finite subset A_i of A, since b_i is a (finite) product of elements of T and A. Put $A' = \cup_{i=1}^{n} A_i$. Hence $\langle T, A'\rangle = \langle T, B\rangle = \langle T, A\rangle$ where A' is a finite subset of A, so for any $a \in A \setminus A'$, we have $a \in \langle T, A'\rangle = \langle T, A \setminus a\rangle$.

Finally, if both A and B are infinite, we can use a similar procedure to obtain a subset A' of A of the same cardinality as B and such that $\langle T, A'\rangle = \langle T, A\rangle$. We then again have $a \in \langle T, A \setminus a\rangle$ for any $a \in A \setminus A'$.

We now introduce some terminology by means of which several results here take on a convenient form.

VIII.5.7 Definition

Let S be an inverse semigroup. For convenience, we include here the empty set $\varnothing$ as an inverse subsemigroup of S. Let T be an inverse subsemigroup of S and let $\varnothing \neq X \subseteq S$. Then X is *T-irredundant* if $\langle T, X\rangle = \langle T, X \setminus x\rangle$ for no $x \in X$; in such a case, X is a *T-basis* for $\langle T, X\rangle$. By definition, the empty set $\varnothing$ is T-irredundant for any such T, and is a *T-basis* for T. We say that S has the *strong basis property* if for any inverse subsemigroups U and T of S, with $U \subseteq T$, any two U-bases for T have the same cardinality.

We say that a subset X of S is *irredundant* if it is $\varnothing$-irredundant, and in such a case X is a *basis* for $\langle X\rangle$. We say that S has the (*weak*) *basis property* if any two bases for any inverse subsemigroup T of S have the same cardinality. If this is the case and an inverse subsemigroup T of S has a basis, call its cardinality the *rank* of T.

Note that if an inverse semigroup has the strong basis property, then it has the (weak) basis property. We can now state the main result of this section by using some of the terminology just introduced.

VIII.5.8 Corollary

Every completely semisimple combinatorial inverse semigroup has the strong basis property.

Proof. Let U and T be inverse subsemigroups of a completely semisimple combinatorial inverse semigroup S such that $U \subseteq T$ and $\langle U, A \rangle = \langle U, B \rangle = T$. If A and B are U-irredundant, then by 5.6 we must have $|A| = |B|$.

We have seen in 1.14 and 1.16 that a free inverse semigroup I_X satisfies the conditions of the above corollary. This result does not yet imply that inverse subsemigroups of I_X have U-bases. We will prove in the next section that this is indeed the case.

In order to study the spread of U-bases relative to the $\mathcal{J}$-classes containing some of their elements, we will use the following symbolism.

VIII.5.9 Notation

For a $\mathcal{J}$-class J of a semigroup S and $A \subseteq S$, set

$$A(J) = \{a \in A \mid J_a > J\}, \qquad A^J = A \cap J,$$

$$A_J = A(J) \cup A^J \quad (= \{a \in A \mid J_a \geq J\}).$$

VIII.5.10 Lemma

Let S be an inverse semigroup. Let U be an inverse subsemigroup of S, $A, B \subseteq S$ be such that $\langle U, A \rangle = \langle U, B \rangle$, and J be a $\mathcal{J}$-class of S. Then

$$\langle U, A(J) \rangle = \langle U, B(J) \rangle, \qquad \langle U, A_J \rangle = \langle U, B_J \rangle.$$

Proof. Let $a \in A(J) \setminus U$. Then a can be expressed as a product of elements from U and $B \cup B^{-1}$ since $\langle U, A \rangle = \langle U, B \rangle$, and involving some $b \in B$. Hence $J < J_a \leq J_b$ by 5.1, so that $b \in B(J)$. Consequently, $A(J) \subseteq \langle U, B(J) \rangle$, and analogously $B(J) \subseteq \langle U, A(J) \rangle$. Therefore, $\langle U, A(J) \rangle = \langle U, B(J) \rangle$, and a similar argument gives $\langle U, A_J \rangle = \langle U, B_J \rangle$.

We are now ready for the desired result.

VIII.5.11 Proposition

Let S be an inverse semigroup with the strong basis property. Let U and T be inverse subsemigroups of S with $U \subseteq T$. Then for every $\mathcal{J}$-class J of S, every U-basis for T has the same number of elements in J.

Proof. Let A and B be two U-bases for T. By 5.10, we have $\langle U, A(J)\rangle = \langle U, B(J)\rangle = T(J)$, say. Further,

$$\langle T(J), A^J\rangle = \langle U, A_J\rangle = \langle U, B_J\rangle = \langle T(J), B^J\rangle$$

again by 5.10. Since A^J and B^J are U-irredundant, they are $T(J)$-bases for $\langle T(J), A^J\rangle$. But then $|A^J| = |B^J|$ by the strong basis property.

VIII.5.12 Exercises

(i) Show that the following conditions on an inverse semigroup S are equivalent.

(α) S is completely semisimple.
(β) For any $a, b \in S$, $ab\mathfrak{D}a$ implies $ab\mathfrak{R}a$.
(γ) For any $a, b \in S$, $ab\mathfrak{D}b$ implies $ab\mathfrak{L}b$.
(δ) For any $a, b \in S$, $ab\mathfrak{D}a\mathfrak{D}b$ implies $a\mathfrak{L}b^{-1}$.

(ii) Let S be an inverse semigroup with the basis property and finite rank n. Show that no generating set of S with more than n elements can be irredundant, and that no irredundant subset of S with less than n elements can generate S.

(iii) Show that condition 5.4(i) on an inverse semigroup S implies that S is completely semisimple and combinatorial.

(iv) Show that the additive group of integers does not have the basis property.

(v) Let S be an inverse semigroup, U an inverse subsemigroup of S, and $A \subseteq S$. Show that A is U-irredundant if and only if A^J is $[U, A(J)]$-irredundant for each $\mathfrak{J}$-class J of S.

All results in this section are due to Jones [3]; this paper contains further relevant material. See also Jones [4], where a complete characterization of inverse semigroups with the strong basis property is given.

VIII.6 THE EXISTENCE OF BASES

The discussion of the preceding section centered around the (strong) basis property, namely around the cardinalities of various (U-)bases for inverse subsemigroups of an inverse semigroup. This, of course, begs the question of the existence of such bases. We provide here a sufficient condition on an inverse semigroup S in order that all inverse subsemigroups of S have a basis. This condition as well as the strong basis property are fulfilled in a free inverse semigroup.

VIII.6.1 Proposition

Let S be an inverse semigroup. If U is an inverse subsemigroup of S and X is a finite subset of S, then X contains a U-basis for $\langle U, X\rangle$. In particular, every finitely generated inverse subsemigroup of S has a basis.

Proof. Let $T = \langle U, X\rangle$. If X is U-irredundant, then X is a basis for T. Otherwise, $\langle U, X\setminus x\rangle = T$ for some $x \in X$. If $X\setminus x$ is not U-irredundant, repeat this process until a U-irredundant subset X' of X is reached. Then X' is a U-basis for T.

VIII.6.2 Corollary

Let S be an inverse semigroup with the strong basis property. Let U and T be inverse subsemigroups of S, A be a finite U-basis for T, and B be a subset of T such that $T = \langle U, B\rangle$. Then $|A| \leq |B|$, and the equality holds if and only if B is a U-basis for T.

Proof. By hypothesis, we have $\langle U, A\rangle = \langle U, B\rangle$. Let $A = \{a_1, a_2, \ldots, a_n\}$. Then for each a_i, we have $a_i \in \langle U, B\rangle$ so that $a_i \in \langle U, B_i\rangle$ for some finite subset B_i of B. Let $B' = \cup_{i=1}^{n} B_i$. Then $T = \langle U, B'\rangle$ and B' is finite. By 6.1, B' contains a U-basis B'' of T. By the strong basis property, we have $|A| = |B''|$ and thus $|A| \leq |B|$. Clearly, $B'' = B$ if and only if B is a U-basis for T.

VIII.6.3 Corollary

Let S be an inverse semigroup with the strong basis property. If T is an inverse subsemigroup of S of finite rank n, then any n generators of T form a basis of T.

Proof. The proof of this corollary is left as an exercise.

Infinitely generated inverse semigroups (even infinitely generated groups) need not have a basis (e.g., the additive group of rationals). A sufficient condition for the existence of bases involves the following concept.

VIII.6.4 Definition

Let P be a partially ordered set and $\mathbb{N}$ be the set of all positive integers. A *depth function* d on P is a mapping $d: P \to \mathbb{N}$ such that $x < y$ in P implies $d(x) > d(y)$ (that is, d is strictly order inverting).

A semigroup S is *layered* if the partially ordered set $S/\mathcal{J}$ of $\mathcal{J}$-classes of S has a depth function. If d is such a depth function, the elements of $S/\mathcal{J}$ with the same image under d are the *layers* of S.

We are now ready for the main result of this section.

VIII.6.5 Theorem

Let S be a layered inverse semigroup with finite $\mathcal{J}$-classes. If U and T are inverse subsemigroups of S with $U \subseteq T$, then T has a U-basis.

Proof. Let d be a depth function for $S/\mathcal{J}$. Since d is strictly order inverting, no two distinct $\mathcal{J}$-classes within the same layer of S are comparable. By putting a well-order on the $\mathcal{J}$-classes within each layer, we can well-order $S/\mathcal{J}$ as $\{J_\alpha | 1 \le \alpha < \lambda\}$ for some ordinal λ in such a way that $\beta > \alpha$ implies $J_\beta \ngeq J_\alpha$.

Let U and T be inverse subsemigroups of S and A be a subset of S such that $\langle U, A\rangle = T$, for example we may take $A = T \setminus U$. For any $\mathcal{J}$-class J_α, define

$$A^\alpha = A \cap J_\alpha, \qquad A_\alpha = \bigcup_{\beta \le \alpha} A^\beta, \qquad T_\alpha = \langle U, A_\alpha \rangle.$$

We now show how we can define a U-basis A'_α for T_α for each α, $1 \le \alpha < \lambda$, by using transfinite induction.

For $\alpha = 1$, we have that $A_1 = A^1$ is finite since each J_α and thus each A^α is finite, and thus by 6.1, T_1 has a U-basis A'_1. Assume that we have defined a U-basis $A'_\beta \subseteq A_\beta$ for all $\beta < \alpha$ and that $A'_\gamma \subseteq A'_\beta$ if $\gamma \le \beta < \alpha$. Set $A''_\alpha = \cup_{\beta<\alpha} A'_\beta$ and suppose that A''_α is not U-irredundant. Then for some $a \in A''_\alpha$, we have $a \in \langle U, A''_\alpha \setminus a\rangle$. There exists $\beta < \alpha$ such that $a \in A'_\beta$ and $\gamma \le \beta$ such that $a \in J_\gamma$. Assume that a can be written as a product involving b, where $b \in (A''_\alpha \setminus a) \cap J_\delta$. Then by 5.1, we have $J_a \le J_b$, that is $J_\gamma \le J_\delta$. Thus $\delta \le \gamma \le \beta$ so that $b \in A'_\beta \setminus a$. But then $a \in \langle U, A'_\beta \setminus a\rangle$ which contradicts U-irredundancy of A'_β.

Consequently, A''_α is U-irredundant. Let $T''_\alpha = \langle U, A''_\alpha\rangle$. Now $\langle U, A_\beta\rangle = \langle U, A'_\beta\rangle$ for all $\beta < \alpha$, so we have $T_\alpha = \langle T''_\alpha, A^\alpha\rangle$. Since A^α is finite, by 6.1, it contains a T''_α-basis $(A^\alpha)'$ for T_α. Set $A'_\alpha = A''_\alpha \cup (A^\alpha)'$. Then

$$\langle U, A'_\alpha\rangle = \langle U, A''_\alpha \cup (A^\alpha)'\rangle = \langle T''_\alpha, (A^\alpha)'\rangle = T_\alpha.$$

Suppose A'_α is not U-irredundant, say $a \in \langle U, A'_\alpha \setminus a\rangle$ for some $a \in A'_\alpha$. Assume that $a \in A''_\alpha$; then $a \in J_\beta$ for some $\beta < \alpha$. Thus $J_\beta \nleq J_\alpha$ and a cannot be a product involving any element of $(A_\alpha)'$. But then $a \in \langle U, A''_\alpha \setminus a\rangle$ contradicting U-irredundancy of A''_α. Otherwise, $a \in (A_\alpha)'$ which implies $\langle T''_\alpha, (A^\alpha)'\rangle = \langle T''_\alpha, (A^\alpha)' \setminus a\rangle$ contradicting T''_α-irredundancy of $(A^\alpha)'$.

Therefore, A'_α is a U-basis for T_α, and $A'_\beta \subseteq A'_\alpha$ if $\beta < a$. By transfinite induction, a U-basis A'_α for S_α is defined for all α, $1 \le \alpha < \lambda$.

Put $A' = \cup_{\alpha<\lambda} A'_\alpha$. Since each A'_α is U-irredundant, we deduce that also A' is U-irredundant, by exactly the same argument as it was shown above that A''_α is U-irredundant when A'_β is U-irredundant for all $\beta < \alpha$. If $a \in A$, then $a \in A_\alpha$ for some α, $1 \le \alpha < \lambda$. Since $\langle U, A_\alpha\rangle = \langle U, A'_\alpha\rangle$, we get $a \in \langle U, A'_\alpha\rangle \subseteq \langle U, A\rangle$. Therefore, $A \subseteq \langle U, A'\rangle$ and thus $T = \langle U, A'\rangle$, that is to say, A' is a U-basis for T.

In order to apply the above theorem to free inverse semigroups, we now show that they are layered.

VIII.6.6 Lemma

Every free inverse semigroup is layered.

Proof. Let X be a nonempty set and consider two elements (A, g) and (B, h) of I_X. Assume that $J_{(A,g)} \geq J_{(B,h)}$. Then

$$(B, h) = (C, s)(A, g)(D, t)$$

for some $(C, s), (D, t) \in I_X$. Hence

$$B = C \cup sA \cup sgD, \qquad h = sgt$$

so that $sA \subseteq B$ and thus $|A| \leq |B|$.

Assume that $|A| = |B|$. Then we must have $sA = B$ so that (A, s^{-1}) is an inverse of (B, s). Thus

$$(A, g)\mathcal{R}(A, s^{-1})\mathcal{L}(B, 1)\mathcal{R}(B, h)$$

which shows that $(A, g)\mathcal{J}(B, h)$. It follows that

$$d: J_{(A,g)} \to |A|$$

is a depth function for $I_X/\mathcal{J}$.

We now collect all the information concerning bases for free inverse semigroups as follows.

VIII.6.7 Corollary

Let X be a nonempty set. The semigroup I_X has the strong basis property. For any inverse subsemigroups U and T of I_X with $U \subseteq T$, T has a U-basis. For any $\mathcal{J}$-class J of I_X, any two U-bases for T have the same number of elements in J.

Proof. It follows from 1.14 and 1.16 that I_X is completely semisimple and combinatorial, so by 5.8 it has the strong basis property. This also implies the last statement of the corollary in view of 5.11. By 6.6, I_X is layered, and by 1.16, all its $\mathcal{J}$-classes are finite, which in view of 6.5 gives the second assertion of the corollary.

While finitely generated inverse semigroups always have a basis, this is not the case for arbitrary inverse semigroups. We exhibit this fact with the following instance.

VIII.6.8 Example

Let Q be the additive group of rationals, and assume that it has a basis X. Let $x \in X$ be arbitrary. Then $X \setminus x \neq \varnothing$; let $y \in X \setminus x$. Then x and y are nonzero, say $x = a/b$ and $y = c/d$ for some nonzero integers a, b, c, d. Let $k = bc$; then

$$kx = bc \cdot \frac{a}{b} = ad \cdot \frac{c}{d} = ady \in \langle X \setminus x \rangle.$$

Since $Q = \langle X \setminus x, x \rangle$, we have $(1/k)x \in \langle X \setminus x, x \rangle$ and thus $(1/k)x = h + nx$ for some $h \in \langle X \setminus x \rangle$ and some integer n. It follows that

$$x = kh + knx \in \langle X \setminus x \rangle.$$

Consequently, $\langle X \setminus x \rangle = \langle X \rangle$ for every $x \in X$, and X cannot be a basis for Q.

VIII.6.9 Exercises

(i) Characterize all bases of a finite primitive combinatorial inverse semigroup S. Show that the rank of S is equal to the number of nonzero idempotents of S if S has more than two elements.

All results in this section stem from Jones [3]; see also Jones [4].

IX

MONOGENIC INVERSE SEMIGROUPS

It is somewhat surprising that free monogenic inverse semigroups are not as simple as are free cyclic groups or semigroups. A representation of these follows, of course, as a special case of the free inverse semigroup on an arbitrary set. There are, however, many features of a free monogenic inverse semigroup worth special attention. For in view of its transparency, one is able to answer many more questions for it than for a general free inverse semigroup.

1. Starting from the characterization of a free inverse semigroup I_x specialized to a single generator, five different representations are derived for it. They are expressed in various ways mainly as triples of integers satisfying certain conditions and under suitable multiplication. One of them is a subdirect product of two copies of the bicyclic semigroup.

2. Congruences can be found on I_x in an explicit way and the several types of these are neatly classified. The determination of congruences on I_x is based on a consideration of the cyclic semigroup generated by x as well as on the behavior of idempotents.

3. The knowledge of congruences makes it possible to describe the entire lattice of congruences on I_x. The homomorphic images can also be explicitly exhibited, thereby providing a classification of the class of all monogenic inverse semigroups.

4. The semigroup I_x admits a presentation as a free semigroup on two generators subject to an infinite number of defining relations. It turns out that no free inverse semigroup (monogenic or not) is finitely related. For every (nonzero) idempotent of a (0)-simple semigroup S which is not completely (0)-simple there exists a bicyclic subsemigroup of S with identity e. This gives a useful criterion for singling out semigroups which are not completely semisimple.

IX.1 FREE MONOGENIC INVERSE SEMIGROUPS

We start with a free monogenic inverse semigroup in its P-representation $I_x = P(Y, G_x; X)$. Here G_x is a free cyclic group on x, which we write as

$$G_x = \{x^k | k \in \mathbb{Z}\};$$

$\mathbb{Z}$ is the set of all integers; Y is the semilattice of all closed finite subsets of G_x containing $x^0 = 1$, which now amounts to finite "intervals" (i.e., consecutive powers of x) containing x^0 and also another power of x; X is the semilattice of all finite "intervals"; with the usual action of G_x on X. Hence the elements of I_x are of the form

$$(\{x^m, x^{m+1}, \ldots, x^0, x, \ldots, x^n\}, x^k)$$

with $m \leq k, 0 \leq n,\ m < n$. It is generated, as an inverse semigroup, by the element $(\{x^0, x\}, x)$.

A free monogenic inverse semigroup admits a number of interesting isomorphic copies. The above is a copy of it in the P-representation, where the elements are determined by three integers. The first isomorphic copy of I_x amounts to taking such triples, with the corresponding restrictions, and giving them suitable multiplication.

IX.1.1 Proposition

Let

$$C_1 = \{(m, k, n) \in \mathbb{Z}^3 | m \leq k, 0 \leq n, m < n\}$$

with multiplication

$$(m, k, n)(m', k', n') = (\min\{m, k + m'\}, k + k', \max\{n, k + n'\}).$$

Then the mapping

$$\varphi : (\{x^m, \ldots, x^0, \ldots, x^n\}, x^k) \to (m, k, n)$$

is an isomorphism of I_x onto C_1.

Proof. The elements of I_x are of the form

$$(\{x^m, \ldots, x^0, \ldots, x^n\}, x^k)$$

with $m, k, n \in \mathbb{Z}, m \leq k, 0 \leq n,\ m < n$ for unique m, n, k so that φ is well defined and evidently a bijection of the sets I_x and C_1. For the product, we obtain

$$\begin{aligned}
&(\{x^m,\ldots,x^0,\ldots,x^n\}, x^k)(\{x^{m'},\ldots,x^0,\ldots,x^{n'}\}, x^{k'}) \\
&\quad = (\{x^m,\ldots,x^n\} \cup x^k\{x^{m'},\ldots,x^{n'}\}, x^k x^{k'}) \\
&\quad = (\{x^m,\ldots,x^n, x^{k+m'},\ldots,x^{k+n'}\}, x^{k+k'}) \\
&\quad = \left(\left\{x^{\min\{m,k+m'\}},\ldots,x^{\max\{n,k+n'\}}\right\}, x^{k+k'}\right)
\end{aligned}$$

which compared with the multiplication in C_1 shows that φ is a homomorphism, and thus an isomorphism of I_x onto C_1.

IX.1.2 Corollary

C_1 is a free monogenic inverse semigroup generated by $(0,1,1)$. Furthermore

$$(m,k,n)^{-1} = (m-k, -k, n-k).$$

Proof. The proof of this corollary is left as an exercise.

The next isomorphic copy of I_x is a subsemigroup of the direct product of two copies of the bicyclic semigroup.

IX.1.3 Proposition

Let

$$C_2 = \{((m,n),(p,q)) \in C \times C \mid m+p = n+q > 0\}$$

with the multiplication of the direct product $C \times C$. Then the mappings

$$\psi : (m,k,n) \to ((n, n-k), (-m, k-m)),$$

$$\theta : ((m,n),(p,q)) \to (-p, m-n, m),$$

are mutually inverse isomorphisms of C_1 and C_2.

Proof. Straightforward verification shows that ψ maps C_1 into C_2, that θ maps C_2 into C_1, and that $\psi\theta$ and $\theta\psi$ are identity mappings on the respective sets. Consequently, ψ and θ are bijections as asserted in the statement of the proposition. As for multiplication, we obtain on the one hand

$$\begin{aligned}
&(m, k, n)\psi(m', k', n')\psi \\
&\quad = ((n, n-k), (-m, k-m))((n', n'-k'), (-m', k'-m')) \\
&\quad = ((n, n-k)(n', n'-k'), (-m, k-m)(-m', k'-m')) \\
&\quad = \big((n+n'-r, n+n'-(k+k')-r), \\
&\qquad (-m-m'-s, k+k'-(m+m')-s)\big) \qquad (1)
\end{aligned}$$

where $r = \min\{n-k, n'\}$, $s = \min\{k-m, -m'\}$, and on the other hand,

$$\begin{aligned}
&((m, k, n)(m', k', n'))\psi \\
&\quad = \big(\min\{m, k+m'\}, k+k', \max\{n, k+n'\}\big)\psi \\
&\quad = \big((\max\{n, k+n'\}, \max\{n, k+n'\}-(k+k')), \\
&\qquad (-\min\{m, k+m'\}, k+k'-\min\{m, k+m'\})\big). \qquad (2)
\end{aligned}$$

In order to show the equality of (1) and (2), we must prove

$$n + n' - \min\{n-k, n'\} = \max\{n, k+n'\}, \qquad (3)$$

$$n + n' - (k+k') - \min\{n-k, n'\} = \max\{n, k+n'\} + (k+k'), \qquad (4)$$

$$-m - m' - \min\{k-m, -m'\} = -\min\{m, k+m'\}, \qquad (5)$$

$$k + k' - (m+m') - \min\{k-m, -m'\} = k + k' - \min\{m, k+m'\}. \qquad (6)$$

Item (3) can be written as

$$-\min\{n-k, n'\} = \max\{-n', k-n\}$$

which evidently holds; item (4) follows from item (3). Item (5) can be written as

$$m + m' + \min\{k-m, -m'\} = \min\{m, k+m'\}$$

which evidently holds; item (6) follows directly from item (5). Consequently, expressions (1) and (2) are equal and ψ is a homomorphism.

IX.1.4 Corollary

C_2 is a free monogenic inverse semigroup generated by $((1,0),(0,1))$.

Proof. The proof of this corollary is left as an exercise.

In the next isomorphic copy of I_x, canonical words in x are given explicit multiplication.

IX.1.5 Proposition

Let

$$C_3 = \{x^{-p}x^{q}x^{-r} \mid 0 \le p, r \le q \ne 0\}$$

(x^0 is an empty symbol) with the multiplication

$$x^{-p_1}x^{q_1}x^{-r_1} \cdot x^{-p_2}x^{q_2}x^{-r_2} = x^{-p}x^{q}x^{-r}$$

where

$$p = p_1 + r_1 + p_2 - a,$$

$$q = q_1 + r_1 + p_2 + q_2 - (a + b),$$

$$r = r_1 + p_2 + r_2 - b,$$

$$a = \min\{q_1, r_1 + p_2\},$$

$$b = \min\{q_2, r_1 + p_2\}.$$

Then the mappings

$$\zeta : ((m, n), (p, q)) \to x^{-p}x^{m+p}x^{-n},$$

$$\tau : x^{-p}x^{q}x^{-r} \to ((q - p, r), (p, q - r))$$

are mutually inverse isomorphisms of C_2 and C_3.

Proof. Straightforward verification shows that ζ maps C_2 into C_3, that τ maps C_3 into C_2, and that $\zeta\tau$ and $\tau\zeta$ are the identity mappings on the respective sets.

For the multiplication, we get, on the one hand,

$$((m_1, n_1),(p_1, q_1))\zeta((m_2, n_2),(p_2, q_2))\zeta$$

$$= x^{-p_1}x^{m_1+p_1}x^{-n_1} \cdot x^{-p_2}x^{m_2+p_2}x^{-n_2} = x^{-p}x^{q}x^{-r} \tag{7}$$

where

$$p = p_1 + n_1 + p_2 - a, \tag{8}$$

$$q = m_1 + p_1 + n_1 + p_2 + m_2 + p_2 - (a + b), \tag{9}$$

$$r = n_1 + p_2 + n_2 - b, \tag{10}$$

$$a = \min\{m_1 + p_1, n_1 + p_2\}, \tag{11}$$

$$b = \min\{m_2 + p_2, n_1 + p_2\}. \tag{12}$$

On the other hand,

$$(((m_1, n_1),(p_1, q_1))((m_2, n_2),(p_2, q_2)))\zeta$$

$$= ((m_1, n_1)(m_2, n_2),(p_1, q_1)(p_2, q_2))\zeta$$

$$= ((m_1 + m_2 - s, n_1 + n_2 - s),(p_1 + p_2 - t, q_1 + q_2 - t))\zeta$$

$$= x^{-(p_1+p_2-t)}x^{m_1+m_2+p_1+p_2-(s+t)}x^{-(n_1+n_2-s)} \tag{13}$$

where $s = \min\{n_1, m_2\}$, $t = \min\{q_1, p_2\}$.

From (11) and (12), we obtain

$$a = \min\{n_1 + q_1, n_1 + p_2\} = n_1 + \min\{q_1, p_2\} = n_1 + t,$$

$$b = p_2 + \min\{m_2, n_1\} = p_2 + s,$$

which substituted in (8), (9), and (10) gives

$$p = p_1 + n_1 + p_2 - n_1 - t = p_1 + p_2 - t,$$

$$q = m_1 + p_1 + n_1 + p_2 + m_2 + p_2 - n_1 - t - p_2 - s$$

$$= m_1 + m_2 + p_1 + p_2 - (s + t),$$

$$r = n_1 + p_2 + n_2 - p_2 - s = n_1 + n_2 - s.$$

This compared with (13) yields that the expressions in (7) and (13) are equal, so that ζ is a homomorphism.

IX.1.6 Corollary

C_3 is a free monogenic inverse semigroup generated by x. Moreover,

$$(x^{-p}x^{q}x^{-r})^{-1} = x^{-(q-r)}x^{q}x^{-(q-p)}.$$

Proof. The proof of this corollary is left as an exercise.

The next isomorphic copy of I_x amounts to a simplified version of the preceding one.

IX.1.7 Proposition

Let

$$C_4 = \{(p, q, r) \in \mathbb{Z}^3 | 0 \leq p, r \leq q \neq 0\}$$

with the multiplication

$$(p, q, r)(p', q', r') = (p + [s - q], [q - s] + s + [q' - s], [s - q'] + r')$$

where $s = r + p'$ and

$$[y] = \begin{cases} y & \text{if } y \geq 0 \\ 0 & \text{if } y < 0. \end{cases}$$

Then the mapping

$$\lambda : x^{-p}x^{q}x^{-r} \to (p, q, r)$$

is an isomorphism of C_3 onto C_4.

Proof. In view of the similar definitions of C_3 and C_4, it suffices to check that λ is a homomorphism. Indeed, on the one hand,

$$\begin{aligned}(x^{-p_1}x^{q_1}x^{-r_1})\lambda(x^{-p_2}x^{q_2}x^{-r_2})\lambda &= (p_1, q_1, r_1)(p_2, q_2, r_2) \\ &= (p_1 + [s - q_1], [q_1 - s] + s + [q_2 - s], [s - q_2] + r_2) \qquad (14)\end{aligned}$$

where $s = r_1 + p_2$, and on the other hand,

$$(x^{-p_1}x^{q_1}x^{-r_1} \cdot x^{-p_2}x^{q_2}x^{-r_2})\lambda = (x^{-p}x^{q}x^{-r})\lambda = (p, q, r) \qquad (15)$$

where p, q, r are as in the statement of 1.5. We first observe that

$$y - \min\{z, y\} = [y - z]$$

for any integers y, z. This gives

$$p = p_1 + r_1 + p_2 - a = p_1 + s - \min\{q_1, s\} = p_1 + [s - q_1],$$

$$q = (q_1 - a) + s + (q_2 - b) = [q_1 - s] + s + [q_2 - s],$$

$$r = r_1 + p_2 + r_2 - b = (s - b) + r_2 = [s - q_2] + r_2$$

whence the expressions (14) and (15) are equal and λ is a homomorphism.

IX.1.8 Corollary

C_4 is a free monogenic inverse semigroup generated by (0, 1, 0). Moreover,

$$(p, q, r)^{-1} = (q - r, q, q - p).$$

Proof. The proof of this corollary is left as an exercise.

Historically the first construction of a free monogenic inverse semigroup is the following.

IX.1.9 Proposition

Let

$$C_5 = \{(p, q, r) \in \mathbb{Z}^3 | p \geq 0, r \geq 0, p + q \geq 0, r + q \geq 0, p + q + r > 0\}$$

with the multiplication

$$(p, q, r)(p', q', r') = (\max\{p, p' - q\}, q + q', \max\{r', r - q'\}).$$

Then the mapping

$$\rho : (p, q, r) \to (p, q - p - r, r)$$

is an isomorphism of C_4 onto C_5.

Proof. It follows easily that ρ is one-to-one and maps C_4 onto C_5. It remains to prove that ρ is a homomorphism. Indeed, on the one hand,

$$\begin{aligned}[(p, q, r)\rho][(p', q', r')\rho] &= (p, q - p - r, r)(p', q' - p' - r', r') \\ &= (\max\{p, p' - q + p + r\}, q + q' - p - p' \\ &\quad -r - r', \max\{r', r - q' + p' + r'\}) \qquad (16)\end{aligned}$$

and on the other hand, letting $s = r + p'$,

$$\begin{aligned}[(p,q,r)(p',q',r')]\rho &= (p+[s-q],[q-s]+s+[q'-s],[s-q']+r')\rho \\ &= (p+[s-q],[q-s]+s+[q'-s]-p-[s-q] \\ &\quad -[s-q']-r',[s-q']+r'). \end{aligned} \tag{17}$$

First

$$\begin{aligned}\max\{p, p'-q+p+r\} &= p+\max\{0, p'+r-q\} \\ &= p+[s-q],\end{aligned}$$

and analogously $\max\{r', r-q'+p'+r'\} = [s-q']+r'$; secondly, using the equality $[a]-[-a]=a$, we get

$$\begin{aligned}&[q-s]+s+[q'-s]-p-[s-q]-[s-q']-r' \\ &\quad = q-s+s+q'-s-p-r' = q+q'-r-p'-p-r'.\end{aligned}$$

This proves that expressions (16) and (17) are equal and hence ρ is an isomorphism of C_4 onto C_5.

IX.1.10 Corollary

C_5 is a free monogenic inverse semigroup generated by $(0,1,0)$. Moreover,

$$(p,q,r)^{-1} = (p+q, -q, q+r).$$

Proof. The proof of this corollary is left as an exercise.

Note that an element of I_x represented by means of a birooted word tree as in VIII.3 can be pictured thus

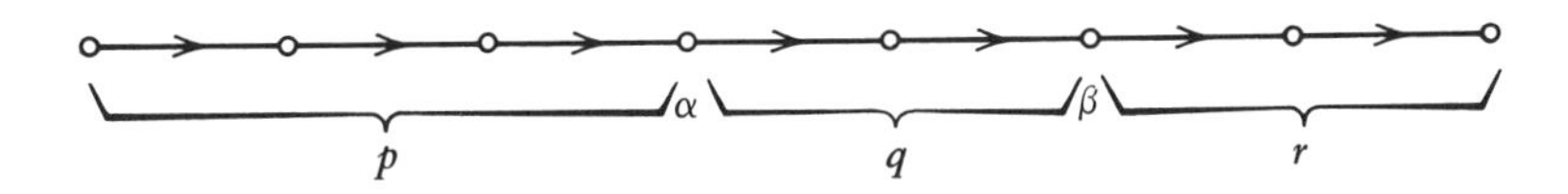

We finally present a realization of I_x as a semigroup of one-to-one partial transformations of a set.

IX.1.11 Proposition

Let $\mathbb{Z}$ be the set of all integers and let

$$\alpha = \begin{pmatrix} \cdots & -3 & -2 & -1 & 0 & 1 & 2 & 3 & \cdots \\ \cdots & -2 & -1 & 0 & & 2 & 3 & 4 & \cdots \end{pmatrix}.$$

Then the inverse subsemigroup $\langle \alpha \rangle$ of $\mathcal{I}(\mathbb{Z})$ generated by α is isomorphic to I_x.

Proof. First observe that

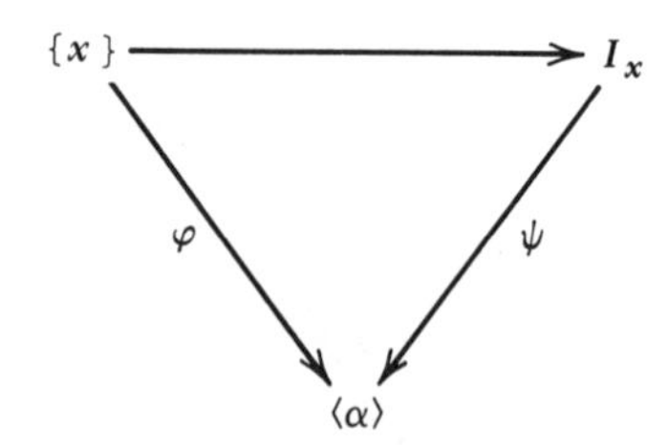

is a commutative diagram with $\varphi : x \to \alpha$, so ψ is a homomorphism of I_x onto $\langle \alpha \rangle$. Furthermore, all elements of I_x can be written uniquely in the form $x^{-q}x^{p+q}x^{-r}$ with $p, q, r \geq 0$ and $p + q \geq r$. Hence

$$\psi : x^{-q}x^{p+q}x^{-r} \to \alpha^{-q}\alpha^{p+q}\alpha^{-r}.$$

It remains to show that ψ is one-to-one. To do this, we must compute the expression $\alpha^{-q}\alpha^{p+q}\alpha^{-r}$. By an obvious induction, we obtain for $n > 0$,

$$\alpha^{n} = \begin{pmatrix} \cdots & -n-1 & -n & -n+1 & \cdots & 0 & 1 & 2 & 3 & \cdots \\ \cdots & -1 & 0 & & & & n+1 & n+2 & n+3 & \cdots \end{pmatrix}$$

$$\alpha^{-n} = \begin{pmatrix} \cdots & -2 & -1 & 0 & 1 & 2 & \cdots & n+1 & n+2 & \cdots \\ \cdots & -n-2 & -n-1 & -n & & & & 1 & 2 & \cdots \end{pmatrix},$$

for $n = 0$, $\alpha^n = \iota_{\mathbb{Z}}$.

Using this we obtain

$$\alpha^{-q}\alpha^{p+q}\alpha^{-r} = \begin{pmatrix} \cdots & -2 & -1 & 0 & 1 & 2 & \cdots & q+1 & q+2 & \cdots \\ \cdots & -q-2 & -q-1 & -q & & & & 1 & 2 & \cdots \end{pmatrix} \overset{q}{\leftarrow}$$

$$\cdot \begin{pmatrix} \cdots & -(p+q) & -(p+q)+1 & \cdots & 0 & 1 & 2 & \cdots \\ \cdots & 0 & & & & (p+q)+1 & (p+q)+2 & \cdots \end{pmatrix} \overset{p+q}{\rightarrow}$$

$$\cdot \begin{pmatrix} \cdots & -2 & -1 & 0 & 1 & 2 & \cdots & r+1 & r+2 & \cdots \\ \cdots & -r-2 & -r-1 & -r & & & & 1 & 2 & \cdots \end{pmatrix} \overset{r}{\leftarrow}$$

$$= \begin{pmatrix} \cdots & -p-2 & -p-1 & -p & -p+1 & \cdots & 0 & 1 & 2 & \cdots & q & q+1 & q+2 & \cdots \\ \cdots & -2 & -1 & 0 & & & & & & & & (p+q)+1 & (p+q)+2 & \cdots \end{pmatrix} \overset{p}{\rightarrow}$$

$$\cdot \begin{pmatrix} \cdots & -2 & -1 & 0 & 1 & 2 & \cdots & r & r+1 & r+2 & \cdots \\ \cdots & -r-2 & -r-1 & -r & & & & & 1 & 2 & \cdots \end{pmatrix} \overset{r}{\leftarrow}$$

$$= \begin{pmatrix} \cdots & -p-2 & -p-1 & -p & -p+1 & \cdots & 0 & 1 & 2 & \cdots & q & q+1 & q+2 & \cdots \\ \cdots & -r-2 & -r-1 & -r & & & & & & & & p+q-r+1 & p+q-r+2 & \cdots \end{pmatrix} \overset{p-r}{\rightarrow}.$$

If we now have

$$\alpha^{-q}\alpha^{p+q}\alpha^{-r} = \alpha^{-b}\alpha^{a+b}\alpha^{-c},$$

then the form of the product we obtained immediately gives $q = b$, $r = c$, $p = a$. Consequently, ψ is one-to-one and thus $\langle\alpha\rangle \cong I_x$.

We now present the first two $\mathfrak{D}$-classes of the semigroup $\langle\alpha\rangle$ in 1.11

$\begin{pmatrix} \cdots & -1 & 0 & 1 & \cdots \\ \cdots & -1 & & 1 & \cdots \end{pmatrix}$	$\begin{pmatrix} \cdots & -1 & 0 & 1 & \cdots \\ \cdots & 0 & & 2 & \cdots \end{pmatrix}$
$\begin{pmatrix} \cdots & -1 & 0 & 1 & 2 & \cdots \\ \cdots & -2 & -1 & & 1 & \cdots \end{pmatrix}$	$\begin{pmatrix} \cdots & -1 & 0 & 1 & 2 & \cdots \\ \cdots & -1 & 0 & & 2 & \cdots \end{pmatrix}$

$\begin{pmatrix} \cdots & -2 & -1 & 0 & 1 & 2 & \cdots \\ \cdots & -2 & & & 1 & 2 & \cdots \end{pmatrix}$	$\begin{pmatrix} \cdots & -2 & -1 & 0 & 1 & 2 & \cdots \\ \cdots & -1 & & & 2 & 3 & \cdots \end{pmatrix}$	$\begin{pmatrix} \cdots & -2 & -1 & 0 & 1 & 2 & \cdots \\ \cdots & 0 & & & 3 & 4 & \cdots \end{pmatrix}$
$\begin{pmatrix} \cdots & -2 & -1 & 0 & 1 & 2 & \cdots \\ \cdots & -3 & -2 & & & 1 & \cdots \end{pmatrix}$	$\begin{pmatrix} \cdots & -2 & -1 & 0 & 1 & 2 & \cdots \\ \cdots & -2 & -1 & & & 2 & \cdots \end{pmatrix}$	$\begin{pmatrix} \cdots & -2 & -1 & 0 & 1 & 2 & \cdots \\ \cdots & -1 & 0 & & & 3 & \cdots \end{pmatrix}$
$\begin{pmatrix} \cdots & -2 & -1 & 0 & 1 & 2 & 3 & \cdots \\ \cdots & -4 & -3 & -2 & & & 1 & \cdots \end{pmatrix}$	$\begin{pmatrix} \cdots & -2 & -1 & 0 & 1 & 2 & 3 & \cdots \\ \cdots & -3 & -2 & -1 & & & 2 & \cdots \end{pmatrix}$	$\begin{pmatrix} \cdots & -2 & -1 & 0 & 1 & 2 & 3 & \cdots \\ \cdots & -2 & -1 & 0 & & & 3 & \cdots \end{pmatrix}$

IX.1.12 Remark

Let N be the set of all nonnegative integers and

$$Y = (N \times N) \setminus \{(0,0)\}$$

and for any $(m, n), (m', n') \in Y$, let

$$(m, n) \wedge (m', n') = \left(\max\{m, m'\}, \max\{n, n'\}\right).$$

It is obvious that Y may be identified with the semilattice of idempotents of I_x by the correspondence

$$\left(\{x^m, \dots, x^0, \dots, x^n\}, 1\right) \to (-m, n).$$

Let k be any element of the additive group of integers $\mathbb{Z}$, and let ψ_k be the element of $\Sigma(Y)$ which is given by

$$\mathbf{d}\psi_k = \{(m, n) \in Y | m \leq -k, 0 \leq n\},$$

$$\mathbf{r}\psi_k = \{(m, n) \in Y | m \leq k, 0 \leq n\},$$

$$\psi_k : (m, n) \to (m + k, n + k) \qquad ((m, n) \in \mathbf{d}\psi_k).$$

One readily checks that $\psi : \mathbb{Z} \to \Sigma(Y)$, $k \to \psi_k$ is a prehomomorphism. Let us consider

$$Q(Y, \mathbb{Z}; \psi) = \{((m, n), k) \in Y \times \mathbb{Z} | (m, n) \in \mathbf{r}\psi_k\},$$

and recall that the multiplication in $Q(Y, \mathbb{Z}; \psi)$ is given by

$$\begin{aligned}
&((m, n), k)((m', n'), k') \\
&\quad = \left(k\left(k^{-1}(m, n) \wedge (m', n')\right), k + k'\right) \\
&\quad = \left(k\left((m - k, n - k) \wedge (m', n')\right), k + k'\right) \\
&\quad = \left(k\left(\max\{m - k, m'\}, \max\{n - k, n'\}\right), k + k'\right) \\
&\quad = \left(\left(\max\{m, k + m'\}, \max\{n, k + n'\}\right), k + k'\right).
\end{aligned}$$

In view of 1.1, the mapping $C_1 \to Q(Y, \mathbb{Z}; \psi)$, given by

$$(m, k, n) \to ((-m, n), k)$$

is an isomorphism. This means that $Q(Y, \mathbb{Z}; \psi)$ is a Q-representation of the free monogenic inverse semigroup.

IX.1.13 Diagram

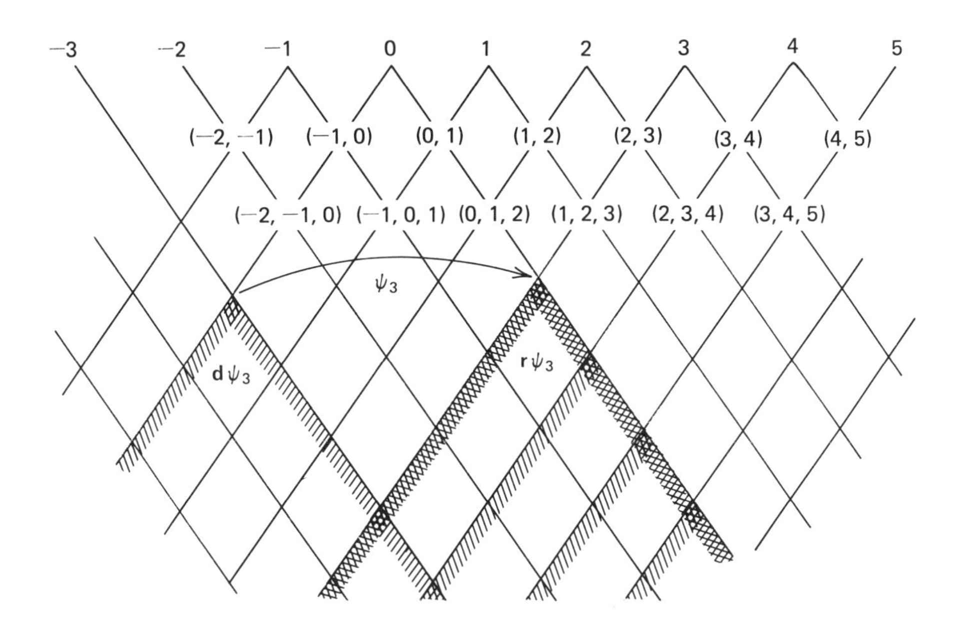

$$\mathbf{d}\psi_3 = \{A \mid 0 \in A \cap 3 + A\}, \qquad \mathbf{r}\psi_3 = \{A \mid 0 \in A \cap -3 + A\},$$

$$\psi_3 A = 3 + A.$$

The above diagram illustrates the Q-representation of I_x. The function ψ_3, corresponding to the group element $3 \in \mathbb{Z}$, is pictured with its domain and range.

The entire picture represents the partially ordered set X for the semigroup $I_x^1 = P(Y, \mathbb{Z}; X)$; omitting the first row, we obtain the set X for I_x. The semilattice Y of idempotents of I_x^1 is represented by the points which lie under 0; again omit the 0 for I_x.

IX.1.14 Diagram

$\mathcal{D}$-class structure of the free monogenic inverse semigroup. See page 406.

IX.1.15 Exercises

(i) Find all principal left and principal right ideals of I_x.

(ii) Construct the P-representation of I_x and the F-representation of I_x^1.

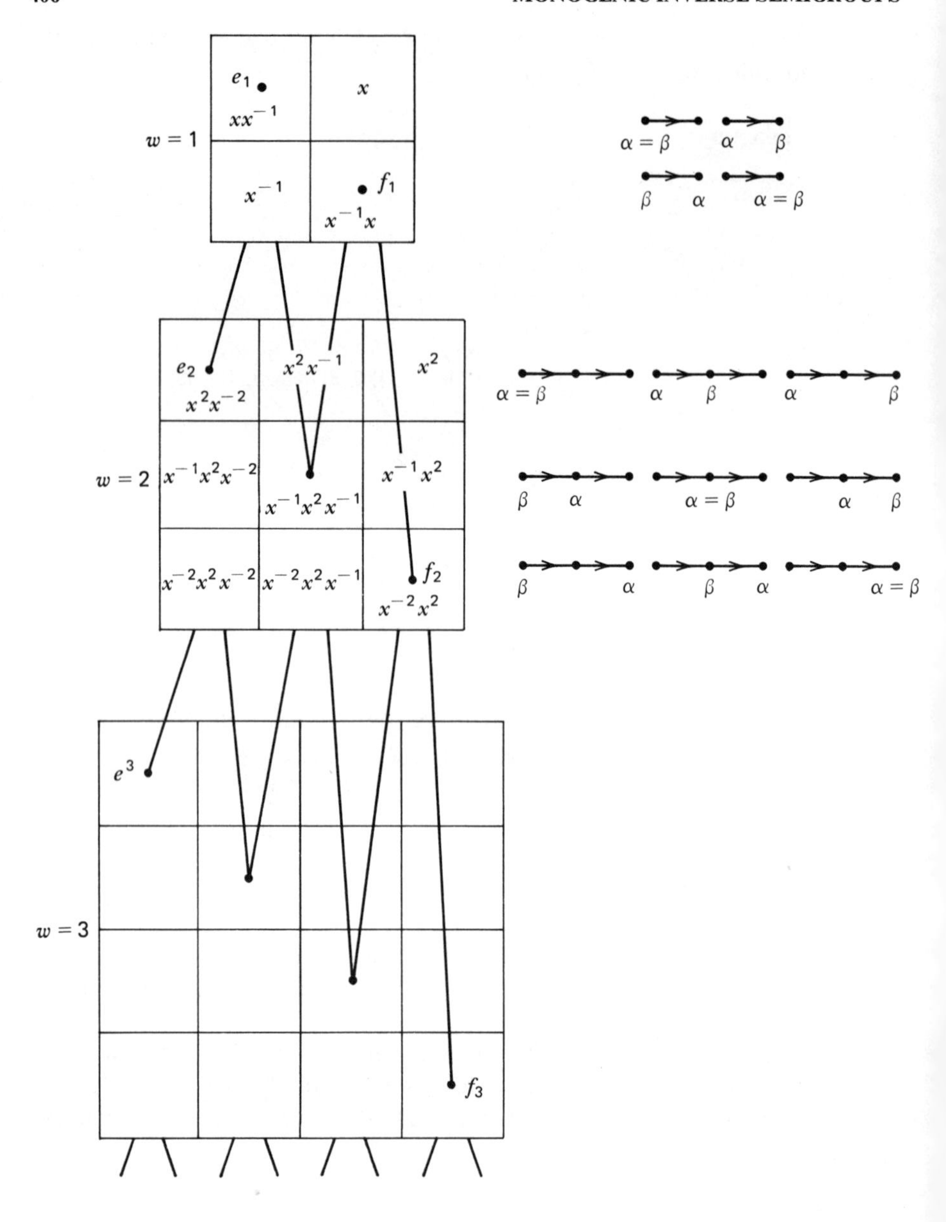

(iii) Find an expression for powers of elements in C_4.

(iv) Show that the multiplication in C_4 can be given the form

$$(p, q, r)(p', q', r')$$
$$= (p + r + p' - s, q + r + p' + q' - (s + s'), r + p' + r' - s'),$$

where $s = \min\{q, r + p'\}$, $s' = \min\{q', r + p'\}$, and also

$$(p, q, r)(p', q', r')$$

$$= \big(p - q + t, t + t' - (r + p'), r' - q' + t'\big),$$

where $t = \max\{q, r + p'\}$, $t' = \max\{q', r + p'\}$.

(v) Show that I_x is a normal extension of its semilattice of idempotents by a bicyclic semigroup.

(vi) Show that I_x is a dense extension of all of its ideals.

(vii) Find a concrete representation of $\Phi(I_x^1)$. Give it the P-, Q-, and F-representations. Also prove that $\Phi(I_x^1)$ is a noncombinatorial bisimple antigroup, and that $\Phi(I_x^1)/\sigma$ is a semidirect product of $\mathbb{Z} \times \mathbb{Z}$ by $\mathbb{Z}/(2)$.

(viii) Find the type $T = T(I_x : E_{I_x})$ and note that $I_x \cong T$. Find an isomorphic copy of T in $\Phi(Y)$ where Y is given in 1.12.

The first copy of a free monogenic inverse semigroup was C_5 constructed by Gluskin [2]; see also Gluskin [3], [4]. C_3 can also be deduced from this paper. Next appeared C_2 due to Scheiblich [2]. Finally, Eberhart–Selden [1] constructed C_4 with the multiplication as in 1.15(iv). Munn [13] deduced the construction of C_5 from his representation of a free inverse semigroup by birooted trees. Isomorphisms between various copies of I_x were computed by Reilly [11]. Note that Gluskin [2] and Scheiblich [2] call monogenic inverse semigroups elementary. An interesting application of monogenic inverse semigroups was found by Meakin [1]. For a different approach to monogenic inverse semigroups, see Preston [12]. Consult also Trueman [1], [2].

IX.2 CONGRUENCES

We first classify the nonidentical congruences on a free monogenic inverse semigroup into four types. Using the isomorphic copy C_2, which we denote by I_x, we prove that for each type, and the parameters figuring in it, there exists on I_x exactly one congruence corresponding to it. For this congruence, we give an explicit expression, thereby providing a complete description of congruences on a free monogenic inverse semigroup. Certain properties of these congruences will be treated in the next section.

We need a considerable amount of notation and several auxiliary results before starting our classification.

IX.2.1 Notation

It will be convenient below to simultaneously use the C_2 and C_3 representations of I_x. Hence we let

$$x = ((1, 0), (0, 1))$$

whence by 1.5,

$$((m, n), (p, q)) = x^{-p}x^{m+p}x^{-n}$$

$$x^{-p}x^{q}x^{-r} = ((q - p, r), (p, q - r)).$$

For any natural number n, let

$$e_n = ((n, n), (0,0)), \qquad f_n = ((0,0), (n, n)),$$

let e_0 and f_0 be empty symbols (see 1.14).

Observe that every idempotent of I_x can be uniquely written as $e_m f_n$ for some m, n such that $m + n > 0$.

IX.2.2 Lemma

For any natural numbers m, k, l,

$$x^m e_k f_l x^{-m} = e_{m+k} f_{l-\min\{m, l\}}, \qquad x^{-m} e_k f_l x^m = e_{k-\min\{m, k\}} f_{m+l},$$

$$e_m = x^m x^{-m}, \qquad f_m = x^{-m} x^m, \qquad x^m = e_m x^m = x^m f_m.$$

Proof. Indeed,

$$\begin{aligned} x^m e_k f_l x^{-m} &= ((m,0), (0, m))((k, k), (l, l))((0, m), (m,0)) \\ &= ((m,0)(k, k)(0, m), (0, m)(l, l)(m,0)) \\ &= ((m + k, k)(0, m), \\ &\qquad (l - \min\{m, l\}, m + l - \min\{m, l\})(m,0)) \\ &= ((m + k, m + k), (l - \min\{m, l\}, l - \min\{m, l\})) \\ &= e_{m+k} f_{l-\min\{m, l\}}. \end{aligned}$$

The second formula follows analogously. The proof of the remaining equalities is left as an exercise.

IX.2.3 Lemma

For any natural number m and any congruence ρ on I_x,

$$e_m \rho e_{m+1} \Leftrightarrow x^m \rho x^{m+1} x^{-1},$$

$$f_m \rho f_{m+1} \Leftrightarrow x^m \rho x^{-1} x^{m+1}.$$

Proof. If $f_m \rho f_{m+1}$, then using 2.2, we obtain

$$x^m = x^m f_m \rho x^m f_{m+1} = (x^m x^{-m})(x^{-1}x)x^m = x^{-1}x(x^m x^{-m} x^m) = x^{-1}x^{m+1}.$$

Conversely, if $x^m \rho x^{-1} x^m$, then

$$f_m = x^{-m} x^m \rho x^{-m}(x^{-1}x^{m+1}) = f_{m+1}.$$

This proves the second equivalence in the statement of the lemma; the first is established analogously.

IX.2.4 Notation

Let ρ be a congruence on I_x. Let $l(\rho)$ [respectively $r(\rho)$] be the least natural number n for which $e_n \rho e_{n+1}$ (respectively $f_n \rho f_{n+1}$); if such n does not exist, we set $l(\rho) = \infty$ [respectively $r(\rho) = \infty$].

IX.2.5 Lemma

If ρ is a congruence on I_x such that $l(\rho) = r(\rho) = \infty$, then $\rho = \varepsilon$.

Proof. Assume that $e \rho f$ for some idempotents e and f in I_x. Since then $e \rho ef$, we may suppose that $e \leq f$. Hence $e = e_m f_n$ and $f = e_k f_l$ with $m \geq k$ and $n \geq l$. By 2.2, we obtain

$$x^{-m} e x^m = x^{-m} e_m f_n x^m = f_{m+n},$$

$$x^{-m} f x^m = x^{-m} e_k f_l x^m = f_{m+l},$$

so that $f_{m+n} \rho f_{m+l}$. The hypothesis implies $f_{m+n} = f_{m+l}$ whence $n = l$. Symmetrically, conjugating with x^{-n}, we get $m = k$. Thus, $e = f$ and ρ is idempotent separating. But then $\rho = \varepsilon$ since $\mathcal{H} = \varepsilon$ on I_x.

IX.2.6 Lemma

If ρ is a congruence on I_x such that $l(\rho) < \infty$ and $r(\rho) < \infty$, then $l(\rho) = r(\rho)$.

Proof. Let $l(\rho) = n \neq \infty$, $r(\rho) = m \neq \infty$. By the definition of m and n, using 2.3, we obtain $x^m \rho x^{-1} x^{m+1}$ and $x^n \rho x^{n+1} x^{-1}$. Assuming that $m < n$, we have

$$\begin{aligned} x^n x^{-1} &= x^m x^{n-m} x^{-1} \rho x^{-1} x^{m+1} x^{n-m} x^{-1} \\ &= x^{-1} x^{n+1} x^{-1} \rho x^{-1} x^n = x^{-1} x^{m+1} x^{n-m-1} \rho x^m x^{n-m-1} = x^{n-1} \end{aligned}$$

whence, by 2.3, we have $e_n \rho e_{n-1}$, which contradicts the minimality of n. Hence $n \leq m$ and analogously $m \leq n$.

Recall from I.3.8 that $[x]$ stands for the cyclic semigroup generated by an element x of some semigroup.

IX.2.7 Definition

A congruence ρ on I_x has *finite index* if $\rho|_{[x]}$ has a finite number of classes, otherwise ρ has *infinite index.*

The *type of* ρ is defined as follows:

(i) (k, l) if $x^{k+l}\rho x^k$ and k, l are least such natural numbers,
(ii) (k, ω) if $k = l(\rho) < \infty$, $r(\rho) < \infty$, and ρ has infinite index,
(iii) (k, ∞^-) if $k = l(\rho) < \infty$, $r(\rho) = \infty$,
(iv) (k, ∞^+) if $l(\rho) = \infty$, $k = r(\rho) < \infty$.

IX.2.8 Definition

For any element $u = ((m, n), (p, q))$ of I_x, the *weight* of u is the number $w(u) = m + p$.

Note that in C_3, $w(x^{-p}x^q x^{-r}) = q$. In fact, two elements are $\mathcal{D}$-equivalent if and only if they have the same weight (see 1.14).

IX.2.9 Lemma

For any element $u = x^{-p}x^q x^{-r}$ of C_3, we have

$$uu^{-1} = e_{q-p}f_p, \qquad u^{-1}u = e_r f_{q-r}.$$

Proof. By 1.5, we get

$$\begin{aligned} uu^{-1} &= ((q-p, r), (p, q-r))((r, q-p), (q-r, p)) \\ &= ((q-p, q-p), (p, p)) = e_{q-p}f_p, \end{aligned}$$

and the second formula follows similarly.

We can finally start our classification.

IX.2.10 Proposition

For any natural number k, the relation ρ defined on I_x by: for $u = ((m, n), (p, q))$, $u' = ((m', n'), (p', q'))$,

$$u\rho u' \Leftrightarrow u = u' \quad \text{or} \quad w(u), w(u') \geq k, \quad (m, n) = (m', n'),$$

is the unique congruence on I_x of type (k, ∞^+).

Proof. We verify first that ρ satisfies the requisite properties. It is clear that ρ is an equivalence relation. With the above notation, we compute

$$\begin{aligned} u((x, y), (w, z)) &= ((m, n)(x, y), (p, q)(w, z)) \\ &= ((m + x - r, n + y - r), (p + w - s, q + z - s)) \end{aligned}$$

where $r = \min\{n, x\}$, $s = \min\{q, w\}$. It follows that

$$(m + x - r) + (p + w - s) \geq m + p \geq k,$$

analogously for the product with u'. Thus ρ is a right congruence, and similarly one shows that ρ is a left congruence. It follows easily from the form of the congruence that it is of type (k, ∞^+).

Conversely, let ρ be a congruence on I_x of type (k, ∞^+). Let $u = x^{-p}x^q x^{-r}$ and $u' = x^{-p'}x^{q'}x^{-r'}$ be distinct elements of I_x in the C_3-representation, and assume that $u \rho u'$. Then $uu^{-1}\rho u'u'^{-1}$ and $u^{-1}u\rho u'^{-1}u'$, which by 2.9 yields

$$e_{q-p}f_p \rho e_{q'-p'}f_{p'}, \qquad e_r f_{q-r} \rho e_{r'} f_{q'-r'}. \tag{18}$$

Let

$$a = \max\{q - p, q' - p'\}, \qquad b = \max\{p, p'\}. \tag{19}$$

Then

$$(e_{q-p}f_p)(e_{q'-p'}f_{p'}) = (e_{q-p}e_{q'-p'})(f_p f_{p'}) = e_a f_b$$

so that, by (18),

$$e_{q'-p'}f_{p'} \rho e_a f_b. \tag{20}$$

By 2.2, $x^{p'}(e_{q'-p'}f_{p'})x^{-p'} = e_{q'}$ and $x^{p'}(e_a f_b)x^{-p'} = e_{a+p'}f_{b-p'}$ which by (20) gives

$$e_{q'} \rho e_{a+p'} f_{b-p'}. \tag{21}$$

Now by (19), we have $q' \leq a + p'$ so that

$$e_{q'} \geq e_{a+p'} \geq e_{a+p'} f_{b-p'},$$

which together with (21) yields $e_{q'} \rho e_{a+p'}$. Since ρ is of type (k, ∞^+), it follows that $q' = a + p'$, which by (19) gives $q - p \leq q' - p'$. By symmetry, we also have $q' - p' \leq q - p$ so that the equality $q - p = q' - p'$ holds. The second part of (18) analogously yields $r = r'$. If $q = q'$, then by $q - p = q' - p'$, we would get $p = p'$, which contradicts the hypothesis that $u \neq u'$. Consequently,

$q \neq q'$. By 2.2, (18) and $q - p = q' - p'$, we get

$$f_q = x^{-(q-p)}\left(e_{q-p}f_p\right)x^{q-p}\rho x^{-(q'-p')}\left(e_{q'-p'}f_{p'}\right)x^{q'-p'} = f_{q'}$$

which then implies that $q, q' \geq k$ since $q \neq q'$ and ρ is of type (k, ∞^+). So far, we have proved

$$q, q' \geq k, \qquad q - p = q' - p', \qquad r = r'. \tag{22}$$

Note that $f_k \rho f_{k+1}$ implies by 2.2,

$$f_{k+1} = x^{-1}f_k x \rho x^{-1}f_{k+1}x = f_{k+2}$$

and in this way, we obtain

$$f_l \rho f_{l'} \qquad (l, l' \geq k). \tag{23}$$

Conversely, assume that (22) holds. Then by 2.2 and (23), we get

$$x^{-p}x^q x^{-r} = x^{q-p}\left(x^{-q}x^q\right)x^{-r} = x^{q-p}f_q x^{-r}\rho x^{q'-p'}f_{q'}x^{-r'}$$

$$= x^{q'-p'}\left(x^{-q'}x^{q'}\right)x^{-r'} = x^{-p'}x^{q'}x^{-r'}.$$

Now in the C_2-representation, we have

$$u = ((q-p, r), (p, q-r)), \qquad u' = \left((q'-p', r'), (p', q'-r')\right)$$

and we have proved that $u\rho u'$ if and only if (22) holds. Consequently ρ has the form as in the statement of the proposition which established uniqueness.

Note that there is an obvious analogue of 2.10 for $\rho_{(k, \infty^-)}$.

IX.2.11 Proposition

For any natural number k, the relation ρ defined on I_x by: for $u = ((m, n), (p, q))$, $u' = ((m', n'), (p', q'))$,

$$u\rho u' \Leftrightarrow u = u' \quad \text{or} \quad w(u), w(u') \geq k, \quad m - n = m' - n',$$

is the unique congruence on I_x of type (k, ω).

Proof. It is again routine to verify that ρ so defined is a congruence. It follows easily that ρ has infinite index. Further, $e_a \rho e_b$ and $f_a \rho f_b$ if and only if $a, b \geq k$. Consequently, ρ is a congruence of type (k, ω).

Conversely, let ρ be a congruence on I_x of type (k, ω), let $u = x^{-p}x^{q}x^{-r}$, $u' = x^{-p'}x^{q'}x^{-r'}$, $u\rho u'$, $u \neq u'$. Let

$$a = \max\{p, p'\}, \qquad c = \max\{r, r'\}.$$

Multiplying $u\rho u'$ by x^a on the left and by x^c on the right, we obtain

$$x^{q+a-p+c-r}\rho x^{q'+a-p'+c-r'}$$

whence $q - p - r = q' - p' - r'$ since ρ has infinite index.

Without loss of generality, we may suppose that $q \geq q'$. As in the proof of 2.10, we let

$$b = \max\{q - p, q' - p'\}$$

and as in the part of that proof following (19), we get $e_{q'}\rho e_{b+p'}f_{a-p'} \leq e_{b+p'} \leq e_{q'}$, whence $e_{q'}\rho e_{b+p'}$. Assume that $q' < k$. Then $e_{q'}\rho e_{b+p'}$ implies $q' = b + p'$, that is $q' - p' = b$. But then, by the definition of b, $q' - p' \geq q - p$. The assumption $q \geq q'$ now implies $p - p' \geq q - q' \geq 0$, so $p \geq p'$. Since $uu^{-1}\rho u'u'^{-1}$ we obtain

$$e_{q-p}f_p\rho e_{q'-p'}f_{p'}$$

which implies

$$x^p e_{q-p}f_p x^{-p}\rho x^p e_{q'-p'}f_{p'}x^{-p}$$

so that $e_q\rho e_{q'+p-p'}$. Also $p - p' \geq 0$ and $q \geq q'$ imply $e_q \leq e_{q'} \leq e_{q'+p-p'}$ whence $e_q\rho e_{q'}$. Since $q' < n$, by hypothesis, we must have $q = q'$. Now $e_q\rho e_{q'+p-p'}$ and $q < k$ imply $q = q' + p - p'$ whence $p = p'$. Finally, $q - p - r = q' - p' - r'$ yields $r = r'$. Consequently, $u = u'$, contradicting the hypothesis $u \neq u'$. Hence $q' \geq k$ and thus also $q \geq k$.

In order to prove that the conditions $w(u), w(u') \geq k$ and $m - n = m' - n'$ imply $u\rho u'$, we first observe that as in the proof of 2.10, the hypothesis on ρ implies $e_s\rho e_t$ and $f_s\rho f_t$ for all $s, t \geq k$. We show first that ρ identifies all $e_m f_n$ with $m + n \geq k$.

Let $m \geq k$. Then $e_m\rho e_k$ implies $x^{-1}e_m x\rho x^{-1}e_k x$ whence $e_{m-1}f_1\rho e_{k-1}f_1$. This in turn implies $x^{-1}e_{m-1}f_1 x\rho x^{-1}e_{k-1}f_1 x$ whence for $k \geq 2$, we get $e_{m-2}f_2\rho e_{k-2}f_2$. We continue this process and at each step assume that k is one larger than in the preceding one. We finish the $e_{m-k}f_k\rho f_k$. Symmetrically, the same type of procedure can be started with $f_m\rho f_k$ and conjugated by x instead of x^{-1}. In this way, we get that ρ identifies all $e_m f_n$ with $m + n \geq k$ and either $m \leq k$ or $n \leq k$. For the remaining $e_m f_n$ with $m + n \geq k$, we have $m \geq k$ and $n \geq k$. In such a case, $e_m f_n\rho e_k f_k$. Consequently, ρ identifies all $e_m f_n$ with $m + n \geq k$.

Now let $u = ((m, n), (p, q))$ and $u' = ((m', n'), (p', q'))$, where $w(u), w(u') \geq k$, $m - n = m' - n'$. Assume first that $m \leq m'$, $p \leq p'$. Then since $m + p, m' + p' \geq k$, by the above, we get

$$\begin{aligned} u = e_m f_p u \rho e_{m'} f_{p'} u &= \big((m', m'), (p', p')\big)\big((m, n), (p, q)\big) \\ &= \big((m', m')(m, n), (p', p')(p, q)\big) \\ &= \big((m', m' + n - m), (p', p' + q - p)\big) \\ &= \big((m', n'), (p', q')\big) = u' \end{aligned}$$

in view of the hypothesis $m - n = m' - n'$. The case $m \geq m'$, $p \geq p'$ is symmetric.

Now assume $m \leq m'$, $p' \leq p$. Then by the above,

$$\begin{aligned} u = u e_n f_q \rho u e_{n'} f_{q'} &= \big((m, n), (p, q)\big)\big((n', n'), (q', q')\big) \\ &= \big((m, n)(n', n'), (p, q)(q', q')\big) \\ &= \big((m + n' - n, n'), (p, q)\big) \\ &= \big((m', n'), (p, q)\big) \end{aligned} \tag{24}$$

since $n \leq n'$ and $q' \leq q$. On the other hand,

$$\begin{aligned} u' = u' e_{n'} f_{q'} \rho u' e_n f_q &= \big((m', n'), (p', q')\big)\big((n, n), (q, q)\big) \\ &= \big((m', n')(n, n), (p', q')(q, q)\big) \\ &= \big((m', n'), (p' + q - q', q)\big) \\ &= \big((m', n'), (p, q)\big) \end{aligned} \tag{25}$$

since $q - p = q' - p'$. From (24) and (25), we deduce that $u \rho u'$. The case $m \geq m'$, $p' \geq p$ is symmetric.

Therefore, $u \rho u'$ in all cases.

IX.2.12 Proposition

For any natural numbers k, l, the relation ρ defined on I_x by: for $u = ((m, n), (p, q))$, $u' = ((m', n'), (p', q'))$,

$$u \rho u' \Leftrightarrow u = u' \quad \text{or} \quad w(u), w(u') \geq k, \quad m - n \equiv m' - n' \pmod{l}$$

is the unique congruence on I_x of type (k, l).

Proof. The argument here is quite analogous to that in 2.11 and is left as an exercise.

Now 2.10 with its dual, 2.11, and 2.12 exhaust all the possible types of nonidentical congruences on I_x. This completes the classification of congruences on I_x.

IX.2.13 Corollary

For any congruence ρ on I_x, we have that I_x/ρ is finite if and only if ρ has finite index.

IX.2.14 Exercises

(i) Call a congruence ρ on a semigroup S *monogenic* if ρ is the congruence on S generated by a single pair. Show that every nonidentical congruence on I_x is monogenic.

(ii) Give an explicit expression for congruences on a free monogenic inverse semigroup in its C_3, C_4, and C_5 representations.

(iii) Prove directly that every nonidempotent of I_x generates a free inverse subsemigroup of I_x. Deduce that an endomorphism of I_x which does not map I_x onto an idempotent must be one-to-one.

Congruences on free monogenic inverse semigroups were found by Eberhart–Selden [1] and Djadčenko–Schein [1]. We have generally followed the latter paper. Congruences on monogenic inverse semigroups were investigated by Gluskin [2].

IX.3 PROPERTIES OF CONGRUENCES

We start by providing an isomorphic copy of the lattice of all congruences on I_x. This is followed by a description of the structure of the quotient semigroup I_x/ρ for each type of congruence ρ on I_x, thereby giving an isomorphic copy of all monogenic inverse semigroups. Among these we single out those which are (relatively) free in the varieties they generate. The section ends with a construction of congruences on I_x in its Q-representation.

IX.3.1 Notation

Let 0 denote the type of the equality relation ε on I_x. If ρ is a congruence on I_x of type (k, u), we write $\rho = \rho_{(k,u)}$.

Let $\mathbb{N}$ be the set of natural numbers. Let $\mathbb{N}^* = \mathbb{N} \cup \{\omega, \infty^-, \infty^+\}$ be ordered as follows:

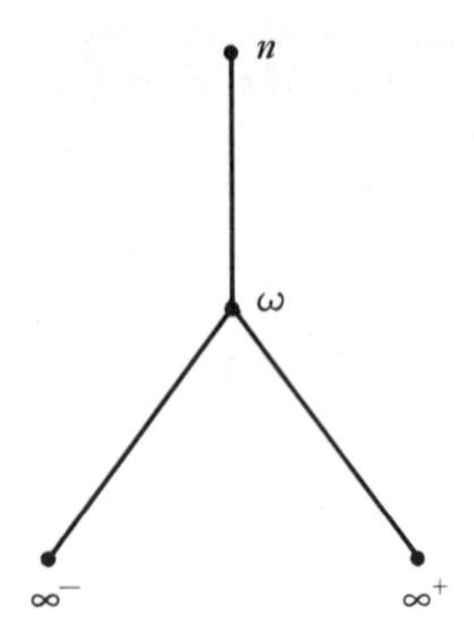

for all $n \in \mathbb{N}$. Write $x \prec y$ for this order, and extend it to $\mathbb{N}$ by letting $m \prec n$ if $n|m$, that is n *divides* m. On $M = (\mathbb{N} \times \mathbb{N}^*) \cup 0$ define a relation $\leq$ by: 0 is the least element of M and

$$(m, x) \leq (n, y) \Leftrightarrow m \geq n, x \prec y.$$

IX.3.2 Theorem

The mapping

$$\tau : \rho \to \text{type } \rho \qquad [\rho \in \mathcal{C}(I_x)]$$

is an isomorphism of $\mathcal{C}(I_x)$ onto M.

Proof. In light of 2.10–2.12, the mapping τ is a bijection of $\mathcal{C}(I_x)$ onto M. Since $(\mathbb{N}, \leq)$ and $(\mathbb{N}, |)$ are lattices, it follows easily that M itself is a lattice. Hence in order to prove that τ is a lattice isomorphism, it suffices to show that both τ and τ^{-1} preserve order (see I.2.8).

In the next table, the elements in the left vertical column are less than or equal, in the order of M, than the elements in the upper horizontal column if and only if $k \geq k'$ and the condition in the corresponding cell is fulfilled. The blank spaces indicate that the corresponding elements are not in the relation $\leq$.

	(k', l')	(k', ω)	(k', ∞^-)	(k', ∞^+)
(k, l)	$l' \mid l$			
(k, ω)	$\omega < l'$	$\omega \leq \omega$		
(k, ∞^-)	$\infty^- < l'$	$\infty^- < \omega$	$\infty^- \leq \infty^-$	
(k, ∞^+)	$\infty^+ < l'$	$\infty^+ < \omega$		$\infty^+ \leq \infty^+$

It remains to prove that the congruences of the respective types are in the same inclusion relation. This is straightforward, so we verify only the first case.

First assume that $\rho_{(k,l)} \subseteq \rho_{(k',l')}$. Since

$$((k + l, k), (0, l))\rho_{(k,l)}((k, k), (0,0)),$$

we obtain

$$((k + l, k), (0, l))\rho_{(k',l')}((k, k), (0,0))$$

and thus $k \geq k'$ and $l \equiv 0 \pmod{l'}$ so that $l'|l$.

Conversely, let $k \geq k'$, $l'|l$. With the notation of 3.1, let $u \neq u'$, $u\rho_{(k,l)}u'$. Then $w(u), w(u') \geq k$ and $m - n \equiv m' - n' \pmod{l}$. It follows that $w(u), w(u') \geq k$ and $m - n \equiv m' - n' \pmod{l'}$, so that $u\rho_{(k',l')}u'$.

Note that congruences of infinite index map onto the part of M of the form

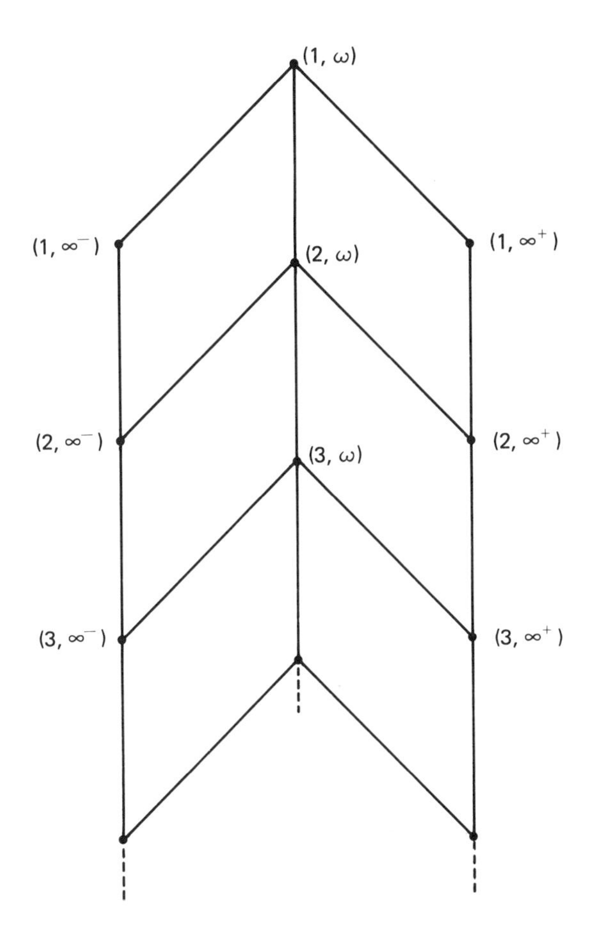

$\mathbb{N}^*$ can be pictured thus

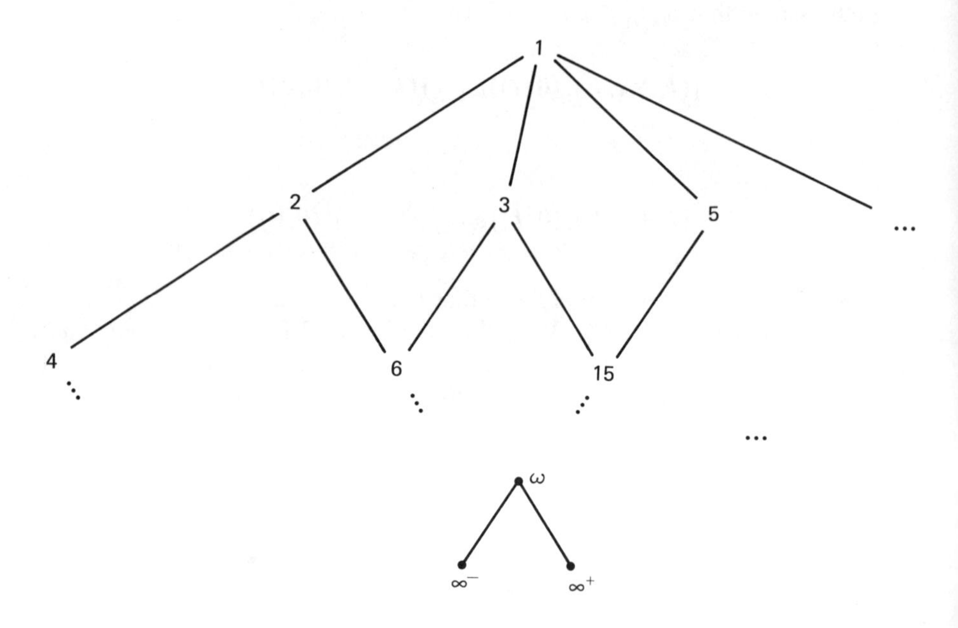

We can now easily classify all monogenic inverse semigroups by means of the quotient semigroups I_x/ρ for all congruences ρ on I_x.

IX.3.3 Notation

For any natural number n, let

$$I_n = \{u \in I_x | w(u) \geq n\}.$$

Then I_n is an ideal of I_x, let $M_n = I_x/I_n$.

IX.3.4 Theorem

(i) $I_x/\rho_{(k,\infty^+)}$ is an ideal extension of the bicyclic semigroup C by M_k. The multiplication in $C \cup M_k^*$ is determined by the partial homomorphism of M_k^* into C:

$$((m,n),(p,q)) \to (m,n).$$

(ii) $I_x/\rho_{(k,\omega)}$ is an ideal extension of the group $\mathbb{Z}$ of integers by M_k. The multiplication in $\mathbb{Z} \cup M_k^*$ is determined by the partial homomorphism of M_k^* into $\mathbb{Z}$:

$$((m,n),(p,q)) \to m - n.$$

(iii) $I_x/\rho_{(k,l)}$ is an ideal extension of the group $\mathbb{Z}/(l)$ by M_k. The multiplication in $\mathbb{Z}/(l) \cup M_k^*$ is determined by the partial homomorphism of M_k^* into $\mathbb{Z}/(l)$:

$$((m,n),(p,q)) \to m - n \ (\text{mod}\ l).$$

Proof. (i) Let $\tau = \rho_{(k,\infty^+)}|_{I_k}$. Then $((0,0),(k,k))\tau$ is the identity of I_k/τ, which is an ideal of $I_x/\rho_{(k,\infty^+)}$. Define ψ on I_k by

$$\psi : ((m,n),(p,q)) \to (m,n).$$

Then ψ is a homomorphism of I_k into C which induces τ. For any $(m,n) \in C$, we have

$$u = ((m,n),(n+k,m+k)) \in I_K$$

and $u\psi = (m,n)$. Consequently, ψ maps I_k onto C which implies that

$$\bar{\psi} : ((m,n),(p,q))\tau \to (m,n)$$

is an isomorphism of I_k/τ onto C.

For $u \in I_x \setminus I_k$, we may identify u and $u\rho_{(k,\infty^+)}$. The multiplication in $I_x/\rho_{(k,\infty^+)}$ is thus determined by the partial homomorphism from M_k^* into I_k/τ given by

$$\theta : ((m,n),(p,q)) \to [((m,n),(p,q))((0,0),(k,k))]\tau$$
$$= ((m,n),(\quad,\quad))\tau.$$

Letting $\varphi : ((m,n),(p,q)) \to (m,n)$ be defined on M_k^*, we deduce that the diagram

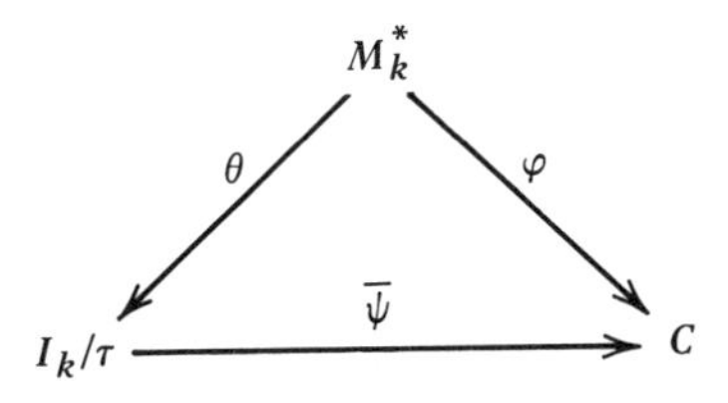

is commutative. It now follows easily that the semigroups $I_x/\rho_{(k,\infty^+)}$ and $C \cup M_k^*$, with the multiplication determined by φ, are isomorphic.

(ii) We can partly reduce this case to the preceding one by observing that $I_x/\rho_{(k,\omega)}$ is a homomorphic image of $I_x/\rho_{(k,\infty^+)}$. It may be obtained by factoring the ideal I_k/τ, in the above notation, through its least group congruence, thereby obtaining an infinite cyclic group.

(iii) In this case, we may identify $(I_k/\tau)/\sigma$ with $\mathbb{Z}$ and further factor it through the subgroup generated by l, thereby obtaining $\mathbb{Z}/(l)$.

The precise argument of the proof of items (ii) and (iii) is left as an exercise.

IX.3.5 Lemma

The automorphism φ of I_x for which $x\varphi = x^{-1}$ is given by

$$\varphi: ((m,n),(p,q)) \to ((p,q),(m,n)).$$

Proof. It is convenient to use C_3 here (see 1.5) since we need powers of generators. Indeed,

$$\begin{aligned}((m,n),(p,q))\varphi &= (x^{-p}x^{m+p}x^{-n})\varphi \\ &= x^{p}x^{-(m+p)}x^{n} \\ &= ((p,0),(0,p))((0,m+p),(m+p,0))((n,0),(0,n)) \\ &= ((p,0)(0,m+p)(n,0),(0,p)(m+p,0)(0,n)) \\ &= ((p,m+p)(n,0),(0,p)(m+p,n)) \\ &= ((p,m+p-n),(m,n)) = ((p,q),(m,n)).\end{aligned}$$

Note that neither $\rho_{(k,\infty^-)}$ nor $\rho_{(k,\infty^+)}$ are invariant under the automorphism φ in 3.5. For example, $e_k\rho_{(k,\infty^-)}e_{k+1}$ but $e_k\varphi = f_k$ and $e_{k+1}\varphi = f_{k+1}$ are not $\rho_{(k,\infty^-)}$-equivalent. However, we have the following statement.

IX.3.6 Lemma

Let φ be as in 3.5. Then the mapping

$$\psi: u\rho_{(k,\infty^+)} \to u\varphi\rho_{(k,\infty^+)} \qquad (u \in I_x)$$

is an isomorphism of $I_x/\rho_{(k,\infty^+)}$ onto $I_x/\rho_{(k,\infty^-)}$.

Proof. It is easy to see that

$$\begin{aligned}&((m,n),(p,q))\rho_{(k,\infty^+)}((m',n'),(p',q')) \\ &\quad\Leftrightarrow ((p,q),(m,n))\rho_{(k,\infty^-)}((p',q'),(m',n')),\end{aligned}$$

and thus for any $u, u' \in I_x$,

$$u\rho_{(k,\infty^+)}u' \Leftrightarrow (u\varphi)\rho_{(k,\infty^-)}(u'\varphi).$$

Consequently, ψ is an isomorphism of $I_x/\rho_{(k,\infty^+)}$ onto $I_x/\rho_{(k,\infty^-)}$.

Besides the monogenic inverse semigroups in 3.4, we still have I_x/ε, that is, the free monogenic inverse semigroup, and $I_x/\rho_{(k,\infty^-)}$. Note that the parameters in the type of a congruence ρ on I_x uniquely determine the quotient semigroup I_x/ρ, and that, except for the isomorphism of $I_x/\rho_{(k,\infty^+)}$ and $I_x/\rho_{(k,\infty^-)}$ (see 3.6), two congruences with different types yield nonisomorphic quotients. This completes our classification of monogenic inverse semigroups.

We now consider certain identities satisfied by various monogenic inverse semigroups.

IX.3.7 Definition

A monogenic inverse semigroup $\langle a \rangle$ is of *type* (k, u) if $\langle a \rangle \cong I_x/\rho_{(k,u)}$ for $k \in \mathbb{N}$ and $u \in \{\infty^+, \infty^-, \omega, l\}$.

IX.3.8 Lemma

Let $((m,n),(p,q)) \in I_x$ be a nonidempotent. Then for every natural number k, we have

$$((m,n),(p,q))^k = \begin{cases} ((m, kn-(k-1)m),(kp-(k-1)q,q)) & \text{if } m < n \\ ((km-(k-1)n, n),(p, kq-(k-1)p)) & \text{if } m > n. \end{cases}$$

Proof. For $k = 1$, this is trivial. Assume true for k, and calculate: for $m < n$, we get $q < p$,

$$\begin{aligned} ((m,n),(p,q))^{k+1} &= ((m, kn-(k-1)m),(kp-(k-1)q,q)) \\ &\quad \cdot ((m,n),(p,q)) \\ &= ((m, kn-(k-1)m)(m,n),(kp-(k-1)q,q) \\ &\quad \cdot (p,q)) \\ &= ((m,(k+1)n-km),((k+1)p-kq,q)); \end{aligned}$$

the case $m > n$ is analogous.

IX.3.9 Proposition

The congruence $\rho_{(k,\omega)}$ is the least congruence ρ on I_x such that I_x/ρ satisfies the identity $y^{-1}y^k = y^k y^{-1}$.

Proof. Let $u = ((m, n), (p, q)) \in I_x$. If $u^2 = u$, then $u = u^{-1}$ so $u^{-1}u = uu^{-1}$ so also $u^{-1}u\rho uu^{-1}$. Assume $u^2 \neq u$. Then either $m < n$ or $m > n$; consider the first case. Then by 3.8, $w(u^k) = m + kp - (k-1)q = (m+p) + (k-1)(p-q)$. Since $m < n$ implies $p > q$, we get that $w(u^k) \geq 1 + k - 1 = k$. Hence $u^k \in I_k$. Since S/ρ is an ideal extension of $\mathbb{Z}$ by I_x/I_k, it follows that the multiplication is determined by a partial homomorphism $\varphi : I_x \setminus I_k \to \mathbb{Z}$. Letting $\bar{u}$ be the image of u in $\bar{I} = I_x/\rho$, we get

$$\bar{u}^{-1}\bar{u}^k = (\bar{u}^{-1}\varphi)\bar{u}^k = (\bar{u}\varphi)^{-1}(\bar{u}\varphi)^k = (\bar{u}\varphi)^k(\bar{u}\varphi)^{-1} = \bar{u}^k\bar{u}^{-1}$$

which proves that $\bar{I}$ satisfies the required identity.

Let τ be any congruence on I_x for which I_x/τ satisfies $u^{-1}u^k = u^k u^{-1}$. In particular, $x^{-1}x^k \tau x^k x^{-1}$. This gives

$$(x^{-1}x^k)^{-1}(x^{-1}x^k)\tau(x^k x^{-1})^{-1}(x^k x^{-1}),$$

$$(x^{-1}x^k)(x^{-1}x^k)^{-1}\tau(x^k x^{-1})(x^k x^{-1})^{-1}$$

whence by 2.2, $f_k \tau e_1 f_{k-1}$ and $e_{k-1} f_1 \tau e_k$. From the first relation, we obtain

$$f_{k+1} = x^{-1} f_k x \tau x^{-1} e_1 f_{k-1} x = f_k$$

and analogously $e_{k+1} \tau e_k$ from the second relation. Hence τ is either of type (k', ω) or of type (k', l') for some $k' \leq k$, $l' \geq 1$. In either case, $\rho \subseteq \tau$, which establishes the minimality of ρ.

IX.3.10 Proposition

The congruence $\rho_{(k,l)}$ is the least congruence ρ on I_x such that I_x/ρ satisfies the identity $y^k = y^{k+l}$.

Proof. With the notation of 3.9, $\bar{u}^k$ is in the group $\mathbb{Z}/(l)$ so that $\bar{u}^k = \bar{u}^{k+l}$. Let τ be such that I_x/τ satisfies $y^k = y^{k+l}$. In particular, $x^k \tau x^{k+l}$ which implies $f_k \tau f_{k+l}$ and $e_k \tau e_{k+l}$. Hence τ must be of type (k', l') where $k \geq k'$ and $l' | l$, so that $\rho \subseteq \tau$.

A precise classification of monogenic inverse semigroups in terms of defining relations is provided by the following result.

IX.3.11 Theorem

Any monogenic inverse semigroup is isomorphic to precisely one of the following. Let $\langle x \rangle$ be a monogenic inverse semigroup with one of the defining relations:

(i) $x^k = x^{-1}x^{k+1}$,
(ii) $x^k x^{-1} = x^{-1}x^k$,
(iii) $x^k = x^{k+l}$,
(iv) $x = x$,

where k and l are natural numbers.

Proof. (i) Let $S = I_x/\rho_{(k,\infty^+)}$ be as in 3.4(i). Then by 2.3, we have $x^k \rho_{(k,\infty^+)} x^{-1}x^{k+1}$. Let ρ be the congruence generated by the pair $(x^k, x^{-1}x^{k+1})$. The foregoing shows that $\rho \subseteq \rho_{(k,\infty^+)}$. Conversely, from $x^k \rho x^{-1}x^{k+1}$ it follows that $f_k \rho f_{k+1}$ by 2.3 and thus $\rho_{(k,\infty^+)} \subseteq \rho$ since $\rho_{(k,\infty^+)}$ is the least congruence on I_x which identifies f_k with f_{k+1}. We conclude that $\rho = \rho_{(k,\infty^+)}$, and hence $x^k = x^{-1}x^{k+1}$ is a defining relation for $I_x/\rho_{(k,\infty^+)}$.

It follows from 3.6 that the defining relation $x^k = x^{-1}x^{k+1}$ can be substituted by $x^k = x^{k+1}x^{-1}$.

(ii) Let ρ be the congruence on I_x generated by the pair $(x^{-1}x^k, x^k x^{-1})$. It follows from 3.9 that $\rho \subseteq \rho_{(k,\omega)}$. On the other hand, using the same argument as in the proof of 3.9, we can show that $x^{-1}x^k \rho x^k x^{-1}$ implies $e_k \rho e_{k+1}$ and $f_k \rho f_{k+1}$. Thus $\rho_{(k,\omega)} \subseteq \rho$, since $\rho_{(k,\omega)}$ is the least congruence on I_x which identifies e_k with e_{k+1} and f_k with f_{k+1}. We conclude that $\rho = \rho_{(k,\omega)}$ and that $x^{-1}x^k = x^k x^{-1}$ is a defining relation for $I_x/\rho_{(k,\omega)}$.

By 3.9, $x^k x^{-1} = x^{-1}x^k$ is not only a defining relation in $I_x/\rho_{(k,\omega)}$ but also an identity.

(iii) The same type of statement is valid here by using 3.10.

(iv) This is the case of a free monogenic inverse semigroup corresponding to I_x/ε.

Conversely, in view of 3.4 any monogenic inverse semigroup is isomorphic to one of the above. It follows easily from the form of the congruences having various types that no two monogenic inverse semigroups with different defining relations can be isomorphic.

IX.3.12 Corollary

A monogenic inverse semigroup $\langle x \rangle$ is

(i) of type (k, ∞^+) if and only if it has the defining relation $x^k = x^{-1}x^{k+1}$,
(ii) of type (k, ω) if and only if it has the defining relation $x^k x^{-1} = x^{-1}x^k$,
(iii) of type (k, l) if and only if it has the defining relation $x^k = x^{k+l}$.

In (i), (k, ∞^-) and $x^k = x^{k+1}x^{-1}$ can each be substituted independently for (k, ∞^+) and $x^k = x^{-1}x^{k+1}$.

Proof. The proof of this corollary is left as an exercise.

After III.2.9, we promised an example showing that in general $\vee_{\rho \in \mathfrak{F}} \rho_{\max} \neq (\vee_{\rho \in \mathfrak{F}} \rho)_{\max}$ for a nonempty family $\mathfrak{F}$ of congruences on an inverse semigroup. We are now ready for it.

IX.3.13 Example

We consider the congruences $\rho = \rho_{(1, \infty^+)}$, $\xi = \rho_{(1, \infty^-)}$, and $\rho_{(1, \omega)}$ on I_x. According to 3.4, ρ is a bicyclic congruence on I_x; in view of 3.6, ξ is also a bicyclic congruence. The definition of $\rho_{(1, \omega)}$ shows immediately that $\sigma = \rho_{(1, \omega)}$. From the results of this section or by direct calculation, one sees that $\rho \vee \xi = \sigma$. In view of III.3.10, we get that $\rho = \rho_{\max}$ and $\xi = \xi_{\max}$ since a bicyclic semigroup is an antigroup. On the other hand, $\sigma_{\max} = \omega$ which shows that $\rho_{\max} \vee \xi_{\max} \neq (\rho \vee \xi)_{\max}$.

IX.3.14 Exercises

(i) Show that I_x is a normal extension of its semilattice of idempotents by an infinite cyclic group.

(ii) Draw the semilattice of idempotents of various types of monogenic inverse semigroups.

(iii) On $\mathcal{C}(I_x)$ find the trace and kernel classes. For each $\rho \in \mathcal{C}(I_x)$, characterize $\rho_{\min}$, $\rho_{\max}$, $\rho^{\min}$, $\rho^{\max}$ in terms of types of congruences on I_x.

(iv) Show that the class of subdirectly irreducible monogenic inverse semigroups consists of

(α) bicyclic semigroups,

(β) cyclic groups of prime order,

(γ) finite monogenic inverse semigroups with zero.

(v) Show explicitly that I_x is a subdirect product of subdirectly irreducible finite semigroups.

(vi) Determine which monogenic inverse semigroups are E-unitary and provide for them the P- and Q-representations.

(vii) Characterize the lattice of E-unitary congruences on I_x.

(viii) Characterize the lattice of antigroup congruences on I_x.

(ix) What are the types of

(α) Rees congruences (relative to ideals),

(β) group congruences,

(γ) bicyclic congruences

on I_x?

(x) Show that every nonidentical congruence on I_x can be uniquely written as the intersection of the Rees congruence relative to some ideal of I_x and either a bicyclic congruence or a cyclic group congruence.

A classification of monogenic inverse semigroups is due to Djadčenko–Schein [1], [2], Eberhart–Selden [1], and Eršova [3]. Identities on these semigroups are discussed in Djadčenko [1], Djadčenko–Schein [2], and Kleiman [4]. An interesting property of monogenic inverse semigroups was established by Eršova [2].

IX.4 PRESENTATIONS

We obtain here a free monogenic inverse semigroup as a quotient of a free semigroup on two generators by certain relations. These relations are countably infinite in number, and we then prove that no finite subcollection is suitable for this representation. Using this, we establish that no free inverse semigroup (monogenic or not) is finitely related in the class of all semigroups. As one of the consequences, we obtain a presentation of the bicyclic semigroup. Generation of semigroups in this section is meant *within the class of all semigroups*.

Let $F = \{u, v\}^+$ be the free semigroup on the set $\{u, v\}$. Let ρ be the congruence on F generated by the set of all pairs of the form

$$(u, uvu), \qquad (v, vuv), \tag{26}$$

$$(u^m v^{m+n} u^n, v^n u^{n+m} v^m) \qquad (m, n \geq 1). \tag{27}$$

Let φ be the unique homomorphism of F into I_x for which

$$u\varphi = x^{-1} = ((0,1),(1,0)),$$

$$v\varphi = x = ((1,0),(0,1)).$$

Since φ maps the set $\{u, v\}$ of generators of F onto the set $\{x^{-1}, x\}$ of generators of I_x, we have that φ maps F onto I_x. Now I_x satisfies the relations

$$x^{-1} = x^{-1}xx^{-1}, \qquad x = xx^{-1}x,$$

$$(x^m x^{-m})(x^{-n}x^n) = (x^{-n}x^n)(x^m x^{-m}),$$

which in view of (26), (27) yields that $\rho \subseteq \overline{\varphi}$, where $\overline{\varphi}$ is the congruence on F induced by φ. In order to establish the opposite inclusion, we need some preparation.

IX.4.1 Lemma

In F/ρ, for all natural numbers m, n, p, we have

$$u^m v^n u^p = \begin{cases} u^{m+p-n} & \text{if } n \leq m, n \leq p \\ u^m v^{n-p} & \text{if } m \geq n \geq p \\ v^{n-m} u^p & \text{if } m \leq n \leq p \\ v^{n-m} u^n v^{n-p} & \text{if } n \geq m, n \geq p. \end{cases} \tag{28}$$

(If in these expressions an exponent is zero, we omit that letter.)

Proof. First let $m \geq n$. Using (26) and (27), we obtain

$$\begin{aligned} u^m v^n u^p &= u^{m-n+1}(u^{n-1}v^{n-1})(vu)u^{p-1} \\ &= u^{m-n+1}(vu)(u^{n-1}v^{n-1})u^{p-1} \\ &= u^{m-n}(uvu)u^{n-1}v^{n-1}u^{p-1} \\ &= u^m v^{n-1} u^{p-1} = \cdots = u^m v^{n-2} u^{p-2} = \cdots . \end{aligned}$$

In the end, we will arrive either at the first or the second equality in (28). One obtains the third equality in (28) analogously.

Next let $n \geq m$, $n \geq p$, $n \leq m + p$. By the hypothesis and the second equality in (28), we get

$$\begin{aligned} u^m v^n u^p &= (u^m v^m)(v^{n-m} u^{n-m}) u^{p+m-n} \\ &= (v^{n-m} u^{n-m})(u^m v^m) u^{p+m-n} \\ &= v^{n-m}(u^n v^m u^{p+m-n}) \\ &= v^{n-m} u^n v^{n-p}. \end{aligned}$$

If $n > m + p$, then $m' = n - m > p$, $p' = n - p < n < m' + p'$. Similarly as in the preceding case, we obtain

$$u^m v^n u^p = u^{n-m'} v^n u^{n-p'} = v^{m'} u^n v^{p'} = v^{n-m} u^n v^{n-p},$$

which proves the fourth equality in (28).

IX.4.2 Lemma

Every element of F/ρ can be written in the form

$$u^m v^n u^p \qquad (0 \leq m \leq n, 0 \leq p \leq n, 0 < n). \tag{29}$$

Proof. First note that 4.1 is valid if we interchange everywhere u and v, and both 4.1 and its dual can be used backward, that is from v^{m+p-n} to $v^m u^n v^p$, and so on. We consider words of the form u^n, v^m, $v^m u^n$, ...

By the first equality in (28), we get $u^n = u^n v^n u^n$, which is of the form (29); v^n is already of that form.

If $n < m$, then $u^m v^n = u^m v^m u^{m-n}$ by the second equality in (28) and $u^m v^m u^{m-n}$ is of the form (29). If $n \geq m$, then $u^m v^n$ is already of the prescribed form. The case $v^n u^p$ is symmetric.

If $p \leq n < m$, then by the preceding case,

$$(u^m v^n) u^p = u^m v^m u^{m-n+p}$$

with $m \geq m - n + p$; the case $m \leq n < p$ is symmetric. If $n < m$ and $n < p$, then by the first equality in (28), and the above,

$$u^m v^n u^p = u^{m+p-n} = u^{m+p-n} v^{m+p-n} u^{m+p-n}$$

which is again of the required form. The dual of 4.1 gives that $v^m u^n v^p$ can be written in the form $u^{m'} v^{n'} u^{p'}$ for some m', n', p' so that this reduces to the case just considered.

Again using the dual of 4.1, we obtain

$$u^q v^m u^n v^p = u^{q+m'} v^{n'} u^{p'},$$

$$v^m u^n v^p u^q = u^{m'} v^{n'} u^{p'+q},$$

and this reduces to the case considered above. In this way, we may reduce any product of powers of u and v to the product of three such and proceed as above.

The first main result of this section can now be proved.

IX.4.3 Theorem

Let S be the semigroup generated by $\{u, v\}$ subject to the defining relations

$$(I) \quad u = uvu, \qquad v = vuv, \tag{30}$$

$$(A_{m,n}) \quad u^m v^{m+n} u^n = v^n u^{n+m} v^m \qquad (m, n \in \mathbb{N}). \tag{31}$$

Then S is a free monogenic inverse semigroup.

Proof. We can consider F/ρ instead of S. In view of the above argument, it remains to take $w\varphi = w'\varphi$ for $w, w' \in F$ and show that $w\rho w'$.

By 4.2, we have $w\rho u^m v^n u^p$ for some m, n, p satisfying the conditions of 4.2. We compute

$$\begin{aligned}(u^m v^n u^p)\varphi &= (u\varphi)^m (v\varphi)^n (u\varphi)^p \\ &= ((0,1),(1,0))^m((1,0),(0,1))^n((0,1),(1,0))^p \\ &= ((0,m)(n,0)(0,p),(m,0)(0,n)(p,0)) \\ &= ((n-m,p),(m,n-p))\end{aligned}$$

and since $\rho \subseteq \overline{\varphi}$, we obtain

$$w\varphi = ((n-m,p),(m,n-p)).$$

Analogously,

$$w'\varphi = ((n'-m',p'),(m',n'-p'))$$

and the hypothesis $w\varphi = w'\varphi$ evidently implies that $m = m'$, $n = n'$, $p = p'$. But then

$$w\rho u^m v^n u^p = u^{m'} v^{n'} u^{p'} \rho w'$$

so that $w\rho w'$. This shows that $\overline{\varphi} \subseteq \rho$ and completes the proof.

IX.4.4 Corollary

Every ρ-class contains exactly one word of the form $u^m v^n u^p$ with $0 \leq m \leq n$, $0 \leq p \leq n$, $0 < n$. If we let $x = v$, $x^{-1} = u$ and define the product of two such words to be equal to the representative of the ρ-class which contains their usual product, then with this multiplication the set of all such words coincides with C_3.

Proof. The first assertion is a direct consequence of 4.3. In order to prove the second assertion, it suffices to verify that the multiplication indicated in the statement of the corollary coincides with that in C_3. Indeed, using the dual of 4.1, we obtain

$$\begin{aligned}&(u^{p_1}v^{q_1}u^{r_1})(u^{p_2}v^{q_2}u^{r_2}) \\ &\quad = u^{p_1}(v^{q_1}u^{r_1+p_2}v^{q_2})u^{r_2} \\ &\quad = \begin{cases} u^{p_1}v^{q_1+q_2-r_1-p_2}u^{r_2} & \text{if } r_1+p_2 \leq q_1,\ r_1+p_2 \leq q_2 \\ u^{p_1}v^{q_1}u^{r_1+p_2-q_2+r_2} & \text{if } q_1 \geq r_1+p_2 \geq q_2 \\ u^{p_1+r_1+p_2-q_1}v^{q_2}u^{r_2} & \text{if } q_1 \leq r_1+p_2 \leq q_2 \\ u^{p_1+r_1+p_2-q_1}v^{r_1+p_2}u^{r_1+p_2-q_2+r_2} & \text{if } r_1+p_2 \geq q_1,\ r_1+p_2 \geq q_2. \end{cases}\end{aligned}$$

Considering these four cases and comparing with the formula for multiplication in 1.5, we see that this product coincides with that in C_3.

We now turn to the question concerning the number of defining relations for a free monogenic inverse semigroup.

IX.4.5 Lemma

The system (30) and (31) of defining relations for the semigroup S is not equivalent to any finite subset of these relations.

Proof. Let $A = \{0, 1, 2, \ldots, n\}$ and u, v be partial transformations of A defined by

$$u = \begin{pmatrix} 0 & 1 & 2 & \cdots & n-1 \\ 1 & 2 & 3 & \cdots & n \end{pmatrix},$$

$$v = \begin{pmatrix} 0 & 1 & 2 & \cdots & n-1 & n \\ 0 & 0 & 1 & \cdots & n-2 & n-1 \end{pmatrix}.$$

We consider the product of partial transformations as written on the right of the argument. One verifies readily that $u = uvu$ and $v = vuv$, and that

$$u^k = \begin{pmatrix} 0 & 1 & 2 & \cdots & n-k \\ k & k+1 & k+2 & \cdots & n \end{pmatrix} \tag{32}$$

if $k \le n$, and $u^k = \varnothing$ if $k > n$; analogously

$$v^k = \begin{pmatrix} 0 & 1 & \cdots & k & k+1 & k+2 & \cdots & n \\ 0 & 0 & \cdots & 0 & 1 & 2 & \cdots & n-k \end{pmatrix} \tag{33}$$

if $k \le n$.

We now consider the validity of the relations $A_{i,j}$ in (31). If $i > n$ or $j > n$, then $u^{i+j} = \varnothing$ and $u^i = \varnothing$ or $u^j = \varnothing$, so that both sides of A_{ij} contain the empty transformation $\varnothing$, and thus $A_{i,j}$ holds. Now let $i \le n$ and $j \le n$. Using (32) and (33), we obtain

$$\begin{aligned} u^i v^i &= \begin{pmatrix} 0 & 1 & \cdots & n-i \\ i & i+1 & \cdots & n \end{pmatrix} \begin{pmatrix} 0 & \cdots & i & i+1 & \cdots & n \\ 0 & \cdots & 0 & 1 & \cdots & n-i \end{pmatrix} \\ &= \begin{pmatrix} 0 & 1 & \cdots & n-i \\ 0 & 1 & \cdots & n-i \end{pmatrix}, \end{aligned} \tag{34}$$

$$\begin{aligned} v^j u^j &= \begin{pmatrix} 0 & \cdots & j & j+1 & \cdots & n \\ 0 & \cdots & 0 & 1 & \cdots & n-j \end{pmatrix} \begin{pmatrix} 0 & 1 & \cdots & n-j \\ j & j+1 & \cdots & n \end{pmatrix} \\ &= \begin{pmatrix} 0 & 1 & \cdots & j & j+1 & \cdots & n \\ j & j & \cdots & j & j+1 & \cdots & n \end{pmatrix}. \end{aligned} \tag{35}$$

Note that the relation $A_{i,j}$ means commutativity of $u^i v^i$ and $v^j u^j$. We are thus interested in the following expressions, which we derive directly from (34) and (35).

$$u^i v^{i+j} u^j = \begin{pmatrix} 0 & \cdots & n-i \\ j & \cdots & j \end{pmatrix} \neq \varnothing = v^j u^{j+i} u^i$$

if $i + j > n$, and

$$u^i v^{i+j} u^j = \begin{pmatrix} 0 & 1 & \cdots & j & j+1 & \cdots & n-i \\ j & j & \cdots & j & j+1 & \cdots & n-i \end{pmatrix} = v^j u^{j+i} u^i$$

if $i + j \leq n$.

Now let S_n be the semigroup generated by $\{u, v\}$. We have just proved that the defining relation $A_{i,j}$ does not hold in S_n if and only if $i, j \leq n < i + j$.

Assume now that the system of defining relations (30) and (31) is equivalent to a finite subset D of these relations. Let $V = \{i + j | A_{i,j} \in D\}$, let $n = \max V$ if $V \neq \varnothing$ and let n be any natural number if $V = \varnothing$. If $A_{i,j} \in D$, then $i + j \leq n$ by the definition of n, and thus $A_{i,j}$ holds in S_n, as we have seen above. Consequently, all defining relations from D hold in S_n. By the hypothesis that D is equivalent to all defining relations in (30) and (31), all the latter relations must hold in S_n. However, we have seen above that $A_{1,n}$ does not hold in S_n, a contradiction. This yields the assertion of the lemma.

IX.4.6 Definition

A semigroup S is said to be *finitely related* if $S \cong X^+/\rho$ for some set X and a congruence ρ on X^+ which is generated by a finite set of relations (that is, pairs of elements of X^+).

The second main result is the following.

IX.4.7 Theorem

No free inverse semigroup is finitely related in the class of all semigroups.

Proof. We consider first the semigroup S in 4.3, and assume that S is finitely related over a set X of generators by means of defining relations R. We may replace X be the set $\{u, v\}$ and every relation from R by a relation resulting from substitution of all occurrences of elements of X by their expressions as products of u and v. Since R is finite, S is definable over the alphabet $\{u, v\}$ by a finite set D of defining relations. Consequently, all the relations from D can be deduced from the defining relations (30) and (31). In such an inference one cannot use but a finite number of relations (30) and (31). Conversely, since D is the set of defining relations for S, all the relations (30) and (31) are deducible from D. It follows that the set (30) and (31) of defining relations is equivalent to some of its finite subsets, which contradicts the assertion of 4.5. We have proved that the assertion of the theorem is valid for a free monogenic inverse semigroup.

For the general case, we let I_X be a free inverse semigroup on a nonempty set X. Assume that I_X is finitely related with a set Y of generators by means of defining relations R. Since we may express each element of Y as a product of elements of X, we may assume that $Y = X$. If X is infinite, then some elements of X do not occur in the defining relations from R since R is finite. In such a case I_X is not an inverse semigroup since if an element x does not occur in R, then $x = xx^{-1}x$ does not hold in I_X. Thus if I_X is finitely related, then X must be finite.

For X a finite set, we add to R the finite set of defining relations of the form $x = y$ for all $x, y \in X$, $x \neq y$. In this way, we obtain an inverse semigroup M which is a homomorphic image of I_X. Since we have identified all the generators of I_X, we must have that M is monogenic. Also, since I_X is free, M must be too. Consequently, M is a free monogenic inverse semigroup which is finitely related (the relations R and $x = y$ for a finite number of $x, y \in X$), which contradicts the first part of the proof. Therefore I_X is not finitely related.

We now derive a presentation of the bicyclic semigroup C.

IX.4.8 Proposition

Let S be a semigroup generated by $\{u, v\}$ subject to the defining relations

$$u = uvu, \qquad v = vuv, \tag{36}$$

$$u = vu^2, \qquad v = v^2u. \tag{37}$$

Then S is a bicyclic semigroup.

Proof. We let $F = \{u, v\}^+$ and let τ be the congruence on F generated by the pairs

$$(u, uvu), \qquad (v, vuv), \qquad (u, vu^2), \qquad (v, v^2u), \tag{38}$$

so that $S \cong F/\tau$. Define φ to be the unique homomorphism of F into C for which $u\varphi = (1,0)$, $v\varphi = (0,1)$. Since φ maps the generators of F onto the generators of C, we deduce that φ maps F onto C. Relations (36) and (37) are valid for $(1,0)$ and $(0,1)$ instead of u and v. Hence $\tau \subseteq \overline{\varphi}$, where $\overline{\varphi}$ is the congruence on F induced by φ.

Relations (38) show that vu is the identity of F/τ. It follows that $v^n u^n = vu$ for any $n > 0$. Consequently, for $m > n$ we get

$$v^m u^n = v^{m-n} v^n u^n = v^{m-n} vu = vuv^{m-n} = v^{m-n};$$

analogously, for $m < n$, we obtain $v^m u^n = u^{n-m}$. Hence every word in F is τ-equivalent to some $u^m v^n$ for $m, n \geq 0$, $m + n > 0$ or to vu.

Now let $w, w' \in F$ be such that $w\varphi = w'\varphi$. Then $w\tau u^m v^n$ or $w\tau vu$ and $w'\tau u^{m'}v^{n'}$ or $w'\tau vu$, so we distinguish four cases.

Case $w\tau u^m v^n$, $w'\tau u^{m'}v^{n'}$. Then $\tau \subseteq \overline{\varphi}$ implies $(u^m v^n)\varphi = (u^{m'}v^{n'})\varphi$ whence $(m, n) = (m', n')$. Consequently, $m = m'$, $n = n'$ so that $w\tau u^m v^n = u^{m'}v^{n'}\tau w'$.

Case $w\tau u^m v^n$, $w'\tau vu$. Then a similar argument leads to $(m, n) = (0,0)$ so that $m = n = 0$, contradicting the requirement $m + n > 0$. Hence this case is impossible. Analogously, $w\tau vu$, $w'\tau u^{m'}v^{n'}$ is impossible.

If $w\tau vu$ and $w'\tau vu$, then trivially $w\tau w'$.

We have proved that $\tau = \overline{\varphi}$ which shows that $F/\tau \cong C$, and S is a bicyclic semigroup.

IX.4.9 Corollary

Let M be a monoid generated by $\{u, v\}$ subject to the defining relation $vu = 1$. Then M is a bicyclic semigroup.

Proof. In the above proof, consider $\{u, v\}^*$ instead of $\{u, v\}^+$. The details are left as an exercise.

For the next corollary we need some information about the congruences on a bicyclic semigroup.

IX.4.10 Lemma

If ρ is a proper congruence on the bicyclic semigroup C, then C/ρ is a group.

Proof. Assume that $(m, n)\rho(p, q)$ where $(m, n) \neq (p, q)$. By taking inverses if necessary, we may assume that $m \neq p$; we may further suppose that $m < p$. It follows that $(m, n)(n, m)\rho(p, q)(q, p)$ so that $(m, m)\rho(p, p)$, and thus

$$(0,0) = (0, m)(m, m)(m,0)\rho(0, m)(p, p)(m,0) = (p - m, p - m).$$

Let $t = p - m$. If $k \leq t$, we get

$$(k, k) = (0,0)(k, k)\rho(t, t)(k, k) = (t, t)$$

and hence $(k, k)\rho(0,0)$. In particular, $(1,1)\rho(0,0)$. Now for any $k \geq 0$, we obtain

$$(k + 1, k + 1) = (k,0)(1,1)(0, k)\rho(k,0)(0,0)(0, k) = (k, k).$$

But then every idempotent of C is ρ-related to $(0,0)$. Thus the quotient C/ρ has only one idempotent, and hence by II.2.10 it must be a group.

IX.4.11 Corollary

Let a, b be elements of any semigroup S and assume that $a = aba$, $b = bab$, $ba < ab$. Then the subsemigroup T of S generated by $\{a, b\}$ is a bicyclic semigroup with identity ab.

Proof. The hypothesis $ab < ba$ gives

$$ba = (ab)(ba) = (ba)(ab)$$

which implies $a = a^2b$, $b = ab^2$, and $ab \neq ba$. In view of 4.8, we see that T must be a homomorphic image of C. By 4.10, proper homomorphic images of C are groups, in which case $a = aba$, $b = bab$, $ab \neq ba$ is impossible. Hence T must be a bicyclic semigroup. Further, the above identities give

$$a = (ab)a = a(ab), \qquad b = (ab)b = b(ab)$$

which implies that ab is the identity of T.

We can now deduce the following important consequence.

IX.4.12 Theorem

If a (0-)simple semigroup is not completely 0-simple, then for every (nonzero) idempotent e, there exists a bicyclic subsemigroup of S with identity e.

Proof. We consider the case with zero; the case without zero has virtually the same proof. By I.6.14, there exists $f \in E_{S^*}$ such that $f < e$ and $e \mathcal{D} f$. Hence I.6.7 provides mutually inverse elements $a, b \in S$ such that $ab = e$ and $ba = f$. Now 4.11 implies that the subsemigroup of S generated by $\{a, b\}$ is a bicyclic semigroup with identity e.

IX.4.13 Corollary

A regular semigroup S is completely semisimple if and only if it does not contain a bicyclic subsemigroup.

Proof. Note first that a bicyclic semigroup being bisimple, it is always contained in some $\mathcal{D}$-class of a semigroup. If S is completely semisimple, then each principal factor of S is completely (0-)simple and hence cannot contain a bicyclic semigroup. If S is not completely semisimple, then some principal factor is (0-)simple but not completely (0-)simple, and 4.12 provides a bicyclic subsemigroup of it.

We are now in a position to establish a partial converse of VIII.5.8.

IX.4.14 Proposition

An inverse semigroup with the basis property is completely semisimple.

Proof. By 4.13, every inverse semigroup which is not completely semisimple contains an isomorphic copy of the bicyclic semigroup C. By contrapositive, it suffices to prove that C does not have the basis property.

First, $(0,3)(2,0) = (0,1)$ shows that

$$C = \langle(0,1)\rangle = \langle(0,2),(0,3)\rangle$$

and thus both $\{(0,1)\}$ and $\{(0,2),(0,3)\}$ generate C. Since $(0,2)(2,0) = (0,0)$, the identity of C, all elements of $\langle(0,2)\rangle$ are of the form $(2,0)^m(0,2)^n = (2m,2n)$, and thus $(0,3) \notin \langle(0,2)\rangle$; analogously $(0,2) \notin \langle(0,3)\rangle$. Hence both $\{(0,1)\}$ and $\{(0,2)(0,3)\}$ are bases of C.

The next result provides a relationship between Green's relations and bicyclic subsemigroups.

IX.4.15 Proposition

Let S be a semigroup with identity e and let $a \in S$. Then $a \in R_e \setminus H_e$ if and only if a has an inverse b such that the subsemigroup of S generated by $\{a, b\}$ is a bicyclic semigroup and $ab = e$.

Proof

Necessity. The hypothesis $a \in R_e$ implies that $ab = e$ for some b, whereas $a \notin L_e$ means that $S \neq Sa$, in particular $e \notin Sa$. Thus $ba \neq e$, and since e is the identity of S, we must have $f = ba < e$. Now we have the setting of 4.11 which then provides the desired bicyclic semigroup.

Sufficiency. Since $ab = e$ and $a = ea$, we have $a\mathcal{R}e$. If $a\mathcal{L}e$, then

$$e = a^{-1}a = a^{-1}aba = eba = ba$$

which is impossible in a bicyclic semigroup. Hence $a \notin L_e$ and thus $a \in R_e \setminus H_e$.

There is an obvious dual statement to 4.15. Also note that b in 4.15 is contained in $L_e \setminus H_e$. Hence the following symmetry is valid in S in 4.15: $R_e \neq H_e$ if and only if $L_e \neq H_e$.

IX.4.16 Exercises

(i) Which monogenic inverse semigroups are finitely related?

(ii) Let a be an element of an inverse semigroup S such that for all $n > 0$, $a^n a^{-n}$ is not comparable with $a^{-n}a^n$. Show that $\langle a \rangle$ is a free inverse semigroup. Also show that the conclusion is false if the above relations hold only for a finite number of integers n.

(iii) Let M be a monoid generated by $\{u, v\}$ subject to the defining relations $uv = vu = 1$. Show that M is an infinite cyclic group.

(iv) Let m be a positive integer and M be a monoid generated by $\{u, v\}$ subject to the defining relations $u^m = uv = vu = 1$. Show that M is a cyclic group of order m.

(v) Give a presentation of a monogenic inverse semigroup S which has a proper ideal M satisfying one of the conditions

(α) M is a bicyclic semigroup,

(β) M is an infinite cyclic group,

(γ) M is a finite cyclic group.

(vi) Characterize all congruences on the bicyclic semigroup C and show that each one of them is induced by a transitive representation of C.

(vii) Let α be an idempotent of $\mathcal{I}(X)$. Show that α is the identity for some bicyclic subsemigroup of $\mathcal{I}(X)$ if and only if the rank of α is infinite. Deduce that X is finite if and only if $\mathcal{I}(X)$ does not contain a bicyclic semigroup.

(viii) An element a of a semigroup S is *left* (respectively *right*) *increasing* if there exists a proper subset T of S such that $aT = S$ (respectively $Ta = S$). In a semigroup S with identity e, show that a is left increasing if and only if $a \in R_e \setminus H_e$.

IX.4.17 Problems

(i) (B. M. Schein [21]) Find an independent set of defining relations for I_X.

All the results up to 4.7 are due to Schein [21], [14]; the first paper contains more material on this subject. Result 4.9 is often given as a definition of a bicyclic semigroup. Consult also Reilly [7]. Further, 4.12 is due to Andersen [1] and 4.14 to Jones [3]. The relationship of bicyclic semigroups with increasing elements is due to Ljapin [2]. Increasing elements in semigroups with one-sided identity were investigated by Schwarz–Jaroker [1]. The position of a bicyclic semigroup as a subsemigroup was also considered by Anderson–Hunter–Koch [1] and Saitô [1].

BISIMPLE INVERSE MONOIDS

1. A bisimple inverse monoid can be constructed from its $\mathcal{R}$-class containing the identity element. This $\mathcal{R}$-class is a right cancellative monoid whose principal left ideals form a semilattice under finite intersection. Conversely, from such a semigroup one can construct a bisimple inverse monoid whose $\mathcal{R}$-class of the identity is isomorphic to the given semigroup. The relationship between these two classes of semigroups can be made into a categorical equivalence if we supply both classes with suitable morphisms. The constructions involved here exhibit some remarkable properties.

2. One can take the subcategory of the category of right cancellative monoids with principal left ideals closed under finite intersection whose objects are positive cones of lattice ordered groups. This category is equivalent to the category of lattice ordered groups and their homomorphisms as well as to a certain category whose objects are combinatorial bisimple inverse semigroups.

3. The lattice of normal congruences on the semilattice E of idempotents of a bisimple inverse monoid S is isomorphic both to the lattice of filters of E self-conjugate in S and to a certain lattice of subsets of the $\mathcal{R}$-class of S containing the identity.

4. The lattice of idempotent separating congruences on a bisimple inverse monoid S is isomorphic both to the lattice of right normal divisors of the $\mathcal{R}$-class of S containing the identity and to a certain lattice of subgroups of the group of units of S. These isomorphisms exhibit several interesting properties.

X.1 CATEGORIES $\mathfrak{B}$ AND $\mathfrak{R}$

We first introduce the two categories in question and the two needed functors. After proving that these are indeed functors of the requisite categories, we introduce the two needed natural equivalences. The four objects so constructed provide the quadruple for the equivalence of the two categories.

The category of principal interest to us here is the following.

X.1.1 Definition

Let a category $\mathfrak{B}$ be given by

Ob $\mathfrak{B}$ are bisimple inverse monoids,

Hom $\mathfrak{B}$ are monoid homomorphisms of Ob $\mathfrak{B}$.

The second category that we consider here is defined thus.

X.1.2 Definition

Let a category $\mathfrak{R}$ be given by

Ob $\mathfrak{R}$ are right cancellative monoids whose principal left ideals are closed under finite intersection,

Hom $\mathfrak{R}$ are homomorphisms of Ob $\mathfrak{R}$, say $\varphi : R \to R'$, such that if $Ra \cap Rb = Rc$ for some $a, b, c \in R$, then $R'(a\varphi) \cap R'(b\varphi) = R'(c\varphi)$.

Note that the morphisms of $\mathfrak{R}$ are also monoid homomorphisms since an object of $\mathfrak{R}$ has the identity element as its sole idempotent, and thus any homomorphism of objects of $\mathfrak{R}$ must map the identity onto the identity. Before we define the first functor, we need an auxiliary result. Recall the notation $a\rho = a^{-1}a$ in II.1.12.

X.1.3 Lemma

Let $S \in$ Ob $\mathfrak{B}$ and R be the $\mathcal{R}$-class of the identity 1 of S.

(i) R is a right cancellative monoid.

(ii) If $a \in R$, then $Sa \cap R = Ra$.

(iii) For any $a, b, c \in R$, $Ra \cap Rb = Rc$ if and only if $(a\rho)(b\rho) = c\rho$.

Proof. (i) If $a, b \in R$, then $a \mathcal{R} b \mathcal{R} 1$ whence $ab \mathcal{R} a1 \mathcal{R} 1$. Hence R is closed under multiplication. For $a, b, x \in R$ such that $ax = bx$, we get $a = axx^{-1} = bxx^{-1} = b$, so R is right cancellative.

(ii) Let $a \in R$. Then by part (i), we have $Ra \subseteq Sa \cap R$. Conversely, let $x \in Sa \cap R$. Then $x = sa$ for some $s \in S$ and

$$ss^{-1} = s(aa^{-1})s^{-1} = (sa)(sa)^{-1} = xx^{-1} = 1$$

which shows that $s \in R$. Consequently $Sa \cap R \subseteq Ra$, and the equality prevails.

(iii) Let $a, b, c \in R$. Assume first that $Ra \cap Rb = Rc$. By part (ii), we have $Sa \cap Sb \cap R = Sc \cap R$ which gives $S(a\rho) \cap S(b\rho) \cap R = S(c\rho) \cap R$. Since clearly $S(a\rho) \cap S(b\rho) = S(a\rho)(b\rho)$, we obtain $S(a\rho)(b\rho) \cap R = S(c\rho) \cap R$. Let $x \in R \cap L_{(a\rho)(b\rho)}$. Then $S(x\rho) = Sx = S(a\rho)(b\rho)$ so that $x\rho = (a\rho)(b\rho)$. Also, using part (ii), we get

$$Rc = Sc \cap R = S(c\rho) \cap R = S(a\rho)(b\rho) \cap R = Sx \cap R = Rx.$$

Consequently, $c\mathcal{L}x$ and thus $c\rho = x\rho$ which yields $(a\rho)(b\rho) = c\rho$.

Conversely, suppose that $(a\rho)(b\rho) = c\rho$. Then

$$Sa \cap Sb = S(a\rho) \cap S(b\rho) = S(a\rho)(b\rho) = S(c\rho) = Sc$$

which by part (ii) gives $Ra \cap Rb = Rc$.

For the first functor, we introduce the following symbolism.

X.1.4 Notation

For every $S \in \text{Ob}\,\mathfrak{B}$, let $U(S)$ be the $\mathcal{R}$-class of the identity of S, and for every $\varphi \in \text{Hom}\,\mathfrak{B}$, say $\varphi : S \to S'$, let $U(\varphi) = \varphi|_{U(S)}$.

X.1.5 Lemma

With the above notation, U is a functor from $\mathfrak{B}$ to $\mathfrak{R}$.

Proof. Let $S \in \text{Ob}\,\mathfrak{B}$ and $R = U(S)$. By 1.3(i), we have that R is a right cancellative monoid. Let $a, b \in R$. In the notation of 1.3, we have

$$Sa \cap Sb = S(a\rho) \cap S(b\rho) = S(a\rho)(b\rho).$$

Letting $c \in R \cap L_{(a\rho)(b\rho)}$ and using 1.3(ii), we obtain

$$Ra \cap Rb = S(a\rho)(b\rho) \cap R = Sc \cap R = Rc$$

which proves that $R \in \text{Ob}\,\mathfrak{R}$.

Next let $\varphi \in \text{Hom}\,\mathfrak{B}$, say $\varphi : S \to S'$ for some $S, S' \in \text{Ob}\,\mathfrak{B}$. Let $\psi = U(\varphi)$, $R = U(S)$, and $R' = U(S')$. Since φ maps the identity of S onto the identity of S', and preserves $\mathcal{R}$-classes, $\psi = \varphi|_{U(S)}$ is a homomorphism of R into R'.

Assume that $Ra \cap Rb = Rc$ for $a, b, c \in R$. By 1.3(iii), we have $(a\rho)(b\rho) = c\rho$ and thus $(a\rho\varphi)(b\rho\varphi) = c\rho\varphi$. Since $x\rho\varphi = x\varphi\rho$ for any $x \in R$, we deduce that $(a\varphi\rho)(b\varphi\rho) = c\varphi\rho$. Now again by 1.3(iii), we get $R'(a\varphi) \cap R'(b\varphi) = R'(c\varphi)$. But then also $R'(a\psi) \cap R'(b\psi) = R'(c\psi)$, which yields $\psi \in \text{Hom}\,\mathfrak{R}$.

The remaining axioms for a functor are verified easily.

X.1.6 Corollary

For any $S \in \text{Ob}\,\mathfrak{B}$ and $R = U(S)$, the mapping

$$Ra \to a\rho \qquad (a \in R)$$

is an isomorphism of the semilattice of principal left ideals of R (under intersection) onto E_S.

Proof. The proof is left as an exercise (bisimplicity implies that the mapping is onto).

In order to construct a functor in the opposite direction, we need some preparation.

X.1.7 Lemma

Let $R \in \text{Ob}\,\mathfrak{R}$ and $a, b \in R$ be $\mathfrak{L}$-related. Then $a = ub$ and $b = u^{-1}a$ for some unit u.

Proof. First $a\mathfrak{L}b$ implies $a = ub$, $b = va$ for some $u, v \in R$. Hence $a = uva$ and $b = vub$ which by right cancellation implies $v = u^{-1}$.

The following construction is of general interest.

X.1.8 Constructions

Let R be a right cancellative monoid, and let $\rho : a \to \rho_a$ be its right regular representation. Since R is right cancellative, each ρ_a is one-to-one, and thus ρ can be considered as a homomorphism of R into the semigroup $\mathcal{I}(R)$ of one-to-one partial transformations of R. Since R has an identity, ρ itself is one-to-one. The inverse subsemigroup of $\mathcal{I}(R)$ generated by $R\rho$ is the *inverse hull* of R, to be denoted by $\Sigma(R)$.

X.1.9 Construction

Now let $R \in \text{Ob}\,\mathfrak{R}$. On $R \times R$ define a relation τ by

$$(a, b)\tau(c, d) \Leftrightarrow a = uc,\ b = ud$$

for some unit u of R. Straightforward verification shows that τ is an equivalence relation. We denote the τ-class containing (a, b) by $[a, b]$, and let

$$R^{-1} \circ R = \{[a, b] \| a, b \in R\}.$$

In order to define a multiplication on $R^{-1} \circ R$, we proceed as follows. In every

$\mathfrak{L}$-class of R choose a representative. For any $a, b \in R$, we have, by hypothesis, $Ra \cap Rb = Rc$ for some $c \in R$. By $a \vee b$ denote the representative of the $\mathfrak{L}$-class L_c. We thus have

$$Ra \cap Rb = R(a \vee b).$$

(Note that in view of 1.3(iii), we have $(a\rho)(b\rho) = (a \vee b)\rho$ if $R = U(S)$ for $S \in \text{Ob}\,\mathfrak{B}$.) Hence $a \vee b = xa$ for some element x of R, which by right cancellation in R is unique. Noting that $a \vee b = b \vee a$, we may introduce the notation $a * b$ and $b * a$ by

$$(a * b)b = a \vee b = (b * a)a.$$

We are now in a position to define a multiplication in $R^{-1} \circ R$ as follows:

$$[a, b][c, d] = [(c * b)a, (b * c)d].$$

X.1.10 Lemma

(i) The multiplication in $R^{-1} \circ R$ is single valued.

(ii) The multiplication in $R^{-1} \circ R$ is independent of the system of representatives chosen in various $\mathfrak{L}$-class.

(iii) $R^{-1} \circ R$ is isomorphic to the inverse hull of R.

Proof. (i) Assume that $[a, b] = [a', b']$ and $[c, d] = [c', d']$. Then

$$a = ua', \qquad b = ub', \qquad c = vc', \qquad d = vd'$$

for some units u and v. It follows that

$$R(b \vee c) = Rb \cap Rc = Rb' \cap Rc' = R(b' \vee c')$$

so that $b \vee c = s(b' \vee c')$ for some unit s by 1.7. Hence

$$(c * b)ub' = (c * b)b = c \vee b = s(c' \vee b') = s(c' * b')b'$$

which in view of right cancellation gives $(c * b)u = s(c' * b')$. Thus

$$(c * b)a = (c * b)ua' = s(c' * b')a'$$

and analogously $(b * c)d = s(b' * c')d'$. Consequently

$$[(c * b)a, (b * c)d] = [(c' * b')a', (b' * c')d']$$

and the multiplication is single valued.

(ii) In 1.9, choose another system of representatives in various $\mathfrak{L}$-classes, and for $Ra \cap Rb = Rc$, denote by $a \vee' b$ the new representative in L_c. Then $a \vee b \mathfrak{L} a \vee' b$ and thus $a \vee' b = u(a \vee b)$ by 1.7 for some unit. Now define $a *' b$ by means of $a \vee' b$ as in 1.9 so that

$$(b *' a)a = b \vee' a = (a *' b)b.$$

Multiplying

$$(b * a)a = b \vee a = (a * b)b$$

on the left by u, we obtain

$$b *' a = u(b * a), \qquad a *' b = u(a * b).$$

Since this is true for all $a, b \in R$, we get

$$[(c *' b)a, (b *' c)d] = [(c * b)a, (b * c)a],$$

as required.

(iii) Now define

$$\theta : [a, b] \to \rho_a^{-1}\rho_b \qquad ([a, b] \in R^{-1} \circ R).$$

In order to see that θ is single valued, let $[a, b] = [c, d]$. Then $a = uc$, $b = ud$ for some unit u. Hence

$$\rho_a^{-1}\rho_b = (\rho_u\rho_c)^{-1}(\rho_u\rho_d) = \rho_c^{-1}(\rho_u^{-1}\rho_u)\rho_d = \rho_c^{-1}\rho_d$$

since ρ_u is a permutation of R. Thus θ is single valued.

Let $a, b, c, d \in R$. Then

$$x \in \mathbf{d}(\rho_b\rho_c^{-1}) \Leftrightarrow xb \in \mathbf{d}\rho_c^{-1} = \mathbf{r}\rho_c = Rc$$

$$\Leftrightarrow xb \in Rb \cap Rc = R(b \vee c) = R(c * b)b$$

$$\Leftrightarrow x \in R(c * b) = \mathbf{r}\rho_{c * b} = \mathbf{d}\rho_{c * b}^{-1} = \mathbf{d}(\rho_{c * b}^{-1}\rho_{b * c})$$

since $\mathbf{d}\rho_{b * c} = R$, and if this is the case, then

$$xb = t(b \vee c) = t(b * c)c = t(c * b)b$$

so that $x = t(c * b)$ for some $t \in R$. It further follows that

$$x\rho_b\rho_c^{-1} = (xb)\rho_c^{-1} = (t(b * c)c)\rho_c^{-1} = t(b * c)$$

$$= t\rho_{b * c} = (t(c * b))\rho_{c * b}^{-1}\rho_{b * c} = x\rho_{c * b}^{-1}\rho_{b * c}$$

which proves that $\rho_b\rho_c^{-1} = \rho_{c*b}^{-1}\rho_{b*c}$. Using this, we obtain

$$
\begin{aligned}
([a,b]\theta)([c,d]\theta) &= (\rho_a^{-1}\rho_b)(\rho_c^{-1}\rho_d) = \rho_a^{-1}(\rho_b\rho_c^{-1})\rho_d \\
&= \rho_a^{-1}\rho_{c*b}^{-1}\rho_{b*c}\rho_d = \rho_{(c*b)a}^{-1}\rho_{(b*c)d} \\
&= ([a,b][c,d])\theta \qquad (1)
\end{aligned}
$$

and θ is a homomorphism.

Now assume that $\rho_a^{-1}\rho_b = \rho_c^{-1}\rho_d$. Note that for any $r \in R$, $\rho_r\rho_r^{-1}$ is the identity transformation on R. Hence

$$
\begin{aligned}
\rho_a^{-1}\rho_a &= \rho_a^{-1}(\rho_b\rho_b^{-1})\rho_a = (\rho_a^{-1}\rho_b)(\rho_a^{-1}\rho_b)^{-1} \\
&= (\rho_c^{-1}\rho_d)(\rho_c^{-1}\rho_d)^{-1} = \rho_c^{-1}(\rho_d\rho_d^{-1})\rho_c = \rho_c^{-1}\rho_c
\end{aligned}
$$

which implies that $Ra = Rc$. Thus $a = uc$ for some unit u by 1.7. Further,

$$\rho_b = \rho_a(\rho_a^{-1}\rho_b) = \rho_a(\rho_c^{-1}\rho_d) = \rho_{uc}\rho_c^{-1}\rho_d = \rho_u(\rho_c\rho_c^{-1})\rho_d = \rho_{ud}$$

and thus $b = ud$. Consequently, $[a, b] = [c, d]$ and θ is one-to-one.

The formula $(\rho_a^{-1}\rho_b)(\rho_c^{-1}\rho_d) = \rho_{(c*b)a}^{-1}\rho_{(b*c)d}$ proved in (1) implies that $\Sigma(R) = \{\rho_a^{-1}\rho_b | a, b \in R\}$ and hence θ is an isomorphism of $R^{-1} \circ R$ onto $\Sigma(R)$.

We are now ready for a functor from $\mathfrak{R}$ to $\mathfrak{B}$.

X.1.11 Notation

For every $R \in \operatorname{Ob} \mathfrak{R}$, let $V(R) = R^{-1} \circ R$. For every $\varphi \in \operatorname{Hom} \mathfrak{R}$, say $\varphi: R \to R'$, let

$$V(\varphi): [a, b] \to [a\varphi, b\varphi] \qquad ([a, b] \in V(R)).$$

X.1.12 Lemma

With the above notation, V is a functor from $\mathfrak{R}$ to $\mathfrak{B}$.

Proof. Let $R \in \operatorname{Ob} \mathfrak{R}$. By 1.10, $V(R) \cong \Sigma(R)$ so that $V(R)$ is an inverse semigroup. Let $[a, b] \in E_{V(R)}$. Then

$$[a,b][a,b] = [(a*b)a, (b*a)b] = [a,b]$$

so $a = u(a*b)a$ and $b = u(b*a)b$ for some unit u. Now right cancellation gives $a*b = b*a = u^{-1}$ whence

$$[a,b] = [(b*a)a, (a*b)b] = [a \vee b, a \vee b].$$

Consequently

$$E_{V(R)} = \{[a, a] \mid a \in R\}.$$

Straightforward verification shows that $[a, b]^{-1} = [b, a]$. Hence, for any idempotents $[a, a]$ and $[b, b]$, the equalities

$$[a, b][b, a] = [a, a], \qquad [b, a][a, b] = [b, b]$$

in view of II.1.8 show that $V(R)$ is bisimple. It follows easily that $[1, 1]$ is the identity of $V(R)$, which proves that $V(R) \in \text{Ob}\,\mathfrak{B}$.

Now let $\varphi \in \text{Hom}\,\mathfrak{R}$, say $\varphi : R \to R'$. Assume that $[a, b] = [a', b']$ so that $a = ua'$, $b = ub'$ for some unit u. Then

$$[a\varphi, b\varphi] = [(ua')\varphi, (ub')\varphi] = [(u\varphi)(a'\varphi), (u\varphi)(b'\varphi)] = [a'\varphi, b'\varphi]$$

since $u\varphi$ is a unit in R'. Hence $\overline{\varphi} = V(\varphi)$ is single valued.

For any $b, c \in R$, we have $Rb \cap Rc = R(b \vee c)$ which by the hypothesis on φ gives $R'(b\varphi) \cap R'(c\varphi) = R'(b \vee c)\varphi$. On the other hand, $R'(b\varphi) \cap R'(c\varphi) = R'(b\varphi \vee c\varphi)$ so that $(b \vee c)\varphi = u'(b\varphi \vee c\varphi)$ for some unit u' in R'. It follows that

$$(b * c)\varphi(c\varphi) = (b \vee c)\varphi = u'(b\varphi \vee c\varphi) = u'(b\varphi * c\varphi)(c\varphi)$$

whence $(b * c)\varphi = u'(b\varphi * c\varphi)$, and analogously $(c * b)\varphi = u'(c\varphi * b\varphi)$.

Using this, we obtain for any $a, b, c, d \in R$,

$$\begin{aligned}
([a, b][c, d])\overline{\varphi} &= [(c * b)a, (b * c)d]\overline{\varphi} = [((c * b)a)\varphi, ((b * c)d)\varphi] \\
&= [(c * b)\varphi(a\varphi), (b * c)\varphi(d\varphi)] \\
&= [u'(c\varphi * b\varphi)(a\varphi), u'(b\varphi * c\varphi)(d\varphi)] \\
&= [a\varphi, b\varphi][c\varphi, d\varphi] \\
&= [a, b]\overline{\varphi}[c, d]\overline{\varphi}
\end{aligned}$$

and $\overline{\varphi}$ is a homomorphism. It is clear that $\overline{\varphi}$ maps the identity of $V(R)$ onto the identity of $V(R')$. Consequently, $\overline{\varphi} \in \text{Hom}\,\mathfrak{B}$.

The remaining axioms for a functor follow without difficulty.

The first natural equivalence follows.

X.1.13 Lemma

For every $S \in \text{Ob}\,\mathfrak{B}$ define a mapping $\xi(S)$ by

$$\xi(S) : a^{-1}b \to [a, b] \qquad [a, b \in U(S)].$$

Then ξ is a natural equivalence of the functors $I_{\mathfrak{B}}$ and VU.

Proof. Fix $S \in \text{Ob}\,\mathfrak{B}$ and let $\theta = \xi(S)$. Let $s \in S$. There exists $x \in S$ such that $ss^{-1} = x^{-1}x$ and $xx^{-1} = 1$ since S is bisimple. Hence $s = ss^{-1}s = x^{-1}(xs)$ where

$$(xs)(xs)^{-1} = x(ss^{-1})x^{-1} = xx^{-1}xx^{-1} = xx^{-1} = 1.$$

Thus $x, xs \in R = U(S)$ so that $s \in R^{-1}R$. Consequently, θ is defined on all of S.

In order to show that ξ is single valued, we let $a, b, c, d \in R$ be such that $a^{-1}b = c^{-1}d$. Then $d = (cc^{-1})d = c(c^{-1}d) = (ca^{-1})b$ and similarly $b = (ac^{-1})d$. Put $v = ca^{-1}$ and $u = ac^{-1}$. Since $d = vb = (vu)d$, we get $vu = vudd^{-1} = dd^{-1} = 1$, and similarly $uv = 1$, and thus u is a unit. Now $b^{-1}a = d^{-1}c$ so that

$$a = bb^{-1}a = bd^{-1}c = udd^{-1}c = uc,$$

$$b = aa^{-1}b = ac^{-1}d = ud.$$

Hence $[a, b] = [c, d]$ and ξ is single valued.

Now let $a, b, c, d \in R$ be arbitrary. Then

$$c * b = (c * b)bb^{-1} = (c \vee b)b^{-1},$$

$$b * c = (b * c)cc^{-1} = (b \vee c)c^{-1},$$

so that by 1.3(iii), we obtain

$$\begin{aligned}(c * b)^{-1}(b * c) &= [(c \vee b)b^{-1}]^{-1}(b \vee c)c^{-1} \\ &= b\big[(b \vee c)^{-1}(b \vee c)\big]c^{-1} \\ &= b(b \vee c)c^{-1} = b(b\rho)(c\rho)c^{-1} = bc^{-1}.\end{aligned}$$

Using this, we get

$$\begin{aligned}(a^{-1}b)\theta(c^{-1}d)\theta &= [a, b][c, d] = [(c * b)a, (b * c)d] \\ &= \big\{[(c * b)a]^{-1}[(b * c)d]\big\}\theta \\ &= \big[a^{-1}(c * b)^{-1}(b * c)d\big]\theta \\ &= \big[a^{-1}(bc^{-1})d\big]\theta \\ &= \big[(a^{-1}b)(c^{-1}d)\big]\theta\end{aligned}$$

and θ is a homomorphism.

Assume that $[a, b] = [c, d]$. Then $a = uc$ and $b = ud$ for some unit u. Hence

$$a^{-1}b = (uc)^{-1}(ud) = c(u^{-1}u)d = c^{-1}d$$

and θ is one-to-one. For any $[a, b] \in R^{-1} \circ R$, we evidently have $(a^{-1}b)\theta = [a, b]$, and thus θ is an isomorphism of S onto $VU(S)$.

To see that the diagram

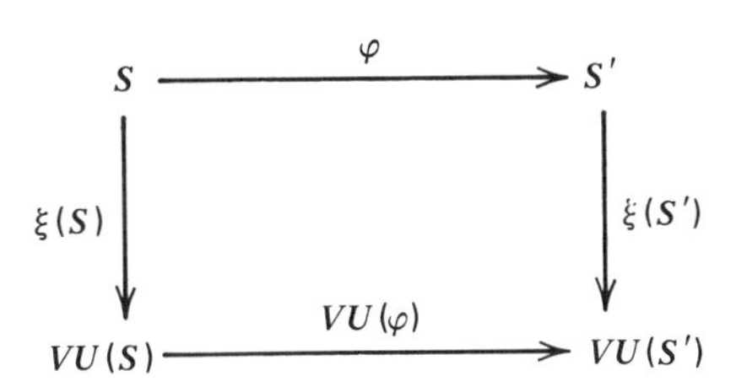

commutes, we let $a, b \in U(S)$ and compute

$$[(a^{-1}b)\xi(S)](VU(\varphi)) = [a\varphi, b\varphi] = (a^{-1}b)\varphi[\xi(S')].$$

Therefore ξ is a natural equivalence of the functors $I_{\mathfrak{B}}$ and VU.

The second natural equivalence follows.

X.1.14 Lemma

For every $R \in \text{Ob}\,\mathfrak{R}$, define a mapping $\eta(R)$ by

$$\eta(R) : r \to [1, r] \qquad (r \in R).$$

Then η is a natural equivalence of the functors $I_{\mathfrak{R}}$ and UV.

Proof. Fix $R \in \text{Ob}\,\mathfrak{R}$ and let $\pi = \eta(R)$. Then for any $r, s \in R$, we have

$$(r\pi)(s\pi) = [1, r][1, s] = [1, rs] = (rs)\pi.$$

If $r\pi = s\pi$, then $[1, r] = [1, s]$ whence $r = s$. If $[a, b]\mathcal{R}[1, 1]$, then $[a, b][b, a] = [1, 1]$ so that $[a, a] = [1, 1]$ and a is a unit, whence $[a, b] = [1, a^{-1}b]$. This proves that π is an isomorphism of R onto $UV(R)$. The diagram

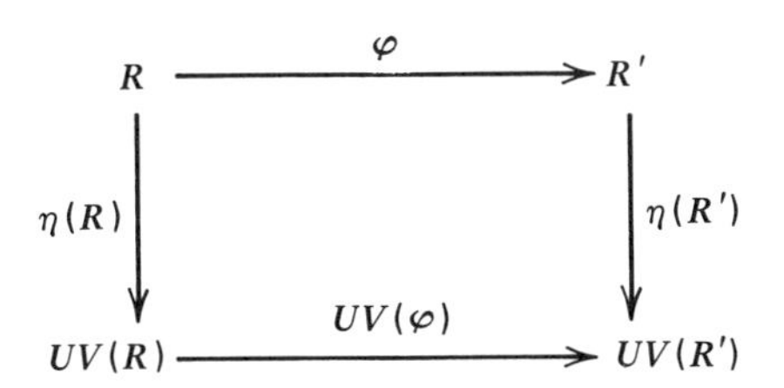

commutes since for any $r \in R$,

$$r\varphi[\eta(R')] = [1, r\varphi] = r[\eta(R)][UV(\varphi)].$$

From 1.5 and 1.12–1.14 we deduce the desired result.

X.1.15 Theorem

The quadruple (U, V, ξ, η) is an equivalence of the categories $\mathfrak{B}$ and $\mathfrak{R}$.

We will now establish two interesting properties of the object $V(R) = R^{-1} \circ R$.

X.1.16 Proposition

Let $R \in \mathrm{Ob}\,\mathfrak{R}$ and let $\eta = \eta(R)$ as in 1.14. Then for any homomorphism φ of R into an inverse semigroup S such that for all $a, b \in R$, $(a\varphi\rho)(b\varphi\rho) = (a \vee b)\varphi\rho$, there exists a unique homomorphism ψ making the diagram

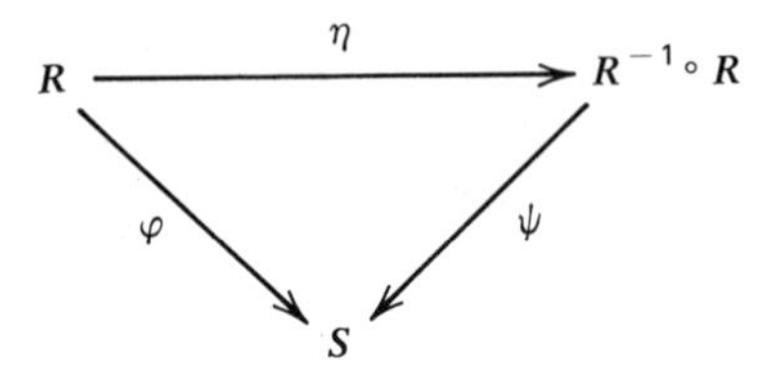

commutative.

Proof. We define ψ by

$$\psi : [a, b] \to (a\varphi)^{-1}(b\varphi) \qquad ([a, b] \in R^{-1} \circ R).$$

In order to see that ψ is single valued, we let $[a, b] = [c, d]$. Then $a = uc$ and $b = ud$ for some unit u. Hence $u\varphi$ is a unit in $R\varphi$ which gives

$$(a\varphi)^{-1}(b\varphi) = ((uc)\varphi)^{-1}(ud)\varphi = (c\varphi)^{-1}(u\varphi)^{-1}(u\varphi)(d\varphi) = (c\varphi)^{-1}(d\varphi)$$

and ψ is single valued.

Let $a, b, c, d \in R$. Then the hypothesis implies that $(b\varphi\rho)(c\varphi\rho) = (b \vee c)\varphi\rho$ which gives

$$(b\varphi)^{-1}(b\varphi)(c\varphi)^{-1}(c\varphi) = ((b \vee c)\varphi)^{-1}(b \vee c)\varphi$$

so that

$$
\begin{aligned}
(b\varphi)(c\varphi)^{-1} &= (b\varphi)[(b \vee c)\varphi]^{-1}(b \vee c)\varphi(c\varphi)^{-1} \\
&= (b\varphi)[((b * c)c)\varphi]^{-1}[(b * c)c]\varphi(c\varphi)^{-1} \\
&= (b\varphi)(c\varphi)^{-1}[(b * c)\varphi]^{-1}[(b * c)\varphi](c\varphi)(c\varphi)^{-1} \\
&= (b\varphi)(c\varphi)^{-1}[(b * c)\varphi]^{-1}[(b * c)\varphi] \\
&= (b\varphi)[((b * c)c)\varphi]^{-1}[(b * c)\varphi] \\
&= (b\varphi)[((c * b)b)\varphi]^{-1}[(b * c)\varphi] \\
&= (b\varphi)(b\varphi)^{-1}[(c * b)\varphi]^{-1}[(b * c)\varphi] \\
&= [(c * b)\varphi]^{-1}[(b * c)\varphi]
\end{aligned}
$$

since $b\mathcal{R}1$ implies $(b\varphi)\mathcal{R}(1\varphi)$, and thus $(b\varphi)(b\varphi)^{-1} = 1\varphi$ and 1φ is the identity of $R\varphi$. Using this, we obtain

$$
\begin{aligned}
[a, b]\psi[c, d]\psi &= (a\varphi)^{-1}\left[(b\varphi)(c\varphi)^{-1}\right](d\varphi) \\
&= (a\varphi)^{-1}((c * b)\varphi)^{-1}(b * c)\varphi(d\varphi) \\
&= [((c * b)a)\varphi]^{-1}[((b * c)d)\varphi] \\
&= [(c * b)a, (b * c)d]\psi \\
&= ([a, b][c, d])\psi
\end{aligned}
$$

and ψ is a homomorphism.

For any $r \in R$, we have

$$r\eta\psi = [1, r]\psi = (1\varphi)^{-1}(r\varphi) = r\varphi$$

and the above diagram commutes. The uniqueness of ψ follows from the fact that the elements of $R^{-1} \circ R$ are of the form $(r\eta)^{-1}(s\eta)$ for $r, s \in R$, so any homomorphism making the above diagram commute must coincide with ψ.

We have defined a one-to-one partial right translation in V.2.1 on any semigroup S; have seen in V.2.2 that they form an inverse subsemigroup of

$\mathcal{I}(S)$ under composition, and denoted this semigroup by $\hat{S}$ in V.2.3. It is easy to see that the elements of $\hat{S}$ whose domain is a principal left ideal form an inverse subsemigroup of $\hat{S}$, to be denoted by $\overline{S}$. We now exhibit an interesting property of $\overline{R}$ when $R \in \operatorname{Ob} \mathfrak{R}$.

X.1.17 Proposition

Let $R \in \operatorname{Ob} \mathfrak{R}$. For each pair $a, b \in R$, define $\rho^{[a,b]}$ by

$$(xa)\rho^{[a,b]} = xb \qquad (x \in R).$$

Then

$$\rho : [a, b] \to \rho^{[a,b]} \qquad ([a, b] \in R^{-1} \circ R)$$

is an isomorphism of $R^{-1} \circ R$ onto $\overline{R}$.

Proof. First note that by right cancellation $\rho^{[a,b]}$ is single valued and one-to-one. Clearly, $\mathbf{d}\rho^{[a,b]} = Ra$ and $\rho^{[a,b]}$ is a partial right translation. Thus $\rho^{[a,b]} \in \overline{R}$.

Assume that $[a, b] = [c, d]$. Then $a = uc$, $b = ud$ for some unit u. Note that then $Ra = Rc$ so that $\mathbf{d}\rho^{[a,b]} = \mathbf{d}\rho^{[c,d]}$. For any $x \in Ra = Rc$, we have $x = ya = yuc$ which implies

$$x\rho^{[a,b]} = (ya)\rho^{[a,b]} = yb = y(ud) = (yu)d = (yuc)\rho^{[c,d]} = x\rho^{[c,d]}$$

and thus $\rho^{[a,b]} = \rho^{[c,d]}$. Hence ρ is indeed defined on $R^{-1} \circ R$.

Assume conversely that $\rho^{[a,b]} = \rho^{[c,d]}$. Then $Ra = Rc$ so $a = uc$ for some unit u. Further, $a\rho^{[a,b]} = a\rho^{[c,d]}$ implies $b = ud$ and hence $[a, b] = [c, d]$. Consequently, ρ is one-to-one. If $\theta \in \overline{R}$, then its domain is a principal left ideal Ra, and for $a\theta = b$, we immediately obtain that $\theta = \rho^{[a,b]}$.

For any $[a, b], [c, d] \in R^{-1} \circ R$, we first have

$$x \in \mathbf{d}(\rho^{[a,b]}\rho^{[c,d]}) \Leftrightarrow x \in Ra, \quad x\rho^{[a,b]} \in Rc$$

$$\Leftrightarrow x = ya, \quad yb = wc \quad \text{for some} \quad y, w \in R.$$

The equation $yb = wc$ implies $yb \in Rb \cap Rc = R(b \vee c) = R(c * b)b$ and $y = u(c * b)$ for some $u \in R$ by right cancellation. Continuing the above equivalences we have

$$\Leftrightarrow x = u(c * b)a \quad \text{for some } u \in R$$

$$\Leftrightarrow x \in R(c * b)a \Leftrightarrow x \in \mathbf{d}\rho^{[(c*b)a,(b*c)d]} = \mathbf{d}\rho^{[a,b][c,d]}.$$

For such x, we get $x = u(c * b)a$,

$$
\begin{aligned}
x\rho^{[a,b]}\rho^{[c,d]} &= (u(c*b)a)\rho^{[a,b]}\rho^{[c,d]} = (u(c*b)b)\rho^{[c,d]} \\
&= (u(b*c)c)\rho^{[c,d]} = u(b*c)d \\
&= (u(c*b)a)\rho^{[(c*b)a,(b*c)d]} \\
&= x\rho^{[a,b][c,d]}
\end{aligned}
$$

which proves that ρ is a homomorphism.

X.1.18 Exercises

(i) Show that both functors U and V preserve direct products, that is,

$$\prod_{\alpha \in A} U(S_\alpha) \cong U\Big(\prod_{\alpha \in A} S_\alpha\Big),$$

$$\prod_{\alpha \in A} V(R_\alpha) \cong V\Big(\prod_{\alpha \in A} R_\alpha\Big).$$

(ii) Show that U maps a bicyclic semigroup onto an infinite cyclic monoid, and that V maps an infinite cyclic monoid onto a bicyclic semigroup.

(iii) Show that the center of a bisimple inverse semigroup is either a group or is empty.

(iv) Characterize Green's relations in $V(R)$ for $R \in \operatorname{Ob} \mathfrak{R}$.

(v) Show that for $R \in \operatorname{Ob} \mathfrak{R}$, the following statements are equivalent.

(α) The group of units of R is trivial.

(β) R is combinatorial.

(γ) $V(R)$ is combinatorial.

(vi) Let $R \in \operatorname{Ob} \mathfrak{R}$ and assume that R is a subsemigroup of a group G for which $R^{-1}R = G$. In 1.16, take $S = G$ and $\varphi = \iota_R$. Show that ψ is a homomorphism of $R^{-1} \circ R$ onto G which induces σ on $R^{-1} \circ R$.

(vii) Let ρ be a congruence on an E-unitary bisimple inverse semigroup. Show that if for some $e \in E_S$, $e\rho \subseteq E_S$, then S/ρ is E-unitary.

(viii) Let ρ be a congruence on an E-unitary bisimple inverse semigroup. For $e, f \in E_S$, show that $e\rho/\sigma$ and $f\rho/\sigma$ are conjugate subgroups of S/σ.

(ix) Show that the inverse hull of an infinite cyclic semigroup is a bicyclic semigroup.

X.1.9 Problems

(i) What can be said about the metacenter of a bisimple inverse semigroup?

The relationship between bisimple inverse monoids and right cancellative monoids whose principal left ideals are closed under finite intersection was established by Clifford [3] (who did not treat their categorical equivalence). Result 1.17 stems from McAlister [10]. The Clifford construction was generalized by Reilly [4] to include bisimple inverse semigroups. This subject was further pursued by Reilly–Clifford [1]. A generalization of Clifford's construction is due to McAlister [2]; see also McAlister [3]. The structure of arbitrary 0-bisimple inverse semigroups was described by McAlister [6] and Munn [11]. E-unitary bisimple inverse semigroups were discussed by McAlister–McFadden [1]. The structure of classes of bisimple inverse semigroups were described by Hickey [3] and Warne [6]. Homomorphisms of bisimple inverse monoids were studied by Warne [1]. Gantos [1] and Iyengar [1] investigated semilattices of bisimple inverse semigroups. See also Pollák [1].

X.2 THE CATEGORY OF l-GROUPS

An interesting full subcategory of $\mathfrak{R}$ is obtained by taking those objects of $\mathfrak{R}$ which form the positive cones of l-groups. We will show that this category is equivalent to the category of l-groups and l-homomorphisms. The corresponding full subcategory of $\mathfrak{B}$ is also brought into play, thereby associating to every l-group a uniquely determined, of course up to an isomorphism, combinatorial bisimple inverse semigroup.

We start with the subcategories of $\mathfrak{B}$ and $\mathfrak{R}$. The following result will be needed presently and is also of independent interest.

X.2.1 Lemma

Let $R \in \operatorname{Ob} \mathfrak{R}$. Then R is left cancellative if and only if $V(R)$ is E-unitary.

Proof. Let R be left cancellative and assume that $[a, b][c, c] = [d, d]$ in $V(R)$. Then

$$(c * b)a = ud, \qquad (b * c)c = ud$$

for some unit u. Hence $(c * b)a = (c * b)b$ which implies $a = b$ by left cancellation. Thus $[a, b]$ is an idempotent and $V(R)$ is E-unitary.

Conversely, let $V(R)$ be E-unitary and assume that $ca = cb$. Then

$$\begin{aligned}[a, b][ca, ca] &= [(ca * b)a, (b * ca)ca] \\ &= [(ca * b)a, (ca * b)b] = [ua, ub]\end{aligned}$$

since

$$R(ca \vee b) = Rca \cap Rb = Rcb \cap Rb = Rcb$$

so that $(ca * b)b = ucb$ for some unit u and thus $ca * b = uc$. Hence $[a, b][ca, ca]$ is an idempotent, and the hypothesis implies that $a = b$.

X.2.2 Definition

Let $\mathfrak{B}_l$ be the full subcategory of $\mathfrak{B}$ for which

$$\text{Ob } \mathfrak{B}_l = \{S \in \text{Ob } \mathfrak{B} \mid S \text{ is } E\text{-unitary, combinatorial, and}$$

$$aR_1 = R_1 a \quad \text{for all } a \in R_1\},$$

where R_1 is the $\mathcal{R}$-class of the identity of S.

Let $\mathfrak{R}_l$ be the full subcategory of $\mathfrak{R}$ for which

$$\text{Ob } \mathfrak{R}_l = \{R \in \text{Ob } \mathfrak{R} \mid R \text{ is left cancellative, combinatorial,}$$

$$\text{and} \quad aR = Ra \quad \text{for all } a \in R\}.$$

Note that a right cancellative monoid is combinatorial if and only if its group of units is trivial. Now 2.1 shows that the restrictions of the functors U and V and the natural equivalences ξ and η to the categories $\mathfrak{B}_l$ and $\mathfrak{R}_l$ provide a categorical equivalence which we state as follows (see 1.15).

X.2.3 Proposition

The quadruple $(U_l, V_l, \xi_l, \eta_l)$ is an equivalence of the categories $\mathfrak{B}_l$ and $\mathfrak{R}_l$.

We now turn to l-groups. We state below only the bare minimum of the material on this subject used in this section, and for an extensive discussion of l-groups refer to the book Birkhoff [1].

X.2.4 Definition

A group G provided with a partial order which satisfies the compatibility condition

$$a \leq b \Rightarrow ca \leq cb, \quad ac \leq bc \qquad (a, b, c \in G)$$

is a *partially ordered group* (briefly *po-group*). If the partial order forms a lattice, G is a *lattice ordered group* (briefly *l-group*).

Note that in an l-group, the compatibility condition for order implies the compatibility condition for lattice operations, that is, for all $a, b, c \in G$,

$$c(a \vee b) = ca \vee cb, \qquad (a \vee b)c = ac \vee bc,$$

$$c(a \wedge b) = ca \wedge cb, \qquad (a \wedge b)c = ac \wedge bc$$

and also for the inversion in G,

$$(a \vee b)^{-1} = a^{-1} \wedge b^{-1}, \qquad (a \wedge b)^{-1} = a^{-1} \vee b^{-1},$$

$$a \leq b \Rightarrow b^{-1} \geq a^{-1}.$$

X.2.5 Definition

Let φ be a homomorphism of a group G into a group H. If G and H are also po-groups, φ is an *o-homomorphism* if it preserves order, that is,

$$a \leq b \Rightarrow a\varphi \leq b\varphi.$$

If G and H are also l-groups, φ is an *l-homomorphism* if it preserves the lattice operations, that is,

$$(a \vee b)\varphi = a\varphi \vee b\varphi, \qquad (a \wedge b)\varphi = a\varphi \wedge b\varphi.$$

For l-groups the o-homomorphisms coincide with the l-homomorphisms.

X.2.6 Definition

Let G be a po-group. Then the set

$$G_{+} = \{g \in G | g \geq 1\}$$

where 1 is the identity of G, is the *positive cone* of G.

We now introduce one more category.

X.2.7 Notation

Let $\mathfrak{G}_l$ be the category of l-groups and l-homomorphisms. For any $G \in \operatorname{Ob} \mathfrak{G}_l$, let $P(G) = G_{+}$ and for $\varphi \in \operatorname{Hom} \mathfrak{G}_l$, say $\varphi : G \to H$, let $P(\varphi) = \varphi|_{G_{+}}$.

X.2.8 Lemma

With the above notation, P is a functor from $\mathfrak{G}_l$ to $\mathfrak{R}_l$.

Proof. Let $G \in \operatorname{Ob} \mathfrak{G}_l$, $R = G_{+}$, and $a, b \in R$. Then

$$a \vee b = (1 \vee ba^{-1})a = (ab^{-1} \vee 1)b$$

so that $R(a \vee b) \subseteq Ra \wedge Rb$. Conversely, let $z = xa = yb$ for some $x, y \in R$. Then $x \wedge y \geq 1$ and

$$x \wedge y = x \wedge xab^{-1} = xa(a^{-1} \wedge b^{-1}) = z(a \vee b)^{-1}$$

whence $z = (x \wedge y)(a \vee b) \in R(a \vee b)$. Consequently, $Ra \cap Rb \subseteq R(a \vee b)$ and the equality prevails. Since R is evidently a cancellative monoid, we conclude that $R \in \operatorname{Ob} \mathfrak{R}$.

Let $a, b, x, y \in R$ be such that $a = xb$ and $b = ya$. Then $a = xya$ so that $y^{-1} = x$. Hence, $y^{-1} \geq 1$ which implies $y \leq 1$ and thus $y = 1$. Consequently, $a = b$ and R is combinatorial.

For any $a, b \in R$, we have $ab = (aba^{-1})a$ where $aba^{-1} \geq aa^{-1} = 1$ so that $aR \subseteq Ra$. Symmetrically $Ra \subseteq aR$ and the equality prevails. Therefore $R \in \operatorname{Ob} \mathfrak{R}_l$.

Let $\varphi \in \operatorname{Hom} \mathfrak{G}_l$, say $\varphi : G \to H$. Evidently φ maps G_+ into H_+. Let $a, b \in R = G_+$. Then $Ra \cap Rb = R(a \vee b)$, whereas in $R' = H_+$ we have

$$R'(a\varphi) \cap R'(b\varphi) = R'(a\varphi \vee b\varphi) = R'(a \vee b)\varphi.$$

Thus $\varphi \in \operatorname{Hom} \mathfrak{R}_l$.

The remaining axioms for a functor clearly hold.

For the next functor, we will need the following auxiliary result of independent interest.

X.2.9 Lemma

Let G be a group and R be a submonoid of G such that $R^{-1}R = G$, R has a trivial group of units, $aR = Ra$ for all $a \in R$, and principal right ideals of R are closed under finite intersection. Define a relation $\leq$ on G by

$$a \leq b \Leftrightarrow a^{-1}b \in R.$$

Then G with this relation as order is an l-group whose positive cone coincides with R.

Proof. Note that $\leq$ is reflexive since $1 \in R$, it is antisymmetric since the group of units of R is trivial, and it is transitive since R is a subsemigroup. By hypothesis $aR = Ra$ for all $a \in R$ which implies that also $a^{-1}R = Ra^{-1}$. This together with the hypothesis that $G = R^{-1}R$ yields that $aR = Ra$ for all $a \in G$.

Now let $a \leq b$ and $c \in G$. Then $a^{-1}b \in R$ whence

$$(ac)^{-1}(bc) = c^{-1}(a^{-1}b)c \in c^{-1}Rc = R,$$

$$(ca)^{-1}(cb) = a^{-1}(c^{-1}c)b = a^{-1}b \in R,$$

and thus $ac \leq bc$, $ca \leq cb$, and G is a po-group. Clearly $R = G_+$.

Let $a, b \in R$. By hypothesis there exists $c \in R$ such that $aR \cap bR = cR$. Hence $c = ar = bs$ for some $r, s \in R$ which implies $a^{-1}c, b^{-1}c \in R$ so that $a \le c$, $b \le c$. Now let $d \in R$ be such that $a \le d$, $b \le d$. Then the elements $r = a^{-1}d$ and $s = b^{-1}d$ are in R and thus $d \in aR \cap bR = cR$ which implies that $d = ct$ for some $t \in R$. Hence $c^{-1}d = t \in R$ which gives $c \le d$. We have proved that $c = a \vee b$.

In order to establish the existence of a join of any pair of elements in G, we let $a, b, c, d \in R$ and will prove that

$$a^{-1}b \vee c^{-1}d = a^{-1}c^{-1}(cb \vee cac^{-1}d). \tag{2}$$

Indeed, by the hypothesis $R^{-1}R = G$, an arbitrary element of G is of the form $a^{-1}b$ with $a, b \in R$. Further, with the above assumptions on a, b, c, d, we have $cac^{-1} \in cRc^{-1} = Rcc^{-1} = R$ in view of the hypothesis $cR = Rc$. Hence $cb, cac^{-1}d \in R$, and the existence of $p = cb \vee cac^{-1}d$ follows by the preceding paragraph. Hence $cac^{-1}d \le p$ which implies that $c^{-1}d \le a^{-1}c^{-1}p$ since G is a po-group. Next let $u, v \in R$ be such that $a^{-1}b \le u^{-1}v$ and $c^{-1}d \le u^{-1}v$. Hence $b^{-1}au^{-1}v, d^{-1}cu^{-1}v \in R$ which implies

$$(cb)(b^{-1}au^{-1}v) = (cac^{-1}d)(d^{-1}cu^{-1}v) \in cbR \cap cac^{-1}dR = pR$$

and thus $p^{-1}(cb)(b^{-1}au^{-1}v) \in R$. It follows that

$$(a^{-1}c^{-1}p)^{-1}u^{-1}v = p^{-1}cau^{-1}v = (p^{-1}cb)(b^{-1}au^{-1}v) \in R$$

and hence $a^{-1}c^{-1}p \le u^{-1}v$. This establishes formula (2).

This suffices to conclude that G is an l-group. For the meet of any two elements $x, y \in G$ is given by

$$x \wedge y = (x^{-1} \vee y^{-1})^{-1}$$

and compatibility of both join and meet with multiplication follows from the fact that G is a po-group.

Our last functor has the following form.

X.2.10 Notation

For every $S \in \operatorname{Ob} \mathfrak{B}_l$, let $Q(S) = S/\sigma$ with the relation

$$a\sigma \le b\sigma \quad \text{if} \quad a^{-1}b\sigma r \quad \text{for some} \quad r\mathcal{R}1.$$

Also for every $\varphi \in \operatorname{Hom} \mathfrak{B}_l$, say $\varphi : S \to T$, let

$$Q(\varphi) : a\sigma \to a\varphi\sigma \qquad (a \in S).$$

X.2.11 Lemma

With the above notation, Q is a functor from $\mathfrak{B}_l$ to $\mathfrak{G}_l$.

Proof. Let $S \in \mathrm{Ob}\,\mathfrak{B}_l$ and let R be the $\mathfrak{R}$-class of the identity of S. In view of 1.15, we may represent S as $R^{-1} \circ R$. The dual of III.7.2 implies that $\sigma|_R = \varepsilon$, the equality relation on R. One verifies directly that for any $a, b \in R$, we have $[1, a][a, b] = [1, b]$ which implies $[a, b]\sigma = ([1, a]\sigma)^{-1}([1, b]\sigma) \in (R\sigma)^{-1}(R\sigma)$.

In view of 2.3, we conclude that $R\sigma$ satisfies all the hypotheses imposed on R in 2.9, and hence $Q(S)$ becomes an l-group under the partial order

$$a\sigma \le b\sigma \Leftrightarrow (a\sigma)^{-1}(b\sigma) \in R\sigma$$

which is evidently equivalent to

$$a\sigma \le b\sigma \Leftrightarrow a^{-1}b\sigma r$$

for some $r \in R$. Therefore $Q(S) \in \mathrm{Ob}\,\mathfrak{G}_l$.

Next let $\varphi \in \mathrm{Hom}\,\mathfrak{B}_l$, say $\varphi : S \to T$. If $a, b \in S$ are such that $a\sigma b$, then $(a\varphi)\sigma(b\varphi)$. This shows that $Q(\varphi)$ is single valued. It follows easily that $Q(\varphi)$ is an o-homomorphism.

The remaining axioms of a functor present no difficulty.

So far we have obtained the following diagram of functors.

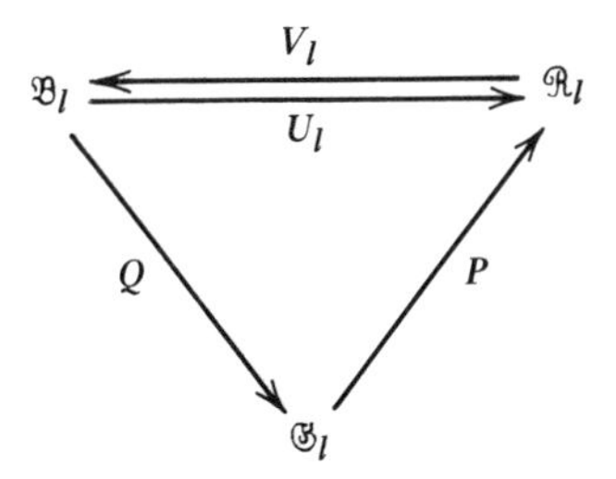

We will show next that the functors $I_{\mathfrak{G}_l}$ and QV_lP and also the functors $I_{\mathfrak{R}_l}$ and PQV_l are equivalent.

X.2.12 Lemma

For every $G \in \mathrm{Ob}\,\mathfrak{G}_l$, define a mapping $\pi(G)$ by

$$\pi(G) : a^{-1}b \to [a, b]\sigma \qquad (a, b \in G_+).$$

Then π is a natural equivalence of the functors $I_{\mathfrak{G}_l}$ and QV_lP.

Proof. Let $G \in \text{Ob}\,\mathfrak{G}_l$ and let $\zeta = \pi(G)$. For any $x \in G$, we have

$$x = (x \wedge 1)(x \vee 1) = (x^{-1} \vee 1)^{-1}(x \vee 1)$$

so that $x = a^{-1}b$ with $a, b \in G_+$. Hence ζ is defined on all of G.

Assume that $a^{-1}b = c^{-1}d$. Let $p = b \vee d$ so that $p \in G_+$. Then

$$\begin{aligned}[a, b][p, p] &= [(p * b)a, (b * p)p] = [(pb^{-1} \vee 1)a, b \vee p] \\ &= [(p \vee b)b^{-1}a, b \vee p] = [(p \vee d)d^{-1}c, d \vee p] \\ &= [(pd^{-1} \vee 1)c, d \vee p] = [(p * d)c, (d * p)p] \\ &= [c, d][p, p]\end{aligned}$$

and thus $[a, b]\sigma[c, d]$, that is $[a, b]\sigma = [c, d]\sigma$. Consequently, ζ is single valued.

Next let $a, b, c, d \in G_+$. First note that

$$c * b = [(c * b)b]b^{-1} = (c \vee b)b^{-1} = cb^{-1} \vee 1$$

and analogously $b * c = bc^{-1} \vee 1$. Taking this into account, we obtain

$$\begin{aligned}(a^{-1}b)\zeta(c^{-1}d)\zeta &= [a, b]\sigma[c, d]\sigma = ([a, b][c, d])\sigma \\ &= [(c * b)a, (b * c)d]\sigma = \left([(c * b)a]^{-1}[(b * c)d]\right)\zeta \\ &= \left[a^{-1}(c * b)^{-1}(b * c)d\right]\zeta \\ &= \left[a^{-1}(bc^{-1} \wedge 1)(bc^{-1} \vee 1)d\right]\zeta \\ &= \left[a^{-1}(bc^{-1})d\right]\zeta = [(a^{-1}b)(c^{-1}d)]\zeta\end{aligned}$$

and ζ is a homomorphism. If $(a^{-1}b)\zeta = 1\zeta$, then $[a, b]\sigma[1, 1]$ which implies that $[a, b]$ is idempotent, so that $a = b$. Thus ζ is one-to-one; it is obvious that it is onto. Therefore ζ is an isomorphism of G onto $QV_lP(G)$.

It remains to establish the commutativity of the diagram

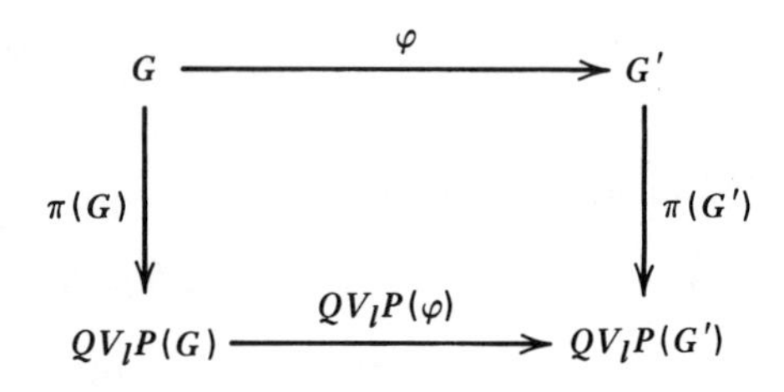

Indeed, for any $a, b \in G_+$,

$$(a^{-1}b)\varphi[\pi(G')] = ((a\varphi)^{-1}(b\varphi))[\pi(G)] = [a\varphi, b\varphi]\sigma$$
$$= [a, b]\sigma[QV_lP(\varphi)] = (a^{-1}b)[\pi(G)][QV_lP(\varphi)].$$

X.2.13 Lemma

For every $R \in \text{Ob}\,\mathfrak{R}_l$, define a mapping $\theta(R)$ by

$$\theta(R) : r \to [1, r]\sigma \qquad (r \in R).$$

Then θ is a natural equivalence of the functors $I_{\mathfrak{R}_l}$ and PQV_l.

Proof. Let $R \in \text{Ob}\,\mathfrak{R}_l$. The mapping $\theta(R)$ is obviously a homomorphism of R onto $PQV_l(R)$, and it is one-to-one by the argument at the beginning of the proof of 2.11. The diagram

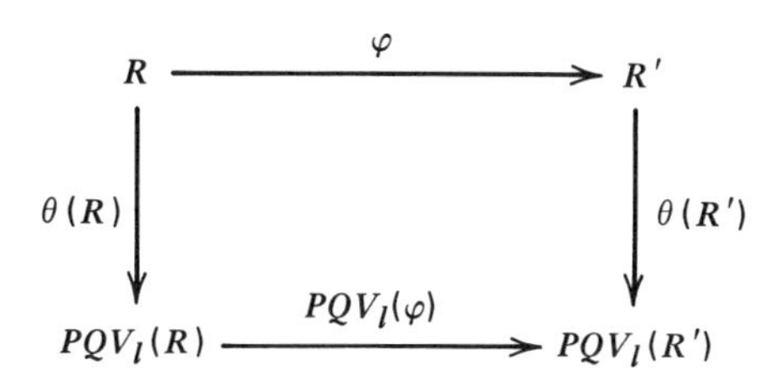

commutes, since for any $r \in R$,

$$r\varphi[\theta(R')] = [1, r\varphi]\sigma = [1, r]\sigma[PQV_l(\varphi)] = r[\theta(R)][PQV_l(\varphi)].$$

From 2.8 and 2.11–2.13 follows the principal result of this section.

X.2.14 Theorem

The quadruple (P, QV_l, π, θ) is an equivalence of the categories $\mathfrak{G}_l$ and $\mathfrak{R}_l$.

Combining this with 2.3 we get the following result.

X.2.15 Corollary

The categories $\mathfrak{V}_l$, $\mathfrak{R}_l$, and $\mathfrak{G}_l$ are equivalent.

Since $V(G_+)$ is an E-unitary inverse semigroup, we can construct the Q-representation for it (see VI.8.14).

X.2.16 Proposition

Let G be an l-group. Denote by $G_{\vee}$ the semilattice under $\vee$ on G_{+}. For any $a, b \in G_{+}$, let

$$\psi_{a^{-1}b}: d \to db^{-1}a \qquad (d \in \mathbf{d}\psi_{a^{-1}b} = \{c \in G_{\vee} | c \geq b\}).$$

Then $V(G_{+}) \cong Q(G_{\vee}, G; \psi)$ where $\psi: g \to \psi_g$ $(g \in G)$.

Proof. Define a function χ by

$$\chi: [a, b] \to (a, a^{-1}b) \qquad ([a, b] \in V(G_{+})).$$

From the proof of VI.7.11, we easily conclude that χ is an isomorphism of $V(G_{+})$ onto $Q(G_{\vee}, G; \psi)$; for we may observe that

$$\chi: [a, b] \to \left([a, b][a, b]^{-1}, [a, b]\sigma\right) = ([a, a], [a, b]\sigma)$$

and the latter can be identified with $(a, a^{-1}b)$. For the corresponding $\psi_{a^{-1}b}$, we have

$$\psi_{a^{-1}b}: [c, d]^{-1}[c, d] \to [c, d][c, d]^{-1}$$

(see VI.7.10), that is, we can write $\psi_{a^{-1}b}: d \to c$ if $[c, d]\sigma[a, b]$. The latter is equivalent to $c^{-1}d = a^{-1}b$ and thus $d = ca^{-1}b$ as in the statement of the proposition.

Note that in

$$S = Q(G_{\vee}, G; \psi) = \{(c, a^{-1}b) \in G_{\vee} \times G | c \geq a\}$$

the subset

$$\{(a, a^{-1}b) | a, b \in G_{\vee}\}$$

coincides with the set of all greatest elements of the different σ-classes. Hence S is an F-inverse semigroup. In the F-representation

$$S \cong F(G_{\vee}, G).$$

X.2.17 Exercises

(i) Let $R \in \operatorname{Ob} \mathfrak{R}$ be combinatorial. For each idempotent e in $S = V(R)$, let

$$P_e = \{a \in S | aa^{-1} = e, a = ae\}.$$

Show that for all $r \in R$, $rR = Rr$ if and only if for all $e \in E_S$ and all $a \in P_e$, $aP_e = P_e a$.

(ii) Show that if $R \in \text{Ob}\,\mathfrak{R}_l$, then R is embeddable into a group.

(iii) Let G and H be l-groups. Show that l-homomorphisms of G into H coincide with the homomorphisms $\varphi : G \to H$ for which $G_+\varphi \subseteq H_+$. Also show that every homomorphism of G_+ into H_+ is uniquely extendable to an l-homomorphism of G into H.

Reilly [5] studied l-bisimple inverse semigroups which arise in a manner that generalizes our procedure in passing from the positive cone of an l-group to a bisimple inverse monoid. This paper contains further results on this subject as well as the proof of 2.1. For a part of 2.8, check with Birkhoff [1].

X.3 NORMAL CONGRUENCES

For an object R of the category $\mathfrak{R}$, we will establish isomorphisms of the lattices of certain subsets of R, called here $*$-filters, and of filters of $E_{V(R)}$ locally self-conjugate in $V(R)$. The latter lattice will then be shown isomorphic to the lattice of normal congruences on $E_{V(R)}$.

With these isomorphisms, the three lattices are then isomorphic. This characterizes the lattice of normal congruences on $E_{V(R)}$ by means of subsets of both R and $E_{V(R)}$.

We need several new concepts. Recall the operation $*$ from 1.9.

X.3.1 Definition

A nonempty subset F of a semigroup S is a *filter* if for all $a, b \in S$, $ab \in F$ if and only if $a, b \in F$. If $R \in \text{Ob}\,\mathfrak{R}$, then a filter Q of R is a *$*$-filter* if for all $q \in Q$, $r \in R$, $q * r \in Q$.

A nonempty subset K of an inverse semigroup S is *locally self-conjugate* if for any $x \in S$, $xKx^{-1} \cap K \neq \varnothing$ implies $xKx^{-1} \subseteq K$.

The main theorem is preceded by several lemmas which also provide the necessary notation.

X.3.2 Lemma

Let $R \in \text{Ob}\,\mathfrak{R}$ and F be a filter of $E_{V(R)}$ locally self-conjugate in $V(R)$. Then

$$Q_F = \{ r \in R | [r, r] \in F \}$$

is a $*$-filter of R.

Proof. If $a, b \in Q = Q_F$, then

$$[ab, ab] = [b, 1][a, a][b, 1]^{-1} \in F$$

since $[b,1][b,1]^{-1} = [b, b] \in F$, and thus $ab \in Q$. Let $ab \in Q$. Then

$$[a, a] = [1,1][a, a][1,1] = [1, b]([b,1][a, a][1, b])[b,1]$$

$$= [1, b][ab, ab][1, b]^{-1} \in F$$

since $[1, b][1, b]^{-1} = [1,1] \in F$ and $[ab, ab] \in F$, and thus $a \in Q$. Further,

$$[b, b][ab, ab] = [b \vee ab, b \vee ab] = [ab, ab] \in F$$

whence $[b, b] \in F$ and $b \in Q$. Thus Q is a filter of R.

Next let $q \in Q$, $r \in R$. Then $[1, r][q, q][1, r]^{-1} \in F$ since $[1, r][1, r]^{-1} \in F$ which gives

$$[1, r][q, q][1, r]^{-1} = [a(q * r), (r \vee q) * r] \in F$$

where $a = r * (r \vee q)$. Hence $a(q * r) \in Q$ and since Q is a filter, we get $q * r \in Q$.

X.3.3 Lemma

Let $R \in \mathrm{Ob}\,\mathfrak{R}$ and Q be a $*$-filter of R. Then

$$F_Q = \{[r, r] \| r \in Q\}$$

is a filter of $E_{V(R)}$ locally self-conjugate in $V(R)$.

Proof. Let $[a, a], [b, b] \in F = F_Q$. Then $a, b \in Q$ whence $a \vee b = (a * b)b \in Q$ which gives

$$[a, a][b, b] = [(b * a)a, (a * b)b] = [a \vee b, a \vee b] \in F.$$

Next let $[a, a][b, b] \in F$. Then $[a \vee b, a \vee b] \in F$ so that $(a * b)b = a \vee b \in Q$ and hence $b \in Q$. Thus $[b, b] \in F$ and analogously $[a, a] \in F$.

Assume that $[a, b][a, b]^{-1}, [c, c] \in F$. Then $[a, b][b, a] = [a, a] \in F$. Thus $a, c \in Q$. We compute

$$[a, b][c, c][b, a] = [(c * b)a, (b * c)c][b, a]$$

$$= [(b * (b \vee c))(c * b)a, ((b \vee c) * b)a]. \qquad (3)$$

Now $(b * (b \vee c))(b \vee c) = b \vee (b \vee c) = b \vee c$ so that $b * (b \vee c) = 1$. Further, $c * b \in Q$ and thus also $(c * b)a \in Q$. Since the element in (3) is idempotent, we obtain that $[a, b][c, c][b, a] \in F$, which was to be proved.

X.3.4 Lemma

Let $R \in \mathrm{Ob}\,\mathfrak{R}$ and Q be a $*$-filter of R. Then τ_Q defined on $E_{V(R)}$ by

$$[a, a]\tau_Q[b, b] \Leftrightarrow Qa \cap Qb \neq \varnothing$$

is a normal congruence on $E_{V(R)}$.

Proof. It is clear that $\tau = \tau_Q$ is reflexive and symmetric. Let $[a, a]\tau[b, b]$ and $[b, b]\tau[c, c]$. Then $pa = qb$ and $rb = sc$ for some $p, q, r, s \in Q$. Let $x = r * q$ and $y = q * r$. Then $x, y \in Q$, $xq = r \vee q = yr$ and thus

$$(xp)a = xqb = yrb = (ys)c$$

with $xp, ys \in Q$. Consequently, $[a, a]\tau[c, c]$, and τ is transitive, and hence is an equivalence relation.

In order to prove normality of τ, we let $[a, a]\tau[b, b]$ and $x, y \in R$. Then

$$\begin{aligned}[y, x][a, a][x, y] &= [(a * x)y, x \vee a][x, y] \\ &= [(x*(x \vee a))(a * x)y, ((x \vee a)* x)y],\end{aligned}$$

where

$$(x*(x \vee a))(x \vee a) = x \vee (x \vee a) = x \vee a$$

whence $x*(x \vee a) = 1$. Hence it suffices to show that

$$[(a * x)y, (a * x)y]\,\tau[(b * x)y, (b * x)y]. \tag{4}$$

Since $[a, a]\tau[b, b]$, we have $pa = qb$ for some $a, b \in Q$. We also have $pa \vee x \in Ra \cap Rx = R(a \vee x)$ so $pa \vee x = p'(a \vee x)$ for some $p' \in R$. To show that $p' \in Q$, let $q = p*(x * a)$ so that $q \in Q$. Now $q(x \vee a) = q(a \vee x) \in Rx$ and

$$\begin{aligned}q(x \vee a) &= (p*(x * a))(x * a)a = (p \vee (x * a))a \\ &= ((x * a) \vee p)a \in Rpa,\end{aligned}$$

so that

$$q(x \vee a) \in Rpa \cap Rx = R(pa \vee x) = Rp'(a \vee x).$$

By right cancellation in R, we get $q = up'$ for some $u \in R$, and therefore $p' \in Q$. Analogously, we obtain $qb \vee x = q'(b \vee x)$ for some $q' \in Q$. It follows that

$$p'(a \vee x) = pa \vee x = qb \vee x = q'(b \vee x)$$

whence $p'(a*x)x = q'(b*x)x$ so that

$$p'(a*x)y = q'(b*x)y \in Q(a*x)y \cap Q(b*x)y.$$

This proves (4), which in turn shows that τ is normal. Note that normality implies that τ is a congruence.

X.3.5 Lemma

Let $R \in \operatorname{Ob} \mathfrak{R}$ and τ be a normal congruence on $E_{V(R)}$. Then

$$Q_\tau = \{r \in R | [r, r]\tau[1,1]\}$$

is a $*$-filter of R.

Proof. Let $a, b \in Q = Q_\tau$. Then $[a, a]\tau[1,1]$, $[b, b]\tau[1,1]$ and hence

$$[b,1][a, a][1, b]\tau[b,1][1, b] = [b, b]\tau[1,1].$$

Since

$$[b,1][a, a][1, b] = [(a*1)b, (1*a)a][1, b]$$

$$= [ab, a][1, b] = [(1*a)ab, (a*1)b]$$

$$= [ab, ab]$$

we deduce that $ab \in Q$.

Now let $ab \in Q$. By the above,

$$[a, a] = [1,1][a, a][1,1] = [1, b]([b,1][a, a][1, b])[b,1]$$

$$= [1, b][ab, ab][b,1]\tau[1, b][b,1] = [1,1]$$

and $a \in Q$. Further,

$$[b, b] = [b,1][1, b]\rho[b,1][a, a][1, b] = [ab, ab]\tau[1,1]$$

and hence also $b \in Q$. Thus Q is a filter of R.

Next let $q \in Q$, $r \in R$. Then, on the one hand,

$$[1, r][q, q][r,1]\tau[1, r][r,1] = [1,1]$$

and on the other hand,

$$[1, r][q, q][r,1] = [a(q*r), (r \vee q)*r]\tau[1,1]$$

where $a = r*(r \vee q)$. But then $[a(q*r), a(q*r)]\tau[1,1]$ whence $a(q*r) \in Q$. Since Q is a filter, we get $q*r \in Q$.

X.3.6 Notation

The order in the following sets is always the inclusion relation.

Fix $R \in \text{Ob}\,\mathfrak{R}$. Let $\mathcal{Q}$ denote the set of all $*$-filters of R. Let $\mathcal{F}$ be the set of all filters of $E_{V(R)}$ locally self-conjugate in $V(R)$. Finally, let $\mathcal{N}$ be the set of all normal congruences on $E_{V(R)}$.

It is verified readily that each of the sets $\mathcal{Q}$, $\mathcal{F}$, $\mathcal{N}$ is closed under intersection and has a greatest element, and is thus a complete lattice by I.2.6.

With the notation introduced in 3.1–3.6, we prove the following result in the proof of which we tacitly use I.2.8.

X.3.7 Theorem

Let $R \in \text{Ob}\,\mathfrak{R}$. The mappings

$$F \to Q_F \qquad (F \in \mathcal{F}), \qquad Q \to F_Q \qquad (Q \in \mathcal{Q})$$

are mutually inverse lattice isomorphisms of $\mathcal{F}$ and $\mathcal{Q}$. The mappings

$$Q \to \tau_Q \qquad (Q \in \mathcal{Q}), \qquad \tau \to Q_\tau \qquad (\tau \in \mathcal{N})$$

are mutually inverse isomorphisms of $\mathcal{Q}$ and $\mathcal{N}$.

Proof. The first part is an obvious consequence of 3.2 and 3.3. For the second part, 3.4 implies that the first function maps $\mathcal{Q}$ into $\mathcal{N}$ and 3.5 implies that the second function maps $\mathcal{N}$ into $\mathcal{Q}$. It remains to establish that they are mutually inverse functions.

Let $\tau \in \mathcal{N}$. Then

$$\begin{aligned} [a, a]\tau_{Q_\tau}[b, b] &\Leftrightarrow Q_\tau a \cap Q_\tau b \neq \varnothing \\ &\Leftrightarrow pa = qb \quad \text{for some } p, q \in Q_\tau \\ &\Leftrightarrow pa = qb \quad \text{for some } [p, p]\tau[q, q]\tau[1,1]. \qquad (5) \end{aligned}$$

If (5) holds, then

$$\begin{aligned} [a, a] &= [a,1][1, a]\tau[a,1][p, p][1, a] = [pa, pa] \\ &= [qb, qb] = [b,1][q, q][1, b]\tau[b,1][1, b] = [b, b]. \end{aligned}$$

Conversely, assume that $[a, a]\tau[b, b]$. Then

$$[1, b][a, a][b,1]\tau[1, b][b, b][b,1] = [1,1] \qquad (6)$$

and

$$[1,b][a,a][b,1] = [a*b, b \vee a][b,1]$$
$$= [(b*(b \vee a))(a*b), (b \vee a)*b]. \qquad (7)$$

Further, $((b \vee a)*b)b = (b \vee a) \vee b = b \vee a = (a*b)b$ so by right cancellation, $(b \vee a)*b = a*b$. In view of (6) and (7), we get $[a*b, a*b]\tau[1,1]$ and thus $a*b \in Q_\tau$. By symmetry, we also have $b*a \in Q$, which together with

$$(a*b)b = a \vee b = (b*a)a$$

implies that (3) holds. Consequently $\tau_{Q_\tau} = \tau$.

Now let $Q \in \mathfrak{Q}$. Then

$$a \in Q_{\tau_Q} \Leftrightarrow [a,a]\tau_Q[1,1] \Leftrightarrow Qa \cap Q \neq \varnothing \Leftrightarrow a \in Q$$

so that $Q_{\tau_Q} = Q$.

X.3.8 Corollary

Let $R \in \text{Ob}\,\mathfrak{R}$. The mappings

$$F \to \tau_{Q_F} \quad (F \in \mathfrak{F}), \qquad \tau \to F_{Q_\tau} \quad (\tau \in \mathfrak{N})$$

are mutually inverse lattice isomorphisms of $\mathfrak{F}$ and $\mathfrak{N}$. For $F \in \mathfrak{F}$, we have

$$[a,a]\tau_{Q_F}[b,b] \Leftrightarrow [a,1]F[1,a] \cap [b,1]F[1,b] \neq \varnothing,$$

and for $\tau \in \mathfrak{N}$,

$$F_{Q_\tau} = \{[a,a] \in E_{V(R)} | [a,a]\tau[1,1]\}.$$

Proof. The first part follows directly from 3.7. The second and third parts amount to a computation of τ_{Q_F} and F_{Q_τ}.

X.3.9 Exercises

(i) Let $R \in \text{Ob}\,\mathfrak{R}$ and Q be a filter of R. Show that Q is a $*$-filter of R if and only if $q \in Q$ and $Rq \cap Rb = Rc$ imply $c \in Qb$.

Most of this section is a simplified version of some of the material in Reilly [5]. Schein [15] proved that any (nonzero) class of a congruence on a (0-)bisimple inverse semigroup uniquely determines the congruence.

X.4 IDEMPOTENT SEPARATING CONGRUENCES

For an object R of the category $\mathfrak{R}$, several isomorphisms of lattices related to R and $S = R^{-1} \circ R$ are established here. The first is an isomorphism of the lattice of left normal divisors of R onto a certain lattice of congruences on R. The second is an isomorphism of the latter lattice onto the lattice of idempotent separating congruences on S. For a congruence τ on R, the quotient semigroup R/τ is compared with the congruence on S which corresponds to τ in the above isomorphisms.

X.4.1 Definition

A subgroup N of the group of units of $R \in \text{Ob}\,\mathfrak{R}$ is a *left normal divisor* of R if $rN \subseteq Nr$ for all $r \in R$.

X.4.2 Notation

Fix $R \in \text{Ob}\,\mathfrak{R}$ and let

$\mathcal{R}_d$ be the lattice of all left normal divisors of R ordered by inclusion,

$\mathcal{R}_c$ be the lattice of all congruences τ on R such that $\tau \subseteq \mathfrak{L}$ and R/τ is right cancellative.

For each $N \in \mathcal{R}_d$, define a relation τ_N on R by

$$a\tau_N b \Leftrightarrow a = hb \quad \text{for some} \quad h \in N.$$

The first pair of isomorphisms is contained in the next statement.

X.4.3 Proposition

The mappings

$$N \to \tau_N \qquad (N \in \mathcal{R}_d), \qquad \tau \to 1\tau \qquad (\tau \in \mathcal{R}_c)$$

are mutually inverse lattice isomorphisms of $\mathcal{R}_d$ and $\mathcal{R}_c$.

Proof. Let $N \in \mathcal{R}_d$. Routine verification shows that τ_N is a right congruence. Let $a\tau_N b$ and $c \in R$. Then $a = hb$ for some $h \in N$, so that $ch = gc$ for some $g \in N$. Hence

$$ca = c(hb) = (ch)b = (gc)b = g(cb)$$

and thus $ca\tau_N cb$. Consequently, τ_N is a congruence on R. Continuing with this notation, $a = hb$ implies $b = h^{-1}a$ which shows that $a\mathfrak{L}b$. Thus $\tau_N \subseteq \mathfrak{L}$. Now assume that $ax\tau_N bx$. Then $ax = ubx$ for some $u \in N$ whence $a = ub$ and $a\tau_N b$. Therefore, $\tau_N \in \mathcal{R}_c$. Clearly, the mapping $N \to \tau_N$ is order preserving.

Now let $\tau \in \mathfrak{R}_c$ and let $N = 1\tau$. Let $a \in N$; then $a\tau 1$ so $a\mathfrak{L}1$ and hence $xa = 1$. But then $axa = a$ so by right cancellation, $ax = 1$. Hence a is a unit, so that $N \subseteq H_1$, the group of units of R. Since $\tau|_{H_1}$ is a congruence, N is a (normal) subgroup of H_1. Let $a \in R$ and $h \in N$. Then $ah\tau a$ so $ah\mathfrak{L}a$ whence $ah = xa$ for some $x \in R$. It follows that $xa\tau a$ so that $x\tau 1$, that is $x \in N$. Consequently, $aN \subseteq Na$ and thus $N \in \mathfrak{R}_d$. Obviously the mapping $\tau \to 1\tau$ is order preserving.

It follows easily that for every $N \in \mathfrak{R}_d$, we have $N = 1\tau_N$. Next let $\tau \in \mathfrak{R}_c$. Assume that $a\tau b$. Then $a\mathfrak{L}b$ so $a = xb$ for some $x \in R$, which implies $xb\tau b$ so that $x\tau 1$. Hence $x \in 1\tau$ which proves that $\tau \subseteq \tau_{1\tau}$. Conversely, let $a\tau_{1\tau}b$. Then $a = hb$ for some $h \in 1\tau$. Hence $h\tau 1$ and thus $a\tau b$. Consequently, $\tau_{1\tau} \subseteq \tau$ and the equality prevails. This proves that the mappings in the statement of the proposition are bijections and thus mutually inverse lattice isomorphisms (see I.2.8).

X.4.4 Notation

Let R be as in 4.2 and $S = V(R)$. Let $\mathfrak{B}_s$ be the lattice of all idempotent separating congruences on S. For every $\rho \in \mathfrak{B}_s$, define $U(\rho)$ on R by

$$aU(\rho)b \Leftrightarrow [1, a]\rho[1, b].$$

For every $\tau \in \mathfrak{R}_c$, define $V(\tau)$ on S by

$$[a, b]V(\tau)[c, d] \Leftrightarrow a = uc,\ b = vd,\ u^{-1}v\tau 1 \text{ for some units } u, v \text{ in } R.$$

X.4.5 Theorem

The mappings

$$\rho \to U(\rho) \qquad (\rho \in \mathfrak{B}_s), \qquad \tau \to V(\tau) \qquad (\tau \in \mathfrak{R}_c)$$

are mutually inverse lattice isomorphisms of $\mathfrak{B}_s$ and $\mathfrak{R}_c$.

Proof. Let $\rho \in \mathfrak{B}_s$. It is clear that $U(\rho)$ is a congruence on R. Let $aU(\rho)b$. Then $[1, a]\rho[1, b]$ and thus $[1, a]\mathfrak{H}[1, b]$ since ρ is idempotent separating. Hence $[1, a] = [u, v][1, b]$ and $[1, b] = [x, y][1, a]$ for some $u, v, x, y \in R$. Consequently,

$$[1, a] = [u, v][1, b] = [u, v][x, y][1, a],$$

and multiplying on the right by $[a, 1]$, we obtain $[1, 1] = [u, v][x, y]$. Analogously, starting with $[1, b]$, we obtain $[1, 1] = [x, y][u, v]$. This means that both

$[u, v]$ and $[x, y]$ are units, and we may write $[u, v] = [1, c]$ and $[x, y] = [1, c^{-1}]$ where c is a unit in R. But then $a = cb$ and $b = c^{-1}a$ so that $a\mathfrak{L}b$. Consequently, $U(\rho) \subseteq \mathfrak{L}$.

Now assume that $axU(\rho)bx$ for some $a, b, x \in R$. Then $[1, a][1, x]$ $\rho[1, b][1, x]$ and multiplying on the right by $[x, 1]$, we get $[1, a]\rho[1, b]$, and thus $aU(\rho)b$. Consequently, $R/U(\rho)$ is right cancellative, which shows that $U(\rho) \in \mathfrak{R}_c$. It is evident that the mapping $\rho \to U(\rho)$ is order preserving.

Now let $\tau \in \mathfrak{R}_c$. We first verify that $V(\tau)$ is well defined. Hence let $[a, b]V(\tau)[c, d]$ and $[a, b] = [a', b']$, $[c, d] = [c', d']$. Then

$$a = uc, \qquad b = vd, \qquad a = sa',$$

$$b = sb', \qquad c = tc', \qquad d = td',$$

for some units u, v, s, t such that $u^{-1}v\tau 1$. Hence

$$a' = (s^{-1}ut)c', \qquad b' = (s^{-1}vt)d',$$

$$(s^{-1}ut)^{-1}(s^{-1}vt) = t^{-1}u^{-1}(ss^{-1})vt = t^{-1}u^{-1}vt\tau t^{-1}t = 1$$

which shows that $[a', b']V(\tau)[c', d']$. Consequently, $V(\tau)$ is well defined.

Clearly, $V(\tau)$ is reflexive, and it follows easily that it is also symmetric. Let $[a, b]V(\tau)[c, d]$ and $[c, d]V(\tau)[e, f]$. Then $a = uc$, $b = vd$, $c = se$, and $d = tf$ for some units u, v, s, t such that $u^{-1}v\tau s^{-1}t\tau 1$. Then $a = use$ and $b = vtf$ with

$$(us)^{-1}(vt) = s^{-1}(u^{-1}v)t\tau s^{-1}t\tau 1.$$

Hence $[a, b]V(\tau)[e, f]$, and thus $V(\tau)$ is also transitive.

Let again $[a, b]V(\tau)[c, d]$ with $a = uc$, $b = vd$, $u^{-1}v\tau 1$. Let $[x, y] \in R^{-1} \circ R$. Then

$$[a, b][x, y] = [(x * b)a, (b * x)y],$$

$$[c, d][x, y] = [(x * d)c, (d * x)y].$$

Further, since v is a unit, we have $Rvd = Rd$ whence $(x * vd)vd = (x * d)d$, so by right cancellation, we get $(x * vd)v = x * d$. In addition, $\tau \in \mathfrak{R}_c$ and in view of 4.3, $N = 1\tau$ is a left normal divisor of R. Since $v^{-1}u \in N$, there exists $w \in N$ such that $(x * d)v^{-1}u = w(x * d)$. Consequently,

$$(x * b)a = (x * vd)uc = (x * d)(v^{-1}u)c = w(x * d)c.$$

Furthermore, $b * x = vd * x = d * x$ and thus

$$[a, b][x, y] = [w(x * d)c, (d * x)y]$$

from which follows

$$[a, b][x, y]V(\tau)[c, d][x, y]$$

since $w\tau 1$. A similar argument can be used to show that

$$[x, y][a, b]V(\tau)[x, y][c, d].$$

Consequently, $V(\tau)$ is a congruence on S. A simple verification shows that $V(\tau) \subseteq \mathcal{H}$. Therefore, $V(\tau) \in \mathfrak{B}_s$. It follows immediately from the definition of $V(\tau)$ that the function $\tau \to V(\tau)$ preserves order.

We let $\rho \in \mathfrak{B}_s$ and will show that $VU(\rho) = \rho$. Let $[a, b]VU(\rho)[c, d]$. Then $a = uc$ and $b = vd$ for some units u and v such that $u^{-1}vU(\rho)1$, that is $[1, u^{-1}v]\rho[1, 1]$. Hence

$$[c, d] = [c, 1][1, d]\rho[c, 1][1, u^{-1}v][1, d] = [c, 1][1, u^{-1}vd]$$

$$= [c, u^{-1}vd] = [uc, vd] = [a, b],$$

which proves that $VU(\rho) \subseteq \rho$.

Conversely, let $[a, b]\rho[c, d]$. Since $\rho \subseteq \mathcal{H}$, we obtain easily that $a = uc$ and $b = vd$ for some units u and v. Since

$$[a, b] = [uc, vd] = [c, u^{-1}vd]$$

we have

$$[1, c][c, u^{-1}vd][d, 1]\rho[1, c][c, d][d, 1] = [1, 1]. \tag{8}$$

The left member is equal to

$$[c * c, (c * c)u^{-1}vd][d, 1] = [1, u^{-1}vd][u^{-1}vd, u^{-1}v] = [1, u^{-1}v]$$

which together with (8) gives $[1, u^{-1}v]\rho[1, 1]$. Consequently, $u^{-1}vU(\rho)1$, which implies that $[a, b]VU(\rho)[c, d]$. Therefore, $\rho \subseteq VU(\rho)$ and the equality prevails.

For $\tau \in \mathfrak{R}_c$, one verifies without difficulty that $UV(\tau) = \tau$. From the established properties for U and V it now follows by I.2.8 that they have all the properties claimed in the statement of the theorem.

X.4.6 Corollary

Let $S \in \text{Ob}\,\mathfrak{B}$ and R be the $\mathcal{R}$-class of S containing the identity. If ρ is an idempotent separating congruence on S, then $\tau = \rho|_R$ is a congruence on R such that (i) $\tau \subseteq \mathcal{L}$ and (ii) R/τ is right cancellative. Conversely, every congruence on R satisfying (i) and (ii) can be uniquely extended to an idempotent separating congruence on S.

Proof. This follows easily from 4.5.

X.4.7 Lemma

Let φ be a homomorphism of a semigroup S onto a semigroup T. Assume that the congruence ρ induced by φ is contained in $\mathcal{K}$, where $\mathcal{K}$ is one of the Green relations $\mathcal{L}, \mathcal{R}, \mathcal{H}$. If $(a\varphi)\mathcal{K}(b\varphi)$, then $a\mathcal{K}b$.

Proof. We prove the lemma for $\mathcal{K} = \mathcal{L}$; the case $\mathcal{K} = \mathcal{R}$ is symmetric; and the case $\mathcal{K} = \mathcal{H}$ follows from the two. Thus assume that $(a\varphi)\mathcal{L}(b\varphi)$. Then $a\varphi = (x\varphi)(b\varphi)$ and $b\varphi = (y\varphi)(a\varphi)$ for some $x, y \in S^1$. Hence $a\varphi = (xb)\varphi$ and $b\varphi = (ya)\varphi$ which says that $a\rho xb$ and $b\rho ya$. The hypothesis implies that $a\mathcal{L}xb$ and $b\mathcal{L}ya$. Thus, $a = wxb$ and $b = zya$ for some $w, z \in S^1$, which evidently implies that $a\mathcal{L}b$.

X.4.8 Proposition

Let $S \in \mathrm{Ob}\,\mathfrak{B}$, R be the $\mathcal{R}$-class of S containing the identity, ρ be an idempotent separating congruence on S, and $U(\rho) = \rho|_R$. Then

$$U(S/\rho) \cong U(S)/U(\rho).$$

Proof. The function $\rho^{\#}$ maps $U(S)$ onto $U(S/\rho)$ by 4.7 with $\mathcal{K} = \mathcal{R}$. The homomorphism $\rho^{\#}|_{U(S)} = (\rho|_{U(S)})^{\#}$ obviously induces $U(\rho)$ on $U(S)$. Thus $U(S/\rho) \cong U(S)/U(\rho)$.

X.4.9 Theorem

Let $R \in \mathrm{Ob}\,\mathfrak{R}$ and $\tau \in \mathcal{R}_c$. Then $R/\tau \in \mathrm{Ob}\,\mathfrak{R}$ and the mapping

$$\theta : [a, b] \to [a\tau, b\tau] \qquad ([a, b] \in V(R))$$

is a homomorphism of $V(R)$ onto $V(R/\tau)$ which induces $V(\tau)$.

Proof. The hypothesis on $\tau \in \mathcal{R}_c$ includes that $\overline{R} = R/\tau$ is right cancellative. Since R has an identity element, so does $\overline{R}$. In order to establish the property of principal left ideals, we will prove

$$\overline{R}(a\tau) \cap \overline{R}(b\tau) = \overline{R}(a \vee b)\tau. \tag{9}$$

First, $Ra \cap Rb = R(a \vee b)$ implies $a \vee b = pa$ for some $p \in R$. Hence $(a \vee b)\tau = (p\tau)(a\tau)$ so that $(a \vee b)\tau \in \overline{R}(a\tau)$. Similarly, $(a \vee b)\tau \in \overline{R}(b\tau)$, which proves one inclusion in (9).

Conversely, let $d\tau \in \overline{R}(a\tau) \cap \overline{R}(b\tau)$, say

$$d\tau = (q\tau)(a\tau) = (r\tau)(b\tau)$$

for some $q, r \in R$. It follows that $d\tau qa\tau rb$ and thus $d\mathcal{L}qa\mathcal{L}rb$ by the assumption on τ. Consequently, for some $u, v \in R$, $d = uqa = vrb$, that is to say, $d \in Ra \cap Rb$. Thus, $d = s(a \vee b)$ for some $s \in R$, and hence $d\tau \in \overline{R}(a \vee b)\tau$. This proves the other inclusion in (9).

In view of 4.7, we may take as a system of representatives of $\mathfrak{L}$-classes of $\overline{R}$ the images under $\tau^{\#}$ of the system of representatives of $\mathfrak{L}$-classes of R. Now (9) makes it possible to define the $\vee$-operation in $\overline{R}$ by

$$a\tau \vee b\tau = (a \vee b)\tau \qquad (a, b \in R). \tag{10}$$

In particular, we have proved that $\overline{R} \in \mathrm{Ob}\,\mathfrak{R}$. It also follows from the uniqueness (by right cancellation) of the element $a * b$ in $(a * b)b = a \vee b$ that (10) implies

$$a\tau * b\tau = (a * b)\tau \qquad (a, b \in R). \tag{11}$$

In order to see that θ is single valued, we first note that $\tau^{\#}$ maps units of R onto units of $\overline{R}$. Hence if $[a, b] = [c, d]$, then $a = uc$, $b = ud$ for some unit u of R, and thus

$$[a\tau, b\tau] = [(u\tau)(c\tau), (u\tau)(d\tau)] = [c\tau, d\tau].$$

That θ is a homomorphism is a consequence of (11); θ obviously maps $V(R)$ onto $V(\overline{R})$.

It remains to prove that θ induces $V(\tau)$. Let $[a, b]\theta = [c, d]\theta$. Then $[a\tau, b\tau] = [c\tau, d\tau]$, and hence

$$a\tau = (w\tau)(c\tau), \qquad b\tau = (w\tau)(d\tau)$$

for some unit $w\tau$ of $\overline{R}$. Thus $a\tau wc$ and $b\tau wd$ which by the hypothesis on τ gives $a\mathfrak{L}wc$ and $b\mathfrak{L}wd$. It follows from 1.8 that $a = xwc$ and $b = ywd$ for some units x, y. Since by 4.7, w also is a unit, and we get that both $u = xw$ and $v = yw$ are units. Further, $wc\tau a = xwc$ implies $x\tau 1$ by right cancellation, and analogously $y\tau 1$. It follows that

$$u^{-1}v = (xw)^{-1}(yw) = w^{-1}x^{-1}yw\tau w^{-1}w = 1,$$

which together with $a = uc$, $b = vd$, u and v units finally yields $[a, b]V(\tau)[c, d]$.

Conversely, assume that $[a, b]V(\tau)[c, d]$, so that $a = uc$, $b = vd$, with u and v units such that $w = u^{-1}v\tau 1$. Since $a\tau = (u\tau)(c\tau)$ and $b\tau = (v\tau)(d\tau) = (uw)\tau(d\tau) = (u\tau)(w\tau)(d\tau) = (u\tau)(d\tau)$, and $u\tau$ is a unit, we obtain $[a\tau, b\tau] = [c\tau, d\tau]$, and thus $[a, b]\theta = [c, d]\theta$.

Therefore, θ induces $V(\tau)$.

X.4.10 Corollary

Let $R \in \mathrm{Ob}\,\mathfrak{R}$. If $\tau \in \mathfrak{R}_c$, then $V(R/\tau) \cong V(R)/V(\tau)$, and if $\rho \in \mathfrak{B}_s$, then $V(R/U(\rho)) \cong V(R)/\rho$.

Proof. The first assertion is an immediate consequence of 4.9. The second assertion follows from the first in view of 4.8.

X.4.11 Exercises

(i) Let $S \in \operatorname{Ob} \mathfrak{B}$ and N be a subgroup of the group of units of S. Show that N is a left normal divisor of the $\mathcal{R}$-class R of the identity of S if and only if $rNr^{-1} \subseteq N$ for all $r \in R$.

(ii) Construct all nongroup congruences on a Reilly semigroup $B(G, \alpha)$.

(iii) Let K be an inverse subsemigroup of an inverse semigroup S. Show that

$$n_s(K) = \{ s \in S \mid s^{-1}K^1 s,\ sK^1 s^{-1} \subseteq K \}$$

is the greatest normal extension of K contained in S.

(iv) For a congruence ρ on a bisimple monoid S, let

$$K_\rho = \{ a \in S \mid a\rho 1 \},$$

$$T_\rho = \{ a \in S \mid aa^{-1}\rho a^{-1}a\rho 1 \},$$

$$Q_\rho = \{ a \in S \mid aa^{-1} = 1,\ a^{-1}a\rho 1 \}.$$

Prove the following statements.

(α) $T_\rho \cap \ker \rho = K_\rho$.

(β) $T_\rho = n_s(K_\rho)$ (see the preceding exercise).

(γ) $T_\rho \cong Q_\rho^{-1} \circ Q_\rho$.

(δ) $E_{K_\rho}\omega = 1\rho_{\min}$.

Now let ξ also be a congruence on S.

(ε) $\operatorname{tr} \rho = \operatorname{tr} \xi$ if and only if $T_\rho = T_\xi$.

(η) $\rho = \xi$ if and only if $K_\rho = K_\xi$.

(v) Let ρ be a congruence on a bisimple inverse semigroup S and let $e \in E_S$, $K = e\rho$. Prove the following statements.

(α) $a \in \ker \rho \Leftrightarrow xax^{-1} \in K$ for some $x \in S$ such that $a \mathcal{L} x \mathcal{R} e$.

(β) For $f, g \in E_S$, $f\rho g$ if and only if $yfy^{-1}, xgx^{-1} \in K$ for some $x, y \in S$ such that $f \mathcal{L} x \mathcal{R} e$, $g \mathcal{L} y \mathcal{R} e$.

X.4.12 Problems

In the following problems S is a bisimple inverse monoid and R is the $\mathcal{R}$-class of its identity.

(i) If a congruence ρ on R can be extended to all of S, is this extension unique?

(ii) Is every congruence on S uniquely determined by the sets

$$N = \{a \in S \mid a\mathcal{H}1, a\rho 1\},$$

$$Q = \{a \in S \mid a\mathcal{R}1, a^{-1}a\rho 1\}?$$

If so, characterize abstractly the pairs (N, Q).

(iii) Let ρ, ξ be congruences on S such that $\rho|_R \subseteq \xi|_R$. Does it follow that $\rho \subseteq \xi$? [A positive answer would solve problem (i).]

Idempotent separating congruences on a bisimple inverse monoid were characterized by Warne [2] in terms of left normal divisors. This was extended to bisimple inverse semigroups by Reilly–Clifford [1] and to regular 0-bisimple semigroups by Munn [8]. General congruences on bisimple inverse semigroups were constructed by Petrich–Reilly [2].

XI

ω-REGULAR SEMIGROUPS

We have seen in II.6 that Reilly semigroups can be faithfully represented as Bruck semigroups over groups. This provided us with a precise structure for bisimple ω-regular semigroups; the goal of this chapter is to give a complete description of the structure of all ω-regular semigroups. The purpose of the extensive discussion of these semigroups is to provide a model for what a structure theory for a class of semigroups may look like and supercedes the importance of specific results proved here.

For a given class $\mathcal{C}$ of semigroups the plan of operation may be the following (not necessarily in this order): (1) a classification of the semigroups in $\mathcal{C}$ into semigroups of possibly simpler structure, (2) a construction of semigroups in each of the classes in terms of sets, functions, groups, semilattices, and so on, (3) the relationships (categorical, varietal, structural, etc.) between the classes of $\mathcal{C}$ and ingredients figuring in these constructions, (4) homomorphisms among special semigroups so constructed, (5) congruences on these special semigroups, (6) the position of semigroups in $\mathcal{C}$ as subsemigroups of arbitrary semigroups, (7) homomorphisms of arbitrary semigroups onto semigroups in $\mathcal{C}$, (8) varieties generated by various subclasses of $\mathcal{C}$, (9) representations of semigroups in $\mathcal{C}$ by one-to-one partial transformations, and (10) amalgamation properties.

We have seen parts of this program carried out for Clifford semigroups in II.2, for Brandt semigroups in II.3, for strict inverse semigroups in II.4, for E-unitary inverse semigroups in VII, for free inverse semigroups in VIII and for monogenic inverse semigroups in IX. But for none of these classes is the theory as complete as for ω-regular semigroups.

1. Homomorphisms of one semigroup in its Reilly representation into another can be constructed explicitly by means of four parameters satisfying a single condition. Using this, the normal hull of a semigroup S in its Reilly representation can be constructed again in terms of a Reilly representation.

2. Four categories may be naturally defined in relation to Reilly semigroups: the category of Reilly semigroups and their monoid homomorphisms,

the category of right cancellative monoids whose principal left ideals form an ω-chain and their homomorphisms, the category whose objects are the ingredients in the Reilly representation with suitable morphisms, and a category whose objects and morphisms are similar to those of the preceding category. These categories are closely related.

3. The position of Reilly semigroups as subsemigroups of arbitrary semigroups is studied in some detail. For any two $\mathcal{D}$-related idempotents e and f of a semigroup S such that $e > f$, all Reilly subsemigroups of S having e and f as its two highest idempotents can be described. In this context, the notion of a normal divisor for an idempotent plays an important role.

4. Simple ω-regular semigroups are proved to be isomorphic to Bruck semigroups over finite chains of groups; the converse follows easily. The number of (maximal) groups in this finite chain is the number of $\mathcal{D}$-classes of the simple ω-regular semigroup, and each of these $\mathcal{D}$-classes is a Reilly semigroup.

5. General ω-regular semigroups can be classified as follows: simple, those with a proper kernel, those without a kernel. In the second case, we are dealing with a special ideal extension of a simple ω-regular semigroup by a finite chain of groups. The third case amounts to an ω-chain of groups, it is thus a very special Clifford semigroup. Isomorphisms of all these types of semigroups have been constructed.

XI.1 HOMOMORPHISMS OF REILLY SEMIGROUPS

We construct here all homomorphisms of one Reilly semigroup $B(G, \alpha)$ into another. This is used to obtain the Reilly representation of the normal hull of $B(G, \alpha)$. Recall the notation $\mathbb{N}$ and N in I.3.1 and ε_z in I.4.1.

XI.1.1 Theorem

Let $S = B(G, \alpha)$ and $T = B(H, \beta)$, where G and H are groups with identities e and f, respectively. Let $\omega : G \to H$ be a homomorphism, $k, l \in N$, and $z \in H$ be such that $\alpha\omega\varepsilon_z = \omega\beta^k$. Define a function $\eta = \eta(\omega, z; k, l)$ by

$$\eta : (m, g, n) \to \left(mk + l, z_m^{-1}(g\omega)z_n, nk + l\right) \qquad [(m, g, n) \in S], \quad (1)$$

where $z_0 = f$ and

$$z_p = z\left(z\beta^k\right)\left(z\beta^{2k}\right) \cdots \left(z\beta^{(p-1)k}\right) \qquad (p \in \mathbb{N}). \quad (2)$$

Then η is a homomorphism of S into T. Conversely, every homomorphism of S into T can be written uniquely in the form $\eta(\omega, z; k, l)$. Moreover, if $k = 0$ then $S\eta$ is a group, otherwise it is a Reilly semigroup.

Proof. In order to prove the direct part, we let $(m, g, n), (p, h, q) \in S$ and $r = \min\{n, p\}$. On the one hand, we have

$$\begin{aligned}
&(m, g, n)\eta(p, h, q)\eta \\
&\quad = \big(mk + l, z_m^{-1}(g\omega)z_n, nk + l\big)\big(pk + l, z_p^{-1}(h\omega)z_q, qk + l\big) \\
&\quad = \Big((m + p - r)k + l, \big[z_m^{-1}(g\omega)z_n\big]\beta^{(p-r)k}\big[z_p^{-1}(h\omega)z_q\big]\beta^{(n-p)k}, \\
&\qquad (n + q - r)k + l\Big) \\
&\quad = \Big((m + p - r)k + l, \big(z_m^{-1}\beta^{(p-r)k}\big)\big(g\omega\beta^{(p-r)k}\big)\big(z_n\beta^{(p-r)k}\big) \\
&\qquad \cdot\big(z_p^{-1}\beta^{(n-r)k}\big)\big(h\omega\beta^{(n-r)k}\big)\big(z_q\beta^{(n-r)k}\big), (n + q - r)k + l\Big) \qquad (3)
\end{aligned}$$

and on the other hand,

$$\begin{aligned}
&[(m, g, n)(p, h, q)]\eta \\
&\quad = \big(m + p - r, (g\alpha^{p-r})(h\alpha^{n-r}), n + q - r\big)\eta \\
&\quad = \Big((m + p - r)k + l, z_{m+p-r}^{-1}\big[(g\alpha^{p-r})(h\alpha^{n-r})\big]\omega z_{n+q-r}, \\
&\qquad (n + q - r)k + l\Big) \\
&\quad = \Big((m + p - r)k + l, z_{m+p-r}^{-1}(g\alpha^{p-r}\omega)(h\alpha^{n-r}\omega)z_{n+q-r}, \\
&\qquad (n + q - r)k + l\Big). \qquad (4)
\end{aligned}$$

We consider the case $n \leq p$; the case $n > p$ is treated similarly. In order to establish the equality of (3) and (4) in this particular case, we must show that

$$\begin{aligned}
&\big(z_m^{-1}\beta^{(p-n)k}\big)\big(g\omega\beta^{(p-n)k}\big)\big(z_n\beta^{(p-n)k}\big)z_p^{-1}(h\omega)z_q \\
&\quad = z_{m+p-n}^{-1}(g\alpha^{p-n}\omega)(h\omega)z_q \qquad (5)
\end{aligned}$$

or, equivalently,

$$z_{m+p-n}\big(z_m^{-1}\beta^{(p-n)k}\big)\big(g\omega\beta^{(p-n)k}\big)\big(z_n\beta^{(p-n)k}\big)z_p^{-1} = g\alpha^{p-n}\omega. \qquad (6)$$

We will need

$$z_{s+t}\big(z_s^{-1}\beta^{tk}\big) = z_t \qquad (s, t \in N), \qquad (7)$$

which follows from $z_t(z_s\beta^{tk}) = z_{t+s}$ which is in turn a direct consequence of the definition. In view of (7) we have that (6) is equivalent to

$$z_t\left(g\omega\beta^{tk}\right)z_t^{-1} = g\alpha^t\omega$$

and thus to

$$\omega\beta^{tk} = \alpha^t\omega\varepsilon_{z_t} \qquad (t \in N, g \in G). \tag{8}$$

We will prove (8) by induction on t. For $t = 0$, (8) clearly holds, and for $t = 1$, it holds by hypothesis. Suppose that (8) is valid for t. Using the hypothesis in the statement of the theorem, we obtain for any $g \in G$,

$$\begin{aligned} g\alpha^{t+1}\omega\varepsilon_{z_{t+1}} &= z_{t+1}^{-1}\left(g\alpha^{t+1}\omega\right)z_{t+1} \\ &= \left(z\beta^{tk}\right)^{-1}\cdots\left(z\beta^k\right)^{-1}\left[z^{-1}\left(g\alpha^{t+1}\omega\right)z\right] \\ &\quad \cdot\left(z\beta^k\right)\cdots\left(z\beta^{tk}\right) \\ &= \left(z\beta^{tk}\right)^{-1}\cdots\left(z\beta^k\right)^{-1}\left[\left(g\alpha^t\right)\alpha\omega\varepsilon_z\right] \\ &\quad \cdot\left(z\beta^k\right)\cdots\left(z\beta^{tk}\right) \\ &= \left(z\beta^{tk}\right)^{-1}\cdots\left(z\beta^k\right)^{-1}\left(g\alpha^t\omega\beta^k\right)\left(z\beta^k\right)\cdots\left(z\beta^{tk}\right) \\ &= \left[\left(z\beta^{(t-1)k}\right)^{-1}\cdots\left(z\beta^k\right)^{-1}z^{-1}\left(g\alpha^t\omega\right)z\left(z\beta^k\right)\right. \\ &\quad \left.\cdots\left(z\beta^{(t-1)k}\right)\right]\beta^k \\ &= \left(g\alpha^t\omega\varepsilon_{z_t}\right)\beta^k = g\omega\beta^{tk}\beta^k = g\omega\beta^{(t+1)k} \end{aligned}$$

proving (8). Consequently (7), (6), and (5) are also valid. Therefore η is a homomorphism.

Conversely, let η be a homomorphism of S into T. Since η preserves Green's relations (see II.5.9), taking into account the form of idempotents in T, we have

$$(0, g, 0) = (l, g\omega, l) \qquad (g \in G) \tag{9}$$

for some $l \in N$ and some function $\omega : G \to H$. It follows at once that ω is a homomorphism. Since $(0, e, 1)\mathcal{R}(0, e, 0)$, we have that $(0, e, 1)\eta\mathcal{R}(l, f, l)$, and thus

$$(0, e, 1)\eta = (l, z, p) \tag{10}$$

for some $z \in H$ and $p \in N$. Further, $(0, e, 1)\mathcal{L}(1, e, 1)$ so that $(l, z, p)\mathcal{L}(1, e, 1)\eta$ which implies $(1, e, 1)\eta = (p, f, p)$. Now $(0, e, 0) > (1, e, 1)$ implies $(l, f, l) \geq (p, f, p)$ which forces $p \geq l$; see II.5.11. Let $k = p - l$.

For any element (m, g, n) in S, we have

$$(m, g, n) = (0, e, 1)^{-m}(0, g, 0)(0, e, 1)^{n} \tag{11}$$

where $x^0 = (0, e, 0)$ for any $x \in S$. Hence we will need an expression for powers of $(0, e, 1)\eta$ $(= (l, z, k + l))$. We claim

$$(l, z, k + l)^{p} = \left(l, z_p, pk + l\right) \qquad (p \in \mathbb{N}). \tag{12}$$

This is trivial for $p = 1$. Assume that (12) holds for p; then

$$\begin{aligned}(l, z, k + l)^{p+1} &= (l, z, k + l)\left(l, z_p, pk + l\right)\\ &= \left(l, z\left(z_p\beta^k\right), (p + 1)k + l\right)\\ &= \left(l, z_{p+1}, (p + 1)k + l\right)\end{aligned}$$

as required. Consequently (12) is true. For $p = 0$, formula (12) yields

$$(k, z, k + l)^{0} = (l, f, l) \tag{13}$$

which can be used in the computation below instead of $(k, z, k + l)^0 = (0, f, 0)$. For $p > 0$, using (12), we have

$$\begin{aligned}(k, z, k + l)^{-p} &= \left[(k, z, k + l)^{p}\right]^{-1} = \left(k, z_p, pk + l\right)^{-1}\\ &= \left(pk + l, z_p^{-1}, l\right).\end{aligned} \tag{14}$$

Using equalities (10)–(14), we finally obtain

$$\begin{aligned}(m, g, n)\eta &= [(0, e, 1)\eta]^{-m}[(0, g, 0)\eta][(0, e, 1)\eta]^{n}\\ &= (k, z, k + l)^{-m}(l, g\omega, l)(l, z, k + l)^{n}\\ &= \left(mk + l, z_m^{-1}, l\right)(l, g\omega, l)\left(l, z_n, nk + l\right)\\ &= \left(mk + l, z_m^{-1}(g\omega)z_n, nk + l\right)\end{aligned} \tag{15}$$

so η has the required form.

For any $g \in G$, we have

$$(0, g, 0)(1, e, 0) = (1, g\alpha, 0).$$

Applying η to both sides and using (15), we obtain

$$(l, g\omega, l)(k + l, z^{-1}, l) = \left(k + l, z^{-1}(g\alpha\omega), l\right) \tag{16}$$

Since the left-hand side is equal to $(k + l, (g\omega\beta^k)z^{-1}, l)$, we deduce from (16) that $(g\omega\beta^k)z^{-1} = z^{-1}(g\alpha\omega)$, which evidently implies $\alpha\omega\varepsilon_z = \omega\beta^k$, as required.

The uniqueness of the parameters ω, z, k, l follows without difficulty.

Let $k = 0$. Then $z_p = z^p$ for all $p > 0$ and

$$(m, g, n)\eta = (l, z^{-m}(g\omega)z^n, l).$$

In particular, η maps S onto a subgroup of T. If $k > 0$, then $\eta|_{E_S}$ is one-to-one and $S\eta$ must be a Reilly semigroup.

We are now ready for the normal hull of a Reilly semigroup $B(G, \alpha)$. Recall the notation $\mathcal{A}(G)$ and ε_z in I.4.1.

XI.1.2 Theorem

Let $S = B(G, \alpha)$ be a Reilly semigroup. Let

$$H = \{(\omega, z) \in \mathcal{A}(G) \times G | \alpha\omega\varepsilon_z = \omega\alpha\}$$

with the multiplication

$$(\omega, z)(\omega', z') = (\omega\omega', (z\omega')z').$$

Define a mapping β by

$$\beta : (\omega, z) \to (\omega\varepsilon_z, z\alpha) \qquad [(\omega, z) \in H].$$

Then H is a group, β is an endomorphism of H and

$$\Phi(S) \cong B(H, \beta).$$

Proof. Let H be as in the statement of the theorem. We verify first that H is a group. For $(\omega, z), (\sigma, v) \in H$, we obtain

$$\begin{aligned}\alpha(\omega\sigma)\varepsilon_{(z\sigma)v} &= (\alpha\omega\varepsilon_z)\varepsilon_{z^{-1}}\sigma\varepsilon_{z\sigma}\varepsilon_v \\ &= (\omega\alpha)\varepsilon_{z^{-1}}\sigma(\sigma^{-1}\varepsilon_z\sigma)\varepsilon_v \\ &= \omega(\alpha\sigma\varepsilon_v) = (\omega\sigma)\alpha\end{aligned}$$

which shows that $(\omega, z)(\sigma, v) \in H$. Easy computation shows that the multiplication on H is associative. Note that $(\iota_G, 1)$ is the identity element of H, where ι_G is the identical automorphism of G and 1 is the identity of G. If

$(\omega, z) \in H$, then

$$\omega^{-1}\alpha = \omega^{-1}(\alpha\omega\varepsilon_z)\varepsilon_{z^{-1}}\omega^{-1} = \omega^{-1}(\omega\alpha)\varepsilon_{z^{-1}}\omega^{-1}$$
$$= \alpha\omega^{-1}(\omega\varepsilon_{z^{-1}}\omega^{-1}) = \alpha\omega^{-1}\varepsilon_{z^{-1}\omega^{-1}}$$

which implies that $(\omega^{-1}, z^{-1}\omega^{-1}) \in H$. Routine verification shows that $(\omega^{-1}, z^{-1}\omega^{-1})$ is the inverse of (ω, z). Consequently, H is a group.

Let β be the mapping defined in the statement of the theorem, and let $(\omega, z), (\sigma, v) \in H$. First note that

$$(\alpha\omega\varepsilon_z)\varepsilon_{z\alpha} = \omega\alpha\varepsilon_{z\alpha} = (\omega\varepsilon_z)\alpha$$

whence $(\omega\varepsilon_z, z\alpha) \in H$, so that β maps H into itself. Furthermore,

$$\begin{aligned}[(\omega, z)\beta][(\sigma, v)\beta] &= (\omega\varepsilon_z, z\alpha)(\sigma\varepsilon_v, v\alpha)\\ &= (\omega\varepsilon_z\sigma\varepsilon_v, (z\alpha\sigma\varepsilon_v)(v\alpha))\\ &= (\omega\sigma\varepsilon_{z\sigma}\varepsilon_v, (z\sigma\alpha)(v\alpha))\\ &= (\omega\sigma\varepsilon_{(z\sigma)v}, [(z\sigma)v]\alpha)\\ &= (\omega\sigma, (z\sigma)v)\beta\\ &= [(\omega, z)(\sigma, v)]\beta\end{aligned}$$

and β is indeed an endomorphism of H. We can thus define the Reilly semigroup $B(H, \beta)$, and it remains to show that $\Phi(S) \cong B(H, \beta)$.

Let $\varphi \in \Phi(S)$. Then φ is an isomorphism of eSe onto fSf for some $e, f \in E_S$. Letting $e = (m, 1, m)$ and $f = (n, 1, n)$, we obtain

$$\mathbf{d}\varphi = \{(m+i, g, m+j) | i, j \in N, g \in G\},$$
$$\mathbf{r}\varphi = \{(n+i, g, n+j) | i, j \in N, g \in G\}.$$

It is clear that both $\mathbf{d}\varphi$ and $\mathbf{r}\varphi$ can be represented as Reilly semigroups. We can thus use 1.1 to obtain the form of φ as follows; the detailed argument is left as an exercise. For some $\omega \in \mathscr{A}(G)$ and $z \in G$ such that $\alpha\omega\varepsilon_z = \omega\alpha$, we have

$$\varphi: (m+i, g, m+j) \to (n+i, z_i^{-1}(g\omega)z_j, n+j) \tag{17}$$

where $z_0 = 1$ and

$$z_p = z(z\alpha)(z\alpha^2)\cdots(z\alpha^{p-1}) \qquad \text{if } p > 0.$$

With this notation, we define a function χ on $\Phi(S)$ by

$$\chi: \varphi \to (m, (\omega, z), n).$$

If also $\chi: \psi \to (p, (\sigma, v), q)$, letting $r = \min\{n, p\}$, routine calculation shows that

$$\mathbf{d}(\varphi\psi) = \{(m + p - r + i, g, m + p - r + j) | i, j \in N, g \in G\},$$

$$\mathbf{r}(\varphi\psi) = \{(n + q - r + i, g, n + q - r + j) | i, j \in N, g \in G\}.$$

Further,

$$\begin{aligned}
&(m + p - r + i, g, m + p - r + j)\varphi\psi \\
&\quad = \left(n + p - r + i, z_{p-r+i}^{-1}(g\omega) z_{p-r+j}, n + p - r + j\right)\psi \\
&\quad = \left(n + q - r + i, v_{n-r+i}^{-1}\left(z_{p-r+i}^{-1}\sigma\right)(g\omega\sigma)\right. \\
&\qquad \left.\cdot(z_{p-r+j}\sigma) v_{n-r+j}, n + q - r + j\right) \\
&\quad = \left(n + q - r + i, \left[(z_{p-r+i}\sigma) v_{n-r+i}\right]^{-1}\right. \\
&\qquad \left.\cdot(g\omega\sigma)\left[(z_{p-r+j}\sigma) v_{n-r+j}\right], n + q - r + j\right).
\end{aligned} \tag{18}$$

Comparing this with (17), we deduce that

$$(\varphi\psi)\chi = (m + p - r, (\theta, u), n + q - r) \tag{19}$$

for some θ and u. Letting $i = j = 0$ in (18), we get for all $g \in G$,

$$g\theta = \left[(z_{p-r}\sigma) v_{n-r}\right]^{-1}(g\omega\sigma)\left[(z_{p-r}\sigma) v_{n-r}\right] \tag{20}$$

and with $i = 0$, $g = 1$, $j = 1$ in (18), we have

$$u = \left[(z_{p-r}\sigma) v_{n-r}\right]^{-1}(z_{p-r+1}\sigma) v_{n-r+1}. \tag{21}$$

It follows from (20) that

$$\theta = \omega\sigma\varepsilon_{(z_{p-r}\sigma)v_{n-r}} \tag{22}$$

and from (21),

$$\begin{aligned}
u &= v_{n-r}^{-1}(z_{p-r}\sigma)^{-1}(z_{p-r+1}\sigma) v_{n-r+1} \\
&= v_{n-r}^{-1}\left(z_{p-r}^{-1} z_{p-r+1}\right)\sigma v_{n-r+1} \\
&= v_{n-r}^{-1}(z\alpha^{p-r}\sigma) v_{n-r+1} \\
&= \left[v_{n-r}^{-1}(z\alpha^{p-r}\sigma) v_{n-r}\right](v\alpha^{n-r}) \\
&= \left(z\alpha^{p-r}\sigma\varepsilon_{v_{n-r}}\right)(v\alpha^{n-r}).
\end{aligned} \tag{23}$$

It now follows from (19), (22), and (23) that

$$(\varphi\psi)\chi = \Big(m + p - r, \Big(\omega\sigma\varepsilon_{(z_{p-r}\sigma)v_{n-r}}, \big(z\alpha^{p-r}\sigma\varepsilon_{v_{n-r}}\big)\big(v\alpha^{n-r}\big)\Big), n + q - r\Big). \tag{24}$$

A simple inductive argument can be used to show that

$$(\omega, z)\beta^k = \big(\omega\varepsilon_{z_k}, z\alpha^k\big) \tag{25}$$

for all $k \geq 0$. Now using (25), we compute

$$\begin{aligned}(\varphi\chi)(\psi\chi) &= (m, (\omega, z), n)(p, (\sigma, v), q)\\ &= (m + p - r, (\omega, z)\beta^{p-r}(\sigma, v)\beta^{n-r}, n + q - r)\\ &= \Big(m + p - r, \big(\omega\varepsilon_{z_{p-r}}, z\alpha^{p-r}\big)\big(\sigma\varepsilon_{v_{n-r}}, v\alpha^{n-r}\big), n + q - r\Big)\\ &= \Big(m + p - r, \Big(\omega\varepsilon_{z_{p-r}}\sigma\varepsilon_{v_{n-r}}, \big(z\alpha^{p-r}\sigma\varepsilon_{v_{n-r}}\big)v\alpha^{n-r}\Big), n + q - r\Big)\\ &= \Big(m + p - r, \Big(\omega\sigma\varepsilon_{(z_{p-r}\sigma)v_{n-r}}, \big(z\alpha^{p-r}\sigma\varepsilon_{v_{n-r}}\big)\big(v\alpha^{n-r}\big)\Big), n + q - r\Big).\end{aligned}$$

This expression coincides with that in (24) which completes the proof that χ is a homomorphism.

Since the parameters m, ω, z, n determine φ uniquely, the homomorphism χ is one-to-one. Conversely, let $(\omega, z) \in \mathcal{Q}(G) \times G$ be such that $\alpha\omega\varepsilon_z = \omega\alpha$. Direct verification shows that the function φ defined by (17) is an element of $\Phi(S)$ for which $\varphi\chi = (\omega, z)$; the details are left as an exercise. Therefore, χ maps $\Phi(S)$ onto $B(H, \beta)$.

XI.1.3 Corollary

With the notation of 1.2, the following assertions are valid.

(i) $\chi\colon \theta^{(m, g, n)} \to (m, (\varepsilon_g, g^{-1}(g\alpha)), n)$ for all $(m, g, n) \in B(G, \alpha)$.

(ii) Let $K = \{(\varepsilon_g, g^{-1}(g\alpha)) | g \in G\}$ and $\gamma = \beta|_K$. Then K is a subgroup of H, γ is an endomorphism of K, and $\Theta(B(G, \alpha)) \cong B(K, \gamma)$.

Proof. (i) One computes without difficulty that

$$\mathbf{d}\theta^{(m, g, n)} = \{(m + i, h, m + j) | i, j \in N, h \in G\},$$

$$\mathbf{r}\theta^{(m, g, n)} = \{(n + i, h, n + j) | i, j \in N, h \in G\},$$

and that for all $h \in G$,

$$(m, g, n)^{-1}(m, h, m)(m, g, n) = (n, h\varepsilon_g, n),$$

$$(m, g, n)^{-1}(m, 1, m + 1)(m, g, n) = (n, g^{-1}(g\alpha), n + 1)$$

which in view of (17) yields the assertion of part (i).

(ii) This follows without difficulty from part (i); the details are left as an exercise.

XI.1.4 Exercises

(i) For a Reilly semigroup $S = B(G, \alpha)$, verify the following statements.

(α) $E_S = \{(m, e, m) | m \geq 0\}$.
(β) $M(S) = \{(m, g, m) \in S | g\alpha = g \in Z(G), m \geq 0\}$.
(γ) $E_S\zeta = \{(m, g, m) \in S | g \in G, m \geq 0\}$.
(δ) $Z(E_S\zeta) = \{(m, g, m) \in S | g\alpha^n \in Z(G)$ for all $n \geq 0; m \geq 0\}$.
(ε) $Z(S) = \{(0, g, 0) \in S | g \in Z(G)\}$.

Also show that all these sets may be different.

(ii) Show that a Reilly semigroup $B(G, \alpha)$ is E-unitary if and only if α is one-to-one. If $B(G, \alpha)$ is E-unitary, is it F-inverse? If so, construct the F-representation for it.

(iii) Construct the conjugate hull of the centralizer of idempotents of a Reilly semigroup.

(iv) Let $S = B(G, \alpha)$ be a Reilly semigroup. Show that α is an automorphism of G if and only if S is E-unitary and for any $a, b \in S$, $a\mathcal{H}b$ implies $a = ub$ for some unit u.

(v) Find necessary and sufficient conditions on the quadruple ω, z, k, l in order for $\eta(\omega, z; k, l)$ to be (α) one-to-one and (β) onto. Deduce necessary and sufficient conditions for isomorphism of Reilly semigroups of the form $B(G, \alpha)$.

(vi) Show directly that a homomorphic image of a Reilly semigroup is either a Reilly semigroup or a group.

XI.1.5 Problems

(i) What is $I(S)$ for S in 1.3 (see VI.4.7 for notation)?

All Reilly semigroups referred to below are meant to be of the form $B(G, \alpha)$. Conditions for isomorphism of Reilly semigroups were established by Reilly [3]; conditions for the existence of a homomorphism of one Reilly semigroup onto another were stated by Munn–Reilly [1]. An expression for all homomorphisms of one Reilly semigroup onto another as well as into a group was stated by Warne [9]. Iyengar [2] computed all homomorphisms of one Reilly semigroup onto another. The lattice of congruences on a Reilly semigroup was investigated by Munn [6]. The normal hull of a Reilly semigroup and of the centralizer of its idempotents were found in Petrich [15].

A construction of the greatest group homomorphic image of, and idempotent separating congruences on, $B(G, \alpha)$ can be found in Munn–Reilly [1]; group congruences on $B(G, \alpha)$ were described by Ault [5].

XI.2 CATEGORICAL TREATMENT

We consider here the subcategory of $\mathfrak{B}$ (see X.1) whose objects have ω-chains as their idempotents and the corresponding subcategory of $\mathfrak{R}$. As a third category, we introduce the one whose objects are groups with an endomorphism, and show that there are full representative functors from this third category into each of the first two categories. We start with an auxiliary result.

XI.2.1 Lemma

Let $R, R' \in \operatorname{Ob} \mathfrak{R}$ and assume that the principal left ideals of R are linearly ordered. Then every homomorphism of R into R' is in $\operatorname{Hom} \mathfrak{R}$.

Proof. Let $a, b \in R$ and consider $Ra \cap Rb$. By the hypothesis, we may assume that $Ra \subseteq Rb$, and thus $Ra \cap Rb = Ra$. Then $a = xb$ for some $x \in R$, whence $a\varphi = (x\varphi)(b\varphi)$. Consequently, $R'(a\varphi) \subseteq R'(b\varphi)$ which implies $R'(a\varphi) \cap R'(b\varphi) = R'(a\varphi)$. Hence $\varphi \in \operatorname{Hom} \mathfrak{R}$.

XI.2.2 Definition

Let $\mathfrak{B}_\omega$ be the full subcategory of $\mathfrak{B}$ whose objects have semilattice of idempotents an ω-chain.

Let $\mathfrak{R}_\omega$ be the full subcategory of $\mathfrak{R}$ whose objects have partially ordered set of principal left ideals an ω-chain.

Note that in view of 2.1, all homomorphisms of objects of $\mathfrak{R}_\omega$ are morphisms in this category. Further, X.1.5 indicates that in the categorical equivalence (U, V, ξ, η) in X.1.15, the objects of $\mathfrak{B}_\omega$ correspond to the objects of $\mathfrak{R}_\omega$. Hence the restriction of (U, V, ξ, η) to these subcategories provides an equivalence of $\mathfrak{B}_\omega$ and $\mathfrak{R}_\omega$, which we formulate as the following statement.

XI.2.3 Proposition

The quadruple $(U_\omega, V_\omega, \xi_\omega, \eta_\omega)$ is an equivalence of the categories $\mathfrak{B}_\omega$ and $\mathfrak{R}_\omega$.

In order to establish the connection of these categories with the Reilly semigroups $B(G, \alpha)$, we introduce the following category.

XI.2.4 Definition

Let $\mathfrak{G}$ be the category with

Ob $\mathfrak{G}$ are pairs (G, α) where G is a group and α is an endomorphism of G,

Hom $\mathfrak{G}$ are triples $(\omega, z, k):(G, \alpha) \to (H, \beta)$ with $\omega : G \to H$ a homomorphism, $z \in H$, $k \in \mathbb{N}$ satisfying $\alpha\omega\varepsilon_z = \omega\beta^k$; these morphisms compose as follows

$$(\omega, z, k)(\omega', z', k') = \big(\omega\omega', (z\omega')z'_k, kk'\big).$$

Recall from 1.1 that $z_0 = f$, the identity of H, and

$$z_p = z\big(z\beta^k\big)\big(z\beta^{2k}\big)\cdots\big(z\beta^{(p-1)k}\big) \qquad \text{if } p > 0;$$

and write $(z, \beta^k)_p$ for z_p.

A long verification shows that $\mathfrak{G}$ satisfies all the axioms for a category. The difficult verification of the associative law for morphisms can actually be deduced from the proof of 2.6 below since the "functor" B in 2.5 is compatible with the composition of morphisms so we can apply associativity of the composition of mappings in $\mathfrak{B}_\omega$.

XI.2.5 Notation

For every $(G, \alpha) \in$ Ob $\mathfrak{G}$, let $B(G, \alpha)$ be the semigroup introduced in II.5.1, and for every $\mathfrak{G}$-morphism $(\omega, z, k):(G, \alpha) \to (H, \beta)$, let

$$B(\omega, z, k):(m, g, n) \to \big(mk, z_m^{-1}(g\omega)z_n, nk\big) \qquad \big[(m, g, n) \in B(G, \alpha)\big].$$

Notice that $B(\omega, z, k) = \eta(\omega, z; k, 0)$ in 1.1.

We are now ready for the principal result of this section.

XI.2.6 Theorem

With the above notation, B is a full representative functor from $\mathfrak{G}$ to $\mathfrak{B}_\omega$.

Proof. For $(G, \alpha) \in$ Ob $\mathfrak{G}$, we get directly from II.6.2 that $B(G, \alpha) \in$ Ob $\mathfrak{B}_\omega$. Analogously, for $(\omega, z, k) \in$ Hom $\mathfrak{G}_\omega$, we have by 1.1 that $B(\omega, z, k) \in$ Hom $\mathfrak{B}_\omega$.

We now check the validity of the composition requirement for morphisms. Let

$$(\omega, z, k):(G, \alpha) \to (H, \beta), \qquad (\omega', z', k'):(H, \beta) \to (K, \gamma)$$

be $\mathfrak{G}$-morphisms. Then for any $(m, g, n) \in B(G, \alpha)$, we obtain

$$\begin{aligned}
&(m, g, n)B(\omega, z, k)B(\omega', z', k') \\
&\quad = \left(mk, (z, \beta^k)_m^{-1}(g\omega)(z, \beta^k)_n, nk\right)B(\omega', z', k') \\
&\quad = \left(mkk', (z', \beta^{k'})_{mk}^{-1}\left[(z, \beta^k)_m^{-1}(g\omega)(z, \beta^k)_n\right]\omega'(z', \beta^{k'})_{nk}^{-1}, nkk'\right) \\
&\quad = \left(mkk', \left[(z, \beta^k)_m\omega'(z', \gamma^{k'})_{mk}\right]^{-1}(g\omega\omega')\right. \\
&\qquad\qquad \left.\cdot\left[(z, \beta^k)_n\omega'(z', \gamma^{k'})_{nk}\right], nkk'\right).
\end{aligned} \tag{26}$$

Formula (8) becomes in the present notation $\omega'\gamma^{tk'} = \beta^t\omega'\varepsilon_{(z', \gamma^{k'})_t}$ for any positive integer t. Using this, we will show

$$\left[(z, \beta^k)_m\omega'\right](z', \gamma^{k'})_{mk} = \left((z\omega')(z', \gamma^{k'})_k, \gamma^{kk'}\right)_m. \tag{27}$$

Indeed

$$\begin{aligned}
&\left((z\omega')(z', \gamma^{k'})_k, \gamma^{kk'}\right)_m \\
&\quad = \left[(z\omega')(z', \gamma^{k'})_k\right]\left[(z\omega')(z', \gamma^{k'})_k\right]\gamma^{kk'} \\
&\qquad \cdots\left[(z\omega')(z', \gamma^{k'})_k\right]\gamma^{(m-1)kk'} \\
&\quad = \left[(z\omega')(z', \gamma^{k'})_k\right]\left[(z\omega'\gamma^{kk'})(z', \gamma^{k'})_k\gamma^{kk'}\right] \\
&\qquad \cdots\left[(z\omega'\gamma^{(m-1)kk'})(z', \gamma^{k'})_k\gamma^{(m-1)kk'}\right] \\
&\quad = \left[(z\omega')(z', \gamma^{k'})_k\right]\left[(z\beta^k\omega'\varepsilon_{(z', \gamma^{k'})_k})(z', \gamma^{k'})_k\gamma^{kk'}\right] \\
&\qquad \cdots\left[(z\beta^{(m-1)k}\omega'\varepsilon_{(z', \gamma^{k'})_{(m-1)k}})(z, \gamma^{k'})\gamma^{(m-1)kk'}\right] \\
&\quad = \left[(z\omega')(z', \gamma^{k'})_k\right]\left[(z', \gamma^{k'})_k^{-1}(z\beta^k\omega')(z', \gamma^{k'})_k(z', \gamma^{k'})\gamma^{kk'}\right] \\
&\qquad \cdots\left[(z', \gamma^{k'})_{(m-1)k}(z\beta^{(m-1)k}\omega')(z', \gamma^{k'})_{(m-1)k}(z', \gamma^{k'})_k\gamma^{(m-1)kk'}\right] \\
&\quad = \left[(z\omega')(z', \gamma^{k'})_k\right]\left[(z', \gamma^{k'})_k^{-1}(z\beta^k\omega')(z', \gamma^{k'})_{k+1}\right] \\
&\qquad \cdots\left[(z', \gamma^{k'})_{(m-1)k}^{-1}(z\beta^{(m-1)k}\omega')(z', \gamma^{k'})_{mk}\right] \\
&\quad = \left[z(z\beta^k)\cdots(z\beta^{(m-1)k})\right]\omega'(z', \gamma^{k'})_{mk} \\
&\quad = \left[(z, \beta^k)_m\omega'\right](z', \gamma^{k'})_{mk}.
\end{aligned}$$

Using (27), we obtain that the expression in (26) is equal to

$$\left(mkk', \left((z\omega')(z', \gamma^{k'})_k, \gamma^{kk'}\right)_m^{-1}(g\omega\omega')\left((z\omega')(z', \gamma^{k'})_k, \gamma^{kk'}\right)_n, nkk'\right)$$

$$= (m, g, n)B\left(\omega\omega', (z\omega')(z', \gamma^{k'})_k, kk'\right)$$

$$= (m, g, n)B((\omega, z, k)(\omega', z', k')).$$

Hence B is compatible with the composition of morphisms. The remaining axioms for a functor are easily verified. That B is full follows from the converse of 1.1, and that it is representative from II.6.2.

We see that the above theorem is but a little more than a categorical formulation of the principal results proved in II.6 and XI.1. We now turn to the corresponding functor concerning $\mathfrak{R}_\omega$.

XI.2.7 Notation

For every $(G, \alpha) \in \mathrm{Ob}\,\mathfrak{G}$, let $R(G, \alpha)$ be the set $G \times N$ together with the multiplication

$$(g, m)(h, n) = (g(h\alpha^m), m + n).$$

If $(\omega, z, k): (G, \alpha) \to (H, \beta)$ is a $\mathfrak{G}$-morphism, define

$$R(\omega, z, k): (g, m) \to ((g\omega)z_m, mk) \qquad [(g, m) \in R(G, \alpha)].$$

Routine modifications of the proofs of II.6.2 and 1.1 provide a proof, entirely analogous to that of 2.6, of the following result. We omit the details.

XI.2.8 Theorem

With the above notation, R is a full representative functor from $\mathfrak{G}$ to $\mathfrak{R}_\omega$.

So far we have obtained the following scheme of functors:

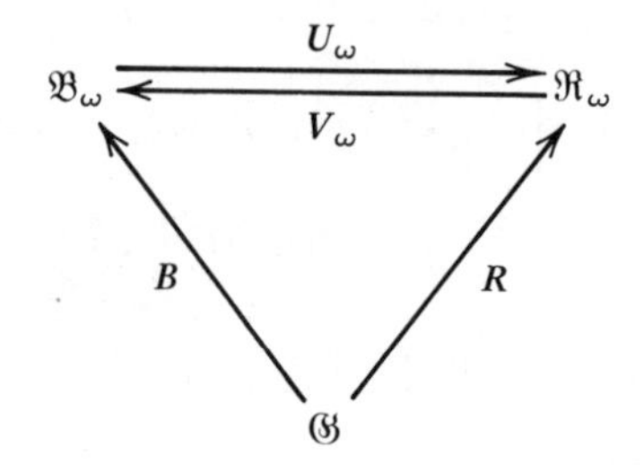

We will prove next that the functors R and $U_\omega B$, and also B and $V_\omega R$ are naturally equivalent.

XI.2.9 Proposition

For every $(G, \alpha) \in \text{Ob}\,\mathfrak{G}$ define a mapping $\lambda(R(G, \alpha))$ by

$$\lambda(R(G,\alpha)) : (g, n) \to (0, g, n) \qquad [(g, n) \in R(G, \alpha)].$$

Then λ is a natural equivalence of the functors R and $U_\omega B$.

Proof. It is evident that $\lambda(R(G, \alpha))$ is an isomorphism of $R(G, \alpha)$ onto $U_\omega B(G, \alpha)$. It remains to prove that the diagram

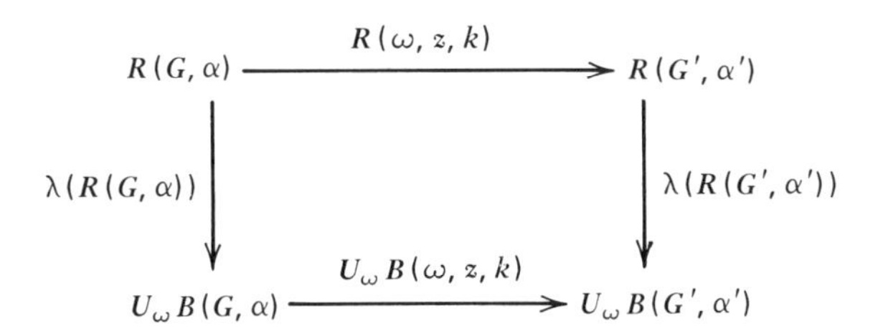

commutes. Indeed, for any $(g, n) \in R(G, \alpha)$, we have

$$\begin{aligned}
&(g, n)[R(\omega, z, k)][\lambda(R(G', \alpha'))] \\
&\quad = ((g\omega)z_n, nk)[\lambda(R(G', \alpha'))] \\
&\quad = (0, (g\omega)z_n, nk) = (0, g, n)[U_\omega B(\omega, z, k)] \\
&\quad = (g, n)[\lambda(R(G, \alpha))][U_\omega B(\omega, z, k)]
\end{aligned}$$

which completes the proof.

XI.2.10 Proposition

For every $(G, \alpha) \in \text{Ob}\,\mathfrak{G}$, define a mapping $\rho(B(G, \alpha))$ by

$$\rho(B(G,\alpha)) : (m, g, n) \to [(e, m), (g, n)] \qquad [(m, g, n) \in B(G, \alpha)].$$

Then ρ is a natural equivalence of the functors B and $V_\omega R$.

Proof. Let $(G, \alpha) \in \text{Ob}\,\mathfrak{G}$ and let $\theta = \rho(B(G, \alpha))$. We show first that θ is an isomorphism of $B(G, \alpha)$ onto $V_\omega R(G, \alpha)$. For any elements of the former, we obtain on the one hand

$$((m, g, n)(p, h, q))\theta = (m + p - r, (g\alpha^{p-r})(h\alpha^{n-r}), n + q - r)\theta$$
$$= [(e, m + p - r), ((g\alpha^{p-r})(h\alpha^{n-r}), n + q - r)], \tag{28}$$

where $r = \min\{n, p\}$, and on the other hand,

$$(m, g, n)\theta(p, h, q)\theta = [(e, m), (g, n)][(e, p), (h, q)]$$
$$= [((e, p)*(g, n))(e, m), ((g, n)*(e, p))(h, q)]. \tag{29}$$

We may take the elements (e, k), with $k \in N$, as representatives in the $\mathfrak{L}$-classes of $R = R(G, \alpha)$. Hence $(g, n) \vee (e, p) = (e, k)$ where evidently $k = \max\{n, p\}$. It follows that

$$((e, p)*(g, n))(g, n) = ((g, n)*(e, p))(e, p) = (e, k) \tag{30}$$

and writing $(e, p)*(g, n) = (a, l)$ and $(g, n)*(e, p) = (b, s)$, we obtain from (30),

$$\left(a(g\alpha^l), l + n\right) = (b, s + p) = (e, k),$$

which yields

$$a = \left(g\alpha^l\right)^{-1}, \qquad l = k - n, \qquad b = e, \qquad s = k - p.$$

We have proved

$$(e, p)*(g, n) = \left(\left(g\alpha^{k-n}\right)^{-1}, k - n\right),$$
$$(g, n)*(e, p) = (e, k - p),$$

which makes it possible to continue (29) as

$$= \left[\left(\left(g\alpha^{k-n}\right)^{-1}, k - n\right)(e, m), (e, k - p)(h, q)\right]$$
$$= \left[\left(\left(g\alpha^{k-n}\right)^{-1}, k - n + m\right), \left(h\alpha^{k-p}, k - p + q\right)\right]$$
$$= \left[\left(\left(g\alpha^{k-n}\right)^{-1}, 0\right)(1, k - n + m),\right.$$
$$\left.\left(\left(g\alpha^{k-n}\right)^{-1}, 0\right)\left(\left(g\alpha^{k-n}\right)\left(h\alpha^{k-p}\right), k - p + q\right)\right]$$
$$= \left[(e, k - n + m), \left(\left(g\alpha^{k-n}\right)\left(h\alpha^{k-p}\right), k - p + q\right)\right]. \tag{31}$$

In order to see that (31) and (28) are equal, it suffices to observe that $k - n = p - r$ and $k - p = n - r$, both of which follow from the trivial equality $k + r = p + n$. Thus expressions (28) and (29) are equal and θ is a homomorphism.

As above, we have

$$\begin{aligned}[(g, m), (h, n)] &= [(g,0)(e, m), (g,0)(g^{-1}h, n)] \\ &= [(e, m), (g^{-1}h, n)] = (m, g^{-1}h, n)\theta\end{aligned}$$

and θ is also onto. If

$$[(e, m), (g, n)] = [(e, p), (h, q)],$$

then

$$(e, m) = (t,0)(e, p), \qquad (g, n) = (t,0)(h, q)$$

for some $t \in G$, which evidently implies $m = p$, $g = h$, and $n = q$. Hence θ is one-to-one and consequently an isomorphism.

It remains to establish the commutativity of the diagram

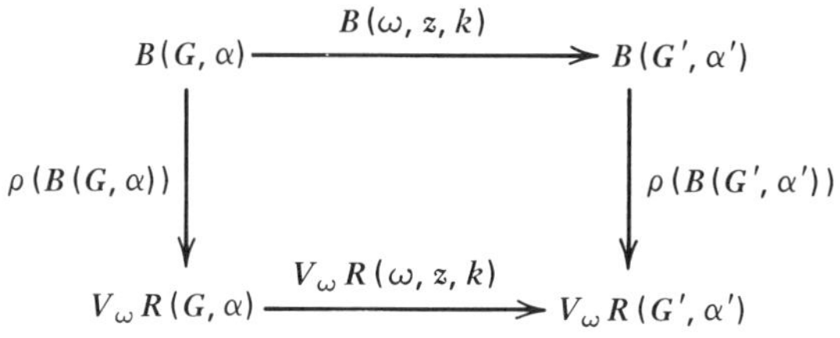

$$\begin{array}{ccc} B(G, \alpha) & \xrightarrow{\quad B(\omega, z, k) \quad} & B(G', \alpha') \\ {\scriptstyle \rho(B(G, \alpha))}\big\downarrow & & \big\downarrow{\scriptstyle \rho(B(G', \alpha'))} \\ V_\omega R(G, \alpha) & \xrightarrow{\quad V_\omega R(\omega, z, k) \quad} & V_\omega R(G', \alpha') \end{array}$$

Indeed, for any $(m, g, n) \in B(G, \alpha)$, we obtain

$$\begin{aligned}(m, g, n)&[B(\omega, z, k)][\rho(B(G', \alpha'))] \\ &= (mk, z_m^{-1}(g\omega)z_n, nk)[\rho(B(G', \alpha'))] \\ &= [(e, mk), (z_m^{-1}(g\omega)z_n, nk)] \\ &= [(z_m^{-1}, 0)(z_m, mk), (z_m^{-1}, 0)((g\omega)z_n, nk)] \\ &= [(z_m, mk), ((g\omega)z_n, nk)] \\ &= [(e, m), (g, n)][V_\omega R(\omega, z, k)] \\ &= (m, g, n)[\rho(B(G, \alpha))][V_\omega R(\omega, z, k)],\end{aligned}$$

which completes the proof.

XI.2.11 Exercises

(i) Let $S = B(G, \alpha)$ be a Reilly semigroup.

(α) Find the product of the endomorphisms $\eta(\omega, z; k, l)$ and $\eta(\omega', z'; k', l')$.

(β) Characterize the group of automorphisms of S.

(ii) Characterize the semigroup of endomorphisms of a bicyclic semigroup.

Part of 2.8 was proved by Rees [1].

XI.3 REILLY SUBSEMIGROUPS

We are interested here in general statements concerning the presence of Reilly semigroups as subsemigroups of a given (not necessarily inverse) semigroup. The position of a bicyclic semigroup as a subsemigroup of an arbitrary semigroup was discussed in IX.4. To facilitate the discussion, we first introduce the following concept.

XI.3.1 Definition

If S is a Reilly semigroup with idempotents $E_S = \{e_0 > e_1 > e_2 > \cdots\}$, then the pair $\{e_0, e_1\}$ is the *initial segment* of S.

Our first task is to characterize Reilly subsemigroups of a given semigroup with specified initial segment.

XI.3.2 Theorem

Let S be a semigroup and let e and f be distinct idempotents of S. If $a, b \in S$ are inverses of each other, $ab = e$, $ba = f$, and G is a subgroup of H_e for which $aG \subseteq Ga$, $Gb \subseteq bG$, then

$$B = \{b^m g a^n | m, n \in N, g \in G\}$$

is a Reilly subsemigroup of S with initial segment $\{e, f\}$, where $a^0 = b^0 = e$. Conversely, any Reilly subsemigroup of S with initial segment $\{e, f\}$ is of the above form for some such a, b, and G.

Moreover, G is the group of units of B, the mapping

$$\alpha : g \to agb \qquad (g \in G)$$

is an endomorphism of G, and

$$\varphi : (m, g, n) \to b^m g a^n \qquad [(m, g, n) \in B(G, \alpha)]$$

is an isomorphism of $B(G, \alpha)$ onto B.

Proof

Necessity. First let $g \in G$. By hypothesis, $ag = g'a$ and $gb = bg''$ for some $g', g'' \in G$. It follows that

$$g' = g'(ab) = (g'a)b = (ag)b = a(gb) = (ab)g'' = g''.$$

Then $g\alpha = agb = g'ab = g' = g''$ so that $ag = (g\alpha)a$ and $gb = b(g\alpha)$. For any $x, y \in G$, we then obtain

$$(xy)\alpha = a(xy)b = (ax)(yb) = (x\alpha)ab(y\alpha) = (x\alpha)(y\alpha),$$

that is, α is an endomorphism of G.

The form of B in the theorem amounts to saying that φ maps $B(G, \alpha)$ onto B. Let $(m, g, n), (p, h, q) \in B(G, \alpha)$. Assume that $n \geq p$. Using the property $ah = (h\alpha)a$, we obtain

$$\begin{aligned}(m, g, n)\varphi(p, h, q)\varphi &= b^m g a^n b^p h a^q = b^m g(a^{n-p}h)a^q \\ &= b^m g(h\alpha^{n-p})a^{n+q-p} = (m, g(h\alpha^{n-p}), n+q-p)\varphi \\ &= ((m, g, n)(p, h, q))\varphi.\end{aligned}$$

The case $n < p$ is treated similarly. Hence φ is a homomorphism.

Let ρ be the congruence on $B(G, \alpha)$ induced by φ. Since $(0, e, 0)\varphi = e$ and $(1, e, 1)\varphi = bea = ba = f$, we have that ρ is not a group congruence. Letting

$$K = \{(m, e, n) | m, n \in N\},$$

we see that K is a full bicyclic subsemigroup of $B(G, \alpha)$. Hence $\rho|_K$ is a nongroup congruence on K and IX.4.10 gives that $\rho|_K = \varepsilon$. Consequently, $\operatorname{tr} \rho = \varepsilon$. In particular, $\rho \subseteq \mu$ and thus $\ker \rho \subseteq E\zeta$ where $E = E_{B(G, \alpha)}$. It is easy to see that

$$E\zeta = \{(m, g, m) | m \in N\}.$$

Now let $(m, g, m) \in \ker \rho$ so that

$$(b^m g a^m)(b^m g a^m) = b^m g a^m.$$

Premultiplying by a^m, postmultiplying by b^m, and taking into account that $ab = e$, we obtain $g^2 = g$ so that $g = e$. This shows that $\ker \rho = E$, which together with $\operatorname{tr} \rho = \varepsilon$ yields $\rho = \varepsilon$. Therefore, φ is one-to-one and thus an isomorphism of $B(G, \alpha)$ onto B. Since $\{(0, g, 0) | g \in G\}$ is the group of units of $B(G, \alpha)$, in view of the isomorphism φ, we deduce that G is the group of units of B.

Sufficiency. Let B be a Reilly subsemigroup of S with initial segment $\{e, f\}$. The proof of II.6.2 clearly indicates that B is of the form as in the statement of the theorem for some $a, b \in S$ which are inverses of each other and satisfy the requirements $ab = e$ and $ba = f$, and some group G with identity e. Since the group of units of a Reilly semigroup is contained in the centralizer of idempotents, for any $g \in G$, we have $gf = fg$, and thus

$$ag = (af)g = a(fg) = a(gf) = ag(ba) = (agb)a. \tag{32}$$

Further,

$$\begin{aligned}(ag^{-1}b)(agb) &= ag^{-1}(ba)gb = ag^{-1}(fg)b \\ &= ag^{-1}g(fb) = afb = ab = e\end{aligned}$$

and analogously $(agb)(ag^{-1}b) = e$. Consequently, $agb \in B \cap H_e = G$, which together with (32) implies $aG \subseteq Ga$. One shows similarly that also $Gb \subseteq bG$.

According to I.6.7, for e, f as in 3.2, one may take any $a \in R_e \cap L_f$ for which $b \in R_f \cap L_e$ is then uniquely determined. (Of course, one may interchange the roles of a and b.) Hence B in 3.2 depends on the choice of a in $R_e \cap L_f$ and G when e, f are given. Alternatively, B is completely determined by G, a, and b. We may thus introduce the following symbolism.

XI.3.3 Notation

We denote the semigroup B defined in 3.2 by $B(G; a, b)$.

The next result expresses the degree of dependence of $B(G; a, b)$ upon the parameters G, a, and b.

XI.3.4 Proposition

Let $U = B(G; a, b)$ and $V = B(H; c, d)$ be subsemigroups of a semigroup S. Then the following statements are equivalent.

(i) $U = V$.
(ii) $G = H$, $ab = cd$, $ba = dc$, $ad, cb \in G$.
(iii) $G = H$, $Ga = Gc$, $bG = dG$.

Proof. (i) *implies* (ii). By 3.2, G and H are groups of units of U and V, respectively, and thus $G = H$. Also, letting $e = ab$ and $f = ba$, we have that $\{e, f\}$ is the initial segment of U. Hence $\{e, f\}$ is the initial segment of V as well, whence $e = cd$ and $f = dc$. But then $ab = cd$ and $ba = dc$. Further,

$$\begin{aligned}ad &= adcd = adab \in Sab, \\ ab &= cd = cdcd = cbad \in Sad, \\ ad &= abad = abS, \\ ab &= abab = adcb \in adS\end{aligned}$$

which shows that $ad\mathcal{H}ab$. Hence

$$ad \in H_e \cap UV \subseteq H_e \cap U = G,$$

as required. A similar argument shows that also $cb \in G$.

(ii) *implies* (iii). Indeed,

$$Ga = Gaba = Gadc = Gc,$$

$$bG = babG = dcbG = dG.$$

(iii) *implies* (i). Indeed, $a \in Ga = Hc \subseteq V$ and similarly $b \in V$. Hence $U \subseteq V$ and by symmetry also $V \subseteq U$.

Observe that in 3.4(ii), the condition $G = H$ says that G and H have the same group of units, and the conditions $ad = bc$ and $ba = dc$ that U and V have the same initial segment.

The next result furnishes the greatest subgroup G of H_{ab} which fits into $B(G; a, b)$ for a given pair a, b.

XI.3.5 Proposition

Let S be a semigroup and $a, b \in S$ be inverses of each other such that $ab > ba$. Then

$$G = \{g \in H_{ab} | gb^n a^n = b^n a^n g \text{ for all } n \in \mathbb{N}\}$$

is the greatest subgroup of H_{ab} for which there exists a Reilly subsemigroup of S with group of units G and initial segment $\{ab, ba\}$.

Proof. First note that

$$a(ab) = a(ba)(ab) = aba = a,$$

$$(ab)b = (ab)(ba)b = bab = b,$$

and thus ab acts as an identity for a and b. Then $ab \in G$ and G is obviously closed under multiplication. If $g \in G$, then for any $n \in \mathbb{N}$,

$$g^{-1}b^n a^n = g^{-1}b^n a^n (ab) = g^{-1}(b^n a^n g)g^{-1} = g^{-1}gb^n a^n g^{-1}$$

$$= (ab)b^n a^n g^{-1} = b^n a^n g^{-1}$$

and hence $g^{-1} \in G$. Therefore, G is a subgroup of H_{ab}. Further, for $g \in G$, we have

$$ag = (aba)g = a(bag) = a(gba) = (agb)a. \tag{33}$$

Now $agb = (ab)(agb) = (agb)(ab)$ and

$$ab = a(ab)b = (aba)(gg^{-1})b = a(bag)g^{-1}b$$
$$= a(gba)g^{-1}b = (agb)(ag^{-1}b)$$

and similarly $ab = (ag^{-1}b)(agb)$ which yields $agb\mathcal{H}ab$, that is to say, $agb \in H_{ab}$. For any $n \in \mathbb{N}$, we obtain

$$(agb)b^n a^n = agb^{n+1}a^n(ab) = a(gb^{n+1}a^{n+1})b$$
$$= a(b^{n+1}a^{n+1}g)b = (ab)b^n a^n(agb) = b^n a^n(agb)$$

which finally shows that $agb \in G$. Hence (33) yields $ag \in Ga$ so that $aG \subseteq Ga$. With $g \in G$, we also have

$$gb = g(bab) = (gba)b = (bag)b = b(agb) \in bG$$

and hence $Gb \subseteq bG$. Note that $ab\mathcal{D}a\mathcal{D}ba$.

In view of 3.2, we have that

$$B = \{b^m g a^n | m, n \in \mathbb{N}, g \in G\}$$

is a Reilly subsemigroup of S with group of units G and initial segment $\{ab, ba\}$. Now let T be a Reilly subsemigroup of S with initial segment $\{ab, ba\}$. Then

$$E_T = \{b^n a^n | n \in N\}$$

where $b^0 a^0 = ab$. Letting H be the group of units of T, we obtain $hb^n a^n = b^n a^n h$ for all $h \in H$ since in a Reilly semigroup the group of units is contained in the centralizer of idempotents. It follows that $H \subseteq G$ which proves maximality of G.

We now consider the following problem: which subgroups G, with identity e, of a semigroup S have the property that for every idempotent f such that $f < e$ and $f\mathcal{D}e$, there is a Reilly subsemigroup of S with group of units G and initial segment $\{e, f\}$? We discuss only the instance when G is a normal subgroup of H_e. For this case, a suitable answer is expressed by means of the following concept (cf. X.4.1).

XI.3.6 Definition

Let G be a subgroup, with identity e, of a semigroup S. Then G is a *normal divisor for e* if $aG \subseteq Ga$ for all $a \in eSe \cap R_e$ and $Gb \subseteq bG$ for all $b \in eSe \cap L_e$.

XI.3.7 Proposition

Let G be a subgroup, with identity e, of a semigroup S. Then G is a normal divisor for e if and only if G is a normal subgroup of H_e and for every $f < e$, $e\mathcal{D}f$, there exists a Reilly subsemigroup of S with group of units G and initial segment $\{e, f\}$.

Proof

Necessity. Let g be a normal divisor for e, and $f < e$, $e\mathcal{D}f$. By I.6.7, there exists $a \in R_e \cap L_f$ and $b \in R_f \cap L_e$ such that $ab = e$, $ba = f$. The hypothesis $f < e$ implies that $a, b \in eSe$, and hence $aG \subseteq Ga$ and $Gb \subseteq bG$. Now 3.2 provides the required Reilly subsemigroup.

Sufficiency. Let $x \in eSe \cap R_e$. If $x \in H_e$, then $xH \subseteq Hx$ since H is a normal subgroup of H_e. Assume next that $x \notin H_e$. By I.6.7, there exists an inverse x' of x such that $xx' = e$. Then ex' is also an inverse of x, and thus $ex'x < e$ with $ex'x\mathcal{D}e$. By hypothesis there exists a Reilly subsemigroup having G as group of units and $\{e, ex'x\}$ as its initial segment. By 3.2, there exist an element a and its inverse b satisfying $ab = e$, $ba = ex'x$, $aG \subseteq Ga$, and $Gb \subseteq bG$. Further, $xb = e(xb) = (xb)e$ and

$$e = xx' = xe(x'xx') = x(ex'x)x' = xbax' \in xbS,$$

$$e = ab = a(ba)b = aex'xb \in Sxb,$$

and thus $xb \in H_e$; also

$$x = xx'x = (xe)x'x = x(ex'x) = xba.$$

Using this and normality of G in H_e, we obtain

$$xG = xbaG \subseteq (xb)Ga \subseteq Gxba = Gx.$$

The second part of the definition of a normal divisor is established analogously.

The next result furnishes the greatest normal divisor for a given idempotent of a semigroup.

XI.3.8 Theorem

For any idempotent e of a semigroup S,

$$N_e = \{g \in H_e | gf = fg \text{ for all } f < e, f\mathcal{D}e\}$$

is, under inclusion, the greatest normal divisor for e.

Proof. It is clear that N_e is closed under multiplication. For $g \in N_e$ and an idempotent $f < e$, $e\mathcal{D}f$, we have $gf = fg$. Hence

$$g^{-1}f = g^{-1}fe = g^{-1}fgg^{-1} = (g^{-1}g)fg^{-1} = efg^{-1} = fg^{-1}$$

which shows that $g^{-1} \in N_e$. Consequently, N_e is a subgroup of H_e.

Next let $x \in eSe \cap R_e$ and x' be an inverse of x such that $xx' = e$ (see I.6.7). Using the hypothesis that $x \in eSe$, we obtain $ex'x \in E_S$, $ex'x \leq e$, and

$$ex'x\mathcal{R}ex'\mathcal{L}x'\mathcal{D}x\mathcal{D}e.$$

Letting $t \in N_e$, we then have

$$xt = xx'xt = (xe)x'xt = x(ex'x)t = xt(ex'x) = (xtx')x. \tag{34}$$

We now show that $xtx' \in N_e$. Indeed,

$$xtx' = (ex)tx' = xt(x'xx') = xt(x'e)$$

$$(xtx')(xt^{-1}x') = x(te)x'xt^{-1}x' = xt(ex'x)t^{-1}x'$$

$$= x(ex'x)tt^{-1}x' = (xe)x'(xe)x' = e$$

and analogously $(xt^{-1}x')(xtx') = e$ so that $xtx' \in H_e$. Next let $f < e$ and $f\mathcal{D}e$. Then

$$(ex'fx)(ex'fx) = ex'f(xe)x'fx = ex'fefx = ex'fx,$$

$$ex'fx\mathcal{L}fx \quad \text{since } x(ex'fx) = xx'fx = efx = fx,$$

$$fx\mathcal{R}f \quad \text{since } (fx)x' = fe = f,$$

so that $ex'fx \in E_{D_e}$ and clearly $ex'fx \leq e$. By the definition of N_e, we then obtain

$$(xtx')f = xtx'(fxx') = xt(x'fx)x' = x(x'fx)tx'$$

$$= (xx')f(xtx') = f(xtx')$$

which gives that $xtx' \in N_e$. Now relation (34) yields that $xt \in N_e x$ so that $xN_e \subseteq N_e x$. The second half of the definition of a normal divisor for e is verified similarly.

Finally, let G be a normal divisor for e and let $f < e$, $f\mathcal{D}e$. By 3.7, there exists a Reilly subsemigroup of S with group of units G and initial segment $\{e, f\}$. Since in a Reilly semigroup the group of units is contained in the centralizer of idempotents, it follows that for any $g \in G$, we have $gf = fg$. Consequently, $G \subseteq N_e$, which proves the maximality of N_e.

The following example shows that a Reilly subsemigroup of a semigroup is not uniquely determined by its group of units and initial segment.

XI.3.9 Example

We characterize here all Reilly subsemigroups B of a Reilly semigroup $S = B(G, \alpha)$ such that B and S have the same identity element e.

Let K be a subgroup of G and let $p \in \mathbb{N}$ and $u \in G$ be such that $K\alpha^p \subseteq u^{-1}Ku$. Then

$$K = \{(0, g, 0) | g \in K\}$$

is a subgroup of the group of units of S, $a = (0, u, p)$ is an element $\mathcal{R}$-equivalent to e, $aH \subseteq Ha$, and $a^{-1}a = (p, 1, p) = f < e$. By 3.2, $B = B(H; a, a^{-1})$ is a Reilly subsemigroup of S with group of units H and initial segment $\{e, f\}$. We may thus write $B = B(K; u, p)$. A simple argument shows that every Reilly subsemigroup of S with identity e is of this form for some K, u, and p.

Now assume that also $B(K'; u', p')$ is given. Using 3.4, we easily obtain that

$$B(K; u, p) = B(K'; u', p') \Leftrightarrow K = K', p = p', uv^{-1} \in K.$$

In fact, $K = K'$ indicates that these subsemigroups have the same group of units and $p = p'$ that they have the same initial segment. It is now easy to construct examples showing that there exist distinct Reilly subsemigroups of S with the same group of units and initial segment. We note that in the case that K is a normal subgroup of G, the above condition becomes $K\alpha^p \subseteq K$, that is, K is α^p invariant, and any u can be used.

Furthermore, it is verified easily that

$$B(K; u, p) \cong B(K, \alpha^p \varepsilon_{u^{-1}}).$$

XI.3.10 Exercises

(i) Let X, A, B be sets such that $B \subset A \subseteq X$ and A and B have the same infinite cardinality and let

$$G = \{\beta \in \mathcal{I}(X) | \mathbf{d}\beta = \mathbf{r}\beta = A, \beta|_B = \iota_B\}.$$

Prove the following statements.

(α) There exists a Reilly subsemigroup of $\mathcal{I}(X)$ with group of units G and initial segment $\{\iota_A, \iota_B\}$.

(β) $\{\iota_A\}$ is the only normal divisor for ι_A.

(γ) There exists no Reilly subsemigroup of $\mathcal{I}(X)$ with group of units H_{ι_A} and initial segment $\{\iota_A, \iota_B\}$.

(ii) Let $Y = N \times N$ with the partial order $\leq$ defined by

$$(m, n) \leq (p, q) \quad \text{if} \quad m \geq p \quad \text{and} \quad n \geq q.$$

Verify that Y is a semilattice and that $\Phi(Y)$ is bisimple. Let $\alpha = \iota_{[(0,0)]}$, $\beta = \iota_{[(1,0)]}$, and $\gamma = \iota_{[(1,1)]}$. Find H_α and show that there exists a Reilly subsemigroup of $\Phi(Y)$ with group of units H_α and initial segment $\{\alpha, \gamma\}$, but no such exists with initial segment $\{\alpha, \beta\}$. (Note that this implies that H_α is not a normal divisor for α.)

(iii) Give an example that fits the following specifications: e is an idempotent of a semigroup S, H is a normal subgroup of H_e, $H \subseteq N_e$, but H is not a normal divisor for e.

(iv) Show that if $\mathcal{H}$ is a congruence on a semigroup S, then every maximal subgroup of S is a normal divisor for its identity.

(v) Let $U = B(G; a, b)$ and $V = B(H; c, d)$ be subsemigroups of a semigroup S. Show that $U \subseteq V$ if and only if $G \subseteq d^m H c^m$, $a \in d^m H c^n$, and $b \in d^n H c^m$ for some $n > m \geq 0$.

(vi) Let S be an inverse semigroup and $e \in E_S$. Show that

$$e\mu = \{x \in H_e | xf = fx \text{ for all } f < e\}.$$

Results 3.2 and 3.7 are due to Ault [4]. A relationship between normal divisors and idempotent separating congruences on a regular 0-bisimple semigroup was established by Munn [5]. Right normal divisors were introduced by Rees [1].

XI.4 SIMPLE ω-REGULAR SEMIGROUPS

The structure theorem for these semigroups says that, up to an isomorphism, they are precisely Bruck semigroups over finite chains of groups. We also prove some auxiliary results which are of independent interest.

The strategy of the proof of the main theorem consists of taking the Munn representation of a sample ω-regular semigroup S and by means of the antigroup S^δ obtaining the structure of S itself. This is made possible by an explicit description of full subtransitive inverse subsemigroups of $\Phi(C_\omega)$ and the fact that $\mu = \mathcal{H}$ in S.

Recall the notation: N is the set of all nonnegative integers, C_ω is the same set with the reverse of the usual order, and C is the bicyclic semigroup. The first lemma describes the structure of $\Phi(C_\omega)$.

XI.4.1 Lemma

For any $m, n \in N$, let

$$\varphi_{m,n}: i \to n - m + i \qquad (i \geq m).$$

Then

$$\Phi(C_\omega) = \{\varphi_{m,n} | m, n \in N\}$$

with multiplication

$$\varphi_{m,n}\varphi_{p,q} = \varphi_{m+p-r,\,n+q-r}$$

where $r = \min\{n, p\}$. In particular, $\Phi(C_\omega) \cong C$.

Proof. From the transparent nature of the ω-chain C_ω, it is clear that the elements of $\Phi(C_\omega)$ are functions of the form

$$\psi_{m,n}: m + k \to n + k \qquad (k \in N)$$

for a fixed pair $m, n \in N$. Writing $m + k = i$, we see that $\psi_{m,n} = \varphi_{m,n}$ as defined in the statement of the lemma. A simple inspection shows that

$$\varphi_{m,n}\varphi_{p,q} = \begin{cases} \varphi_{m+p-n,\,q} & \text{if } n \le p \\ \varphi_{m,\,n+q-p} & \text{if } n > p, \end{cases}$$

which can be written compactly as in the statement of the lemma. It is now obvious that the mapping

$$\varphi_{m,n} \to (m, n) \qquad (m, n \in N)$$

is an isomorphism of $\Phi(C_\omega)$ onto C.

The second lemma concerns full subtransitive inverse subsemigroups of $\Phi(C_\omega)$.

XI.4.2 Lemma

For every natural number d, let

$$B_d = \{\varphi_{m,n} | m \equiv n \ (\mathrm{mod}\ d)\}.$$

Then, the semigroups B_d $(d = 1, 2, \ldots)$ are precisely all the full subtransitive inverse subsemigroups of $\Phi(C_\omega)$.

Proof. Since the idempotents of $\Phi(C_\omega)$ are of the form $\varphi_{m,m}$, it is clear that B_d is full. Let $m, n \in N$; then for any p such that $p \ge n$ and $p \equiv m \ (\mathrm{mod}\ d)$, we have $m\varphi_{m,p} = p$ and $[p] \subseteq [n]$, where, as usual, $[p]$ is the principal ideal of C_ω generated by p. Thus B_d is also subtransitive. Simple verification shows that B_d is an inverse subsemigroup of $\Phi(C_\omega)$.

Conversely, let S be a full subtransitive inverse subsemigroup of $\Phi(C_\omega)$. There exists $\varphi \in S$ such that $[0\varphi] \subseteq [1]$ and thus $0\varphi = d \ge 1$, and hence

$\varphi_{0,d} \in S$. Let d be the least natural number such that $\varphi_{0,d} \in S$. By iteration of $\varphi_{0,d}$, we obtain $\varphi_{0,d}^k = \varphi_{0,kd}$ for $k = 1, 2, \ldots$. Since S is full, we also have $\varphi_{m,m} \in S$ for any $m \in N$, and thus $\varphi_{m,m}\varphi_{0,kd} \in S$, that is to say, $\varphi_{m,m+kd} \in S$. Taking inverses, we also have $\varphi_{m+kd,m} \in S$, which shows that $B_d \subseteq S$. Let $\varphi_{m,n} \in S$. We write $n = m + kd + r$ with $0 \leq r < d$, to obtain

$$\varphi_{m,m+kd+r}\varphi_{m+kd,m} = \varphi_{m,m+r} \in S.$$

If $m > 0$, we further get

$$\varphi_{m-1,m-1+d}\varphi_{m,m+r}\varphi_{m-1+d,m-1} = \varphi_{m-1,m-1+r} \in S.$$

Continuing this procedure, we eventually get that $\varphi_{0,r} \in S$. By the minimality of d, we obtain $r = 0$. Consequently $B_d = S$.

The third lemma describes the $\mathcal{D}$-structure of B_d.

XI.4.3 Lemma

For every natural number d and $i = 0, 1, \ldots, d-1$,

$$B_{d,i} = \{\varphi_{m,n} | m \equiv n \equiv i \ (\mathrm{mod}\ d)\}$$

is a $\mathcal{D}$-class of B_d; it is a bicyclic semigroup.

Proof. For any $\varphi_{m,n}, \varphi_{p,q} \in B_d$, we have

$$\varphi_{m,n}\mathcal{D}\varphi_{p,q} \Leftrightarrow \varphi_{m,n}\mathcal{R}\varphi_{u,v}\mathcal{L}\varphi_{p,q} \quad \text{for some} \quad \varphi_{u,v} \in B_d$$

$$\Leftrightarrow m = u, v = q, u \equiv v \ (\mathrm{mod}\ d)$$

$$\Leftrightarrow m \equiv q \ (\mathrm{mod}\ d).$$

Now fix $0 \leq i < d$. It follows that

$$D_{\varphi_{i,i}} = \{\varphi_{m,n} \in B_d | \varphi_{m,n}\mathcal{D}\varphi_{i,i}\}$$

$$= \{\varphi_{m,n} | m \equiv n \ (\mathrm{mod}\ d), m \equiv i \ (\mathrm{mod}\ d)\} = B_{d,i}.$$

Hence the sets $B_{d,i}$ for $i = 0, 1, \ldots, d-1$ exhaust all $\mathcal{D}$-classes of B_d. Straightforward verification shows that the mapping

$$\varphi_{md+i,nd+i} \to (m, n) \qquad (m, n \in N)$$

is an isomorphism of $B_{d,i}$ onto C.

The fourth lemma concerns inverse semigroups with a restriction on idempotents.

XI.4.4 Lemma

Let S be an inverse semigroup with the property that for any $e, f \in E = E_S$, there is a unique $\varphi \in \Phi(E)$ such that $\mathbf{d}\varphi = [e]$ and $\mathbf{r}\varphi = [f]$. Then $\mu = \mathcal{H}$ in S.

Proof. Let $a, b \in S$ be such that $a\mathcal{H}b$. Then $aa^{-1} = bb^{-1}$ and $a^{-1}a = b^{-1}b$ so that

$$\mathbf{d}\delta^a = \mathbf{d}\delta^{aa^{-1}} = \mathbf{d}\delta^{bb^{-1}} = \mathbf{d}\delta^b$$

and analogously $\mathbf{r}\delta^a = \mathbf{r}\delta^b$. Now the hypothesis implies that $\delta^a = \delta^b$ whence $a\mu b$ by IV.2.3. Hence $\mathcal{H} \subseteq \mu$ and the opposite inclusion always holds.

In the following theorem there appears a *finite chain of groups* $T = G_0 \cup G_1 \cup \cdots \cup G_{d-1}$ with the ordering $G_0 > G_1 > \cdots > G_{d-1}$. Note that "finite" modifies the word "chain" only, so that the groups G_i are assumed to be arbitrary. If e_i is the identity of G_i for $0 \leq i < d$, then e_0 is the identity of T and G_0 is its group of units. We are finally ready for the structure theorem for simple ω-regular semigroups.

XI.4.5 Theorem

A semigroup S is simple ω-regular if and only if it is isomorphic to a Bruck semigroup over a finite chain of groups.

Proof

Necessity. The Munn representation δ, see IV.2.3, is a homomorphism of S onto a full inverse subsemigroup S^δ of $\Phi(C_\omega)$ inducing μ on S. In view of IV.2.7, S^δ is subtransitive, which together with its fullness, by 4.2, implies that $S^\delta = B_d$ for some natural number d. It is clear that S satisfies the hypothesis of 4.4 which gives that $\mu = \mathcal{H}$ in S. Combining these findings, we see that δ is a homomorphism of S onto B_d which induces $\mathcal{H}$ on S. For any $m, n \in N$ such that $m \equiv n \pmod{d}$, we let

$$H_{m,n} = \{s \in S \mid \delta^s = \varphi_{m,n}\}.$$

Hence the sets $H_{m,n}$ are precisely the $\mathcal{H}$-classes of S and

$$H_{m,n}H_{p,q} \subseteq H_{m+p-r,\,n+q-r}$$

where $r = \min\{n, p\}$ since $S/\mu \cong B_d$.

This already describes a good deal of the structure of S. The $\mathcal{D}$-structure of S follows directly from the $\mathcal{D}$-structure of B_d since $S/\mathcal{H} \cong B_d$. Hence by 4.3, the $\mathcal{D}$-classes of S are

$$D_i = \cup\{H_{m,n} \mid m \equiv n \equiv i \pmod{d}\}$$

for $i = 0, 1, \ldots, d-1$, and $D_i/\mathcal{H}$ is a bicyclic semigroup. It follows that each D_i is a Reilly semigroup; its structure is given by II.6.2 as a Bruck semigroup over a group.

Idempotents of S form an ω-chain $e_0 > e_1 > e_2 > \cdots$. For $i = 0, 1, \ldots, d-1$, let $G_i = H_{i,i}$ and let $T = \bigcup_{i=0}^{d-1} G_i$. Then T is the complete inverse image of the set $\{\varphi_{0,0}, \varphi_{1,1}, \ldots, \varphi_{d-1,d-1}\}$ under δ and is thus a subsemigroup of S. Hence T is a finite chain of groups with idempotents $e_0 > e_1 > \cdots > e_{d-1}$. Note that $e_i \in H_{i,i}$ for all $i \in N$. Hence, by the above description of $\mathcal{D}$-classes of S, we conclude that for $0 \leq i < d$, e_i is the highest idempotent of D_i and is thus its identity. It follows that G_i is the group of units of D_i for $0 \leq i < d$.

Recall that in the proof of II.6.2 we chose an element in the $\mathcal{H}$-class "next" to the group of units in the $\mathcal{R}$-class of the identity, and that this element gave rise to an endomorphism of the group of units which was later used to describe the multiplication in the Reilly semigroup. We could do the same here for each $\mathcal{D}$-class D_i. However, in order to achieve the desired form of multiplication, the choice of elements in various $\mathcal{D}$-classes D_i cannot be totally arbitrary.

Let a be any element of $H_{0,d}$. For $i = 0, 1, \ldots, d-1$, let $a_i = e_i a$. Then $a_i \in H_{i,i} H_{0,d} \subseteq H_{i,d+i}$ so that a_i is in the "right" $\mathcal{H}$-class in comparison with the proof of II.6.2. This is our "choice" of the element in the Reilly semigroups $D_0, D_1, \ldots, D_{d-1}$. We have remarked above that e_i is the identity of D_i for $0 \leq i < d$. Since $a_i \in D_i$, we obtain $a_i e_i = a_i$ whence, for any natural number k, we have $a_i^k = (e_i a)^k = e_i a^k$. According to the proof of II.6.2, every element of D_i can be uniquely written in the form

$$a_i^{-m} g a_i^n = (a_i^m)^{-1} g a_i^n = (e_i a^m)^{-1} g (e_i a^n) = a^{-m}(e_i g e_i) a^n = a^{-m} g a^n$$

for some $g \in G_i$, $m, n \in N$. Recall that a^0 here means the empty symbol. It follows that the mapping

$$\chi \colon a^{-m} g a^n \to (m, g, n) \qquad (m, n \in N, g \in T)$$

is a bijection of S onto $N \times T \times N$. We still need a homomorphism $\alpha \colon T \to G_0$ such that χ is an isomorphism of S onto $B(T, \alpha)$.

For any $g \in G_i$, we have $ag \in H_{0,d} H_{i,i} \subseteq H_{0,d}$ and hence, by the unique representation of elements of S obtained above, we have $ag = ha$ for a unique element h in G_0. Now writing $h = g\alpha$, we obtain a function $\alpha \colon T \to G_0$. The same argument as in the last two parts of the proof of II.6.2 shows that α is a homomorphism and χ is an isomorphism of S onto $B(T, \alpha)$.

Sufficiency. This follows easily from II.5.2, II.5.7, and II.5.13.

XI.4.6 Exercises

(i) Which Green's relations on a simple ω-regular semigroup are congruences?

(ii) For which sets X does $\mathcal{J}(X)$ contain a simple ω-semigroup which is not a Reilly semigroup?

(iii) Show that every simple ω-semigroup can be embedded into a Reilly semigroup.

(iv) Construct all homomorphisms of one Bruck semigroup over a finite chain into another.

(v) Let $T = \{G_0 > G_1 > \cdots > G_{n-1}\}$ be a finite chain of groups determined by homomorphisms φ_i: $G_i \to G_{i+1}$ for $i = 0, 1, \ldots, n-2$, let φ_{n-1}: $G_{n-1} \to G_0$ be a homomorphism, and let $S = B(T, \alpha)$, where α is determined by the homomorphisms φ_i. Show that S is E-unitary if and only if all φ_i are one-to-one.

(vi) Let S be an inverse semigroup. For any equivalence relation ρ on S, let ρ^0 denote the greatest congruence on S contained in ρ. Show that

$$\mathcal{H}^0 = \mathcal{L}^0 = \mathcal{R}^0 \subseteq \mathcal{D}^0 \subseteq \mathcal{J}^0,$$

and that in general $\mathcal{H}^0 \neq \mathcal{D}^0 \neq \mathcal{J}^0$.

XI.4.7 Problems

(i) Let T be a monoid and α be a homomorphism of T into its group of units. What are necessary and sufficient conditions on T and α in order that $B(T, \alpha)$ be E-unitary?

The theorem of this section is due to Kočin [1]; the proof here generally follows Howie [4]. An alternative construction of simple ω-regular semigroup was given independently by Munn [9]. Ault [4] investigated simple ω-regular subsemigroups of a given semigroup. Baird [1], [2] studied certain sublattices of the congruence lattice on simple ω-regular semigroups; all congruences on them were described by Petrich [12].

XI.5 GENERAL ω-REGULAR SEMIGROUPS

We classify ω-regular semigroups as follows:

(α) those which coincide with their kernel,

(β) those with a proper kernel,

(γ) those without a kernel.

Those of type (α) are simple, for they cannot have a proper ideal, and hence their structure is given in 4.5. For the remaining two types, the following simple statement will be useful.

XI.5.1 Lemma

A regular semigroup whose idempotents form a finite chain is a finite chain of groups.

Proof. Let S be a regular semigroup with E_S consisting of the chain $e_1 > e_2 > \cdots > e_n$. Note that S is an inverse semigroup. Let $a \in S$; then $e_m \mathcal{R} a \mathcal{L} e_k$ for some $1 \leq m, k \leq n$. In the Munn representation, δ^a is an isomorphism of $[aa^{-1}] = [e_m]$ onto $[a^{-1}a] = [e_k]$, which implies that $e_m = e_k$ since both chains $[e_m]$ and $[e_k]$ are finite. But then $a \mathcal{H} e_m$ which proves that S is a finite chain of groups.

The next theorem gives the global structure of ω-regular semigroups with proper kernel.

XI.5.2 Theorem

A semigroup S is an ω-regular semigroup with a proper kernel if and only if S is an ideal extension of a simple ω-regular semigroup K by a finite chain of groups with a zero adjoined Q determined by a homomorphism ψ of Q^* into the group of units of K.

Proof

Necessity. Let S be an ω-regular semigroup with proper kernel K. Then K is a simple semigroup by I.5.7, and hence it is isomorphic to a Bruck semigroup over a finite chain of groups in view of 4.5 since it is obviously an ω-regular semigroup.

Let $Q = S/K$. Since K is proper, Q has at least two elements. Thus Q contains nonzero idempotents. Let E_S consist of the chain $e_1 > e_2 > e_3 > \cdots$. It follows that for some $n \geq 1$, we have $e_i \in K$ if and only if $i > n$, and thus $E_{Q^*} = \{e_1, e_2, \ldots, e_n\}$. Now 5.1 asserts that Q is a chain of groups with idempotents $e_1, e_2, \ldots, e_n, 0$, so that Q is a chain of groups with a zero adjoined.

Since K is a monoid, it is a retract ideal of S by I.9.16. By I.9.14, the extension S of K by Q is determined by a partial homomorphism $\psi: Q^* \to K$. Let e be the identity of K and let $f \in E_{Q^*}$. Then $f > e$ and thus $e = ef = e(f\psi)$ so that $e \leq f\psi$. But then $e = f\psi$ since e is the highest idempotent in K. Let $x \in Q^*$. Then $x \mathcal{H} f$ for some $f \in E_{Q^*}$ and thus $(x\psi) \mathcal{H} (f\psi)$ since ψ is a homomorphism. But then $(x\psi) \mathcal{H} e$. Consequently, ψ maps Q^* into the group of units of K.

Sufficiency. This is verified without difficulty.

The definition of a Bruck semigroup over a finite chain of groups as well as the situation in the above theorem involve a homomorphism of a finite chain of groups into a group. We know by II.2.6 that every semilattice of groups is strong. If that semilattice is either a finite chain or an ω-chain, it is easily seen that the required system of homomorphisms among these groups can be reduced to the following:

$$G_1 \xrightarrow{\varphi_1} G_2 \xrightarrow{\varphi_2} G_3 \longrightarrow \quad \cdots \xrightarrow{\varphi_{n-1}} G_n \tag{35}$$

$$G_1 \xrightarrow{\varphi_1} G_2 \xrightarrow{\varphi_2} G_3 \longrightarrow \cdots \quad \xrightarrow{\varphi_{n-1}} G_n \xrightarrow{\varphi_n} \cdots \tag{36}$$

respectively. For we can define φ_{ij} by: $\varphi_{ii} = \iota_{G_i}$ and $\varphi_{ij} = \varphi_i \varphi_{i+1} \cdots \varphi_{j-1}$ if $i > j$ so that the multiplication is given by

$$g * h = (g\varphi_{ik})(h\varphi_{jk})$$

if $g \in G_i$, $h \in G_j$, and $k = \max\{i, j\}$. We will always assume the existence of homomorphisms φ_i and call them *connecting homomorphisms* for the semigroup. We can express homomorphisms of finite chains of groups into a group as follows.

XI.5.3 Lemma

Let T be a finite chain of groups with connecting homomorphisms as in (35) and let G be a group. Let φ_n: $G_n \to G$ be a homomorphism, and on T define a function φ by

$$\varphi\colon g \to g\varphi_i \varphi_{i+1} \cdots \varphi_{n-1}\varphi_n \quad \text{if } g \in G_i.$$

Then φ is a homomorphism of T into G. Conversely, every homomorphism of T into G can be so obtained. Moreover, φ_n and φ determine each other uniquely and $\phi|_{G_n} = \varphi_n$.

Proof. Let φ be as defined above. If $g \in G_i$, $h \in G_j$, and $i < j$, then

$$\begin{aligned}(g\varphi)(h\varphi) &= (g\varphi_i \cdots \varphi_n)(h\varphi_j \cdots \varphi_n) = [(g\varphi_i \cdots \varphi_{j-1})h]\varphi_j \cdots \varphi_n \\ &= (gh)\varphi_j \cdots \varphi_n = (gh)\varphi;\end{aligned}$$

the case $i = j$ is trivial, and the case $i > j$ is analogous. Consequently, φ is a homomorphism.

Conversely, let φ: $T \to G$ be a homomorphism and let $\varphi_n = \varphi|_{G_n}$. Let e_i be the identity of G_i for $i = 1, 2, \ldots, n$ and e be the identity of G. Note that $g\varphi_i = ge_{i+1} = e_{i+1}g$ for any $g \in G_i$ and $1 \le i < n$. Letting $g \in G_i$ and $1 \le i < n$, we have

$$\begin{aligned}g\varphi &= (g\varphi)e = (g\varphi)(e_n\varphi) = (ge_n)\varphi = (ge_{i+1} \cdots e_n)\varphi \\ &= g\varphi_i\varphi_{i+1} \cdots \varphi_{n-1}\varphi_n,\end{aligned}$$

as required.

The last two assertions of the lemma are obvious.

The lemma indicates that any homomorphism of T into G is in fact uniquely determined by its restriction to the lowest maximal subgroup of T, that is, the kernel of T. We can apply this to the following situations.

XI.5.4 Remark

Let $S = B(T, \alpha)$ where T is a finite chain of groups as given in (35). Letting $\varphi_n = \alpha|_{G_n}$, we obtain a circular diagram of homomorphisms:

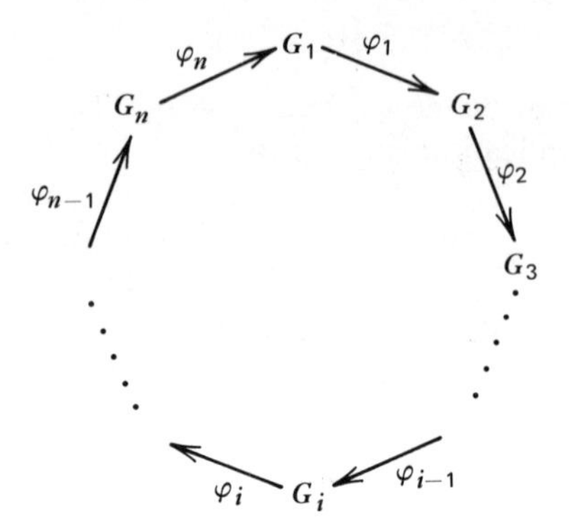

According to 5.3, such a diagram uniquely determines α and thus also S.

XI.5.5 Remark

For the semigroups in 5.2, we have the circular diagram above and for the finite chain of groups Q^*, the homomorphisms

$$H_1 \xrightarrow{\psi_1} H_2 \xrightarrow{\psi_2} \cdots \xrightarrow{\psi_{m-1}} H_m.$$

For the homomorphism ψ of Q^* into the group of units G_1 of K, letting $\psi_m = \psi|_{H_m}$, we have the "big dipper" diagram of homomorphisms:

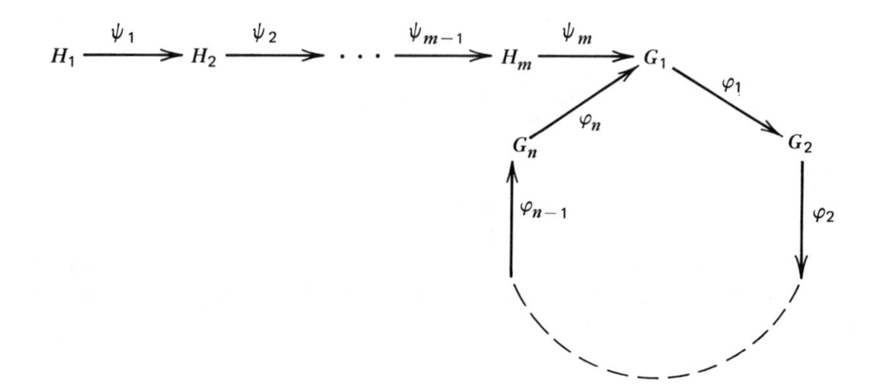

According to 4.5, 5.2, and 5.3, such diagrams determine ω-regular semigroups with proper kernel.

Finally, we can take care of ω-regular semigroups without kernel. For it, we will need the following concept.

XI.5.6 Definition

A semigroup is an *ω-chain of groups* if it is a semilattice of groups indexed by an ω-chain.

XI.5.7 Theorem

A semigroup S is an ω-regular semigroup without kernel if and only if S is an ω-chain of groups.

Proof

Necessity. Let S be as stipulated, and let I be a proper ideal of S. Then S/I is a regular semigroup whose idempotents form a finite chain. According to 5.1, we have that S/I is a finite chain of groups. It follows that S is an ω-chain of these groups since it has no kernel.

Sufficiency. Let S be given by connecting homomorphisms as in (36). Then for any $n \in \mathbb{N}$, $\bigcup_{i=n}^{\infty} G_i$ is an ideal of S, and the intersection of all such ideals is empty. Hence S has no kernel. It is obvious that S is an ω-regular semigroup.

At the beginning of this section, we distinguished three types of ω-regular semigroups, namely (α), (β), and (γ). It is clear that if two ω-regular semigroups are isomorphic, they must be of the same type. We now discuss conditions and expressions for isomorphisms for each type of ω-regular semigroup.

For simple ω-regular semigroups, we may use the same notation as for the bisimple case treated in 1.1. In view of the fact that we are interested here only in isomorphisms, we have that ω is an isomorphism, $k = 1$ and $l = 0$, so that the notation $\eta(\omega, z)$ may be used.

XI.5.8 Proposition

Let $S = B(G, \alpha)$ and $T = B(H, \beta)$, where G and H are finite chains of groups. Let ω be an isomorphism of G onto H and z be an element of the group of units of H satisfying $\alpha\omega\varepsilon_z = \omega\beta$. Define a function $\eta = \eta(\omega, z)$ on S by

$$\eta\colon (m, g, n) \to \left(m, z_m^{-1}(g\omega)z_n, n\right)$$

where z_0 is the identity of H and

$$z_p = z(z\beta)(z\beta^2)\cdots(z\beta^{p-1}) \quad \text{if } p > 0.$$

Then η is an isomorphism of S onto T. Conversely, every isomorphism of S onto T can be written uniquely in the form $\eta(\omega, z)$.

Proof. The argument here goes along the same lines as in the case of a Reilly semigroup in 1.1 with the simplification arising from the fact that $k = 1$ and $l = 0$. The reason for this is that z is in the group of units of D and thus manipulation with z is the same as in the group case. The verification of these assertions is left as an exercise.

We can now obtain conditions for isomorphism of two Bruck semigroups over finite chains of groups in a more explicit form by elaborating upon the isomorphisms of finite chains of groups.

XI.5.9 Corollary

Let $S = B(G, \alpha)$ and $T = B(H, \beta)$ where G is a chain of groups $G_1 > G_2 > \cdots > G_m$ with connecting homomorphisms φ_i and H is a chain of groups $H_1 > H_2 > \cdots > H_n$ with connecting homomorphisms ψ_i. Let α and β be determined by homomorphisms $\varphi_m: G_m \to G_1$ and $\psi_n: H_n \to H_1$ as in 5.3. Then $S \cong T$ if and only if

(i) $m = n$,
(ii) there exist isomorphisms ω_i of G_i onto H_i for $i = 1, 2, \ldots, n$ satisfying

$$\varphi_i\omega_{i+1} = \omega_i\psi_i \quad \text{for } 1 \leq i < n, \qquad \varphi_n\omega_1\varepsilon_z = \omega_n\psi_n \tag{37}$$

for some $z \in H_1$ with ε_z an inner automorphism of H_1.

Proof

Necessity. Assume that $S \cong T$. Then 5.8 implies the existence of an isomorphism ω of G onto H satisfying $\alpha\omega\varepsilon_z = \omega\beta$ for some $z \in H_1$. Now II.2.8 provides the isomorphisms ω_i satisfying the first part of (37) since $G \cong H$ obviously forces $m = n$. The second part of (37) follows from $\alpha\omega\varepsilon_z = \omega\beta$ by restriction to G_n.

Sufficiency. In view of 5.8, it remains to show that conditions (37) imply $\alpha\omega\varepsilon_z = \omega\beta$ in the obvious notation. Indeed, for $g \in G_i$, we obtain

$$\begin{aligned} g\alpha\omega\varepsilon_z &= g\varphi_i \cdots \varphi_n\omega_1\varepsilon_z = g\varphi_i \cdots \varphi_{n-1}\omega_n\psi_n \\ &= g\varphi_i \cdots \varphi_{n-2}\omega_{n-1}\psi_{n-1}\psi_n \\ &= \cdots = g\omega_i\psi_i \cdots \psi_n = g\omega\beta, \end{aligned}$$

as required.

We can illustrate the situation in the above corollary by commutativity of the following diagram:

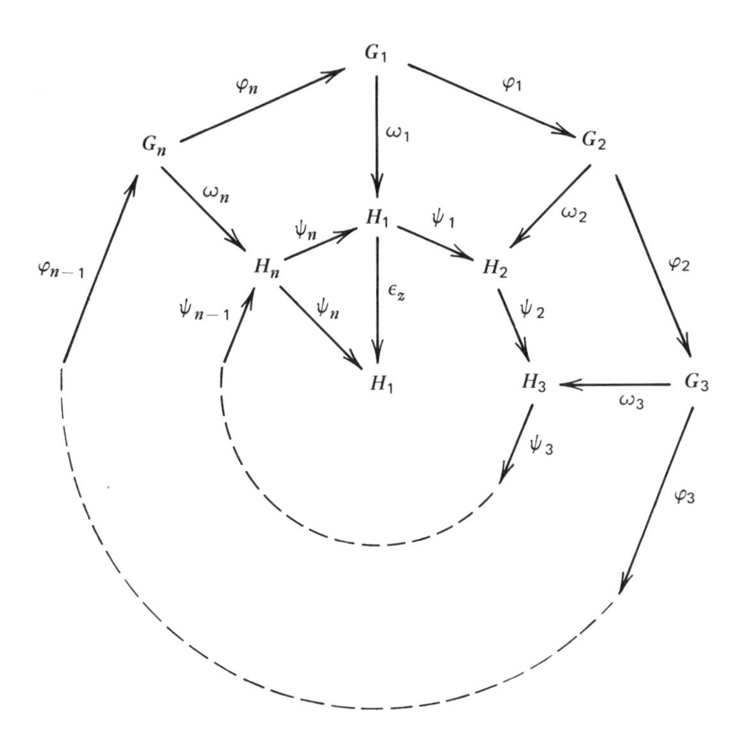

We consider next isomorphisms of ω-regular semigroups with proper kernel and start with an auxiliary statement.

XI.5.10 Lemma

Let S and T be ω-regular semigroups with proper kernels K and L, respectively. Let $Q = S/K$ and $R = T/L$ and let φ: $Q^* \to K$ and ψ: $R^* \to L$ be homomorphisms which determine the multiplications in S and T, respectively. Further, let ξ and η be isomorphisms of K onto L and of Q onto R, respectively, satisfying the condition $\varphi\xi = (\eta|_{Q^*})\psi$. On S define a mapping χ by the requirement that $\chi|_K = \xi$, $\chi|_{Q^*} = \eta|_{Q^*}$. Then χ is an isomorphism of S onto T. Conversely, every isomorphism of S onto T can be so obtained.

Proof. The proof of the direct part reduces to a simple verification that $(xy)\chi = (x\chi)(y\chi)$ for $x \in K$, $y \in Q^*$ and for $x \in Q^*$, $y \in K$, which presents no difficulty. For the converse, one sets $\xi = \chi|_K$, $0\eta = 0$, $\eta|_{Q^*} = \chi|_{Q^*}$ and verifies easily that ξ and η fulfill all the requirements.

We thus arrive at the following isomorphism criterion for ω-regular semigroups with proper kernel.

XI.5.11 Proposition

Let S and T be ω-regular semigroups represented as in 5.5 by

$$G_1 \xrightarrow{\varphi_1} G_2 \longrightarrow \cdots \longrightarrow G_m \xrightarrow{\varphi_m} G_{m+1} \longrightarrow \cdots \longrightarrow G_{m+n} \xrightarrow{\varphi_{m+n}} G_{m+1},$$

$$H_1 \xrightarrow{\psi_1} H_2 \longrightarrow \cdots \longrightarrow H_k \xrightarrow{\psi_k} H_{k+1} \longrightarrow \cdots \longrightarrow H_{k+l} \xrightarrow{\psi_{k+1}} H_{k+1},$$

respectively. Then $S \cong T$ if and only if

(i) $m = k$, $n = l$,
(ii) there exist isomorphisms ω_i of G_i onto H_i for $i = 1, 2, \ldots, m + n$ and $z \in H_{m+1}$ satisfying

$$\varphi_i \omega_{i+1} = \omega_i \psi_i \quad \text{for } 1 \leq i \leq m + n, \qquad \varphi_{m+n} \omega_{m+1} \varepsilon_z = \omega_{m+n} \psi_{m+n}.$$

Proof. This follows by a straightforward argument from 5.5, II.2.8, 5.9, and 5.10; the details are left as an exercise.

The conditions of the above theorem can be represented by commutativity of the following diagram:

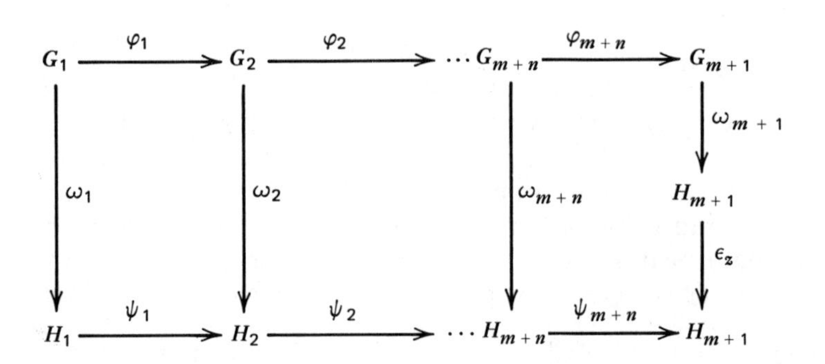

For ω-regular semigroups without kernel, we have the following simple result.

XI.5.12 Proposition

Let S and T be ω-chains of groups represented by connecting homomorphisms

$$G_1 \xrightarrow{\varphi_1} G_2 \xrightarrow{\varphi_2} G_3 \longrightarrow \cdots,$$

$$H_1 \xrightarrow{\psi_1} H_2 \xrightarrow{\psi_2} H_3 \longrightarrow \cdots,$$

respectively. For each positive integer i, let ω_i be an isomorphism of G_i onto H_i, and assume that $\varphi_i \omega_{i+1} = \omega_i \psi_i$ for $i = 1, 2, \ldots$. On S define a function ω by the requirement that $\omega|_{G_i} = \omega_i$ for all $i \in \mathbb{N}$. Then ω is an isomorphism of S onto T. Conversely, every isomorphism of S onto T can be obtained in this fashion.

Proof. This follows easily from II.2.8.

Finally, the conditions of the above theorem amount to commutativity of the diagram

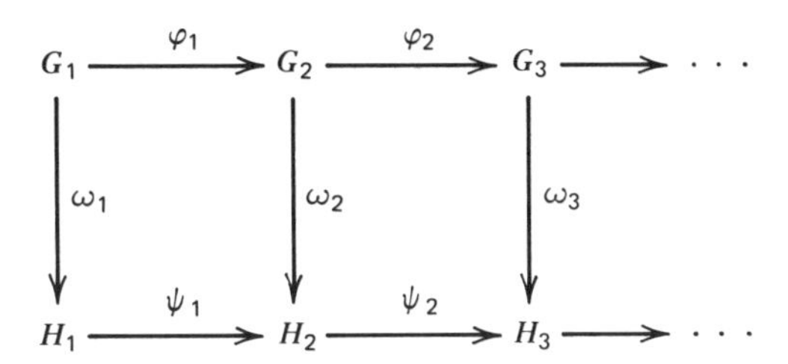

XI.5.13 Exercises

(i) Which Green's relations on an ω-regular semigroup are congruences?

(ii) Prove that every ω-regular semigroup can be embedded into a Reilly semigroup.

(iii) Construct all homomorphisms of an ω-chain of groups into another.

(iv) Characterize the semigroup of endomorphisms of an ω-chain.

(v) Show that the following conditions on a semigroup S are equivalent.

(α) S is a semilattice of simple semigroups.

(β) S is *intraregular* (that is, $a \in Sa^2S$ for all $a \in S$).

(γ) $\eta = \mathcal{J}$.

(vi) Show that a semigroup S is a semilattice of groups and simple ω-regular semigroups if and only if S is regular, intraregular, and for every $e \in E_S$, the partially ordered set

$$\{f \in E_S | f \geq e, f \mathcal{J} e\}$$

is a finite chain.

(vii) Construct explicitly all congruences on an ω-chain of groups.

The structure theory for ω-regular semigroups was completed by Munn [9]. His work, as well as that of Kočin [1], was obviously inspired by the Reilly theorem proved in II.6.2. This theorem created an impetus for much work in the direction of establishing structure theorems for regular semigroups whose idempotents involve, but are more complex than, ω-chains. This work includes papers by Ault–Petrich [1], Feller–Gantos [1], Hogan [1]–[4], Lallement [3], McLean [1], Mills [1], Nivat–Perrot [1] (for a correct version of their theorem see Lallement [4]), Perrot [1], Warne [7]–[14], Warne–Hogan [1], and White [1]. The isomorphism theorem 5.8 is due to Kočin [1]. The explicit form for isomorphisms of all types of ω-regular semigroups in terms of group isomorphisms is due to Munn [9].

XII

VARIETIES

Inverse semigroups considered as algebras with the binary operation of multiplication and the unary operation of inversion form a variety. The study of inverse semigroup varieties, and of the lattice of all such varieties, contributes in an essential way to a better understanding of the structure of inverse semigroups as well as provides a frame for a possible classification. For some special varieties, we are able to make the structure of inverse semigroups quite transparent by the knowledge of the identities they satisfy.

1. The groundwork for the study of these varieties is laid by providing a base for the identities of the variety $\mathcal{I}$ of all inverse semigroups. The cardinality of the lattice $\mathcal{L}(\mathcal{I})$ of varieties of inverse semigroups is that of the continuum. The variety $\mathcal{I}$ is not a join of a finite number of its proper subvarieties. It is generated by its finite combinatorial members.

2. The mapping $\mathcal{V} \to \mathcal{V} \vee \mathcal{G}$, where $\mathcal{G}$ is the variety of all groups, is a homomorphism of the lattice $\mathcal{L}(\mathcal{I})$ onto its interval $[\mathcal{G}, \mathcal{I}]$. The congruence induced on $\mathcal{L}(\mathcal{I})$ by this homomorphism exhibits some interesting properties and admits a characterization in terms of antigroups.

3. Analogously, the mapping $\mathcal{V} \to \mathcal{V} \cap \mathcal{G}$ is a homomorphism of $\mathcal{L}(\mathcal{I})$ onto $\mathcal{L}(\mathcal{G})$. The congruence induced by it also merits special attention. The intersection of the two congruences mentioned is not the equality relation.

4. The lowest three levels of the lattice $\mathcal{L}(\mathcal{I})$ consist of varieties of groups, Clifford semigroups, and strict inverse semigroups. The structure of these semigroups is well understood and all varieties of these are characterized in terms of identities (everything modulo group varieties).

5. The varieties among those just discussed whose intersection with groups is a variety of abelian groups are called here quasiabelian strict varieties. Using the description of the lattice of all abelian group varieties, one can give a concrete isomorphic copy of the lattice of all these varieties.

6. Small varieties are those which are generated by groups, semilattices, and monogenic inverse semigroups. For the lattice of these varieties one also has a suitable concrete isomorphic copy.

7. Varieties of completely semisimple inverse semigroups can be characterized by identities. This holds also for those among them all of whose members are cryptic ($\mathcal{H}$ is a congruence). There are interesting homomorphisms among various lattices of completely semisimple varieties.

8. For cryptic completely semisimple varieties, the classes of the congruence which is induced by the homomorphism $\mathcal{V} \to (\mathcal{V} \vee \mathcal{G}, \mathcal{V} \cap \mathcal{G})$ turn out to be intervals of the lattice $\mathcal{L}(\mathcal{I})$. Its least and greatest elements can be expressed by means of groups and antigroups in $\mathcal{V}$, the latter by using the Malcev product of varieties.

9. Varieties $\mathcal{V}$ of inverse semigroups with the property that each member of $\mathcal{V}$ has an E-unitary cover in $\mathcal{V}$ admit several interesting characterizations. Inverse semigroup varieties $\mathcal{V}$ having E-unitary covers for its members over groups in a fixed group variety $\mathcal{U}$ can be characterized in terms of the Malcev product.

XII.1 GENERALITIES

We have discussed varieties of semigroups in I.11. The same definitions carry over to any species of algebras and in our case to the class of inverse semigroups. In this case, the variables, figuring in an identity, may appear with an inverse. *We will use the definitions and results in* I.11 *as if they were stated for inverse semigroups.*

We start by proving that inverse semigroups, as algebras with the operations of multiplication and inversion, form a variety. Some useful lemmas are proved here which will be needed later and are of independent interest.

XII.1.1 Theorem

Considered as algebras with the two operations of multiplication and inversion, inverse semigroups form a variety determined by associativity and by either of the following systems of identities:

$$\text{(I1)} \quad x = xx^{-1}x, \qquad (x^{-1})^{-1} = x, \qquad x^{-1}xy^{-1}y = y^{-1}yx^{-1}x,$$

$$\text{(I2)} \quad x = xx^{-1}x, \qquad (x^{-1})^{-1} = x, \qquad (xy)^{-1} = y^{-1}x^{-1},$$

$$xx^{-1}x^{-1}x = x^{-1}xxx^{-1}.$$

Moreover, every (multiplicative) homomorphism of, and congruence on, inverse semigroups preserves inverses.

Proof. We have defined an inverse semigroup S in II.1.1 by the property that the system of equations $a = axa$, $x = xax$ has a unique solution for any $a \in S$, denoted by a^{-1}. Hence S satisfies $x = xx^{-1}x$ and by II.1.3 also $(x^{-1})^{-1} = x$. By II.1.2, idempotents of S commute, so S also satisfies $x^{-1}xy^{-1}y = y^{-1}yx^{-1}x$. This establishes the validity of system (I1) for inverse semigroups.

Let S satisfy (I1). Then S is regular, and in order to prove that S is an inverse semigroup, in view of II.1.2, it suffices to show that its idempotents commute. Because of the identity $x^{-1}xy^{-1}y = y^{-1}yx^{-1}x$, it remains to show that every idempotent of S is of the form $x^{-1}x$. Hence let $e \in E_S$. Then

$$e^{-1} = e^{-1}ee^{-1} = (e^{-1}e)(ee^{-1}) = (ee^{-1})(e^{-1}e)$$

whence $ee^{-1} = e^{-1}$ which implies $e = e^{-1}e$.

If S is an inverse semigroup, then it satisifies system (I2) in view of II.1.3.

Conversely, let S satisfy (I2). Then S is regular, and again by II.1.2, it remains to prove that its idempotents commute. Let $e, f \in E_S$. As above, we obtain $e = e^{-1}e$ whence

$$e^{-1} = (e^{-1}e)^{-1} = e^{-1}e = e.$$

Hence

$$ef = (ef)(ef)^{-1}(ef) = (ef)(f^{-1}e^{-1})(ef) = (ef)(fe)(ef) = (ef)^2$$

and thus $fe = f^{-1}e^{-1} = (ef)^{-1} = ef$, as required.

For the last assertion of the theorem, see III.1.1 and its proof.

We will need the following result; for a proof, we refer to Olšanskiĭ [1] or Vaughan-Lee [1].

XII.1.2 Lemma

The cardinality of the lattice of all varieties of groups is that of the continuum.

Everything we have seen in I.10 concerning defining relations and in I.11 concerning identities and varieties carries over verbatim to the situation where we consider inverse semigroups. The identities may now involve inverses, varieties are closed under taking of inverse subsemigroups, and so on, and the free object is I_X. We have the following result concerning the entire lattice of varieties of inverse semigroups.

XII.1.3 Proposition

The set of all varieties of inverse semigroups is a complete lattice under inclusion and has the cardinality of the continuum.

Proof. The argument for the upper bound for the cardinality of this lattice is the same as in I.11.5. The lower bound is given by 1.2, and the conclusion follows. The intersection of a nonempty set of varieties is evidently a variety, and hence by I.2.6, we have a complete lattice.

XII.1.4 Notation

We will consistently use the following notation:

(i) $\mathcal{I}$—the variety of inverse semigroups,
(ii) $\mathcal{L}(\mathcal{V})$—the lattice of all subvarieties of a variety $\mathcal{V}$ of inverse semigroups,
(iii) $\mathcal{G}$—the variety of groups,
(iv) $\mathcal{S}$—the variety of semilattices,
(v) $\mathcal{T}$—the variety of trivial inverse semigroups,
(vi) $[u_\alpha = v_\alpha]_{\alpha \in A}$—the variety of inverse semigroups satisfying $u_\alpha = v_\alpha$ for all $\alpha \in A$,
(vii) $\langle \mathcal{C} \rangle$—the variety of inverse semigroups generated by a nonempty class $\mathcal{C}$ of inverse semigroups; if $\mathcal{C} = \{S\}$, we write $\langle S \rangle$ instead of $\langle \mathcal{C} \rangle$.
(viii) $\mathcal{C}_n = [x^n = x^{n+1}]$ for every natural number n.
(ix) 2—a two-element set.

From IX.3.9 and IX.3.10, we immediately have the following consequence.

XII.1.5 Corollary

For any natural numbers k and l,

(i) $I_x/\rho_{(k,\omega)}$ is $[y^{-1}y^k = y^k y^{-1}]$-free,
(ii) $I_x/\rho_{(k,l)}$ is $[y^k = y^{k+l}]$-free.

An interesting property of the variety $\mathcal{I}$ is given in the following result.

XII.1.6 Theorem

The variety $\mathcal{I}$ is not a join of a finite number of its proper subvarieties.

Proof. It evidently suffices to show that the join of two proper subvarieties of $\mathcal{I}$ is again a proper subvariety of $\mathcal{I}$. Hence let $\mathcal{U}$ and $\mathcal{V}$ be proper subvarieties of $\mathcal{I}$. Let X be a countably infinite set and (F, φ) be a $\mathcal{U}$-free inverse semigroup on X. Then $X\varphi$ does not generate a free inverse semigroup since $\mathcal{U}$ is a proper subvariety of $\mathcal{I}$. Hence either VIII.4.4(i) or (ii) fails to hold. If VIII.4.4(i) fails, then $X \cap X^{-1} \neq \varnothing$, say $x = y^{-1}$ for some $x, y \in K$. It follows that $y = y^{-1}$ is a law for $\mathcal{U}$ so that $\mathcal{U} \subseteq [y = y^{-1}]$.

Now suppose that VIII.4.4(ii) fails. There exist $y, y_{ij} \in X\varphi \cup (X\varphi)^{-1}$ satisfying all the properties in VIII.4.4(ii) except that $y \neq y_{i1}$ for $i = 1, 2, \ldots, n$. Letting $e = e_1 e_2 \cdots e_n$, we get $\lambda y \geq e$ which we write as $yy^{-1}e = e$. The equality $yy^{-1}e = e$ in F means that $yy^{-1}e = e$ is a law in $\mathcal{U}$, and thus $\mathcal{U} \subseteq [yy^{-1}e = e]$. Recall from VIII.4.4 that $e_i = \lambda(y_{i1} y_{i2} \cdots y_{in_i})$ with $y_{ij} \neq y_{i(j+1)}$ for $j = 1, 2, \ldots, n_i - 1$, $i = 1, 2, \ldots, n$. Since in the present case $y, y_{ij} \in X\varphi \cup (X\varphi)^{-1}$, in the expression for e, the various y_{ij} and y are either variables or their inverses are variables.

The same type of argument is applicable to the variety $\mathcal{V}$. We thus arrive at three possible cases.

Case 1

$$\mathcal{U} \vee \mathcal{V} \subseteq [yy^{-1}e = e] \vee [yy^{-1}f = f] \subseteq [yy^{-1}ef = ef].$$

We must show that $[yy^{-1}ef = ef]$ is a proper subvariety of $\mathcal{I}$, or equivalently, that $yy^{-1}ef = ef$ does not hold in I_X.

We adopt the notation of the necessity part of the proof of VIII.4.4. The expression for $\Delta(e)$ is obtained by taking the union over i of the expression in formula (1) in the proof of VIII.4.4. The only elements of $X\varphi \cup (X\varphi)^{-1}$ in $\Delta(e)$ are the elements $z_{11}, z_{21}, \ldots, z_{n1}$. We have seen above that $y \neq y_{i1}$ for $i = 1, 2, \ldots, n$. Hence $z \neq z_{i1}$ for $i = 1, 2, \ldots, n$. Since $z \in X \cup X^{-1}$, we conclude that $z \notin \Delta(e)$. An analogous argument shows that also $z \notin \Delta(f)$ and hence $z \notin \Delta(e) \cup \Delta(f) = \Delta(ef)$. Consequently, $\lambda y \ngeq ef$ which implies that $yy^{-1}ef \neq ef$, as required.

Case 2

$$\mathcal{U} \vee \mathcal{V} \subseteq [z = z^{-1}].$$

It is obvious that $z = z^{-1}$ is not a law in I_X.

Case 3

$$\mathcal{U} \vee \mathcal{V} \subseteq [yy^{-1}e = e] \vee [z = z^{-1}] \subseteq [(yy^{-1})(y^{-1}y)e = y^{-1}ye].$$

We have seen above that

$$\Delta(e) \cap (X \cup X^{-1}) = \{z_{11}, z_{21}, \ldots, z_{n1}\}$$

which implies

$$\Delta(y^{-1}ye) \cap (X \cup X^{-1}) = \{z^{-1}, z_{11}, z_{21}, \ldots, z_{n1}\}$$

whereas

$$\Delta(yy^{-1}y^{-1}ye) \cap (X \cup X^{-1}) = \{z, z^{-1}, z_{11}, \ldots, z_{n1}\}.$$

It follows that $(yy^{-1})(y^{-1}y)e \neq y^{-1}ye$ and hence $[(yy^{-1})(y^{-1}y)e = y^{-1}ye] \neq \mathcal{I}$.

Therefore in all cases $\mathcal{U} \vee \mathcal{V} \neq \mathcal{I}$.

The following result will be useful in the next section and in the next chapter. It is also of independent interest.

XII.1.7 Lemma

The variety of all inverse semigroups is generated by its finite combinatorial members.

Proof. Let X be a nonempty set. Fix any element $(A, g) \in I_X$, and let

$$F_{(A,g)} = \{(B, h) \in I_X | J_{(A,g)} \leq J_{(B,h)}\}.$$

We now compute

$$\begin{aligned} J_{(A,g)} \leq J_{(B,h)} &\Leftrightarrow J((A, g)) \subseteq J((B, h)) \\ &\Leftrightarrow (A, g) = (C, s)(B, h)(D, t) \\ &\Leftrightarrow (A, g) = (C \cup sB \cup shD, sht) \\ &\Leftrightarrow A = C \cup sB \cup shD, \quad g = sht \end{aligned}$$

for some $(C, s), (D, t) \in I_X$. If this is the case, $sB \subseteq A$, $s \in C \subseteq A$, and $h \in B$. Since A is finite, there is only a finite number of choices for B and hence also for h. It follows that the set $F_{(A,g)}$ is finite. It is clear that $F_{(A,g)}$ has the following property: $xy \in F_{(A,g)}$ implies $x, y \in F_{(A,g)}$ for any $x, y \in S$. Consequently, the set $I_{(A,g)} = I_X \setminus F_{(A,g)}$ is an ideal of I_X if nonempty. Let $\varphi_{I_{(A,g)}}: I_X \to I_X/I_{(A,g)}$ be the natural homomorphism (we formally set $I_X/\varnothing = I_X$).

In this way, for every $\mathcal{J}$-class J of I_X, we obtain an ideal $\bar{J}$ ($= I_{(A,g)}$ if $J = J_{(A,g)}$) such that $I_X/\bar{J}$ is finite, with $\varphi_{\bar{J}}: I_X \to I_X/\bar{J}$ the natural homomorphism. Let x and y be two distinct elements of I_X. If $J_x = J_y$, then $x\varphi_{\bar{J}_x} \neq y\varphi_{\bar{J}_x}$; if $J_x \nleq J_y$, then $x\varphi_{\bar{J}_x} \neq 0 = y\varphi_{\bar{J}_x}$, and if $J_x < J_y$, then $x\varphi_{\bar{J}_y} = 0 \neq y\varphi_{\bar{J}_y}$. It follows that the mapping

$$\varphi: x \to (x\varphi_{\bar{J}})_{J \in I_X/\mathcal{J}}$$

is one-to-one, and it is obviously a homomorphism. Thus $\varphi: I_X \to \prod_{J \in S/\mathcal{J}} I_X/\bar{J}$ is an embedding.

We now let $\mathcal{V}$ be a variety of inverse semigroups which contains all finite combinatorial inverse semigroups. Then $I_X/\bar{J} \in \mathcal{V}$ and thus also $\prod_{J \in I_X/\mathcal{J}} I_X/\bar{J}$ and finally also I_X. But then $\mathcal{V}$ contains all free inverse semigroups, and thus also all inverse semigroups.

XII.1.8 Proposition

$$\bigvee_{n=1}^{\infty} \mathcal{C}_n = \mathcal{I}.$$

Proof. First let S be a finite combinatorial inverse semigroup. For every $s \in S$, the cyclic semigroup $[s]$ generated by s is finite, hence $s^n = s^{n+m}$ for some natural numbers n and m. If m were greater than 1, $[s]$ would have a nontrivial subgroup, contradicting the hypothesis that S is combinatorial. Thus $s^n = s^{n+1}$, where $n = n(s)$. Let $S = \{s_1, s_2, \ldots, s_k\}$. Then for $n = \sum_{i=1}^{k} n(s_i)$ we immediately obtain that $s_i^n = s_i^{n+1}$ for $i = 1, 2, \ldots, k$. Thus $S \in \mathcal{C}_n$.

By 1.7, the variety $\mathcal{I}$ is generated by its finite combinatorial members, and hence by the above $\mathcal{I} \subseteq \vee_{n=1}^{\infty} \mathcal{C}_n$. Since the opposite inclusion is trivial, the assertion of the proposition follows.

The following concept will be used extensively.

XII.1.9 Definition

A variety $\mathcal{V}$ of inverse semigroups is *combinatorial* if all its members are combinatorial semigroups.

These varieties admit simple characterizations as follows.

XII.1.10 Lemma

The following conditions on a variety $\mathcal{V}$ of inverse semigroups are equivalent.

(i) $\mathcal{V}$ is combinatorial.
(ii) $\mathcal{V} \subseteq \mathcal{C}_n$ for some n.
(iii) $\mathcal{V} \cap \mathcal{G} = \mathcal{T}$.

Proof. (i) *implies* (ii). Suppose $\mathcal{V} \nsubseteq \mathcal{C}_n$ for all positive integers n. Then for each n, there exists S_n in $\mathcal{V}$ and an element $a_n \in S_n$ such that $a_n^{n+1} \neq a_n^n$. Letting $\langle a_n \rangle$ be the monogenic inverse semigroup generated by a_n, we consider $T = \prod_{n=1}^{\infty} \langle a_n \rangle$. Let $a = (a_1, a_2, a_3, \ldots)$. Then a is of infinite order. Hence the monogenic inverse semigroup $\langle a \rangle$ generated by a is isomorphic to $I_X/\rho_{(k,\omega)}$ or to $I_X/\rho_{(k,\infty^-)}$ or to $I_X/\rho_{(k,\infty^+)}$. In any case, using IX.3.4, $\langle a \rangle \in \mathcal{V}$ implies that $\mathbb{Z} \in \mathcal{V}$, and thus $\mathcal{V}$ is not combinatorial. By contrapositive, if $\mathcal{V}$ is combinatorial, then $\mathcal{V} \subseteq \mathcal{C}_n$ for some n.

(ii) *implies* (iii). If $G \in \mathcal{C}_n \cap \mathcal{G}$, then for any element a of G, $a^{n+1} = a^n$ and thus $a^2 = 1$. Hence $\mathcal{V} \cap \mathcal{G} \subseteq \mathcal{C}_n \cap \mathcal{G} = \mathcal{T}$.

(iii) *implies* (i). For any $S \in \mathcal{V}$, all maximal subgroups of S must be trivial; hence $\mathcal{H} = \varepsilon$ in S, and S is combinatorial.

XII.1.11 Exercises

(i) Prove that the varieties of Clifford semigroups and the variety of all inverse semigroups are the only varieties of inverse semigroups closed under forming of dense ideal extensions (equivalently under taking of the translational hull).

(ii) Prove that (α) trivial semigroups, (β) groups, (γ) semilattices, and (δ) inverse semigroups are the only varieties of inverse semigroups closed under forming of normal extensions.

(iii) Show that if ρ is a fully invariant congruence on an inverse semigroup, so is $\rho_{\min}$.

(iv) Give an example of an inverse semigroup in which μ is not fully invariant. (Hence the preceding exercise is not true for $\rho_{\max}$ instead of $\rho_{\min}$.)

(v) Show that for any $\mathcal{V} \in \mathcal{L}(\mathcal{I})$, the interval $[\mathcal{V}, \mathcal{V} \vee \mathcal{G}]$ is a complete modular sublattice of $\mathcal{L}(\mathcal{I})$.

(vi) Let $\mathcal{U} \in \mathcal{L}(\mathcal{G})$ and $\mathcal{V} \in \mathcal{L}(\mathcal{I})$ be such that $\mathcal{V} \subseteq \mathcal{C}_n$. Show that $(\mathcal{U} \vee \mathcal{V}) \cap \mathcal{C}_n = \mathcal{V}$.

XII.1.12 Problems

(i) (P. R. Jones) Is every relatively free inverse semigroup completely semisimple?

(ii) Characterize varieties of inverse semigroups generated by their anti-groups.

Systems of axioms for inverse semigroups were investigated by Schein [9], [10]. Results 1.6 and 1.8 are due to Kleiman [3]; for 1.7, see Munn [13] and Hall [6]. A characterization of semigroup varieties consisting solely of inverse semigroups was given by Klun [1]. For interesting properties of inverse semigroup varieties consult Martinov–Martinova [1] and Rasin [1], and for further properties of combinatorial varieties Kleiman [7]. See also Perkins [1].

XII.2 JOINS WITH GROUPS

On the lattice of all varieties of inverse semigroups we define the mapping $\varphi_1 : \mathcal{U} \to \mathcal{U} \vee \mathcal{G}$, and prove that it is a lattice homomorphism. The congruence ν_1 induced by it is of considerable interest since it provides the first decomposition of the lattice $\mathcal{L}(\mathcal{I})$ and because it has nice properties: its classes are complete modular sublattices of $\mathcal{L}(\mathcal{I})$.

XII.2.1 Lemma

Let ρ and ξ be congruences on an inverse semigroup S. If $\rho \cap \sigma = \xi \cap \sigma$, then $\operatorname{tr} \rho = \operatorname{tr} \xi$. The converse holds if S is E-unitary.

Proof. If $\rho \cap \sigma = \xi \cap \sigma$, then by III.2.5,

$$\operatorname{tr} \rho = \operatorname{tr} \rho \cap \operatorname{tr} \sigma = \operatorname{tr}(\rho \cap \sigma) = \operatorname{tr}(\xi \cap \sigma) = \operatorname{tr} \xi \cap \operatorname{tr} \sigma = \operatorname{tr} \xi.$$

Conversely, assume that S is E-unitary and $\operatorname{tr} \rho = \operatorname{tr} \xi$. Using III.2.5 and III.4.8, we get

$$\operatorname{tr}(\rho \cap \sigma) = \operatorname{tr} \rho \cap \operatorname{tr} \sigma = \operatorname{tr} \rho = \operatorname{tr} \xi = \operatorname{tr} \xi \cap \operatorname{tr} \sigma = \operatorname{tr}(\xi \cap \sigma),$$

$$\ker(\rho \cap \sigma) = \ker \rho \cap \ker \sigma = E_S = \ker \xi \cap \ker \sigma = \ker(\xi \cap \sigma)$$

and thus by III.1.5, we have $\rho \cap \sigma = \xi \cap \sigma$.

XII.2.2 Corollary

For any $\mathcal{U}, \mathcal{V} \in \mathcal{L}(\mathcal{I})$, we have

$$\mathcal{U} \vee \mathcal{G} = \mathcal{V} \vee \mathcal{G} \Leftrightarrow \operatorname{tr} \rho(\mathcal{U}) = \operatorname{tr} \rho(\mathcal{V}).$$

Proof. This follows directly from I.11.11 and 2.1 since a free inverse semigroup I_X is E-unitary and an antiisomorphism of lattices maps joins onto meets.

XII.2.3 Notation

Let $\mathcal{Q}$ denote the class of all antigroups.

The next result provides a connection between antigroups and joins with groups.

XII.2.4 Theorem

For any $\mathcal{U}, \mathcal{V} \in \mathcal{L}(\mathcal{I})$,

$$\mathcal{U} \vee \mathcal{G} \subseteq \mathcal{V} \vee \mathcal{G} \Leftrightarrow \mathcal{U} \cap \mathcal{Q} \subseteq \mathcal{V} \cap \mathcal{Q}.$$

Proof. Assume that $\mathcal{U} \vee \mathcal{G} \subseteq \mathcal{V} \vee \mathcal{G}$, and let $S \in \mathcal{U} \cap \mathcal{Q}$. Then $S \in \mathcal{V} \vee \mathcal{G}$ and thus, in view of I.11.9, there exist $V \in \mathcal{V}$, $G \in \mathcal{G}$, a subdirect product $T \subseteq V \times G$, and a homomorphism φ of T onto S. Let π be the projection of T onto V. If $e, f \in E_T$ are such that $e\pi = f\pi$, then $e = f$ since the projection onto G identifies all idempotents. Hence π is idempotent separating. Also, π induces a homomorphism of T/μ onto V/μ, which is then one-to-one by VII.4.7. Hence $T/\mu \in \mathcal{V}$ and since φ induces a homomorphism of T/μ onto S/μ, we also have $S/\mu \in \mathcal{V}$. But S is an antigroup so that $S \in \mathcal{V}$, which proves that $\mathcal{U} \cap \mathcal{Q} \subseteq \mathcal{V} \cap \mathcal{Q}$.

Conversely, assume that $\mathcal{U} \cap \mathcal{Q} \subseteq \mathcal{V} \cap \mathcal{Q}$, and let $S \in \mathcal{U} \vee \mathcal{G}$. By the first part of the proof, we have $(\mathcal{U} \vee \mathcal{G}) \cap \mathcal{Q} \subseteq \mathcal{U} \cap \mathcal{Q}$. Hence

$$S/\mu \in (\mathcal{U} \vee \mathcal{G}) \cap \mathcal{Q} \subseteq \mathcal{U} \cap \mathcal{Q} \subseteq \mathcal{V} \cap \mathcal{Q} \subseteq \mathcal{V}.$$

By VII.4.8, S is a homomorphic image of a subdirect product of S/μ with some group H, so that $S \in \mathcal{V} \vee \mathcal{G}$. Consequently, $\mathcal{U} \vee \mathcal{G} \subseteq \mathcal{V} \vee \mathcal{G}$.

The above theorem has very useful consequences as follows.

XII.2.5 Corollary

For any $\mathcal{U}, \mathcal{V} \in \mathcal{L}(\mathcal{I})$,

$$\mathcal{U} \vee \mathcal{G} = \mathcal{V} \vee \mathcal{G} \Leftrightarrow \mathcal{U} \cap \mathcal{Q} = \mathcal{V} \cap \mathcal{Q}.$$

XII.2.6 Corollary

For any $\mathcal{V} \in \mathcal{L}(\mathcal{I})$, we have

$$\mathcal{V} \vee \mathcal{G} = \{S \in \mathcal{I} \mid S/\mu \in \mathcal{V}\}.$$

Proof. If $S \in \mathcal{V} \vee \mathcal{G}$, then by 2.5,

$$S/\mu \in (\mathcal{V} \vee \mathcal{G}) \cap \mathcal{Q} = \mathcal{V} \cap \mathcal{Q} \subseteq \mathcal{V}.$$

Conversely, if $S/\mu \in \mathcal{V}$, then VII.4.8 implies (as in the proof of 2.4) that $S \in \mathcal{V} \vee \mathcal{G}$.

Recall the notation $c(w)$ in VIII.2.2.

XII.2.7 Corollary

If $\mathcal{V} = [u_\alpha = v_\alpha]_{\alpha \in A}$, then

$$\mathcal{V} \vee \mathcal{G} = \left[u_\alpha^{-1} t_\alpha^{-1} t_\alpha u_\alpha = v_\alpha^{-1} t_\alpha^{-1} t_\alpha v_\alpha\right]_{\alpha \in A} \tag{1}$$

where $t_\alpha \notin c(u_\alpha) \cup c(v_\alpha)$.

Proof. Let S be an inverse semigroup. Assume first that $S \in \mathcal{V} \vee \mathcal{G}$. Then by 2.6, $S/\mu \in \mathcal{V}$ and hence S/μ satisfies $u_\alpha = v_\alpha$ for all $\alpha \in A$. For any substitution for u_α and v_α in S, we then have $u_\alpha \mu = v_\alpha \mu$; this means that $u_\alpha^{-1} t^{-1} t u_\alpha = v_\alpha^{-1} t^{-1} t v_\alpha$ for all $t \in S$. But then S satisfies all identities $u_\alpha^{-1} t_\alpha^{-1} t_\alpha u_\alpha = v_\alpha^{-1} t_\alpha^{-1} t_\alpha v_\alpha$ with $t_\alpha \notin c(u_\alpha) \cup c(v_\alpha)$.

Conversely, let S be a member of the variety on the right-hand side of (1). Then S satisfies $u_\alpha^{-1} t_\alpha^{-1} t_\alpha u_\alpha = v_\alpha^{-1} t_\alpha^{-1} t_\alpha v_\alpha$ and hence for any substitution in S, $u_\alpha^{-1} t_\alpha^{-1} t_\alpha u_\alpha = v_\alpha^{-1} t_\alpha^{-1} t_\alpha v_\alpha$ and thus $u_\alpha \mu v_\alpha$. Hence S/μ satisfies $u_\alpha = v_\alpha$ for all $\alpha \in A$, and thus $S/\mu \in \mathcal{V}$. By 2.6, we then have $S \in \mathcal{V} \vee \mathcal{G}$.

We are now able to prove the following interesting result.

XII.2.8 Theorem

The mapping

$$\varphi_1 : \mathcal{V} \to \mathcal{V} \vee \mathcal{G} \qquad [\mathcal{V} \in \mathcal{L}(\mathcal{I})]$$

is a complete lattice homomorphism of $\mathcal{L}(\mathcal{I})$ onto $[\mathcal{G}, \mathcal{I}]$. Denote by ν_1 the congruence on $\mathcal{L}(\mathcal{I})$ induced by φ_1. Then for every $\mathcal{V} \in \mathcal{L}(\mathcal{I})$,

$$\mathcal{V}\nu_1 = [\langle \mathcal{V} \cap \mathcal{A} \rangle, \mathcal{V} \vee \mathcal{G}]$$

and is a complete modular sublattice of $\mathcal{L}(\mathcal{I})$.

Proof. Using the antiisomorphism ρ in I.11.11, we may consider the mapping

$$\rho(\mathcal{V}) \to \rho(\mathcal{V}) \cap \rho(\mathcal{G}) = \rho(\mathcal{V}) \cap \sigma.$$

We have seen in III.7.9(iii) that this is a complete homomorphism. It follows that φ_1 is a complete homomorphism, and it obviously maps $\mathcal{L}(\mathcal{I})$ onto $[\mathcal{G}, \mathcal{I}]$.

Let $\mathcal{N}$ be the ν_1-class of $\mathcal{V}$. It is clear that $\mathcal{V} \vee \mathcal{G}$ is the greatest element of $\mathcal{N}$. Since

$$\langle \mathcal{V} \cap \mathcal{A} \rangle \cap \mathcal{A} = \mathcal{V} \cap \mathcal{A},$$

2.5 implies that $\langle \mathcal{V} \cap \mathcal{A} \rangle \in \mathcal{N}$. For any $\mathcal{U} \in \mathcal{N}$, we have by 2.5 that $\mathcal{V} \cap \mathcal{A} = \mathcal{U} \cap \mathcal{A}$ which implies

$$\langle \mathcal{V} \cap \mathcal{A} \rangle = \langle \mathcal{U} \cap \mathcal{A} \rangle \subseteq \mathcal{U}.$$

Hence $\mathcal{N} \subseteq [\langle \mathcal{V} \cap \mathcal{A} \rangle, \mathcal{V} \vee \mathcal{G}]$. Conversely, let $\mathcal{U}$ be a variety such that $\langle \mathcal{V} \cap \mathcal{A} \rangle \subseteq \mathcal{U} \subseteq \mathcal{V} \vee \mathcal{G}$. Then

$$\langle \mathcal{V} \cap \mathcal{A} \rangle \vee \mathcal{G} \subseteq \mathcal{U} \vee \mathcal{G} \subseteq \mathcal{V} \vee \mathcal{G}$$

and since the first variety here is equal to the last, we obtain $\mathcal{U} \vee \mathcal{G} = \mathcal{V} \vee \mathcal{G}$, that is to say, $\mathcal{U}\nu_1\mathcal{V}$, and thus $\mathcal{U} \in \mathcal{N}$. Therefore, $\mathcal{N} = [\langle \mathcal{V} \cap \mathcal{C} \rangle, \mathcal{V} \vee \mathcal{G}]$.

Since $\mathcal{N}$ is an interval of the complete lattice $\mathcal{L}(\mathcal{I})$, it is a complete sublattice of $\mathcal{L}(\mathcal{I})$. In view of 2.2, the antiisomorphism ρ in I.11.11 maps $\mathcal{N}$ into a single trace class T of the lattice of congruences on I_X. By III.2.5, T is a modular lattice, and hence the image of $\mathcal{N}$ under ρ is modular. Since a lattice antiisomorphic to a modular lattice is itself modular, it follows that $\mathcal{N}$ also is modular.

Note that also 2.5 implies that ν_1 is a congruence.

We have seen in III.7.9(ii) that for any congruence ρ on an E-unitary inverse semigroup, we have $\rho_{\min} = \rho \cap \sigma$. Applying this to the free inverse semigroup I_X on a countably infinite set X and ρ a fully invariant congruence ρ on I_X, we obtain

$$\mathcal{V}(\rho) \vee \mathcal{G} = \mathcal{V}(\rho) \vee \mathcal{V}(\sigma) = \mathcal{V}(\rho \cap \sigma) = \mathcal{V}(\rho_{\min}).$$

Thus, $\rho_{\min}$ is itself fully invariant and corresponds to the greatest element of $\mathcal{V}(\rho)\nu_1$. In order to get a similar interpretation for $\rho_{\max}$, we first need the following result, which is evidently of independent interest.

XII.2.9 Lemma

Let S be a relatively free inverse semigroup on an infinite set X. Then μ is a fully invariant congruence on S.

Proof. Let $u, v \in S$ be written as products of some elements of X and their inverses, say

$$u = u(x_1, x_2, \ldots, x_n), \qquad v = v(x_1, x_2, \ldots, x_n)$$

where not all x_i need occur in both u and v. Assume that $u\mu v$. By III.3.3, we have $u^{-1}eu = v^{-1}ev$ for all $e \in E_S$, and thus in particular

$$u^{-1}x_{n+1}x_{n+1}^{-1}u = v^{-1}x_{n+1}x_{n+1}^{-1}v,$$

where x_{n+1} is any element of $X \setminus \{x_1, x_2, \ldots, x_n\}$. Since X is a relatively free set of generators for S, the equality

$$[u(x_1, x_2, \ldots, x_n)]^{-1}x_{n+1}x_{n+1}^{-1}[u(x_1, x_2, \ldots, x_n)]$$
$$= [v(x_1, x_2, \ldots, x_n)]^{-1}x_{n+1}x_{n+1}^{-1}[v(x_1, x_2, \ldots, x_n)] \qquad (2)$$

is actually an identity satisfied by S.

Now let φ be any endomorphism of S. For any $e \in E_S$, we obtain by (2)

$$\begin{aligned}(u\varphi)^{-1}e(u\varphi) &= [u(x_1\varphi, x_2\varphi,\dots,x_n\varphi)]^{-1}ee^{-1}[u(x_1\varphi, x_2\varphi,\dots,x_n\varphi)] \\ &= [v(x_1\varphi, x_2\varphi,\dots,x_n\varphi)]^{-1}ee^{-1}[v(x_1\varphi, x_2\varphi,\dots,x_n\varphi)] \\ &= (v\varphi)^{-1}e(v\varphi),\end{aligned}$$

which of course yields $(u\varphi)\mu(v\varphi)$. Consequently, μ is fully invariant.

The desired result can now be established.

XII.2.10 Proposition

Let X be an infinite set. If ρ is a fully invariant congruence on I_X, then so is ρ_{max}.

Proof. Let ρ be a fully invariant congruence on I_X. The conclusion is trivial for $\rho = \omega$. Assume that ρ is proper. Then S/ρ is a relatively free inverse semigroup on an infinite set of generators.

Let φ be an endomorphism of I_X. If $a\rho b$, then $(a\varphi)\rho(b\varphi)$ since ρ is fully invariant. The mapping

$$\bar{\varphi}: a\rho \to a\varphi\rho \qquad (a \in S)$$

is thus an endomorphism of S/ρ. Let $a\rho_{max}b$. Then $(a\rho)\mu(b\rho)$ and thus $((a\rho)\bar{\varphi})\mu((b\rho)\bar{\varphi})$ by 2.9. But then $((a\varphi)\rho)\mu((b\varphi)\rho)$ so that $(a\varphi)\rho_{max}(b\varphi)$. Consequently, ρ_{max} is fully invariant.

In view of 2.2, 2.8, and 2.10, we see that for X countably infinite and ρ a fully invariant congruence on I_X, to ρ_{max} corresponds the variety $\langle \mathcal{V} \cap \mathcal{A} \rangle$. Combining this with the remark following 2.8, we obtain the following correspondence:

$$\begin{array}{ccccc} \rho_{min} & \subseteq & \rho & \subseteq & \rho_{max} \\ \downarrow & & \downarrow & & \downarrow \\ \mathcal{V} \vee \mathcal{G} & \supseteq & \mathcal{V} & \supseteq & \langle \mathcal{V} \cap \mathcal{A} \rangle \end{array}$$

In particular $\mathcal{V}(\rho_{max}) = \langle \mathcal{V} \cap \mathcal{A} \rangle$.

XII.2.11 Exercises

(i) Let ρ be a congruence on an inverse semigroup S and let $\mathcal{V} \in \mathcal{L}(\mathcal{I})$ be such that $S/\rho \in \mathcal{V}$. Show that $S/\rho_{min} \in \mathcal{V} \vee \mathcal{G}$ and $S/\rho_{max} \in \mathcal{V} \cap \mathcal{A}$.

(ii) Let I be an infinite set, $S = B(1, I)$, $\mathcal{V} = \langle \Omega(S) \rangle$. Show that $\mathcal{V}\nu_1$ contains only one element.

(iii) For a group G, prove

$$\langle B_2^1 \rangle \vee \langle G \rangle = \langle B(G,2)^1 \rangle.$$

(iv) For a group variety $\mathcal{U}$, prove

$$\langle B_2^1 \rangle \vee \mathcal{U} = \langle B(G,2)^1 | G \in \mathcal{U} \rangle.$$

XII.2.12 Problems

(i) (P. R. Jones) Which varieties $\mathcal{V}$ of inverse semigroups have the property that every member of $\mathcal{V}$ can be embedded into an antigroup in $\mathcal{V}$? ($\mathcal{I}$ and combinatorial varieties are such.)

(ii) Let $\mathcal{V} = [u_\alpha = v_\alpha]_{\alpha \in A}$. Find a system of defining identities for $\langle \mathcal{V} \cap \mathcal{C} \rangle$.

Kleiman [3] introduced the mapping φ_1 and the congruence ν_1, results 2.5–2.8 are due to him; in connection with 2.8, see also Žitomirskiĭ [5]. Result 2.9 was mentioned in Kleiman [3]; it and 2.10 were proved by T. E. Hall. Congruence ν_1 was further investigated by Reilly [16]. For more information on joins of group varieties and combinatorial varieties consult Kleiman [6].

XII.3 MEETS WITH GROUPS

We introduce the mapping $\varphi_2 : \mathcal{V} \to \mathcal{V} \cap \mathcal{G}$ on the lattice of varieties of inverse semigroups and prove that it is a homomorphism. Letting ν_2 be the congruence induced by φ_2, and $\nu = \nu_1 \cap \nu_2$, we obtain two further congruences on $\mathcal{L}(\mathcal{I})$ which exhibit some interesting properties.

In Sections 7 and 8, we will study congruences ν_1, ν_2, and ν on completely semisimple varieties, where more precise statements can be made.

XII.3.1 Definition

For every variety $\mathcal{V}$ of inverse semigroups, the intersection $\mathcal{V} \cap \mathcal{G}$ is the *group part* of $\mathcal{V}$. For every group variety $\mathcal{U}$, let

$$Q(\mathcal{U}) = \{ \mathcal{V} \in \mathcal{L}(\mathcal{I}) | \mathcal{V} \cap \mathcal{G} = \mathcal{U} \}.$$

Hence $Q(\mathcal{U})$ is the set of all varieties of inverse semigroups with group part $\mathcal{U}$. The next result is an analogue of 2.8.

XII.3.2 Theorem

The mapping

$$\varphi_2 : \mathcal{V} \to \mathcal{V} \cap \mathcal{G} \qquad [\mathcal{V} \in \mathcal{L}(\mathcal{I})]$$

is a homomorphism of $\mathcal{L}(\mathcal{I})$ onto $\mathcal{L}(\mathcal{G})$. Denote by ν_2 the congruence on $\mathcal{L}(\mathcal{I})$ induced by φ_2. Then for every $\mathcal{V} \in \mathcal{L}(\mathcal{I})$, $\mathcal{V} \cap \mathcal{G}$ is the least element of $\mathcal{V}\nu_2$ and $\mathcal{I}$ is the join of all members of $\mathcal{V}\nu_2$. Hence only $\mathcal{G}\nu_2$ has a greatest element.

Proof. Using the antiisomorphism ρ in I.11.11, we may consider the mapping

$$\rho(\mathcal{V}) \to \rho(\mathcal{V}) \vee \rho(\mathcal{G}) = \rho(\mathcal{V}) \vee \sigma.$$

We have seen in III.5.6 that this is a homomorphism. It follows that φ_2 is a homomorphism, and it obviously maps $\mathcal{L}(\mathcal{I})$ onto $\mathcal{L}(\mathcal{G})$.

For any $\mathcal{V} \in \mathcal{L}(\mathcal{I})$, $\mathcal{V} \cap \mathcal{G}$ is obviously the least element of $\mathcal{V}\nu_2$. In addition, we may consider only ν_2-classes of group varieties. But these are precisely $Q(\mathcal{U})$ for all $\mathcal{U} \in \mathcal{L}(\mathcal{G})$.

Now let $\mathcal{U} \in \mathcal{L}(\mathcal{G})$. For every natural number n, let $\mathcal{U}_n = \mathcal{U} \vee \mathcal{C}_n$. Then the first assertion of the theorem gives $\mathcal{U}_n \cap \mathcal{G} = \mathcal{U}$ and thus $\mathcal{U}_n \in Q(\mathcal{U})$. Using 1.8, we obtain

$$\bigvee_{\mathcal{V} \in Q(\mathcal{U})} \mathcal{V} \supseteq \bigvee_{n=1}^{\infty} \mathcal{U}_n \supseteq \bigvee_{n=1}^{\infty} \mathcal{C}_n = \mathcal{I}$$

and thus $\vee_{\mathcal{V} \in Q(\mathcal{U})} \mathcal{V} = \mathcal{I}$, as asserted. Since $\mathcal{I} \in Q(\mathcal{G})$, the remaining ν_2-classes do not have a greatest element.

As an analogue of 2.2, we have the following result.

XII.3.3 Proposition

For any $\mathcal{U}, \mathcal{V} \in \mathcal{L}(\mathcal{I})$, we have

$$\mathcal{U} \cap \mathcal{G} = \mathcal{V} \cap \mathcal{G} \Leftrightarrow (\ker \rho(\mathcal{U}))\omega = (\ker \rho(\mathcal{V}))\omega.$$

Proof. Since $\rho(\mathcal{G}) = \sigma$ and ρ is a lattice antiisomorphism, we obtain

$$\mathcal{U} \cap \mathcal{G} = \mathcal{V} \cap \mathcal{G} \Leftrightarrow \rho(\mathcal{U}) \vee \sigma = \rho(\mathcal{V}) \vee \sigma.$$

By III.5.5, we have that $\ker(\rho(\mathcal{U}) \vee \sigma) = (\ker \rho(\mathcal{U}))\omega$; whence follows the assertion of the proposition.

We now introduce the third congruence as follows.

XII.3.4 Notation

Let $\nu = \nu_1 \cap \nu_2$.

We can picture the ν_1-, ν_2-, and ν-classes of a variety $\mathcal{V}$ as follows:

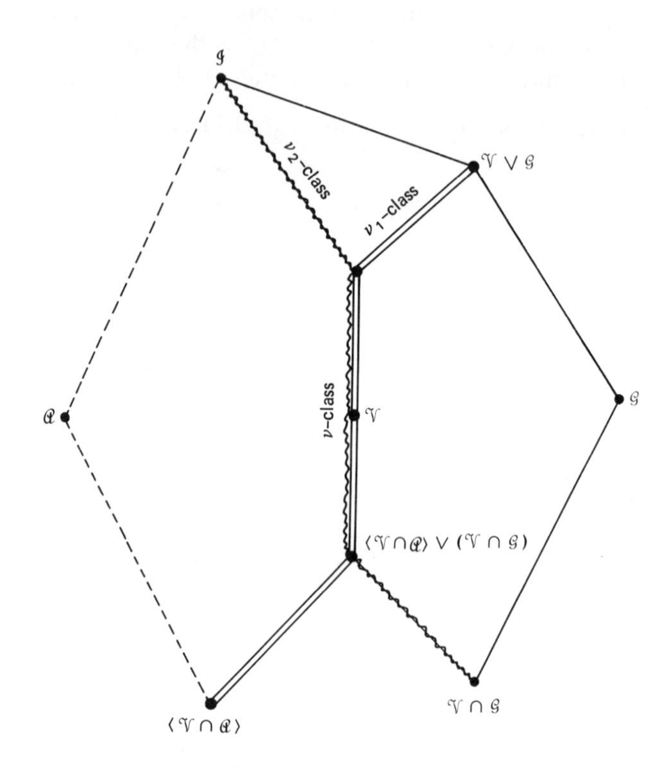

The following example shows that ν is not the equality relation on $\mathcal{L}(\mathcal{I})$. Recall from II.3.3 that B_2 denotes a five-element combinatorial Brandt semigroup.

XII.3.6 Example

Let $\mathbb{Z}$ be the group of additive integers, $G = \mathbb{Z} \times \mathbb{Z}$, $I = \{1, 2\}$, and $S = B(G, I)$. Let $\alpha_1 = \iota_G$ and α_2 be the automorphism of G given by $(a, b)\alpha_2 = (b, a)$. Let $T = S \cup G$ with the multiplication in S and G as given, and for $g \in G$, $(i, a, j) \in S$, let

$$g(i, h, j) = (i, (g\alpha_i)h, j), \qquad (i, h, j)g = (i, h(g\alpha_j), j).$$

Routine verification shows that this multiplication is associative, with $(0,0)$ the identity of T, so that T is an inverse semigroup. It is just as easy to verify that $\mathcal{H}$ is a congruence on T and that $T/\mathcal{H} \cong B_2^1$.

It follows from 2.6 that $T \in \langle B_2^1 \rangle \vee \mathcal{G}$. Let $\mathcal{V} = \langle T \rangle$. Since B_2^1 satisfies the identity $x^2 = x^3$, each semigroup in $\langle B_2^1 \rangle$ is combinatorial and thus is an antigroup. This together with 2.5 gives

$$\langle B_2^1 \rangle = \langle B_2^1 \rangle \cap \mathcal{A} = \big(\langle B_2^1 \rangle \vee \mathcal{G} \big) \cap \mathcal{A}.$$

Hence

$$\mathcal{V} \cap \mathcal{A} = \langle B_2^1 \rangle. \tag{3}$$

Since T contains a copy of $\mathbb{Z}$, it follows that $\mathcal{V}$ contains the variety $\mathcal{AG}$ of abelian groups, and thus $\mathcal{V} \cap \mathcal{G} \supseteq \mathcal{AG}$. Denoting the word $xx^{-1}x^{-1}xyy^{-1}y^{-1}y$ by e, simple inspection shows that T satisfies the identity

$$\big(exeyex^{-1}ey^{-1}e \big)^2 = exeyex^{-1}ey^{-1}e.$$

Hence this is a law in $\mathcal{V}$. This law in a group amounts to the commutator of x and y being equal to the identity element. Consequently, $\mathcal{V} \cap \mathcal{G} \subseteq \mathcal{AG}$, which together with the other inclusion established above gives

$$\mathcal{V} \cap \mathcal{G} = \mathcal{AG}. \tag{4}$$

Write $[x, y] = xyx^{-1}y^{-1}$. Since $x^3 = x^2$ is a law in $\langle B_2^1 \rangle$, so is $[x, y]^3 = [x, y]^2$. The latter is obviously a law in an abelian group, and thus $[x, y]^3 = [x, y]^2$ is a law in $\mathcal{U} = \langle B_2^1 \rangle \vee \mathcal{AG}$. However, by letting $x = (2,(1,0),1) \in S$ and $y = (1,0) \in G$, and computing

$$\begin{aligned}
[x, y] &= (2,(1,0),1)(1,0)(1,(-1,0),2)(-1,0) \\
&= (2,(2,0),1)(1,(-1,-1),2) \\
&= (2,(1,-1),2)
\end{aligned}$$

we see that $[x, y]^3 \neq [x, y]^2$ in T. Thus $[x, y]^3 = [x, y]^2$ is not a law in $\mathcal{V}$, which shows that $\mathcal{U} \neq \mathcal{V}$. By (3) and (4), we have

$$\mathcal{U} = (\mathcal{V} \cap \mathcal{A}) \vee (\mathcal{V} \cap \mathcal{G}).$$

Hence $\mathcal{U} \subseteq \mathcal{V}$ and in view of 2.8, $\mathcal{U}$ is the least element of the ν-class containing $\mathcal{V}$. Hence this ν-class is nontrivial, and ν is not the equality relation on $\mathcal{L}(\mathcal{I})$.

This example shows somewhat more, namely the following statement is true.

XII.3.7 Corollary

The lattice $\mathcal{L}(\langle B_2^1 \rangle \vee \mathcal{G})$ is not modular.

Proof. For $\mathcal{V}$ in 3.6, the right-hand side of 3.5 becomes

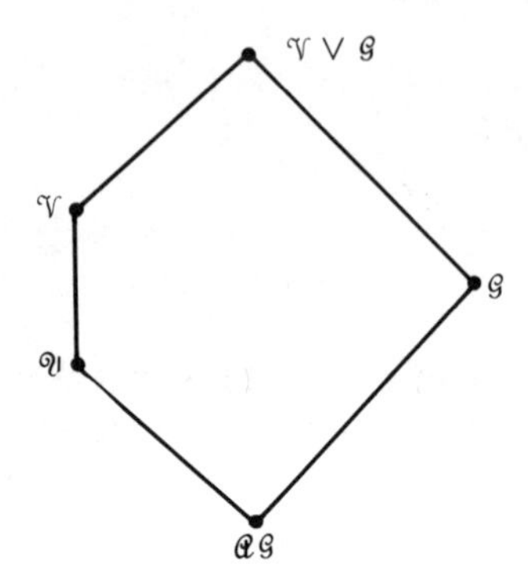

We have seen in 3.6 that $\mathcal{U} \neq \mathcal{V}$. That all other vertices in the diagram are distinct is a matter of simple inspection. We have also seen in 3.6 that $\mathcal{V} \subseteq \langle B_2^1 \rangle \vee \mathcal{G}$. Consequently, $\mathcal{L}(\langle B_2^1 \rangle \vee \mathcal{G})$ is nonmodular.

We will see in the next section that $\mathcal{L}(\langle B_2 \rangle \vee \mathcal{G})$ is modular.

XII.3.8 Exercises

(i) Show that the mappings defined on $\mathcal{L}(\mathcal{I})$ by

$$\mathcal{V} \to \mathcal{V} \cap \mathcal{S}, \qquad \mathcal{V} \to \mathcal{V} \vee \mathcal{S}$$

are homomorphisms.

(ii) Let $\mathcal{V} \in Q(\mathcal{T})$ and $\mathcal{U}, \mathcal{U}' \in \mathcal{L}(\mathcal{G})$. Show that if $\mathcal{V} \vee \mathcal{U} = \mathcal{V} \vee \mathcal{U}'$, then $\mathcal{U} = \mathcal{U}'$.

(iii) Show that T in 3.6 is isomorphic to a subsemigroup of $\Omega(S)$ contained in $\Pi(S) \cup \Sigma(S)$, where $\Sigma(S)$ is the group of units of $\Omega(S)$.

(iv) Show that the homomorphism φ_2 in 3.2 is not complete.

(v) Give a direct proof of the first part of 3.2.

XII.3.9 Problems

(i) If $\mathcal{V} \in \mathcal{L}(\mathcal{I})$, what can be said about $(\rho(\mathcal{V}))^{\min}$, $(\rho(\mathcal{V}))^{\max}$?

Kleiman [3] established 3.2 and introduced the objects φ_2, ν_2, and ν. Reilly [15] constructed Example 3.6 and deduced 3.7; Kleiman ([3], correction) provided another example. Congruences ν_2 and ν were further investigated by Reilly [16].

XII.4 THE FIRST THREE LAYERS

The sublattice of $\mathfrak{L}(\mathcal{I})$ consisting of group varieties forms an ideal of $\mathfrak{L}(\mathcal{I})$. The lattice of all varieties of Clifford semigroups which do not entirely consist of groups forms a sublattice of $\mathfrak{L}(\mathcal{I})$ which is an isomorphic copy of $\mathfrak{L}(\mathcal{G})$. These varieties coincide with those of the form $\mathfrak{U} \vee \mathcal{S}$ where $\mathfrak{U}$ runs over all group varieties. A second sublattice of $\mathfrak{L}(\mathcal{I})$ isomorphic to $\mathfrak{L}(\mathcal{G})$ is obtained by taking all varieties of strict inverse semigroups which do not entirely consist of Clifford semigroups. These varieties coincide with those of the form $\mathfrak{U} \vee \mathcal{B}$, where $\mathfrak{U}$ runs over all group varieties and $\mathcal{B} = \langle B_2 \rangle$. The set of all these varieties forms an ideal of $\mathfrak{L}(\mathcal{I})$.

This is what we term the first three layers of the lattice $\mathfrak{L}(\mathcal{I})$: (i) $\mathfrak{U}$, (ii) $\mathfrak{U} \vee \mathcal{S}$, and (iii) $\mathfrak{U} \vee \mathcal{B}$, where $\mathfrak{U}$ runs over all group varieties. We prove that the lattice of all varieties of strict inverse semigroups is isomorphic to the direct product of a three-element chain and $\mathfrak{L}(\mathcal{G})$, and establish several interesting properties of the varieties under study.

We now introduce the varieties which form the subject of this section.

XII.4.1 Definition

A variety consisting solely of Clifford (respectively strict inverse) semigroups is a *Clifford* (respectively *strict*) *variety*. We denote by $\mathcal{S}\mathcal{G}$ the class of all Clifford semigroups (semilattices of groups), by $\mathcal{S}\mathcal{I}$ the class of all strict inverse semigroups, and by $\mathcal{B}$ the class of all combinatorial strict inverse semigroups.

We will see below that all these classes are actually varieties.

For the remainder of this chapter, we will use the following abbreviated notation. We remark first that an element a of an inverse semigroup S is contained in a subgroup of S if and only if a commutes with its inverse. Hence we may introduce the following concepts.

XII.4.2 Definition

The fact that an inverse semigroup S satisfies an identity of the form $uu^{-1} = u^{-1}u$, where u is an element of a free inverse semigroup, will be expressed by saying that "*S satisfies the identity $u \in G$*." Similarly, "*S satisfies the identity $u \in E$*" means that S satisfies the identity $u^2 = u$. We can thus speak of the varieties $[u \in G]$ and $[u \in E]$.

All identities occurring in our discussion will be inverse semigroup identities.

XII.4.3 Theorem

If $\mathcal{U} = [u_\alpha = v_\alpha]_{\alpha \in A}$ is a group variety, then

$$\mathcal{U} \vee \mathcal{S} = \left[x \in G, u_\alpha v_\alpha^{-1} \in E\right]_{\alpha \in A}$$

$$= \{(\text{strong}) \text{ semilattices of groups in } \mathcal{U}\}$$

$$= \{S \in \mathcal{I} \mid S \text{ is a subdirect product of groups in } \mathcal{U}$$

$$\text{with a zero possibly adjoined}\}.$$

Proof. Denote the four classes in the statement of the theorem by A, B, C, and D, respectively.

It is clear that $\mathcal{U} \subseteq B$. Also $\mathcal{S} \subseteq B$ since all identities figuring in B are homotypical. Thus $A \subseteq B$.

Let $S \in B$. Then S satisfies the identity $x \in G$, that is, is a union of its subgroups, and is a semilattice Y of groups G_β. Each of these groups G_β satisfies the identity $u_\alpha v_\alpha^{-1} \in E$ for each $\alpha \in A$, which in a group can be written as $u_\alpha = v_\alpha$. This shows that all the groups G_β are in $\mathcal{U}$. Consequently, $B \subseteq C$; the inclusion $C \subseteq D$ follows directly from II.2.6.

Finally, let $S \in D$, say $S \subseteq \prod_{\delta \in \Delta} T_\delta$ where $T_\delta \in \mathcal{U}$ or $T_\delta = G_\delta^0$ with $G_\delta \in \mathcal{U}$. For any group G, we have

$$G^0 \cong (G \times Y_2)/(G \times \{0\}).$$

It follows that $G_\delta^0 \in \mathcal{U} \vee \mathcal{S}$, and hence $T_\delta \in \mathcal{U} \vee \mathcal{S}$ for all $\delta \in \Delta$. But then also $S \in \mathcal{U} \vee \mathcal{S}$, and hence $D \subseteq A$.

It follows from 4.3 that $\mathcal{S}\mathcal{G}$ is a variety. The next lemma shows that the varieties $\mathcal{U}$ and $\mathcal{U} \vee \mathcal{S}$, where $\mathcal{U}$ ranges over $\mathcal{L}(\mathcal{G})$, actually exhaust the Clifford varieties.

XII.4.4 Lemma

Let $\mathcal{V}$ be a Clifford variety which is not a group variety. Then $\mathcal{V} = (\mathcal{V} \cap \mathcal{G}) \vee \mathcal{S}$.

Proof. Since not all semigroups in $\mathcal{V}$ are groups, $\mathcal{V}$ must contain an inverse semigroup with at least two idempotents. But then $Y_2 \in \mathcal{V}$ and hence $\mathcal{S} \subseteq \mathcal{V}$. Thus $(\mathcal{V} \cap \mathcal{G}) \vee \mathcal{S} \subseteq \mathcal{V}$. Conversely, let $S \in \mathcal{V}$. Then S is a subdirect product of groups G_α in $\mathcal{V} \cap \mathcal{G}$ and of these with a zero adjoined by II.2.6. As in the proof of 4.3, we obtain $G_\alpha^0 \in (\mathcal{V} \cap \mathcal{G}) \vee \mathcal{S}$. But then also $S \in (\mathcal{V} \cap \mathcal{G}) \vee \mathcal{S}$ being a subdirect product of semigroups G_α^0 and G_α.

XII.4.5 Corollary

$$\mathcal{L}(\mathcal{S}\mathcal{G}) = \mathcal{L}(\mathcal{G}) \cup \{\mathcal{U} \vee \mathcal{S} \mid \mathcal{U} \in \mathcal{L}(\mathcal{G})\}.$$

Proof. The proof of this corollary is left as an exercise.

We have thus completed the second level and start treating the third. The result below shows that $\mathcal{S}\mathcal{I}$ is a variety and provides a basis for its identities.

XII.4.6 Proposition

$\mathcal{S}\mathcal{I} = [yxy^{-1} \in G]$.

Proof. Let B be a Brandt semigroup, and assume that for $x, y \in B$, $yxy^{-1} \neq 0$. Then

$$yxy^{-1} = (i, g, j)(j, h, j)(j, g^{-1}, i) = (i, ghg^{-1}, i) \in G.$$

Hence a Brandt semigroup satisfies $yxy^{-1} \in G$, and thus all strict inverse semigroups satisfy this identity.

Conversely, let S be an inverse semigroup which satisfies the identity $yxy^{-1} \in G$. Let $e, f, g \in E_S$ be such that $e > f$, $e > g$, $f \mathcal{D} g$. Then $f = a^{-1}a$, $g = aa^{-1}$ for some $a \in S$ by II.1.7. Hence $ae = a(a^{-1}a)e = afe = af = a$ and similarly $a = ea$. By the hypothesis,

$$(eae)(eae)^{-1} = (eae)^{-1}(eae)$$

whence $aa^{-1} = a^{-1}a$ since $a = ae = ea$. But then $f = g$ and II.4.5 gives that S is strict.

XII.4.7 Lemma

Let $B_\alpha = B(G_\alpha, I_\alpha)$ be a Brandt semigroup for $\alpha \in A$. Then

$$\Big(\prod_{\alpha \in A} B_\alpha\Big)/I \cong B\Big(\prod_{\alpha \in A} G_\alpha, \prod_{\alpha \in A} I_\alpha\Big)$$

where

$$I = \Big\{(a_\alpha) \in \prod_{\alpha \in A} B_\alpha \mid \text{at least one } a_\alpha = 0_\alpha\Big\}.$$

Proof. Routine verification shows that

$$\chi : \begin{cases} (i_\alpha, g_\alpha, j_\alpha)_{\alpha \in A} \to ((i_\alpha)_{\alpha \in A}, (g_\alpha)_{\alpha \in A}, (j_\alpha)_{\alpha \in A}) \\ 0 \to 0 \end{cases}$$

is the required isomorphism. The details are left as an exercise.

XII.4.8 Proposition

$$\mathfrak{B} = [yxy^{-1} \in E] = \{S \in \mathfrak{I} \mid S \text{ is a subdirect product of combinatorial Brandt semigroups if } |S| > 1\}$$
$$= \langle B_2 \rangle.$$

Proof. Denote by C and D the second and the third classes in the statement of the proposition.

Let $S \in \mathfrak{B}$. Then S is strict, so by 4.6, it satisfies the identity $yxy^{-1} \in G$. But S is also combinatorial, so that $yxy^{-1} \in G$ implies $yxy^{-1} \in E$. Hence $\mathfrak{B} \subseteq C$.

Next let $S \in C$ be nontrivial. Then S satisfies the identity $yxy^{-1} \in G$ and is thus strict by 4.6. Hence S is a subdirect product of semigroups B_α which are either groups or Brandt semigroups. The identity $yxy^{-1} \in E$ in a group G implies that G is trivial. Consequently, all B_α are combinatorial. This proves that $C \subseteq D$.

In order to establish that $D \subseteq \langle B_2 \rangle$, it suffices to show that $\langle B_2 \rangle$ contains every combinatorial Brandt semigroup. By 4.7, for any set J, $\langle B_2 \rangle$ contains $B(1, 2^J)$ and hence it contains every combinatorial Brandt semigroup. Consequently, $D \subseteq \langle B_2 \rangle$.

Any semigroup in C must be strict by 4.6, and thus combinatorial because of the identity $yxy^{-1} \in E$. Hence $C \subseteq \mathfrak{B}$ and the equality prevails. In particular, $\mathfrak{B}$ is a variety, which together with the obvious fact that $B_2 \in \mathfrak{B}$ yields that $\langle B_2 \rangle \subseteq \mathfrak{B}$.

XII.4.9 Lemma

For any nonempty set I and group G,

$$B(G, I) \cong (B(1, I) \times G)/(\{0\} \times G).$$

Proof. The proof of this lemma is left as an exercise.

XII.4.10 Theorem

If $\mathfrak{U} = [u_\alpha(x_i) = v_\alpha(x_j)]_{\alpha \in A}$ is a group variety, then

$$\mathfrak{U}\mathfrak{B} = \left[yxy^{-1} \in G,\, u_\alpha\left(y_i x_i y_i^{-1}\right)\left[v_\alpha\left(y_j x_j y_j^{-1}\right)\right]^{-1} \in E\right]_{\alpha \in A}$$
$$= \{\text{strict inverse semigroups whose subgroups belong to } \mathfrak{U}\}$$
$$= \{S \in \mathfrak{I} \mid S \text{ is a subdirect product of groups in } \mathfrak{U} \text{ and Brandt semigroups over groups in } \mathfrak{U}\}.$$

Proof. Denote the four classes in the statement of the theorem by A, B, C, and D, respectively.

Clearly, $\mathcal{U} \subseteq B$. Further, B_2 satisfies the identity $yxy^{-1} \in E$ by 4.8, so it satisfies $yxy^{-1} \in G$, and each $u_\alpha(y_i x_i y_i^{-1})$, $v_\alpha(y_j x_j y_j^{-1})$ is idempotent. It then follows that $B_2 \in B$, and by 4.8, $\mathcal{B} \subseteq B$. Since B is a variety, we obtain $A \subseteq B$.

Any semigroup in B is strict by 4.6. Any group satisfying the identities in B must be in $\mathcal{U}$. This shows that $B \subseteq C$.

Let $S \in C$. By the proof of II.4.5, S is a subdirect product of its principal factors B_α which are Brandt semigroups with one possible exception which is a group. Consequently, $C \subseteq D$.

In order to show that $D \subseteq A$, it suffices to show that a Brandt semigroup S over a group in $\mathcal{U}$ is in A. So let $S = B(G, I)$ where $G \in \mathcal{U}$. Now 4.9 gives $S \in \mathcal{U} \vee \mathcal{B}$. Consequently, $D \subseteq A$.

The following simple result is not surprising.

XII.4.11 Lemma

$$\langle I_x \rangle = \langle C \rangle.$$

Proof. By IX.1.3, I_x is a subdirect product of two copies of C, which proves that $I_x \in \langle C \rangle$ and thus $\langle I_x \rangle \subseteq \langle C \rangle$. Conversely, $C \cong I_x/\rho_{(1,\infty^+)}$ by IX.3.4, so that $C \in \langle I_x \rangle$ and hence $\langle C \rangle \subseteq \langle I_x \rangle$.

XII.4.12 Notation

We denote $\mathcal{C} = \langle C \rangle = \langle I_x \rangle$.

XII.4.13 Proposition

The following statements concerning a variety $\mathcal{V}$ of inverse semigroups are true.

(i) $\mathcal{V}$ is a group variety if and only if $Y_2 \notin \mathcal{V}$.
(ii) $\mathcal{V}$ is a Clifford variety if and only if $B_2 \notin \mathcal{V}$.
(iii) $\mathcal{V}$ is a strict variety if and only if $B_2^1 \notin \mathcal{V}$.

Proof. The direct part is trivial for all the statements.

(i) If $Y_2 \notin \mathcal{V}$, then every semigroup in $\mathcal{V}$ contains exactly one idempotent and is thus a group by II.2.10.

(ii) Assume that $B_2 \notin \mathcal{V}$. Since B_2 is a homomorphic image of I_x, we get $I_x \notin \mathcal{V}$. Hence $C \notin \mathcal{V}$ by 4.11. Let S be a monogenic inverse semigroup in $\mathcal{V}$. By IX.3.4, $S \cong I_x/\rho$ where ρ cannot be of type (n, ∞^-) or (n, ∞^+) by the above, and it cannot be of type (n, m) or (n, ω) for $n > 1$ since for $n > 1$, S would have a homomorphic image isomorphic to B_2. But then S must be a cyclic group. Consequently, any semigroup in $\mathcal{V}$ is a union of groups, and hence a Clifford semigroup.

(iii) Assume that $B_2^1 \notin \mathcal{V}$. Then as in the proof of part (ii), using I_x^1 instead of I_x, we conclude that $C \notin \mathcal{V}$. Thus, every semigroup in $\mathcal{V}$ is completely semisimple.

By contradiction suppose that $\mathcal{V}$ contains an inverse semigroup S which is not strict. Then by II.4.5, there are idempotents in S such that $e > f$, $e > g$, $f \neq g$, $f \mathcal{D} g$. Let $Q = S/I(f)$. Then $I(f)$ cannot be empty since $f \neq g$. Thus, Q has an ideal B which contains f and g and is a Brandt semigroup since S is completely semisimple by the above. By II.1.7, there exists $a \in B$ such that $aa^{-1} = f$ and $a^{-1}a = g$; we get a Brandt semigroup $T = \{f, g, a, a^{-1}, 0\}$. But $T \cup \{e\} \cong B_2^1$ and we have arrived at the contradiction that $B_2^1 \in \mathcal{V}$. Consequently, any $S \in \mathcal{V}$ must be strict.

XII.4.14 Corollary

The following statements hold in $\mathcal{L}(\mathcal{I})$.

(i) $\mathcal{S}$ is the least nongroup variety.
(ii) $\mathcal{B}$ is the least non-Clifford variety.
(iii) $\langle B_2^1 \rangle$ is the least nonstrict variety.
(iv) $\mathcal{T} \prec \mathcal{S} \prec \mathcal{B} \prec \langle B_2^1 \rangle$.

Proof. The proof of this corollary is left as an exercise.

We construct next a simple isomorphic copy of $\mathcal{L}(\mathcal{S}\mathcal{I})$. We will need the following result, which is an analogue of 4.4.

XII.4.15 Lemma

Let $\mathcal{V}$ be a strict variety of inverse semigroups which is not a Clifford variety. Then $\mathcal{V} = (\mathcal{V} \cap \mathcal{G}) \vee \mathcal{B}$.

Proof. By hypothesis, $\mathcal{V}$ contains a strict inverse semigroup S which is not a Clifford semigroup. Then S is a subdirect product of groups and Brandt semigroups B_α, one of which, say B, is not a group or a group with zero. Hence $B/\mathcal{H}$ is a combinatorial Brandt semigroup which is not a two-element chain. Such a semigroup clearly contains a copy of B_2. Thus, $B_2 \in \mathcal{V}$ whence by 4.8, $\mathcal{B} \subseteq \mathcal{V}$. Consequently, $(\mathcal{V} \cap \mathcal{G}) \vee \mathcal{B} \subseteq \mathcal{V}$.

Conversely, let $S \in \mathcal{V}$. Then S is a subdirect product of groups (in $\mathcal{V}$) and of Brandt semigroups (over groups in $\mathcal{V}$). The last part of the proof of 4.10 now yields that $S \in (\mathcal{V} \cap \mathcal{G}) \vee \mathcal{B}$.

XII.4.16 Theorem

The function

$$\chi: \mathcal{V} \to (\mathcal{V} \cap \mathcal{B}, \mathcal{V} \cap \mathcal{G}) \qquad [\mathcal{V} \in \mathcal{L}(\mathcal{S}\mathcal{I})]$$

is a lattice isomorphism of $\mathcal{L}(\mathcal{S}\mathcal{I})$ onto $\mathcal{L}(\mathcal{B}) \times \mathcal{L}(\mathcal{G})$.

Proof. In the following, we use 4.14(iv). Let $\mathcal{U}, \mathcal{V} \in \mathcal{L}(\mathcal{S}\mathcal{I})$ and assume that $\mathcal{U}\chi = \mathcal{V}\chi$. If $\mathcal{U} \cap \mathcal{B} = \mathcal{T}$, then $\mathcal{U} = \mathcal{U} \cap \mathcal{G} = \mathcal{V} \cap \mathcal{G} = \mathcal{V}$. If $\mathcal{U} \cap \mathcal{B} = \mathcal{S}$, then by 4.4, we have

$$\mathcal{U} = (\mathcal{U} \cap \mathcal{G}) \vee \mathcal{S} = (\mathcal{V} \cap \mathcal{G}) \vee \mathcal{S} = \mathcal{V}.$$

Finally, if $\mathcal{U} \cap \mathcal{B} = \mathcal{B}$, then by 4.15, the same type of argument gives $\mathcal{U} = \mathcal{V}$. Consequently, χ is one-to-one.

Now let $\mathcal{U} \in \mathcal{L}(\mathcal{G})$. Then $\mathcal{U}\chi = (\mathcal{T}, \mathcal{U})$, $(\mathcal{S} \vee \mathcal{U})\chi = (\mathcal{S}, \mathcal{U})$ by 4.3 and $(\mathcal{B} \vee \mathcal{U})\chi = (\mathcal{B}, \mathcal{U})$ by 4.10. Consequently, χ maps $\mathcal{L}(\mathcal{S}\mathcal{I})$ onto $\mathcal{L}(\mathcal{B}) \times \mathcal{L}(\mathcal{G})$.

It is clear that χ preserves inclusion; a simple inspection shows that also χ^{-1} is order preserving. Now I.2.8 gives that χ is a lattice isomorphism.

Note that by 4.14(iv), $\mathcal{L}(\mathcal{B})$ is a three-element chain. We also deduce that $\mathcal{L}(\mathcal{B} \vee \mathcal{G}) \cong \mathcal{L}(\mathcal{B}) \times \mathcal{L}(\mathcal{G})$. For another property of this lattice, we first need a familiar statement.

XII.4.17 Lemma

Let G be a group, $\mathcal{N}$ be the lattice of its normal subgroups, and $\mathcal{F}$ be the lattice of its fully invariant subgroups. Then $\mathcal{N}$ is modular and $\mathcal{F}$ is a sublattice of $\mathcal{N}$.

Proof. The first assertion is a well-known result in group theory and its proof is left as an exercise. The verification of the second assertion is immediate.

XII.4.18 Corollary

The lattice $\mathcal{L}(\mathcal{S}\mathcal{I})$ is modular.

Proof. The lattice $\mathcal{L}(\mathcal{G})$ is antiisomorphic to the lattice $\mathcal{F}\mathcal{I}(\mathcal{G}_X)$ of fully invariant congruences on a free group on a countably infinite set of generators, as in I.11.11. The lattice $\mathcal{F}\mathcal{I}(\mathcal{G}_X)$ is isomorphic to the lattice of fully invariant subgroups in $\mathcal{G}_X$ by the mapping $\rho \to \ker \rho$. It follows from 4.17 that the latter lattice is modular, and thus so is $\mathcal{L}(\mathcal{G})$. The assertion of the corollary now follows from 4.16.

XII.4.19 Exercises

(i) Define a function χ by

$$\chi: \mathcal{U} \to \mathcal{U} \vee \mathcal{S} \qquad [\mathcal{U} \in \mathcal{L}(\mathcal{G})].$$

Show that χ is an isomorphism of $\mathcal{L}(\mathcal{G})$ onto the interval $[\mathcal{S}, \mathcal{S}\mathcal{G}]$.

(ii) Define a function τ by

$$\tau: \mathcal{U} \to \mathcal{U} \vee \mathcal{B} \qquad [\mathcal{U} \in \mathcal{L}(\mathcal{G})].$$

Show that τ is an isomorphism of $\mathcal{L}(\mathcal{G})$ onto the interval $[\mathcal{B}, \mathcal{S}\mathcal{I}]$.

(iii) Show that $\langle B(G,I)\rangle = \langle G\rangle \vee \mathcal{B}$ if $|I| > 1$.

(iv) Let $\mathcal{V} \in \mathcal{L}(\mathcal{I})$. Show that $\mathcal{V} \in \mathcal{L}(\mathcal{G})$ if and only if some (respectively every) system of defining identities for $\mathcal{V}$ contains at least one heterotypical identity.

(v) Show that

$$\mathcal{S}\mathcal{I} = [yxy^2 \in G] = \left[(yx^2y^{-1})(yx^2y^{-1})^{-1} = (yxy^{-1})(yxy^{-1})^{-1}\right].$$

(vi) Show that $\mathcal{B} = [yxy^{-1} = yx^2y^{-1}]$.

(vii) Show that $\mathcal{S}\mathcal{G} = [x^2x^{-2} = xx^{-1}]$.

(viii) Show that the following conditions on an inverse semigroup S are equivalent.

(α) S is not a Clifford semigroup.

(β) S contains an element a with $a^2 < a$.

(γ) S contains elements a, b such that $ab \in E_S$, $ba \notin E_S$.

(ix) Let S be an inverse semigroup and $\mathcal{U}$ be a group variety. Show that $eSe \in \mathcal{S} \vee \mathcal{U}$ for all $e \in E_S$ if and only if $S \in \mathcal{B} \vee \mathcal{U}$.

(x) For any Clifford variety $\mathcal{V}$ and nonempty set X, construct a $\mathcal{V}$-free Clifford semigroup on X.

(xi) Characterize the free monogenic strict inverse semigroup.

(xii) Characterize the congruences ν_1 and ν_2 on the lattice of all strict varieties.

(xii) Let $\mathcal{V}$ be a variety of inverse semigroups. Show that every normal subsemigroup of every semigroup in S is a kernel in S if and only if $\mathcal{V}$ is a Clifford variety.

(xiv) For a group variety $\mathcal{U} = [u_\alpha = 1]$, prove

(α) $\mathcal{U} \vee \mathcal{S} = [x \in G, u_\alpha^2 = u_\alpha]$,

(β) $\mathcal{U} \vee \mathcal{B} = [xyx^{-1} \in G, u_\alpha^3 = u_\alpha^2]$.

(xv) Show that any lattice of varieties of inverse semigroups which properly contains $\mathcal{S}\mathcal{I}$ is nonmodular.

XII.4.20 Problems

(i) Let $K_n = \{1, 2, \ldots, n\}$. Note that $\mathcal{I}(\varnothing)$ is a trivial semigroup, $\mathcal{I}(K_1) \cong Y_2$, and it can be shown that $\langle \mathcal{I}(\mathbb{N})\rangle = \mathcal{I}$, the variety of all inverse semigroups. We thus have

$$\langle I(\varnothing)\rangle = \mathcal{T} \subset \langle \mathcal{I}(K_1)\rangle = \mathcal{S} \subset \langle \mathcal{I}(K_2)\rangle \subseteq \langle \mathcal{I}(K_3)\rangle \subseteq \cdots$$

$$\subseteq \langle \mathcal{I}(K_n)\rangle \subseteq \cdots \subseteq \langle \mathcal{I}(\mathbb{N})\rangle = \mathcal{I}.$$

Characterize the remaining $\langle \mathcal{I}(K_n) \rangle$. [Recall that $\mathcal{I}(X)$ is the symmetric inverse semigroup on X.]

(ii) Construct a suitable isomorphic copy of the free strict inverse semigroup on a nonempty set.

(iii) In any inverse semigroup denote by β the least $\mathcal{B}$-congruence. Is there a simple way to obtain β from η?

Result 4.5 is due to Petrich [10] and Djadčenko [1]; 4.6 and 4.8 to Kleiman [2] and Reilly [15]; 4.10 and 4.13 to Kleiman [2], see also Djadčenko [1]; 4.16 to Djadčenko [1] and Kleiman [2]; 4.14(iv) to Kleiman [2]. The absence of finite bases for varieties belonging to certain intervals was investigated in Kleiman [4]. See Maševickiĭ [1] for identities satisfied by Brandt semigroups. Also consult Trachtman [1].

XII.5 QUASIABELIAN STRICT VARIETIES

We call an inverse semigroup S quasiabelian if $xyx^{-1}y^{-1} \in E_S$ for all $x, y \in S$ and consider here varieties consisting of quasiabelian strict inverse semigroups. The variety of all quasiabelian strict inverse semigroups coincides with the join of the variety of abelian groups with $\mathcal{B}$. Using the fact that the lattice of varieties of abelian groups can be easily described, we give an isomorphic copy of the lattice of all quasiabelian strict varieties. We consider the joins of varieties of abelian groups, and their Burnside subvarieties, with $\mathcal{S}$ and $\mathcal{B}$.

XII.5.1 Definition

For elements x and y of an inverse semigroup S, we write $[x, y] = xyx^{-1}y^{-1}$ and call it the *commutator* of x and y. If S satisfies the identity $[x, y] \in E$, we call it *quasiabelian*.

XII.5.2 Notation

We will use the following notation:

$\mathcal{QA} = [[x, y] \in E]$ quasiabelian inverse semigroups

$\mathcal{AG}$—abelian groups,

$\mathcal{AG}_n = \mathcal{AG} \cap [x^{n+1} = x]$ for $n \geq 1$.

Note that if $S \in \mathcal{QA}$ then all subgroups of S are abelian.
The first principal result is the following.

XII.5.3 Theorem

(i) $\mathcal{AG} = [x = yxy^{-1}] = \mathcal{G} \cap \mathcal{QA} = \langle \mathbb{Z} \rangle$.

(ii) $\mathcal{AG} \vee \mathcal{S} = [xy = yx] = \mathcal{SG} \cap \mathcal{QA} = \langle \mathbb{Z}^0 \rangle$.

(iii) $\mathcal{AG} \vee \mathcal{B} = [(zxz^{-1})(zyz^{-1}) = (zyz^{-1})(zxz^{-1})]$

$= \mathcal{SI} \cap \mathcal{QA} = \langle B(\mathbb{Z}, 2) \rangle$

$=$ {strict inverse semigroups with abelian subgroups}.

Proof. (i) Any abelian group satisfies the identity $x = yxy^{-1}$. Let S be an inverse semigroup satisfying the identity $x = yxy^{-1}$. For any $e, f \in E_S$, we have $e = fef = ef$ and analogously $f = ef$ so that $e = f$. Hence, S is a group by II.2.10 and thus an abelian group. This establishes the first equality in (i); the second is obvious. The third equality follows from the well-known property that every abelian group is a homomorphic image of a free abelian group, which is a direct sum of infinite cyclic groups.

(ii) Since $(xy)(yx)^{-1} = xyx^{-1}y^{-1} = [x, y]$, 4.3 implies the equality $\mathcal{AG} \vee \mathcal{S} = \mathcal{SG} \cap \mathcal{QA}$. The last equality in (ii) follows from 4.3 and the last equality in (i). The inclusions

$$\mathcal{AG} \vee \mathcal{S} \subseteq [xy = yx] \subseteq \mathcal{SG} \cap \mathcal{QA}$$

are obvious.

(iii) Denote by A, B, C, D, and E the classes figuring in (iii). As in part (ii), now using 4.9, we deduce that $A = C$. Obviously, $\mathcal{AG} \subseteq B$, and by 4.8, $\mathcal{B} = [yxy^{-1} \in E]$ so that also $\mathcal{B} \subseteq B$, and thus $A \subseteq B$.

Let $S \in B$. Then S satisfies the identity

$$(yxy^{-1})(yxy^{-1})^{-1} = (yxy^{-1})^{-1}(yxy^{-1})$$

which shows that S is strict. Thus, $S \in \mathcal{SI}$ and is a subdirect product of groups and Brandt semigroups B_α. The identity in B immediately implies that subgroups of B_α are abelian. Consider $B_\alpha = B(G, I)$ over an abelian group G. Then

$$(i, g, j)(k, h, l)(j, g^{-1}, i)(l, h^{-1}, k) \neq 0$$

if and only if $j = k = l = i$, and if this is the case, then this expression is equal to $(i, 1, i)$ since G is abelian. Consequently, each B_α satisfies the identity $[x, y] \in E$, and thus also S satisfies this identity. Therefore, $S \in \mathcal{QA}$, which proves that $B \subseteq C$. This establishes the first two equalities in (iii).

In particular, $B(\mathbb{Z}, 2) \in \mathcal{SI} \cap \mathcal{QA}$ which implies that $D \subseteq C$. By part (i), we obtain $\mathcal{AG} = \langle \mathbb{Z} \rangle \subseteq \langle B(\mathbb{Z}, 2) \rangle$ and by 4.8,

$$\mathcal{B} = \langle B_2 \rangle = \langle B(1, 2) \rangle \subseteq \langle B(\mathbb{Z}, 2) \rangle$$

which implies that $A \subseteq D$. This proves the fourth equality in (iii).

It is clear that any semigroup in $\mathcal{Q}\mathcal{A}$ has all subgroups abelian. Hence $C \subseteq E$. Let $S \in E$. Then S is a subdirect product of abelian groups and Brandt semigroups B_α over abelian groups. Such a semigroup B_α is in $\mathcal{Q}\mathcal{A}$. Consequently, $E \subseteq C$.

The next result is analogous to the preceding one but includes the hypothesis that $x^{n+1} = x$ for all x in these semigroups.

XII.5.4 Theorem

Let n be a natural number.

(i) $\mathcal{A}\mathcal{G}_n = [x = yx^{n+1}y^{-1}] = \langle \mathbb{Z}/(n) \rangle$.
(ii) $\mathcal{A}\mathcal{G}_n \vee \mathcal{S} = [xy = yx^{n+1}] = \langle (\mathbb{Z}/(n))^0 \rangle$.
(iii) $\mathcal{A}\mathcal{G}_n \vee \mathcal{B} = [(zxz^{-1})(zyz^{-1}) = (zyz^{-1})(zxz^{-1})^{n+1}]$
$= \langle B(\mathbb{Z}/(n), 2) \rangle$.

Proof. (i) The argument here amounts to a modification of that in the proof of 5.3(i).

(ii) The proof of the equality of the first and the third classes follows along the same lines as the corresponding items in 5.3(ii). It is clear that the first class is contained in the second. Let S be an inverse semigroup satisfying the identity $xy = yx^{n+1}$. Letting x be an idempotent, we deduce that S is a Clifford semigroup. Letting $y = xx^{-1} = x^{-1}x$, we get $x = x^{n+1}$ and hence S also satisfies $xy = yx$. Thus $S \in \mathcal{S}\mathcal{G} \cap \mathcal{Q}\mathcal{A}$ and all subgroups of S are in $\mathcal{A}\mathcal{G}_n$. It then follows from 4.3 that $S \in \mathcal{A}\mathcal{G}_n \vee \mathcal{S}$. Consequently, the second class in (ii) is contained in the first.

(iii) The argument here is similar to that for part (ii) using 5.3(iii). We prove only that the second class is contained in the first. Hence let S be an inverse semigroup satisfying the identity in the second class. Let $e, f, g \in E_S$ be such that $e > f$, $e > g$, and $f \mathcal{D} g$. Then $f = a^{-1}a$ and $g = aa^{-1}$ for some $a \in S$, by II.1.7, and by hypothesis

$$(eae)(ea^{-1}e) = (ea^{-1}e)(eae)^{n+1}.$$

Since $a = ea = ae$, we obtain

$$g = aa^{-1} = a^{-1}a^{n+1} = (a^{-1}a)a^n(a^{-1}a) = fa^nf$$

so that $g \le f$. Symmetrically, $f \le g$ which implies that $f = g$. By II.4.5, S is a subdirect product of groups and Brandt semigroups B_α. Let $B_\alpha = B(G, I)$. The identity in the second class in (iii) is valid in G, which easily implies that $G \in \mathcal{A}\mathcal{G}_n$. But then 4.10 implies that $S \in \mathcal{A}\mathcal{G}_n \vee \mathcal{B}$. It follows that the second class in (iii) is contained in the first.

The remaining details of this proof are left as an exercise.

We can now describe the lattice of all quasiabelian strict varieties. To start with, we consider the varieties of abelian groups.

XII.5.5 Lemma

The mapping

$$\chi : \mathcal{AG}_n \to n \qquad (n \in \mathbb{N})$$

is an isomorphism of the lattice of all proper varieties of abelian groups onto the lattice of natural numbers $\mathbb{N}$ ordered by division.

Proof. Let $\mathcal{V}$ be a proper variety of abelian groups and $\mathcal{V} \neq \mathcal{T}$. Suppose that the orders of elements in groups in $\mathcal{V}$ are unbounded. Then $\mathcal{V}$ contains $\mathbb{Z}/(n)$ for all positive integers n. For each positive integer n, let $\varphi_n : \mathbb{Z} \to \mathbb{Z}/(n)$ be the natural homomorphism. Routine verification shows that the mapping $m \to (m\varphi_n)$ is an embedding of $\mathbb{Z}$ into $\prod_{n=1}^{\infty} \mathbb{Z}/(n)$. But then $\mathbb{Z} \in \mathcal{V}$ which implies $\mathcal{AG} = \mathcal{V}$ by 5.3(i), contradicting the hypothesis that $\mathcal{V}$ is a proper subvariety of $\mathcal{AG}$.

Let n be the greatest positive integer n such that $\mathbb{Z}/(n) \in \mathcal{V}$. In light of 5.4(i), we obtain $\mathcal{AG}_n \subseteq \mathcal{V}$. Let x be an element of a semigroup S in $\mathcal{V}$. Then x cannot be of infinite order since in that case, $\mathbb{Z}$ would be in $\mathcal{V}$. Let m be the order of x. Since $\mathbb{Z}/(n) \in \mathcal{V}$, there is a cyclic group $\langle y \rangle$ in $\mathcal{V}$ for which the order of y is equal to n. Further, the direct product $G = \langle x \rangle \times \langle y \rangle$ is in $\mathcal{V}$. Consider the element $a = (x, y)$ of G. For any natural number k, we have

$$a^k = (x,1)^k(1, y)^k = (x^k,1)(1, y^k).$$

It follows that the order q of a is the least common multiple of m and n. Hence $\mathbb{Z}/(q) \in \mathcal{V}$, and by the maximality of n, we also must have $q \leq n$. But then $q = n$ which proves that m divides n. This shows that $S \in \mathcal{AG}_n$ and proves that $\mathcal{AG}_n = \mathcal{V}$.

Consequently, χ is a one-to-one mapping of the set of proper subvarieties of $\mathcal{L}(\mathcal{AG})$ onto $\mathbb{N}$. It is easy to see that $\mathcal{AG}_m \subseteq \mathcal{AG}_n$ if and only if m divides n, which implies that χ is an isomorphism.

XII.5.6 Corollary

Let $\mathbb{N}^1$ stand for the set of natural numbers ordered by division with a greatest element ∞ adjoined.

(i) $\mathcal{L}(\mathcal{AG}) \cong \mathbb{N}^1$.
(ii) $\mathcal{L}(\mathcal{SG} \cap \mathcal{QA}) \cong Y_2 \times \mathbb{N}^1$.
(iii) $\mathcal{L}(\mathcal{SI} \cap \mathcal{QA}) \cong Y_3 \times \mathbb{N}^1$, Y_3 is a three-element chain.

Proof. This follows immediately from 5.5, 4.14(iv), and 4.16.

The lattice $\mathbb{N}$ has the following interesting property.

XII.5.7 Lemma

The lattice $\mathbb{N}$ of natural numbers under division is distributive.

Proof. Let K be the set of all nonnegative integers with the usual order. Then K is a chain, so it is a distributive lattice. Hence $K^{\mathbb{N}}$, the Cartesian product of $\mathbb{N}$ copies of K with coordinatewise operations, is also distributive. Define a mapping θ by

$$\theta : n = p_{i_1}^{n_{i1}} p_{i_2}^{n_{i2}} \cdots p_{i_m}^{n_{im}} \to (n_i) \qquad (n \in \mathbb{N})$$

where p_{i_j} are distinct primes ordered as in K and

$$n_i = \begin{cases} n_{i_r} & \text{if } i = i_r \text{ and } p_{i_r} \text{ appears in the decomposition of } n \\ 0 & \text{otherwise.} \end{cases}$$

A simple verification shows that θ is an embedding of $\mathbb{N}$ into $K^{\mathbb{N}}$ and thus $\mathbb{N}$ must itself be distributive.

XII.5.8 Corollary

The lattices $\mathfrak{L}(\mathcal{AG})$, $\mathfrak{L}(\mathcal{SG} \cap \mathcal{QA})$, and $\mathfrak{L}(\mathcal{SI} \cap \mathcal{QA})$ are complete and distributive.

Proof. Completeness follows from I.2.6 since these lattices have a greatest element and are closed under intersection. For distributivity, by 5.6, it suffices to check that $\mathbb{N}^1$ is distributive. This follows by 5.7 and the simple inspection of the distributive law for triples involving the greatest element of $\mathbb{N}^1$.

XII.5.9 Exercises

(i) Show that $\mathcal{AG} \vee \mathcal{S} = \langle \overline{\mathbb{Z}} \rangle = \langle \mathbb{Z} \times Y_2 \rangle$, where $\overline{\mathbb{Z}}$ is the group $\mathbb{Z}$ with an identity adjoined.

(ii) Show that $\mathcal{AG} \vee \mathcal{B} = \langle I_x / \rho_{(2, \omega)} \rangle$.

(iii) Show that a Brandt semigroup satisfies the identity $[x, y] \in E$ if and only if its subgroups are abelian.

The results in this section concerning groups and the lattice $\mathbb{N}$ are folklore.

XII.6 SMALL VARIETIES

A variety of inverse semigroups is said to be small if it is generated by groups, semilattices, and/or monogenic inverse semigroups. We will characterize small varieties and prove that they form a lattice isomorphic to $(C_\omega \times \mathbb{N}^1)^1$ where C_ω is an ω-chain, $\mathbb{N}$ is the lattice of all natural numbers under division, and $(\ )^1$ means addition of a greatest element. The lattice of all small varieties contains the first three layers of $\mathcal{L}(\mathcal{I})$ considered in Section 4, and hence also the quasiabelian strict varieties considered in Section 5.

We also characterize those small varieties which are generated by monogenic inverse semigroups, as well as the intersections $\mathcal{C} \cap \mathcal{G}$, $\mathcal{C} \cap \mathcal{SG}$, and $\mathcal{C} \cap \mathcal{SI}$, where $\mathcal{C}$ is the variety generated by the bicyclic semigroup.

XII.6.1 Definition

A variety of inverse semigroups is *small* if it is generated by groups, semilattices, and/or monogenic inverse semigroups.

XII.6.2 Notation

Let M_0 be a trivial semigroup, $M_1 = Y_2$, and for $n > 1$, let M_n have the meaning given to it in IX.3.3. (Note: M_1 there is isomorphic to M_0 here.) For $n \geq 0$, let $\mathcal{M}_n = \langle M_n \rangle$. Recall that C stands for the bicyclic semigroup and C_ω for the ω-chain of nonnegative integers.

XII.6.3 Definition

For any variety $\mathcal{V}$ of inverse semigroups, we define the *index* of $\mathcal{V}$ by

$$\operatorname{ind} \mathcal{V} = \sup\{ n | M_n \in \mathcal{V} \}.$$

A variety $\mathcal{V}$ is of *finite index* if $\operatorname{ind} \mathcal{V} < \infty$.

For the description of the lattice of small varieties, we first prove the following auxiliary result.

XII.6.4 Lemma

If $\mathcal{V}$ is a small variety of finite index, then $\mathcal{V} = \langle M_{\operatorname{ind} \mathcal{V}} \rangle \vee (\mathcal{V} \cap \mathcal{G})$.

Proof. Let $\mathcal{V}$ be a small variety of finite index n. If $n = 0$, then $\mathcal{V}$ consists of groups. If $n = 1$, then $\mathcal{V}$ consists of Clifford semigroups, hence it is generated by $M_1 = Y_2$ and $\mathcal{V} \cap \mathcal{G}$. For $n > 1$, $\mathcal{V}$ is generated by a set M of monogenic inverse semigroups and a set of groups. Each semigroup in M is of the form $I_x/\rho_{(k,\omega)}$ or $I_x/\rho_{(k,l)}$ with $k \leq n$, since the presence of other types would contradict the hypothesis that $\operatorname{ind} \mathcal{V} = n < \infty$. Each of these semigroups is an ideal extension of a cyclic group G by a semigroup M_k with $k \leq n$, and hence a subdirect product of G and M_k by I.9.15 and I.9.16. On the other hand, if $k \leq n$, then M_k is a homomorphic image of M_n. Consequently, $\mathcal{V}$ is generated by M_n and $\mathcal{V} \cap \mathcal{G}$.

The desired isomorphic copy of the lattice of small varieties of finite index is provided by the following result.

XII.6.5 Theorem

The mapping

$$\chi : \mathcal{V} \to (\operatorname{ind} \mathcal{V}, \mathcal{V} \cap \mathcal{G})$$

is an isomorphism of the lattice of small varieties of finite index onto the lattice $C_\omega \times \mathcal{L}(\mathcal{G})$.

Proof. If $\mathcal{U}\chi = \mathcal{V}\chi$, then $\operatorname{ind} \mathcal{U} = \operatorname{ind} \mathcal{V}$ and $\mathcal{U} \cap \mathcal{G} = \mathcal{V} \cap \mathcal{G}$ which in view of 6.4 implies that both $\mathcal{U}$ and $\mathcal{V}$ are generated by $M_{\operatorname{ind} \mathcal{V}}$ and $\mathcal{V} \cap \mathcal{G}$. Consequently $\mathcal{U} = \mathcal{V}$ and χ is one-to-one.

Let $n \in C_\omega$ and $\mathcal{U} \in \mathcal{L}(\mathcal{G})$, and set $\mathcal{V} = \langle M_n \rangle \vee \mathcal{U}$. Now using 3.2, we get

$$\mathcal{V} \cap \mathcal{G} = (\langle M_n \rangle \vee \mathcal{U}) \cap \mathcal{G} = (\langle M_n \rangle \cap \mathcal{G}) \vee (\mathcal{U} \cap \mathcal{G}) = \mathcal{U}$$

since by IX.3.8, M_n satisfies $x^{n+1} = x^n$ so $\langle M_n \rangle \cap \mathcal{G} = \mathcal{T}$.

Since $M_n \in \mathcal{V}$, we have $\operatorname{ind} \mathcal{V} \geq n$. By IX.3.9, M_n satisfies the identity $x^{-1}x^n = x^n x^{-1}$. This identity is satisfied in any group, so that it is a law in $\mathcal{V}$. But M_{n+1} does not satisfy the identity $x^{-1}x^n = x^n x^{-1}$ [take the generator $x = ((1,0),(0,1))$ of M_n, for example, and use IX.3.8]. Thus, $M_{n+1} \notin \mathcal{V}$ which shows that $\operatorname{ind} \mathcal{V} = n$. Consequently, $\mathcal{V}\chi = (n, \mathcal{U})$ and χ is onto.

A verification that both χ and χ^{-1} preserve order is left as an exercise. The desired conclusion now follows from I.2.8.

XII.6.6 Definition

For any variety $\mathcal{V}$ of inverse semigroups we say that $\mathcal{V} \cap \mathcal{G}$ is *abelian* if it consists solely of abelian groups.

XII.6.7 Proposition

A small variety of index greater than 1 is generated by monogenic inverse semigroups if and only if its group part is abelian. Each such variety is generated by a unique monogenic inverse semigroup.

Proof. Let $\mathcal{V}$ be a small variety of index greater than 1 which is generated by monogenic inverse semigroups. Hence by 4.11, $\mathcal{V} \subseteq \langle I_x \rangle = \mathcal{C}$. Straightforward verification shows that the bicyclic semigroup C satisfies the identity

$[x, y]^2 = [x, y]$. Hence any group in $\mathcal{C} \cap \mathcal{G}$ must satisfy this identity and is thus abelian. Consequently, $\mathcal{V} \cap \mathcal{G} \subseteq \mathcal{V} \cap \mathcal{C} \subseteq \mathcal{AG}$.

Conversely, assume that $\mathcal{V} \cap \mathcal{G}$ consists of abelian groups. Then $\mathcal{V} \cap \mathcal{G}$ is generated by some cyclic group G in view of 5.5. So by 6.4, $\mathcal{V}$ itself is generated by M_n and G if $n = \operatorname{ind} \mathcal{V} < \infty$, and by C otherwise. In the first case, $\mathcal{V}$ is generated by $I_x/\rho_{(n,\omega)}$ if G is infinite and by $I_x/\rho_{(n,m)}$ if G is finite of order m. In any case, $\mathcal{V}$ is generated by a unique monogenic inverse semigroup.

XII.6.8 Theorem

The lattice L of all varieties generated by monogenic inverse semigroups is a complete distributive lattice isomorphic to $(C_\omega \times \mathbb{N}^1)^1$.

Proof. Let $\mathcal{V}$ be a small variety of infinite index. Then $\mathcal{V}$ contains M_n for all $n \geq 2$. Let ρ_n be the congruence on I_x of type $(n, 1)$ for $n \geq 2$. It is clear that $\cap_{n \geq 2} \rho_n = \varepsilon$ so that I_x is a subdirect product of the semigroups M_n, $n \geq 2$, by I.4.18. Hence $I_x \in \mathcal{V}$, and the hypothesis that $\mathcal{V}$ is generated by monogenic inverse semigroups gives $\mathcal{V} = \mathcal{C}$. Since $\mathcal{C}$ is the greatest element of L, we obtain by 6.5 and 6.7 that $L \cong (C_\omega \times \mathcal{L}(\mathcal{AG}))^1$, and thus by 5.6, $L \cong (C_\omega \times \mathbb{N}^1)^1$. Distributivity of the last lattice follows by an argument similar to that in the proof of 5.8.

Now let $\{\mathcal{V}_\alpha\}_{\alpha \in A} \subseteq L$. If one of the $\mathcal{V}_\alpha$ is equal to $\mathcal{C}$, then $\mathcal{C} = \vee_{\alpha \in A} \mathcal{V}_\alpha$. Hence assume that all $\mathcal{V}_\alpha$ are of finite index. In view of the proof of 6.7, $\mathcal{V}_\alpha = \langle I_x/\rho_{(m_\alpha, n_\alpha)} \rangle$ for some $2 \leq m_\alpha < \omega$, $1 \leq n_\alpha \leq \omega$. If $\sup\{m_\alpha\} = \omega$, then $\vee_{\alpha \in A} \mathcal{V}_\alpha = \mathcal{C}$, and if $\sup\{m_\alpha\} = m$, then $\vee_{\alpha \in A} \mathcal{V}_\alpha = \langle I_x/\rho_{(m,n)} \rangle$ where $n = \sup\{n_\alpha\}$. Consequently, L is a complete lattice.

XII.6.9 Theorem

The following statements hold.

(i) $\mathcal{C} \cap \mathcal{G} = \mathcal{AG}$.

(ii) $\mathcal{C} \cap \mathcal{SG} = \mathcal{S} \vee \mathcal{AG}$.

(iii) $\mathcal{C} \cap \mathcal{SI} = \mathcal{B} \vee \mathcal{AG}$.

Proof. (i) We have seen in the proof of 6.7 that $\mathcal{C} \cap \mathcal{G} \subseteq \mathcal{AG}$. Since $C/\sigma \cong \mathbb{Z}$, we get that $\mathbb{Z} \in \mathcal{C}$. Hence by 5.3, we have $\mathcal{AG} = \langle \mathbb{Z} \rangle \subseteq \mathcal{C} \cap \mathcal{G}$.

(ii) Let $S \in \mathcal{C} \cap \mathcal{SG}$. Then S is a semilattice of groups G_α each of which is contained in $\mathcal{C} \cap \mathcal{G} = \mathcal{AG}$, using part (i). Thus $S \in \mathcal{S} \vee \mathcal{AG}$, and hence $\mathcal{C} \cap \mathcal{SG} \subseteq \mathcal{S} \vee \mathcal{AG}$. The opposite inclusion follows from $\mathcal{AG} \subseteq \mathcal{C}$ in part (i).

(iii) Let $S \in \mathcal{C} \cap \mathcal{SI}$. Then S is a subdirect product of groups and/or Brandt semigroups B_α. Then $B_\alpha \in \mathcal{C}$, and by part (i), the maximal subgroups of B_α must be abelian. It follows from 4.7 that

$$B_\alpha = B(G_\alpha, I_\alpha) \in \langle B(1, I_\alpha) \rangle \vee \langle G_\alpha \rangle$$

which gives $B_\alpha \in \mathcal{B} \vee \mathcal{AG}$. But then also $S \in \mathcal{B} \vee \mathcal{AG}$, which proves that

$\mathcal{C} \cap \mathcal{S}\mathcal{I} \subseteq \mathcal{B} \vee \mathcal{A}\mathcal{G}$. The opposite inclusion follows from $\mathcal{B} \subseteq \mathcal{C}$ and $\mathcal{A}\mathcal{G} \subseteq \mathcal{C}$ in part (i).

XII.6.10 Diagram: small varieties

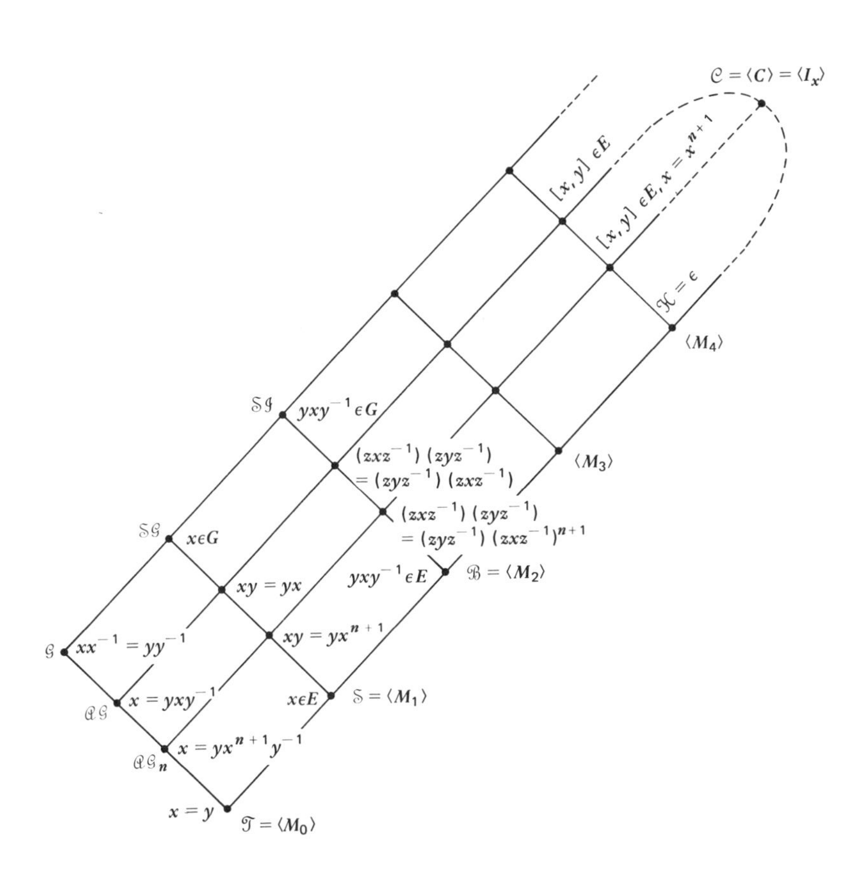

XII.6.11 Exercises

(i) Characterize the restriction of the congruences ν_1 and ν_2 to the lattice of small varieties of finite index.

(ii) Show that the bicyclic semigroup satisfies the identities

(α) $xy^2x^2yxy^2x = xy^2xyx^2y^2x$,

(β) $xy^2x^2y^2x^2y = xy^2xyxyx^2y$,

(γ) $[x, y] \in E$.

(iii) Show that for any abelian group variety $\mathcal{U}$,

$$\bigvee_{n=1}^{\infty} (\mathcal{U} \vee \langle M_n \rangle) = \bigvee_{n=1}^{\infty} \langle M_n \rangle = \mathcal{C}.$$

XII.6.12 Problems

(i) Find a basis for the identities of $\mathcal{C}$. Characterize the semigroups in $\mathcal{C}$.

(ii) (E. I. Kleiman) Is the set of varieties generated by monogenic inverse semigroups a sublattice of $\mathcal{L}(\mathcal{I})$?

(iii) For $n \geq 0$, let $\mathcal{M}_n = \langle M_n \rangle$. Is it true that

$$\mathcal{C} \cap (\mathcal{M}_n \vee \mathcal{G}) = \mathcal{M}_n \vee \mathcal{AG}?$$

(The inclusion $\supseteq$ is easy to establish.)

(iv) Characterize $\mathcal{C} \vee \mathcal{G}$.

(v) (P. R. Jones) Is there a variety in $\mathcal{C}$ which is not small?

Results 6.5 and 6.8 are due to Djadčenko [1] who also introduced all relevant concepts; result 6.9(i) was proved by Kleiman [1]. The paper of Kleiman [4] contains many interesting facts concerning the nonexistence of finite bases for identities of many varieties, including $\langle B_2^1 \rangle$ and $\mathcal{C}$.

XII.7 COMPLETELY SEMISIMPLE VARIETIES

These are the varieties all of whose members are completely semisimple inverse semigroups. They are characterized here in several ways including by the identities they satisfy. A similar analysis is performed for completely semisimple varieties whose members are cryptic (that is, $\mathcal{H}$ is a congruence). The lattice of all combinatorial varieties is proved to be isomorphic to the lattice of all cryptic completely semisimple varieties containing all groups. A homomorphism inducing ν on a sublattice is also constructed.

The following concepts are basic for our discussion.

XII.7.1 Definition

A semigroup in which $\mathcal{H}$ is a congruence is *cryptic*. A variety is *completely semisimple* (respectively *cryptic*) if all its members are completely semisimple (respectively cryptic).

For any inverse semigroup variety $\mathcal{V}$, let

$$T(\mathcal{V}) = \{S \in \mathcal{V} \mid \langle S \rangle \text{ is combinatorial}\}.$$

The first major result here characterizes completely semisimple varieties in several ways.

XII.7.2 Theorem

The following conditions concerning a variety $\mathcal{V}$ of inverse semigroups are equivalent.

(i) $\mathcal{V}$ is a completely semisimple variety.
(ii) $\mathcal{V} \subseteq [x^n \in G]$ for some natural number n.
(iii) $T(\mathcal{V}) \subseteq \mathcal{C}_n$ for some natural number n.
(iv) $T(\mathcal{V})$ is a variety.

Moreover, the least positive integer satisfying (ii) coincides with the least positive integer satisfying (iii). If n is this integer, then $T(\mathcal{V}) = \mathcal{V} \cap \mathcal{C}_n$.

Proof. (i) *implies* (ii). Assume that $\mathcal{V} \nsubseteq [x^n \in G]$ for all positive integers n. Hence for every n, there exists a semigroup $S_n \in \mathcal{V}$ and an element $a_n \in S_n$ such that $a^n a^{-n} \neq a^{-n} a^n$. Let $T = \prod_{n=1}^{\infty} \langle a_n \rangle$ and let $a = (a_1, a_2, a_3, \ldots)$. Then $\langle a \rangle$ is not of type (n, ω) for any n since $x^n x^{-1} = x^{-1} x^n$ implies $a^n a^{-n} = a^{-n} a^n$, see IX.3.12(ii). Also, $\langle a \rangle$ is not of type (n, m), for if so, then a^n would be contained in a subgroup and thus $a^n a^{-n} = a^{-n} a^n$, a contradiction. Thus $\langle a \rangle$ is either of type (n, ∞^-) or (n, ∞^+). In either case, $\langle a \rangle \in \mathcal{V}$ implies that $C \in \mathcal{V}$, and thus $\mathcal{V}$ is not a completely semisimple variety.

(ii) *implies* (iii). Let $S \in T(\mathcal{V})$ and let $a \in S$. Then $\langle a \rangle \in \langle S \rangle$ which is combinatorial by hypothesis. By 1.10, we have $\langle S \rangle \subseteq \mathcal{C}_k$ for some positive integer k, which implies that $\langle a \rangle$ satisfies the identity $x^{k+1} = x^k$. Then $\langle a \rangle \cong M_k$ for some integer k. Since $\langle a \rangle \in \mathcal{V}$, we also have $a^n a^{-n} = a^{-n} a^n$. Now the generator $x = ((1,0),(0,1))$ of M_k must satisfy $x^n x^{-n} = x^{-n} x^n$. But $x^n x^{-n} = ((n,n),(0,0))$ and $x^{-n} x^n = ((0,0),(n,n))$ so that $n \geq k$. It follows that $T(\mathcal{V}) \subseteq \mathcal{C}_n$.

(iii) *implies* (iv). Let $S \in \mathcal{V} \cap \mathcal{C}_n$, where n is the smallest positive integer satisfying (iii). Then $\langle S \rangle \subseteq \mathcal{C}_n$ and so $\langle S \rangle$ is combinatorial. It follows that $S \in T(\mathcal{V})$ whence $\mathcal{V} \cap \mathcal{C}_n \subseteq T(\mathcal{V})$; the opposite inclusion is obvious.

(iv) *implies* (i). Assume that $T(\mathcal{V})$ is a variety but $\mathcal{V}$ is not completely semisimple. Then $C \in \mathcal{V}$ and thus also I_x and M_n are in $\mathcal{V}$ for all positive integers n. It follows that $M_n \in T(\mathcal{V})$ for all n, so $\prod_{n=1}^{\infty} M_n \in T(\mathcal{V})$ since $T(\mathcal{V})$ is a variety. But C is a homomorphic image of I_x which is a subdirect product of the semigroups M_n, so $C \in T(\mathcal{V})$, which is impossible since then also $\mathbb{Z} \in T(\mathcal{V})$. Consequently, $\mathcal{V}$ must be completely semisimple.

Now assume that $T(\mathcal{V}) \subseteq [x^{n+1} = x^n]$ where n is least such. By (ii) we know that $\mathcal{V} \subseteq [x^m \in G]$ for some positive integer m; let m be the least such. Then there exists $S \in \mathcal{V}$ and $a \in S$ such that $a^m a^{-m} = a^{-m} a^m$ but $a^k a^{-k} \neq a^{-k} a^k$ for all $k < m$. Then $\langle a \rangle / J(a^m) \cong M_m$. Since $\langle M_m \rangle$ is combinatorial, we have $M_m \in T(\mathcal{V})$. Hence

$$\langle M_m \rangle \subseteq T(\mathcal{V}) \subseteq [x^{n+1} = x^n],$$

and thus $m \leq n$. On the other hand, $T(\mathcal{V}) \subseteq [x^m \in G]$ by part (ii), and since $T(\mathcal{V})$ is combinatorial, we also have $T(\mathcal{V}) \subseteq [x^m = x^{m+1}]$. By the minimality of n, we must have $n = m$.

The second major result characterizes completely semisimple cryptic varieties in a similar fashion.

XII.7.3 Theorem

The following conditions concerning a variety $\mathcal{V}$ of inverse semigroups are equivalent.

(i) $\mathcal{V}$ is a completely semisimple cryptic variety.
(ii) $\mathcal{V} \subseteq \mathcal{C}_n \vee \mathcal{G}$ for some natural number n.
(iii) $\mathcal{V} \subseteq [x^{n+1}yy^{-1}x^{-n-1} = x^n yy^{-1}x^{-n}]$ for some natural number n.
(iv) $\mathcal{V} \cap \mathcal{A} \subseteq \mathcal{C}_n$ for some natural number n.
(v) $T(\mathcal{V}) = \mathcal{V} \cap \mathcal{A}$.
(vi) $\mathcal{V} \cap \mathcal{A}$ is a variety.

Proof. We note first that by 2.7,

$$\mathcal{C}_n \vee \mathcal{G} = \left[x^{n+1}yy^{-1}x^{-n-1} = x^n yy^{-1}x^{-n}\right]$$

and hence, (ii) and (iii) are equivalent.

(i) *implies* (iii). In view of 7.2, we have $\mathcal{V} \subseteq [x^n \in G]$ for some natural number n. Let $S \in \mathcal{V}$ and $x, y \in S$. Then x^n is in a subgroup of S and thus

$$x^n = (x^n x^{-n})x^n = x^{2n}x^{-n} = x^{2n}x^{-n}xx^{-1} = x^n xx^{-1} = x^{n+1}x^{-1}$$

which implies $x^n \mathcal{R} x^{n+1}$ and analogously $x^n \mathcal{L} x^{n+1}$. Hence $x^n \mathcal{H} x^{n+1}$ which implies $x^n y \mathcal{H} x^{n+1}y$ since $\mathcal{H}$ is a congruence on S. But then $x^n yy^{-1}x^{-n} = x^{n+1}yy^{-1}x^{-n-1}$, as required.

(ii) *implies* (iv). Using 2.5, we obtain

$$\mathcal{V} \cap \mathcal{A} \subseteq (\mathcal{C}_n \vee \mathcal{G}) \cap \mathcal{A} = \mathcal{C}_n \cap \mathcal{A} = \mathcal{C}_n.$$

(iv) *implies* (v). We always have $T(\mathcal{V}) \subseteq \mathcal{V} \cap \mathcal{A}$ since each semigroup in $T(\mathcal{V})$ is an antigroup. If $S \in \mathcal{V} \cap \mathcal{A}$, the hypothesis implies that $S \in \mathcal{C}_n$. Hence $\langle S \rangle$ must be combinatorial, and we get $S \in T(\mathcal{V})$.

(v) *implies* (vi). Since $T(\mathcal{V})$ is closed under taking subsemigroups and homomorphic images, and $\mathcal{V} \cap \mathcal{A}$ is closed under taking of direct products, the hypothesis implies that $\mathcal{V} \cap \mathcal{A}$ is a variety.

(vi) *implies* (i). Since $(\mathcal{V} \cap \mathcal{A}) \cap \mathcal{G} = \mathcal{T}$, 1.10 implies that $\mathcal{V} \cap \mathcal{A}$ is combinatorial and $\mathcal{V} \cap \mathcal{A} \subseteq \mathcal{C}_n$ for some n. We always have $T(\mathcal{V}) \subseteq \mathcal{V} \cap \mathcal{A}$ so that

$T(\mathcal{V}) \subseteq \mathcal{C}_n$ for some n. By 7.2, we have that $\mathcal{V}$ is a completely semisimple variety. Let $S \in \mathcal{V}$. Then $S/\mu \in \mathcal{V} \cap \mathcal{A}$ and hence S/μ is combinatorial. But then $\mu = \mathcal{H}$, and S is cryptic.

XII.7.4 Notation

Let $\mathcal{CS}$ and $\mathcal{CSH}$ denote the classes of completely semisimple inverse semigroups and such cryptic semigroups, respectively. (Since C belongs to the variety generated by the semigroups M_n, $n \geq 1$, neither of these classes is a variety.) We denote by $\mathcal{L}(\mathcal{CS})$ and $\mathcal{L}(\mathcal{CSH})$ the set of all varieties contained in $\mathcal{CS}$ and $\mathcal{CSH}$, respectively.

XII.7.5 Corollary

Both $\mathcal{L}(\mathcal{CS})$ and $\mathcal{L}(\mathcal{CSH})$ are ideals of $\mathcal{L}(\mathcal{I})$.

Proof. If $\mathcal{V} \in \mathcal{L}(\mathcal{CS})$ and $\mathcal{U} \in \mathcal{L}(\mathcal{I})$, then by 7.2, $\mathcal{V} \subseteq [x^n \in G]$ for some natural number n, and thus $\mathcal{V} \cap \mathcal{U} \subseteq [x^n \in G]$, which by 7.2 shows that $\mathcal{V} \cap \mathcal{U} \in \mathcal{L}(\mathcal{CS})$. If also $\mathcal{V}' \in \mathcal{L}(\mathcal{CS})$, then for some n', $\mathcal{V}' \subseteq [x^{n'} \in G]$ and thus $\mathcal{V} \vee \mathcal{V}' \subseteq [x^k \in G]$ for $k = \max\{n, n'\}$, which gives that $\mathcal{V} \vee \mathcal{V}' \in \mathcal{L}(\mathcal{CS})$. Consequently, $\mathcal{L}(\mathcal{CS})$ is an ideal of $\mathcal{L}(\mathcal{I})$.

Using 7.3, a similar argument proves the second assertion of the corollary.

XII.7.6 Remark

Let $\mathcal{V}$ be the variety in 3.6. Since $\mathcal{V} \cap \mathcal{A} = \langle B_2^1 \rangle$ and $\langle B_2^1 \rangle$ is a combinatorial variety (satisfying the identity $x^3 = x^2$), by 7.3 we conclude that $\mathcal{V} \in \mathcal{CSH}$. Also $T(\mathcal{V}) = \langle B_2^1 \rangle$ is the greatest combinatorial variety contained in $\mathcal{V}$, $\mathcal{V} \cap \mathcal{G} = \mathcal{AG}$ is the greatest group variety contained in $\mathcal{V}$, and $\mathcal{V} \neq \langle B_2^1 \rangle \vee \mathcal{AG}$. We deduce that not every variety in $\mathcal{CSH}$ is the join of a combinatorial and a group variety.

We will next establish some interesting relationships among various lattices under study. To this end, we first prove two auxiliary results.

XII.7.7 Lemma

If $\mathcal{U} \in \mathcal{L}(\mathcal{G})$ and $\mathcal{V} \in Q(\mathcal{T})$, then

$$(\mathcal{U} \vee \mathcal{V}) \cap \mathcal{A} = \mathcal{V}.$$

Proof. Obviously, $(\mathcal{U} \vee \mathcal{V})\nu_1\mathcal{V}$ which by 2.5 yields

$$(\mathcal{U} \vee \mathcal{V}) \cap \mathcal{A} = \mathcal{V} \cap \mathcal{A} = \mathcal{V},$$

where the last equality holds since $\mathcal{V} \in Q(\mathcal{T})$.

XII.7.8 Lemma

If $\mathcal{U} \in Q(\mathcal{G})$, then

$$\langle \mathcal{U} \cap \mathcal{Q} \rangle \vee \mathcal{G} = \mathcal{U}.$$

Proof. By 2.8, we have

$$\langle \mathcal{U} \cap \mathcal{Q} \rangle \vee \mathcal{G} = \mathcal{U} \vee \mathcal{G} = \mathcal{U}.$$

The third major result of this section can now be established.

XII.7.9 Theorem

The mappings

$$\theta : \mathcal{V} \to \mathcal{V} \vee \mathcal{G} \qquad [\mathcal{V} \in Q(\mathcal{T})],$$

$$\tau : \mathcal{U} \to \mathcal{U} \cap \mathcal{Q} \qquad [\mathcal{U} \in Q(\mathcal{G}) \cap \mathcal{L}(\mathcal{CSH})],$$

are mutually inverse isomorphisms of $Q(\mathcal{T})$ and $Q(\mathcal{G}) \cap \mathcal{L}(\mathcal{CSH})$.

Proof. If $\mathcal{V} \in Q(\mathcal{T})$, then $\mathcal{V}\theta = \mathcal{V} \vee \mathcal{G} \subseteq \mathcal{C}_n \vee \mathcal{G}$ for some n so that $\mathcal{V}\theta \in \mathcal{L}(\mathcal{CSH})$ by 7.3; $\mathcal{V}\theta \in Q(\mathcal{G})$ trivially. If $\mathcal{U} \in Q(\mathcal{G}) \cap \mathcal{L}(\mathcal{CSH})$, then $\mathcal{U} \cap \mathcal{Q}$ is a variety by 7.3(vi), and since $(\mathcal{U} \cap \mathcal{Q}) \cap \mathcal{G} = \mathcal{T}$, we get $\mathcal{U}\tau \in Q(\mathcal{T})$.

Applying 7.7 to the case $\mathcal{U} = \mathcal{G}$, we immediately get $\theta\tau = \iota_{Q(\mathcal{T})}$. On the other hand, 7.8 yields $\tau\theta = \iota_{Q(\mathcal{G}) \cap \mathcal{L}(\mathcal{CSH})}$. This gives that the functions θ and τ are one-to-one correspondences of the sets indicated in the statement of the theorem. It is obvious that both θ and τ are order preserving. But then both are lattice isomorphisms by I.2.8.

XII.7.10 Proposition

The mapping

$$\chi : \mathcal{V} \to (\mathcal{V} \cap \mathcal{Q}, \mathcal{V} \cap \mathcal{G}) \qquad [\mathcal{V} \in \mathcal{L}(\mathcal{CSH})]$$

is a homomorphism of $\mathcal{L}(\mathcal{CSH})$ onto $Q(\mathcal{T}) \times \mathcal{L}(\mathcal{G})$ which induces ν on $\mathcal{L}(\mathcal{CSH})$.

Proof. We have seen in 3.2 that $\mathcal{V} \to \mathcal{V} \cap \mathcal{G}$ is in fact a homomorphism on all of $\mathcal{L}(\mathcal{I})$. The mapping

$$\mathcal{V} \to \mathcal{V} \vee \mathcal{G} \to (\mathcal{V} \vee \mathcal{G}) \cap \mathcal{Q}$$

is the composition of the homomorphism introduced in 2.8 on all of $\mathcal{L}(\mathcal{I})$ and the homomorphism τ defined in 7.9 only on $Q(\mathcal{G}) \cap \mathcal{L}(\mathcal{CSH})$. It follows directly from 2.5 that $(\mathcal{V} \vee \mathcal{G}) \cap \mathcal{C} = \mathcal{V} \cap \mathcal{C}$. By 7.3, we have that $\mathcal{V} \cap \mathcal{C}$ is a variety and thus clearly $\mathcal{V} \cap \mathcal{C} \in Q(\mathcal{T})$. Consequently, χ is a homomorphism. By the definition of ν and 2.5, we have that χ induces ν on $\mathcal{L}(\mathcal{CSH})$.

To see that χ is onto, we let $\mathcal{U} \in Q(\mathcal{T})$ and $\mathcal{U}' \in \mathcal{L}(\mathcal{G})$. Let $\mathcal{V} = \mathcal{U} \vee \mathcal{U}'$. Then $\mathcal{V} \in \mathcal{L}(\mathcal{CSH})$ by 7.3,

$$\mathcal{V} \cap \mathcal{C} = (\mathcal{U} \vee \mathcal{U}') \cap \mathcal{C} = \mathcal{U}$$

by 7.7, and

$$\mathcal{V} \cap \mathcal{G} = (\mathcal{U} \vee \mathcal{U}') \cap \mathcal{G} = (\mathcal{U} \cap \mathcal{G}) \vee (\mathcal{U}' \cap \mathcal{G}) = \mathcal{U}'$$

by 3.2. Consequently, $\mathcal{V}\chi = (\mathcal{U}, \mathcal{U}')$ and χ is onto.

XII.7.11 Diagram: completely semisimple varieties

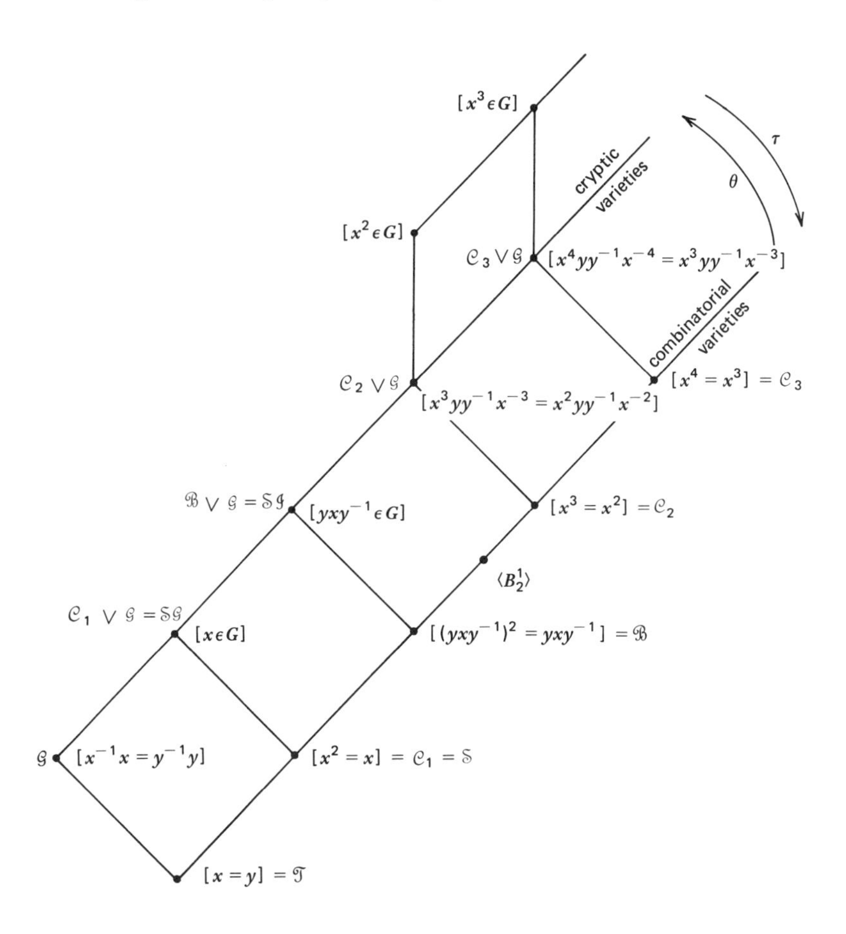

XII.7.12 Exercises

(i) Find a semilattice Y of smallest order for which $\Phi(Y)$ satisfies the conditions of 7.2 but not the conditions of 7.3.

(ii) Show that for every natural number n,

$$[x^n \in G] = [x^n = x^{-1}x^{n+1}] = [x^n = x^{n+1}x^{-1}]$$
$$= [x^{n+1}x^{-n-1} = x^n x^{-n}].$$

(iii) Give an example of an inverse semigroup all of whose homomorphic images are completely semisimple but which does not generate a completely semisimple variety.

(iv) Show that there exists a cryptic variety which is not completely semisimple if and only if $\mathcal{C}$ is cryptic.

XII.7.13 Problems

(i) Let $\mathcal{V}$ be a variety of inverse semigroups. What are necessary and sufficient conditions on some (respectively every) system of defining identities in order for $\mathcal{V}$ to be a Clifford variety?

(ii) (P. R. Jones) Does there exist a cryptic variety which is not completely semisimple?

All theorems and the proposition in this section are due to Reilly [15]. This paper contains further interesting results on the subject.

XII.8 THE MALCEV PRODUCT

This product has proved useful in many considerations concerning the lattice of subvarieties of a variety of algebras. Here it is used to obtain the greatest element of the ν-class of a cryptic completely semisimple variety of inverse semigroups. In fact, for such a variety, $\mathcal{V}$, the ν-class coincides with the interval $[(\mathcal{V} \cap \mathcal{G}) \vee (\mathcal{V} \cap \mathcal{C}), (\mathcal{V} \cap \mathcal{G}) \circ (\mathcal{V} \cap \mathcal{C})]$ where $\circ$ stands for the Malcev product.

XII.8.1 Definition

Let $\mathcal{U}$ and $\mathcal{V}$ be any classes of inverse semigroups. The class of all inverse semigroups S for which there exists a congruence ρ on S with the property that all idempotent ρ-classes are in $\mathcal{U}$ and $S/\rho \in \mathcal{V}$ is the *Malcev product* of $\mathcal{U}$ and $\mathcal{V}$, denoted by $\mathcal{U} \circ \mathcal{V}$.

The Malcev product $\mathcal{U} \circ \mathcal{V}$ of varieties need not be a variety. However, it does have the following property.

XII.8.2 Lemma

If $\mathcal{U}, \mathcal{V} \in \mathcal{L}(\mathcal{I})$, then $\mathcal{U} \circ \mathcal{V}$ is closed under direct products and subsemigroups.

Proof. Let $S_\alpha \in \mathcal{U} \circ \mathcal{V}$ for $\alpha \in A$, and for each $\alpha \in A$, let ρ_α be a congruence on S_α figuring in the definition of the Malcev product. On $S = \prod_{\alpha \in A} S_\alpha$ define a relation ρ by

$$(a_\alpha)\rho(b_\alpha) \quad \text{if} \quad a_\alpha \rho_\alpha b_\alpha \quad \text{for all } \alpha \in A.$$

Then ρ is a congruence on S such that for all $(e_\alpha) \in E_S$, $(e_\alpha)\rho \cong \prod_{\alpha \in A} e_\alpha \rho_\alpha$ and $S/\rho \cong \prod_{\alpha \in A} S_\alpha/\rho_\alpha$, so that $S \in \mathcal{U} \circ \mathcal{V}$.

Next let $S \in \mathcal{U} \circ \mathcal{V}$ with the corresponding congruence ρ, and let T be an inverse subsemigroup of S. Then $\rho' = \rho|_T$ is a congruence on T which gives $T \in \mathcal{U} \circ \mathcal{V}$.

XII.8.3 Proposition

If $\mathcal{U} \in \mathcal{L}(\mathcal{G})$ and $\mathcal{V} \in \mathcal{L}(\mathcal{I})$, then $\mathcal{U} \circ \mathcal{V}$ is an inverse semigroup variety.

Proof. We have by 8.2 that $\mathcal{U} \circ \mathcal{V}$ is closed under direct products and inverse subsemigroups. Let $S \in \mathcal{U} \circ \mathcal{V}$ and ρ be a corresponding congruence on S. Since $\mathcal{U}$ is a group variety, in light of III.3.2, we have that ρ is idempotent separating.

Let φ be a homomorphism of S onto T. Note that $K = \ker \rho$ is a semilattice of groups in $\mathcal{U}$. Hence $A = K\varphi$ is also a semilattice of groups in $\mathcal{U}$. Now K is a normal subsemigroup of S which evidently implies that A is a normal subsemigroup of T. Since A is also a semilattice of groups and is full in T, in view of III.3.6, we obtain that A is the kernel of the congruence τ on T defined by

$$a \tau b \Leftrightarrow a \mathcal{H} b, \quad ab^{-1} \in A. \tag{5}$$

We now define

$$\psi : s\rho \to (s\varphi)\tau \qquad (s \in S).$$

In order to see that ψ is single valued, we let $a, b \in S$ be such that $a \rho b$. Since $\rho \subseteq \mathcal{H}$, we have $a \mathcal{H} b$ so that $(a\varphi)\mathcal{H}(b\varphi)$. Also $ab^{-1}\rho bb^{-1}$ implies $ab^{-1} \in K$ so that $(a\varphi)(b\varphi)^{-1} \in A$. By (5), we obtain $(a\varphi)\tau(b\varphi)$ which establishes that ψ is single valued. Hence ψ is a homomorphism of S/ρ onto T/τ and thus $T/\tau \in \mathcal{V}$.

We have seen above that A is a semilattice of groups in $\mathcal{U}$. Since τ is idempotent separating and $A = \ker \tau$, it follows that all idempotent τ-classes are in $\mathcal{U}$. Therefore, $T \in \mathcal{U} \circ \mathcal{V}$ and $\mathcal{U} \circ \mathcal{V}$ is a variety.

The principal result of this section is the following.

XII.8.4 Theorem

For any $\mathcal{V} \in \mathcal{L}(\mathcal{CSH})$, the ν-class of $\mathcal{V}$ coincides with the interval

$$[(\mathcal{V} \cap \mathcal{G}) \vee (\mathcal{V} \cap \mathcal{C}),\ (\mathcal{V} \cap \mathcal{G}) \circ (\mathcal{V} \cap \mathcal{C})]$$

and is a complete modular lattice.

Proof. Let $\mathcal{N}$ be the ν-class containing $\mathcal{V}$. It follows from 2.8 that without any restriction on the variety $\mathcal{V}$, $\mathcal{N}$ is modular and has as a least element $(\mathcal{V} \cap \mathcal{G}) \vee \langle \mathcal{V} \cap \mathcal{C} \rangle$. By 7.3, $\mathcal{V} \cap \mathcal{C}$ is a variety, so the least element of $\mathcal{N}$ is $(\mathcal{V} \cap \mathcal{G}) \vee (\mathcal{V} \cap \mathcal{C})$.

Let $\mathcal{U} = (\mathcal{V} \cap \mathcal{G}) \circ (\mathcal{V} \cap \mathcal{C})$. Then $\mathcal{U}$ is a variety by 8.3 since $\mathcal{V} \cap \mathcal{G}$ is a group variety. If $G \in \mathcal{V} \cap \mathcal{G}$, then using the universal congruence on G, we see that $G \in \mathcal{U}$, whence $\mathcal{V} \cap \mathcal{G} \subseteq \mathcal{U} \cap \mathcal{G}$. Conversely, let $G \in \mathcal{U} \cap \mathcal{G}$. Then there exists a congruence ρ on G such that $1\rho \in \mathcal{V} \cap \mathcal{G}$ and $G/\rho \in \mathcal{V} \cap \mathcal{C}$. The latter forces ρ to be the universal congruence. But then the former says that $G = 1\rho \in \mathcal{V}$. Consequently, $\mathcal{U} \cap \mathcal{G} \subseteq \mathcal{V} \cap \mathcal{G}$ and the equality prevails.

If $A \in \mathcal{V} \cap \mathcal{C}$, then using the identity congruence we obtain that $A \in \mathcal{U}$, and thus $\mathcal{V} \cap \mathcal{C} \subseteq \mathcal{U} \cap \mathcal{C}$. Conversely, let $A \in \mathcal{U} \cap \mathcal{C}$. Then there exists a congruence ρ on A such that all idempotent ρ-classes are in $\mathcal{V} \cap \mathcal{G}$ and $A/\rho \in \mathcal{V} \cap \mathcal{C}$. The former implies that ρ is idempotent separating by III.3.2 and the latter that $\rho = \rho_{max}$ by III.3.10. But then $\rho = \mu = \varepsilon$ and hence $A \in \mathcal{V}$. Consequently, $\mathcal{U} \cap \mathcal{C} \subseteq \mathcal{V} \cap \mathcal{C}$ and the equality prevails.

We have proved that $\mathcal{U} \nu \mathcal{V}$. Next let $\mathcal{W}$ be any variety ν-related to $\mathcal{V}$. Then $\mathcal{W} \cap \mathcal{C} = \mathcal{V} \cap \mathcal{C}$ by 2.5 and the latter is a variety by 7.3. Hence $\mathcal{W} \cap \mathcal{C}$ is also a variety which in view of 7.3 yields that $\mathcal{W} \in \mathcal{L}(\mathcal{CSH})$. Since $\mathcal{W} \nu \mathcal{V}$, it follows that

$$(\mathcal{W} \cap \mathcal{G}) \circ (\mathcal{W} \cap \mathcal{C}) = (\mathcal{V} \cap \mathcal{G}) \circ (\mathcal{V} \cap \mathcal{C}).$$

Now observing that for any $S \in \mathcal{W}$, μ is a congruence on S such that all idempotent classes are in $\mathcal{W} \cap \mathcal{G}$ and $S/\mu \in \mathcal{W} \cap \mathcal{C}$, we obtain $S \in (\mathcal{W} \cap \mathcal{G}) \circ (\mathcal{W} \cap \mathcal{C})$. This proves the maximality of $(\mathcal{V} \cap \mathcal{G}) \circ (\mathcal{V} \cap \mathcal{C})$.

Since $\mathcal{N}$ is clearly convex, we deduce that

$$\mathcal{V}\nu = [(\mathcal{V} \cap \mathcal{G}) \vee (\mathcal{V} \cap \mathcal{C}),\ (\mathcal{V} \cap \mathcal{G}) \circ (\mathcal{V} \cap \mathcal{C})].$$

Each interval of a complete lattice is itself complete, so $\mathcal{N}$ is complete.

XII.8.5 Corollary

If $\mathcal{U} \in \mathcal{L}(\mathcal{G})$ and $\mathcal{V} \in Q(\mathcal{T})$, then $[\mathcal{U} \vee \mathcal{V}, \mathcal{U} \circ \mathcal{V}]$ is a ν-class.

Proof. The proof of this corollary is left as an exercise.

It also follows from the above proof that a ν-class intersecting $\mathcal{L}(\mathcal{CSH})$ is entirely contained in $\mathcal{L}(\mathcal{CSH})$.

We have seen that for an inverse semigroup variety $\mathcal{V}$, $\langle \mathcal{V} \cap \mathcal{Q} \rangle \vee (\mathcal{V} \cap \mathcal{G})$ is the least element of $\mathcal{V}\nu$. According to an example of Reilly [16], the class $\mathcal{V}\nu$ need not have a greatest element. It is easy to see that $\mathcal{G}\nu$ consists of $\mathcal{G}$ alone: 1.6 implies that $\mathcal{I}\nu_1$ consists of $\mathcal{I}$ alone. Another example follows.

XII.8.6 Example

Let X be a countably infinite set, $T = B(1, G_X)$ and S be an ideal extension of T by G_X^0 with the mixed products given by

$$g(h,1,k) = (gh,1,k), \qquad (h,1,k)g = (h,1,g^{-1}k).$$

One verifies directly that S is an antigroup. Letting $\mathcal{V} = \langle S \rangle$, we get $S \in \mathcal{V} \cap \mathcal{Q}$ and hence $S \in \langle \mathcal{V} \cap \mathcal{Q} \rangle$ so that $\mathcal{V} = \langle \mathcal{V} \cap \mathcal{Q} \rangle$. On the other hand, S contains G_X, and since X is countably infinite, $\mathcal{V}$ must contain all groups. Thus $\mathcal{G} \subseteq \mathcal{V}$ so that $\mathcal{V} = \mathcal{V} \vee \mathcal{G}$. By 2.8, $\mathcal{V}\nu_1 = [\langle \mathcal{V} \cap \mathcal{Q} \rangle, \mathcal{V} \vee \mathcal{G}]$ and hence $\mathcal{V}\nu_1$ contains only $\mathcal{V}$.

XII.8.7 Exercises

(i) Show that $\mathcal{S} \circ \mathcal{G}$ coincides with the class of E-unitary inverse semigroups. (Note that $\mathcal{G} \circ \mathcal{S} = \mathcal{S}\mathcal{G}$.)

(ii) Show that for any $\mathcal{U} \in \mathcal{L}(\mathcal{G})$, we have $\mathcal{U} \circ \mathcal{S} = \mathcal{U} \vee \mathcal{S}$.

(iii) Show that for any variety $\mathcal{V}$ of inverse semigroups, we have $\mathcal{G} \circ \mathcal{V} = \mathcal{G} \vee \mathcal{V}$.

(iv) Verify all the statements in 8.6; show that S is isomorphic to a subsemigroup of the translational hull of T and that it satisfies the identity $x^3x^{-3} = x^2x^{-2}$.

XII.8.8 Problems

(i) For the variety $\mathcal{V}$ in 3.6, characterize $\mathcal{Q}\mathcal{G} \circ \langle B_2^1 \rangle$ (the greatest element of $\mathcal{V}\nu$).

Result 8.3 is due to Bales [1], and what follows it is due to Reilly [16]. These papers contain a wealth of information concerning the Malcev product. Results concerning finite bases of the Malcev product were established by Kleiman [4]. For the source of the Malcev product consult Malcev [1].

XII.9 VARIETIES AND E-UNITARY COVERS

We have defined E-unitary covers in VII.4.5. Consider the following problem: which varieties $\mathcal{V}$ of inverse semigroups have the property that every semigroup

S in $\mathcal{V}$ has an E-unitary cover in $\mathcal{V}$? We have seen in VII.4.6 and VIII.1.10 that every inverse semigroup has an E-unitary cover and thus the above property holds for the variety $\mathcal{I}$ of all inverse semigroups.

The varieties having the above property are characterized here in several ways. We then refine the problem by considering varieties $\mathcal{V}$ with the property that all semigroups in $\mathcal{V}$ have an E-unitary cover over some group G in a fixed group variety $\mathcal{U}$. This leads to a characterization in terms of Malcev products.

It is convenient to introduce the following concept.

XII.9.1 Definition

A variety $\mathcal{V}$ of inverse semigroups *has E-unitary covers* if every S in $\mathcal{V}$ has an E-unitary cover in $\mathcal{V}$.

The notation $\rho(\mathcal{V})$ introduced in I.11.11 will now be used for varieties of inverse semigroups.

XII.9.2 Theorem

The following conditions on a nontrivial variety $\mathcal{V}$ of inverse semigroups are equivalent

(i) $\mathcal{V}$ has E-unitary covers.

(ii) The free objects in $\mathcal{V}$ are E-unitary.

(iii) The free object in $\mathcal{V}$ on a countably infinite set of generators is E-unitary.

(iv) $\mathcal{V}$ is generated by its E-unitary members.

(v) $\ker \rho(\mathcal{V})$ is closed.

Proof. (i) *implies* (ii). Let S be a $\mathcal{V}$-free inverse semigroup, T be an E-unitary inverse semigroup in $\mathcal{V}$, and $\varphi : T \to S$ be an idempotent separating epimorphism. Let $X \subseteq S$ be a set of $\mathcal{V}$-free generators of S and let C be a cross section of the congruence on T induced by φ. Define ψ on X by: $x\psi = c$ if $c\varphi = x$ and $c \in C$. Then ψ extends uniquely to a homomorphism ψ on S into T. Hence $\psi\varphi$ is an endomorphism of S whose restriction to X is the identity mapping on X. Since X is a set of $\mathcal{V}$-free generators of S, we must have $\psi\varphi = \iota_S$. But then ψ is one-to-one and is thus a monomorphism of S into T. Since T is E-unitary so is $S\psi$ and thus also S.

(ii) *implies* (iii). This is trivial.

(iii) *implies* (iv). This follows directly from I.11.11.

(iv) *implies* (i). Let $S \in \mathcal{V}$. In view of I.11.7 and the hypothesis, there exist E-unitary inverse semigroups T_α in $\mathcal{V}$, an inverse semigroup T which is a subdirect product of T_α's and an epimorphism $\varphi : T \to S$. Letting ρ be the congruence on T induced by φ, we obtain the following diagram of epimorphisms:

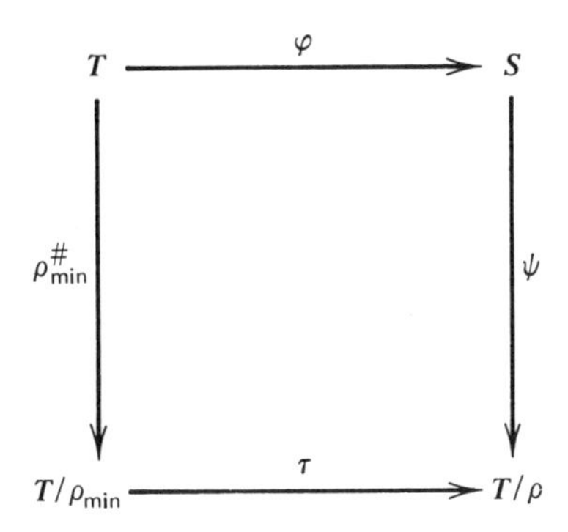

where $\tau: t\rho_{\min} \to t\rho$ ($t \in T$), and ψ is an isomorphism. Since $\operatorname{tr}\rho = \operatorname{tr}\rho_{\min}$, τ is one-to-one on idempotents and is thus idempotent separating. We know that $\sigma = \omega_{\min}$, which then implies that $\sigma \supseteq \rho_{\min}$. This together with the fact that T is E-unitary gives $\ker\rho_{\min} \subseteq \ker\sigma = E_T$ so that $\ker\rho_{\min} = E_T = E_T\omega$. It follows that $T/\rho_{\min}$ is E-unitary by III.7.3. Since $T/\rho_{\min} \in \mathcal{V}$, we have proved that S has an E-unitary cover in $\mathcal{V}$.

(iii) *is equivalent to* (v) by III.7.3.

We now consider E-unitary covers over groups belonging to a fixed group variety.

XII.9.3 Definition

Let $\mathcal{U}$ be a variety of groups. An inverse semigroup S *has an E-unitary cover over* $\mathcal{U}$ if S has an E-unitary cover T over a group G in $\mathcal{U}$. A variety $\mathcal{V}$ of inverse semigroups *has E-unitary covers over* $\mathcal{U}$ if every semigroup S in $\mathcal{V}$ has an E-unitary cover over $\mathcal{U}$.

It follows that $\mathcal{V}$ has E-unitary covers if and only if it has E-unitary covers over its group part $\mathcal{V} \cap \mathcal{G}$. Note that in view of VII.4.2, an E-unitary cover for an inverse semigroup S over a group variety $\mathcal{U}$ is a subdirect product of S with a group in $\mathcal{U}$. The next result gives a simple criterion for an inverse semigroup to have an E-unitary cover over a group variety.

XII.9.4 Theorem

Let S be an inverse semigroup and $\mathcal{U}$ be a variety of groups. Then S has an E-unitary cover over $\mathcal{U}$ if and only if $u^2 = u$ a law in $\mathcal{U}$ implies $u^2 = u$ is a law in S.

Proof

Necessity. In view of the hypothesis and VII.4.2, there exists a group G in $\mathcal{U}$ and a full prehomomorphism $\theta: G \to C(S)$, in notation $\theta: g \to \theta_g$. Let $u = u(x_1, x_2, \ldots, x_n)$ be a (inverse semigroup) word and assume that $u^2 = u$ is a law in $\mathcal{U}$. Let $s_1, s_2, \ldots, s_n \in S$. Fullness of θ implies that for each $i = 1, 2, \ldots, n$, there exists $g_i \in G$ such that $s_i \in \theta_{g_i}$. The definition of a prehomomorphism implies that $u(s_1, s_2, \ldots, s_n) \in \theta_{u(g_1, g_2, \ldots, g_n)}$. Since $u^2 = u$ is a law in G, we must have $u(g_1, g_2, \ldots, g_n) = 1$, the identity of G. But $\theta_1 = E_S$, as we have seen at the beginning of the proof of VII.4.2. Thus, $u(s_1, s_2, \ldots, s_n) \in E_S$ and hence $u^2 = u$ is a law in S.

Sufficiency. Let $\alpha: I_S \to S$ be the homomorphism which extends the identity map on S. Let G be the $\mathcal{U}$-free group on the set S and let $\beta: I_S \to G$ be the natural homomorphism. For every $g \in G$, let

$$\theta_g = \{s \in S \mid s = u\alpha,\ u\beta = g \text{ for some } u \in I_S\}.$$

Since β is an epimorphism, $\theta_g \neq \varnothing$ for every $g \in G$. Let $a \in S$ and $b \in \theta_g$ be such that $a \leq b$. Hence $a = be$, $b = u\alpha$, $u\beta = g$ for some $e \in E_S$ and $u \in I_S$. Considering e as an element of I_S, we have $a = be = (ue)\alpha$, $(ue)\alpha = (u\alpha)1 = g$, and thus $a \in \theta_g$. Hence θ_g is an order ideal of S.

Next let $a, b \in \theta_g$, say $a = u\alpha$, $b = v\alpha$, $u\beta = v\beta = g$ for some $u, v \in I_S$. Then $a^{-1}b = (u^{-1}v)\alpha$, $(u^{-1}v)\beta = 1$, and the hypothesis implies that $a^{-1}b \in E_S$; analogously, we have $ab^{-1} \in E_S$. Hence θ_g is permissible.

Let $a \in \theta_g$, $b \in \theta_h$, say $a = u\alpha$, $u\beta = g$, $b = v\alpha$, $v\beta = h$. Then $ab = (uv)\alpha$, $(uv)\beta = gh$, which shows that $ab \in \theta_{gh}$. Finally, θ is full since α is an epimorphism. Consequently, the function $\theta: g \to \theta_g$ $(g \in G)$ satisfies all the conditions of VII.4.2 and thus S has an E-unitary cover over G.

We may let S range over a variety $\mathcal{V}$ in 9.4 which then yields the following consequence.

XII.9.5 Corollary

Let $\mathcal{U}$ be a variety of groups and $\mathcal{V}$ be a variety of inverse semigroups. Then $\mathcal{V}$ has E-unitary covers over $\mathcal{U}$ if and only if $u^2 = u$ a law in $\mathcal{U}$ implies $u^2 = u$ is a law in $\mathcal{V}$.

XII.9.6 Corollary

If an inverse semigroup S has an E-unitary cover over a group variety $\mathcal{U}$, then $\langle S \rangle$ has E-unitary covers over $\mathcal{U}$.

Proof. Let $u^2 = u$ be a law in $\mathcal{U}$. Then $u^2 = u$ is a law in S by 9.4. But then $u^2 = u$ is also a law in $\langle S \rangle$ and the desired conclusion follows from 9.5.

It is of interest to notice the following simple statement.

XII.9.7 Proposition

No combinatorial variety of inverse semigroups different from $\mathcal{T}$ and $\mathcal{S}$ has E-unitary covers.

Proof. Let $\mathcal{V}$ be a combinatorial variety. For any S in $\mathcal{V}$, the Green relation $\mathcal{H}$ is the equality relation. Thus, any idempotent separating epimorphism of S is an isomorphism. Hence S has an E-unitary cover in $\mathcal{V}$ if and only if it is itself E-unitary. Now assume that $\mathcal{V} \neq \mathcal{T}, \mathcal{S}$. It follows that $B_2 \in \mathcal{V}$. Since B_2 has a zero and is not a semilattice, it is not E-unitary. Therefore B_2 has no E-unitary cover over $\mathcal{V}$.

Since a semilattice is necessarily E-unitary, $\mathcal{S}$ has E-unitary covers over any variety of groups. In contrast, 9.7 shows that $\mathcal{B}$ does not have this property. However, it is very close to having it, as the next result indicates.

XII.9.8 Proposition

The variety $\mathcal{B}$ has E-unitary covers over every nontrivial variety of groups.

Proof. Let $n > 1$ be an integer. Setting $S = Q = B_2$ and $K = \mathbb{Z}/(n)$ in VII.4.14, we easily see that B_2 has an E-unitary cover over $\mathbb{Z}/(n)$. Hence B_2 has an E-unitary cover over $\mathcal{AG}_n = \langle \mathbb{Z}/(n) \rangle$. Thus 9.6 implies that $\mathcal{B} = \langle B_2 \rangle$ has E-unitary covers over $\mathcal{AG}_n$. This evidently gives the desired result.

The next result indicates that $\mathcal{B}$ represents an exception among the strict varieties in the present context.

XII.9.9 Proposition

Every strict variety, except $\mathcal{B}$, has E-unitary covers.

Proof. By 4.16, every strict variety has one of the forms $\mathcal{U}$, $\mathcal{U} \vee \mathcal{S}$, $\mathcal{U} \vee \mathcal{B}$ for some group variety $\mathcal{U}$. We first consider the case $\mathcal{V} = \mathcal{U} \vee \mathcal{B}$ where $\mathcal{U}$ is a nontrivial group variety.

Let $S \in \mathcal{V}$. By I.11.9, there exist $G \in \mathcal{U}$, $B \in \mathcal{B}$, a subdirect product T of G and B, and an epimorphism $\varphi: T \to S$. Since $\mathcal{U} \neq \mathcal{T}$, 9.8 provides an E-unitary cover P of B over $\mathcal{U}$. Now VII.4.2 yields that $P \in \mathcal{U} \vee \mathcal{B} = \mathcal{V}$. Let $\psi: P \to B$ be an idempotent separating epimorphism and let

$$T' = \{(g, p) \in G \times P \mid (g, p\psi) \in T\}.$$

Then T' is clearly an inverse subsemigroup of $G \times P$. Since G is a group and P is E-unitary, $G \times P$ is E-unitary, and thus so is T'. Define a mapping χ by

$$\chi: (g, p) \to (g, p\psi)\varphi \qquad [(g, p) \in T'].$$

It follows easily that χ is a homomorphism of T' into S. Let $s \in S$. Then $t\varphi = s$ for some $t \in T$ since φ is an epimorphism. Then $t = (g, b)$ for some $g \in G$ and $b \in B$. Hence $p\psi = b$ for some $p \in P$ since ψ is an epimorphism. It follows that $(g, p\psi) \in T$ and thus $(g, p) \in T$ and clearly $(g, p)\chi = s$. Thus $\chi : T' \to S$ is an epimorphism. Hence $\mathfrak{V}$ must be generated by its E-unitary members and 9.2 asserts that $\mathfrak{V}$ has E-unitary covers.

The case of $\mathfrak{V} = \mathfrak{U} \vee \mathfrak{S}$ is treated similarly. By 9.7 we know that $\mathfrak{B}$ has no E-unitary covers.

XII.9.10 Notation

For any property P of identities on inverse semigroups, we let

$$[u = v \mid u = v \text{ has property } P]$$

denote the class of all inverse semigroups satisfying all the identities $u = v$ where $u = v$ has property P.

The third principal result of this section follows.

XII.9.11 Theorem

The following statements are valid for any variety $\mathfrak{U}$ of groups.

(i) $\langle \mathfrak{S} \circ \mathfrak{U} \rangle = [u^2 = u \mid u^2 = u \text{ is a law in } \mathfrak{U}]$
$= \{S \in \mathfrak{I} \mid S \text{ has an } E\text{-unitary cover over } \mathfrak{U}\}$.

(ii) $\langle \mathfrak{S} \circ \mathfrak{U} \rangle$ is the largest variety of inverse semigroups having E-unitary covers over $\mathfrak{U}$.

(iii) $\mathfrak{U}$ is the smallest variety of groups over which $\langle \mathfrak{S} \circ \mathfrak{U} \rangle$ has E-unitary covers.

(iv) $\langle \mathfrak{S} \circ \mathfrak{U} \rangle \cap \mathfrak{G} = \mathfrak{U}, \quad \langle \mathfrak{S} \circ \mathfrak{U} \rangle \cap \operatorname{Ob} \mathfrak{E} = \mathfrak{S} \circ \mathfrak{U}$.

Proof. (i) Let $\mathfrak{V}$, $\mathfrak{W}$, and $\mathfrak{Z}$ be the three classes in part (i). First let $S \in \mathfrak{S} \circ \mathfrak{U}$ and let $u^2 = u$ be a law in $\mathfrak{U}$. It follows from the definition of $\mathfrak{S} \circ \mathfrak{U}$ that $S/\sigma \in \mathfrak{U}$ and thus $u^2 = u$ is a law in S/σ. Hence S satisfies $u^2 \sigma u$ and we may consider $u \in \ker \sigma = E_S$. Thus $u^2 = u$ is a law in S. Consequently, $\mathfrak{S} \circ \mathfrak{U} \subseteq \mathfrak{W}$ and thus $\mathfrak{V} = \langle \mathfrak{S} \circ \mathfrak{U} \rangle \subseteq \mathfrak{W}$.

The inclusion $\mathfrak{W} \subseteq \mathfrak{Z}$ follows directly from 9.4. Next let $S \in \mathfrak{Z}$. Then S has an E-unitary cover T over G for some $G \in \mathfrak{U}$. It follows that $T \in \mathfrak{S} \circ \mathfrak{U}$ and hence $S \in \langle \mathfrak{S} \circ \mathfrak{U} \rangle = \mathfrak{V}$. Therefore $\mathfrak{Z} \subseteq \mathfrak{V}$.

(ii) This is an obvious consequence of part (i).

(iii) Let $\mathfrak{V}$ be a variety of groups over which $\langle \mathfrak{S} \circ \mathfrak{U} \rangle$ has E-unitary covers and let $G \in \mathfrak{U}$. Then $G \in \langle \mathfrak{S} \circ \mathfrak{U} \rangle$ and hence has an E-unitary cover T over $\mathfrak{V}$. But then T must itself be a group. Since G is a homomorphic image of T, we obtain $G \in \mathfrak{V}$. Consequently $\mathfrak{U} \subseteq \mathfrak{V}$, as required.

(iv) Let $G \in \langle \mathfrak{S} \circ \mathfrak{U} \rangle \cap \mathfrak{G}$ and let $u^2 = u$ be a law in $\mathfrak{U}$. Then $u^2 = u$ is also a law in G by 9.4. Hence $G \in \mathfrak{U}$ since every law in $\mathfrak{U}$, except $xx^{-1} = yy^{-1}$, can be written in the form $u^2 = u$. Consequently $\langle \mathfrak{S} \circ \mathfrak{U} \rangle \cap \mathfrak{G} \subseteq \mathfrak{U}$; the opposite inclusion is obvious.

Next let $S \in \langle \mathfrak{S} \circ \mathfrak{U} \rangle \cap \text{Ob}\, \mathfrak{E}$. Then $S/\sigma \in \langle \mathfrak{S} \circ \mathfrak{U} \rangle \cap \mathfrak{G} = \mathfrak{U}$ by the first formula in part (iv). Since S is E-unitary, we get $S \in \mathfrak{S} \circ \mathfrak{U}$. Thus $\langle \mathfrak{S} \circ \mathfrak{U} \rangle \cap \text{Ob}\, \mathfrak{E} \subseteq \mathfrak{S} \circ \mathfrak{U}$; the opposite inclusion is trivial.

Going back to the free inverse semigroup on a countably infinite set of generators, we have the following result.

XII.9.12 Proposition

For any variety $\mathfrak{U}$ of groups and any variety $\mathfrak{V}$ of inverse semigroups, we have

$$\ker \rho(\mathfrak{U}) = \ker \rho(\mathfrak{V}) \Leftrightarrow \mathfrak{U} \subseteq \mathfrak{V} \subseteq \langle \mathfrak{S} \circ \mathfrak{U} \rangle.$$

Proof. First assume that $\ker \rho(\mathfrak{U}) = \ker \rho(\mathfrak{V})$. This means that $w^2 = w$ is a law in $\mathfrak{U}$ if and only if $w^2 = w$ is a law in $\mathfrak{V}$. It follows from 9.11(i) that $\mathfrak{V} \subseteq \langle \mathfrak{S} \circ \mathfrak{U} \rangle$. Since $\mathfrak{U}$ is a group variety, we have $\operatorname{tr} \rho(\mathfrak{U}) = \omega$ and thus $\operatorname{tr} \rho(\mathfrak{U}) \supseteq \operatorname{tr} \rho(\mathfrak{V})$. This together with the hypothesis that $\ker \rho(\mathfrak{U}) = \ker \rho(\mathfrak{V})$ implies that $\rho(\mathfrak{U}) \supseteq \rho(\mathfrak{V})$ and thus $\mathfrak{U} \subseteq \mathfrak{V}$.

Conversely, suppose that $\mathfrak{U} \subseteq \mathfrak{V} \subseteq \langle \mathfrak{S} \circ \mathfrak{U} \rangle$. The first inclusion implies that $\rho(\mathfrak{U}) \supseteq \rho(\mathfrak{V})$ and thus $\ker \rho(\mathfrak{U}) \supseteq \ker \rho(\mathfrak{V})$. The second inclusion implies that $\ker \rho(\mathfrak{U}) \subseteq \ker \rho(\mathfrak{V})$ by 9.11(i), as above. Therefore $\ker \rho(\mathfrak{U}) = \ker \rho(\mathfrak{V})$.

XII.9.13 Exercises

(i) For any variety $\mathfrak{V}$ of inverse semigroups, show that

  (α) $\langle \mathfrak{V} \cap \text{Ob}\, \mathfrak{E} \rangle$ is the largest variety contained in $\mathfrak{V}$ having E-unitary covers,

  (β) $\langle \mathfrak{V} \cap \text{Ob}\, \mathfrak{E} \rangle = \{S \in \mathfrak{I} \mid S \text{ has an } E\text{-unitary cover in } \mathfrak{V}\}$.

(ii) Let $\mathfrak{V}$ be a variety of inverse semigroups. Show that each of the following statements implies the next.

  (α) $\mathfrak{V}$ has E-unitary covers.

  (β) For every $S \in \mathfrak{V}$, there exists $G \in \mathfrak{V} \cap \mathfrak{G}$, an inverse semigroup T which is a subdirect product of S/μ and G, and an idempotent separating epimorphism $\varphi : T \to S$.

  (γ) $\mathfrak{V}$ is the minimum element of its ν-class.

  Also show that (β) does not imply (α) in general.

(iii) Show that every Clifford variety has E-unitary covers. (Two different proofs are possible: the direct proof, and passing through the free objects and using 9.4.)

(iv) On the lattice $\mathfrak{L}(\mathfrak{I})$ define a relation ξ by

$$\mathfrak{U}\xi\mathfrak{V} \Leftrightarrow \mathfrak{U} \cap \text{Ob}\,\mathfrak{E} = \mathfrak{V} \cap \text{Ob}\,\mathfrak{E}.$$

Show that ξ is an $\cap$-congruence on $\mathfrak{L}(\mathfrak{I})$ but not a congruence. For every $\mathfrak{V} \in \mathfrak{L}(\mathfrak{I})$, show that $\langle \mathfrak{V} \cap \text{Ob}\,\mathfrak{E} \rangle$ is the least member of the ξ-class containing $\mathfrak{V}$. Also show that $\nu \subseteq \xi \subseteq \nu_2$.

(v) Show that every inverse semigroup free in a variety of inverse semigroups containing all groups is E-unitary.

XII.9.14 Problems

(i) Does (γ) imply (β) in 9.13(ii)?
(ii) Which varieties $\langle B(G, \alpha) \rangle$ have E-unitary covers?

This entire section is due to Petrich–Reilly [4]; a part of 9.2 was also established by Pastijn [4].

XIII

AMALGAMATION

Amalgamation has its origin in a very natural setting. Given two algebras of the same kind having a nonempty intersection, does there exist an algebra, of the same kind, which contains them both as subalgebras? This problem setting can be immediately generalized to an arbitrary number of algebras. The situation is abstracted by requiring that the algebras be pairwise disjoint but that each contain an isomorphic copy of some fixed algebra.

1. The class of inverse semigroups has the representation extension property and the free representation extension property. These two in general imply the strong representation property which in turn can be used to prove the strong amalgamation property for inverse semigroups within the class of all semigroups. Using this one proves that the class of all inverse semigroups has the strong amalgamation property (within inverse semigroups).

2. The variety of all groups as well as any inverse semigroup variety consisting entirely of commutative semigroups have the strong amalgamation property. For groups, this can be derived from the inverse semigroup case, but historically represents the first significant result in the theory of amalgamation. The class of finite inverse semigroups does not have the weak amalgamation property, whereas the class of E-unitary inverse semigroups does not have the special amalgamation property.

3. Conversely, it turns out that any proper inverse semigroup variety which is not a group variety and has the weak amalgamation property consists entirely of commutative semigroups. This gives a complete answer to the question which inverse semigroup varieties have the (weak, strong) amalgamation property (modulo group varieties).

4. One defines the free product and the amalgamated free product of inverse semigroups in a natural way and proves that the results concerning semigroups in general carry over to this case as well. The same type of thing can be done for any variety of inverse semigroups.

5. The strong amalgamation property for the class of inverse semigroups can be used to show that if an inverse semigroup has a maximal conjugate

(respectively normal) extension then it must have idempotent metacenter. This completes some of the results in VI.2.

XIII.1 STRONG AMALGAMATION FOR INVERSE SEMIGROUPS

Our purpose here is to prove the strong amalgamation property for the class of all inverse semigroups. To this end, we first introduce several concepts and prove a sequence of lemmas. The path goes through the representation of inverse semigroups by transformations of a set and ends with an embedding into the symmetric inverse semigroup on a set.

XIII.1.1 Construction

Let U be an inverse subsemigroup of a semigroup S, and let $\rho\colon U \to \mathfrak{T}(X)$ be a representation. Write $\rho\colon u \to \rho^u$ for all $u \in U$. We adjoin a common identity 1 to both S and U (even if they already have an identity) and denote the resulting semigroups by $\overline{S}$ and $\overline{U}$, respectively. Extend the representation ρ to $\overline{U}$ by letting $\rho^1 = \iota_X$.

On $X \times \overline{S}$ define a relation δ_1 by

$$(x, a)\delta_1(y, b) \Leftrightarrow a = us, \quad b = vs, \quad x\rho^u = y\rho^v$$

$$\text{for some } u, v \in \overline{U}, s \in \overline{S}.$$

It follows easily that δ_1 is reflexive, and it is obvious that it is symmetric. Let δ be the transitive closure of δ_1. Then δ is an equivalence relation on $X \times \overline{S}$; we denote its class containing (x, a) by $[x, a]$.

For every $c \in S$, define α^c by

$$[x, a]\alpha^c = [x, ac] \qquad ([x, a] \in (X \times \overline{S})/\delta).$$

It follows from δ_1 being compatible with $(x, s) \to (x, sc)$ that α^c is single valued. We now define a function

$$\alpha\colon c \to \alpha^c \qquad (c \in S).$$

It is then clear that $\alpha\colon S \to \mathfrak{T}[(X \times \overline{S})/\delta]$ is a representation. Define a map f by

$$f\colon x \to [x, 1] \qquad (x \in X).$$

The ordered pair (α, f) constructed above is the *free extension of ρ to S*.

Henceforth we will write xu instead of $x\rho^u$ for all $x \in X$, $u \in \overline{U}$. For any (x, s), $(y, t) \in X \times \overline{S}$, $(x, s)\delta(y, t)$ is equivalent to the following condition. There exist a natural number n and elements $x_i \in X$, $1 \leq i \leq n + 1$, u_i, $v_i \in \overline{U}$

and $s_i \in \overline{S}$, $1 \leq i \leq n$ such that

$$x = x_1, \qquad s = u_1 s_1, \qquad x_{n+1} = y, \qquad v_n s_n = t, \tag{1}$$

$$x_1 u_1 = x_2 v_1, \qquad x_2 u_2 = x_3 v_2, \ldots, x_n u_n = x_{n+1} v_n, \tag{2}$$

$$v_1 s_1 = u_2 s_2, \qquad v_2 s_2 = u_3 s_3, \ldots, v_{n-1} s_{n-1} = u_n s_n. \tag{3}$$

In the case $s = t = 1$, we have $u_1 = s_1 = v_n = s_n = 1$.

XIII.1.2 Lemma

Assume that conditions (1)–(3) hold and let

$$w_i = u_1 v_1^{-1} u_2 v_2^{-1} \cdots u_i v_i^{-1} \qquad (i = 1, 2, \ldots, n).$$

(i) $w_n^{-1} w_n v_n s_n = w_n^{-1} u_1 s_1$.
(ii) $x_{n+1} w_n^{-1} = x_1 w_n w_n^{-1}$.

Proof. (i) Indeed,

$$\begin{aligned}
w_n^{-1} w_n v_n s_n &= v_n \big(u_n^{-1} w_{n-1}^{-1} w_{n-1} u_n \big) \big(v_n^{-1} v_n \big) s_n \\
&= v_n \big(v_n^{-1} v_n \big) \big(u_n^{-1} w_{n-1}^{-1} w_{n-1} u_n \big) s_n && \text{commuting of idempotents} \\
&= w_n^{-1} w_{n-1} v_{n-1} s_{n-1} && \text{by (3)} \\
&= \cdots = w_n^{-1} w_1 v_1 s_1 \\
&= v_n u_n^{-1} \cdots v_2 u_2^{-1} v_1 \big(u_1^{-1} u_1 \big) \big(v_1^{-1} v_1 \big) s_1 && \text{definition of } w_n, w_1 \\
&= v_n u_n^{-1} \cdots v_2 u_2^{-1} v_1 \big(v_1^{-1} v_1 \big) \big(u_1^{-1} u_1 \big) s_1 && \text{commuting of idempotents} \\
&= w_n^{-1} u_1 s_1 && \text{definition of } w_n.
\end{aligned}$$

(ii) Indeed,

$$\begin{aligned}
x_{n+1} w_n^{-1} &= x_{n+1} v_n v_n^{-1} w_n^{-1} && \text{definition of } w_n \\
&= x_n u_n v_n^{-1} w_n^{-1} && \text{by (2)} \\
&= x_n (u_n v_n^{-1} v_n u_n^{-1})(v_{n-1} v_{n-1}^{-1}) w_{n-1}^{-1} && \text{definition of } w_n \\
&= x_n (v_{n-1} v_{n-1}^{-1})(u_n v_n^{-1} v_n u_n^{-1}) w_{n-1}^{-1} && \text{commuting of idempotents} \\
&= x_{n-1} u_{n-1} v_{n-1}^{-1} u_n v_n^{-1} w_n^{-1} && \text{by (2), definition of } w_n \\
&= \cdots = x_1 u_1 v_1^{-1} \cdots u_n v_n^{-1} w_n^{-1} \\
&= x_1 w_n w_n^{-1}. && \text{definition of } w_n.
\end{aligned}$$

XIII.1.3 Definition

A subsemigroup U of a semigroup S has the *representation extension property in* S if for any set X and any representation $\rho\colon U \to \mathfrak{T}(X)$, there exists a set Y disjoint from X and a representation $\alpha\colon S \to \mathfrak{T}(X \cup Y)$ such that $\alpha_X = \rho$.

Recall from IV.5.6 that $\alpha_X = \rho$ means that $\alpha^u|_X = \rho^u$ for all $u \in U$. Compare the next lemma with IV.5.7.

XIII.1.4 Lemma

If U is an inverse subsemigroup of a semigroup S, then U has the representation extension property in S.

Proof. Let $\rho\colon U \to \mathfrak{T}(X)$ be a representation of U and define δ_1 and δ as above. Take $x, y \in X$ such that $(x,1)\delta = (y,1)\delta$; we show first that $x = y$.

By the definition of δ, there exist a positive integer n and elements $x_i \in X$ for $1 \le i \le n+1$, $u_i, v_i \in \overline{U}$, $s_i \in \overline{S}$ for $1 \le i \le n$ satisfying equations (1)–(3) with $s = t = 1$.

If $(x,1)\delta_1(y,1)$, then $x = y$ since $u_1 s_1 = v_n s_n = 1$ implies $u_1 = v_1 = 1$; this takes care of the case $n = 1$. Assuming that $n > 1$, we have

$x = x_1 = x_2 v_1 = x_2(w_1^{-1}w_1)(v_1 s_1)$	$u_1 = 1 = s_1$ and $w_1 = v_1^{-1}$
$= x_2(w_1^{-1}w_1)(u_2 s_2)$	by (3)
$= x_2(u_2 u_2^{-1})(w_1^{-1}w_1)(u_2 s_2)$	commuting of idempotents
$= x_3 v_2 u_2^{-1}(w_1^{-1}w_1)(u_2 s_2)$	by (2)
$= x_3 v_2(v_2^{-1}v_2)(u_2^{-1}w_1^{-1}w_1 u_2)(v_2^{-1}v_2)s_2$	commuting of idempotents
$= x_3(w_2^{-1}w_2 v_2 s_2)$	
$= x_3(w_2^{-1}w_2)u_3 s_3$	by (3)
$= \cdots = x_{n+1}w_n^{-1}w_n v_n s_n$	
$= x_{n+1}w_n^{-1}w_n$	since $v_n s_n = 1$
$= x_{n+1}w_n^{-1}$	by 1.2(i)
$= x_1 w_n w_n^{-1}$	by 1.2(ii)
$= y$	by symmetry and $x = x_{n+1}w_n^{-1}w_n$.

Now let $x \in X$ and $u \in U$. In the definition of δ_1, letting $u, 1, 1$ stand for u, v, s, respectively, we obtain that $(x, u)\delta_1(xu, 1)$, which implies

$$[x, 1]\alpha^u = [x, u] = [xu, 1].$$

In view of the above, we may identify each $[x, 1]$ with the element x, so that the last equation gives $\alpha_X = \rho$. Finally, put $Y = (X \times \overline{S})/\delta \setminus Xf$.

XIII.1.5 Definition

A subsemigroup U of a semigroup S has the *free representation extension property in S* if for any sets $X' \subseteq X$ and any representations

$$\rho\colon \overline{U} \to \mathfrak{T}(X), \qquad \rho'\colon \overline{U} \to \mathfrak{T}(X'),$$

such that $\rho_{X'} = \rho'$ and $\rho^1 = \iota_X$, we have $\delta|_{(X' \times \overline{S})} = \delta'$ where δ is defined for ρ and δ' for ρ' as in 1.1.

XIII.1.6 Lemma

If U is an inverse subsemigroup of a semigroup S, then U has the free representation extension property in S.

Proof. Let $\rho\colon \overline{U} \to \mathfrak{T}(X)$ and $\rho'\colon \overline{U} \to \mathfrak{T}(X')$ be representations of $\overline{U}$ with $X' \subseteq X$, $\rho_{X'} = \rho'$, and $\rho^1 = \iota_X$. Define δ_1 and δ on $X \times \overline{S}$, δ_1' and δ' on $X' \times \overline{S}$ as in 1.1. Take $x, y \in X'$ and $s, t \in \overline{S}$ such that $(x, s)\delta(y, t)$; it suffices to show that $(x, s)\delta'(y, t)$.

By the definition of δ, there exists a positive integer n and elements $x_i \in X$ for $1 \leq i \leq n + 1$, $u_i, v_i \in \overline{U}$ and $s_i \in \overline{S}$ for $1 \leq i \leq n$ such that (1), (2), and (3) hold.

We will need the following statement:

$$(z, up)\delta_1'(zu, p) \qquad (z \in X', u \in \overline{U}, p \in \overline{S}), \tag{4}$$

which follows directly from the definition of δ_1'. If also $u^2 = u$, then (4) implies

$$(zu, p)\delta_1'(zu, up) \qquad (z \in X', u \in \overline{U}, p \in \overline{S}). \tag{5}$$

We now compute

$$(x, s) = (x_1, u_1 s_1)\delta_1'(x_1 u_1, s_1) \qquad \text{by (4)}$$

$$= (x_2 v_1 u_1^{-1} u_1, s_1) = (x_2 v_1 (v_1^{-1} v_1)(u_1^{-1} u_1), s_1) \qquad \text{by (2)}$$

$$\delta_1'(x_2 v_1, (v_1^{-1} v_1)(u_1^{-1} u_1) s_1) \qquad \text{by (4)}$$

$$= (x_2v_1, (u_1^{-1}u_1)(v_1^{-1}v_1)s_1) \qquad \text{commuting of idempotents}$$

$$= (x_2v_1, u_1^{-1}u_1v_1^{-1}(u_2s_2)) \qquad \text{by (3)}$$

$$\delta_1'(x_2v_1u_1^{-1}u_1v_1^{-1}u_2, s_2) \qquad \text{by (4)}$$

$$= (x_2w_1^{-1}w_1u_2, s_2) \qquad \text{definition of } w_1$$

$$= (x_2(w_1^{-1}w_1)(u_2u_2^{-1})u_2, s_2)$$

$$= (x_2(u_2u_2^{-1})(w_1^{-1}w_2)u_2, s_2) \qquad \text{commuting of idempotents}$$

$$= (x_3v_2u_2^{-1}w_1^{-1}w_1u_2, s_2) \qquad \text{by (2)}$$

$$= (x_3v_2u_2^{-1}w_1^{-1}w_1u_2v_2^{-1}v_2, s_2) \qquad \text{commuting of idempotents}$$

$$\delta_1'(x_3v_2u_2^{-1}w_1^{-1}w_1u_2, v_2^{-1}v_2s_2) \qquad \text{by (5)}$$

$$= (x_3v_2u_2^{-1}w_1^{-1}w_1u_2, v_2^{-1}(u_3s_3)) \qquad \text{by (2)}$$

$$\delta_1'(x_3v_2u_2^{-1}w_1^{-1}w_1u_2v_2^{-1}u_3, s_3) \qquad \text{by (4)}$$

$$= (x_3w_2^{-1}w_2u_3, s_3) \qquad \text{definition of } w_2$$

$$= \cdots = (x_nw_{n-1}^{-1}w_{n-1}u_n, s_n)$$

$$= (x_n(u_nu_n^{-1})(w_{n-1}^{-1}w_{n-1})u_n, s_n) \qquad \text{commuting of idempotents}$$

$$= (x_{n+1}v_nu_n^{-1}w_{n-1}^{-1}w_{n-1}u_n, s_n) \qquad \text{by (2)}$$

$$\delta_1'(x_{n+1}, v_n(v_n^{-1}v_n)(u_n^{-1}w_{n-1}^{-1}w_{n-1}u_n)s_n) \qquad \text{by (4)}$$

$$= (x_{n+1}, v_n(u_n^{-1}w_{n-1}^{-1}w_{n-1}u_n)(v_n^{-1}v_n)s_n) \qquad \text{commuting of idempotents}$$

$$= (x_{n+1}, w_n^{-1}w_nv_ns_n) \qquad \text{definition of } w_n$$

$$= (x_{n+1}, w_n^{-1}u_1s_1) \qquad \text{by 1.2(i)}$$

$$\delta_1'(x_{n+1}w_n^{-1}, u_1s_1) \qquad \text{by (4)}$$

$$= (x_1w_nw_n^{-1}, u_1s_1) \qquad \text{by 1.2(iii)}$$

$$\delta_1'(x_1, w_nw_n^{-1}u_1s_1) \qquad \text{by (4)}$$

$$\delta'(y, t) \qquad \text{by symmetry and } (x, s)\delta_1'(x_{n+1}, w_n^{-1}w_nv_ns_n) \text{ above.}$$

XIII.1.7 Definition

Let U be a subsemigroup of a semigroup S. Let X and Y be sets, $\rho\colon U \to \mathfrak{T}(X)$ and $\beta\colon S \to \mathfrak{T}(Y)$ be representations, and $g\colon X \to Y$ be a function. The ordered pair (β, g) is an *extension of* ρ *to* S if

(i) $\rho^u g = g\beta^u \quad (u \in U)$,
(ii) $\{xg | x \in X\} \cup \{xg\beta^s | x \in X, s \in S\} = Y$.

Condition 1.7(i) means commutativity of the diagram

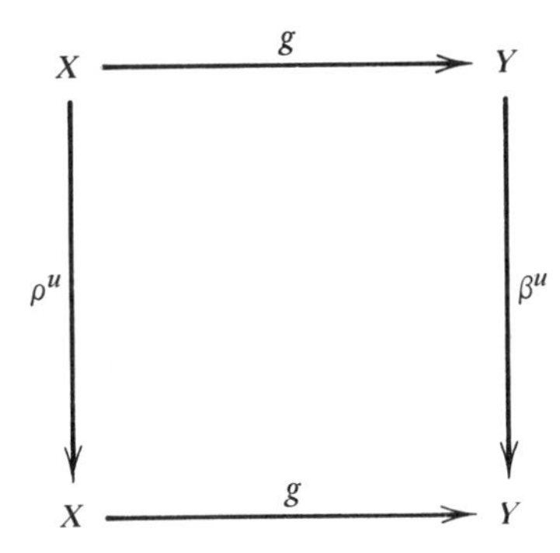

for all $u \in U$.

XIII.1.8 Lemma

Let the notation be as in 1.7 and let $\beta^1 = \iota_Y$. Then there exists a surjection $\varphi\colon (X \times \bar{S})/\delta \to Y$ such that

$$f\varphi = g, \qquad \varphi\beta^s = \alpha^s\varphi \qquad (s \in S).$$

Proof. Define φ by

$$\varphi\colon (x, s)\delta \to xg\beta^s \qquad [(x, s)\delta \in (X \times \bar{S})/\delta].$$

In order to show that φ is single valued, we let $(x, s)\delta(y, t)$, and assume that equations (1)–(3) are satisfied. Hence

$$x_i\rho^{u_i} = x_{i+1}\rho^{v_i} \qquad (i = 1, 2, \ldots, n)$$

which by the hypothesis $\rho^u g = g\beta^u$ implies, for any $s \in S$,

$$x_{i+1}g\beta^{v_i s} = x_{i+1}g\beta^{v_i}\beta^s = x_{i+1}\rho^{v_i}g\beta^s = x_i\rho^{u_i}g\beta^s = x_i g\beta^{u_i}\beta^s = x_i g\beta^{u_i s}.$$

Hence, using (1)–(3), we get

$$xg\beta^s = x_1 g\beta^{u_1 s_1} = x_2 g\beta^{v_1 s_1} = x_2 g\beta^{u_2 s_2} = \cdots$$

$$= x_n g\beta^{u_n s_n} = x_{n+1} g\beta^{v_n s_n} = yg\beta^t,$$

as required. Hence φ is single valued.

For any $x \in X$, we have

$$xf\varphi = (x,1)\delta\varphi = xg\beta^1 = xg$$

which gives $f\varphi = g$. If also $s, t \in \bar{S}$, we obtain

$$(x,t)\delta\varphi\beta^s = (xg\beta^t)\beta^s = xg\beta^{ts} = (x,ts)\delta\varphi = (x,t)\delta\alpha^s\varphi$$

so that $\varphi\beta^s = \alpha^s\varphi$ for all $s \in S$. If $y \in Y$, then $y = xg$ for some $x \in X$ or $y = xg\beta^s$ for some $x \in X$ and $s \in S$, which implies that φ maps $(X \times \bar{S})/\delta$ onto Y.

XIII.1.9 Definition

A subsemigroup U of a semigroup S has the *strong representation extension property in* S if for any sets $X' \subseteq X$ and any representations

$$\rho: U \to \mathfrak{T}(X), \qquad \beta: S \to \mathfrak{T}(X')$$

such that $\rho_{X'} = \beta|_U$, there exists a set Y disjoint from X and a representation $\alpha: S \to \mathfrak{T}(X \cup Y)$ such that $(\alpha|_U)_X = \rho$ and $\alpha_{X'} = \beta$.

XIII.1.10 Lemma

Let U be a subsemigroup of a semigroup S having both the representation extension property and the free representation extension property in S. Then U also has the strong representation property in S.

Proof. Let $\rho: U \to \mathfrak{T}(X)$ and $\beta: S \to \mathfrak{T}(X')$ be representations with $X' \subseteq X$ and $\rho_{X'} = \beta|_U$. Construct δ and the free extension (α, f) of ρ to S as in 1.1. By the free representation extension property, the pair $(\alpha_{(X' \times \bar{S})\delta^\#}, f|_{X'})$ is the free extension of $\rho_{X'}$ to S. It is easily seen that $(\beta, \iota_{X'})$ is an extension of $\rho_{X'}$ to S. Hence by 1.8, there exists a mapping φ of $(X' \times \bar{S})\delta^\#$ onto X' such that $(f|_{X'})\varphi = \iota_{X'}$ and $\varphi\beta^s = (\alpha^s|_{(X' \times \bar{S})\delta^\#})\varphi$ for all $s \in S$.

Put $\theta = (\varphi \circ \varphi^{-1}) \cup \varepsilon_0$ where ε_0 is the identical relation on $(X \times \bar{S})/\delta$. Clearly θ is an equivalence relation on $(X \times \bar{S})/\delta$. For every $s \in S$, define

$$\gamma^s: p\theta \to p\alpha^s\theta \qquad [p \in (X \times \bar{S})/\delta].$$

To see that γ^s is single valued, let $p\theta = q\theta$. Then $p\theta\beta_s = q\theta\beta_s$ and $\varphi\beta^s = \alpha^s\varphi$ yield $p\alpha^s\theta = q\alpha^s\theta$, as required. Also $(x, t)\delta\alpha^s = (x, ts)\delta$ implies that $(X' \times \overline{S})\delta^{\#}\alpha^s \subseteq (X' \times \overline{S})\delta^{\#}$ and thus $\gamma^s \in \mathfrak{T}((X \times \overline{S})\delta^{\#}\theta^{\#})$. It is clear that the function

$$\gamma: s \to \gamma^s \qquad (s \in S)$$

is a representation of S.

The equation $f\varphi = \iota_{X'}$ yields that for any $x' \in X'$, $x' = x'f\varphi = (x', 1)\delta\varphi$. Hence the mapping

$$(x', 1)\delta \to (x', 1)\delta\varphi = x' \qquad (x' \in X')$$

is one-to-one. Also the mapping

$$x \to (x, 1)\delta \qquad (x \in X)$$

is one-to-one by the proof 1.4.

We prove next that for any $x \in X \setminus X'$, we have $(x, 1)\delta \notin (X' \times \overline{S})\delta^{\#}$. We argue by contradiction assuming that there exist $x \in X \setminus X'$, $x' \in X'$, and $t \in S$ such that $(x, 1)\delta(x', t)$.

Take a set Z disjoint from X and of the same cardinality as $X \setminus X'$ and let a map $x \to x^*$ be a bijection from $X \setminus X'$ to Z. We extend, for each $u \in U$, the function $\rho^u \in \mathfrak{T}(X)$ to a function $\tau^u \in \mathfrak{T}(X \cup Z)$ by defining

$$x^*\tau^u = \begin{cases} (x\rho^u)^* & \text{if} \quad x\rho^u \in X \setminus X' \\ x\rho^u & \text{if} \quad x\rho^u \in X'. \end{cases}$$

We define $\tau: u \to \tau^u \quad (u \in U)$ and verify easily that $\tau: U \to \mathfrak{T}(X \cup Z)$ is a representation of U. Consider the free extension of τ to S, denoting by δ^* the associated equivalence relation on $(X \cup Z) \times \overline{S}$.

It follows from the definition of the relation δ_1 that $\delta_1^*|_{(X \times \overline{S})} = \delta_1$ since $\tau_X = \rho$. Hence the hypothesis $(x, 1)\delta(x', t)$ implies $(x, 1)\delta^*(x', t)$. Furthermore, the hypothesis implies the existence of $x_2, \ldots, x_{n+1} \in X$, $u_2, \ldots, u_n \in \overline{U}$, $v_1, \ldots, v_n \in \overline{U}$, and $s_2, \ldots, s_n \in \overline{S}$ such that

$$v_n s_n = t, \tag{6}$$

$$x_1 = x_2\rho^{v_2}, \quad x_2\rho^{u_2} = x_3\rho^{v_3}, \ldots, x_n\rho^{u_n} = x'\rho^{v_n}, \tag{7}$$

$$v_1 = u_2 s_2, \quad v_2 s_2 = u_3 s_3, \ldots, v_{n-1}s_{n-1} = u_n s_n \tag{8}$$

by (1), (2), and (3) with $x = x_1$, $s = u_1 = s_1 = 1$, $x_{n+1} = x'$. Now $x = x_2\rho^{v_2} \in X \setminus X'$ in (7) implies that $x_2 \in X \setminus X'$ and thus $x^* = (x_2\rho^{v_2})^*$ implies

$$x^* = x_2^*\tau^{v_2} \tag{9}$$

by the definition of τ. Now we take the second equation in (7), namely $x_2\rho^{u_2} = x_3\rho^{v_3}$, where $x_2 \in X \setminus X'$, as we have just seen.

Case 1. $x_2\rho^{u_2} \in X \setminus X'$. Then $x_3 \in X \setminus X'$ and $(x_2\rho^{u_2})^* = (x_3\rho^{v_3})^*$ implies

$$x_2^*\tau^{u_2} = x_3^*\tau^{v_3} \tag{10}$$

by the definition of τ.

Case 2. $x_2\rho^{u_2} \in X'$. Then $x_2^*\tau^{u_2} = (x_2\rho^{u_2})^* = x_3\rho^{v_3} = x_3\tau^{v_3}$ by the definition of τ, and thus

$$x_2^*\tau^{u_2} = x_3\tau^{v_3}. \tag{11}$$

We may continue this process with the next equation in (7), namely $x_3\rho^{u_3} = x_4\rho^{v_3}$, and in each of the above two cases consider the two cases: $x_3\rho^{u_3} \in X \setminus X'$ or $x_3\rho^{u_3} \in X'$. The same type of argument will yield in the first case

$$x_3^*\tau^{u_3} = x_4^*\tau^{v_3} \tag{12}$$

or

$$x_3^*\tau^{u_3} = x_4\tau^{v_3} \tag{13}$$

and in the second case; we may take

$$x_3\tau^{u_3} = x_4\tau^{v_3}. \tag{14}$$

We continue this process thus obtaining one of the sequences

(9), (10), (12),

(9), (10), (13),

(9), (11), (14).

As soon as an unstarred x_i makes appearance in continuing this procedure, we complete the sequence on the pattern of (7). Since the last equation in (7) is of the form $x_n\rho^{u_n} = x'\rho^{v_n}$ with $x' \in X'$ and thus $x_n\rho^{u_n} \in X'$, in the above procedure there will occur a change of the form x_i starred to x_{i+1} unstarred somewhere along the way. We thus obtain a sequence of the form (7) with starred x_i up to a certain place and τ instead of ρ. This together with (6) and (8) shows that $(x^*, 1)\delta^*(x', t)$.

We have seen above that $(x, 1)\delta^*(x', t)$ and we now conclude that $(x^*, 1)\delta^*(x, 1)$. The proof of 1.4 applied to δ^* yields that the mapping

$$f^*: X \cup Z \to \big((X \cup Z) \times \overline{S}\big)/\delta^*$$

defined by $f^*: y \to (y, 1)\delta^*$ is one-to-one. But then

$$xf^* = (x, 1)\delta = (x^*, 1)\delta = x^*f^*$$

implies that $x = x^*$ contradicting the hypothesis that X and Z are disjoint.

We deduce that the mapping

$$x \to (x, 1)\delta\theta \qquad (x \in X)$$

is one-to-one, and may identify X with its image under this map.

For any $x' \in X'$, $x \in X$, $u \in U$, $s \in S$, using (4), we get

$$x\gamma^u = (x, 1)\delta\theta\gamma^u = (x, 1)\delta\alpha^u\theta = (x, u)\delta\theta = (x\rho^u, 1)\delta\theta = x\rho^u,$$

$$\begin{aligned} x'\gamma^s &= (x', 1)\delta\theta\gamma^s = (x', 1)\delta\alpha^s\theta = (x', 1)\delta\alpha^s\theta\theta^{-1} \\ &= (x', 1)\delta\varphi\beta^s\varphi^{-1} = x'\beta^s\varphi^{-1} = (x\beta^s, 1)\delta\theta = x'\beta^s. \end{aligned}$$

This proves that $(\gamma|_U)_X = \rho$ and $\gamma_{X'} = \beta$.

XIII.1.11 Lemma

An amalgam $[S_i; U]_{i \in I}$ of semigroups is strongly embeddable if U has the strong representation extension property in each S_i.

Proof. In view of I.12.4, we may consider only the case $|I| = 2$. Hence we let $I = \{1, 2\}$ and $S = S_1$, $T = S_2$. Without loss of generality, we assume that $S \cap T = U$. As before, we adjoin a common identity 1 to S and T to obtain $\overline{S}$, $\overline{T}$, and $\overline{U}$. Put

$$X_0 = Y_0 = \overline{U}, \qquad X_1 = S \setminus U, \qquad Y_1 = T \setminus U,$$

and let

$$\alpha_1: S \to \mathfrak{T}(X_1 \cup X_0), \qquad \beta_1: T \to \mathfrak{T}(Y_0 \cup Y_1), \qquad \gamma_0: U \to \mathfrak{T}(X_0)$$

be the restrictions to S, T, and U of the right regular representations of $\overline{S}$, $\overline{T}$, and $\overline{U}$, respectively. Then for all $u \in U$, α_1^u and β_1^u agree on their common domain X_0 with γ_0^u so $\alpha_1^u \cup \beta_1^u$ is a function and the map

$$\gamma_1: u \to \alpha_1^u \cup \beta_1^u \qquad (u \in U)$$

is a representation $\gamma_1: U \to \mathfrak{T}(X_1 \cup X_0 \cup Y_1)$.

By the strong representation extension property of U in S, there exist a set X_2 and a representation

$$\alpha_2: S \to \mathfrak{T}(X_2 \cup X_1 \cup X_0 \cup Y_1)$$

such that

$$(\alpha_2^s)_{X_1 \cup X_0} = \alpha_1^s, \qquad (\alpha_2^u)_{X_1 \cup X_0 \cup Y_1} = \gamma_1^u \qquad (s \in S, u \in U).$$

Similarly, there exist a set Y_2 and a representation

$$\beta_2 \colon T \to \mathfrak{T}(X_1 \cup Y_0 \cup Y_1 \cup Y_2)$$

such that

$$(\beta_2^t)_{Y_0 \cup Y_1} = \beta_1^t, \qquad (\beta_2^u)_{X_1 \cup Y_0 \cup Y_1} = \gamma_1^u \qquad (t \in T, u \in U).$$

Once again, for all $u \in U$, α_2^u and β_2^u agree on their common domain $X_1 \cup X_0 \cup Y_1$, this time agreeing with γ_1^u. We now define

$$\gamma_2 \colon u \to \alpha_2^u \cup \beta_2^u \qquad (u \in U)$$

to obtain a representation $\gamma_2 \colon U \to \mathfrak{T}(X_2 \cup X_1 \cup X_0 \cup Y_1 \cup Y_2)$.

Continuing this process, we obtain disjoint sets

$$\ldots, X_2, X_1, X_0, Y_1, Y_2, \ldots,$$

representations $\alpha_1, \alpha_2, \ldots$ of S, $\beta_1, \beta_2, \ldots$ of T, $\gamma_0, \gamma_1, \ldots$ of U such that, for $n = 1, 2, \ldots,$

$$\alpha_n \colon S \to \mathfrak{T}(X_n \cup \cdots \cup X_1 \cup X_0 \cup Y_1 \cup \cdots \cup Y_{n-1}),$$

$$\beta_n \colon T \to \mathfrak{T}(X_{n-1} \cup \cdots \cup X_1 \cup X_0 \cup Y_1 \cup \cdots \cup Y_n),$$

$$\gamma_{n-1} \colon U \to \mathfrak{T}(X_{n-1} \cup \cdots \cup X_1 \cup X_0 \cup Y_1 \cup \cdots \cup Y_{n-1}),$$

$$(\alpha_n^u)_{X_{n-1} \cup \cdots \cup X_1 \cup X_0 \cup Y_1 \cup \cdots \cup Y_{n-1}}$$

$$= (\beta_n^u)_{X_{n-1} \cup \cdots \cup X_1 \cup X_0 \cup Y_1 \cup \cdots \cup Y_{n-1}} \qquad (u \in U),$$

$$\alpha_1^s \subseteq \alpha_2^s \subseteq \alpha_3^s \subseteq \cdots \qquad (s \in S),$$

$$\beta_1^t \subseteq \beta_2^t \subseteq \beta_3^t \subseteq \cdots \qquad (t \in T).$$

We thus may define

$$\alpha \colon s \to \alpha_1^s \cup \alpha_2^s \cup \cdots \qquad (s \in S),$$

$$\beta \colon t \to \beta_1^t \cup \beta_2^t \cup \cdots \qquad (t \in T)$$

to obtain representations

$$\alpha: S \to \mathfrak{T}(\cdots \cup X_2 \cup X_1 \cup X_0 \cup Y_1 \cup Y_2 \cup \cdots),$$

$$\beta: T \to \mathfrak{T}(\cdots \cup X_2 \cup X_1 \cup X_0 \cup Y_1 \cup Y_2 \cup \cdots).$$

Now α and β are faithful since α_1 and β_1 are faithful. It follows easily that $\alpha^u = \beta^u$ for all $u \in U$. Assume that $\alpha^s = \beta^t$ for some $s \in S$, $t \in T$. Then $\alpha_1^s = \beta_1^t$ so that, for the element $1 \in X_0 = Y_0 = \overline{U}$, we have $s = 1\alpha_1^s = 1\beta_1^t = t$ whence $s = t \in S \cap T = U$, giving $S\alpha \cap T\beta = U\alpha$ $(= U\beta)$.

XIII.1.12 Corollary

An amalgam $[S_i, U]_{i \in I}$ where S_i are semigroups and U is an inverse semigroup is strongly embeddable.

Proof. This follows directly from 1.4, 1.6, 1.10, and 1.11.

The long and winding path to the proof of strong amalgamation for inverse semigroups comes to an end with the following result.

XIII.1.13 Theorem

The class of inverse semigroups has the strong amalgamation property.

Proof. In view of I.12.4, it suffices to consider two inverse semigroups S and T with a common inverse subsemigroup U. Without loss of generality, we assume that $S \cap T = U$. By 1.12, there exists a semigroup V containing S and T as subsemigroups. Adjoin an identity to V to obtain a semigroup $\overline{V}$. Let ρ be the right regular representation of $\overline{V}$ restricted to V. Further let

$$\alpha = \rho|_S, \qquad \beta = \rho|_T, \qquad \gamma = \rho|_U.$$

Then $\alpha: S \to \mathfrak{T}(V)$, $\beta: T \to \mathfrak{T}(V)$, $\gamma: U \to \mathfrak{T}(V)$ are faithful representations of S, T, and U, respectively, since ρ is faithful. For each $s \in S$, $t \in T$, $u \in U$, let

$$\overline{\alpha_s} = \alpha_s|_{Vs^{-1}}, \qquad \overline{\beta_t} = \beta_t|_{Vt^{-1}}, \qquad \overline{\gamma_u} = \gamma_u|_{Vu^{-1}}.$$

According to IV.5.9, the maps

$$\overline{\alpha}: s \to \overline{\alpha_s}, \qquad \overline{\beta}: t \to \overline{\beta_t}, \qquad \overline{\gamma}: u \to \overline{\gamma_u}$$

are faithful representations

$$\overline{\alpha}: S \to \mathfrak{I}(V), \qquad \overline{\beta}: T \to \mathfrak{I}(V), \qquad \overline{\gamma}: U \to \mathfrak{I}(V)$$

and it is clear that $\bar{\alpha}|_U = \bar{\beta}|_U = \bar{\gamma}$. (This gives the weak amalgamation property for inverse semigroups.)

Let V' be the set V with the opposite multiplication, namely $a * b = ba$ in V'. Let ρ' be the right regular representation of $\bar{V}'$ restricted to V'. We can now repeat the above definitions with primes attached to obtain $S', T', U', \alpha', \beta', \gamma', \alpha'_s, \beta'_t, \gamma'_u, \overline{\alpha'_s}, \overline{\beta'_t}, \overline{\gamma'_u}, \overline{\alpha'}, \overline{\beta'}, \overline{\gamma'}$, respectively. Now combine these representations as follows:

$$\left(\overline{\alpha'}, \bar{\alpha}\right): s \to \left(\overline{\alpha'_s}, \overline{\alpha_s}\right) \qquad (s \in S),$$

$$\left(\overline{\beta'}, \bar{\beta}\right): t \to \left(\overline{\beta'_t}, \overline{\beta_t}\right) \qquad (t \in T),$$

$$\left(\overline{\gamma'}, \bar{\gamma}\right): u \to \left(\overline{\gamma'_u}, \overline{\gamma_u}\right) \qquad (u \in U).$$

In this way, we obtain the representations

$$\left(\overline{\alpha'}, \bar{\alpha}\right): S \to P, \qquad \left(\overline{\beta'}, \bar{\beta}\right): T \to P, \qquad \left(\overline{\gamma'}, \bar{\gamma}\right): U \to P$$

where $P = (\mathcal{I}(V))' \times \mathcal{I}(V)$, and clearly

$$\left(\overline{\alpha'}, \bar{\alpha}\right)|_U = \left(\overline{\beta'}, \bar{\beta}\right)|_U = \left(\overline{\gamma'}, \bar{\gamma}\right).$$

In order to prove that the function $(\overline{\alpha'}, \bar{\alpha}) \cup (\overline{\beta'}, \bar{\beta})$ is one-to-one, we let $s \in S \setminus T$ and $t \in T \setminus S$ and distinguish three cases.

Case 1. ss^{-1} and tt^{-1} are not $\mathcal{L}_V$-related. Then

$$\mathbf{d}\overline{\alpha_s} = Vs^{-1} = Vss^{-1} \neq Vtt^{-1} = Vt^{-1} = \mathbf{d}\overline{\beta_t}$$

and $\overline{\alpha_s} \neq \overline{\beta_t}$.

Case 2. ss^{-1} and tt^{-1} are not $\mathcal{R}_V$-related. Then they are not $\mathcal{L}_{V'}$-related and Case 1 gives that $\overline{\alpha'_s} \neq \overline{\beta'_t}$.

Case 3. $ss^{-1}(\mathcal{L} \cap \mathcal{R})tt^{-1}$. Then $ss^{-1} = tt^{-1}$ and thus

$$(ss^{-1})\overline{\alpha_s} = (ss^{-1})\rho_s = s \neq t = (tt^{-1})\rho_t = (tt^{-1})\overline{\beta_t}$$

and thus $\overline{\alpha_s} \neq \overline{\beta_t}$.

We have proved that $s \in S \setminus U$ and $t \in T \setminus U$ implies $(\overline{\alpha'_s}, \overline{\alpha_s}) \neq (\overline{\beta'_t}, \overline{\beta_t})$. By contrapositive, if $\delta \in S(\overline{\alpha'}, \bar{\alpha}) \cap T(\overline{\beta'}, \bar{\beta})$, then $\delta = (\overline{\alpha'_s}, \alpha_s) = (\overline{\beta'_t}, \overline{\beta_t})$ for some $s \in S$ and $t \in T$, and thus either $s \in U$ or $t \in U$, and in either case $\delta \in U(\overline{\gamma'}, \bar{\gamma})$. Therefore

$$S\left(\overline{\alpha'}, \bar{\alpha}\right) \cap T\left(\overline{\beta'}, \bar{\beta}\right) = U\left(\overline{\gamma'}, \bar{\gamma}\right)$$

in the inverse semigroup P, as required.

XIII.1.14 Exercises

(i) Show that the free extension (α, f) of ρ to S constructed in 1.1 is indeed an extension of ρ to S according to 1.7.

(ii) Show that φ in 1.8 is unique relative to the properties stated there.

(iii) Let S be an inverse semigroup. For $U = E_S$ and ρ the right regular representation of E_S, define δ as in 1.1. Show that $(x, s)\delta(y, t)$ if and only if $xs = yt$. Next let

$$T = \{(ss^{-1}, s)\delta | s \in S\}.$$

For every $s \in S$, let $\hat{\alpha}^s = \alpha^s|_T$. Show that the mapping $\hat{\alpha} : s \to \hat{\alpha}^s$ $(s \in S)$ is a representation of S equivalent to the right regular representation of S.

(iv) Let the notation be as in 1.1 and let

$$\theta = \{((x, us), (xu, s)) | x \in X, u \in \overline{U}, s \in \overline{S}\}.$$

Show that δ is the equivalence relation generated by θ.

(v) Let S be the semilattice of two groups G and H determined by a homomorphism $\varphi : G \to H$. For $U = G$ and ρ the right regular representation of G, let (α, f) be the free extension of ρ to S. Show that α is equivalent to ξ_G where ξ is the right regular representation of S. Also prove that for $U = H$ we get in this way a representation equivalent to ξ.

(vi) Show that if an amalgam of inverse semigroups is strongly embeddable into a finite semigroup, then it is also strongly embeddable into a finite inverse semigroup.

(vii) Show by an example that in the proof of 1.13 consideration of representations $\overline{\alpha}$, $\overline{\beta}$, $\overline{\gamma}$ may not lead to the desired conclusion concerning injectivity of the mapping $\overline{\alpha} \cup \overline{\beta}$.

XIII.1.15 Problems

(i) (T. E. Hall). Is there a short proof of the special amalgamation property for inverse semigroups, directly, or from the free representation extension property for inverse semigroups?

The proof of strong amalgamation for inverse semigroup carried out in this section can be extracted from Hall [5] (see also Hall [9]); another proof can be found in Hall [4]. Result 1.12 was first established by Howie [3] using different methods.

The problem of amalgamation of inverse semigroups was posed by Preston [9]. Hall [4], [5], [7], [8], [9] and Howie [2] contain further properties of inverse semigroups in the context of amalgamation. In particular, Hall [8] is replete with applications of amalgamation of inverse semigroups. An application of amalgamation will be treated in XIII.5.

XIII.2 VARIETIES WITH THE STRONG AMALGAMATION PROPERTY

From the preceding section we derive here that both the class of groups and the class of semilattices have the strong amalgamation property. The latter is used to prove that any inverse semigroup variety all of whose members are commutative also has the strong amalgamation property. We end the section with two negative results.

In the proofs below we will not mention explicitly that for the desired conclusions one actually uses I.12.4.

XIII.2.1 Lemma

The variety $\mathcal{G}$ of all groups has the strong amalgamation property.

Proof. Let $[G, H; U; \varphi, \psi]$ be a group amalgam. By 1.13, it is embeddable into an inverse semigroup, S say, according to the following commutative diagram:

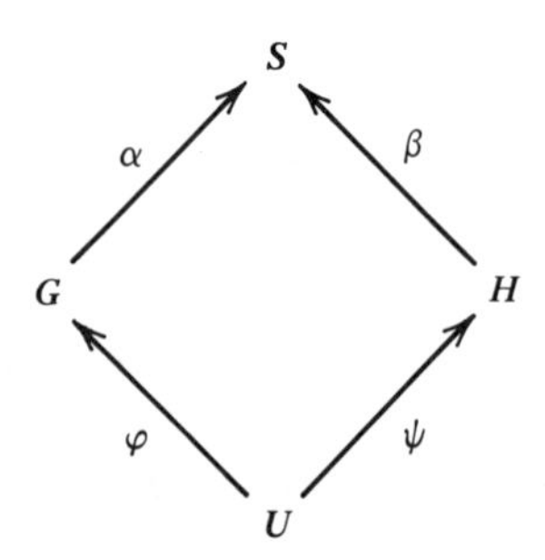

Then $G\alpha$ and $H\beta$ are subgroups of S with intersection $U\varphi\alpha$. But then both $G\alpha$ and $H\beta$ are contained in a maximal subgroup K of S. Hence the above amalgam is embeddable into the group K.

XIII.2.2 Lemma

The variety $\mathcal{S}$ of semilattices has the strong amalgamation property.

Proof. Let $[X, Y; U; \varphi, \psi]$ be a semilattice amalgam. By 1.13 it is embeddable into an inverse semigroup S. But then this amalgam is strongly embeddable into the semilattice E_S.

XIII.2.3 Theorem

Every variety of commutative inverse semigroups has the strong amalgamation property.

Proof. First let S and T be commutative inverse semigroups, and let U be their common inverse subsemigroup. We may suppose that $U = S \cap T$. To each semigroup S, T, U we may adjoin a common identity 1 (even if such already exists) and denote the resulting semigroups by $\bar{S}, \bar{T}, \bar{U}$. Since $E_{\bar{S}} \cap E_{\bar{T}} = E_{\bar{U}}$, by 1.2, there exists a semilattice Z and monomorphisms $\zeta : E_{\bar{S}} \to Z$ and $\pi : E_{\bar{T}} \to Z$ such that $\zeta|_{E_{\bar{U}}} = \pi|_{E_{\bar{U}}}$ and

$$E_{\bar{S}}\zeta \cap E_{\bar{T}}\pi = E_{\bar{U}}\zeta = E_{\bar{U}}\pi.$$

Put

$$X = E_S\zeta, \qquad \bar{X} = E_{\bar{S}}\zeta, \qquad Y = E_T\pi, \qquad \bar{Y} = E_{\bar{T}}\pi.$$

We can write S and T as Clifford semigroups $S = [X; G_\alpha, \varphi_{\alpha,\beta}]$ and $T = [Y; H_\alpha, \psi_{\alpha,\beta}]$. Then $\bar{S} = [\bar{X}; G_\alpha, \varphi_{\alpha,\beta}]$ and $\bar{T} = [\bar{Y}; H_\alpha, \psi_{\alpha,\beta}]$ in the obvious notation. If an element x of S is in G_α, we will write x_α for x; denote by 1_α the identity of G_α; similarly for T.

Now let $V = \bar{S} \times \bar{T} \setminus \{(1,1)\}$, and on V define a relation θ by

$$(x_\alpha, y_\beta)\theta(w_\gamma, z_\delta) \Leftrightarrow \text{there exists } u_\sigma \in \bar{U} \text{ for which}$$

$$\sigma \geq \alpha\beta = \gamma\delta \quad \text{in} \quad Z, \qquad u_\sigma x_\alpha = 1_\sigma w_\gamma, \qquad 1_\sigma y_\beta = u_\sigma z_\delta.$$

Routine verification shows that θ is a congruence on V. (It is in fact the congruence generated by the set

$$\{((u_\sigma, 1_\sigma), (1_\sigma, u_\sigma)) | u_\sigma \in \bar{U}\},$$

which we do not actually need.)

We define the functions

$$\varphi : s \to (s,1)\theta \qquad (s \in S),$$

$$\psi : t \to (1,t)\theta \qquad (t \in T).$$

We will now verify that the functions φ and ψ satisfy the conditions for strong embedding of the amalgam $[S, T; U]$.

First it is clear that φ and ψ are homomorphisms. Let $x_\alpha, y_\beta \in S$ and assume that $x_\alpha\varphi = y_\beta\varphi$. Then $(x_\alpha, 1)\theta(y_\beta, 1)$. So by the definition of θ, $\alpha = \beta$ and there exists $u_\sigma \in \bar{U}$ such that

$$\sigma \geq \alpha \quad \text{in} \quad Z, \qquad u_\sigma x_\alpha = 1_\sigma w_\gamma, \qquad 1_\sigma 1 = u_\sigma 1.$$

Hence $u_\sigma = 1_\sigma$ and thus $x_\alpha = y_\beta$. Consequently, φ is one-to-one; the proof for ψ is symmetric.

Next let $u_\alpha \in U$. Since $u_\alpha^{-1}u_\alpha = 1_\alpha 1$ and $1_\alpha 1 = u_\alpha^{-1}u_\alpha$, we obtain

$$u_\alpha\varphi = (u_\alpha, 1)\theta = (1, u_\alpha)\theta = u_\alpha\psi$$

which proves that $\varphi|_U = \psi|_U$.

Finally, let $x_\alpha \in S$, $y_\beta \in T$ and assume that $x_\alpha\varphi = y_\beta\psi$. By the definition of θ, we get $\alpha = \beta$ and the existence of $u_\sigma \in \overline{U}$ for which

$$\sigma \geq \alpha \quad \text{in} \quad Z, \qquad u_\sigma x_\alpha = 1_\sigma 1, \qquad 1_\sigma 1 = u_\sigma y_\alpha.$$

It follows that $\sigma = \alpha$ and $x_\alpha = u_\sigma^{-1} = y_\alpha \in U$. We have proved that $S\varphi \cap T\psi \subseteq U\varphi$; the opposite inclusion is trivial. Therefore, the amalgam $[S, T; U]$ is strongly embedded into the inverse semigroup V.

Now let $\mathcal{V}$ be a variety of commutative inverse semigroups and S, T above be in $\mathcal{V}$. Then $\overline{S}, \overline{T}, \overline{U} \in \mathcal{V} \vee \mathbb{S}$ and hence also $V \in \mathcal{V} \vee \mathbb{S}$. If $\mathcal{V} \supseteq \mathbb{S}$, then the above says that $\mathcal{V}$ has the strong amalgamation property. If $\mathcal{V} \not\supseteq \mathbb{S}$, then $\mathcal{V}$ is a group variety, and by the above, $[S, T; U]$ can be strongly embedded into a subgroup K of V as in the proof of 2.1, with $K \in \mathcal{V}$ (consult XII.4.5 for Clifford varieties).

There are, however, plenty of classes of inverse semigroups which do not have the strong amalgamation property. As an illustration, we prove the following negative results.

XIII.2.4 Proposition

The class of finite inverse semigroups does not have the weak amalgamation property.

Proof. Let $S = \{x, x^{-1}, e, f, 0\}$ and $S' = \{y, y^{-1}, e, g, 0\}$ be combinatorial Brandt semigroups with common idempotents e and 0. Let T be the retract extension of S' by the two-element chain $\{f, 0'\}$ determined by the homomorphism $f \to g$. Hence

$$T = \{y, y^{-1}, e, g, 0, f\}.$$

Take $U = \{e, f, 0\}$ and assume that there exist a semigroup W and embeddings $\varphi : S \to W$ and $\psi : T \to W$ which agree on U. We have $e\mathcal{J}f$ in S, and $e\mathcal{J}g < f$ in T and thus

$$(g\psi)\mathcal{J}(e\psi) = (e\varphi)\mathcal{J}(f\varphi) = f\psi, \qquad g\psi < f\psi$$

which in view of IX.4.12 implies that W contains a copy of the bicyclic semigroup, and is thus infinite. We illustrate this example by the following diagram:

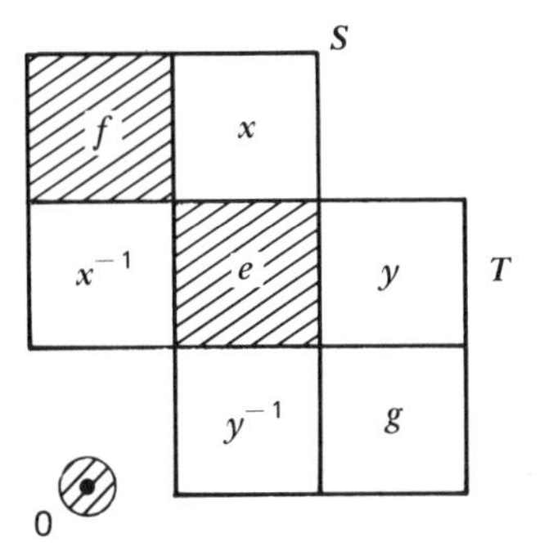

XIII.2.5 Proposition

The class of E-unitary inverse semigroups does not have the special amalgamation property.

Proof. Let G be a nontrivial finite group with identity e, and let $Y_2 = \{0, 1\}$ be a semilattice with $1 > 0$. Put $S = Y \times G$ and

$$U = (\{0\} \times G) \cup \{(1, e)\}.$$

Then both S and U are E-unitary inverse semigroups. Let ψ be an isomorphism of S onto a semigroup S' disjoint from S. We will show that the amalgam $\mathfrak{A} = [S, S'; U; \iota_U, \psi|_U]$ is not strongly embeddable in any E-unitary inverse semigroup.

We identify U with its image under ψ and may picture the situation as follows.

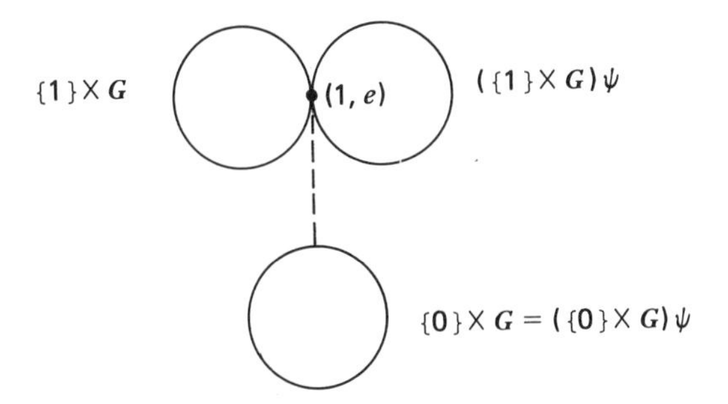

Suppose that the amalgam is strongly embedded into a semigroup H. Without loss of generality, we may assume that T is a semilattice Y of two groups H_0 and H_1, where in fact $H_0 = \{0\} \times G = (\{0\} \times G)\psi$. Since H_1 must contain copies of the groups $\{1\} \times G$ and $(\{1\} \times G)\psi$ whose intersection is the identity, we must have $|H_1| > |H_0|$. But then the homomorphism $H_1 \to H_0$ determining the multiplication in H cannot be one-to-one, and thus H is not E-unitary.

The following observation will be useful.

XIII.2.6 Lemma

For any set X, $\mathscr{I}(X)$ is an antigroup.

Proof. Let $\alpha \in E_{\mathscr{I}(X)}\zeta$ and $x \in \mathbf{d}\alpha$. Then $\varepsilon_{\{x\}} = \alpha\varepsilon_{\{x\}}$ which implies that $x\alpha = x\alpha\varepsilon_{\{x\}}$ so that $x\alpha = x$. Hence α is an idempotent. This shows that $E_{\mathscr{I}(X)} = E_{\mathscr{I}(X)}\zeta$ and IV.2.5 yields the desired conclusion.

XIII.2.7 Exercises

(i) Show that the class of antigroups has the strong amalgamation property.

(ii) Show that the class of combinatorial inverse semigroups does not have the special amalgamation property.

XIII.2.8 Problems

(i) Does the class of E-unitary inverse semigroups have the weak amalgamation property?

Schreier [1] initiated the theory of amalgamation for groups and established 2.1 directly. Weak amalgamation for semilattices was deduced by Horn–Kimura [1]. Hall [4] noted 2.1 and 2.2 as a consequence of the strong amalgamation property for inverse semigroups. Result 2.3 is due to Taĭclin [1] and Imaoka [1], [2]; see also Howie [2]. Result 2.4 can be found in Hall [4] (credited to C. J. Ash), whereas 2.5 is due to T. E. Hall. Crawley [1] deduced the special amalgamation property for Clifford semigroups.

XIII.3 THE CONVERSE

The aim of this section is to establish that the only proper inverse semigroup varieties which are not group varieties and have the weak amalgamation property are the varieties of commutative inverse semigroups. The proof of this is effected by a long sequence of lemmas and an intermediate result.

XIII.3.1 Lemma

Let S be an inverse semigroup and E be a subsemilattice of E_S. Then

$$E^c = \{ s \in S | sEs^{-1}, \quad s^{-1}Es \subseteq E, \quad ss^{-1}, s^{-1}s \in E \}$$

is the greatest inverse subsemigroup of S having E as its semilattice of idempotents.

Proof. For any $s, t \in E^c$, we get

$$(st)E(st)^{-1} = s(tEt^{-1})s^{-1} \subseteq sEs^{-1} \subseteq E,$$

$$(st)(st)^{-1} = s(tt^{-1})s^{-1} \in sEs^{-1} \subseteq E,$$

and symmetrically $(st)^{-1}E(st) \subseteq E$, $(st)^{-1}(st) \in E$. It then follows that E^c is an inverse subsemigroup of S whose semilattice of idempotents coincides with E. If T is any inverse subsemigroup of S with the semilattice of idempotents E, then for any $t \in T$, we clearly have $t \in E^c$. Consequently, $T \subseteq E^c$, which proves the assertion concerning E^c.

XIII.3.2 Lemma

Let E be a semilattice and identify E with $E_{\Phi(E)}$. Then the Munn representation of $\Phi(E)$ is the identical transformation on $\Phi(E)$.

Proof. Let $\alpha \in \Phi(E)$ and denote by δ^α its image in $\Phi(E)$ under the Munn representation. Let $\mathbf{d}\alpha = [e]$ and $\mathbf{r}\alpha = [f]$. In view of the identification of E and $E_{\Phi(E)}$, that is to say, of each $g \in E$ with $\iota_{[g]}$, we obtain

$$\begin{aligned} \mathbf{d}\theta_\alpha &= \{ \iota_{[g]} | g \in E, \quad \iota_{[g]} \leq \alpha\alpha^{-1} = \iota_{[e]} \quad \text{in} \quad \Phi(E) \} \\ &= \{ g \in E | g \leq e \quad \text{in} \quad E \} = [e] = \mathbf{d}\alpha. \end{aligned}$$

For $g \in \mathbf{d}\alpha$, we get

$$\begin{aligned} \mathbf{d}\left(\alpha^{-1}\iota_{[g]}\alpha\right) &= \{ x \in \mathbf{r}\alpha | x\alpha^{-1} \in \mathbf{d}\iota_{[g]} \} \\ &= \{ x \leq f | x\alpha^{-1} \leq g \} \\ &= \{ x \in E | x \leq g\alpha \} = [g\alpha], \end{aligned}$$

which gives

$$g\theta_\alpha = \iota_{[g]}\theta_\alpha = \alpha^{-1}\iota_{[g]}\alpha = \iota_{[g\alpha]} = g\alpha,$$

and hence $\theta_\alpha = \alpha$.

XIII.3.3 Lemma

Let X be a finite nonempty set, and denote by $P(X)$ the semilattice of all subsets of X under intersection. Then $\mathcal{I}(X) \cong \Phi(P(X))$.

Proof. We may identify $E_{\mathcal{I}(X)}$ with $P(X)$. In such a case, the Munn representation of $\mathcal{I}(X)$ maps $\mathcal{I}(X)$ into $\Phi(P(Y))$ and has the form

$$\delta : \alpha \to \bar{\alpha} \qquad [\alpha \in \mathcal{I}(X)]$$

where $\bar{\alpha} : Y \to Y\alpha$ if $Y \subseteq \mathbf{d}\alpha$. By 2.6, $\mathcal{I}(X)$ is an antigroup which implies that δ is a monomorphism.

Now let $\beta \in \Phi(P(X))$, and define α as follows:

$$\mathbf{d}\alpha = \{x \in X | \{x\} \in \mathbf{d}\beta\} \cup \{\varnothing\},$$

$$\{x\alpha\} = \{x\}\beta \quad \text{if } \{x\} \in \mathbf{d}\beta, \qquad \varnothing\alpha = \varnothing.$$

This definition is meaningful and thus β maps one-element subsets onto one-element subsets. The properties of β immediately give that $\alpha \in \mathcal{I}(X)$ and $\mathbf{d}\bar{\alpha} = \mathbf{d}\beta$. Furthermore, for any $Y \in \mathbf{d}\beta$, taking into account that β also preserves joins in $P(X)$, we get

$$\begin{aligned} x \in Y\bar{\alpha} &\Leftrightarrow x = y\alpha \quad \text{for some} \quad y \in Y \\ &\Leftrightarrow \{x\} = \{y\}\beta \quad \text{for some} \quad y \in Y \\ &\Leftrightarrow x \in Y\beta \end{aligned}$$

which shows that $\bar{\alpha} = \beta$. Consequently, ψ maps $\mathcal{I}(X)$ onto $\Phi(P(X))$.

XIII.3.4 Notation

Let B_2 be as in II.3.3. Letting $x = (1, 1, 2)$ and $e = xx^{-1}, f = x^{-1}x$, we have

$$B_2 = \{x, x^{-1}, e, f, 0\}.$$

Also let X be a nonempty set, and consider the set $P(X)$ of all subsets of X as a semilattice under intersection. Let

$$S = (P(X) \times B_2)/(P(X) \times \{0\})$$

and denote its semilattice of idempotents by E. Also let

$$C_u = (P(X) \times \{u\}) \cup \{0\}$$

for $u \in \{e, f\}$ so that E is the orthogonal sum of C_e and C_f.

We will identify E with $E_{\Phi(E)}$ throughout. Let $\{\pi_j | j \in J\}$ be the set of all permutations of X, where J is some index set. For each $j \in J$, define a mapping α_j on C_e by

$$\alpha_j : (Y, e) \to (Y\pi_j, f), \qquad 0 \to 0,$$

where $Y\pi_j = \{y\pi_j | y \in Y\}$. Then α_j is an isomorphism of C_e onto C_f, and hence $\alpha_j \in \Phi(E)$. Let S_j be the inverse subsemigroup of $\Phi(E)$ generated by the set $E \cup \{\alpha_j\}$.

XIII.3.5 Lemma

Let the notation be as in 3.4. Then $S \cong S_j$ for any $j \in J$.

Proof. Since S is combinatorial, it is an antigroup, and thus the Munn representation $\delta : S \to \Phi(E)$ is one-to-one. The image S^δ of S is evidently the inverse subsemigroup of $\Phi(E)$ generated by $E = E_{\Phi(E)}$ and the isomorphism

$$\alpha_1 : (Y, e) \to (Y, f), \qquad 0 \to 0.$$

Hence $S^\delta = S_1$ in the notation introduced above since π_1 is then the identity permutation of X. Consequently, $S \cong S_1$.

Now let $j \in J$ be arbitrary, and define a function β on C_f by

$$\beta : (Y, f) \to (Y\pi_j, f), \qquad 0 \to 0.$$

Clearly β is an automorphism of C_f and $\alpha_1\beta = \alpha_j$. Then $\gamma = \iota_{C_e} \cup \beta$ is an automorphism of E, which then induces an automorphism δ of $\Phi(E)$ by the usual rule:

$$\delta : \varphi \to \gamma^{-1}\varphi\gamma \qquad [\varphi \in \Phi(E)].$$

A simple inspection shows that $\alpha_1\delta = \alpha_j$ and since δ maps idempotents of S_1 onto idempotents of S_j, $\delta|_{S_1}$ is an isomorphism of S_1 onto S_j. Therefore $S_1 \cong S_j$ and thus also $S \cong S_j$.

We can now prove the first principal result of this section.

XIII.3.6 Theorem

Let $\mathcal{V}$ be a variety of inverse semigroups with the weak amalgamation property. If $\mathcal{V}$ is not a Clifford variety, then $\mathcal{V} = \mathcal{I}$.

Proof. Let $\mathcal{V}$ be as in the hypothesis part of the theorem. By XII.4.13, we know that $\mathcal{V}$ must contain B_2. We now use all the notation introduced in 3.4 and assume that X is finite. Hence $\mathcal{V}$ contains the chain $\{e, 0\}$, and it thus contains all semilattices. In particular $P(X) \in \mathcal{V}$ and thus also $S \in \mathcal{V}$. According to 3.5, for every $j \in J$, $S \cong S_j$ and hence $S_j \in \mathcal{V}$.

As before, we identify the idempotents of S_j with the semilattice E of idempotents of S (note $S = S_1$ with π_1 the identity mapping on X). By the hypothesis on $\mathcal{V}$, there exists an inverse semigroup T in $\mathcal{V}$ and for each $j \in J$, an isomorphism φ_j of S_j into T such that all φ_j agree on E. We may take that $\cup_{j \in J} S_j\varphi_j$ generates T as an inverse semigroup. Also, we identify each $i \in E$ with $i\varphi_1$ (which is also equal to $i\varphi_j$ for all $j \in J$) thus identifying E with $E\varphi_1$ ($= E\varphi_j$). By 3.1, there exists a greatest inverse subsemigroup E^c of T having E as its semilattice of idempotents. But then $S_j\varphi_j \subseteq E^c$ for all $j \in J$, and thus also

$$T = \bigcup_{j \in J} S_j\varphi_j \subseteq E^c \subseteq T$$

and consequently $T = E^c$. Hence E is the entire semilattice of idempotents of T. Let $\delta : T \to \Phi(E)$ be the Munn representation; we will show that $T^\delta = \Phi(E)$.

Fix $j \in J$. Since E is the common semilattice of idempotents of both S_j and T, and since φ_j is a monomorphism of S_j into T which leaves every element of E fixed, we have $\delta^{\alpha_j\varphi_j} = \delta^{\alpha_j}$; here δ^{α_j} denotes the element which corresponds to α_j under the Munn representation of S_j. Since $S_j \subseteq \Phi(E)$ we have by 3.2 that $\delta_j^{\alpha_j} = \alpha_j$. Thus $\delta^{\alpha_j\varphi_j} = \alpha_j$, and $\alpha_j \in T^\delta$ for all $j \in J$.

Now let $\alpha \in \Phi(E)$. If α is an isomorphism of $[(Y, e)]$ onto $[(Y', f)]$, then $|Y| = |Y'|$ and by finiteness of X, there exists $j \in J$ such that $\alpha = \iota_{[(Y, e)]}\alpha_j$. If α is an isomorphism of $[(Y, e)]$ onto $[(Y', e)]$, then α is the product of an isomorphism of $[(Y, e)]$ onto $[(Y', f)]$ and of an isomorphism of $[(Y', f)]$ onto $[(Y', e)]$, and thus of the form $\alpha = \iota_{[(Y, e)]}\alpha_j\alpha_1^{-1}$ (recall that π_1 is the identity mapping on X). The remaining cases, namely $\alpha : [(Y, f)] \to [(Y', e)]$ and $\alpha : [(Y, f)] \to [(Y', f)]$ are treated similarly. Consequently, α is in the inverse semigroup generated by $\cup_{j \in J} S_j\varphi_j$, whence $T^\delta = \Phi(E)$. It follows that $\Phi(E) \in \mathcal{V}$.

Recall that $C_e = (P(X) \times \{e\}) \cup \{0\} = [(X, e)]$. Then $\Phi(C_e)$ is a subsemigroup of $\Phi(E)$ and hence $\Phi(C_e) \in \mathcal{V}$. Furthermore, $P(X) \cong P(X) \times \{e\} = C_e \setminus \{0\}$ and hence $\Phi(P(X))$ is isomorphic to a subsemigroup of $\Phi(C_e)$. It follows that $\Phi(P(X)) \in \mathcal{V}$. By 3.3, $\mathcal{I}(X) \cong \Phi(P(X))$ which then yields $\mathcal{I}(X) \in \mathcal{V}$.

If now Q is any finite inverse semigroup, by means of the Wagner representation, see IV.1.6, it is isomorphic to a subsemigroup of $\mathcal{I}(Q)$. By the finiteness of Q, according to the above, we must have $\mathcal{I}(Q) \in \mathcal{V}$ and thus also $Q \in \mathcal{V}$. This means that $\mathcal{V}$ contains all finite inverse semigroups, which in view of XII.1.7 yields that $\mathcal{V}$ contains all inverse semigroups.

It remains to consider the varieties of Clifford semigroups. We will need a universal algebraic concept.

XIII.3.7 Definition

A variety $\mathcal{V}$ of inverse semigroups has the *congruence extension property* if for every S in $\mathcal{V}$, every congruence on any subsemigroup of S can be extended to a congruence on S.

XIII.3.8 Lemma

Let $\mathcal{V}$ be a Clifford variety with the weak amalgamation property. If $\mathcal{V}$ is not a group variety, then $\mathcal{V} \cap \mathcal{G}$ has the congruence extension property.

Proof. Let $\mathcal{V}$ be as in the hypothesis part of the lemma. Let $G \in \mathcal{V} \cap \mathcal{G}$, U be a subgroup of G and ρ be a congruence on U. Form the semilattice of groups $S = U \cup U/\rho$ with the connecting homomorphism $\rho^{\#} : U \to U/\rho$. We consider the amalgam $[G, S; U]$ and observe that both G and S are in $\mathcal{V}$. Assuming the weak amalgamation property for $\mathcal{V}$, there exists $W \in \mathcal{V}$ and isomorphisms $\varphi : G \to W$ and $\psi : S \to W$ which agree on U.

Denote by 1 the identity of U (which is also the identity of G and S). Then $1 \geq 1\rho$ and hence $1\psi \geq 1\rho\psi$ in W. Since W is a Clifford semigroup, the mapping θ defined by

$$\theta : x \to (1\rho\psi)x \qquad (x \in H_{1\psi})$$

is a homomorphism of $H_{1\psi}$ into $H_{1\rho\psi}$. Since G is a group, the composition $\varphi\theta$ is a homomorphism of G into $H_{1\rho\psi}$, and for all $u \in U$, we also have

$$u\varphi\theta = (1\rho\psi)(u\varphi) = (1\rho\psi)(u\psi) = ((1\rho)u)\psi = (u\rho)\psi = u\rho^{\#}\psi$$

so that $(\varphi\theta)|_U = \rho^{\#}\psi$. Consequently, the congruence on G induced by $\varphi\theta$ extends the given congruence ρ. Therefore $\mathcal{V} \cap \mathcal{G}$ has the congruence extension property.

This proof can be illustrated by the following diagram.

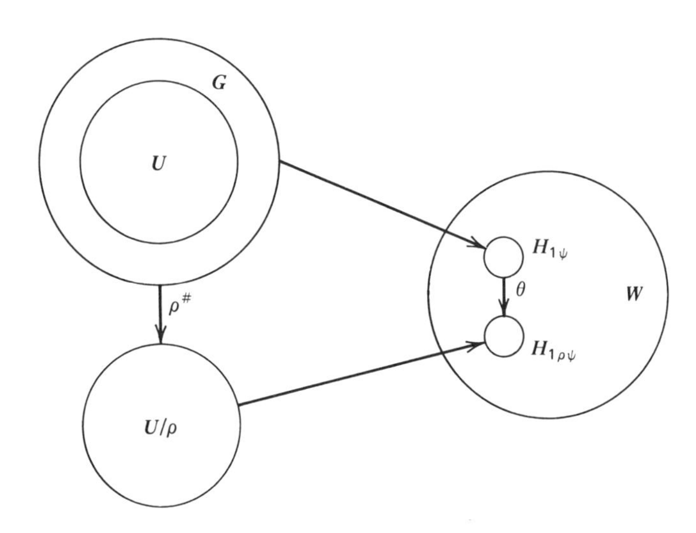

XIII.3.9 Lemma

A group variety $\mathcal{V}$ with the congruence extension property consists of abelian groups.

Proof. Let $\mathcal{V}$ have the congruence extension property and let $G \in \mathcal{V}$. We write commutators in G as $[x, y] = x^{-1}y^{-1}xy$, and for subgroups H and K of G, let $[H, K]$ be the subgroup of G generated by the elements $[h, k]$, where $h \in H, k \in K$.

1. We first claim that for any abelian subgroup A of G,

$$[A, G] \cap A = \{1\}, \tag{15}$$

where 1 is the identity of G. To prove this, we first note that $G \times G$ is in $\mathcal{V}$ and thus has the congruence extension property. Further, $A \times A$ is a subgroup of $G \times G$, and

$$\Delta_A = \{(a, a) | a \in A\}$$

is a normal subgroup of $A \times A$ since the latter is abelian. By the congruence extension property there exists a normal subgroup N of $G \times G$ such that $N \cap (A \times A) = \Delta_A$. For any $a \in A$ and $g \in G$, we obtain

$$([a, g], 1) = (a^{-1}, a^{-1})[(g^{-1}, 1)(a, a)(g, 1)]$$

$$\in \Delta_A[(g, 1)^{-1}\Delta_A(g, 1)] \subseteq N.$$

Hence $[A, G] \times \{1\}$ is a subgroup of N which implies

$$\Delta_A = N \cap (A \times A) \supseteq ([A, G] \times \{1\}) \cap (A \times A)$$

$$= ([A, G] \cap A) \times \{1\}$$

which evidently yields (15).

2. Next let G be a subdirectly irreducible group in $\mathcal{V}$. Denote by M the unique minimal nontrivial normal subgroup of G. We now claim that

$$M \subseteq Z(G). \tag{16}$$

Indeed, let A be a nontrivial abelian subgroup of M. One verifies easily that $[A, G]$ is a normal subgroup of G. By minimality of M, either $[A, G] \supseteq M$ or $[A, G] = \{1\}$. In the first case, we have

$$[A, G] \cap A \supseteq M \cap A = A \neq \{1\}$$

which contradicts relation (15). Thus $[A, G] = \{1\}$. Hence $A \subseteq Z(G)$, and since A is any abelian subgroup of M, we deduce that (16) holds.

Continuing with this notation, we next let $g \in G$. Then the subgroup $A = \langle g, M \rangle$ generated by g and M is abelian in view of (16). Again either

$[A, G] \supseteq M$ or $[A, G] = \{1\}$. In the first case, we get

$$[A, G] \cap A \supseteq M \neq \{1\}$$

contradicting (15). Thus we must have $[A, G] = \{1\}$ which implies that $g \in A \subseteq Z(G)$. This being the case for all $g \in G$, we conclude that G is abelian.

Since every group in $\mathcal{V}$ is a subdirect product of subdirectly irreducible groups in $\mathcal{V}$, all groups in $\mathcal{V}$ must be abelian.

We are finally ready for the main result of this section.

XIII.3.10 Theorem

Let $\mathcal{V}$ be a proper variety of inverse semigroups with the weak amalgamation property. If $\mathcal{V}$ is not a group variety, then all semigroups in $\mathcal{V}$ are commutative.

Proof. If $\mathcal{V}$ is not a Clifford variety, then $\mathcal{V} = \mathcal{I}$ by 3.6. Let $\mathcal{V}$ be a Clifford variety. Then by 3.8, $\mathcal{V} \cap \mathcal{G}$ has the congruence extension property and thus by 3.9 all its members are abelian groups. By XII.4.4, we have $\mathcal{V} = (\mathcal{V} \cap \mathcal{G}) \vee \mathcal{S}$ which implies that all semigroups in $\mathcal{V}$ are commutative.

We can summarize the highlights of this and the preceding two sections as follows.

XIII.3.11 Theorem

(i) The variety $\mathcal{G}$ of all groups,
(ii) the variety $\mathcal{I}$ of all inverse semigroups,
(iii) all varieties of commutative inverse semigroups,

have the strong amalgamation property, and apart from group varieties there are no other varieties of inverse semigroups with the weak amalgamation property.

Proof. The strong amalgamation property for groups is proved in 2.1, for inverse semigroups in 1.13, and for the varieties of commutative inverse semigroups in 2.3. The last assertion is the content of 3.10.

Note that the abelian group varieties are characterized in XII.5.5, and an isomorphic copy of the lattice of all varieties of commutative inverse semigroup is given in XII.5.6(ii).

XIII.3.12 Exercises

(i) Prove directly that the variety of all Clifford semigroups does not have the weak amalgamation property.

(ii) Show that the class of combinatorial inverse semigroups does not have the weak amalgamation property.

XIII.3.13 Problems

(i) Do there exist proper non-abelian group varieties which have the weak or the strong amalgamation property?

(ii) (T. E. Hall) Which varieties of inverse semigroups have the special amalgamation property?

Result 3.1 is due to Reilly–Scheiblich [1]; 3.3 stems from Howie [4]; 3.9 was proved by Biró–Kiss–Palfy [1]. The rest of the section makes up the content of Hall [6].

XIII.4 FREE PRODUCTS OF INVERSE SEMIGROUPS

A construction similar to that of a free inverse semigroup can be used to define an inverse free product of inverse semigroups. This can be carried further to varieties of inverse semigroups by defining the relatively free product for each variety. We characterize such a product for the varieties of groups, commutative inverse semigroups, and Clifford semigroups.

In order to adapt the concepts of a free product and an amalgamated free product of semigroups to inverse semigroups, we must account for the additional unary operation of inversion. A natural way of going about this is as follows. Recall the notation and results in I.13.

XIII.4.1 Lemma

Let $\{S_i\}_{i \in I}$ be a family of pairwise disjoint inverse semigroups. Define a unary operation on $F = \prod^* S_i$ by

$$(s_1 s_2 \cdots s_n)^{-1} = s_n^{-1} s_{n-1}^{-1} \cdots s_1^{-1}.$$

Let τ be the congruence on F generated by the set

$$T = \{(u, uu^{-1}u) | u \in F\} \cup \{(u^{-1}uv^{-1}v, v^{-1}vu^{-1}u) | u, v \in F\}.$$

Then F/τ is an inverse semigroup and $\theta_i \tau^{\#}$ is one-to-one for all $i \in I$, where $\theta_i : S_i \to F$ is the natural embedding.

Proof. It follows at once that the unary operation $u \to u^{-1}$ is an involution on F. The proof that F/τ is an inverse semigroup goes along the same lines as that in VIII.1.1 when showing that Z/ρ is an inverse semigroup. Indeed, F/τ is obviously regular, and the only argument is that of proving that all idempotents of F/τ are of the form $(u^{-1}u)\tau$ which can be shown as in the proof of VIII.1.1.

Now let $s, t \in S_i$ be such that $s\theta_i\tau^{\#} = t\theta_i\tau^{\#}$. Then $s\tau t$; so by I.4.13, there exists a sequence $s = a_1, a_2, \ldots, a_n = t$ such that $a_i = u_i p_i v_i$, $a_{i+1} = u_i q_i v_i$ with $(p_i, q_i) \in T \cup T^{-1}$, $u_i, v_i \in F^1$ for $i = 1, 2, \ldots, n-1$. It follows that $u_1, v_1 \in S_i^1$ and $p_1 \in S_i^1$ and $p_1 \in S_i$ whence $q_1 \in S_i$, which yields $a_2 \in S_i$. In n steps we get that $u_i, v_i \in S_i^1$ and $p_i, q_i \in S_i$ for $i = 1, 2, \ldots, n$. Since S_i is an inverse semigroup, we conclude that $a_1 = a_2 = \cdots = a_n$, and thus $s = t$. Consequently, $\theta_i\tau^{\#}$ is one-to-one.

We may thus introduce the following concepts.

XIII.4.2 Definition

The semigroup $(\prod{}^* S_i)/\tau$ in 4.1 is the *inverse free product* of the inverse semigroups S_i, to be denoted by $\prod_{\mathrm{inv}}^* S_i$. When $I = \{1, 2\}$, we will write $S_1 {}^*_{\mathrm{inv}} S_2$ instead of $\prod_{\mathrm{inv}}^* S_i$. The homomorphism $\nu_i = \theta_i\tau^{\#}$ is the *natural embedding* of S_i into $\prod_{\mathrm{inv}}^* S_i$ for all $i \in I$.

The situation we have arrived at for the class of inverse semigroups is analogous to that for the class of all semigroups in I.13. Indeed, using $\prod_{\mathrm{inv}}^* S_i$ and ν_i instead of $\prod{}^* S_i$ and θ_i, respectively, we may redo everything from I.13.3 to I.13.7 for the class of inverse semigroups. The verification of this assertion is left as a useful exercise.

We can further modify the above product to an arbitrary variety $\mathcal{V}$ of inverse semigroups as follows.

XIII.4.3 Definition

Let $\mathcal{V}$ be a variety of inverse semigroups and $\{S_i\}_{i \in I}$ be a family of pairwise disjoint semigroups in $\mathcal{V}$. Let $\tau_{\mathcal{V}}$ be the least $\mathcal{V}$-congruence on $\prod_{\mathrm{inv}}^* S_i$. Then $(\prod_{\mathrm{inv}}^* S_i)/\tau_{\mathcal{V}}$ is the $\mathcal{V}$-*free product* of semigroups S_i, to be denoted by $\prod_{\mathcal{V}}^* S_i$. The mapping $\nu_i\tau_{\mathcal{V}}^{\#}$ is the *natural embedding* of S_i into $\prod_{\mathcal{V}}^* S_i$ for every $i \in I$ (the one-to-oneness of $\nu_i\tau_{\mathcal{V}}^{\#}$ follows as in the proof of 4.1).

We consider first the $\mathcal{G}$-free products which we call the *group free products* and denote by $\prod_g^* G_i$. Continue the notation of 4.1 and 4.2.

XIII.4.4 Theorem

Let $\{G_i\}_{i \in I}$ be a family of pairwise disjoint groups, and $\{1\}$ be a one-element group. Then

$$\prod_g{}^* G_i = \prod_{\mathrm{inv}}{}^*_{\{1\}} G_i \cong \prod{}^*_{\{1\}} G_i.$$

Proof. Let $F = \prod{}^* G_i$ and $Q = \prod_{\mathrm{inv}}^* G_i$.

Recall that, by definition, $\prod_g^* G_i = Q/\sigma$, where σ is as usual the least group congruence, and $\prod_{\mathrm{inv}\{1\}}^* G_i = Q/\rho$ where ρ is the congruence generated by the set

$$R = \{(1\varphi_i\nu_i, 1\varphi_j\nu_j) \| i, j \in I\}.$$

In order to show the equality in the assertion of the theorem, we must show that $\sigma = \rho$. To this end, we introduce a relation ρ' on Q by

$$a'\rho'b' \Leftrightarrow a' = u'(1_i\tau)v', \quad b' = u'(1_j\tau)v' \quad \text{for some}$$

$$u', v' \in Q^1, i, j \in I.$$

(Note that ρ is the transitive closure of ρ'.) Lifting the right-hand side of this definition to F, we get

$$(a\tau)\rho'(b\tau) \Leftrightarrow a\tau u1_iv, \quad b\tau u1_jv \quad \text{for some} \quad u, v \in F^1, i, j \in I. \tag{17}$$

We show next that any two idempotents of Q are ρ'-related. First note that $(1_i\tau)\rho'(1_j\tau)$ for any $i, j \in I$. We have seen in the proof of 4.1 that any idempotent of Q is of the form $(w^{-1}w)\tau$ for some $w \in F$. Let $w = w_1w_2 \cdots w_n$ where we may assume that $w_i \in G_i$, $i = 1, 2, \ldots, n$. Then

$$\begin{aligned}(w^{-1}w)\tau &= \left(w_n^{-1} \cdots w_2^{-1}(w_1^{-1}w_1)w_2 \cdots w_n\right)\tau \\ &= \left(w_n^{-1} \cdots w_2^{-1}1_1w_2 \cdots w_n\right)\tau \\ &\rho'\left(w_n^{-1} \cdots w_2^{-1}1_2w_2 \cdots w_n\right)\tau \\ &= \left(w_n^{-1} \cdots w_3^{-1}1_2w_3 \cdots w_n\right)\tau \cdots \rho'1_n\tau\end{aligned}$$

so that $[(w^{-1}w)\tau]\rho[1_n\tau]$. For $u = u_1u_2 \cdots u_m \in F$, we similarly obtain $[(u^{-1}u)\tau]\rho[1_m\tau]$ and hence $[(w^{-1}w)\tau]\rho[(u^{-1}u)\tau]$. It follows that ρ is a group congruence, and thus $\sigma \subseteq \rho$.

Conversely, assume that (17) holds. Then

$$\left[a\left(v^{-1}1_i1_jv\right)\right]\tau\left[u1_ivv^{-1}1_i1_jv\right]\tau\left[u1_jvv^{-1}1_i1_jv\right]\tau\left[b\left(v^{-1}1_i1_jv\right)\right]$$

which shows that $(a\tau)\sigma(b\tau)$. Hence $\rho' \subseteq \sigma$ and thus $\rho \subseteq \sigma$. Therefore $\rho = \sigma$ as required.

Note next that $\Pi^*_{\{1\}}G_i = F/\xi$ where ξ is the congruence generated by the relation

$$T = \{(1\varphi_i\theta_i, 1\varphi_j\theta_j) \| i, j \in I\}.$$

In order to establish the isomorphism in the assertion of the theorem, it suffices to show that for any $u, v \in F$,

$$u \xi v \Leftrightarrow (u\tau)\rho(v\tau). \tag{18}$$

Since ξ identifies the identities of $G_i \theta_i$, it follows easily that $u \xi u u^{-1} u$ and $u^{-1} u v^{-1} v \xi v^{-1} v u^{-1} u$ for all $u \in F$ and thus $\tau \subseteq \xi$ by definition. Now ξ/τ identifies all the idempotents of $F/\tau = Q$ and thus $\sigma \subseteq \xi/\tau$, and by the above $\rho \subseteq \xi/\tau$. A comparison of the definitions of ξ and ρ easily gives the inclusion $\xi/\tau \subseteq \rho$. Hence $\rho = \xi/\tau$ which proves (18), and establishes the desired isomorphism.

For the variety of commutative inverse semigroups, we write $\Pi^*_{\text{csg}} S_i$ for the (relatively) free product, and have the following simple result.

XIII.4.5 Proposition

Let $\{S_i\}_{i \in I}$ be a family of pairwise disjoint commutative inverse semigroups. To every S_i adjoin an identity 1_i and denote the new semigroup $\bar{S}_i$. Then $\Pi^*_{\text{csg}} S_i$ is isomorphic to the subsemigroup of $\Pi \bar{S}_i$ consisting of all elements (s_i) with at most a finite number of $s_i \neq 1_i$ and not all $s_i = 1_i$.

Proof. The proof of this proposition is left as an exercise.

XIII.4.6 Corollary

Let S_1 and S_2 be commutative inverse semigroups and denote by $S_1 \overset{*}{_{\text{csg}}} S_2$ their (relatively) free product. Then $S_1 \overset{*}{_{\text{csg}}} S_2 \cong \bar{S}_1 \times \bar{S}_2 \setminus (1_1, 1_2)$.

Proof. The proof is left as an exercise.

Note that these semigroups were used in the proof of 2.3.

We will construct next the Clifford representation of the free product of two Clifford semigroups in the class of Clifford semigroups; the general case is handled along the same lines.

XIII.4.7 Notation

For disjoint semilattices Y and Y', we denote their free product in the class of semilattices by $Y \overset{*}{_s} Y'$. As in 4.6, it follows without difficulty that

$$Y \overset{*}{_s} Y' \cong \bar{Y} \times \overline{Y'} \setminus \{(1, 1')\} \tag{19}$$

where we have adjoined the identities 1 and 1′ to Y and Y', respectively (getting the semilattices $\bar{Y}$ and $\overline{Y'}$). In order to simplify the notation, we will identify the two semigroups in (19), and use the shorter notation $Y \overset{*}{_s} Y'$.

As before, $G \overset{*}{_g} G'$ denotes the group free product of the disjoint groups G

and G'. Let $\varphi : G \to H$ and $\varphi' : G' \to H'$ be homomorphisms of groups. By the group analogue of I.13.5(i), there exists a unique homomorphism, which we denote by $\varphi * \varphi'$, making the following diagram

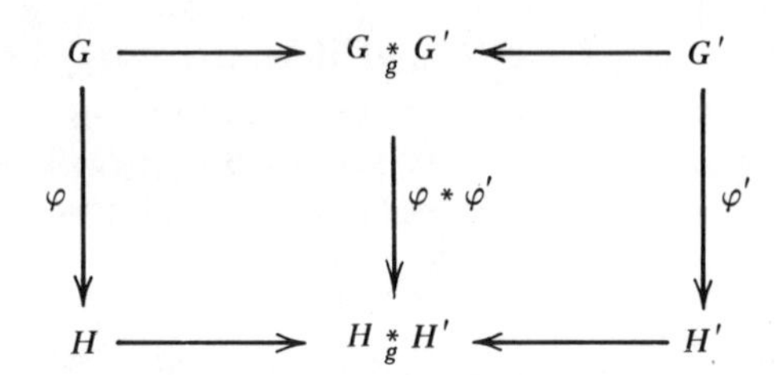

commutative (the unlabeled arrows are the natural embeddings).

For $1 \in \overline{Y}$ and $1' \in \overline{Y'}$, we let G_1 and $G_{1'}$ be the trivial groups, and for $\alpha \in Y$, $\alpha' \in Y'$, $\varphi_{1,1}$, $\varphi_{1,\alpha}$, $\varphi'_{1',1'}$, $\varphi'_{1',\alpha'}$ be the trivial homomorphisms.

We are now able to provide the desired construction.

XIII.4.8 Theorem

Let $S = [Y; G_\alpha, \varphi_{\alpha,\beta}]$ and $S' = [Y'; G'_{\alpha'}, \varphi'_{\alpha',\beta'}]$ be Clifford semigroups. Then

$$\left[Y \underset{s}{*} Y'; G_\alpha \underset{g}{*} G'_{\alpha'}, \varphi_{\alpha,\beta} * \varphi'_{\alpha',\beta'}\right] \tag{20}$$

is the free product of S and S' in the class of Clifford semigroups.

Proof. We verify first that the system of homomorphisms given in the above representation is transitive. Thus let $(\alpha, \alpha') \geq (\beta, \beta') \geq (\gamma, \gamma')$, and consider the diagram

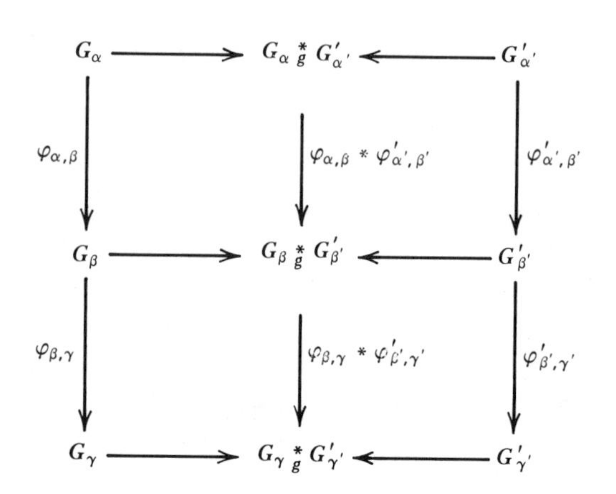

In addition, we have

$$\varphi_{\alpha,\gamma} * \varphi'_{\alpha',\gamma'} : G_\alpha \underset{g}{*} G'_{\alpha'} \to G_\gamma \underset{g}{*} G'_{\gamma'}.$$

By the transitivity of the systems $\{\varphi_{\alpha,\beta}\}$ and $\{\varphi'_{\alpha',\beta'}\}$ and the uniqueness assertion mentioned in 4.7, we deduce that

$$\left(\varphi_{\alpha,\beta} * \varphi'_{\alpha',\beta'}\right)\left(\varphi_{\beta,\gamma} * \varphi'_{\beta',\gamma'}\right) = \varphi_{\alpha,\gamma} * \varphi'_{\alpha',\gamma'}.$$

Consequently, the semigroup (20) is defined; denote it by F.

Let $T = [Z; H_\alpha, \pi_{\alpha,\beta}]$ be a Clifford semigroup and $\theta : S \to T$ and $\theta' : S' \to T$ be homomorphisms. As in II.2.8, θ induces a homomorphism $\psi : Y \to Z$ and θ' induces $\psi' : Y' \to Z$. We denote by $\psi * \psi'$ the unique homomorphism of $Y \underset{s}{*} Y'$ into Z induced by ψ and ψ', and by $\kappa : S \to F$ and $\kappa' : S' \to F$ the natural embeddings. We thus arrive at the following diagram

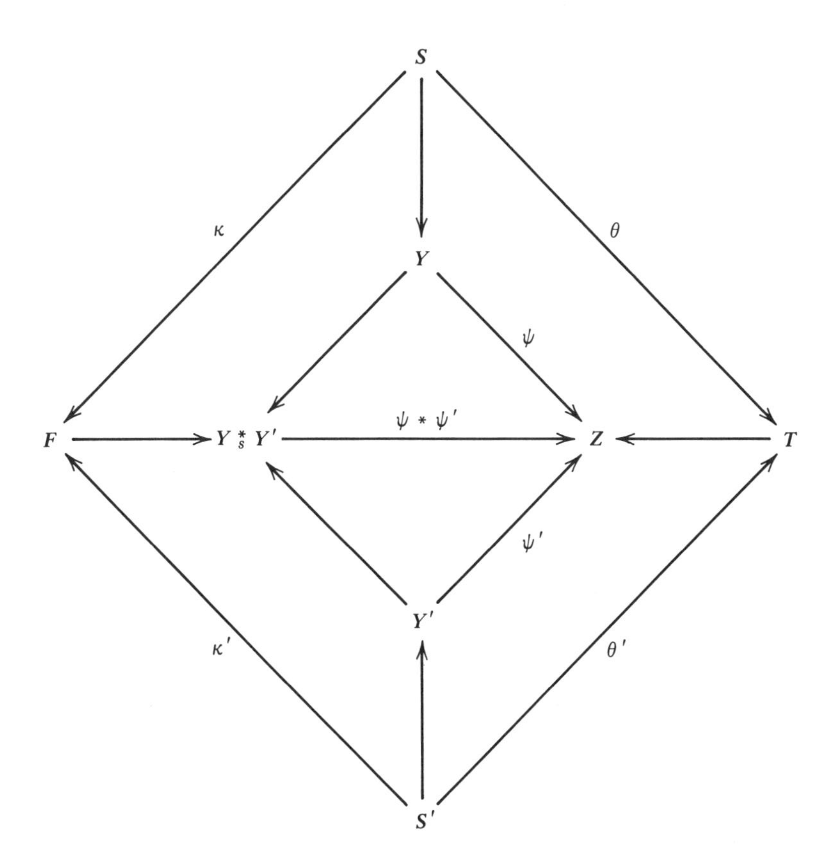

where the unlabeled arrows are the natural homomorphisms. To construct a homomorphism from F to T, we use the homomorphism $\psi * \psi' : Y \underset{s}{*} Y' \to Z$ and for each $(\alpha, \alpha') \in Y \underset{s}{*} Y'$, define a mapping on $G_\alpha \underset{g}{*} G'_{\alpha'}$ by

$$\xi_{(\alpha, \alpha')} : g_1 g_1' \cdots g_n g_n' \to (g_1\theta)(g_1'\theta') \cdots (g_n\theta)(g_n'\theta').$$

This is actually the unique homomorphism of $G_\alpha \underset{g}{*} G'_{\alpha'}$ into $H_{(\alpha\psi)(\alpha'\psi')}$ induced by $\theta|_{G_\alpha}$ and $\theta'|_{G'_{\alpha'}}$ and satisfying the commutativity of the corresponding diagram. We have actually adjoined an identity to Y, Y', and Z and extended the mappings ψ and ψ' in order to simplify the notation. To show that the aggregate $\{\psi, \xi_{(\alpha, \alpha')}\}$ determines a homomorphism of F to T, as in II.2.8, it remains to show that for any $(\alpha, \alpha') \geq (\beta, \beta')$, the following diagram commutes:

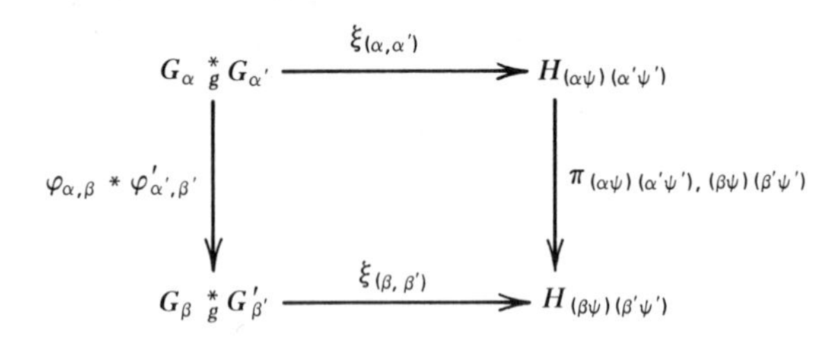

Indeed,

$$\begin{array}{ccc}
g_1 g_1' \cdots g_n g_n' & \xrightarrow{\xi_{(\alpha,\alpha')}} & (g_1\theta)(g_1'\theta') \cdots (g_n\theta)(g_n'\theta') \\
\Big\downarrow \varphi_{\alpha,\beta} * \varphi'_{\alpha',\beta'} & & \\
(g_1\varphi_{\alpha,\beta})(g_1'\varphi'_{\alpha',\beta'}) \cdots (g_n\varphi_{\alpha,\beta})(g_n'\varphi'_{\alpha',\beta'}) & \xrightarrow{\xi_{(\beta,\beta')}} & (g_1\varphi_{\alpha,\beta}\theta) \cdots (g_n'\varphi'_{\alpha',\beta'}\theta')
\end{array}$$

and

$$\begin{aligned}
&((g_1\theta)(g_1'\theta')) \cdots (g_n\theta)(g_n'\theta'))\pi_{(\alpha\psi)(\alpha'\psi'),(\beta\psi)(\beta'\psi')} \\
&= (g_1\theta\pi_{\alpha\psi,(\beta\psi)(\beta'\psi')}) \cdots (g_n'\theta'\pi_{\alpha'\psi',(\beta\psi)(\beta'\psi')}) \\
&= (g_1\varphi_{\alpha,\beta}\theta\pi_{\beta\psi,(\beta\psi)(\beta'\psi')}) \cdots (g_n'\varphi'_{\alpha',\beta'}\theta'\pi_{\beta'\psi',(\beta\psi)(\beta'\psi')}) \\
&= (g_1\varphi_{\alpha,\beta}\theta) \cdots (g_n'\varphi'_{\alpha',\beta'}\theta')
\end{aligned}$$

which establishes commutativity of the above diagram. This defines a homomorphism $\xi : F \to T$, as in 4.7, which evidently makes the corresponding diagrams, involving the natural embeddings, commutative.

It is easy to see that $S\kappa$ and $S'\kappa'$ generate F, which then implies the uniqueness of ξ with the requisite properties. Therefore, by the Clifford semigroup version of I.13.7, F is the free product of S and S' in the class of Clifford semigroups.

XIII.4.9 Notation

For the free product F of S and S' constructed above, we write $S *_{sg} S'$. The free product of abelian groups, in the class of abelian groups, reduces evidently to the direct product, so we may use the notation $G \times G'$ and $\varphi \times \varphi'$.

XIII.4.10 Corollary

Let S and S' be as in 4.8 and also commutative. Then their free product in the class of commutative inverse semigroups is isomorphic to

$$\left[Y *_s Y'; G_\alpha \times G'_{\alpha'}, \varphi_\alpha \times \varphi'_{\alpha'}\right].$$

Proof. The proof of this corollary is left as an exercise.

XIII.4.11 Example of a Free Product

$G * H$ where $G = \{e, a\}$, $H = \{\alpha\}$ are groups.

In the class of inverse semigroups:

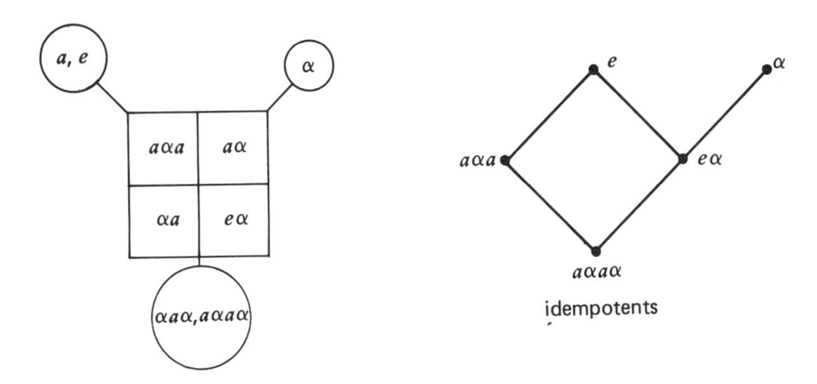

In the class of Clifford semigroups:

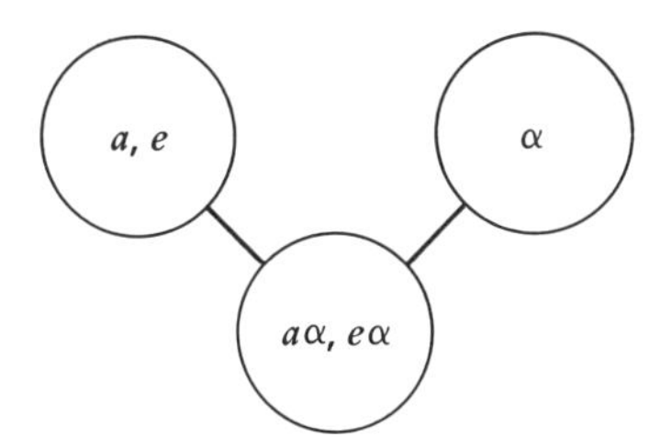

In the class of groups:

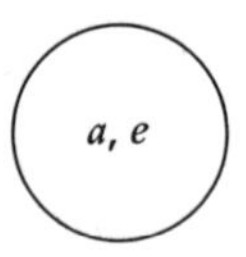

XIII.4.12 Exercises

(i) Let $U = \{e\}$ and $Y_2 = \{0,1\}$ be the semilattices. Find their free product in the following classes: (α) inverse semigroups, (β) Clifford semigroups, (γ) commutative inverse semigroups, and (δ) semilattices.

(ii) Let $\{X_i\}_{i\in I}$ be a family of pairwise disjoint nonempty sets. Show that $\prod^*_{\text{inv}} I_{X_i} \cong I_{\cup X_i}$. Deduce that for any nonempty set X, $\prod^*_{\text{inv}} I_x \cong I_X$ where x runs through all elements of X.

(iii) Let $\{S_i\}_{i\in I}$ be a family of pairwise disjoint inverse semigroups. Show that $\prod^*_{\text{inv}} S_i$ is isomorphic to $(I_{\cup S_i})/\rho$ where ρ is the congruence generated by the relations valid in various S_i.

(iv) Generalize 4.8 to an arbitrary family of Clifford semigroups.

(v) Let $\{G_i\}_{i\in I}$ be a family of pairwise disjoint groups. Show that $\prod^*_{sg} G_i$ can be embedded into $Y_I \times \prod^*_g G_i$. Derive the form of a free Clifford semigroup on I in the special case when all G_i are infinite cyclic groups.

(vi) Let B_2' be an isomorphic copy of B_2 disjoint from B_2. Show that the free product of B_2 and B_2' amalgamating their idempotents is isomorphic to $B(\mathbb{Z}, 2)$.

Free products of inverse semigroups were studied by Preston [9]. Knox [1] and McAlister [7] investigated the free product of groups in the class of inverse semigroups. The Hopf property for free products was investigated by Jones [8], [11]. For more discussion on free products, consult Howie [4] and Jones [14].

XIII.5 AN APPLICATION TO ESSENTIAL CONJUGATE EXTENSIONS

We have seen in VI.2.12 that an inverse semigroup S with idempotent metacenter has a maximal essential extension, and that any such is isomorphic to $\Psi(S)$. We will use the strong amalgamation property of inverse semigroups to establish a converse of this result, namely that an inverse semigroup having a maximal essential conjugate extension must necessarily have idempotent metacenter. We then deduce analogous results for essential normal extensions.

The proof of the main result is by contrapositive, namely for every essential conjugate extension C of an inverse semigroup A with nonidempotent metacenter, there exists an essential conjugate extension D of A properly containing C. The argument goes as follows. We construct an essential normal extension B of S, with $B \cap C = S$, S being a certain inverse subsemigroup of C which contains A and which has nonidempotent metacenter. We then amalgamate B and C with S as a core and use this to produce an essential conjugate extension of A properly containing C.

First let S be an inverse semigroup and let $c \in M(S) \setminus E_S$. By VI.2.4, c is contained in a maximal subgroup of S, say with identity e, and $c \in Z(eSe)$. Let $\langle c \rangle$ be the cyclic group generated by c. We let u and v be elements generating cyclic groups $\langle u \rangle$ and $\langle v \rangle$, respectively. Assume that the mappings $c \to u$ and $c \to v$ extend to isomorphisms of $\langle c \rangle$ onto $\langle u \rangle$ and $\langle v \rangle$, respectively, and that the semigroups S, $\langle u \rangle$, and $\langle v \rangle$ are pairwise disjoint.

We will need the set

$$\hat{S} = \langle u \rangle \times SeS \times \langle v \rangle,$$

and the notation

$$\bar{t} = (u^0, t, v^0) \qquad (t \in SeS)$$

where u^0 and v^0 are the identities of $\langle u \rangle$ and $\langle v \rangle$, respectively. We can find a set $\bar{S} = \{\bar{s} | s \in S\}$ such that the mapping $s \to \bar{s}$ is a bijection of S and $\bar{S}$ and

$$\bar{S} \cap \hat{S} = \{\bar{t} | t \in SeS\}.$$

Note that $t \to \bar{t}$ $(t \in SeS)$ is a bijection of SeS onto $\bar{S} \cap \hat{S}$. We put

$$S' = \bar{S} \cup \hat{S}.$$

In order to define a multiplication on S', we must first prove the following statement.

XIII.5.1 Lemma

If $xey = x'ey'$ for some $x, y, x', y' \in S$, then $xc^k y = x'c^k y'$ for all $k \in \mathbb{Z}$.

Proof. Since $c \in Z(eSe)$, we get

$$\begin{aligned} xc^k y &= x(ex^{-1}xe)c^k y = xc^k ex^{-1}(xey) = xc^k ex^{-1}(x'ey') \\ &= xc^k(ex^{-1}x'e)y' = (xex^{-1})(x'c^k y') \end{aligned}$$

and analogously $x'c^k y' = (xc^k y)(y'^{-1}ey')$. Consequently,

$$\begin{aligned} x'c^k y' &= (xc^k y)(y'^{-1}ey') = (xex^{-1})(x'c^k y')(y'^{-1}ey') \\ &= (xex^{-1})x'c^k(ey')(ey')^{-1}(ey') = (xex^{-1})(x'c^k y') \\ &= xc^k y. \end{aligned}$$

We define a multiplication on $\hat{S}$ by

$$(u^m, xey, v^n)(u^{m'}, x'ey', v^{n'}) = (u^{m+m'}, xc^{nm'}yx'ey', v^{n+n'}).$$

Now 5.1 together with the fact that the mappings $c \to u$ and $c \to v$ extend to isomorphisms of $\langle c \rangle$ onto $\langle u \rangle$ and $\langle v \rangle$, respectively, guarantee single-valuedness of this operation.

A simple verification also shows that for any $s, t \in SeS$ and $m, n, m', n' \in \mathbb{Z}$,

$$(u^m, s, v^n)\bar{t} = (u^m, st, v^n) = \bar{s}(u^m, t, v^n),$$

and in particular $\bar{s}\bar{t} = \overline{st}$. This makes it possible to extend the multiplication in $\hat{S}$ to all of S' as follows: for any $r, s \in S$, $t \in SeS$, $m, n \in \mathbb{Z}$, we set

$$\bar{r}\bar{s} = \overline{rs},$$

$$\bar{s}(u^m, t, v^n) = (u^m, st, v^n),$$

$$(u^m, t, v^n)\bar{s} = (u^m, ts, v^n).$$

The first formula indicates that the mapping $s \to \bar{s}$ is an isomorphism of S onto $\bar{S}$.

XIII.5.2 Lemma

With the above multiplication, S' is an inverse semigroup and $\bar{S}$ is a full inverse subsemigroup of S'.

Proof. The verification of associativity is long but straightforward; we check only the most tedious case. Indeed,

$$\begin{aligned}
&[(u^{m_1}, x_1ey_1, v^{n_1})(u^{m_2}, x_2ey_2, v^{n_2})](u^{m_3}, x_3ey_3, v^{n_3}) \\
&= (u^{m_1+m_2}, x_1c^{n_1m_2}ey_1x_2ey_2, v^{n_1+n_2})(u^{m_3}, x_3ey_3, v^{n_3}) \\
&= (u^{m_1+m_2+m_3}, x_1c^{n_1m_2+(n_1+n_2)m_3}(ey_1x_2e)y_2x_3ey_3, v^{n_1+n_2+n_3}) \\
&= (u^{m_1+m_2+m_3}, x_1c^{n_1(m_2+m_3)}y_1x_2c^{n_2m_3}y_2x_3ey_3, v^{n_1+n_2+n_3}) \\
&= (u^{m_1}, x_1ey_1, v^{n_1})(u^{m_2+m_3}, x_2c^{n_2m_3}y_2x_3ey_3, v^{n_2+n_3}) \\
&= (u^{m_1}, x_1ey_1, v^{n_1})[(u^{m_2}, x_2ey_2, v^{n_2})(u^{m_3}, x_3ey_3, v^{n_3})],
\end{aligned}$$

as required.

If (u^m, t, v^n) is idempotent, then $u^{2m} = u^m$ so $u^m = u^0$ and analogously $v^n = v^0$ so $t^2 = t$, and thus $(u^m, t, v^n) = \bar{t}$. Hence all idempotents of S' are of the form $\bar{f}$ for some idempotent f in S, which implies that $\bar{S}$ is full in S'.

Routine verification shows that for any $(u^m, xey, v^n) \in \hat{S}$, $(u^{-m}, y^{-1}c^{nm}x^{-1}, v^{-n})$ is an inverse of (u^m, xey, v^n). It follows that S' is regular, and since $\bar{S}$ is a full inverse subsemigroup of S', we obtain that S' is an inverse semigroup.

We deduce from the proof that

$$(u^m, xey, v^n)^{-1} = (u^{-m}, y^{-1}c^{nm}x^{-1}, v^{-n})$$

for any element of $\hat{S}$. Also note that $\hat{S}$ is an ideal of S'. We now put

$$\bar{u} = (u, e, v^0), \qquad \bar{v} = (u^0, e, v).$$

Routine calculation shows that

$$\bar{u}^{-1} = (u^{-1}, e, v^0), \qquad \bar{v}^{-1} = (u^0, e, v^{-1}),$$

$$\bar{u}\bar{u}^{-1} = \bar{u}^{-1}\bar{u} = \bar{v}\bar{v}^{-1} = \bar{v}^{-1}\bar{v} = \bar{c}\bar{c}^{-1} = \bar{c}^{-1}\bar{c} = \bar{e}$$

so that $\bar{u}$, $\bar{v}$, $\bar{c}$, and $\bar{e}$ belong to the same $\mathcal{H}$-class of S'. We can picture the situation as follows:

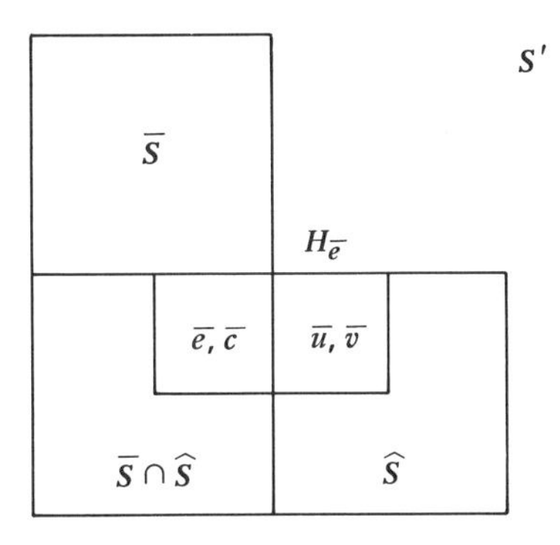

XIII.5.3 Lemma

The semigroup S' is generated (as an inverse semigroup) by the set $\bar{S} \cup \{\bar{u}, \bar{v}\}$. For every $\bar{x} \in \bar{S}$, we have

$$\bar{u}^{-1}\bar{x}\bar{u} = \bar{u}\bar{x}\bar{u}^{-1} = \bar{v}\bar{x}\bar{v}^{-1} = \bar{v}^{-1}\bar{x}\bar{v} = \bar{e}\bar{x}\bar{e} = \overline{exe}.$$

Moreover, $\bar{S}$ is a normal subsemigroup of S', and $\bar{c} \in M(S')$.

Proof. Routine calculation shows that for any element of $\hat{S}$ we have $(u^m, xey, v^n) = \overline{xe}\,\bar{u}^m\bar{v}^n\overline{ey}$, which proves the first assertion of the lemma. For any $\bar{x} \in \bar{S}$, we obtain

$$\bar{u}^{-1}\bar{x}\bar{u} = (u^{-1}, e, v^0)\bar{x}(u, e, v^0) = (u^0, exe, v^0) = \overline{exe} = \bar{e}\bar{x}\bar{e},$$

and the remaining equalities are checked just as easily. Since S' is generated by the set $\bar{S} \cup \{\bar{u}, \bar{v}\}$, it follows that $\bar{S}$ is a self-conjugate inverse subsemigroup of S'. By 5.2, $\bar{S}$ is full in S', so that $\bar{S}$ is a normal subsemigroup of S'.

We have remarked above that $\bar{c}$ is contained in $H_{\bar{e}}$. It is easily verified that every element of $\bar{e}S'\bar{e}$ is of the form (u^m, t, v^n) for some $t \in eSe$, $m, n \in \mathbb{Z}$. Since $c \in Z(eSe)$, we obtain

$$\bar{c}(u^m, t, v^n) = (u^m, ct, v^n) = (u^m, tc, v^n) = (u^m, t, v^n)\bar{c}$$

which again by VI.2.4 yields that $\bar{c} \in M(S')$.

XIII.5.4 Lemma

The semigroup $\bar{S}$ intersects every $\mathcal{H}$-class of S'. Moreover,

$$H_{\bar{e}} = \{(u^m, x, v^n) | x\mathcal{H}e \text{ in } S, m, n \in \mathbb{Z}\}.$$

Proof. For any $(u^m, t, v^n) \in \hat{S}$, easy calculation shows that

$$(u^m, t, v^n)(u^m, t, v^n)^{-1} = \bar{t}\bar{t}^{-1} = \overline{tt^{-1}},$$

$$(u^m, t, v^n)^{-1}(u^m, t, v^n) = \bar{t}^{-1}\bar{t} = \overline{t^{-1}t},$$

so that $(u^m, t, v^n)\mathcal{H}\bar{t}$ in S'. It follows that for $t \in SeS$, $H_{\bar{t}}$ in S' consists of the elements (u^m, s, v^n) with $s\mathcal{H}t$ in S and $m, n \in \mathbb{Z}$, whence follow both assertions of the lemma.

XIII.5.5 Lemma

Let ρ be a congruence on S'. If ρ restricted to $\bar{S}$ is the equality relation, then also ρ restricted to $\bar{S} \cup H_{\bar{e}}$ is the equality relation.

Proof. Assume that the restriction of ρ to $\bar{S}$ is the equality relation. Since $\bar{S}$ is full in S', it follows that ρ is idempotent separating, and thus $\rho \subseteq \mathcal{H}$. Hence an element of $H_{\bar{e}}$ can only be ρ-related to an element of $H_{\bar{e}}$. In order to prove the lemma, it thus suffices to show that the ρ-class $\bar{e}\rho$ is trivial. Assume that $(u^m, x, v^n)\rho\bar{e}$. By 5.4, we have $x\mathcal{H}e$ in S. An easy calculation shows that

$$\overline{c^{-m}}(u^m, x, v^n) = \bar{v}^{-1}(u^m, x, v^n)\bar{v} \in \bar{e}\rho,$$

$$\overline{c^n}(u^m, x, v^n) = \bar{u}^{-1}(u^m, x, v^n)\bar{u} \in \bar{e}\rho,$$

and thus $\overline{c^{-m}}, \overline{c^n} \in \bar{e}\rho$. Since $\rho|_{\bar{S}}$ is the equality relation, we obtain $\overline{c^{-m}} = \overline{c^n} = \bar{e}$ and thus $u^m = u^0$, $v^n = v^0$ in view of the isomorphisms $\langle u \rangle \cong \langle c \rangle \cong \langle v \rangle$. Consequently $(u^m, x, v^n) = \bar{x} \in \bar{S}$. By the hypothesis $\rho|_{\bar{S}} = \varepsilon$, we conclude that $(u^m, x, v^n) = \bar{e}$. Thus $\bar{e}\rho = \{\bar{e}\}$, and ρ restricted to $\bar{S} \cup H_{\bar{e}}$ is the equality relation.

We are now ready to complete the first step of the proof of the principal result of this section.

XIII.5.6 Theorem

Let S be an inverse semigroup and $c \in M(S) \setminus E_S$. Then c is contained in the $\mathcal{H}$-class of some idempotent e. There exists an essential normal extension B of S with the following properties.

(i) B is generated as an inverse semigroup by the set $S \cup \{a, b\}$ where $a, b \in B \setminus S$ and $a \neq b$.

(ii) The elements a, b, c are $\mathcal{H}$-related to e in B.

(iii) For all $s \in S$, we have $a^{-1}sa = asa^{-1} = b^{-1}sb = bsb^{-1} = ese$.

(iv) The element c is in the metacenter of B.

Proof. Let the notation be as introduced above. We consider the partially ordered set $\mathcal{P}$ of congruences on S' under inclusion whose restriction to $\bar{S}$ is the equality relation. By Zorn's lemma, $\mathcal{P}$ has a maximal element, say ρ. In view of 5.3, we conclude that S'/ρ is an essential normal extension of $\bar{S}\rho^{\#}$. We identify $\bar{S}$ with $\bar{S}\rho^{\#}$ via the isomorphism $s \to \bar{s}\rho^{\#}$, and set $B = S'/\rho$, $a = \bar{u}\rho^{\#}$, $b = \bar{v}\rho^{\#}$. Now item (i) follows from 5.3 and 5.5, (ii) follows from 5.4, and (iii) and (iv) follow from 5.3.

We may now turn to the second step. The following diagram illustrates the situation discussed in the proof of the next theorem.

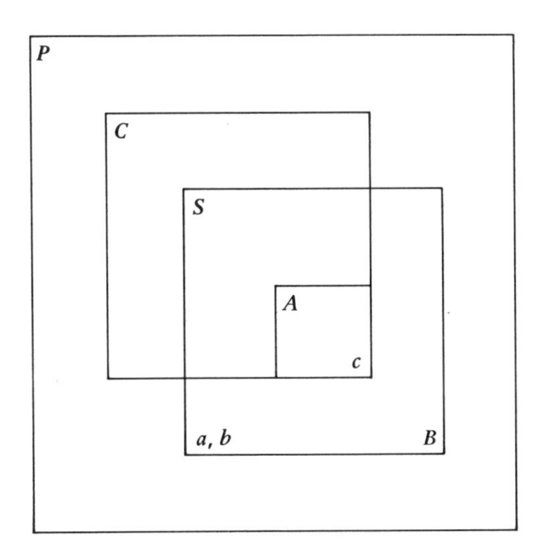

XIII.5.7 Theorem

Let C be an essential conjugate extension of an inverse semigroup A with nonidempotent metacenter. Then there exists an essential conjugate extension of A which contains C properly.

Proof. Let $c \in M(A) \setminus E_A$. Let $\mathfrak{X}$ be the partially ordered set under inclusion of all inverse subsemigroups of C which are essential conjugate extensions of A and have c in their metacenter. Note that $A \in \mathfrak{X}$. By Zorn's lemma, $\mathfrak{X}$ has a maximal element, say S. Let e, a, b, and B be as in 5.6; in addition, we assume that $B \cap C = S$. By 5.6(iv), c is in the metacenter of B. For all $x \in A$, by 5.6(iii), we have

$$a^{-1}xa = axa^{-1} = b^{-1}xb = bxb^{-1} = exe \in A,$$

and for all $s \in S$, $s^{-1}xs \in A$ since A is self-conjugate in S. Since B is generated by the set $S \cup \{a, b\}$, we conclude that B is a conjugate extension of A.

Let ρ be a congruence on B whose restriction to A is the equality relation, and let $\tau = \rho|_S$. Since $\tau|_A$ is the equality relation and S is an essential conjugate extension of A, it follows that τ is the equality relation. But B is an essential normal extension of S, and thus ρ must itself be the equality relation. We conclude that B is an essential conjugate extension of A.

We now consider the amalgam $[B, C; S]$ with the inclusion mappings. In view of 1.13, the class of inverse semigroups has the strong amalgamation property. Hence there exists an inverse semigroup P containing B and C as inverse subsemigroups with the property $B \cap C = S$. We may assume that P is generated as an inverse semigroup by the set $B \cup C$. By 5.6(i), P is generated as an inverse semigroup by the set $\{a, b\} \cup C$. Since both B and C are conjugate extensions of A, it follows that also P is a conjugate extension of A.

Let $\mathfrak{Y}$ be the partially ordered set under inclusion of all congruences on P whose restriction to C is the equality relation. By Zorn's lemma, $\mathfrak{Y}$ has a maximal element, say ξ. Let $\theta = \xi|_B$. Then θ is a congruence on B whose restriction to A is the equality relation. Since B is an essential normal extension of A, see 5.6, θ must be the equality relation. Let $D = P/\xi$. It follows that the restrictions of $\xi^{\#}$ to B and C are monomorphisms of B and C, respectively, into D.

We will show next that D is an essential conjugate extension of $A\xi^{\#}$. Since P is a conjugate extension of A, it is clear that D is a conjugate extension of $A\xi^{\#}$. Let ν be a congruence on D whose restriction to $A\xi^{\#}$ is the equality relation. We lift ν to P by defining ν^* on P by

$$x\nu^*y \Leftrightarrow (x\xi^{\#})\nu(y\xi^{\#}).$$

It follows easily that ν^* is a congruence on P which contains ξ and whose restriction to A is the equality relation. Since C is an essential conjugate extension of A, the restriction of ν^* to C must be the equality relation. This implies that $\nu^* \in \mathcal{Y}$. By the maximality of ξ, we must have $\xi = \nu^*$ which implies that ν is the equality relation. Consequently D is an essential conjugate extension of $A\xi^{\#}$ which contains $C\xi^{\#}$, where $A\xi^{\#} \cong A$ and $C\xi^{\#} \cong C$.

It remains to prove that D contains $C\xi^{\#}$ properly. By contradiction, suppose that $D = C\xi^{\#}$. Clearly $B\xi^{\#}$ is an essential normal extension of $A\xi^{\#}$ which has $c\xi^{\#}$ in its metacenter. The hypothesis $D = C\xi^{\#}$ implies that $B\xi^{\#}$ is an inverse subsemigroup of $C\xi^{\#}$ which contains $S\xi^{\#}$. By the maximality of S in $\mathcal{X}$, we must have $S\xi^{\#} = B\xi^{\#}$, which is impossible since S is properly contained in B and $\xi^{\#}$ is one-to-one on B. Thus the assumption $D = C\xi^{\#}$ must be false, and hence $C\xi^{\#}$ must be properly contained in D.

XIII.5.8 Corollary

An inverse semigroup has a maximal essential conjugate extension if and only if it has idempotent metacenter.

Proof. This follows directly from VI.2.12 and 5.7.

In view of the statements in VI.2.12, we may even speak of the "greatest" essential conjugate extension in the case of idempotent metacenter.

We now consider essential normal extensions. The following statements are faithful analogues of VI.2.9, VI.2.10, and VI.2.12; their proofs can be deduced from the proofs of the cited statements by trivial modifications, which is left as an exercise.

XIII.5.9 Proposition

Let S be a normal extension of an inverse semigroup K. Then $\theta(S:K)$ is one-to-one if and only if $M(K) = E_K$ and S is an essential extension of K.

XIII.5.10 Theorem

If S is an inverse semigroup with idempotent metacenter, then $\Phi(S)$ is a maximal essential normal extension of $\Theta(S)$.

XIII.5.11 Corollary

Let S be an inverse semigroup with idempotent metacenter. Then S has a maximal essential normal extension. Every essential normal extension of S is isomorphic to its type in $\Phi(S)$, and every maximal such is isomorphic to $\Phi(S)$. Every inverse subsemigroup of $\Phi(S)$ containing $\Theta(S)$ is the type of some essential normal extension of S.

We also have the analogue of 5.7 as follows.

XIII.5.12 Theorem

Let C be an essential normal extension of an inverse semigroup A with nonidempotent metacenter. Then there exists an essential normal extension of A which properly contains C.

Proof. First note that B in 5.6 is an essential normal extension of S. Since C is now a normal extension of A, S in the proof of 5.7 is also a normal extension of A. The proof of 5.7 goes through to the appearance and properties of P. Letting $F = E_A^c$ in 3.1 relative to P, we obtain the greatest inverse subsemigroup of P having E_A for its semilattice of idempotents. It follows that F is a normal extension of A which contains both B and C. The argument already used in the proof of 5.7 can be applied here to show that F is actually an essential extension of A. It follows that $F\xi^{\#}$ is an essential normal extension of $A\xi^{\#}$ containing $C\xi^{\#}$. That $C\xi^{\#} \neq F\xi^{\#}$ follows along the lines of the last paragraph of the proof of 5.7. Consequently $F\xi^{\#}$ contains $C\xi^{\#}$ properly.

XIII.5.13 Corollary

An inverse semigroup has a maximal essential normal extension if and only if it has idempotent metacenter.

Proof. This is a direct consequence of 5.10 and 5.12.

Note that 5.7 and 5.12 prove somewhat more than is used in the corresponding parts of 5.8 and 5.13, respectively. Indeed, if the metacenter of S is not idempotent, then to every essential conjugate (normal) extension of S there is one such which is larger.

XIII.5.14 Exercises

(i) State and prove the analogues of 5.7–5.11 for groups.

(ii) Let U, T, V be semigroups and $[\ ,\ ]: V \times U \to Z(T)$ be a function satisfying

$$[v_1, u_2][v_1v_2, u_3] = [v_1, u_2u_3][v_2, u_3] \quad (v_i \in V, u_i \in U, i = 1,2,3).$$

On $S = U \times T \times V$ define a multiplication by

$$(u, t, v)(u', t', v') = (uu', t[v, u']t', vv').$$

Show that S is a semigroup and that it generalizes the semigroup $\hat{S}$ at the beginning of this section.

(iii) Continue the notation of the preceding exercise and assume that U, T, V are inverse semigroups, T is a monoid with identity e, $[v, u] \in$

H_e for all $v \in V$, $u \in U$, and $[e', e''] = e$ for all $e' \in E_U$, $e'' \in E_V$.

(α) Show that S is an inverse semigroup.

(β) Find the inverse of an arbitrary element of S.

(γ) Show that U, T, and V can be embedded into S.

(δ) Show that if either U or V has a zero, then $S = U \times T \times V$.

XIII.5.15 Problems

(i) What are the analogues of results 5.7–5.11 for the proper inverse semigroup varieties with the strong amalgamation property? [VI.2.13(xii) takes care of semilattices, 5.14(i) of groups; Gluskin [5] of abelian groups; Gluskin [7] of commutative inverse semigroups.]

All the results of this section, except those already contained in VI.2, are due to Pastijn [5]. The corresponding problem for the variety of groups was previously solved by Gluskin [5]. Gluskin's proof was incorrect, but its underlying ideas, with nontrivial adaptations, carried over to the case of inverse semigroups.

XIV
THE TRACE

One can try to construct inverse semigroups from some more special classes such as semilattices and groups. Indeed, such an attempt, for example, succeeded via P-semigroups and idempotent separating homomorphic images. Other building blocks may be used such as Brandt semigroups and semilattices. The present approach provides certain constructions based on these building blocks from which any inverse semigroup may be obtained. The importance of such devices is theoretical because it makes the structure of a general inverse semigroup more transparent, but concretely it may be very difficult to actually construct a new semigroup out of the given ingredients.

1. One defines a Brandt groupoid as a partial groupoid satisfying some fairly restrictive axioms. Adjoining a new element to a Brandt groupoid which acts as the zero as well as the product for the elements with undefined product yields a Brandt semigroup, and conversely. The same type of procedure is valid for Croisot groupoids versus primitive inverse semigroups.

2. An inverse triple is defined as an abstraction of the trace, the semilattice of idempotents and certain functions among $\mathcal{R}$-classes of the trace of an inverse semigroup. From this abstraction a unique inverse semigroup can be constructed, certain of whose parameters coincide with the given inverse triple.

3. A different set of ingredients consisting of an abstraction of the trace, the natural order, and some products of idempotents and elements may also be used to construct inverse semigroups. The inverse semigroup so obtained is again unique relative to the requirement that certain of its parameters coincide with the given ingredients.

4. Strict inverse semigroups admit a construction based on the knowledge of their semilattice of idempotents and certain mappings among $\mathcal{J}$-classes of its trace. The aim here is more restricted than in the cases discussed above, but one does get a somewhat more elaborate construction.

XIV.1 BRANDT AND CROISOT GROUPOIDS

We define these groupoids abstractly and then establish their close relationship with the familiar classes of Brandt and primitive inverse semigroups. These groupoids play a decisive role in this chapter where we consider the trace of an arbitrary and of some special inverse semigroups.

XIV.1.1 Definition

A *partial groupoid* is a nonempty set G together with a multiplication which may not be defined for all pairs of elements of G.

We consider the following axioms, which a partial groupoid G is susceptible to satisfy.

(A) For any $a, b, c \in G$, ab and $(ab)c$ are defined if and only if bc and $a(bc)$ are defined, and if so then $(ab)c = a(bc)$.

(B) For every $e, f \in G$ such that $e^2 = e$, $f^2 = f$, there exists an element $a \in G$ such that $a = ea = af$.

(C) For any $a \in G$, there exists a unique $a^{-1} \in G$ such that aa^{-1} and $(aa^{-1})a$ are defined and $a = (aa^{-1})a$.

The following terminology will simplify our statements.

XIV.1.2 Definition

A partial groupoid G is a *Croisot groupoid* if it satisfies axioms (A) and (C); it is a *Brandt groupoid* if it satisfies axioms (A), (B), and (C).

We will need a number of simpler properties of these groupoids. In view of (A), in a product, we may omit all the parentheses.

XIV.1.3 Lemma

Let G be a Croisot groupoid. Then for every $a \in G$, there exist unique e and f such that $a = ea = af$, and a^{-1} is the unique element such that $a^{-1}a = f$.

Proof. For $a \in G$, by (A) and (C), we have $a = ea = af$ with $e = aa^{-1}$, $f = a^{-1}a$. Assume that also $a = ga$. Then

$$a = aa^{-1}a = aa^{-1}(ga) = a(a^{-1}g)a$$

which gives $a^{-1} = a^{-1}g$. Multiplying by a on the left, we get $e = eg$, whence $e = ege = eee$ so that $g = e$. Hence in $a = ea$, e is unique; one shows symmetrically that f is unique with the property $a = af$. If $a'a = f$, then

$$a = af = aa'a = aa^{-1}a$$

immediately implies that $a' = a^{-1}$.

XIV.1.4 Definition

For an element a of a Croisot groupoid G, we call e and f in 1.3 the *left* and the *right identity* of a, to be denoted by λa and $a\rho$, respectively; the element a^{-1} is the *inverse* of a. If $a^2 = a$, then a is an *idempotent*.

Note that for $a \in G$,

$$\lambda a = aa^{-1}, \qquad a\rho = a^{-1}a,$$

and that axiom (C) simply assures the unique existence of a^{-1} such that $a = aa^{-1}a$. Further properties of these groupoids follow.

XIV.1.5 Lemma

The following statements are true in a Croisot groupoid G.

(i) If e and f are idempotents and ef is defined, then $e = f$.
(ii) The product ab is defined if and only if $a\rho = \lambda b$.
(iii) If ab and bc are defined, so is abc.
(iv) $(a^{-1})^{-1} = a$.
(v) $\lambda a^{-1} = a\rho, \quad \lambda a = a^{-1}\rho$.

Proof. (i) For any e, f as in (i), letting $x = (ef)^{-1}$, we obtain

$$ef = (ef)x(ef) = (ef)(fx)(ef) = (ef)(xe)(ef) = (ef)(xefx)(ef)$$

which by uniqueness implies

$$x = fx = xe = (xe)(fx) = x^2 = xex = xfx$$

so that $e = f$.

(ii) Let ab be defined. Then $ab = (a(a\rho))((\lambda b)b)$ so that $(a\rho)(\lambda b)$ is defined. Since $a\rho$ and λb are idempotents, part (i) yields that $a\rho = \lambda b$. Conversely, if $a\rho = \lambda b$, then $a\rho = (a\rho)^2 = (a^{-1}a)(bb^{-1})$ and ab is defined.

(iii) Let ab and bc be defined. Then $(ab)\rho = b\rho$ and $b\rho = \lambda c$ by part (ii). Hence again by part (ii), $(ab)c$ is defined. By axiom (A), we get $(ab)c = a(bc)$.

(iv) Let $a \in G$. Since aa^{-1} exists, $(a(a^{-1}a))a^{-1} = a((a^{-1}a)a^{-1})$ exists, and thus also $a^{-1}aa^{-1}$ exists. Further,

$$a = aa^{-1}a = (aa^{-1}a)a^{-1}a = a(a^{-1}aa^{-1})a$$

implies $a^{-1} = a^{-1}aa^{-1}$ by the uniqueness condition in (C), whence $(a^{-1})^{-1} = a$.

(v) By part (iv),

$$\lambda a^{-1} = a^{-1}(a^{-1})^{-1} = a^{-1}a = a\rho$$

and similarly $\lambda a = a^{-1}\rho$.

In order to establish the precise relationship between Croisot and Brandt groupoids, we need the following concept.

XIV.1.6 Definition

Let $\{G_\alpha\}_{\alpha \in A}$ be a family of pairwise disjoint partial groupoids. On the set $G = \cup_{\alpha \in A} G_\alpha$ we define an operation $*$ as follows: for $a \in G_\alpha$, $b \in G_\beta$, $a * b$ is defined if and only if $\alpha = \beta$ and ab is defined in G_α, and if so, then $a * b = ab$. This provides G with the structure of a partial groupoid, and we write $G = \Sigma_{\alpha \in A} G_\alpha$ and call G an *orthogonal sum* of the groupoids G_α, which are then the *components* of G.

The sought for relationship between the Brandt and the Croisot groupoids can now be made explicit.

XIV.1.7 Theorem

An orthogonal sum of Brandt groupoids is a Croisot groupoid. Conversely, every Croisot groupoid is uniquely an orthogonal sum of Brandt groupoids.

Proof. The direct part follows from the even more general statement that an orthogonal sum of Croisot groupoids is again a Croisot groupoid. Indeed, axiom (A) holds since the product of elements in distinct components is undefined, and (C) holds automatically in an orthogonal sum of groupoids satisfying (C).

Conversely, let G be a Croisot groupoid. Define a relation ξ on G by

$$a \xi b \Leftrightarrow \text{there exists } c \in G \text{ such that } ac \text{ and } cb \text{ are defined.}$$

For any $a \in G$, with the notation of 1.3,

$$a = af = a(a^{-1}a) = aa^{-1}a$$

which we can now write without parentheses, so that both aa^{-1} and $a^{-1}a$ are defined which gives that ξ is reflexive. Assume next that $a\xi b$, so that ac and cb are defined for some c. Then by 1.5(iii), acb is defined. Hence $(acb)(acb)^{-1}(acb) = acb$ implies that both $b(acb)^{-1}$ and $(acb)^{-1}a$ are defined. Hence ξ is symmetric. Now assume that $a\xi b$ and $b\xi c$ so that ap, pb, bq, and qc are defined for some p, q. Using 1.5(iii) several times, we deduce that $apbqc$ and thus $a(pbq)$ and $(pbq)c$ are defined so that $a\xi c$. Thus ξ is also transitive, and is hence an equivalence relation.

Let $e^2 = e\xi f = f^2$. Then for some a, both ea and af are defined, and hence so is eaf by 1.5(iii). It follows that $e(eaf) = (eaf)f = eaf$, which verifies axiom (B) in the ξ-class of e and f. If ab is defined, then by 1.3, the right identity e of a is equal to the left identity of b, so ae and eb are defined and thus $a\xi b$. By contrapositive, if $a\not\xi b$, then ab is not defined. If ab is defined, then similarly as just above, using (A), (C), and 1.5(iii), we get that aa^{-1} and $a^{-1}(ab)$ are defined so that $a\xi ab$. Consequently, G is an orthogonal sum of the ξ-classes B_α each of which is a Brandt groupoid.

Let G be also an orthogonal sum of Brandt groupoids C_α, and denote by τ the equivalence relation induced by this decomposition. If $a\tau b$, then by (B) there exists $c \in G$ such that $(a\rho)c(\lambda b) = c$; this implies by 1.5(iii) that acb is defined, and so $a\xi b$. Thus $\tau \subseteq \xi$. If $a\xi b$, then there exists $c \in G$ such that ac and cb are defined, and hence a, b and c must belong to some C_α, that is to say $a\tau c\tau b$. Consequently, $\xi \subseteq \tau$, and the equality prevails.

XIV.1.8 Notation

Let G be a partial groupoid. We adjoin an element 0 to G and define an operation on $G^0 = G \cup \{0\}$ by

$$a * b = \begin{cases} ab & \text{if } ab \text{ is defined in } G \\ 0 & \text{otherwise,} \end{cases}$$

$$a * 0 = 0 * a = 0 * 0 = 0.$$

The following result establishes a link between the concepts discussed above and the familiar ones from semigroups.

XIV.1.9 Theorem

The following statements are valid for any partial groupoid G.

(i) G satisfies (A) if and only if G^0 is a semigroup.
(ii) G is a Croisot groupoid if and only if G^0 is a primitive inverse semigroup.
(iii) G is a Brandt groupoid if and only if G^0 is a Brandt semigroup.

Proof. (i) Assume that G satisfies (A). Let $a, b, c \in G^0$ and suppose that $(a * b) * c \neq 0$. Then $a * b \neq 0$ so ab is defined in G and $a * b = ab$, so $(ab) * c \neq 0$ implies that $(ab)c$ is defined in G. By (A), both bc and $a(bc)$ are defined in G so that $(ab)c = a(bc)$ again by (A). But then

$$(a * b) * c = (ab)c = a(bc) = a * (b * c).$$

By symmetry, we conclude that

$$(a * b) * c \neq 0 \Leftrightarrow a * (b * c) \neq 0$$

and the associative law holds in G^0.

Conversely, assume that G^0 is a semigroup. Suppose that ab and $(ab)c$ are defined in G. This means that in G^0, $(a * b) * c \neq 0$, so that $a * (b * c) \neq 0$ whence both bc and $a(bc)$ are defined and $(ab)c = a(bc)$ holds by the associative law in G^0. By symmetry, we conclude that (A) holds.

(ii) Let G be a Croisot groupoid. Then by part (i), G^0 is a semigroup. By axiom (C), G^0 is a regular semigroup. By 1.5(i), the product of any two distinct nonzero idempotents is equal to zero, which evidently means that G^0 is a primitive inverse semigroup.

Conversely, assume that G^0 is a primitive inverse semigroup. Then G satisfies (A) by part (i). By II.4.3, G^0 is an orthogonal sum of Brandt semigroups. Using II.3.5, one sees easily that the equation $a = axa$, with $a \neq 0$, has exactly one solution in a Brandt semigroup, and hence also in G^0. Consequently, G satisfies axiom (C).

(iii) In view of the proof of part (ii) and 1.7, it suffices to observe that an orthogonal sum $\Sigma_{\alpha \in A} B_\alpha$ of Brandt semigroups has the property that $(\Sigma_{\alpha \in A} B_\alpha)^*$ satisfies axiom (B) if and only if $|A| = 1$, that is to say, $\Sigma_{\alpha \in A} B_\alpha$ has a single component.

Note that axiom (A) (associativity) gives an abstract characterization of the groupoid S^* where S is a nontrivial semigroup with zero. Observe also that in a Croisot groupoid G, if ab is defined, then

$$(ab)^{-1} = b^{-1}a^{-1}, \qquad a = (a^{-1})^{-1}$$

as follows from 1.9(ii). From the proof of 1.9(ii) and (iii), we also deduce the following consequence.

XIV.1.10 Corollary

Let C be a Croisot groupoid. Then $C = \Sigma_{\alpha \in A} B_\alpha$, an orthogonal sum of Brandt groupoids, if and only if $C^0 = \Sigma_{\alpha \in A} B_\alpha^0$, an orthogonal sum of Brandt semigroups.

Hence the correspondence of Croisot groupoids [axioms (A) and (C)] and Brandt groupoids [axioms (A), (B), and (C)] is concordant with the correspondence of primitive inverse semigroups and Brandt semigroups.

XIV.1.11 Exercises

(i) Show that the conclusion in 1.3 together with axiom (A) implies axiom (C).

Brandt [1], see also [2], introduced the groupoid which now carries his name. Its relationship with what is now called a Brandt semigroup was clarified by Clifford [1]. Croisot [1] investigated more general groupoids; see also Clifford [4], Hoehnke [1]–[4], Schein [6], Stolt [1], Tamari [1].

XIV.2 STRUCTURE MAPPINGS

One might ask whether the elegant construction theorem for Clifford semigroups has a suitable analogue for the class of all inverse semigroups. Among the parameters figuring in the construction of a (strong) semilattice of groups there are three basic ingredients: partial products (within each maximal subgroup), the order of $\mathcal{R}$-classes (that is, the underlying semilattice) and the structure mappings (among some of the maximal subgroups). In the generalization we discuss here, the three ingredients just mentioned translate into the trace, the semilattice of idempotents, and the structure mappings of an inverse semigroup. The principal result here asserts that these three parameters indeed determine the inverse semigroup uniquely.

We start our discussion by introducing the relevant concepts.

XIV.2.1 Definition

The *trace* $\operatorname{tr}(S)$ of an inverse semigroup S is the set S together with the partial multiplication: $a * b$ is defined precisely when $ab \in R_a \cap L_b$ and is then equal to ab. We will usually denote the product in $\operatorname{tr}(S)$ again by juxtaposition.

It will be convenient to have the following simple statement.

XIV.2.2 Lemma

For any inverse semigroup S, the partial product in $\operatorname{tr}(S)$ can be given by

$$a * b = ab \quad \text{if} \quad a\rho = \lambda b$$

and is undefined otherwise.

Proof. This is an easy consequence of I.6.6.

XIV.2.3 Definition

Let S be an inverse semigroup. For any pair $e, f \in E_S$, $e \geq f$, define $\varphi_{e,f}$ by

$$\varphi_{e,f}\colon x \to fx \qquad (x \in R_e).$$

Then $\varphi_{e,f}\colon R_e \to R_f$, and all such $\varphi_{e,f}$ are the *structure mappings* of S.

It is easy to check that $b = fx$ is the unique element in R_f with the property $x \geq b$. In view of 1.9, we may speak of Green's relations in a Croisot groupoid in the obvious way. In order to simplify our statements, we introduce the following concept.

XIV.2.4 Definition

Let T be a Croisot groupoid, and E be the set of its idempotents provided with a semilattice structure, written $\wedge$. For each pair $e, f \in E$, $e \geq f$, let $\varphi_{e,f}: R_e \to R_f$ be a function, and assume that for all $e, f, g \in E$, the following axioms are fulfilled.

(S1) $\varphi_{e,e} = \iota_{R_e}$;
(S2) $\varphi_{e,f}\varphi_{f,g} = \varphi_{e,g}$ if $e \geq f \geq g$;
(S3) $e\varphi_{e,f} = f$ if $e \geq f$;
(S4) if $a\rho = \lambda b = e \geq f$, let $\bar{f} = (a^{-1}\varphi_{e,f})\rho$, then $\lambda a \geq \bar{f}$ and

$$(ab)\varphi_{\lambda a, \bar{f}} = \left(a^{-1}\varphi_{e,f}\right)^{-1}\left(b\varphi_{e,f}\right).$$

Under these assumptions, we call $(T, E; \varphi_{e,f})$ an *inverse triple*.

Note that the product in (S4) is indeed defined in T. In the main theorem below, the greatest difficulty arises in the proof of the associative law. For a given inverse triple $(T, E; \varphi_{e,f})$ we will first prove several lemmas needed for establishing the associative law. It will be convenient to introduce the following set of functions.

For any $e, f \in E$, $e \geq f$, define $\psi_{e,f}$ by

$$\psi_{e,f}: a \to \left(a^{-1}\varphi_{e,f}\right)^{-1} \qquad (a \in L_e).$$

The functions $\psi_{e,f}$ are obviously "structure mappings" dual to the given ones $\varphi_{e,f}$; in the second lemma below, we will see that they satisfy the conditions dual to (S1)–(S4). Furthermore, the conclusion of (S4) can now be written symmetrically:

$$(ab)\varphi_{\lambda a, \bar{f}} = \left(a\psi_{e,f}\right)\left(b\varphi_{e,f}\right).$$

We will use this and (S1)–(S3) below without express reference.

XIV.2.5 Lemma

If $\lambda a = e \geq f$, $\bar{e} = a\rho$ and $\bar{f} = (a\varphi_{e,f})\rho$, then $\bar{e} \geq \bar{f}$ and $a\varphi_{e,f} = a\psi_{\bar{e},\bar{f}}$.

Proof. The present a substituted for a^{-1} in (S4) gives that $\bar{e} \geq \bar{f}$. Applying (S4) to the product $a^{-1} = a^{-1}e$, we obtain

$$a^{-1}\varphi_{\bar{e},\bar{f}} = (a^{-1}e)\varphi_{\bar{e},\bar{f}} = (a^{-1}\psi_{e,f})(e\varphi_{e,f}) = (a^{-1}\psi_{e,f})f = a^{-1}\psi_{e,f}$$

and taking inverses gives the desired equality.

XIV.2.6 Lemma

The mappings $\psi_{e,f}$ satisfy the duals of (S1)–(S4).

Proof. The duals of (S1)–(S3) follow without difficulty. The dual of (S4): if $a\rho = \lambda b = e \geq f$ and $\bar{f} = (b\varphi_{e,f})\rho$, then $b\rho \geq \bar{f}$ and

$$(ab)\psi_{b\rho,\bar{f}} = (a\psi_{e,f})(b\varphi_{e,f}). \tag{1}$$

First note that $(a^{-1}\varphi_{e,f})\rho = \lambda(a\psi_{e,f})$. By (S4), we then have

$$(ab)\varphi_{\lambda a,\lambda(a\psi_{e,f})} = (a\psi_{e,f})(b\varphi_{e,f})$$

which gives

$$\big((ab)\varphi_{\lambda a,\lambda(a\psi_{e,f})}\big)\rho = (b\varphi_{e,f})\rho = \bar{f}.$$

Using this and 2.5, we obtain

$$(ab)\varphi_{\lambda a,\lambda(a\psi_{e,f})} = (ab)\psi_{b\rho,((ab)\varphi_{\lambda a,\lambda(a\psi_{e,f})})\rho} = (ab)\varphi_{b\rho,\bar{f}}$$

and (1) is proved.

XIV.2.7 Lemma

Let $e = a\rho$ and $f = \lambda a$. Then the mapping

$$\sigma_a : t \to (a^{-1}\varphi_{e,t})\rho \qquad (t \in [e])$$

is an isomorphism of $[e]$ onto $[f]$.

Proof. That σ_a maps $[e]$ into $[f]$ follows from (S4). Symmetrically $\sigma_{a^{-1}}$ maps $[f]$ into $[e]$. For any $t \in [e]$, we obtain

$$t\sigma_a\sigma_{a^{-1}} = (a^{-1}\varphi_{e,t})\rho\sigma_{a^{-1}} = \Big[\big(a\varphi_{f,(a^{-1}\varphi_{e,t})\rho}\big)\rho\Big]\mathcal{R}\big[(a^{-1}\varphi_{e,t})\rho\big]\mathcal{R}t$$

which implies that $t\sigma_a\sigma_{a^{-1}} = t$. It follows that $\sigma_a\sigma_{a^{-1}} = \iota_{[e]}$ and symmetrically $\sigma_{a^{-1}}\sigma_a = \iota_{[f]}$. Thus $\sigma_{a^{-1}}$ is the inverse of σ_a so that σ_a is a bijection of $[e]$ and $[f]$. If $s \leq t \leq e$, then

$$s\sigma_a = (a^{-1}\varphi_{e,s})\rho = (a^{-1}\varphi_{e,t}\varphi_{t,s})\rho \leq (a^{-1}\varphi_{e,t})\rho = t\sigma_a$$

again using (S4). Hence σ_a is order preserving, and by symmetry, so is $\sigma_{a^{-1}}$. The proof of I.2.8 yields that σ_a is an isomorphism.

XIV.2.8 Lemma

Let $e, f, g, h \in E$. If α is an isomorphism of $[f]$ onto $[g]$, then

$$(e \wedge f)\alpha \wedge h = \left[e \wedge (g \wedge h)\alpha^{-1}\right]\alpha.$$

Proof. Note that

$$(e \wedge f)\alpha \wedge h = (e \wedge f)\alpha \wedge (g \wedge h) \tag{2}$$

since $(e \wedge f)\alpha \leq g$, and

$$e \wedge (g \wedge h)\alpha^{-1} = (e \wedge f) \wedge (g \wedge h)\alpha^{-1} \tag{3}$$

since $(g \wedge h)\alpha^{-1} \leq f$. Applying α to (3) and comparing with (2) gives the desired equality.

We are now ready for the principal result of this section.

XIV.2.9 Theorem

Let $(S, E; \varphi_{e,f})$ be an inverse triple. A product can be defined on S by

$$a \circ b = \left(a^{-1}\varphi_{a\rho, e}\right)^{-1}\left(b\varphi_{\lambda b, e}\right) \tag{4}$$

where $e = a\rho \wedge \lambda b$. Then $(S, \circ)$ is the unique inverse semigroup with trace S, semilattice of idempotents E and structure mappings $\varphi_{e,f}$.

Conversely, every inverse semigroup can be so constructed.

Proof. Let $(S, E; \varphi_{e,f})$ be an inverse triple. With the notation as in the statement of the theorem, we have

$$\left(a^{-1}\varphi_{a\rho, e}\right)^{-1}\rho = \lambda\left(a^{-1}\varphi_{a\rho, e}\right) = e = \lambda\left(b\varphi_{\lambda b, e}\right)$$

which shows that the product on the right-hand side in (4) is performed in the Croisot groupoid S.

We verify the associative law next. Thus let $a, b, c \in S$ and $e = a\rho$, $f = \lambda b$, $g = b\rho$, $h = \lambda c$. Let $\alpha = \sigma_{b^{-1}}$, as defined in 2.7, so that α is an isomorphism of $[f]$ onto $[g]$. Further,

$$\begin{aligned}(a \circ b)\rho &= \left(b\varphi_{f, e \wedge f}\right)^{-1}\left(a^{-1}\varphi_{e, e \wedge f}\right)\left(a^{-1}\varphi_{e, e \wedge f}\right)^{-1}\left(b\varphi_{f, e \wedge f}\right) \\ &= \left(b\varphi_{f, e \wedge f}\right)^{-1}\left(b\varphi_{f, e \wedge f}\right) = \left(b\varphi_{f, e \wedge f}\right)\rho = (e \wedge f)\alpha \end{aligned} \tag{5}$$

where the second equality follows from properties of a Croisot groupoid. This locates the product $a \circ b$. By 2.7 we obtain

$$\left(b\varphi_{f,[(e \wedge f)\alpha \wedge h]\alpha^{-1}}\right)\rho = (e \wedge f)\alpha \wedge h. \tag{6}$$

Now

$$\begin{aligned}
(a \circ b) \circ c &= \left[\left(a\psi_{e,e \wedge f}\right)\left(b\varphi_{f,e \wedge f}\right)\right] \circ c \\
&= \left[\left(a\psi_{e,e \wedge f}\right)\left(b\varphi_{f,e \wedge f}\right)\right] \psi_{(e \wedge f)\alpha,(e \wedge f)\alpha \wedge h}\left(c\varphi_{h,(e \wedge f)\alpha \wedge h}\right) \\
&= \left(a\psi_{e,e \wedge f}\psi_{e \wedge f,[(e \wedge f)\alpha \wedge h]\alpha^{-1}}\right) \\
&\quad \cdot\left(b\varphi_{f,e \wedge f}\psi_{e \wedge f,[(e \wedge f)\alpha \wedge h]\alpha^{-1}}\right)\left(c\varphi_{h,(e \wedge f)\alpha \wedge h}\right) \\
&\qquad\qquad \text{by (1), (5), (6)} \\
&= \left(a\psi_{e,e \wedge (g \wedge h)\alpha^{-1}}\right)\left(b\varphi_{f,e \wedge (g \wedge h)\alpha^{-1}}\right)\left(c\varphi_{h,(e \wedge f)\alpha \wedge h}\right) \\
&\qquad\qquad \text{by 2.8,} \quad (7)
\end{aligned}$$

and similarly

$$\begin{aligned}
a \circ (b \circ c) &= a \circ \left[\left(b\psi_{g,g \wedge h}\right)\left(c\varphi_{h,g \wedge h}\right)\right] \\
&= \left(a\psi_{e,e \wedge (g \wedge h)\alpha^{-1}}\right) \\
&\quad \cdot\left[\left(b\psi_{g,g \wedge h}\right)\left(c\varphi_{h,g \wedge h}\right)\right]\varphi_{(g \wedge h)\alpha^{-1},e \wedge (g \wedge h)\alpha^{-1}} \\
&= \left(a\psi_{e,e \wedge (g \wedge h)\alpha^{-1}}\right)\left(b\psi_{g,g \wedge h}\psi_{g \wedge h,(e \wedge f)\alpha \wedge h}\right) \\
&\quad \cdot\left(c\varphi_{h,g \wedge h}\varphi_{g \wedge h,(e \wedge f)\alpha \wedge h}\right) \\
&= \left(a\psi_{e,e \wedge (g \wedge h)\alpha^{-1}}\right)\left(b\psi_{g,(e \wedge f)\alpha \wedge h}\right)\left(c\varphi_{h,(e \wedge f)\alpha \wedge h}\right).
\end{aligned} \tag{8}$$

Using 2.8 and 2.6, we further obtain

$$b\varphi_{f,e \wedge (g \wedge h)\alpha^{-1}} = b\psi_{g,(e \wedge f)\alpha \wedge h}$$

which together with (7) and (8) establishes the associative law.

If ab is defined in S, then $a\rho = \lambda b$ whence $a \circ b = ab$. For any $a \in S$, we have $a = aa^{-1}a$, which then implies that $a = a \circ a^{-1} \circ a$, and thus $(S, \circ)$ is a regular semigroup. Also, for $e \in E$, we obtain $e \circ e = e$. Conversely, let $a \circ a = a$. Then letting $e = a\rho \wedge \lambda a$, we have

$$\begin{aligned}
\lambda a = \lambda(a \circ a) &= \left(a^{-1}\varphi_{a\rho,e}\right)^{-1}\left(a\varphi_{a\lambda,e}\right)\left(a\varphi_{a\lambda,e}\right)^{-1}\left(a^{-1}\varphi_{a\rho,e}\right) \\
&= \left(a^{-1}\varphi_{a\rho,e}\right)\rho
\end{aligned}$$

which implies that $\lambda a = e$. It follows dually that $a\rho = e$. Hence

$$a = \left(a^{-1}\varphi_{e,e}\right)^{-1}\left(a\varphi_{e,e}\right) = a^2$$

and thus $a \in E$. Consequently, E coincides with the set of idempotents of $(S, \circ)$. It follows easily that for any $e, f \in E$, $e \circ f = e \wedge f$. This together with regularity of $(S, \circ)$ shows that it is an inverse semigroup.

By 1.5(ii) and 2.2, we have $\operatorname{tr}((S, \circ)) = S$. We have already remarked that $E_{(S, \circ)} = E$. If $\lambda a = e \geq f$, then

$$f \circ a = \left(f\varphi_{e,f}\right)^{-1}\left(a\varphi_{e,f}\right) = f\left(a\varphi_{e,f}\right) = a\varphi_{e,f},$$

that is, $\varphi_{e,f}$ are the structure mappings for $(S, \circ)$. The uniqueness of the inverse semigroup $(S, \circ)$ with trace S, semilattice of idempotents E, and structure mappings $\varphi_{e,f}$ follows directly from the form of the product in (4).

For the converse, let S be an inverse semigroup. We must show that $(\operatorname{tr}(S), E_S, \varphi_{e,f})$ is an inverse triple and that the associated multiplication $\circ$ coincides with the multiplication in S.

We verify first that $\operatorname{tr}(S)$ is a Croisot groupoid. Let $a, b, c \in S$. Assume first that ab and $(ab)c$ are defined in $\operatorname{tr}(S)$. By 2.2, we then have $bb^{-1} = a^{-1}a$ and $cc^{-1} = b^{-1}a^{-1}ab$ whence

$$cc^{-1} = b^{-1}(a^{-1}a)b = b^{-1}(bb^{-1})b = b^{-1}b,$$

$$a^{-1}a = bb^{-1}(a^{-1}a)bb^{-1} = b(b^{-1}a^{-1}ab)b^{-1} = bcc^{-1}b^{-1}$$

so again by 2.2, we have that bc and $a(bc)$ are defined. The proof of the converse is analogous, and if these statements hold, then $(ab)c = a(bc)$ in $\operatorname{tr}(S)$. Thus axiom (A) in 1.1 holds.

It follows immediately from 2.2 that both aa^{-1} and $(aa^{-1})a$ are defined and $a = (aa^{-1})a$ in $\operatorname{tr}(S)$. In order to prove uniqueness, we assume that ab and $(ab)a$ are defined in $\operatorname{tr}(S)$ and $a = (ab)a$. By 2.2, we have $a^{-1}a = bb^{-1}$ and $b^{-1}a^{-1}ab = aa^{-1}$, whence

$$bab = b(ab)(ab)^{-1}(ab) = bab(aa^{-1}) = b(aba)a^{-1}$$

$$= b(aa^{-1}) = bb^{-1}a^{-1}ab = a^{-1}ab = bb^{-1}b = b.$$

By uniqueness of inverses, we get $b = a^{-1}$. Hence axiom (C) in 1.1 is satisfied as well, and thus $\operatorname{tr}(S)$ is a Croisot groupoid.

It is obvious that the idempotents of S and of $\operatorname{tr}(S)$ coincide. The structure mappings evidently satisfy axioms (S1)–(S3) in 2.4. Assume that $a\rho = \lambda b = e \geq f$ and let $\bar{f} = (a^{-1}\varphi_{e,f})\rho$. Then

$$\bar{f} = \left(fa^{-1}\right)^{-1}\left(fa^{-1}\right) = afa^{-1}$$

and hence $\lambda a = aa^{-1} \geq afa^{-1} = \bar{f}$. Further, on the one hand,

$$(ab)\varphi_{\lambda a, \bar{f}} = \bar{f}ab = afa^{-1}ab = afb,$$

and on the other hand,

$$\left(a^{-1}\varphi_{e,f}\right)^{-1}\left(b\varphi_{e,f}\right) = \left(fa^{-1}\right)^{-1}(fb) = afb$$

which verifies axiom (S4). We have proved that $(\mathrm{tr}(S), E_S, \varphi_{e,f})$ is an inverse triple. Finally, letting $e = (a\rho)(\lambda b)$, we obtain

$$\begin{aligned} a \circ b &= \left(a^{-1}\varphi_{a\rho, e}\right)^{-1}\left(b\varphi_{\lambda b, e}\right) = \left(ea^{-1}\right)^{-1}(eb) = aeb \\ &= \left[a(a\rho)\right]\left[(\lambda b)b\right] = ab, \end{aligned}$$

and the two multiplications coincide.

XIV.2.10 Remark

We can formulate the above theorem more explicitly by using the representation $B(G, I)$ for the Brandt semigroups which arise from the Brandt groupoids in the orthogonal sum representation of the Croisot groupoid. We now outline this procedure.

Let $S = \Sigma_{\alpha \in A} B_\alpha$ be an orthogonal sum where each B_α is a Brandt groupoid. Then B_α^0 is a Brandt semigroup and we may write $B_\alpha^0 = B(G_\alpha, I_\alpha)$. Assume that $I_\alpha \cap I_\beta = \varnothing$ if $\alpha \neq \beta$, and let $I = \cup_{\alpha \in A} I_\alpha$. The assumption that the set E of idempotents of S carries a semilattice structure can be transferred to I by

$$i \leq j \Leftrightarrow (i, e_\alpha, i) \leq (j, e_\beta, j)$$

where $i \in I_\alpha$, $j \in I_\beta$. We denote the product so defined in I by juxtaposition. The $\mathcal{R}$-classes of S are the sets

$$R_i = \{(i, g, j) | g \in G_\alpha, j \in I_\alpha\}$$

if $i \in I_\alpha$, for all $i \in I$. The system of functions $\varphi_{e,f}$ satisfying the requisite properties in 2.4 can now be labeled as $\varphi_{i,j}$ with $i \geq j$, with the axioms (S1)–(S4) translated into this notation. The product in this notation has the form:

$$(i, g, j) \circ (k, h, l) = \left[(j, g^{-1}, i)\varphi_{j, jk}\right]^{-1}\left[(k, h, l)\varphi_{k, jk}\right].$$

XIV.2.11 Example

We show on a simple example how the constructions in this section can be used to obtain a concrete, and well-known, structure theorem. In fact, the construction of an inverse triple can be regarded as a generalization of the (strong) semilattice of groups to be now constructed.

Let $(S, E, \varphi_{e,f})$ be an inverse triple, where $S = \Sigma_{\alpha \in Y} G_\alpha$, each G_α is a group with identity e_α. Since each G_α has only one idempotent, Y inherits the semilattice structure from E in the obvious way, and we may label the functions $\varphi_{\alpha,\beta}$ with $\alpha, \beta \in Y$, $\alpha \geq \beta$. The axioms for an inverse triple yield

$$\begin{aligned}
&\text{(S1)} \quad \varphi_{\alpha,\alpha} = \iota_{G_\alpha}, \\
&\text{(S2)} \quad \varphi_{\alpha,\beta}\varphi_{\beta,\gamma} = \varphi_{\alpha,\gamma} \quad \text{if} \quad \alpha \geq \beta \geq \gamma, \\
&\text{(S3)} \quad e_\alpha \varphi_{\alpha,\beta} = e_\beta \quad \text{if} \quad \alpha \geq \beta, \\
&\text{(S4)} \quad \text{for } a, b \in G_\alpha, \alpha \geq \beta, (ab)\varphi_{\alpha,\beta} = (a^{-1}\varphi_{\alpha,\beta})^{-1}(b\varphi_{\alpha,\beta})
\end{aligned} \tag{9}$$

and for $b = e_\alpha$, using (S3), we obtain

$$a\varphi_{\alpha,\beta} = (ae_\alpha)\varphi_{\alpha,\beta} = \left(a^{-1}\varphi_{\alpha,\beta}\right)^{-1} e_\beta = \left(a^{-1}\varphi_{\alpha,\beta}\right)^{-1}$$

so that (9) means that $\varphi_{\alpha,\beta}$ is a homomorphism.

Furthermore, the multiplication is given by

$$a \circ b = \left(a\varphi_{\alpha,\alpha\beta}\right)\left(b\varphi_{\beta,\alpha\beta}\right)$$

if $a \in G_\alpha$, $b \in G_\beta$. We have thus obtained a strong semilattice of groups. Conversely, every semilattice of groups can be so constructed.

The next two propositions show how certain restrictions on structure mappings are reflected as properties of the multiplication of the semigroup.

XIV.2.12 Proposition

Let S be an inverse semigroup. Then S is E-unitary if and only if the structure mappings of S are one-to-one.

Proof. First assume that S is E-unitary. Let $e = xx^{-1} = yy^{-1}$, $f \leq e$, $x\varphi_{e,f} = y\varphi_{e,f}$. Then $fx = fy$ and thus $x\sigma y$. But then $x(\sigma \cap \mathcal{R})y$ which implies that $x = y$ since S is E-unitary (see III.7.2). Consequently $\varphi_{e,f}$ is one-to-one.

Conversely, suppose that the structure mappings are one-to-one. Let $a \in S$, $e = aa^{-1}$, $f \in E_S$ be such that $fa = f$. Then

$$fe = f(aa^{-1}) = (fa)a^{-1} = fa^{-1} = (af)^{-1} = f$$

so that

$$a\varphi_{e,f} = fa = fe = e\varphi_{e,f}.$$

The hypothesis now implies that $a = e \in E_S$, and hence S is E-unitary.

XIV.2.13 Proposition

Let S be an inverse semigroup. Then S is a semidirect product of a semilattice and a group if and only if the structure mappings of S are bijections.

Proof. Let S be a semidirect product of a semilattice Y and a group G. By VII.5.24, S can be taken to be equal to $P(Y, G; Y)$. Since S is then E-unitary, 2.12 gives that the structure mappings are one-to-one. For any idempotents $(\alpha, 1) \geq (\beta, 1)$ and any $g \in G$, we have $(\beta, g) = (\beta, 1)(\alpha, g)$ which says that $(\alpha, g)\varphi_{(\alpha,1),(\beta,1)} = (\beta, g)$. Consequently $\varphi_{(\alpha,1),(\beta,1)}$ maps $R_{(\alpha,1)}$ onto $R_{(\beta,1)}$.

Conversely, assume that the structure mappings are bijections. By 2.12, S is E-unitary, so we may take $S = P(Y, G; X)$. Let $\alpha \in Y$ and $g \in G$. By the construction of $P(Y, G; X)$, there exists $\beta \in Y$ such that $g^{-1}\beta \in Y$. Hence $(\beta, g) \in S$ and thus also $(\alpha, 1)(\beta, g) \in S$. But then $(\alpha \wedge \beta, g) \in S$ and hence $g^{-1}(\alpha \wedge \beta) \in Y$. Now

$$R_{(\alpha,1)} = \{(\alpha, h) | h^{-1}\alpha \in Y\},$$

$$R_{(\alpha \wedge \beta,1)} = \{(\alpha \wedge \beta, h) | h^{-1}(\alpha \wedge \beta) \in Y\}.$$

By hypothesis, there exists $h \in G$ such that $h^{-1}\alpha \in Y$ and

$$(\alpha, h)\varphi_{(\alpha,1),(\alpha \wedge \beta,1)} = (\alpha \wedge \beta, g).$$

It follows that $(\alpha \wedge \beta, 1)(\alpha, h) = (\alpha \wedge \beta, g)$ so that $h = g$. Consequently, $g^{-1}\alpha \in Y$, that is to say, $(\alpha, g) \in S$. We have proved that the product on S is defined on all of $Y \times G$ and is thus a semidirect product of Y and G.

XIV.2.14 Exercises

(i) Let C be the bicyclic semigroup, and let $B = B(1, N)$ where $N = \{0, 1, 2, \dots\}$. Show that $\operatorname{tr}(C) = B^*$ and that

$$(m, n)\varphi_{(m,m),(p,p)} = (p, p - m + n) \quad \text{if} \quad m \leq p$$

is a structure mapping for C. Also show directly that the system so arising satisfies axioms (S1)–(S4) and write down the corresponding multiplication.

(ii) Do all the items in the preceding exercise for a Reilly semigroup $B(G, \alpha)$.

(iii) Let $N = \{0, 1, 2, \dots\}$ and G be a group. Let $T = B(G, N)^*$ with the ordering of idempotents

$$(m, 1, m) \leq (n, 1, n) \Leftrightarrow m \geq n.$$

Let the structure mappings be defined by

$$(m, g, n)\varphi_{(m,m),(p,p)} = (p, g\alpha^{p-m}, p - m + n) \quad \text{if} \quad m \geq p.$$

Show that the resulting system satisfies axioms (S1)–(S4) and write down the corresponding multiplication.

(iv) Construct all inverse semigroups (up to isomorphism) of orders 2, 3, and 4.

(v) Is there a connection between the functions σ_a in 2.7, the function σ in I.6.11 and the Munn representation?

(vi) Let S be an inverse semigroup. Show that for any $a, b \in S$,

$$a\mu b \Leftrightarrow a\mathcal{H}b \quad \text{and} \quad a\varphi_{e,f}\mathcal{H}b\varphi_{e,f} \quad \text{for all} \quad f \leq e = \lambda a.$$

All results in this section are due to Meakin [2], see also [3]. Some ramifications concerning the existence of the systems discussed here were investigated by Byleen [1] and Byleen–Komjáth [1].

XIV.3 INDUCTIVE GROUPOIDS

We consider here the following problem: to what extent do the trace and the natural partial order of an inverse semigroup S determine the multiplication of S? We will see that with only "a few" partial products of elements and idempotents, the multiplication of S is uniquely determined. The trace, the natural order, and the partial multiplication of an inverse semigroup will be abstractly characterized as a triple.

The following concept is basic for our discussion.

XIV.3.1 Definition

Let T be a Croisot groupoid on which a partial order $\leq$ is defined. Assume that for all $a, b, c, d \in T$, the following axioms are fulfilled.

(I1) If $a \leq b$ and $c \leq d$, and the products ac and bd are defined, then $ac \leq bd$.

(I2) If $a \leq b$, then $a^{-1} \leq b^{-1}$.

(I3) If e and f are idempotents in T, then they have a greatest lower bound $e \wedge f$.

(I4) If e is an idempotent and $e \leq \lambda a$, then there exists a unique element $e * a \in T$ such that $e * a \leq a$ and $\lambda(e * a) = e$.

Then the triple $(T, \leq, *)$ is an *inductive groupoid*.

We will first explore the relationship between inductive groupoids and inverse triples and then prove an analogue of 2.9 for inductive groupoids.

XIV.3.2 Proposition

Let $(S, \leq, *)$ be an inductive groupoid. Let E be the set of idempotents of S with the induced semilattice structure. For any $e, f \in E$, $e \geq f$, let

$$\varphi_{e,f}: a \to f * a \qquad (a \in R_e).$$

Then $(S, E; \varphi_{e,f})$ is an inverse triple.

Proof. First observe that $\varphi_{e,f}$ is well defined, for $a \in R_e$ and $e \geq f$ gives $f \leq \lambda a$, so $f * a$ is defined and $\lambda(f * a) = f$ shows that $f * a \in R_f$.

We now verify axioms (S1)–(S4) in 2.4. Let $e \in E$, $a \in R_e$. Then $a\varphi_{e,e} = e * a = a$ by the uniqueness requirement in (I4). Hence $\varphi_{e,e} = \iota_{R_e}$. Next let $e, f, g \in E$ be such that $e \geq f \geq g$ and let $a \in R_e$. Using (I4) twice, we obtain

$$g * (f * a) \leq f * a \leq a.$$

Since $\lambda(g * (f * a)) = g$, we get $g * (f * a) = g * a$ by (I4). But then

$$a\varphi_{e,f}\varphi_{f,g} = g * (f * a) = g * a = a\varphi_{e,g}$$

which verifies (S2). Let $e, f \in E$ be such that $e \geq f$. Again (I4) yields $f = f * e = e\varphi_{e,f}$, and (S3) is verified as well.

Now let $a\rho = b\lambda = e \geq f$ and $\bar{f} = (a^{-1}\varphi_{e,f})\rho$. By (I4), we have $f * a^{-1} \leq a^{-1}$ so by (I2), $(f * a^{-1})^{-1} \leq a$, and hence by (I1),

$$\bar{f} = \left(a^{-1}\varphi_{e,f}\right)\rho = (f * a^{-1})\rho \leq a^{-1}\rho = \lambda a,$$

as required. Further, $f * a^{-1} \leq a^{-1}$ implies $(f * a^{-1})^{-1} \leq a$ by (I2) and $f * b \leq b$. Since

$$(f * a^{-1})^{-1}\rho = \lambda(f * a^{-1}) = f = \lambda(f * b),$$

axiom (I1) gives

$$(f * a^{-1})^{-1}(f * b) \leq ab. \tag{10}$$

Further,

$$\lambda\left[(f * a^{-1})^{-1}(f * b)\right] = \lambda(f * a^{-1})^{-1} = (f * a^{-1})\rho = \bar{f}$$

which together with (10), by (I4), gives

$$(f * a^{-1})^{-1}(f * b) = \bar{f} * ab.$$

This means that

$$(ab)\varphi_{\lambda a, \bar{f}} = \left(a^{-1}\varphi_{e,f}\right)^{-1}\left(b\varphi_{e,f}\right)$$

and (S4) holds.

Therefore $(S, E; \varphi_{e,f})$ is an inverse triple.

The relationship in the opposite direction is provided by the following result.

XIV.3.3 Proposition

Let $(S, E; \varphi_{e,f})$ be an inverse triple. On S define a relation $\leq$ by

$$a \leq b \Leftrightarrow \lambda a \leq \lambda b, \quad a = b\varphi_{\lambda b, \lambda a}, \tag{11}$$

and a partial operation $*$ by

$$e * a = a\varphi_{\lambda a, e} \quad \text{if} \quad e \leq \lambda a. \tag{12}$$

Then $(S, \leq, *)$ is an inductive groupoid.

Proof. We could verify the axioms for an inductive groupoid directly. Instead, we use 2.9 which gives us an inverse semigroup $(S, \circ)$ with requisite properties. Now write the product in $(S, \circ)$ by juxtaposition. We assert that in $(S, \circ)$,

$$a \leq b \Leftrightarrow \lambda a \leq \lambda b, \quad a = (\lambda a)b. \tag{13}$$

Indeed, if $a \leq b$, then $\lambda a \leq \lambda b$ and $a = eb$ for some idempotent e, which gives

$$a = eb = (\lambda a)eb = e(\lambda a)b = (ea)a^{-1}b = aa^{-1}b = (\lambda a)b.$$

This proves one implication in (13); the opposite one is trivial. Now (13) shows that the order in $(S, \circ)$ agrees with the order in (11). The $*$-operation in (12) obviously coincides, whenever it is defined, with the product in $(S, \circ)$.

We now verify axioms (I1)–(I4) in 3.1. Let $a \leq b$ and $c \leq d$. Then $a = eb$ and $c = fd$ for some $e, f \in E$. Hence $ac = ebfd = (ebfb^{-1})bd$ so that $ac \leq bd$. Further, $a^{-1} = b^{-1}e = (b^{-1}eb)b^{-1}$ and thus $a^{-1} \leq b^{-1}$. This verifies axioms (I1) and (I2); (I3) follows directly from the fact that E_S is a semilattice under its usual order. Next let $e \in E_S$, $e \leq aa^{-1}$. Then $ea \leq a$ and $\lambda(ea) = e$. Suppose that $b \leq a$ and $\lambda b = e$. Then $b = fa$ for some $f \in E_S$ and

$$b = (bb^{-1})b = eb = efa$$

and from $\lambda(efa) = e$, we obtain $ef = e$ so that $b = ea$. This verifies axiom (I4), and shows that $(S, \leq, *)$ is an inductive groupoid.

From the preceding two propositions we see that the relationship between inverse triples and inductive groupoids is quite close. Indeed, the semilattice and the structure mappings are traded off for the partial order and the partial multiplication.

In order to simplify the statements below, we introduce a new concept.

XIV.3.4 Definition

For any inverse semigroup S, the *imprint* of S, $\mathrm{im}(S)$, is the triple $(\mathrm{tr}(S), \leq, *)$ where $\leq$ is the natural partial order in S, and $*$ is a partial product defined by: for $e \in E_S$, $a \in S$, $e \leq \lambda a$, $e * a = ea$.

We now prove the main result of this section.

XIV.3.5 Theorem

Let $(S, \leq, *)$ be an inductive groupoid. A product can be defined in S by

$$a \circ b = (e * a^{-1})^{-1}(e * b) \tag{14}$$

where $e = a\rho \wedge \lambda b$. Then $(S, \circ)$ is the unique inverse semigroup on S with imprint $(S, \leq, *)$.

Conversely, every inverse semigroup can be so constructed.

Proof. That $(S, \circ)$ is an inverse semigroup with trace S follows directly from 3.2 and 2.9. We will need the following statement.

$$\text{If } a \leq b, \text{ then } a = \lambda a * b. \tag{15}$$

Indeed, $a \leq b$ implies $a^{-1} \leq b^{-1}$ by (I2) so $aa^{-1} \leq bb^{-1}$ by (I1), and thus $\lambda a \leq \lambda b$. Then $\lambda a * b \leq b$ and $\lambda(\lambda a * b) = \lambda a$ by (I4), and a has the same properties as $\lambda a * b$ so that by the uniqueness part of (I4), we must have $a = \lambda a * b$.

Let $e \in E$, $a \in S$ and assume that $e \leq \lambda a$. Note that $e^{-1} = e$ in view of axiom (C) in 1.1. Then $e\rho \wedge \lambda a = e \wedge \lambda a = e$ whence in view of (15), we have $e * e = e$. Thus

$$e \circ a = (e * e)(e * a) = e(e * a) = e * a \tag{16}$$

where the last equality holds since $e = \lambda(e * a)$.

Let $a \leq b$. Then $a = \lambda a * b = \lambda a \circ b \in E \circ b$ by (15) and (16). Conversely, assume that $a = e \circ b$ for some $e \in E$. Then $a = (f * e)^{-1}(f * b)$ where $f = e \wedge \lambda b$. Now $f \leq e$ and $\lambda f = f$ imply $f = f * e$ by the uniqueness in (I4). Thus

$a = f(f * b) = f * b \leq b$ where the second equality follows similarly as above. Consequently

$$a \leq b \Leftrightarrow a \in E \circ b$$

and the natural partial order of $(S, \circ)$ coincides with $\leq$.

Therefore $(S, \leq, *)$ is the imprint of $(S, \circ)$. Let $\cdot$ be another product on S having the same imprint as $(S, \circ)$. Let $a, b \in S$, $e = a\rho \wedge \lambda b$. Then

$$\begin{aligned} a \cdot b &= a \cdot (a^{-1} \cdot a \cdot b \cdot b^{-1}) \cdot b = a \cdot (a\rho \cdot \lambda b) \cdot b \\ &= a \cdot (a\rho \wedge \lambda b) \cdot b = a \cdot e \cdot b = (a \cdot e) \cdot (e \cdot b) \\ &= (a \cdot e)(e \cdot b) = (e \cdot a^{-1})^{-1}(e \cdot b) = (e * a^{-1})^{-1}(e * b) \\ &= a \circ b \end{aligned}$$

where we have exhausted the hypothesis $\operatorname{im}(S, \circ) = \operatorname{im}(S, \cdot)$. Therefore $(S, \circ)$ and $(S, \cdot)$ coincide.

Conversely, let S be an inverse semigroup. We have verified in the proof of 2.9 that $\operatorname{tr}(S)$ is a Croisot groupoid, and in the proof of 3.3 that $(\operatorname{tr}(S), E_S, *)$ is an inductive groupoid.

Let $a, b \in S$ and $e = a\rho \wedge \lambda b$. Then $e \leq a\rho = \lambda a^{-1}$ so by (I4), $e * a^{-1}$ exists, and further $e \leq \lambda b$ so

$$((e * a^{-1})^{-1})\rho = \lambda(e * a^{-1}) = e = \lambda(e * b)$$

which by 1.5(iii) implies that $(e * a^{-1})^{-1}(e * b)$ exists in $\operatorname{tr}(S)$. Furthermore,

$$\begin{aligned} a \circ b &= (e * a^{-1})^{-1}(e * b) = (ea^{-1})^{-1}(eb) \\ &= (ae)(eb) = a(a^{-1}a)(bb^{-1})b = ab. \end{aligned}$$

Note the similarity of the product formula (4) in 2.9 and (14) in 3.5.

XIV.3.6 Exercises

(i) Show that in the bicyclic semigroup C,

(α) $(m, n) \leq (p, q) \Leftrightarrow m - n = p - q$, $p \leq m$ is the natural partial ordering,

(β) $(p, p)*(m, n) = (p, p - m + n)$ if $m \leq p$ is the $*$-operation.

Also show directly that this system satisfies axioms (I1)–(I4) and write down the corresponding multiplication.

(ii) Do all the items in the preceding exercise for a Reilly semigroup $B(G, \alpha)$.

(iii) Let the notation be as in 2.14(iii). Define an ordering on T by

$$(m, g, n) \leq (p, h, q) \Leftrightarrow q \leq n, \quad h\alpha^{n-q} = g, \quad m - n = p - q,$$

and the $*$-operation by

$$(m, 1, m) * (p, h, q) = (m, h\alpha^{m-p}, m - p + q)$$

$$\text{if} \quad (m, 1, m) \leq (p, 1, p).$$

Show that this system satisfies axioms (I1)–(I4) and write down the corresponding multiplication.

(iv) Construct the inductive groupoid for a Clifford semigroup $[Y; G_\alpha, \varphi_{\alpha,\beta}]$.

(v) Let ψ be a bijection of an inverse semigroup S onto an inverse semigroup T. Show that ψ is an isomorphism if and only if

(α) ψ is an isomorphism of traces (for the definition, see 3.7 below),
(β) ψ is an order isomorphism for natural partial orders.

(vi) Show that there exist nonisomorphic inverse semigroups with isomorphic traces and isomorphic associated partially ordered sets.

(vii) Prove 3.3 directly (that is, without using 2.9).

XIV.3.7 Problems

(i) What are necessary and sufficient conditions on inverse semigroups S and T in order that their traces (respectively, their traces and idempotents) be isomorphic? For the definition of isomorphism φ of the partial groupoids G and H: ab defined in G if and only if $(a\varphi)(b\varphi)$ defined in H and then $(ab)\varphi = (a\varphi)(b\varphi)$.

Result 3.3 can be found in Meakin [2], whereas 3.5 is due to Schein [10] with a direct proof. Structures related to inductive groupoids were studied by Dubikajtis [1], [2], Dubikajtis–Jarek [1], Ehresmann [1], [2]. For another kind of construction of inverse semigroups, consult Grillet [2].

XIV.4 A CONSTRUCTION OF STRICT INVERSE SEMIGROUPS

We have seen in II.4.5 that for a strict inverse semigroup there exist functions among some $\mathcal{D}$-classes which preserve certain products; see also II.4.6. Based on these functions, we give below a structure theorem for strict inverse semigroups. It is convenient to introduce the following concept.

XIV.4.1 Definition

For a strict inverse semigroup S, we call the functions $\varphi: J_b \to J_a$ appearing in II.4.5 the *$\mathcal{J}$-structure mappings* for S.

Further properties of these functions can be found in IV.4.6.

XIV.4.2 Construction

Let $S = \Sigma_{\alpha \in P} B_\alpha$ be a Croisot groupoid, where each B_α is a Brandt groupoid and P is a partially ordered set. Let E be the set of idempotents of S and assume that E is a semilattice under an operation $\wedge$ with no two distinct idempotents in the same B_α comparable. For each pair $\alpha, \beta \in P$, $\alpha \geq \beta$, let $\chi_{\alpha,\beta}: B_\alpha \to B_\beta$ be a partial homomorphism. Assume that for any $\alpha, \beta, \gamma \in P$ the following conditions hold.

(C1) $\chi_{\alpha,\alpha} = \iota_{B_\alpha}$.
(C2) $\chi_{\alpha,\beta}\chi_{\beta,\gamma} = \chi_{\alpha,\gamma}$ if $\alpha > \beta > \gamma$.
(C3) $\alpha \geq \beta$, $e\chi_{\alpha,\beta} = f \Leftrightarrow e \geq f$ where $e \in E_{B_\alpha}, f \in E_{B_\beta}$.

On S define a multiplication $\circ$ by: for $a \in B_\alpha$, $b \in B_\beta$,

$$a \circ b = (a\chi_{\alpha,\gamma})(b\chi_{\beta,\gamma}) \tag{17}$$

where $a\rho \wedge \lambda b \in B_\gamma$.

XIV.4.3 Theorem

The operation $\circ$ defined in 4.2 is the unique multiplication on S for which $(S, \circ)$ is a strict inverse semigroup with trace S, semilattice of idempotents E, and $\mathcal{J}$-structure mappings $\varphi_{\alpha,\beta}$.

Conversely, every strict inverse semigroup can be so constructed.

Proof. Let S be as in 4.2. For any $e, f \in E_S$ such that $e \geq f$, define

$$\varphi_{e,f} = \chi_{\alpha,\beta}|_{R_e} \quad \text{if} \quad e \in B_\alpha, f \in B_\beta.$$

Conditions (S1), (S2), and (S3) in 2.4 follow directly from (C1), (C2), and (C3) in 4.2, respectively. In order to verify condition (S4), we let $a\rho = \lambda b = e \geq f$ and $\bar{f} = (a^{-1}\varphi_{e,f})\rho$, where $e \in B_\alpha$ and $f \in B_\beta$. Since $\chi_{\alpha,\beta}$ is a partial homomorphism, we have

$$\begin{aligned}(\lambda a)\chi_{\alpha,\beta} &= (aa^{-1})\chi_{\alpha,\beta} = (a\chi_{\alpha,\beta})(a^{-1}\chi_{\alpha,\beta}) = (a\chi_{\alpha,\beta})(a\chi_{\alpha,\beta})^{-1} \\ &= \lambda(a\chi_{\alpha,\beta}).\end{aligned}$$

Now condition (C3) with $e = \lambda a$ and $f = \lambda(a\chi_{\alpha,\beta})$ gives $\lambda a \geq \lambda(a\chi_{\alpha,\beta})$. This can be written as $\lambda a \geq (a^{-1}\varphi_{e,f})\rho = \bar{f}$ giving the first assertion in (S4). Furthermore,

$$(ab)\varphi_{\lambda a,\bar{f}} = (ab)\chi_{\alpha,\beta} = (a\chi_{\alpha,\beta})(b\chi_{\alpha,\beta})$$

$$= \left(a^{-1}\chi_{\alpha,\beta}\right)^{-1}(b\chi_{\alpha,\beta}) = \left(a^{-1}\varphi_{e,f}\right)^{-1}(b\varphi_{e,f}),$$

as required.

We have proved that $(S, E; \varphi_{e,f})$ is an inverse triple. The product formula (17) evidently coincides with the multiplication (4) in 2.9. It follows from 2.9 that $(S, \circ)$ is the unique inverse semigroup with trace S, semilattice of idempotents E, and structure mappings $\varphi_{e,f}$. Let $e, f, g \in E$ be such that $e \geq f$, $e \geq g$, $f \mathcal{D} g$. Then $e \in B_\alpha$ and $f, g \in B_\beta$ for some $\alpha, \beta \in P$. Condition (C3) implies that $f = g$. Now II.4.5 yields that S is strict. The given mappings $\chi_{\alpha,\beta}$ are easily seen to be the $\mathcal{J}$-structure mappings for $(S, \circ)$. The uniqueness of $(S, \circ)$ as asserted in the statement of the theorem now follows from the uniqueness of $(S, \circ)$ invoked above.

Conversely, let S be a strict inverse semigroup. Its trace $S = \Sigma_{\alpha \in P} B_\alpha$ is a Croisot groupoid, where the B_α are the Brandt groupoids which are the $\mathcal{J}$-classes of S, and $P = S/\mathcal{J}$ with the induced partial order. No two distinct $\mathcal{J}$-related idempotents of S are comparable since S is completely semisimple. The existence of $\mathcal{J}$-structure mappings is guaranteed by II.4.5; they are partial homomorphisms by II.4.6(iii). Condition (C1) follows from the proof of II.4.5, condition (C2) from II.4.6(v), and the product formula (17) from II.4.6(iv).

Let $e \in E_{B_\alpha}$ and $f \in E_{B_\beta}$. If $\alpha \geq \beta$ and $e\chi_{\alpha,\beta} = f$, then

$$f = e\chi_{\alpha,\beta} = (e\chi_{\alpha,\beta})(e\chi_{\alpha,\beta}) \in E_{B_\beta}$$

which by II.4.6(i) and (ii) implies $f = e(e\chi_{\alpha,\beta}) = ef$ and thus $e \geq f$. Conversely, assume that $e \geq f$. Then $\alpha \geq \beta$ and by II.4.5(iv) (β) and II.4.6(i), we have $f = ef = (e\chi_{\alpha,\beta})f$ which implies that $f = e\chi_{\alpha,\beta}$ since $e\chi_{\alpha,\beta}$ and f are nonzero idempotents of a Brandt semigroup. This verifies condition (C3).

The $\mathcal{J}$-structure mappings in 4.3 are partial homomorphisms of Brandt groupoids. They can be constructed as follows.

XIV.4.4 Lemma

Let $S = B(G, I)$ and $T = B(H, J)$ be Brandt semigroups, $\xi: I \to J$ and $u: I \to H$ be functions, in notation $u: i \to u_i$, and $\omega: G \to H$ be a homomorphism. Define a mapping χ by

$$\chi: (i, g, j) \to \left(i\xi, u_i^{-1}(g\omega)u_j, j\xi\right) \qquad [(i, g, j) \in S^*].$$

Then χ is a partial homomorphism of S^* into T^*, and conversely, every partial homomorphism of S^* into T^* can be so constructed.

Proof. The argument here is very close to that in II.3.7 and VII.4.9 and is left as an exercise.

XIV.4.5 Exercises

(i) Let S be a strict inverse semigroup. Show that the $\mathcal{J}$-structure mappings of S are one-to-one if and only if S is E-unitary and satisfies: $e \geq g, f \geq g, e\mathcal{D}f$ implies $e = f$.

(ii) Specialize 4.2 to obtain the Clifford representation of a Clifford semigroup.

BIBLIOGRAPHY

Adjan, S. I.

[1] Identities in special semigroups, *Doklady Akad. Nauk SSSR* **143** (1962), 499–502 (Russian).

[2] Defining relations and algorithmic problems for groups and semigroups, *Trudy Matem. Inst. Steklova* **85** (1966), 1–123 (Russian).

Aĭzenštat, A. Ja. and B. K. Boguta

[1] On the lattice of varieties of semigroups, in *Semigroup Varieties and Endomorphism Semigroups*, Leningrad Ped. Inst. (1979), 3–46 (Russian).

Alimpić, B. P. and D. N. Krgović

[1] Idempotent pure congruences on Clifford semigroups, in *Algebra Conf.*, Novi Sad (1981), 13–18.

Allouch, D.

[1] Extensions de demi-groupes inverses, *Semigroup Forum* **16** (1978), 111–116.

[2] Sur les extensions de demi-groupes strictement réguliers, Doctoral Dissertation, University of Montpellier, 1979.

Andersen, O.

[1] Ein Bericht über die Struktur abstrakter Halbgruppen, Staatsexamensarbeit, Hamburg, 1952.

Anderson, J. A.

[1] Characters of commutative semigroups, *Math. Sem. Notes*, Kobe University **7** (1979), 301–308.

Anderson, L. W., R. P. Hunter and R. J. Koch

[1] Some results on stability in semigroups, *Trans. Amer. Math. Soc.* **117** (1965), 521–529.

Applebaum, C. H.

[1] Some structure theorems for inverse ω-semigroups, *Fund. Math.* **99** (1978), 79–91.

Ash, C. J.

[1] Dense, uniform and densely subuniform chains, *J. Austral. Math. Soc.* **23** (1977), 1–8.

[2] A uniform chain whose inverse semigroup has no chart, *Semigroup Forum* **18** (1979), 1–4.

[3] Uniform labelled semilattices, *J. Austral. Math. Soc.* **28** (1979), 385–397.

[4] The $\mathcal{J}$-classes of an inverse semigroup, *J. Austral. Math. Soc.* **28** (1979), 427–432.

Ash, C. J. and T. E. Hall

[1] Inverse semigroups on graphs, *Semigroup Forum* **11** (1975), 140–145.

Ault, J. E.

[1] Extensions of primitive inverse semigroups, Doctoral Dissertation, Pennsylvania State University, 1970.

[2] Extensions of one primitive inverse semigroup by another, *Canad. J. Math.* **24** (1972), 270–278.

[3] The translational hull of an inverse semigroup, *Glasgow Math. J.* **14** (1973), 56–64; announced in *Semigroup Forum* **4** (1972), 165–168.

[4] Semigroups with bisimple and simple ω-subsemigroups, *Semigroup Forum* **9** (1975), 318–333.

[5] Group congruences on a bisimple ω-semigroup, *Semigroup Forum* **10** (1975), 351–366.

Ault, J. E. and M. Petrich

[1] The structure of ω-regular semigroups, *J. Reine Angew. Math.* **251** (1971), 110–141; announced in *Bull. Amer. Math. Soc.* **77** (1971), 196–199.

Baird, B. B.

[1] Inverse semigroups of homeomorphisms between open subsets, *J. Austral. Math. Soc.* **24** (1977), 92–102.

[2] Inverse semigroups of homeomorphisms are Hopfian, *Canad. J. Math.* **31** (1979), 800–807.

[3] The group of units as a maximal inverse subsemigroup, *Semigroup Forum* **25** (1982), 381–382.

[4] Maximal inverse subsemigroups of $S(X)$, *Glasgow Math. J.* **24** (1983), 53–64.

Baird, G. R.

[1] On a sublattice of the lattice of congruences on a simple regular ω-semigroup, *J. Austral. Math. Soc.* **13** (1972), 461–471.

[2] Congruences on simple regular ω-semigroups, *J. Austral. Math. Soc.* **14** (1972), 155–167.

[3] On 0-simple semigroups, *Semigroup Forum* **5** (1973), 270–272.

[4] Congruence-free inverse semigroups with zero, *J. Austral. Math. Soc.* **20** (1975), 110–114.

Bales, J. L.

[1] On product varieties of inverse semigroups, *J. Austral. Math. Soc.* **28** (1979), 107–119.

Barnes, B. A.

[1] Representations of the l^1-algebra of an inverse semigroup, *Trans. Amer. Math. Soc.* **218** (1976), 361–396.

Batbedat, A.

[1] Le produit demi-direct pour les demi-groupes inverses, *Semigroup Forum* **17** (1979), 283–305.

Beĭda, A. A.

[1] Elementary inverse semigroups, *Vescī Akad. Navuk BSSR, Ser. Fiz.-Matem. Navuk issue* **6** (1981), 58–63 (Russian; English summary).

Birkhoff, G.

[1] *Lattice Theory*, 3rd ed., Amer. Math. Soc. Coll. Publ., Providence, 1967.

Biró, B., E. W. Kiss and P. P. Pálfy

[1] On the congruence extension property, in *Universal Algebra*, North Holland, Amsterdam (1981), 129–151 (Proc. Conf. Esztergom 1977); announced in *Semigroup Forum* **15** (1977), 183–184.

Blyth, T. S.

[1] On the greatest isotone homomorphic group image of an inverse semigroup, *J. London Math. Soc.* (2)**1** (1969), 260–264.

[2] Residuated inverse semigroups, *J. London Math. Soc.* (2)**1** (1969), 243–248; correction: ibid. **2** (1970), 592.

Bonzini, C.

[1] Sulla somma di reticoli, *Rend. Ist. Lombardo, Accad. Sci. Lett.* **A103** (1969), 871–885.

Brandt, H.

[1] Über eine Verallgemeinerung des Gruppenbegriffs, *Math. Ann.* **96** (1927), 360–365.

[2] Über die Axiome des Gruppoids, *Vierteljsch. Naturforsch. Ges. Zürich* **85** (1940), 95–104.

Brashear, P. W.

[1] An embedding for inverse Clifford semigroup, *J. Natur. Sci. Math.* **16** (1976), 87–90.

Bredihin, D. A.

[1] Inverse semigroups of automorphisms of universal algebras, *Sibir. Matem. Ž.* **17** (1976), 499–507 (Russian).

[2] Inverse semigroups of local automorphisms, *Uspehi Matem. Nauk* **34** (208) (1979), 181–182 (Russian).

Bruck, R. H.

[1] *A survey of binary systems*, Ergeb. Math., Heft 20, Springer, Berlin, 1958.

Bruns, G. and H. Lakser

[1] Injective hulls of semilattices, *Canad. Math. Bull.* **13** (1970), 115–118.

Bullman-Fleming, S. and K. McDowell

[1] Flatness and projectivity in commutative regular semigroups, *Semigroup Forum* **19** (1980), 247–259.

Burgess, D. C. J.

[1] Note on the preceding paper, *Proc. Cambridge Phil. Soc.* **57** (1961), 237–238.

Burgess, W. D.

[1] The injective hull of S-sets, S a semilattice of groups, *Semigroup Forum* **23** (1981), 241–246.

Byleen, K. E.

[1] The structure of regular and inverse semigroups, Doctoral Dissertation, University of Nebraska, 1977.

Byleen, K. and P. Komjáth

[1] The admissible trace problem for E-unitary inverse semigroups, *Semigroup Forum* **15** (1978), 235–246.

Cameron, P. J., and M. Deza

[1] On permutation geometries, *J. London Math. Soc.* (2) **20** (1979), 373–386.

Ceresa-Genet, G.

[1] Schemi in semigruppi inversi, *Rend. Ist. Lombardo, Accad. Sci. Lett.* **A109** (1975), 201–212.

Chen, S. Y. and S. C. Hsieh

[1] Factorizable inverse semigroups, *Semigroup Forum* **8** (1974), 283–297.

Clement, A. and F. Pastijn

[1] Inverse semigroups with idempotents dually well-ordered, *J. Austral. Math. Soc.* **35** (1983), 373–385.

Clifford, A. H.

[1] Semigroups admitting relative inverses, *Annals Math.* **42** (1941), 1037–1049.

[2] Matrix representations of completely simple semigroups, *Amer. J. Math.* **64** (1942), 327–342.

[3] A class of d-simple semigroups, *Amer. J. Math.* **15** (1953), 541–556.

[4] Partial homomorphic images of Brandt groupoids, *Proc. Amer. Math. Soc.* **16** (1965), 538–544.

Clifford, A. H. and G. B. Preston

[1] *The Algebraic Theory of Semigroups*, Math. Surveys No. 7, Amer. Math. Soc., Providence, Vol. I (1961), Vol. II (1967).

Coudron, A.

[1] Sur les extensions de demi-groupes réciproques, *Bull. Soc. Roy. Sci. Liège* **31** (1968), 409–419.

Crawley, J. W.

[1] Epimorphisms and semilattices of groups, *Semigroup Forum* **21** (1980), 95–112.

Croisot, R.

[1] Une interprétation des relations d'équivalences dans un ensemble, *C. R. Acad. Sci. Paris* **226** (1948), 616–617.

[2] Equivalences principales bilatères définies dans un demi-groupe, *J. Math. Pures Appl.* **36** (1957), 373–417.

D'Alarcao, H.

[1] Idempotent-separating extensions of inverse semigroups, *J. Austral. Math. Soc.* **9** (1969), 211–217.

Dean, R. A. and R. H. Oehmke

[1] Idempotent semigroups with distributive right congruence lattices, *Pacific J. Math* **14** (1964), 1187–1209.

Djadčenko, G. G.

[1] On identities in monogenic inverse semigroups, in *Algebra and Number Theory*, Nalčik **2** (1977), 57–77 (Russian).

[2] Construction of monogenic inverse semigroups, *Zap. Naučn. Sem. Leningrad. Otdel. Matem. Inst. Steklova* **103** (1980), 66–75 (Russian).

Djadčenko, G. G. and B. M. Schein

[1] Monogenic inverse semigroups, in *Algebra and Number Theory*, Nalčik **1** (1973), 3–26 (Russian).

[2] The structure of finite monogenic inverse semigroups, *Matem. Zap. Ural. Univ.* **9** (1974), 15–21 (Russian).

Domanov, O. I.

[1] Semigroups of all partial automorphisms of universal algebras, *Izv. Vysš. Učebn. Zav. Matem.* issue **8** (1971), 52–58 (Russian).

[2] The semisimplicity and identities of the semigroup algebra of inverse semigroups, *Matem. Issled.* **38** (1976), 123–131 (Russian).

Dubikajtis, L.

[1] Groupoïde inductif élémentaire et pseudogroupe élémentaire, *Bull. Acad. Pol. Sci., Série Math.* **11** (1963), 469–472.

[2] Certaines extensions de la notion de groupoïde inductif et de celle de pseudogroupe, *Colloq. Math.* **12** (1964), 163–185.

Dubikajtis, L. and P. Jarek

[1] Pseudogroupe élémentaire commutatif et semi-groupe régulier commutatif, *Colloq. Math.* **12** (1964), 187–193.

Eberhart, C. and J. Selden

[1] One parameter inverse semigroups, *Trans. Amer. Math. Soc.* **168** (1972), 53–66.

Edwards, C. E.

[1] A simplification of D'Alarcao's idempotent separating extensions of inverse semigroups, *Internat. J. Math. Sci.* **1** (1978), 393–396.

Ehresmann, C.

[1] Gattungen von lokalen Strukturen, *Jahresb. Dtsch. Math.-Ver.* **60** (1957), 49–77.

[2] Catégories inductives et pseudogroupes, *Ann. Inst. Fourier* **10** (1960), 307–332.

Eršova, T. I.

[1] Inverse semigroups with a certain type of lattice of inverse subsemigroups, *Matem. Zap. Ural. Univ.* **7** (1969), 62–76 (Russian).

[2] Determinability of monogenic inverse semigroups by the lattice of inverse subsemigroups, *Matem. Zap. Ural. Univ.* **8** (1971), 34–49 (Russian).

[3] On monogenic inverse semigroups, *Matem. Zap. Ural. Univ.* **8** (1971), 30–33 (Russian).

[4] Lattice isomorphisms of inverse semigroups, *Izv. Vysš. Učebn. Zav. Matem.* issue **9** (1972), 25–32 (Russian).

[5] Inverse semigroups with a finiteness condition, *Izv. Vysš. Učebn. Zav. Matem.* issue **11** (1977), 7–14 (Russian).

[6] Lattice isomorphisms of Brandt semigroups, *Uč. Zap. Ural. Univ.* **13** (1982), 27–39 (Russian).

Evans, T.

[1] The lattice of semigroup varieties, *Semigroup Forum* **2** (1971), 1–43.

Feller, E. H. and R. L. Gantos

[1] Completely injective semigroups, *Pacific J. Math.* **31** (1969), 359–366.

Fortunatov, V. A.

[1] On semidirect products of a semilattice and a group, in *Semigroup Theory and its Appl.*, Saratov University **3** (1974), 129–139 (Russian).

[2] Perfect congruences on monogenic inverse semigroups, in *Algebra and Number Theory*, Nalčik **4** (1978), 93–98 (Russian).

Fountain, J. B. and P. Lockley

[1] Semilattices of groups with distributive congruence lattices, *Semigroup Forum* **14** (1977), 81–91.

Fotedar, G. L.

[1] On a class of simple inverse semigroup, *Riv. Mat. Univ. Parma* **4** (1978), 49–53.

Freese, R. S. and J. B. Nation

[1] Congruence lattices of semilattices, *Pacific J. Math.* **49** (1973), 51–58.

Fulp, R. O.

[1] On homomorphisms of commutative inverse semigroups, *Czechoslovak Math. J.* **16** (1966), 72–75.

Gantos, R. L.

[1] Semilattices of bisimple inverse semigroups, *Quart. J. Math. Oxford* (2)**22** (1971), 379–393.

Gelbaum, B. R. and S. A. Schanuel

[1] Characterization of the semigroup of matrix units, *J. Austral. Math. Soc.* **29** (1980), 291–296.

Gluskin, L. M.

[1] Simple semigroups with zero, *Doklady Akad. Nauk SSSR* **103** (1955), 5–8 (Russian).

[2] Elementary generalized groups, *Matem. Sbornik* **41** (1957), 23–36 (Russian).

[3] On inverse semigroups, *Zap. Mat.-Meh. Fak. i Harkov. Mat. Obšč.* (4)**28** (1961), 103–110 (Russian).

[4] On the question of elementary inverse semigroups, *Zap. Meh.-Matem. Fak. Harkov. Univ.* **29** (1963), 143–144 (Russian).

[5] On dense embeddings, *Trudy Mosk. Matem. Obšč.* **29** (1973), 119–131 (Russian).

[6] Extensions of commutative semigroups, *Izv. Vysš. Učebn. Zav. Matem.* issue **11** (1979), 3–12; announced in *Doklady Akad. Nauk SSSR* **230** (1976), 757–760 (Russian).

[7] Dense extensions of commutative semigroups, in *Algebraic Theory of Semigroups*, North Holland, Amsterdam (1976), 155–171 (Proc. Conf. Szeged University 1976).

Goberstein, S. M.

[1] Φ-simple inverse semigroups, in *Semigroup Theory and its Appl.*, Saratov University **4** (1978), 11–18 (Russian).

[2] Fundamental order relations on right inverse semigroups, in *Contemporary Algebra*, Leningrad Ped. Inst. **5** (1976), 20–39 (Russian).

[3] Fundamental order relations on inverse semigroups and on their generalizations, *Semigroup Forum* **21** (1980), 317–328.

[4] On the modularity of the lattice of fundamental orders on a semilattice, *Semigroup Forum* **24** (1982), 83–86.

Golubov, E. A.

[1] On residual finiteness of regular semigroups, *Matem. Zametki* **17** (1975), 423–432 (Russian).

Goralčik, P.

[1] One remarkable property of the bicyclic semigroup, *Comment. Math. Univ. Carolinae* **12** (1971), 503–518.

Gould, M.

[1] An easy proof of Ponizovski's theorem, *Semigroup Forum* **15** (1977), 181–182.

Grätzer, G.

[1] *Universal Algebra*, 2nd ed., Springer, New York, Heidelberg, Berlin (1979).

[2] *General Lattice Theory*, Academic Press, New York (1978).

Green, D. G.

[1] Extensions of a semilattice by an inverse semigroup, *Bull. Austral. Math. Soc.* **9** (1973), 21–31.

[2] The lattice of congruences on an inverse semigroup, *Pacific J. Math.* **57** (1975), 141–152.

Grillet, P. A.

[1] Left coset extensions, *Semigroup Forum* **7** (1974), 200–263.

[2] A construction of inverse semigroups, *Semigroup Forum* **8** (1974), 169–176.

Grossman, P. A. and H. Lausch

[1] Interpolation on semilattices, in *Semigroups*, Academic Press, New York (1980), 57–65 (Proc. Conf. Monash University 1979).

Hall, T. E.

[1] On the natural order of $\mathcal{J}$-classes and of idempotents in a regular semigroup, *Glasgow Math. J.* **11** (1970), 167–168.

[2] On the lattice of congruences on a semilattice, *J. Austral. Math. Soc.* **12** (1971), 456–460.

[3] The partially ordered set of all $\mathcal{J}$-classes of a finite semigroup, *Semigroup Forum* **6** (1973), 263–264.

[4] Free products with amalgamation of inverse semigroups, *J. Algebra* **34** (1975), 375–385.

[5] Representation extension and amalgamation for semigroups, *Quart. J. Math. Oxford* (2)**29** (1978), 309–334.

[6] Inverse semigroup varieties with the amalgamation properties, *Semigroup Forum* **16** (1978), 37–51.

[7] Amalgamation and inverse and regular semigroups, *Trans. Amer. Math. Soc.* **246** (1978), 395–406.

[8] Inverse and regular semigroups and amalgamation, in *Proc. Symp. Regular Semigroups*, Northern Illinois University (1979), 50–79.

[9] Generalized inverse semigroups and amalgamation, in *Semigroups*, Academic Press, New York (1980), 145–157 (Proc. Conf. Monash University 1979).

Halmos, P. R.

[1] *Finite-dimensional Vector Spaces*, Van Nostrand, Princeton, Toronto, London (1958).

Hamilton, H. B.

[1] Semilattices whose structure lattice is distributive, *Semigroup Forum* **8** (1974), 245–253.

Hamilton, H. B. and T. Tamura

[1] Finite inverse perfect semigroups and their congruences, *J. Austral. Math. Soc.* **32** (1982), 114–128.

Hardy, D. W. and Y. Tirasupa

[1] Semilattices of proper inverse semigroups, *Semigroup Forum* **13** (1976), 29–36.

Head, T. and J. A. Anderson

[1] On injective and flat commutative regular semigroups, *Semigroup Forum* **21** (1980), 283–284.

Hickey, J. B.

[1] Bisimple inverse semigroups and uniform semilattices, Doctoral Dissertation, University of Glasgow, 1970.

[2] A type of product for uniform semilattices, *Semigroup Forum* **6** (1973), 198–215.

[3] A family of bisimple inverse semigroups, *Semigroup Forum* **11** (1975), 30–48.

Hinkle, C. V.

[1] Semigroups of right quotients of a semigroup which is a semilattice of groups, *J. Algebra* **31** (1974), 276–286; announced in *Semigroup Forum* **5** (1972/73), 167–173.

Hoehnke, H.-J.

[1] Über die Erzeugung von monomialen Gruppendarstellungen durch Brandtsche Gruppoide, *Math. Zeitschr.* **77** (1961), 68–80.

[2] Zur Theorie der Gruppoide. I., *Math. Nachr.* **24** (1962), 137–168.

[3] Zur Theorie der Gruppoide. VI, *Math. Nachr.* **25** (1963), 191–198.

[4] Zur Theorie der Gruppoide. VII, *Math. Nachr.* **27** (1963/64), 289–298.

[5] Über antiautomorphe und involutorische primitive Halbgruppen, *Czechoslovak J. Math.* **15** (1965), 50–63.

Hogan, J. W.

[1] Homomorphisms of ω^n-bisimple semigroups, *J. Natur. Sci. Math.* **12** (1972), 159–167.

[2] Homomorphisms of certain right cancellative semigroups, *J. Natur. Sci. Math.* **12** (1972), 160–167.

[3] Homomorphisms and congruences on ω^α-bisimple semigroups, *J. Austral. Math. Soc.* **15** (1973), 441–460.

[4] Bisimple semigroups with idempotents well-ordered, *Semigroup Forum* **6** (1973), 298–316.

Horn, A. and N. Kimura

[1] The category of semilattices, *Algebra Univ.* **1** (1971), 26–38.

Houghton, C. H.

[1] Embedding inverse semigroups in wreath products, *Glasgow Math J.* **17** (1976), 77–82.

Howie, J. M.

[1] The maximum idempotent separating congruence on an inverse semigroup, *Proc. Edinburgh Math. Soc.* (2)**14** (1964), 71–79.

[2] Commutative semigroup amalgams, *J. Austral. Math. Soc.* **8** (1968), 609–630.

[3] Semigroup amalgams whose cores are inverse semigroups, *Quart J. Math. Oxford* (2)**26** (1975), 23–45.

[4] *An Introduction to Semigroup Theory*, Academic Press, London (1976).

[5] A congruence-free inverse semigroup associated with a pair of infinite cardinals, *J. Austral. Math. Soc.* **31** (1981), 337–342.

Howie, J. M. and G. Lallement

[1] Certain fundamental congruences on a regular semigroup, *Proc. Glasgow Math. Assoc.* **7** (1966), 145–159.

Howie, J. M. and B. M. Schein

[1] Anti-uniform semilattices, *Bull. Austral. Math. Soc.* **1** (1969), 263–268.

Hsieh, S. C.

[1] Contributions to the theory of semilattices, Doctoral Dissertation, University of South Carolina, 1970.

[2] An embedding theorem for semilattices, *Tamkang J. Math.* **2** (1971), 177–180.

Imaoka, T.

[1] Free products with amalgamation of semigroups, Doctoral Dissertation, Monash University, 1978.

[2] Free products with amalgamation of commutative inverse semigroups, *J. Austral. Math. Soc.* **22** (1976), 246–251.

Inasaridze, H. N.

[1] On some questions in the theory of semigroups, *Works of Stalin University*, Tbilis **76** (1959), 247–260 (Russian).

[2] On simple semigroups, *Matem. Sbornik* **57** (1962), 225–232 (Russian); *Amer. Math. Soc. Transl.* (2)**45** (1965), 147–155.

[3] Extensions of commutative inverse semigroups, *Sakharth. SSR Mecn. Akad. Moambe* **46** (1967), 11–18 (Russian).

Iyengar, H. R. K.
[1] Semilattices of bisimple inverse semigroups, *Quart. J. Math. Oxford* (2)**23** (1972), 299–318; announced in *Semigroup Forum* **2** (1971), 44–48.
[2] Homomorphisms of ω-regular and I-regular semigroups, *Math. Japon.* **17** (1972), 85–103.

Jacobson, N.
[1] *Structure of Rings*, Amer. Math. Soc. Coll. Publ. Vol. 37, Providence (1956).

Jarek, P.
[1] Commutative regular semigroups, *Colloq. Math.* **12** (1964), 195–208.

Johnson, C. S. and F. R. McMorris
[1] Completely cyclic injective semilattices, *Proc. Amer. Math. Soc.* **36** (1972), 385–388.
[2] Injective hulls of certain S-systems over a semilattice, *Proc. Amer. Math. Soc.* **32** (1972), 371–375.
[3] Weakly self-injective semilattices, *Semigroup Forum* **6** (1973), 3–11.
[4] Nonsingular semilattices and semigroups, *Czechoslovak Math. J.* **26** (1976), 280–282.

Jones, P. R.
[1] Inverse subsemigroups of free inverse semigroups, Doctoral Dissertation, Monash University, 1975.
[2] The lattice of inverse subsemigroups of a reduced inverse semigroup, *Glasgow Math. J.* **17** (1976), 161–172.
[3] A basis theorem for free inverse semigroups, *J. Algebra* **49** (1977), 172–190.
[4] Basis properties for inverse semigroups, *J. Algebra* **50** (1978), 135–152.
[5] Semimodular inverse semigroups, *J. London Math. Soc.* **17** (1978), 446–456.
[6] Distributive inverse semigroups, *J. London Math. Soc.* **17** (1978), 457–466.
[7] Lattice isomorphisms of inverse semigroups, *Proc. Edinburgh Math. Soc.* **21** (1978), 149–157.
[8] The Hopf property and free products of semigroups, *J. Austral. Math. Soc.* **27** (1979), 358–364.
[9] Lattice isomorphisms of distributive inverse semigroups, *Quart. J. Math. Oxford* (2)**30** (1979), 301–314.
[10] Congruences contained in an equivalence relation on a semigroup, *J. Austral. Math. Soc.* **29** (1980), 162–176.
[11] The Hopf property and $\underset{\sim}{K}$-free products of semigroups, *Semigroup Forum* **20** (1980), 343–368.
[12] Inverse semigroups determined by their lattices of inverse subsemigroups, *J. Austral. Math. Soc.* **30** (1981), 321–346.
[13] Inverse semigroups whose full inverse subsemigroups form a chain, *Glasgow Math. J.* **22** (1981), 159–165.
[14] A graphical representation for the free product of E-unitary inverse semigroups, *Semigroup Forum* **24** (1982), 195–221.

Jónsson, B.
[1] Extensions of relational structures, in *The Theory of Models*, North Holland, Amsterdam (1965), 146–157 (Proc. Int. Symp. Berkeley 1963).

Jordan, M. J.
[1] Inverse H-semigroups and t-semisimple inverse H-semigroups, *Trans. Amer. Math. Soc.* **163** (1972), 75–84.

Justin, J.
[1] Sur une construction de Bruck et Reilly, *Semigroup Forum* **3** (1971), 148–155.

Jürgensen, H.
[1] Inf-Halbverbände als syntaktische Halbgruppen, *Acta Math. Acad. Sci. Hung.* **31** (1978), 37–41.
[2] Disjunktive Teilmengen inverser Halbgruppen, *Archiv Math.*, Brno **15** (1979), 205–208.

Karvellas, P. H.

[1] Inversive semirings, *J. Austral. Math. Soc.* **18** (1974), 277–288.

Kastl, J.

[1] One method of representation of inverse semigroups, in *Seminar on Semigroup Theory*, Japan (1976), 73–89.

Keenan, M. and G. Lallement

[1] On certain codes admitting inverse semigroups as syntactic monoids, *Semigroup Forum* **8** (1974), 312–331.

Kimura, N.

[1] On semigroups, Doctoral Dissertation, Tulane University, 1957.

Kleiman, E. I.

[1] On free inverse semigroups, *Matem. Zap. Ural. Univ.* **8** (1972), 49–72 (Russian).

[2] On basis of identities of Brandt semigroups, *Semigroup Forum* **13** (1977), 209–218.

[3] Some properties of the lattice of varieties of inverse semigroups, in *Research in Modern Algebra*, Sverdlovsk University (1977), 56–72; correction: ibid. (1979), 207; announced in On the lattice of varieties of inverse semigroups, *Izv. Vysš. Učebn. Zav. Matem.* issue **7** (1976), 106–109 (Russian).

[4] On basis of identities of inverse semigroup varieties, *Sibir. Matem. Ž.* **20** (1979), 760–777 (Russian).

[5] On a covering condition in the lattice of varieties of inverse semigroups, in *Investigations of Algebraic Systems by Properties of Their Subsystems*, Sverdlovsk University (1980) 76–91 (Russian).

[6] On the join of group varieties and combinatorial varieties of inverse semigroups, *Sverdlovsk Ped. Inst.* (VINITI Publ.) (1980), 13 pp. (Russian).

[7] Two theorems on combinatorial varieties of inverse semigroups, *Sverdlovsk Ped. Inst.* (VINITI Publ.) (1980), 12 pp. (Russian).

[8] On a condition for covering in the lattice of varieties of inverse semigroups, *Uč. Zap. Ural. Univ.* **12** (1980), 76–91 (Russian).

[9] On a pseudo-variety generated by a finite semigroup, *Uč. Zap. Ural. Univ.* **13** (1982), 40–42 (Russian).

Klein-Barmen, F.

[1] Über gewisse Halbverbände und kommutative Semigruppen, I, *Math. Zeitschr.* **48** (1942), 275–288.

[2] Über gewisse Halbverbände und kommutative Semigruppen, II, *Math. Zeitschr.* **41** (1948), 355–366.

Klun, M. J.

[1] A characterization of varieties of inverse semigroups, *Semigroup Forum* **10** (1975), 1–7.

Knox, N.

[1] The inverse semigroup coproduct of an arbitrary non-empty collection of groups, Doctoral Dissertation, Tennessee State University, 1974.

[2] The inverse hull of the free semigroup on a set X, *Semigroup Forum* **16** (1978), 345–354.

Kočin, B. P.

[1] The structure of inverse ideally-simple ω-semigroups, *Vestnik Leningrad. Univ.* **23**, issue 7 (1968), 41–50 (Russian).

Köhler, P.

[1] Quasi-decompositions of semigroups, *Houston J. Math.* **5** (1979), 525–542.

Kowol, G. and H. Mitsch

[1] Nilpotent inverse semigroups with central idempotents, *Trans. Amer. Math. Soc.* **271** (1982), 437–449.

Koževnikov, O. B.

[1] Generalized Brandt semigroups, in *Modern Algebra*, Leningrad Ped. Inst. (1974), 57–69 (Russian).

Kožuhov, I. B.

[1] Self-injective semigroup rings of inverse semigroups, *Izv. Vysš. Učebn. Zav. Matem.* issue **2** (1981), 46–51 (Russian).

Kupcov, A. I.

[1] Certain properties of endomorphism semigroups of commutative regular semigroups, in *Semigroup Varieties and Endomorphism Semigroups*, Leningrad Ped. Inst. (1979), 67–79 (Russian).

[2] On a property of endomorphism semigroups of commutative regular semigroups, in *Semigroup Varieties and Endomorphism Semigroups*, Leningrad Ped. Inst. (1979), 80 (Russian).

[3] On periodicity of the semigroup of homomorphisms of commutative regular semigroups, in *Contemporary Algebra, Groupoids and Their Homomorphisms*, Leningrad Ped. Inst. (1980), 89–98 (Russian).

[4] On certain properties of functors h^A and h_C in the category of commutative regular semigroups, in *Modern Algebra, Semigroup Constructions*, Leningrad Ped. Inst. (1981), 17–22 (Russian).

Lajos, S.

[1] On semilattices of groups, *Proc. Japan Acad.* **45** (1969), 383–384.

[2] Characterization of completely regular inverse semigroups, *Acta Sci. Math.*, Szeged **31** (1970), 229–231.

[3] Semigroups that are semilattices of groups, *Math. Japon.* **16** (1971), 25–34.

[4] On semilattices of groups. II, *Proc. Japan Acad.* **47** (1971), 36–37.

[5] A remark on semilattices of groups, *Ann. Univ. Sci.*, Budapest **15** (1972), 121–122.

[6] A note on semilattices of groups, *Acta Sci. Math.*, Szeged **33** (1972), 315–317.

Lallement, G.

[1] Congruences et équivalences de Green sur un demi-groupe régulier, *C. R. Acad. Sci. Paris* **262** (1966), 613–616.

[2] Demi-groupes réguliers, *Anali Mat. Pura Apl.* **77** (1967), 47–129.

[3] Structure d'une classe de demi-groupes inverses 0-simples, *C. R. Acad. Sci. Paris* **271** (1970), 8–11.

[4] Structure theorems for regular semigroups, *Semigroup Forum* **4** (1972), 95–123.

Lallement, G. and E. Milito

[1] Recognizable languages and finite semilattices of groups, *Semigroup Forum* **11** (1975), 181–184.

Lallement, G. and M. Petrich

[1] Décompositions I-matricielles d'un demi-groupe, *J. Math. Pures Appl.* **45** (1966), 67–117; announced in Some results concerning completely 0-simple semigroups, *Bull. Amer. Math. Soc.* **70** (1964), 777–778.

[2] Structure d'une classe de demi-groupes réguliers, *J. Math. Pures Appl.* **48** (1969), 345–397; announced in Structure of a class of regular semigroups and rings, *Bull. Amer. Math. Soc.* **73** (1967), 419–422.

[3] A generalization of the Rees theorem in semigroups, *Acta Sci. Math.*, Szeged **30** (1969), 113–132.

[4] Extensions of a Brandt semigroup by another, *Canad. J. Math.* **22** (1970), 974–983.

LaTorre, D. R.

[1] On semigroups that are semilattices of groups, *Czechoslovak J. Math.* **21** (1971), 369–370.

Lausch, H.

[1] Cohomology of inverse semigroups, *J. Algebra* **35** (1975), 273–303.

[2] Inverse semigroups with certain universal properties, *J. Austral. Math. Soc.* **21** (1976), 166–170.

[3] Relative cohomology of groups, in *Group Theory*, Lecture Notes Math No. 573, Springer, Berlin, Heidelberg, New York (1977), 66–72 (Canberra Conf. 1975).

Lavrov, B. N.

[1] Maximal semigroups, *Izv. Vysš. Učebn. Zav. Matem.* **5** (1972), 66–73 (Russian).

Leemans, H. and F. Pastijn

[1] Embedding inverse semigroups in bisimple congruence-free inverse semigroups, *Quart. J. Math. Oxford* (2)**34** (1983), 455–458.

Lesohin, M. M.

[1] Homomorphisms into regular semigroups, *Sibir. Matem. Ž.* **4** (1963), 1431–1432 (Russian).

[2] Additive semigroups of homomorphisms of regular semigroups, *Trudy Naučn. Obed. Prep. Fiz.-Mat. Fak. Ped. Inst. Daln. Vostoka* **7** (1966), 54–67 (Russian).

[3] The approximation of semigroups, *Uč. Zap. Leningrad. Ped. Inst.* **328** (1967), 147–171 (Russian).

L'Heureux, J. E.

[1] A note on principal normal inverse sub-semigroups, *J. Natur. Sci. Math.* **6** (1966), 231–233.

Li, I.

[1] The Burnside algebra of a finite inverse semigroup, *Zap. Naučn. Sem. Leningrad. Otdel. Matem. Inst. Steklova* **46** (1974), 41–52 (Russian).

Liber, A. E.

[1] On symmetric generalized groups, *Matem. Sbornik* **33** (1953), 531–544 (Russian).

[2] On the theory of generalized groups, *Doklady Akad. Nauk. SSSR* **97** (1954), 25–28 (Russian).

Liber, S. A.

[1] On free algebras of normal closures of varieties, in *Ordered Sets and Lattices*, Saratov University **2** (1974), 51–53 (Russian).

Libih, A. L.

[1] Inverse semigroups of local automorphisms of abelian groups, in *Research in Algebra*, Saratov University **3** (1973), 25–33 (Russian).

[2] Local automorphisms of commutative monomorphic inverse semigroups, in *Studies in Algebra*, Saratov University **4** (1974), 56–69 (Russian).

[3] On the theory of inverse semigroups of local automorphisms, in *Semigroup Theory and its Appl.*, Saratov University **3** (1974), 46–59 (Russian).

[4] On determinability of an abelian group by its inverse semigroup of local automorphisms, *Izv. Vysš. Učebn. Zav. Matem.* issue **6** (1976), 62–65 (Russian); English transl. *Soviet. Math.* (Iz. VUZ) **20** (1976), 53–55.

[5] Local automorphisms of monogenic inverse semigroups, in *Semigroup Theory and its Appl.*, Saratov University **4** (1978), 11–18 (Russian).

Ljapin, E. S.

[1] Canonical form of elements of an associative system given by defining relations, *Uč. Zap. Leningrad. Ped. Inst.* **89** (1953), 45–54 (Russian).

[2] Increasing elements of associative systems, *Uč. Zap. Leningrad. Ped. Inst.* **89** (1953), 55–65 (Russian).

[3] Associative systems of all partial transformations, *Doklady Akad. Nauk SSSR* **88** (1953), 13–15; correction: ibid. **92** (1953), 692 (Russian).

[4] Abstract characterization of certain semigroups of transformations, *Uč. Zap. Leningrad. Ped. Inst.* **103** (1955), 5–29 (Russian).

[5] On the existence and uniqueness of the solution of an equation of general type in connection with invertibility in semigroups of transformations, *Doklady Akad. Nauk SSSR* **116** (1957), 552–555 (Russian).

[6] Invertibility of elements in semigroups, *Uč. Zap. Leningrad. Ped. Inst.* **166** (1958), 65–74 (Russian).

[7] *Semigroups*, Fizmatgiz, Moscow (1960) (Russian); (2nd edition), English transl. by Amer. Math. Soc. (1968).

McAlister, D. B.

[1] A homomorphism theorem for semigroups, *J. London Math. Soc.* **43** (1968), 355–366.

[2] Inverse semigroups separated over a subsemigroup, *Trans. Amer. Math. Soc.* **182** (1973), 85–117.

[3] On 0-simple inverse semigroups, *Semigroup Forum* **8** (1974), 347–360.

[4] Groups, semilattices and inverse semigroups, *Trans. Amer. Math. Soc.* **192** (1974), 227–244.

[5] Groups, semilattices and inverse semigroups, II, *Trans. Amer. Math. Soc.* **196** (1974), 351–370.

[6] 0-bisimple inverse semigroups, *Proc. London Math. Soc.* (3)**28** (1974), 193–221.

[7] Inverse semigroups generated by a pair of subgroups, *Proc. Roy. Soc. Edinburgh* **A77** (1976), 9–22.

[8] Some covering and embedding theorems for inverse semigroups, *J. Austral. Math. Soc.* **22** (1976), 188–211.

[9] v-prehomomorphisms on inverse semigroups, *Pacific J. Math.* **67** (1976), 215–231.

[10] One-to-one partial right translations of a right cancellative semigroup, *J. Algebra* **43** (1976), 231–251.

[11] E-unitary inverse semigroups over semilattices, *Glasgow Math. J.* **19** (1978), 1–12.

[12] Embedding inverse semigroups in coset semigroups, *Semigroup Forum* **20** (1980), 255–267.

[13] Amenably ordered inverse semigroups, *J. Algebra* **65** (1980), 118–146.

McAlister, D. B. and R. McFadden

[1] Zig-zag representations and inverse semigroups, *J. Algebra* **32** (1974), 178–206.

[2] The free inverse semigroup on two commuting generators, *J. Algebra* **32** (1974), 215–233.

McAlister, D. B. and N. R. Reilly

[1] E-unitary covers for inverse semigroups, *Pacific J. Math.* **68** (1977), 161–174.

McFadden, R.

[1] Proper Dubreil-Jacotin inverse semigroups, *Glasgow Math. J.* **16** (1975), 40–51.

McFadden, R. and L. O'Carroll

[1] F-inverse semigroups, *Proc. London Math. Soc.* **22** (1971), 652–666.

McFadden, R. and H. Schneider

[1] Completely simple and inverse semigroups, *Proc. Cambridge Phil. Soc.* **57** (1961), 234–236.

MacLane, S.

[1] *Categories for the Working Mathematician*, Grad. Texts Math. No. 5, Springer, New York (1971).

McLean, P.

[1] Contributions to the theory of 0-simple inverse semigroups, Doctoral Dissertation, University of Stirling, 1973.

McMorris, F. R.

[1] The quotient semigroup of a semigroup that is a semilattice of groups, *Glasgow Math. J.* **12** (1971), 18–23.

Mahmood, S. J., J. D. P. Meldrum and L. O'Carroll

[1] Inverse semigroups and near-rings, *J. London Math. Soc.* **23** (1981), 45–60.

Majumdar, S.

[1] Homomorphisms of an inverse semigroup, *J. Natur. Sci. Math.* **4** (1964), 125–132.

Malcev, A. I.

[1] Multiplication of classes of algebraic systems, *Sibir. Matem. Ž.* **8** (1967), 346–365 (Russian).

[2] *Algebraic Systems*, Grundl. Math. Wiss., Bd. 192, Springer, New York, Heidelberg, Berlin (1973).

Maniakowski, F.

[1] Sur les axiomes du pseudogroupe, *Bull. Acad. Pol. Sci., Série Math.* **12** (1964), 197–201.

Manukjanc, M. G.

[1] A construction of the endomorphism semigroup of certain commutative regular semigroups, *Izv. Vysš. Učebn. Zav. Matem.* issue **8** (1974), 65–71 (Russian); English transl. *Soviet Math. (Iz. VUZ)* **18** (1974), 51–55.

Martinov, L. M. and T. A. Martinova

[1] Ideally reachable varieties of inverse semigroups, *Izv. Vysš. Učebn. Zav. Matem.* issue **8** (1978), 67–73 (Russian).

Masat, F. E.

[1] A note on ideal decompositions of inverse semigroups, *Nanta Math.* **10** (1977), 40.

Maševickiĭ, G. I.

[1] Identities in Brandt semigroups, in *Semigroup Varieties and Semigroups of Endomorphisms*, Leningrad Ped. Inst. (1979), 126–137 (Russian).

Maxson, C. J. and K. C. Smith

[1] Centralizer near-rings determined by completely regular inverse semigroups, *Semigroup Forum* **22** (1981), 47–58.

Meakin, J.

[1] One-sided congruences on inverse semigroups, *Trans. Amer. Math. Soc.* **206** (1975), 67–82.

[2] On the structure of inverse semigroups, *Semigroup Forum* **12** (1976), 6–14.

[3] Coextensions of inverse semigroups, *J. Algebra* **46** (1977), 315–333.

[4] The partially ordered set of $\mathcal{J}$-classes of an inverse semigroup, *J. London. Math. Soc.* (2)**21** (1980), 244–256.

Melnik, I. I.

[1] On the theory of generalized groups, in *Semigroup Theory and Its Appl.*, Saratov University **2** (1971), 51–59 (Russian).

Mills, J. E.

[1] The idempotents of a class of 0-simple inverse semigroups, *Pacific J. Math.* **81** (1979), 159–166.

Mitchell, J. M. O.

[1] Counting word trees, *Glasgow Math. J.* **15** (1974), 148–149.

Mitsch, H.

[1] Inverse semigroups and their natural order, *Bull. Austral. Math. Soc.* **19** (1978), 59–65.

[2] Congruences on N-simple inverse semigroups, *Math. Japon.* **4** (1981), 349–358.

Mogilevskiĭ, M. G.

[1] Order relations on the symmetric inverse semigroup, in *Semigroup Theory and its Appl.*, Saratov University **3** (1974), 63–71 (Russian).

Munn, W. D.

[1] Matrix representations of semigroups, *Proc. Cambridge Phil. Soc.* **53** (1957), 5–12.

[2] The characters of the symmetric inverse semigroup, *Proc. Cambridge Phil. Soc.* **53** (1957), 13–18.

[3] A class of irreducible matrix representations of an arbitrary inverse semigroup, *Proc. Glasgow Math. Assoc.* **5** (1961), 41–48.

[4] Brandt congruences on inverse semigroups, *Proc. London Math. Soc.* **14** (1964), 154–164.

[5] A certain sublattice of the lattice of congruences on a regular semigroup. *Proc. Cambridge Phil. Soc.* **60** (1964), 385–391.

[6] The lattice of congruences on a bisimple ω-semigroup, *Proc. Roy. Soc. Edinburgh* **A67** (1965/67), 175–184.

[7] Uniform sublattices and bisimple inverse semigroups, *Quart. J. Math. Oxford* (2)**17** (1966), 151–159.
[8] The idempotent separating congruences on a regular 0-bisimple semigroup, *Proc. Edinburgh Math. Soc.* (2)**15** (1967), 233–240.
[9] Regular ω-semigroups, *Glasgow Math. J.* **9** (1968), 46–66.
[10] Fundamental inverse semigroups, *Quart. J. Math. Oxford* (2)**21** (1970), 157–170.
[11] 0-bisimple inverse semigroups, *J. Algebra* **15** (1970), 570–588.
[12] On simple inverse semigroups, *Semigroup Forum* **1** (1970), 63–74.
[13] Free inverse semigroups, *Proc. London Math. Soc.* (3)**29** (1974), 385–404; announced in *Semigroup Forum* **5** (1973), 262–269.
[14] Congruence-free inverse semigroups, *Quart. J. Math. Oxford* (2)**25** (1974), 463–484.
[15] A note on congruence-free inverse semigroups, *Quart. J. Math. Oxford* (2)**26** (1975), 385–387.
[16] A note on E-unitary inverse semigroups, *Bull. London Math. Soc.* **8** (1976), 71–76.
[17] Semiunitary representations of inverse semigroups, *J. London Math. Soc.* (2)**18** (1978), 75–80.
[18] Ideals of free inverse semigroups, *J. Austral. Math. Soc.* **30** (1980), 157–167.
[19] An embedding theorem for free inverse semigroups, *Glasgow Math. J.* **22** (1981), 217–227.
[20] Semiprimitivity of inverse semigroup algebras, *Proc. Roy. Soc. Edinburgh* **A93** (1982/83), 83–98.
[21] The algebra of a combinatorial inverse semigroup, *J. London Math. Soc.* (2)**27** (1983), 35–38.
[22] On the singular ideals of inverse semigroup algebras, *Semigroup Forum* **26** (1983), 375–377.

Munn, W. D. and R. Penrose
[1] A note on inverse semigroups, *Proc. Cambridge Phil. Soc.* **51** (1955), 396–399.

Munn, W. D. and N. R. Reilly
[1] Congruences on a bisimple ω-semigroup, *Proc. Glasgow Math. Assoc.* **7** (1966), 184–192.

Neumann, B. H.
[1] *Universal algebra*, Lecture Notes, New York University, 1962.

Neumann, H.
[1] *Varieties of Groups*, Ergeb. Math. Grenzgeb., Springer, Berlin (1967).

Nguen, K.
[1] Structure of certain inverse semigroups of transformations of a finite set, *Vestnik Leningrad. Univ.* (13)**3** (1973), 48–55 (Russian).
[2] Certain inverse semigroups of transformations of a finite set, *Vestnik Leningrad. Univ.* (7)**2** (1974), 48–54 (Russian; English summary).

Nichols, J. W.
[1] A class of maximal inverse subsemigroups of T_X, *Semigroup Forum* **13** (1976), 187–188.

Nivat, M. and J.-F. Perrot
[1] Une généralisation du monoïde bicyclique, *C. R. Acad. Sci. Paris* **271** (1970), 824–827.

O'Carroll, L.
[1] Reduced inverse semigroups and partially ordered semigroups, *J. London Math. Soc.* (2)**9** (1974), 293–301.
[2] Quasi-reduced inverse semigroups, *J. London Math. Soc.* **9** (1974), 142–150.
[3] A note on free inverse semigroups, *Proc. Edinburgh Math. Soc.* (2)**19** (1974), 17–23.
[4] Reduced inverse semigroups, *Semigroup Forum* **8** (1984), 270–276.
[5] Inverse semigroups as extensions of semilattices, *Glasgow Math. J.* **16** (1975), 12–21.
[6] Embedding theorems for proper inverse semigroups, *J. Algebra* **42** (1976), 26–40.
[7] Idempotent-determined congruences on inverse semigroups, *Semigroup Forum* **12** (1976), 233–243.

[8] Strongly E-reflexive inverse semigroups, *Proc. Edinburgh Math. Soc.* (2)**20** (1976/77), 339–354.

[9] Strongly E-reflexive inverse semigroups. II, *Proc. Edinburgh Math. Soc.* (2)**21** (1978), 1–10.

[10] A note on strongly E-reflexive inverse semigroups, *Proc. Amer. Math. Soc.* **79** (1980), 352–354.

[11] On some embedding theorems for inverse semigroups, *Proc. Amer. Math. Soc.* **83** (1981), 8–10.

Olšanskiĭ, A. Ju.

[1] On the problem of a finite basis for the identities of groups, *Izv. Akad. Nauk SSSR* **34** (1970), 376–384 (Russian); English transl. *Math. USSR-Izvestija* **4** (1970), 381–389.

Osondu, K. E.

[1] Homomorphisms of semilattices of semigroups, *Semigroup Forum* **19** (1980), 133–138.

Papert, D.

[1] Congruence relations in semilattices, *J. London Math. Soc.* **39** (1964), 723–729.

Pastijn, F.

[1] On completely regular inverse semigroups, *Simon Stevin* **37** (1973/74), 135–138.

[2] Semilattices with transitive automorphism group, *J. Austral. Math. Soc.* **29** (1980), 29–34.

[3] Division theorems for inverse and pseudo-inverse semigroups, *J. Austral. Math. Soc.* **31** (1981), 415–420.

[4] Inverse semigroup varieties generated by E-unitary inverse semigroups, *Semigroup Forum* **24** (1982), 87–88.

[5] Essential normal and conjugate extensions of inverse semigroups, *Glasgow Math. J.* **23** (1982), 123–130.

[6] L'enveloppe conjuguée d'un demi-groupe inverse à métacentre idempotent, *Canad. J. Math.* **34** (1982), 900–909.

Perkins, P.

[1] Bases for equational theories of semigroups, *J. Algebra* **11** (1969), 298–314.

Perrot, J. F.

[1] Contribution à l'étude des monoïdes sytactiques et de certains groupes associés aux automates finis, Doctoral Dissertation, University of Paris, 1972.

[2] Une famille de monoïdes inversifs 0-bisimples généralisant le monoïde bicyclique, *Sém. Dubreil, University of Paris* **3** (1973), 15 pp.

Petrich, M.

[1] Translational hull and semigroups of binary relations, *Glasgow Math. J.* **9** (1968), 12–21.

[2] Representations of semigroups and the translational hull of a regular Rees matrix semigroup, *Trans. Amer. Math. Soc.* **143** (1969), 303–318.

[3] On a class of completely semisimple inverse semigroups, *Proc. Amer. Math. Soc.* **24** (1970), 671–679.

[4] On transitive representations of semigroups by 1-1 partial transformations, *Duke Math. J.* **38** (1971), 305–316.

[5] On ideals of a semilattice, *Czechoslovak Math. J.* **22** (1972), 361–367.

[6] Regular semigroups satisfying certain conditions on idempotents and ideals, *Trans. Amer. Math. Soc.* **170** (1972), 245–269.

[7] L'enveloppe de translations d'un demi-treillis de groupes, *Canad. J. Math.* **21** (1973), 164–177.

[8] *Introduction to Semigroups*, Merrill, Columbus (1973).

[9] The translational hull of a semilattice of weakly reductive semigroups, *Canad. J. Math.* **26** (1974), 1520–1536.

[10] Varieties of orthodox bands of groups, *Pacific J. Math.* **58** (1975), 209–217.

[11] Congruences on inverse semigroups, *J. Algebra* **55** (1978), 231–256.

[12] Congruences on simple ω-semigroups, *Glasgow Math. J.* **20** (1979), 87–101.
[13] The conjugate hull of an inverse semigroup, *Glasgow Math. J.* **21** (1980), 103–124.
[14] Some categorical equivalences for E-unitary inverse semigroups, *Trans. Amer. Math. Soc.* **259** (1980), 493–503.
[15] Extensions normales de demi-groupes inverses, *Fund. Math.* **112** (1981), 187–203.
[16] Free inverse semigroups, *Colloq. Math.* (to appear).

Petrich, M. and N. R. Reilly
[1] A representation of E-unitary inverse semigroups, *Quart. J. Math. Oxford* (2)**30** (1979), 339–350.
[2] Congruences on bisimple inverse semigroups, *J. London Math. Soc.* (2)**22** (1980), 251–262.
[3] A network of congruences on an inverse semigroup, *Trans. Amer. Math. Soc.* **270** (1982), 309–325.
[4] E-unitary covers and varieties of inverse semigroups, *Acta Sci. Math.*, Szeged (to appear).

Plahotnik V. V.
[1] Inverse semigroups which are unions of two abelian groups, in *Theory and Appl. Probl. Diff. Eq. and Algebra*, Kiev (1978), 200–203 (Russian).

Płonka, J.
[1] On a method of construction of abstract algebras, *Fund. Math.* **61** (1967), 183–189.

Pollák, G.
[1] Construction of bisimple inverse semigroups from right cancellative semigroups, *Math. Semin. Notes*, Kobe University **7** (1979), 421–426.

Pondelíček, B.
[1] On the characters of the semigroups whose idempotents form a chain, *Čas. Pěst. Mat.* **91** (1966), 4–7 (Czech; Russian and English summaries).

Ponizovskiĭ, I. S.
[1] On homomorphisms of finite inverse semigroups, *Doklady Akad. Nauk SSSR* **142** (1962), 1258–1260 (Russian).
[2] Inverse semigroups with a finite number of idempotents, *Doklady Akad. Nauk SSSR* **143** (1972), 1282–1285 (Russian).
[3] On homomorphisms of finite inverse semigroups, *Uspehi Matem. Nauk.* (110)**18** (1963), 151–153 (Russian).
[4] On representations of inverse semigroups by partial one-to-one transformations, *Izv. Akad. Nauk SSSR, Ser. Matem.* **28** (1964), 989–1002 (Russian).
[5] A remark on inverse semigroups, *Uspehi Matem. Nauk* (126)**20** (1965), 147–148 (Russian).
[6] On the radical of the Burnside algebra of a finite inverse semigroup, in *Algebraic Theory of Semigroups*, North Holland, Amsterdam (1976), 463–478 (Proc. Conf. Szeged University 1966).

Preston, G. B.
[1] Inverse semi-groups, *J. London Math. Soc.* **29** (1954), 396–403.
[2] Inverse semi-groups with minimal right ideals, *J. London Math. Soc.* **29** (1954), 404–411.
[3] Representations of inverse semi-groups, *J. London Math. Soc.* **29** (1954), 411–419.
[4] The structure of normal inverse semigroups, *Proc. Glasgow Math. Assoc.* **3** (1956), 1–9.
[5] A note on representations of inverse semigroups, *Proc. Amer. Math. Soc.* **8** (1957), 1144–1147.
[6] Congruences on Brandt semigroups, *Math. Ann.* **139** (1959), 91–94; correction: ibid. **143** (1961), 465.
[7] Matrix representations of inverse semigroups, *J. Austral. Math. Soc.* **9** (1969), 29–61.
[8] Free inverse semigroups, *J. Austral. Math. Soc.* **16** (1973), 443–453.
[9] Inverse semigroups: some open questions, in *Proc. Symp. Inverse Semigroups*, Northern Illinois University (1973), 122–139.

[10] Representations of inverse semigroups by one-to-one partial transformations of a set, *Semigroup Forum* **6** (1973), 240–245; addendum: ibid. **8** (1974), 277.

[11] Semigroups and graphs, in *Semigroups*, Academic Press, New York (1980), 225–237 (Proc. Conf. Monash University 1979).

[12] Monogenic inverse semigroups, in *Séminaire d'Algèbre*, University of Paris (1980), 25 pp.

Rasin, V. V.

[1] Finitely reachable varieties of inverse semigroups, *Izv. Vysš. Učebn. Zav. Matem.* issue **8** (1978), 80–87 (Russian).

Rees, D.

[1] On the ideal structure of a semigroup satisfying a cancellation law, *Quart. J. Math. Oxford* **19** (1948), 101–108.

Reilly, N. R.

[1] Contributions to the theory of inverse semigroups, Doctoral Dissertation, University of Glasgow, 1965.

[2] Embedding inverse semigroups in bisimple inverse semigroups, *Quart. J. Math. Oxford* (2)**16** (1965), 183–187.

[3] Bisimple ω-semigroups, *Proc. Glasgow Math. Assoc.* **7** (1966), 160–167.

[4] Bisimple inverse semigroups, *Trans. Amer. Math. Soc.* **132** (1968), 101–114.

[5] Congruences on a bisimple inverse semigroup in terms of RP-systems, *Proc. London Math. Soc.* (3)**23** (1971), 99–127.

[6] Inverse semigroups of partial transformations and θ-classes, *Pacific J. Math.* **41** (1972), 215–235.

[7] Free generators in free inverse semigroups, *Bull. Austral. Math. Soc.* **7** (1972), 407–424; correction: ibid. **9** (1973), 479.

[8] Semigroups of order preserving partial transformations of a totally ordered set, *J. Algebra* **22** (1972), 409–427.

[9] The translational hull of an inverse semigroup, *Canad. J. Math.* **26** (1974), 1050–1068.

[10] Congruence-free inverse semigroups, *Proc. London Math. Soc.* (3)**33**, (1976), 497–514.

[11] Free inverse semigroups, in *Algebraic Theory of Semigroups*, North Holland, Amsterdam (1976), 479–508 (Proc. Conf. Szeged University 1976).

[12] Not all fundamental 0-bisimple inverse semigroups contain charts, *Semigroup Forum* **12** (1976), 176–179.

[13] Maximal inverse subsemigroups of T_X, *Semigroup Forum* **15** (1977), 319–326.

[14] Enlarging the Munn representation of inverse semigroups, *J. Austral. Math. Soc.* **23** (1977), 28–41.

[15] Varieties of completely semisimple inverse semigroups, *J. Algebra* **65** (1980), 427–444.

[16] Modular sublattices of the lattice of varieties of inverse semigroups, *Pacific J. Math.* **89** (1980), 405–417.

[17] The breadth of the lattice of those varieties of inverse semigroups which contain the variety of groups, *Proc. Roy. Soc. Edinburgh* **A93** (1982/83), 319–325.

Reilly, N. R. and A. H. Clifford

[1] Bisimple inverse semigroups as semigroups of ordered triples, *Canad. J. Math.* **20** (1968), 25–39.

Reilly, N. R. and W. D. Munn

[1] *E*-unitary congruences on inverse semigroups, *Glasgow Math. J.* **17** (1976), 57–75.

Reilly, N. R. and H. E. Scheiblich

[1] Congruences on regular semigroups, *Pacific J. Math* **23** (1967), 349–360.

Rodriquez, G.

[1] La determinazione dei semigruppi che risultano unione di un gruppo e di un semireticolo, *Rend. Ist. Lombardo, Accad. Sci. Lett.* **A103** (1969), 199–221.

[2] La determinazione dei semigruppi che sono unione di un semireticolo e di un semigruppo inverso unione di gruppi, *Rend. Ist. Lombardo, Accad. Sci. Lett.*, **A103** (1969), 886–920.

[3] Semigruppi inversi generati da copie di semigruppi inversi assegnati, *Rend. Ist. Lombardo, Accad. Sci. Lett.* **A104** (1970), 653–679.

Rozenblat, B. V.

[1] On positive theories of free inverse semigroups, *Sibir. Matem. Ž.* **20** (1979), 1282–1293 (Russian).

[2] Equations on a free inverse semigroup, *Uč. Zap. Ural. Univ.* **13** (1982), 117–120 (Russian).

Rukolaĭne, A. V.

[1] Characters of finite inverse semigroups, *Zap. Naučn. Sem. Leningrad. Otdel. Matem. Inst. Steklova* **71** (1977), 207–215 (Russian).

[2] The center of the semigroup algebra of a finite inverse semigroup over the field of complex numbers, *Zap. Naučn. Sem. Leningrad. Otdel. Matem. Inst. Steklova* **75** (1978), 154–159 (Russian).

[3] Semigroup algebras of finite inverse semigroups over arbitrary fields, *Zap. Naučn. Sem. Leningrad. Otdel. Matem. Inst. Steklova* **103** (1980), 117–123 (Russian).

Saitô, T.

[1] Note on semigroups having no minimal ideals, *Bull. Tokyo Gakugei Univ.* **9** (1958), 13–16.

[2] Proper ordered inverse semigroups, *Pacific J. Math.* **15** (1965), 649–666.

[3] Ordered inverse semigroups, *Trans. Amer. Math. Soc.* **153** (1971), 99–138.

[4] No free inverse semigroup is orderable, *Proc. Japan. Acad.* **50** (1974), 837–838.

Saliĭ, V. N.

[1] On proper representations of generalized groups by matrices, in *Trudy Mol. Uč.*, Saratov University (1964), 88–92 (Russian).

[2] A theorem on homomorphisms of strong commutative bands of semigroups, in *Semigroup Theory and its Appl.*, Saratov University **2** (1971), 69–74 (Russian).

Scheiblich, H. E.

[1] Semimodularity and bisimple ω-semigroups, *Proc. Edinburgh Math. Soc.* (2)**17** (1970), 79–81.

[2] A characterization of a free elementary inverse semigroup, *Semigroup Forum* **2** (1971), 76–79.

[3] Concerning congruences on symmetric inverse semigroups, *Czechoslovak Math. J.* **23** (1973), 1–10.

[4] Free inverse semigroups, *Proc. Amer. Math. Soc.* **38** (1973), 1–7; announced in *Semigroup Forum* **4** (1972), 351–359.

[5] Kernels of inverse semigroup homomorphisms, *J. Austral. Math. Soc.* **18** (1974), 289–292.

Schein, B. M.

[1] A system of axioms for semigroups embeddable into generalized groups, *Doklady Akad. Nauk SSSR* **134** (1960), 1030–1033 (Russian); English transl. *Soviet Math. Doklady* **1** (1961), 1180–1183.

[2] Embedding of a semigroup into a generalized group, *Uspehi Matem. Nauk* (98)**16** (1961), 218–219 (Russian).

[3] Embedding semigroups into generalized groups, *Matem. Sbornik* **55** (1961), 379–400 (Russian).

[4] On subdirectly irreducible semigroups, *Doklady Akad. Nauk SSSR* **144** (1962), 999–1002 (Russian).

[5] Representations of generalized groups, *Izv. Vysš. Učebn. Zav. Matem.* issue **3** (1962), 164–176 (Russian).

[6] On the theory of generalized groups, *Doklady Akad. Nauk SSSR* **153** (1963), 296–299 (Russian).

[7] Generalized groups with well-ordered set of idempotents, *Mat.-Fyz. Časopis* **14** (1964), 259–262.

[8] Involuted semigroups of full binary relations, *Doklady Akad. Nauk SSSR* **156** (1964), 1300–1303 (Russian).

[9] Positive axioms for generalized groups, *Revue Roumaine Math. Pures Appl.* **10** (1965), 1–2.

[10] On the theory of generalized groups and generalized heaps, in *Semigroup Theory and its Appl.*, Saratov University **1** (1965), 286–324 (Russian); *Amer. Math. Soc. Transl.* (2)**113** (1979), 89–122.

[11] Homomorphisms and subdirect decompositions of semigroups, *Pacific J. Math.* **17** (1966), 529–547.

[12] *o*-rings and LA-rings, *Izv. Vysš. Učebn. Zav. Matem.* **2** (1966), 111–122 (Russian); *Amer. Math. Soc. Transl.* (2)**96** (1970), 137–152.

[13] Semigroups of strong subsets, *Volžskiĭ Matem. Sbornik* **4** (1966), 180–186 (Russian).

[14] The symmetric transformation semigroup is covered by its inverse subsemigroups, *Acta Math. Acad. Sci. Hung.* **22** (1971), 163–170 (Russian).

[15] A remark concerning congruences on (0-) bisimple inverse semigroups, *Semigroup Forum* **3** (1971), 80–83.

[16] Homomorphisms of symmetric semigroups into inverse semigroups, in *Semigroup Theory and its Appl.*, Saratov University **2** (1971), 90–93 (Russian).

[17] Completions, translational hulls and ideal extensions of inverse semigroups, *Czechoslovak Math. J.* **23** (1973), 575–610.

[18] Inverse semigroups that do not admit isomorphic representations by partial transformations of their proper subsets, in *Semigroup Theory and its Appl.*, Saratov University **3** (1974), 139–148 (Russian).

[19] Injectives in certain classes of semigroups, *Semigroup Forum* **9** (1974), 159–171.

[20] A new proof for the McAlister *P*-theorem, *Semigroup Forum* **10** (1975), 185–188.

[21] Free inverse semigroups are not finitely presentable, *Acta Math. Acad. Sci. Hung.* **26** (1975), 41–52.

[22] Injective commutative semigroups, *Algebra University* **6** (1976), 395–397.

[23] Injective monars over inverse semigroup, in *Algebraic Theory of Semigroups*, North Holland, Amsterdam (1976), 519–544 (Proc. Conf. Szeged University 1976).

[24] Embedding semigroups into inverse semigroups, in *Algebra and Number Theory*, Nalčik **2** (1977), 147–163 (Russian).

Schreier, O.

[1] Die Untergruppen der freien Gruppen, *Abh. Math. Semin. Hamburg* **5** (1927), 161–183.

Schwab, E.

[1] On the equation $ixi = i$ on regular and inverse semigroups, *An. Univ. Timişoara, Ser. Şti. Mat.* **7** (1969), 253–256 (Rumanian; German summary).

[2] On characterizations of inverse semigroups, *Magyar Tud. Akad. Mat. Fiz Oszt. Közl.* **21** (1973), 201–204 (Hungarian; English summary).

Schwarz, V. Ja. and I. Š. Jaroker

[1] Increasing elements of semigroups with one-sided identity, *Uspehi Matem. Nauk* (118)**19** (1964), 209–214 (Russian).

Shoji, K. and M. Yamada

[1] D*-simple inverse semigroups, *J. Algebra* **65** (1980), 317–327.

Sribala, S.

[1] On Σ-ordered inverse semigroups, *Acta Sci. Math.*, Szeged **35** (1973), 207–210.

[2] Cohomology and extensions of inverse semigroups, *J. Algebra* **47** (1977), 1–17.

Steinfeld, O.

[1] On semigroups which are unions of completely 0-simple semigroups, *Czechoslovak Math. J.* **16** (1966), 63–69.

Stolt, B.

[1] Zur Axiomatik des Brandtschen Gruppoids, *Math. Zeitschr.* **70** (1958), 156–164.

Sullivan, R. P.
[1] Partial translations of semigroups, *Acta Sci. Math.*, Szeged **41** (1979), 221–225.

Szendrei, M. B.
[1] On an extension of semigroups, *Acta Sci. Math.*, Szeged **39** (1977), 367–389.

Šaronova, T. N.
[1] Elementary inverse subsemigroups of the semigroup of linear transformations, in *Research in Algebra*, Saratov University **4** (1974), 115–118 (Russian).

Ševrin, L. N.
[1] Basic problems in the theory of projections of semilattices, *Matem. Sbornik* **66**(108) (1965), 568–597 (Russian); *Amer. Math. Soc. Transl.* **96**(2) (1970), 1–35.

Šimelfenig, O. V.
[1] Representation of commutative idempotent semigroups, *Izv. Vysš. Učebn. Zav. Matem.* issue **7** (1971), 99–107 (Russian).

Širjajev, V. M.
[1] On inverse semigroups with given *G*-radical, *Doklady Akad. Nauk BSSR* **14** (1970), 782–785 (Russian).
[2] On a class of inverse semigroups, *Vestnik Belorus. Univ.*, *Ser. I*, issue **2** (1970), 10–13 (Russian).
[3] On certain extension in the class of inverse semigroups, *Vestnik Belorusk. Univ. Ser. I*, issue **1** (1971), 15–23 (Russian).
[4] Semilattices of width 2, *Semigroup Forum* **13** (1976), 149–177.
[5] Certain properties of the category of inverse semigroups, *Izv. Vysš. Učebn. Zav. Matem.* issue **5** (1977), 125–136 (Russian).
[6] Semilattices of width 3 whose indecomposable elements form a chain, *Semigroup Forum* **17** (1979), 201–240.

Šneperman, L. B.
[1] Maximal inverse subsemigroups of the semigroup of linear transformations, *Izv. Vysš. Učebn. Zav. Matem.* issue **11** (1974), 93–100 (Russian).
[2] Periodic linear inverse semigroups, *Vescī Akad. Navuk BSSR*, *Ser. Fiz. Matem.* **4** (1976), 22–28 (Russian).

Taĭclin, M. A.
[1] Existentially closed regular commutative semigroups, *Algebra i Logika* **12** (1973), 689–703 (Russian).

Tamari, D.
[1] Les images homomorphes des groupoïdes de Brandt et l'immersion des semi-groupes, *C. R. Acad. Sci. Paris* **229** (1949), 1291–1293.

Tichy, T. and J. Vinarek
[1] On the algebraic characterization of system of 1-1 partial mappings, *Comment. Math. Univ. Carolinae* **13** (1972), 711–720.

Tirasupa, Y.
[1] Weakly factorizable inverse semigroups, *Semigroup Forum* **18** (1979), 283–291.

Todorov, K.
[1] Über die inversen Unterhalbgruppen der endlichen symmetrischen Halbgruppe, *Arch. Math.*, Basel **33** (1979), 23–28.

Trachtman, A. N.
[1] A basis for identities of the five-element Brandt semigroup, *Matem. Zap. Ural. Univ.* **12** (1981), 117–132 (Russian).

Trotter, P. G.
[1] Congruence-free inverse semigroups, *Semigroup Forum* **9** (1974), 109–116.
[2] Projectives on inverse semigroups, *Semigroup Forum* **26** (1983), 167–190.

Trotter, P. G. and T. Tamura

[1] Completely semisimple inverse Δ-semigroups admitting principal series, *Pacific J. Math.* **68** (1977), 515–525.

Trueman, D. C.

[1] Direct products of monogenic semigroups, Doctoral Dissertation, Monash University, 1981; announced in *Bull. Austral. Math. Soc.* **24** (1981), 475–480.

[2] Direct products of cyclic semigroups, in *Semigroups*, Academic Press, New York (1980), 103–110 (Proc. Conf. Monash University 1979).

[3] Direct product of finite monogenic inverse semigroups, *Proc. Roy. Soc. Edinburgh* **A92** (1982), 301–317.

Varlet, J. C.

[1] On separation properties of semilattices, *Semigroup Forum* **10** (1975), 220–228.

Vaughan-Lee, M. R.

[1] Uncountably many varieties of groups, *Bull. London Math Soc.* **2** (1970), 280–286.

Važenin, Ju. M.

[1] On the elementary theory of free inverse semigroups, *Semigroup Forum* **9** (1974), 185–195.

Venkatesan, P. S.

[1] On a class of inverse semigroups, *Amer. J. Math.* **84** (1962), 578–582.

[2] On decomposition of semigroups with zero, *Math. Zeitschr.* **92** (1966), 164–174.

Wagner, V. V.

[1] On the theory of partial transformations, *Doklady Akad. Nauk SSSR* **84** (1952), 653–656 (Russian).

[2] Generalized groups, *Doklady Akad. Nauk SSSR* **84** (1952), 1119–1122 (Russian).

[3] The theory of generalized heaps and generalized groups, *Matem. Sbornik* **32** (1953), 545–632 (Russian).

[4] Generalized heaps reducible to generalized groups, in *Sci. Yearbook*, Saratov University (1954/55), 668–669 (Russian).

[5] Represenation of ordered semigroups, *Matem. Sbornik* **38** (1956), 203–240 (Russian); *Amer. Math. Soc. Transl.* (2)36 (1964), 295–336.

[6] Generalized heaps reducible to generalized groups, *Ukrain. Matem. Ž.* **8** (1956), 235–253 (Russian).

[7] Semigroups associated with a generalized heap, *Matem. Sbornik* **52** (1960), 597–628 (Russian).

[8] Generalized heaps and generalized groups with a transitive compatibility relation, *Uč. Zap. Saratov. Univ. Meh.-Matem.* **70** (1961), 25–39 (Russian).

[9] Foundation of differential geometry and modern algebra, in *Works 4th All-Union Math. Conf. 1961*, Vol. 1, *Akad. Nauk SSSR* (1963), 17–29 (Russian).

[10] On the theory of antigroups, *Izv. Vysš. Učebn. Zav. Matem.* issue **4** (1971), 3–15 (Russian).

[11] *t*-simple representations of antigroups, *Izv. Vysš. Učebn. Zav. Matem.* issue **9** (1971), 18–29 (Russian).

[12] On the theory of involuted semigroups, *Izv. Vysš. Učebn. Zav. Matem.* issue **10** (1971), 24–35 (Russian).

Warne, R. J.

[1] Homomorphisms of *d*-simple semigroups with identity, *Pacific J. Math.* **14** (1964), 1111–1122.

[2] The idempotent-separating congruences of a bisimple inverse semigroup, *Publ. Math.*, Debrecen **13** (1966), 203–206.

[3] On certain bisimple inverse semigroups, *Bull. Amer. Math. Soc.* **72** (1966), 679–682.

[4] A class of bisimple inverse semigroups, *Pacific J. Math.* **18** (1966), 563–577.

[5] Extensions of Brandt semigroups and applications, *Illinois J. Math.* **10** (1966), 652–660;

announced in Extension of Brandt semigroups, *Bull. Amer. Math. Soc.* **72** (1966), 683–684.
[6] Bisimple inverse semigroups mod groups, *Duke Math. J.* **34** (1967), 787–811.
[7] I-bisimple semigroups, *Trans. Amer. Math. Soc.* **130** (1968), 367–386.
[8] Extensions of I-bisimple semigroups, *Canad. J. Math.* **19** (1967), 419–426; correction: ibid. **20** (1968), 511–512.
[9] Extensions of $\omega^n I$-bisimple semigroups, *Math. Japon.* **13** (1968), 105–121.
[10] Congruences on ω^n-bisimple semigroups, *J. Austral. Math. Soc.* **9** (1969), 257–274.
[11] E-bisimple semigroups, *J. Natur. Sci. Math.* **10** (1970), 51–81.
[12] I-regular semigroups, *Math. Japon.* **15** (1970), 91–100.
[13] Some properties of simple I-regular semigroups, *Comp. Math.* **22** (1970), 181–195.
[14] $\omega^n I$-bisimple semigroups, *Acta Math. Acad. Sci. Hung.* **21** (1970), 121–150.

Warne, R. J. and J. W. Hogan
[1] Homomorphisms of ω^n-right cancellative semigroups, *J. Natur. Sci. Math.* **7** (1967), 227–235.

Warne, R. J. and L. K. Williams
[1] Characters on inverse semigroups, *Czechoslovak Math. J.* **11** (1961), 150–155.

Wenger, R.
[1] Self-injective semigroup rings for finite inverse semigroups, *Proc. Amer. Math. Soc.* **20** (1969), 213–216.

White, G. L.
[1] The dual ordinal of a bisimple inverse semigroup, *Semigroup Forum* **6** (1973), 295–297.

Yamada, M.
[1] H-compatible orthodox semigroups, in *Algebraic Theory of Semigroups*, North Holland, Amsterdam (1976), 721–748 (Proc. Conf. Szeged University 1976).
[2] J-compatible orthodox semigroups, *Proc. Japan Acad.* **55** (1979), 301–305.

Yoshida, R.
[1] McAlister representation of proper simple inverse semigroups, *Proc. 2nd Symp. on Semigroups*, Tokyo Gakugei University (1978), 17–24.

Yusuf, S. M.
[1] Inner transformations of an inverse semigroup, *J. Natur. Sci. Math.* **2** (1962), 101–109.
[2] Semigroups with operators, *J. Natur. Sci. Math.* **3** (1963), 57–72.
[3] A Jordan-Hölder theorem for inverse semigroups with operators, *J. Natur. Sci. Math.* **3** (1963), 129–137.
[4] A structure theorem for inverse semigroups, *J. Natur. Sci. Math.* **4** (1964), 103–108.
[5] A structure theorem for an inverse semigroup which has a given inverse semigroup as its principal normal inverse subsemigroup, *J. Natur. Sci. Math.* **5** (1965), 231–236.
[6] Inverse semimodules, *J. Natur. Sci. Math.* **6** (1966), 111–117.

Zapletal, J.
[1] Distinguishing subsets in semilattices, *Arch. Math.*, Brno **9** (1973), 73–82.
[2] On the characterization of semilattices satisfying the descending chain conditions and some remarks on distinguishing subsets, *Arch. Math.*, Brno **10** (1974), 123–128.

Zitarelli, D. E.
[1] Subdirectly irreducible finite inverse semigroups, Doctoral Dissertation, Pennsylvania State University, 1970.
[2] Inverse subsemigroups of Rees matrix semigroups, *Bull. Austral. Math. Soc.* **9** (1973), 445–463.
[3] Compatible extensions and subdirectly irreducible semigroups, *Semigroup Forum* **9** (1974), 241–252.

Žitomirskiĭ, G. I.
[1] On homomorphisms of generalized heaps, in *Semigroup Theory and Its Appl.*, Saratov University **1** (1965), 220–237 (Russian).

[2] On the congruence lattice in a generalized heap, *Izv. Vysš. Učebn. Zav. Matem.* issue **1** (1965), 56–61 (Russian).

[3] On imprimitivity relations of generalized groups of transformations, *Matem. Sbornik* **73** (1967), 500–512 (Russian).

[4] On regular and stable relations on generalized heaps, *Izv. Vysš. Učebn. Zav. Matem.* issue **1** (1968), 64–77 (Russian).

[5] Varieties of generalized heaps, in *Semigroup Theory and Its Appl.*, Saratov University **2** (1971), 27–35 (Russian).

[6] Generalized heaps and generalized groups with a modular lattice of congruence relations, *Izv. Vysš. Učebn. Zav. Matem.* issue **6** (1972), 26–35 (Russian).

[7] Generalized groups with a Boolean lattice of congruence relations, in *Ordered Sets and Lattices*, Saratov University **2** (1974), 18–27 (Russian).

[8] Extensions of universal algebras, in *Semigroup Theory and Its Appl.*, Saratov University **4** (1978), 19–40 (Russian).

SYMBOLS

(A)	associativity for partial groupoids,
$\|A\|$	cardinality of a set A
$\langle A \rangle$	inverse subsemigroup generated by A
$A(J) = \{a \in A \mid J_a > J\}, A \subseteq S, J$ is a $\mathcal{J}$-class	
$A^J = A \cap J$	
$A_J = A(J) \cup A$	
$A_1 A_2 \cdots A_n$	complex multiplication
A^n	n^{th} power of a set
(B)	one of the axioms for Brandt groupoids
$B_d = \{\varphi_{m,n} \mid m \equiv n \pmod{d}\}$	
$B_{d,i} = \{\varphi_{m,n} \mid m \equiv n \equiv i \pmod{d}\}$	
B_2	five-element combinatorial Brandt semigroup
B_X	birooted word trees on X
$B(G, I)$	Brandt semigroup
$B(T, \alpha)$	Bruck semigroup over T
C	bicyclic semigroup
(C)	one of the axioms for Croisot groupoids
C_1	free monogenic inverse semigroup
C_2	free monogenic inverse semigroup
C_3	free monogenic inverse semigroup
C_4	free monogenic inverse semigroup
C_5	free monogenic inverse semigroup
C_X	free Clifford semigroup
C_ω	ω-chain
$c_S(A)$	centralizer of A in S
$(C1)$–$(C3)$	axioms for the construction of strict inverse semigroups
$C(S)$	permissible subsets of an inverse semigroup S
D_a	$\mathcal{D}$-class of a
E	idempotents of a free inverse semigroup

E_A	idempotents in a set A
$F_Q = \{[r, r] \mid r \in Q\}$	
(F1), (F2), (R3)	axioms for the construction of R-inverse semigroups
(F1)–(F3)	axioms for an F-pair
${}^I G$	functions from subsets of I into G (on the right)
G^I	functions from subsets of I into G (on the left)
G^0	group with zero
G^0	partial groupoid with zero
G_+	positive cone of an l-group
$\lvert G \rvert$	cardinality of the set of vertices of a graph G
G_X	free group on X
$G_\vee$	semilattice on G_+ under $\vee$
H_a	$\mathcal{H}$-class of a
I_C	identity functor on C
$I_n = \{u \in I_x \mid w(u) \geq n\}$	
I_x	free monogenic inverse semigroup on x
I_X	free inverse semigroup on X
$I(a) = \{x \in J(a) \mid J(x) \neq J(a)\}$	
$i_S(T)$	idealizer of T in S
$I(S) = \{\psi \in \Psi(S) \mid \tilde{\psi} \in \mathcal{IA}(S/\sigma)\}$	
(I1), (I2)	axioms for inverse semigroups
(I1)–(I4)	axioms for an inductive groupoid
$J(s)$	ideal generated by s
J_a	$\mathcal{J}$-class of a
K^{-1}	inverse of a subset KH
$L(s)$	left ideal generated by s
L_a	$\mathcal{L}$-class of a
$l(T)$	letters labeling edges of T
(M)	condition related to congruence-freeness
$M = (\mathbb{N} \times \mathbb{N}^*) \cup 0$	
$M_n = I_x/I_n$	
$M_e = \{f \in E_S \mid f \leq e\}$	
$M(S)$	metacenter of an inverse semigroup S
M_X	free inverse monoid on X
N	nonnegative integers
N_e	greatest normal divisor for e
$\mathbb{N}$	natural numbers $1, 2, 3, \ldots$
$\mathbb{N}^* = \mathbb{N} \cup \{\omega, \infty^-, \infty^+\}$	
$\mathbb{N}^1$	$\mathbb{N}$ with the greatest element ∞ adjoined
P_X	G_X together with a partial order
$P(X)$	subsets of X under intersection
$Q_F = \{r \in R \mid [r, r] \in F\}$	

$Q_\tau = \{r \in R \mid [r, r]\tau[1,1]\}$

$Q(\mathcal{U}) = \{\mathcal{V} \in \mathcal{L}(\mathcal{I}) \mid \mathcal{V} \cap \mathcal{G} = \mathcal{U}\}$

$R(s)$	right ideal generated by s
R_a	$\mathcal{R}$-class of a
$\langle S \rangle$	variety generated by an inverse semigroup S
S^1	semigroup with an identity possibly adjoined
$\hat{S}$	one-to-one partial right translations of S
S^K	functions from K to S
S^w	image under the Wagner representation of S
S^δ	image under the Munn representation of S
S^*	semigroup with zero removed
(S1)–(S4)	axioms for structure mappings
$\bar{S}$	semigroup with an identity adjoined

$T(S) = \{\tau_{(\lambda,\rho)} \mid (\lambda, \rho) \in \Omega(S)\}$

$T(\mathcal{V}) = \{S \in \mathcal{V} \mid \langle S \rangle$ is combinatorial$\}$

$U(\rho)$	$aU(\rho)b \Leftrightarrow [1, a]\rho[1, b]$
$V(\tau)$	$[a, b]V(\tau)[c, d] \Leftrightarrow a = uc,\ b = vd,\ u^{-1}v\tau 1$
$V(G)$	vertices of a graph G
X^1	X with the greatest element adjoined
X^+	free semigroup on X
X^*	free monoid on X
Y_X	free semilattice on X
Y_2	two-element semilattice
Z	free monoid with involution on X
$\mathbb{Z}$	integers
$Z(S)$	center of a semigroup S
$c(w)$	content of a word w
$c(A)$	content of a set A

$[d] = d$ if $d \geq 0$, $[d] = 0$ if $d < 0$

$e_n = ((n, n), (0, 0))$

$f_n = ((0, 0), (n, n))$

$l(\rho)$	least n such that $e_n \rho e_{n+1}$
m_g	greatest element of the σ-class g
$r(w)$	reduced of a word w
$r(A)$	reduced of a set A
$r(\rho)$	least n such that $f_n \rho f_{n+1}$
s^{-1}	inverse of an element s
$[s]$	cyclic semigroup generated by s
w	Wagner representation
w^a	image of a under the Wagner representation
$\hat{w}$	left factors of a word w
$w(u)$	weight of a word u

$|w| = |x_1||x_2| \cdots |x_n|$ if $w = x_1 x_2 \cdots x_n$

$|x| = x$ if $x \in X$, $|x| = x^{-1}$ if $x \in X'$, $|\varnothing| = \varnothing$

$\{x\}$ singleton x

$z_p = z(z\beta^k)(z\beta^{2k}) \cdots (z\beta^{(p-1)k})$, $z_0 = f$

$z_p = (z, \beta^k)_p$

SCRIPT LETTERS

$\mathcal{A}$ class of all antigroups

$\mathcal{AG}$ variety of abelian groups

$\mathcal{AG}_n = \mathcal{AG} \cap [x^{n+1} = x]$

$\mathcal{AG}(S)$ lattice of antigroup congruences on S

$\mathcal{A}(S)$ automorphism group of S

$\mathcal{B}$ variety of combinatorial strict inverse semigroups

$\mathcal{B}(X)$ binary relations on a set X

$\mathcal{B}_s$ idempotent separating congruences on $V(R)$

$\mathcal{C}$ $a\mathcal{C}b \Leftrightarrow a^{-1}b,\ ab^{-1} \in E_S$

$\mathcal{C}$ variety generated by a bicyclic semigroup

$\langle \mathcal{C} \rangle$ variety generated by a class $\mathcal{C}$

$\mathcal{C}_n = [x^n = x^{n+1}]$

$\mathcal{C}(S)$ lattice of congruences on S

$\mathcal{CP}(S)$ congruence pairs for S

$\mathcal{CS}$ completely semisimple varieties

$\mathcal{CSH}$ completely semisimple cryptic varieties

$\mathcal{D}$ Green's relation

$\mathcal{E}(S)$ semigroup of endomorphisms of S

$\mathcal{F}$ filters of $E_{V(R)}$ locally self-conjugate in $V(R)$

$\mathcal{F}$ $a\mathcal{F}b \Leftrightarrow a^{-1}b \in E_S$

$[\mathcal{F}]$ variety determined by the family of identities $\mathcal{F}$

$\mathcal{F}(X)$ partial transformations of X (on the right)

$\mathcal{F}'(X)$ partial transformations of X (on the left)

$\mathcal{F}_0(X) = \{\alpha \in \mathcal{F}(X) | \text{rank } \alpha \leq 1\}$

$\mathcal{F}_0'(X) = \{\alpha \in \mathcal{F}'(X) | \text{rank } \alpha \leq 1\}$

$\mathcal{FI}(S)$ fully invariant congruences on S

$\mathcal{G}$ variety of groups

$\mathcal{H}$ Green's relation

$\mathcal{I}$ variety of inverse semigroups

$\mathcal{I}(X)$ symmetric inverse semigroup on X (on the right)

$\mathcal{I}'(X)$ symmetric inverse semigroup on X (on the left)

$\mathcal{I}_0(X) = \{\alpha \in \mathcal{I}(X) | \text{rank } \alpha \leq 1\}$

$\mathcal{I}_0'(X) = \{\alpha \in \mathcal{I}(X) | \text{rank } \alpha \leq 1\}$

$\mathcal{I}_Y$	ideals of a semilattice Y
$\mathcal{I}\mathcal{A}(G)$	group of inner automorphisms of G
$\mathcal{J}$	Green's relation
$\mathcal{K}(S)$	kernels in an inverse semigroup S
$\mathcal{K}(\rho)$	kernel normal system of ρ
$\mathcal{L}$	Green's relation
$\mathcal{L}(G)$	lattice of subgroups of a group G
$\mathcal{L}(\mathcal{V})$	varieties contained in a variety $\mathcal{V}$
$\mathcal{L}(\mathcal{C}\mathcal{S})$	lattice of completely semisimple varieties
$\mathcal{L}(\mathcal{C}\mathcal{S}\mathcal{K})$	lattice of completely semisimple cryptic varieties
$\mathcal{M}_n = \langle M_n \rangle$	
$\mathcal{N}$	normal congruences on $E_{V(R)}$
$\mathcal{N}(E_S)$	normal congruences on E_S
$\mathcal{P}_Y$	principal ideals of a semilattice Y
$\mathcal{P}_\nu$	subsets of cardinality ν
$\mathcal{P}_S = \{[a] \mid a \in S\}$	
$\mathcal{P}(W)$	all subsets of W
$\mathcal{Q}$	$*$-filters of R
$\mathcal{Q}\mathcal{A}$	variety of quasiabelian inverse semigroups
$\mathcal{R}$	Green's relation
$\mathcal{R}_c$	right cancellative congruences on R contained in $\mathcal{L}$
$\mathcal{R}_d$	left normal divisors of R
$\mathcal{R}(S)$	E-reflexive congruences on S
$\mathcal{R}_Y$	retract ideals of a semilattice Y
$\mathcal{S}$	variety of semilattices
$\mathcal{S}(X)$	symmetric group on X (on the right)
$\mathcal{S}'(X)$	symmetric group on X (on the left)
$\mathcal{S}\mathcal{G}$	variety of Clifford semigroups
$\mathcal{S}\mathcal{I}$	variety of strict inverse semigroups
$\mathcal{T}$	trivial variety
$\mathcal{T}_X$	cross section of isomorphism classes of word trees
$\mathcal{T}(X)$	semigroup of transformations on X (on the right)
$\mathcal{T}'(X)$	semigroup of transformations on X (on the left)
$\mathcal{U}(S)$	E-unitary congruences on S
$\mathcal{V}(\rho)$	variety corresponding to ρ

GREEK LETTERS

$\Gamma(S)$	inner left translations of S
$\Delta(S)$	inner right translations of S
$\Delta(u) = A \quad \text{if} \quad u = (A, g)$	

$\Theta(S)$	inner part of the conjugate hull $\Psi(S)$
$\Lambda(S)$	left translations of S
$\Pi(S)$	inner bitranslations of S
$\mathrm{P}(S)$	right translations of S
$\Sigma(R)$	inverse hull of R
$\Sigma(Y)$	isomorphisms among ideals of Y
Φ	isomorphism of $\hat{S}$ onto $C(S)$
$\Phi(S)$	normal hull of S
$\Phi(Y)$	isomorphisms among principal ideals of Y
$\Psi(S)$	conjugate hull of S
$\Omega(S)$	translational hull of S
$[\alpha]$	principal ideal generated by α
δ	Munn representation
δ_1	relation used for strong amalgamation
δ^a	image of a under the Munn representation
$\varepsilon = \varepsilon_X$	equality relation on X
ε_g	inner automorphism induced by g
$\eta = \eta_S$	least semilattice congruence on S
θ	congruence of trace classes
$\theta^a : k \to a^{-1}ka \quad (k \in aKa^{-1})$	
$\theta = \theta_S : S \to \Psi(S)$	canonical homomorphism
$\iota = \iota_X$	identity mapping on X
κ	$\cap$ -congruence of kernel classes
$\bar{\kappa} : \Psi(S) \to \Psi(T)$	isomorphism induced by κ
$\lambda = \lambda_S$	least E-reflexive congruence on S
λ_s	inner left translation induced by s
$\lambda^v : s \to vs \quad (s \in S)$	
$\mu = \mu_S$	greatest idempotent separating congruence on S
$\nu = \nu_S$	least Clifford congruence on S
$\nu_K : \Psi(K) \to \Psi(K\pi)$	isomorphism
$\nu = \nu_1 \cap \nu_2$	
ν_1	$\mathcal{U}\nu_1\mathcal{V} \Leftrightarrow \mathcal{U} \vee \mathcal{G} = \mathcal{V} \vee \mathcal{G}$
ν_2	$\mathcal{U}\nu_2\mathcal{V} \Leftrightarrow \mathcal{U} \cap \mathcal{G} = \mathcal{V} \cap \mathcal{G}$
$\xi : (\lambda, \rho) \to \xi_{(\lambda, \rho)} \quad ((\lambda, \rho) \in \Omega(S))$	
$\xi_{\mathcal{K}}$	congruence associated with $\mathcal{K}$
$\xi_{(\lambda, \rho)} : e \to \lambda^{-1}e\rho \quad (e \in \lambda E_S \rho^{-1})$	
$\pi : S \to \Omega(S)$	canonical homomorphism
$\pi = \pi_S$	least E-unitary congruence on S
π_s	inner bitranslation induced by s
π_ρ	least E-unitary congruence containing ρ
ρ_s	inner right translation induced by s

ρ_I	Rees congruence relative to the ideal I
ρ^e	relation of elementary ρ-transition
ρ^t	transitive closure of ρ
ρ^{-1}	inverse of a binary relation ρ
$\rho^{\#} : S \to S/\rho$	natural homomorphism
ρ^*	congruence generated by ρ
$\rho^v : s \to sv \quad (s \in S)$	
$\rho_{(k,u)}$	congruence of type (k, u)
$\rho_{(K,\tau)}$	congruence with kernel K and trace τ
$\rho(\mathcal{V})$	fully invariant congruence corresponding to $\mathcal{V}$
$\rho_{\min}$	least congruence with trace $\operatorname{tr}\rho$
$\rho_{\max}$	greatest congruence with trace $\operatorname{tr}\rho$
$\rho^{\min}$	least congruence with kernel $\ker\rho$
$\rho^{\max}$	greatest congruence with kernel $\ker\rho$
$\sigma = \sigma_S$	least group congruence on S
$\sigma_a : t \to (a^{-1}\varphi_{e,t})\rho \quad (t \in [e])$	
$\tau = \tau_S$	greatest idempotent pure congruence on S
τ_H	transitivity relation of H
τ^H	syntactic congruence relative to H
τ_N	$a\tau_N b \Leftrightarrow a = hb$ for some $h \in N$
τ_Q	$[a, a]\tau_Q[b, b] \Leftrightarrow Qa \cap Qb \neq \varnothing$
$\tau_{(\lambda,\rho)} : s \to \lambda^{-1}s\rho \quad (s \in \lambda S\rho^{-1})$	
φ_H	transitive representation induced by H
$\varphi_H^s : (Ha)\omega \to (Has)\omega \quad \text{if} \quad \lambda(as) \in H$	
φ_g	relation on an inverse semigroup
$\overline{\varphi}$	congruence induced by φ
$\overline{\varphi^s}$	φ^s cut to $\mathbf{r}\varphi^{s^{-1}}$
$\varphi^\alpha : i \to \varphi\alpha i \quad (i \in \mathbf{d}\alpha, \alpha i \in \mathbf{d}\varphi)$	
$\varphi_{e,f}$	structure mapping
$\varphi_{m,n} : i \to n - m + i \quad (i \geq m)$	
φ_0	completion to a representation by transformations
$\varphi_1 : \mathcal{V} \to \mathcal{V} \vee \mathcal{G} \quad [\mathcal{V} \in \mathcal{L}(\mathcal{I})]$	
$\varphi_2 : \mathcal{V} \to \mathcal{V} \cap \mathcal{G} \quad [\mathcal{V} \in \mathcal{L}(\mathcal{I})]$	
$\varphi_Y : s \to \varphi^s\vert_Y$	
$\chi_{\alpha,\beta}$	$\mathcal{J}$-structure mapping
$\tilde{\psi} : a\sigma \to a\psi\sigma \quad (a \in \lambda S\rho)$	
$\psi_{e,f}$	dual of a structure mapping
ψ_G	transitive representation induced by G
$\psi_G^s : Ga \to Gas \quad \text{if} \quad Gas \subseteq R$	
${}^{\beta}\psi : \mu \to \mu\beta\psi \quad (\mu \in \mathbf{d}\beta, \mu\beta \in \mathbf{d}\psi)$	
ω	universal relation

MIXED LETTERS

(F, φ)	free $\mathcal{C}$-algebra
(S, ρ)	normal extension with $K = \ker \rho$, $\operatorname{tr} \pi = \operatorname{tr} \rho$
(Y, G)	F-pair
(ξ, S, η)	normal extension of K by Q along π
$(Y, G; \psi)$	unitary triple
$(Y, G; X)$	McAlister triple
(K, π, Q)	normal extension triple
$\langle A, B \rangle$	inverse semigroup generated by $A \cup B$
$\langle A, x, y \rangle$	inverse semigroup generated by $A \cup \{x, y\}$
$\langle S, Q; \varphi \rangle$	ideal extension of S by Q
$[u \in E] = [u^2 = u]$	
$[u \in G] = [uu^{-1} = u^{-1}u]$	
$[Y, G; \eta]$	subdirect product of Y and G
$[Y; S_\alpha, \varphi_{\alpha,\beta}]$	strong semilattice Y of semigroups S_α
$[S_i; U; \varphi_i]_{i \in I}$	amalgam of semigroups S_i with core U
$F(Y, G)$	F-representation
$R(Y, G)$	R-inverse semigroup
$P(Y, G; X)$	P-representation
$Q(Y, G; \psi)$	Q-representation
$Q(Y, \pi, P; \psi)$	normal extension of Y by P
$T(S : K)$	type of a conjugate extension S of K
$T(V : S)$	type of an ideal extension V of S
$\mathcal{T}(V : S)$	congruence induced by $\tau(V : S)$
$\theta(S : K) : S \to \Psi(K)$	canonical homomorphism
$\tau(V : S) : V \to \Omega(S)$	canonical homomorphism
$B(G; a, b) = \{b^m g a^n \mid m, n \in \mathbb{N}, g \in G\}$	
$\Pi(\alpha, \beta)$	(α, β)-path on a tree
$A \setminus B = \{a \in A \mid a \notin B\}$	
$E\zeta = E_S\zeta$	centralizer of idempotents
$\mathcal{F} \Rightarrow u = v$	implication
$gY = \{g\alpha \mid \alpha \in Y\}$	
$GY = \{g\alpha \mid g \in G, \alpha \in Y\}$	
$H\omega$	closure of H
$\operatorname{Ob} C$	objects of C
$\operatorname{Hom} C$	morphisms of C
$\operatorname{Hom}_C(a, b) = \operatorname{Hom}(a, b)$	
$\alpha : a \to b$	
$\operatorname{im}(S)$	imprint of S
$\operatorname{ind} \mathcal{V} = \sup\{n \mid M_n \in \mathcal{V}\}$	
$\operatorname{Inv\,lim}\{G_\alpha\}_{\alpha \in I}$	inverse limit of groups

$\operatorname{Inv}\lim\{G_\alpha\}_{\alpha\in Y}$	inverse limit of groups over a semilattice
$R^{-1}\circ R = \{[a,b] \mid a,b\in R\}$	
S/I	Rees quotient semigroup
S/ρ	quotient semigroup
$\operatorname{tr}(S)$	trace of S
$(T,\le,*)$	inductive groupoid
$(T,E;\varphi_{e,f})$	inverse triple
$[u=v \mid u=v$ has property $P]$	variety
HK	complex product
$\lambda a = aa^{-1}$	
$a\rho = a^{-1}a$	
$s\rho$	ρ-class containing s
$\alpha\beta$	product of binary relations α and β
$\alpha\beta$	composition of morphisms α and β
$\alpha\beta$	oriented edge
$\varphi\cdot\varphi' : i\to(\varphi i)(\varphi' i)$	$(i\in \mathbf{d}\varphi\cap\mathbf{d}\varphi')$
$\psi\cdot\psi' : \mu\to(\mu\psi)(\mu\psi')$	$(\mu\in \mathbf{d}\psi\cap\mathbf{d}\psi')$
$u=v$	identity
$a\to b$	$a^2=a$ implies $b^2=b$
$a\vee b$	$Ra\cap Rb = R(a\vee b)$
$a*b$	$(a*b)b = a\vee b$
τ/ρ	$(a\rho)(t/\rho)(b\rho)\Leftrightarrow a\tau b$
(k,u)	type of a monogenic inverse semigroup
(k,l)	type of a congruence
(k,ω)	type of a congruence
(k,∞^-)	type of a congruence
(k,∞^+)	type of a congruence
$[a,b]$	class of an equivalence relation containing (a,b)
$[u=v]$	variety determined by $u=v$
$[x,y]=xyx^{-1}y^{-1}$	commutator of x and y
$[u_\alpha=v_\alpha]_{\alpha\in A}$	variety determined by $\{u_\alpha=v_\alpha\}_{\alpha\in A}$
$[\alpha,\beta]$	interval from α to β
$\alpha\to\beta$	oriented edge
$\ker:\rho\to\ker\rho\quad[\rho\in\mathcal{C}(S)]$	
$\ker\rho$	kernel of ρ
$\ker\varphi$	kernel of the congruence induced by φ
$\operatorname{tr}:\rho\to\operatorname{tr}\rho\quad[\rho\in\mathcal{C}(S)]$	
$\operatorname{tr}\rho$	trace of ρ
$\operatorname{tr}\varphi$	trace of the congruence induced by φ
$\operatorname{rank}\alpha$	rank of α
$\eta(\omega,z;k,l)$	homomorphism of Reilly semigroups
$\eta(\omega,z)=\eta(\omega,z;1,0)$	

type ρ	type of ρ
$\cong$	isomorphism
$\leq$	partial order
$\leq$	natural partial order
$\vee$, $\vee Y$, $\vee_{\alpha \in y}\alpha$	join
$\wedge$, $\wedge Y$, $\wedge_{\alpha \in y}\alpha$	meet
$\prec$	covering
$\prec$	order on $\mathbb{N}^* = \mathbb{N} \cup \{\omega, \infty^-, \infty^+\}$
$\varnothing$	empty set
$\varnothing$	empty word
$\mid$	division of integers
2	two-element set
1_a	identity morphism on a

BOLDFACE

$\mathbf{d}\alpha$	domain of α
$\mathbf{r}\alpha$	range of α

PRODUCTS AND SUMS

$\prod_{\alpha \in A} S_\alpha$	direct product
$\prod^*_{i \in I} S_i = \prod^* S_i$	free product
$\prod^*_U S_i$	amalgamated free product
$\prod^*_{\text{inv}} S_i$	inverse free product
$\prod^*_{\text{csg}} S_i$	commutative inverse free product
$\prod^*_g S_i$	group free product
$\prod^*_{\mathcal{V}} S_i$	$\mathcal{V}$-free product
$\prod_{i=1}^n a_i = a_1 a_2 \cdots a_n$	
$P \circ Q$	ordinal product
$S_1 \times S_2 \times \cdots \times S_m$	direct product
$S_1 * S_2 * \cdots * S_n$	free product
$S_1 *_U S_2$	amalgamated free product
$S_1 *_{\text{inv}} S_2$	inverse free product
$S_1 *_{\text{csg}} S_2$	commutative inverse free product
$S_1 *_g S_2$	group free product
$Y *_s Y'$	semilattice free product
$S *_{sg} S'$	Clifford free product
$G \operatorname{wr} Q$	right wreath product
$P \operatorname{wl} G$	left wreath product

$\mathfrak{U} \circ \mathfrak{V}$	Malcev product
$\Sigma_{\alpha \in A} G_\alpha$	orthogonal sum of groupoids
$\Sigma_{\alpha \in A} S_\alpha$	orthogonal sum of semigroups
$\oplus_{\alpha \in A} \varphi_\alpha$	sum of representations
$\varphi_1 \oplus \varphi_2 \oplus \cdots \oplus \varphi_n$	sum of representations

GERMAN

$\mathfrak{a}$	isomorphism of $\Lambda(S)$ and $\mathfrak{F}'(I) \operatorname{wl} G$
$\mathfrak{b}$	isomorphism of $\mathrm{P}(S)$ and $G \operatorname{wr} \mathfrak{F}(I)$

CATEGORIES

We state here only a brief description of the objects of these categories.

$\mathfrak{B}$	bisimple inverse monoids
$\mathfrak{B}_l$	Ob $\mathfrak{B}$, E-unitary, combinatorial, $aR_1 = R_1 a$ for all $a \in R_1$
$\mathfrak{B}_\omega$	Ob $\mathfrak{B}$, E_S is an ω-chain
$\mathfrak{D}$	$(Y, \pi, P; \psi)$
$\mathfrak{E}$	E-unitary inverse semigroups
$\mathfrak{F}$	F-inverse semigroups
$\mathfrak{G}$	(G, α), α is an endomorphism of the group G
$\mathfrak{G}_l$	lattice ordered groups
$\mathfrak{M}$	McAlister triples $(Y, G; X)$
$\mathfrak{P}$	(G, I), G is a group, I is a nonempty set
$\mathfrak{Q}$	Brandt semigroups
$\mathfrak{R}$	right cancellative monoids with condition on principal left ideals
$\mathfrak{R}_l$	Ob $\mathfrak{R}$, left cancellative, combinatorial, $aR = Ra$ for all $a \in R$
$\mathfrak{R}_\omega$	Ob $\mathfrak{R}$, principal left ideals form an ω-chain
$\mathfrak{S}$	inverse semigroups
$\mathfrak{T}$	F-pairs
$\mathfrak{U}$	unitary triples $(Y, G; \psi)$

FUNCTORS

$B : \mathfrak{P} \to \mathfrak{Q}$

$B : \mathfrak{G} \to \mathfrak{B}_\omega$

$F : \mathfrak{T} \to \mathfrak{F}$

$K : \mathfrak{F} \to \mathfrak{T}$

$P : \mathfrak{M} \to \mathfrak{E}$

$P : \mathfrak{G}_l \to \mathfrak{R}_l$

$Q : \mathfrak{D} \to \mathfrak{S}$

$Q : \mathfrak{B}_l \to \mathfrak{G}_l$

$R : \mathfrak{E} \to \mathfrak{M}$

$R : \mathfrak{G} \to \mathfrak{R}_\omega$

$T : \mathfrak{S} \to \mathfrak{D}$

$U : \mathfrak{B} \to \mathfrak{R}$

$U_l : \mathfrak{B}_l \to \mathfrak{R}_l$

$U_\omega : \mathfrak{B}_\omega \to \mathfrak{R}_\omega$

$V : \mathfrak{R} \to \mathfrak{B}$

$V_l : \mathfrak{R}_l \to \mathfrak{B}_l$

$V_\omega : \mathfrak{R}_\omega \to \mathfrak{B}_\omega$

INDEX